D0078837

# ELECTRIC CIRCUITS AND NETWORKS

## Second Edition

# ELECTRIC CIRCUITS AND NETWORKS
## Second Edition

**R. D. STRUM**
**J. R. WARD**

Professors of Electrical and Computer Engineering
U.S. Naval Postgraduate School, Monterey, California

**Prentice-Hall, Inc., Englewood Cliffs, New Jersey 07632**

**Library of Congress Cataloging in Publication Data**

Strum, Robert D.
Electric circuits and networks.

Includes index.
1. Electric circuits. 2. Electric Networks.
I. Ward, John Robert. II. Title.
TK454.S84 1985 621.319'2 84-16070
ISBN 0-13-248170-7

Editorial/production supervision: Alberta Boddy
and Barbara Palumbo
Manufacturing buyer: Anthony Caruso



Printed in the United States of America

10 9 8 7 6 5 4 3 2 1

ISBN 0-13-248170-7 01

Prentice-Hall International, Inc., *London*
Prentice-Hall of Australia Pty. Limited, *Sydney*
Editora Prentice-Hall do Brasil, Ltda., *Rio de Janeiro*
Prentice-Hall Canada Inc., *Toronto*
Prentice-Hall of India Private Limited, *New Delhi*
Prentice-Hall of Japan, Inc., *Tokyo*
Prentice-Hall of Southeast Asia Pte. Ltd., *Singapore*
Whitehall Books Limited, *Wellington, New Zealand*

# CONTENTS

# PREFACE

*Electric Circuits and Networks* has been designed as a textbook for sophomore and junior courses and contains a comprehensive explanation of the theory, as well as a variety of worked and annotated problems illustrating applications of the theory. It assumes no prior knowledge of circuit analysis and can be used as a class text, a supplemental text, or as the basis for self-study. Further, parts of the later chapters (for example, convolution and Fourier transform) will be helpful in subsequent and/or more advanced circuits or systems courses. State variable techniques have been introduced in several places throughout the text, but not at the expense of the normal development of classical circuit analysis.

Even though most students will have encountered electric circuits in a prior physics course, we have found that a review of the basic principles is valuable; thus the two-terminal elements and the operational amplifier are discussed in Chapter 1, followed by the Kirchhoff constraints in Chapter 2. This permits us to establish systematic procedures for writing loop, node and state equations, to introduce the Thévenin equivalent, and to develop these procedures into a facility that many students lack. In Chapters 3 and 4, the classical solution of homogeneous and nonhomogeneous circuit differential equations is treated. If desired, these two chapters may be covered quite briefly, but they should not be omitted since they introduce the important concept of the characteristic or natural modes of response. In Chapter 5, we apply the method of the Laplace transform to the solution of ordinary differential equations. The concept of impedance follows naturally.

Chapter 6 establishes the foundation for the operational or transform methods of circuit analysis. The impedance concept is extended to include the network function, and poles, zeros, and stability are defined.

Sinusoidal steady-state solutions are treated in Chapter 7 with the introduction of the sinusoidal network function and the phasor. This is applied to frequency response calculations, frequency response plots, and filters in Chapter 8, and to ac power, transformers, and polyphase circuits in Chapter 9.

Network theorems are developed in Chapter 10, and the characteristics of two-port networks follow in Chapter 11. This would be a good point to complete in a two-quarter course or a one-semester course if the first four (review) chapters were covered briefly.

We regard the material of Chapters 1 to 11 as a solid foundation for the engineering analysis of electric and electronic circuits and as the point of departure for advanced circuits or systems courses. In such a course, Chapters 12 (Fourier methods), 13 (convolution), and 14 (matrix methods of analysis) can be utilized in any order.

This book differs from the usual text in quite a fundamental way. Many years of teaching undergraduate circuits and systems courses has shown that students appreciate some kind of guide for their own study. Extensive experimentation with computer-assisted instruction, programmed learning, and the use of sets of solved problems has convinced us that the illustrative problem format provides the support that students need. The introductory material in each chapter supplies a concise but complete outline of the objectives; students can be encouraged to participate in the subject matter development as well as in problem solving, and the instructor can bring difficulties to the attention of the students and can discuss the students' questions *as they arise*. This creates a harmonious and efficient teaching–learning environment.

As a result of class testing, we are satisfied that we have been able to answer—in the text—almost all of the questions that students ask. This book will thus be as valuable for self-study as it has proved to be for classwork. It is our suggestion that students attempt to solve the worked problems for themselves before looking at our solutions.

ROBERT D. STRUM
JOHN R. WARD

# ELECTRIC CIRCUITS AND NETWORKS
## Second Edition

## Chapter 1

# THE BASIC CIRCUIT ELEMENTS AND WAVEFORMS

Circuit analysis is the process of investigating the behavior of any given electrical circuit. That is, given the particular set of circuit *elements* (resistors, capacitors, sources, op amps, etc.) and a drawing which shows how they are connected together (the circuit *diagram*), we proceed to compute the various voltages and currents within the circuit. But first the behavior of the individual elements must be established—in mathematical terms.

### QUANTITIES AND UNITS

The physical quantities which will be most commonly encountered, together with their corresponding SI units (*Système International d'Unités*) are listed in Table 1-1.

It is convenient to indicate large submultiples and large multiples of the basic units in terms of the prefixes listed in Table 1-2.

For example, $5 \times 10^{-10}$ F may be written either as 0.5 nF or as 500 pF.

The units and standards used in this book correspond to "*IEEE* Recommended Practice: Rules for the Use of Units of the International System of Units," *IEEE Spectrum,* March 1971.

**Table 1-1 Some Quantities and Units.**

| Quantity | | Symbol | SI unit and abbreviation |
|---|---|---|---|
| Fundamental quantities | Mass | $m$, $M$ | kilogram, kg |
| | Length | $l$, $L$ | meter, m |
| | Time | $t$ | second, s |
| | Current | $i$, $I$ | ampere, A |
| Charge | | $q$, $Q$ | coulomb, C |
| Flux linkage | | $\lambda$ | weber, Wb |
| Voltage | | $v$, $V$ | volt, V |
| Power | | $p$, $P$ | watt, W |
| Energy, work | | $w$, $W$ | joule, J |
| Resistance | | $R$ | ohm, Ω |
| Conductance | | $G$ | siemen, S |
| Capacitance | | $C$ | farad, F |
| Inductance | | $L$ | henry, H |
| Frequency | | $f$ | hertz, Hz |

**Table 1-2 Prefixes.**

| Factor | Prefix and abbreviation |
|---|---|
| $10^{-12}$ | pico, p |
| $10^{-9}$ | nano, n |
| $10^{-6}$ | micro, $\mu$ |
| $10^{-3}$ | milli, m |
| $10^{3}$ | kilo, k |
| $10^{6}$ | mega, M |
| $10^{9}$ | giga, G |

## CHARGE, CURRENT, WORK, VOLTAGE, AND POWER

Charge is moved through a circuit as current, which is simply a measure of the charge transferred per second. In other words,

$$i = \frac{dq}{dt} \tag{1.1}$$

If a charge $q$ is moved through a potential difference (or voltage) $v$, then the work done in moving the charge is

$$w = qv \tag{1.2}$$

This is in fact the *definition* of voltage, which is the work done *per unit charge* in moving a charge from one point to another in an electric field. Thus

$$v = \frac{w}{q} \tag{1.3}$$

In terms of an elementary charge $dq$, $v = dw/dq$ and, multiplying by equation (*1.1*),

$$vi = \frac{dw}{dt} = p(t) \tag{1.4}$$

since by definition the time rate-of-change of energy is the power $p(t)$.

Alternatively we may write

$$w(t) = \int_0^t p(\tau)\, d\tau + w(0) \tag{1.5}$$

where $w(0)$ is the energy stored at time $t = 0$, or the work done prior to $t = 0$. The integral represents the work done, or the energy transferred, between time $t = 0$ and the current time $t$. Note that the variable of integration, $\tau$, is a *dummy*. The symbol $t$ is often used, but it can be confused with the upper limit $t$.

## LUMPED, TWO-TERMINAL ELEMENTS

Each elementary circuit element has two terminals. It will be assumed that the current into one such terminal *exactly* equals the current out of the other at the same instant. That is, the element must be small enough for the time of propagation of an electrical effect to be practically negligible. The element is then said to be *lumped*. This is not a valid approximation in a wave-guide or antenna, which is said to be *distributed*.

## SOURCES AND SIGNALS

Circuits contain *active* and *passive* elements. Here we will describe the ideal *active* elements—the sources of electrical energy.

An *independent* voltage source, which is represented by the symbol of Fig. 1-1*a*, is defined to have a prespecified terminal voltage, $v(t)$, *regardless* of the circuit into which it is connected. That is, the voltage $v(t)$ is *independent* of the current flowing. The voltage $v(t)$ is often called an input *signal*, and may have any waveform—sinusoid, squarewave, exponential, speech, noise, constant, etc. Similarly, an *inde-*

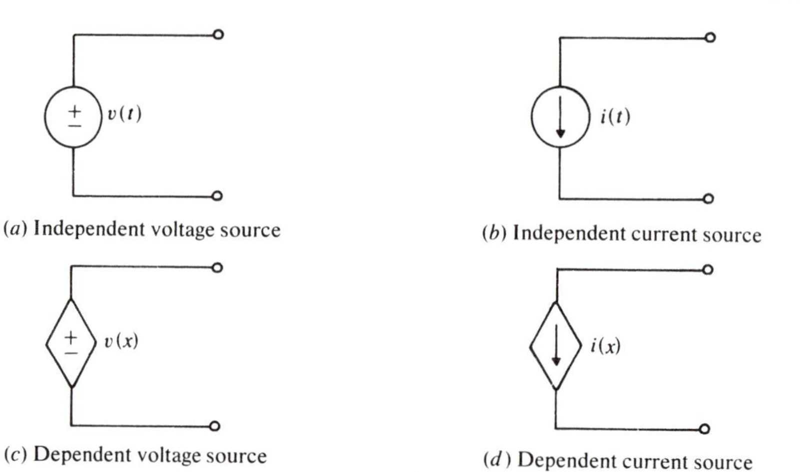


(a) Independent voltage source

(b) Independent current source

(c) Dependent voltage source

(d) Dependent current source

**Fig. 1-1**

*pendent* current source (Fig. 1-1*b*) delivers a prespecified current, $i(t)$, regardless of the terminal voltage.

The terminal voltage of a *dependent* or *controlled* voltage source (Fig. 1-1*c*) is a specified function of some variable $x$, usually a current or voltage in the circuit. That is, $v = v(x)$, regardless of the current flowing through the source. Similarly, the current supplied by a *dependent* or *controlled* current source (Fig. 1-1*d*) will be $i = i(x)$, regardless of the terminal voltage.

Independent source voltages and/or currents (signals) are typically generated in batteries, power supplies, oscillators, tape recorder heads, solar cells, and the like. Controlled sources play a central role in the mathematical representation of transistors and other active electronic devices.

## LINEAR PASSIVE ELEMENTS

Often the *passive* elements (resistors, capacitors, and inductors) may be characterized by *constant* parameter values (resistance, capacitance, or inductance). They are then said to be *linear* and *time-invariant*.

From materials obeying Ohm's law, i.e. linear resistive materials, we obtain resistors whose terminal current and voltage are related linearly through the resistance $R$,

$$v_R(t) = Ri_R(t), \qquad i_R(t) = \frac{1}{R}v_R(t) \tag{1.6}$$

For inductors utilizing linear magnetic materials, the terminal relations, in terms of the inductance $L$ are

$$v_L(t) = L\frac{di_L(t)}{dt}, \qquad i_L(t) = \frac{1}{L}\int_0^t v_L(\tau)\,d\tau + i_L(0) \tag{1.7}$$

And for capacitors, constructed from linear dielectric material, we find that in terms of the capacitance $C$,

$$v_C(t) = \frac{1}{C}\int_0^t i_C(\tau)\,d\tau + v_C(0), \qquad i_C(t) = C\frac{dv_C(t)}{dt} \tag{1.8}$$

In each case the circuit parameter ($R$, $L$, or $C$) may be regarded as merely the constant of proportionality in the mathematical statement of a (linear) physical relationship.

Some notes on equations (*1.7*) and (*1.8*) are in order. The variable of integration $\tau$ is a *dummy*. Both $t$ and $x$ are often used instead, but if $t$ is the variable of integration it must not be confused with the upper limit $t$. Sometimes $t_0$ is used for the lower limit, together with the initial values $i_L(t_0)$ and $v_C(t_0)$. Indefinite integrals will occasionally be encountered. This is less desirable usage, since the constants of integration or initial values do not then appear explicitly.

## TIME-VARYING PASSIVE ELEMENTS

More generally than before, capacitance $C$ and inductance $L$ are defined by

$$q = Cv \tag{1.9}$$

and

$$\lambda = Li \tag{1.10}$$

where the flux linkage $\lambda$ is related to voltage through Faraday's law, $v = d\lambda/dt$. Equations (*1.9*) and (*1.10*) are illustrated graphically in Fig. 1-2.

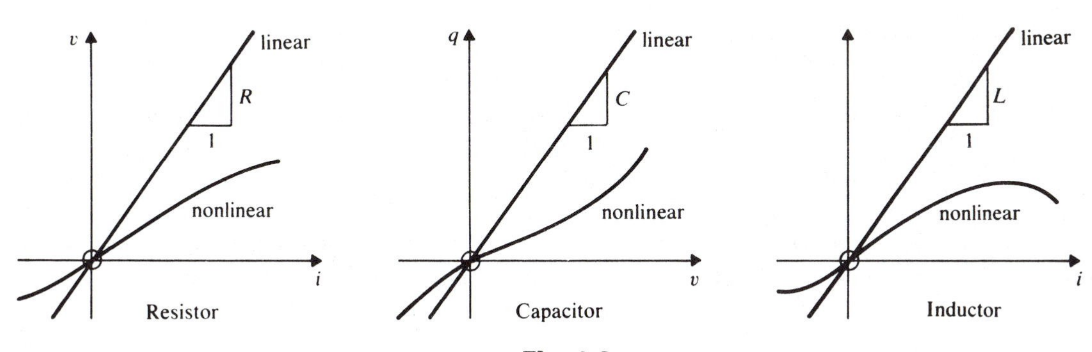


**Fig. 1-2**

If $C$ and $L$ are time-dependent, we write

$$i = \frac{dq}{dt} = \frac{d}{dt}(Cv) \tag{1.11}$$

and

$$v = \frac{d\lambda}{dt} = \frac{d}{dt}(Li) \tag{1.12}$$

Of course resistance may also be time-dependent, but this does not affect equation (*1.6*).

## NONLINEAR PASSIVE ELEMENTS

For a capacitor there is often a nonlinear relationship between $q$ and $v$ (see Fig. 1-2). Then

$$i = \frac{dq(v)}{dt} = \frac{dq(v)}{dv}\frac{dv}{dt} = C(v)\frac{dv}{dt} \tag{1.13}$$

Similarly for an inductor,

$$v = \frac{d\lambda(i)}{dt} = \frac{d\lambda(i)}{di}\frac{di}{dt} = L(i)\frac{di}{dt} \tag{1.14}$$

Here $C(v)$ and $L(i)$ are called *incremental* capacitance and inductance, and they should be interpreted as the *local* slope of the $q - v$ or the $\lambda - i$ curve which describes the element.

A nonlinear resistor is simply described by

$$v = R(i)i \tag{1.15}$$

## MUTUAL INDUCTANCE

If two coils are magnetically coupled, as in a transformer, a changing current in one will induce a voltage in the other (Faraday's law). In a circuit diagram we must indicate the corresponding polarities, which will depend on the geometrical form of and relationship between the coils. By convention, a dot is placed on each coil such that an *increasing* current into one dot will induce a *positive* voltage at the other. Thus with reference to Fig. 1-3 and assuming linearity and time-invariance,

$$v_2 = +M\frac{di_1}{dt} \tag{1.16}$$

where $M$ is called the *mutual inductance* between the two coils.

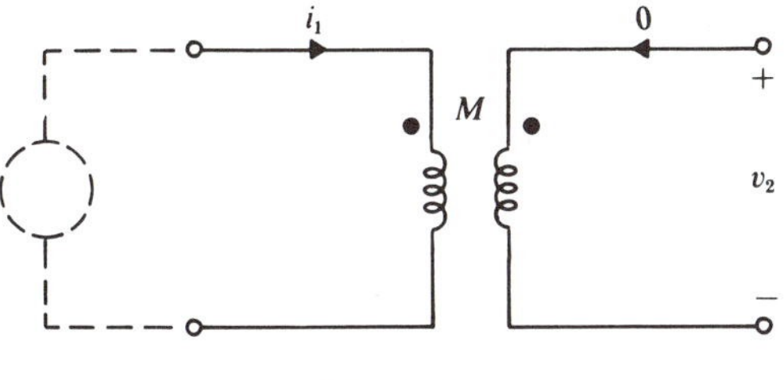


Fig. 1-3

## THE OPERATIONAL AMPLIFIER

The operational amplifier or op amp of Fig. 1-4*a* is an active electronic element that draws power from either one or two power supplies (dc voltage sources). However, the power sources are usually omitted from circuit diagrams, as in Fig. 1-4. (The manufacturer's data sheets will specify the necessary power connections.) This device is *operational* in the sense that it can be used, with appropriate passive elements, to accurately perform the *operations* of addition, subtraction, integration, etc. Op amps

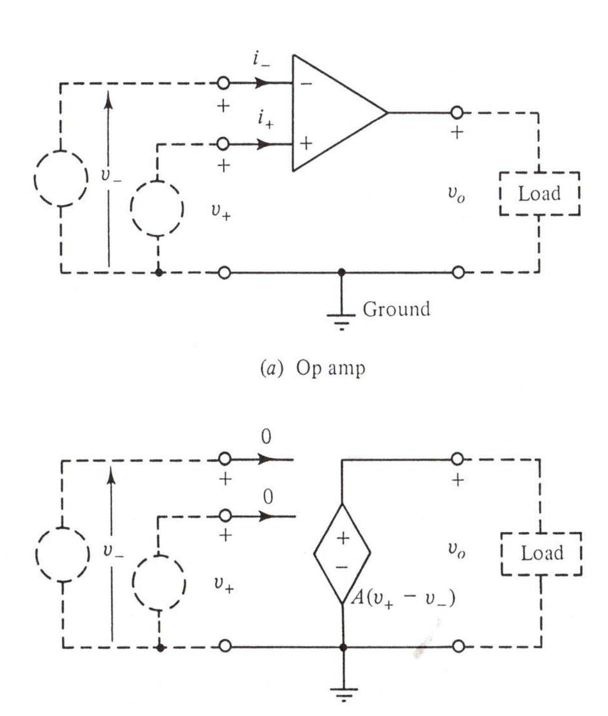


(*a*) Op amp

(*b*) Model of ideal op amp ($A \to \infty$)

Fig. 1-4

are available in the form of low-cost integrated circuits on silicon chips, and are commonly used in a wide variety of circuits and systems.

The *ideal* op amp is defined by

1. $v_o = A(v_+ - v_-) \qquad A \to \infty$
   which implies, for finite $v_o$, that $v_+ = v_-$.
2. $i_+ = i_- = 0$

where $v_+$ and $v_-$ are the voltages at the op amp's + and − input terminals (so marked on the manufacturer's data sheets) relative to *ground,* the zero reference of potential. Thus the ideal op amp can be modeled or represented by the equivalent circuit of Fig. 1-4*b*.

## REFERENCE POLARITY AND DIRECTION

The *reference* voltage polarity symbol on the element in Fig. 1-5 does *not* necessarily mean that the top terminal is positive! What it does mean is that *if* $v$ is positive, *then* the top terminal is positive. If $v$ is negative, then the top terminal will be negative—with respect to the bottom terminal.

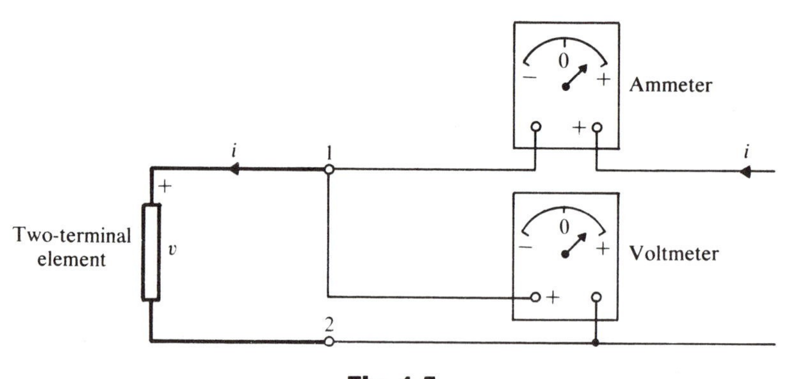


**Fig. 1-5**

Put another way, if a voltmeter were connected across the element, with the positive voltmeter lead connected to the plus-marked end of the element, then the voltmeter would read positively when $v$ is positive, and negatively whenever $v$ becomes negative.

Note that $v$ is strictly a potential *difference,* and that if $v_1$ and $v_2$ are the voltages of the terminals relative to ground or other reference, then $v = v_1 - v_2$; that is, the voltmeter measures the potential *difference* between its terminals.

Similarly, the *reference* directional arrow for current does *not* necessarily mean that current is flowing down through the element in Fig. 1-5. This would be true only when $i$ is positive. Current would flow upward whenever $i$ is negative.

## ASSOCIATED SIGN CONVENTION

It is necessary to *relate* or *associate* the reference direction of current in an element to the reference polarity of voltage. The + polarity symbol for the voltage reference *must* (in the usual associated convention) be placed on the end of the element *entered* by the reference current (see Fig. 1-5). This convention has been assumed in equations (*1.6*)–(*1.16*).

In equation (*1.4*) power has been defined as $p = vi$. Hence when $v$ and $i$ (see Fig. 1-5) are of the same sign, $p$ will be positive, corresponding to the delivery of power *to* the element. Electrical energy will be given up by the element when $p$ is negative—that is, when $v$ and $i$ are of opposite sign.

## CHARACTERISTICS OF WAVEFORMS

The word *waveform* describes the mathematical "shape" of a signal. We are familiar with a number of possibilities, such as $v_1(t) = 2e^{-3t}$, $i(t) = -4$, and $v_2(t) = 117\sqrt{2}\cos 377t$, but we must be prepared to meet some less common functions. For that matter, we are not yet equipped to recognize $v_2(t)$, above, as the mathematical representation of ordinary household utility voltage.

One common function is the simple constant, representing as it does a battery or an electronic constant-voltage or constant-current power supply. The terms *dc voltage* and *dc current* are often used in this connection, where *dc* stands for direct (i.e. constant) "current." This function is graphed in Fig. 1-6*a*, and is written $v(t) = K$.

Probably the most common electrical waveform is the sinusoid $v(t) = A\cos(\omega t + \phi)$. (We could equally well work with the sine function.) In Fig. 1-6*b*,

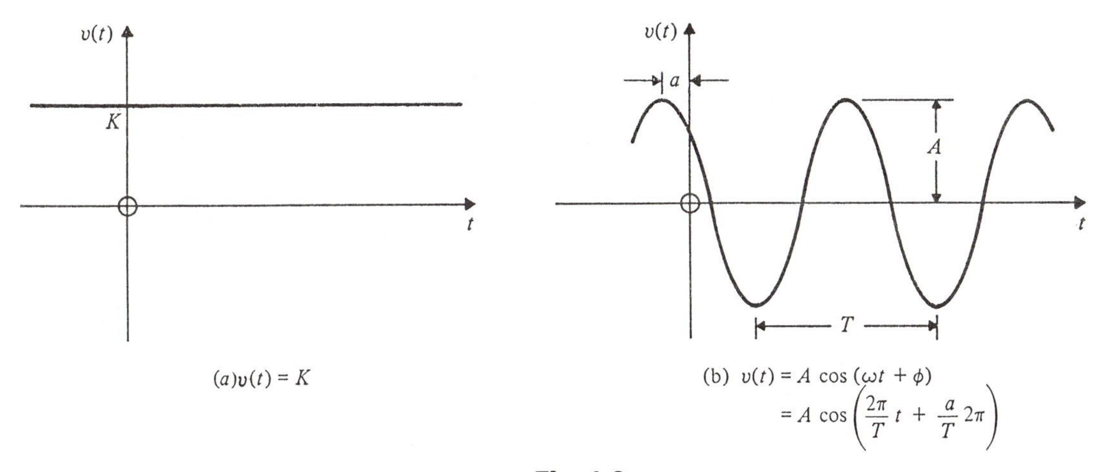


(*a*) $v(t) = K$

(b) $v(t) = A\cos(\omega t + \phi) = A\cos\left(\frac{2\pi}{T}t + \frac{a}{T}2\pi\right)$

**Fig. 1-6**

$A$ is seen to be the zero-to-peak *amplitude*, $T$ is the *period*, $f = 1/T$ is the *true frequency*, $\omega = 2\pi/T$ is the *angular frequency*, and $\phi = 2\pi a/T$ is the phase angle. $T$, $f$, and $\omega$ are related by

$$T = \frac{1}{f} = \frac{2\pi}{\omega}$$

Since the vast majority of electrical power in the civilized world is generated, transmitted, and used in sinusoidal form, the ubiquity of the sinusoid in electrical engineering will be appreciated. The terms *ac voltage* and *ac current* are common, where *ac* stands for alternating (i.e. sinusoidal) "current."

### Example 1.1

Suppose we are looking at an oscilloscope which is simultaneously displaying two sinusoids of the same frequency, as in Fig. 1-7. The period—the time of one complete *cycle*—can easily be read off the display: $T$ = 2 milliseconds = $2 \times 10^{-3}$ second. Then the (true) frequency is $f = 1/T$ = 500 hertz and the angular frequency is $\omega = 2\pi f$ = 3141.6 radians/second = 3.1416 kiloradians/second. The zero-to-peak amplitudes can also be read off from the oscilloscope: $A_1 = 0.2$ and $A_2 = 12$ volts.

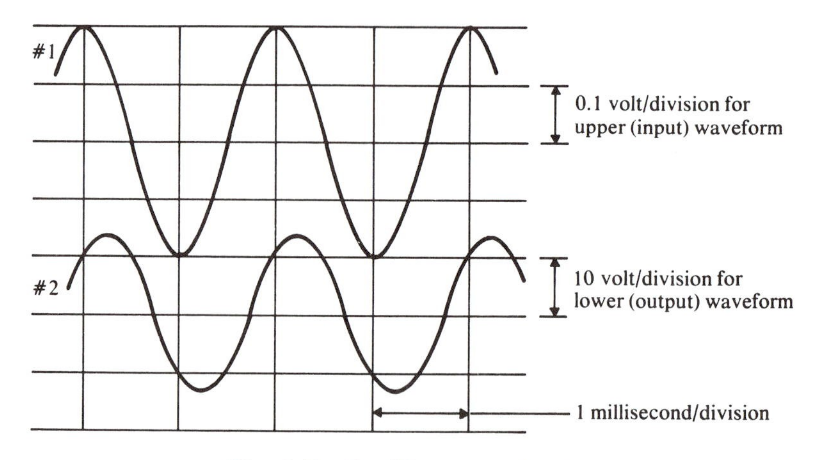


**Fig. 1-7 Oscilloscope display** ■

The situation is less clear in phase. Is the upper waveform in Fig. 1-7 $0.2 \cos 3141.6t$, or $0.2 \sin 3141.6t$, or $0.2 \cos (3141.6t + \phi)$, where $\phi$ is still to be found? Usually we have no way of telling, since we would have to relate the waveform to the origin of time (see Fig. 1-6*b*) and this origin, as it concerns the oscilloscope display, is usually an unknown distance into the past.

Fortunately, we seldom need to find the phase of a single sinusoid, but determine instead the phase of one sinusoid *relative* to another, which we designate as the *reference sinusoid*. For example, if the upper waveform in Fig. 1-7 were the input to an electronic amplifier and the lower waveform were the output, our interest would center on the *phase shift* caused by the amplifier—the amount by which the lower (output) waveform *leads* or *lags* the upper.

### Example 1.2

Notice, first, that a given event (say a peak or a zero-crossing) occurs *later,* i.e. to the right, in the lower sinusoid of Fig. 1-7. We say that the lower waveform is *lagging* the upper waveform in phase. The magnitude of the shift is about one-eighth of a cycle, or an angle of one-eighth of $2\pi$ radians, i.e. $2\pi/8 = \pi/4$ radian. Thus, designating the upper waveform as the reference,

$$v_1(t) = 0.2 \cos 3141.6t$$

and

$$v_2(t) = 12 \cos (3141.6t - \pi/4)$$

■

Next to be considered is the family of *singularity functions*. The first is the *unit function* or the *unit step function,* $\mathbb{1}(t)$. (The symbol $u(t)$ is often used for the unit function, but since $u(t)$ is also used to represent any general input, the notation $\mathbb{1}(t)$ is preferred.) By definition,

$$\mathbb{1}(t) = \begin{cases} 0 & \text{for } t < 0 \\ 1 & \text{for } t > 0 \end{cases} \tag{1.17}$$

This function is not defined (or is defined in various ways) at $t = 0$. It is graphed in Fig. 1-8.

If we integrate the unit step, we obtain the *unit ramp function*

$$r(t) = \begin{cases} 0 & \text{for } t < 0 \\ t & \text{for } t > 0 \end{cases} \tag{1.18}$$

which is graphed in Fig. 1-9. (Its derivative is undefined at $t = 0$.)

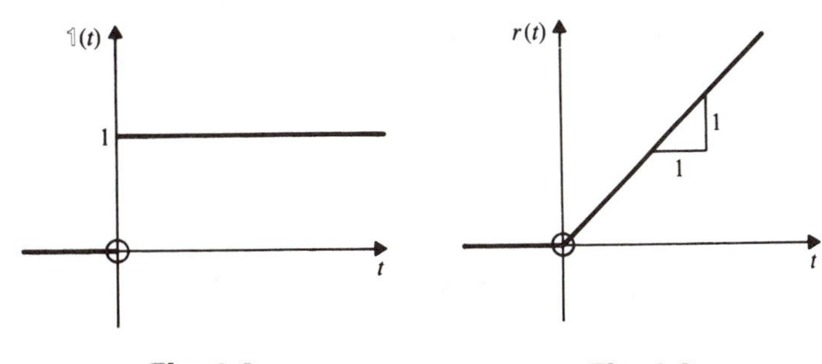


**Fig. 1-8** **Fig. 1-9**

The *unit rectangular pulse* is defined by

$$p_a(t) = \begin{cases} 0 & \text{for } t < 0 \\ 1/a & \text{for } 0 < t < a \\ 0 & \text{for } t > a \end{cases} \tag{1.19}$$

This function is undefined at $t = 0$ and $t = a$. It is clear from its graph (see Fig. 1-10) that the area under the pulse, between $t = 0$ and $t = a$, is unity.

If we now allow the width $a$ of the unit rectangular pulse to approach zero, we obtain in the limit the *unit impulse* or the *unit delta function*. That is,

$$\delta(t) = \begin{cases} 0 & \text{for } t \neq 0 \\ \text{undefined} & \text{for } t = 0 \end{cases} \tag{1.20}$$

and

$$\int_{-\xi}^{\xi} \delta(t)\, dt = 1 \qquad \text{for any } \xi > 0 \tag{1.21}$$

where equation (*1.21*) simply specifies that the area under the curve is still unity. We cannot of course graph this function directly, since its amplitude approaches infinity. But by convention a delta function is represented by an arrow, as in Fig. 1-11, where it must be understood that the amplitude of the delta function refers to its *area*.

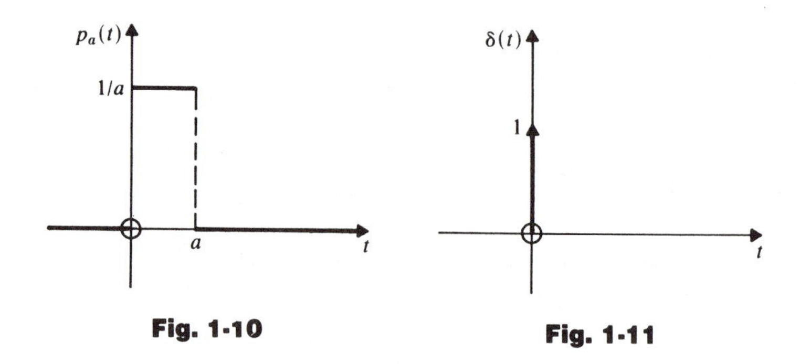


**Fig. 1-10** **Fig. 1-11**

It can easily be shown, heuristically, that the unit step is the integral of the unit impulse, although a rigorous proof is difficult. Thus the sequence of functions $r(t)$, $\mathbb{1}(t)$, $\delta(t)$ are related to each other by differentiation as we move to the right in the sequence, or by integration if we move to the left. This sequence of the singularity functions can be, and often is, extended in both directions.

## THE SHIFTED FUNCTIONS

The function $f(t - a)$ is delayed in time by $a$ as compared with $f(t)$. The graph of $f(t - a)$ is *shifted a* to the *right* compared with the graph of $f(t)$. This is illustrated in Fig. 1-12*a*, together with a simple example in Fig. 1-12*b*. (Another example of time-shift can be seen in Fig. 1-7. The second sinusoid is delayed 0.25 millisecond relative to the first.)

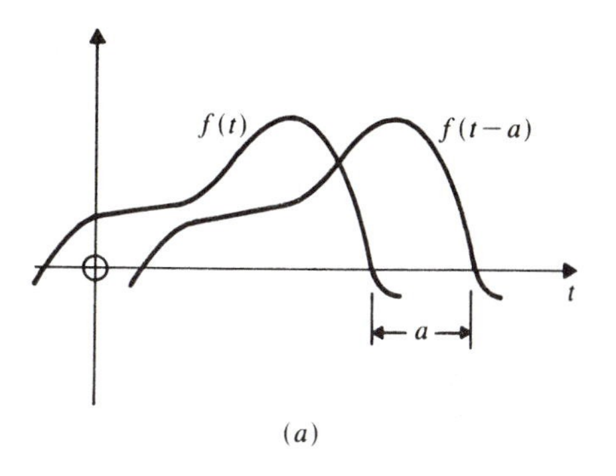


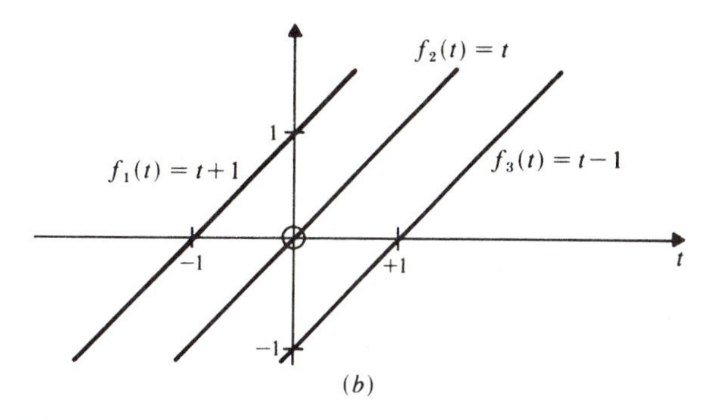


**Fig. 1-12**

### Example 1.3

Suppose that the waveform of Fig. 1-13*a* is the voltage across a 2-farad capacitor, and that we wish to find the current. The first step is necessarily to express $v_C(t)$ mathematically. Comparing Fig. 1-13*a* and *b*, it can be seen that

$$v_C(t) = \tfrac{3}{2}r(t) - \tfrac{3}{2}r(t-2)$$

where $r(t)$ is the unit ramp. Then from equation (*1.8*),

$$i_C(t) = C\frac{dv_C}{dt} = 2\frac{d}{dt}\{\tfrac{3}{2}r(t) - \tfrac{3}{2}r(t-2)\} = 3\mathbb{1}(t) - 3\mathbb{1}(t-2)$$

since $dr(t)/dt = \mathbb{1}(t)$. The function $i_C(t)$ is shown in Fig. 1-14.

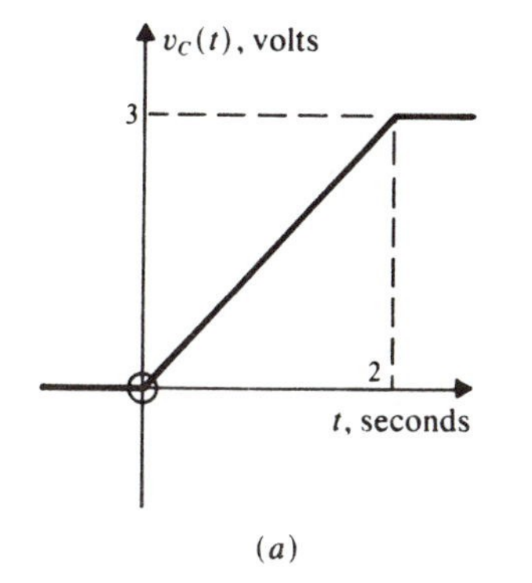


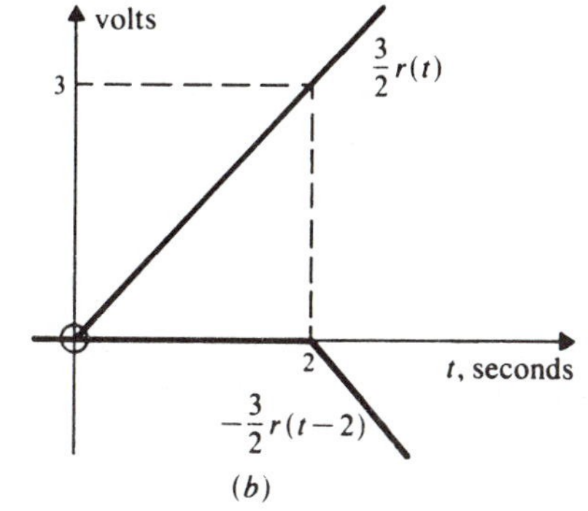


**Fig. 1-13**

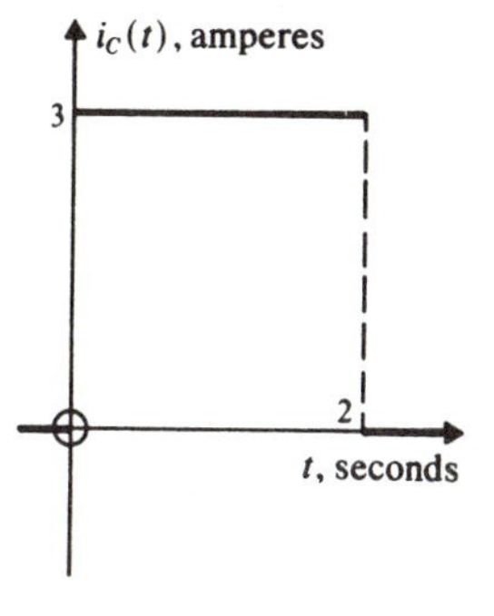


**Fig. 1-14**

Alternatively, since $i_C(t) = 2\,dv_C/dt$, we could have obtained Fig. 1-14 by differentiating Fig. 1-13*a* graphically (i.e. by reading off the slope of each segment) and multiplying by two. ■

## CHARACTERISTICS AND MEASUREMENT OF REPETITIVE WAVEFORMS

Often in electrical engineering we deal with waveforms, $x(t)$, which are *repetitive,* i.e. functions for which

$$x(t + T) = x(t) \qquad \text{for all } t$$

We have already discussed one such function, the sinusoid.

The *period* of the waveform is the smallest (positive) value of $T$ for which the above relation is valid (see Fig. 1-15). The *frequency* is $1/T$.

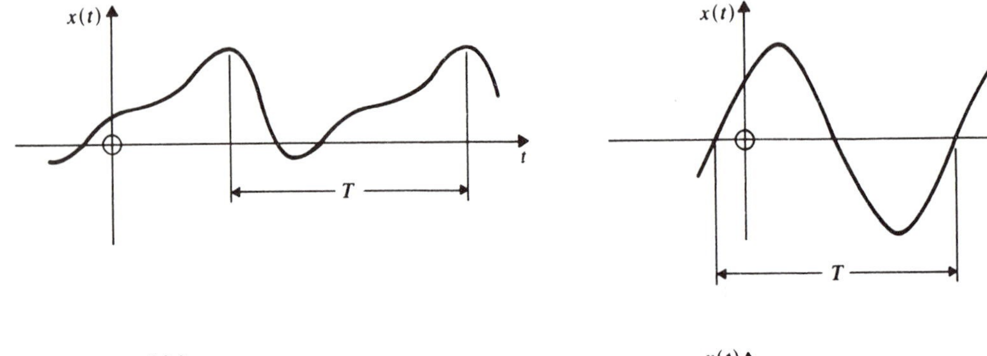


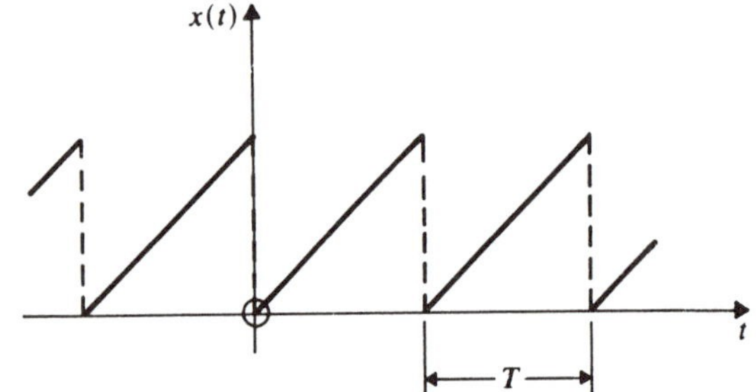


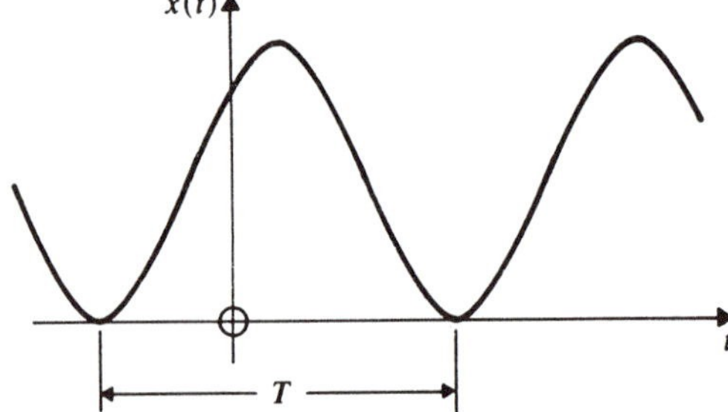


**Fig. 1-15 Repetitive waveforms**

Given an oscilloscope—or, for "slow" waveforms, a pen-recorder—the "shape" of the waveform can be observed, sketched, or photographed. The value of $x(t)$ can then be determined for any and all values of $t$.

If a waveform is repetitive, we may attach meaning to the *average value,*

$$X_{av} = \frac{1}{T}\int_{t_0}^{t_0+T} x(\tau)\, d\tau \qquad (1.22)$$

which is by definition the average value of $x$, taken over one complete cycle of the waveform. This is the characteristic measured by an ordinary moving-coil (D'Arsonval) meter, provided that the waveform's frequency is high compared with the "natural" mechanical frequency of the meter. (If it is not, the meter will either "follow" the waveform as it varies with time—as in a pen-recorder—or the meter needle will vibrate about the average value.)

We might also ask how "effectively" some repetitive current waveform heats up a given resistance element—in a stove, for example. To quantify the concept of "effectiveness" we look for the value of the dc current, $I_{\text{eff}}$, which would generate the *same* amount of heat in a resistor $R$ (see Fig. 1-16). Now

$$\begin{aligned}\text{Power dissipated by the dc current} &= I_{\text{eff}} V_{\text{eff}} && \text{(from equation (1.4))}\\ &= I_{\text{eff}}(RI_{\text{eff}}) && \text{(from equation (1.6))}\\ &= RI_{\text{eff}}^2\end{aligned}$$

and

$$\textit{Average} \text{ power dissipated by } i(t) = \frac{1}{T}\int_{t_0}^{t_0+T} i(\tau)v(\tau)\,d\tau = \frac{1}{T}\int_{t_0}^{t_0+T} Ri^2(\tau)\,d\tau$$

Equating these powers,

$$RI_{\text{eff}}^2 = \frac{1}{T}\int_{t_0}^{t_0+T} Ri^2(\tau)\,d\tau$$

or

$$I_{\text{eff}} = \sqrt{\frac{1}{T}\int_{t_0}^{t_0+T} i^2(\tau)\,d\tau} = I_{\text{RMS}} \tag{1.23}$$

where $I_{\text{RMS}}$ (root-mean-square current) is synonymous with $I_{\text{eff}}$. Similarly,

$$V_{\text{eff}} = \sqrt{\frac{1}{T}\int_{t_0}^{t_0+T} v^2(\tau)\,d\tau} = V_{\text{RMS}} \tag{1.24}$$

A number of special instruments are available which read "true" RMS or effective values (of a current or voltage waveform); for example, the hot-wire or thermocouple meter, the dynamometer, and some electronic instruments.

Sometimes a moving-coil meter with a rectifier is *calibrated* to read RMS values *for a particular waveform,* usually a sinusoid. It will *not* correctly read the RMS values of other waveforms, but a waveform correction can be computed and then applied to the reading.

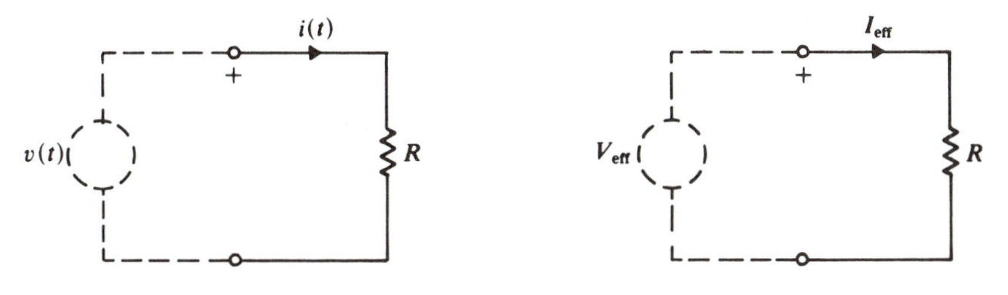


**Fig. 1-16**

# ILLUSTRATIVE PROBLEMS

*Note:* In this chapter all units will be given in full, often with their standard abbreviation, as listed in Tables 1-1 and 1-2. In subsequent chapters only the abbreviations will be used.

## *Charge, Current, Work, Voltage, and Power*

**1.1** The charge on one plate of a capacitor varies as

$$q(t) = 10 \sin 2t \text{ coulombs, C}$$

Find the equation for the current flowing into the capacitor.

From equation (*1.1*),

$$i(t) = \frac{dq(t)}{dt} = 20 \cos 2t \text{ amperes, A}$$

**1.2** The current in a wire varies according to the equation $i(t) = 2t$, for $t \geq 0$. Find the total charge that passes through the wire during the period between $t = 0$ and $t = 10$ seconds.

From equation (*1.1*), $dq = i\,dt$ or

$$\int_{q(0)}^{q(10)} dq = \int_0^{10} i(\tau)\,d\tau$$

$$q\Big|_{q(0)}^{q(10)} = \int_0^{10} 2\tau\,d\tau$$

$$q(10) - q(0) = \tau^2\Big|_0^{10} = 100 \text{ coulombs, C}$$

Here $q(10)$ is the charge that has moved along the wire up to the time $t = 10$ seconds, and $q(0)$ is that which has passed up to time $t = 0$. The *difference* is therefore the charge that has moved in the 10-second period of interest. (It is generally preferable to work with definite integrals, thereby avoiding difficulties with the constant of integration.)

**1.3** A constant current of 2 amperes flows for 5 seconds into an initially uncharged capacitor, raising the potential between its plates from 0 to 100 volts at a constant rate. Calculate the work done during this process.

As in Problem 1.2,

$$q(5) - q(0) = \int_0^5 2\,d\tau = 2\tau\Big|_0^5 = 10 \text{ coulombs, C}$$

or

$$q(5) = 10 \text{ coulombs, C}$$

Before we can compute the work done during the charging process, we must express $q$ in terms of $v$. Thus from the data, $v \propto q$, or $q = 0.1v$. (The constant of proportionality, 0.1, is by definition the capacitance of this capacitor; see equation (*1.9*).)

Equation (*1.3*) can be written $v = dw/dq$, or $dw = v\,dq$. But $v = 10q$; so, integrating with respect to the dummy variable $x$,

$$\int_{w(0)}^{w(5)} dw = \int_{q(0)}^{q(5)} 10x\,dx = 5x^2\Big|_0^{10} = 500 \text{ joules, J}$$

Exactly the same amount of work would be recovered from the capacitor if the current were reversed for 5 seconds. In other words, no energy is dissipated; it is merely *stored* in the capacitor.

### *Alternative Solution*

From equation (*1.4*), $p = vi = (20t)(2) = 40t$. Then from equation (*1.5*),

$$w(t) = \int_0^t p(\tau)\,d\tau + w(0)$$

or

$$w(5) = \int_0^5 40\tau\,d\tau = 20\tau^2\Big|_0^5 = 500 \text{ joules, J}$$

**1.4** A certain element has a potential difference of 100 volts while drawing a current of 5 amperes. Calculate the rate at which energy is changing in this element.

The time rate-of-change of energy is power. Then by equation (*1.4*),

$$p = vi = (100)(5) = 500 \text{ watts, W}$$

In the absence of a directional relationship between $v$ and $i$ there is no way of telling whether the power is flowing into or out of the element. This will be taken up in Problem 1.13.

**1.5** Given $w(t)$ and $v(t)$ as in Fig. 1-17, sketch and label the curves for $p(t)$ and $i(t)$.

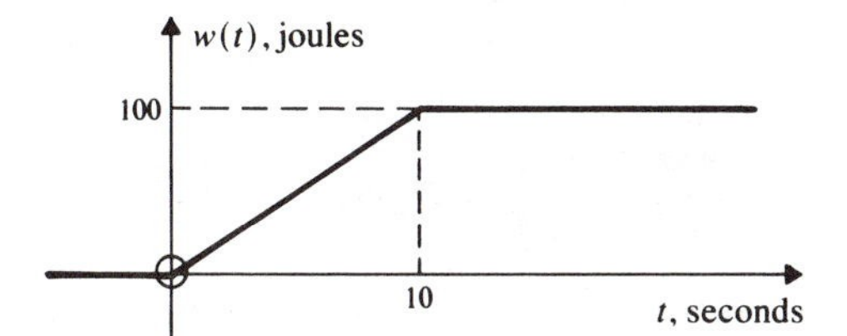


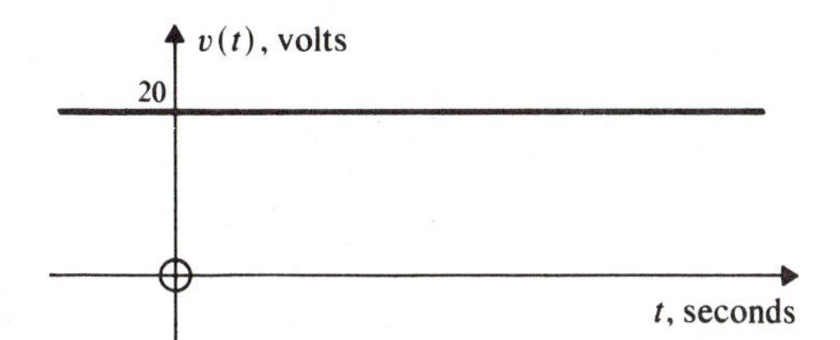


**Fig. 1-17**

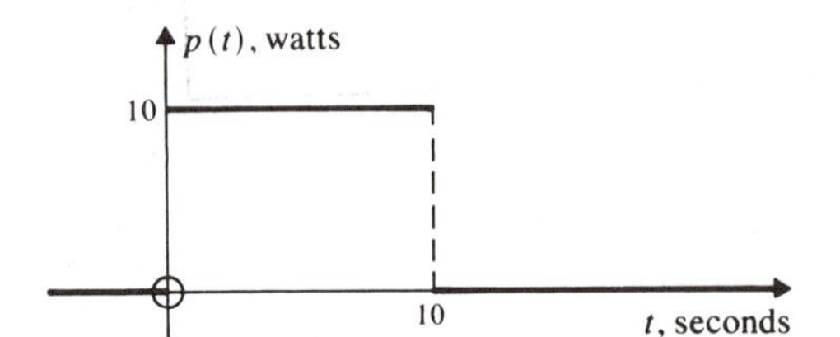


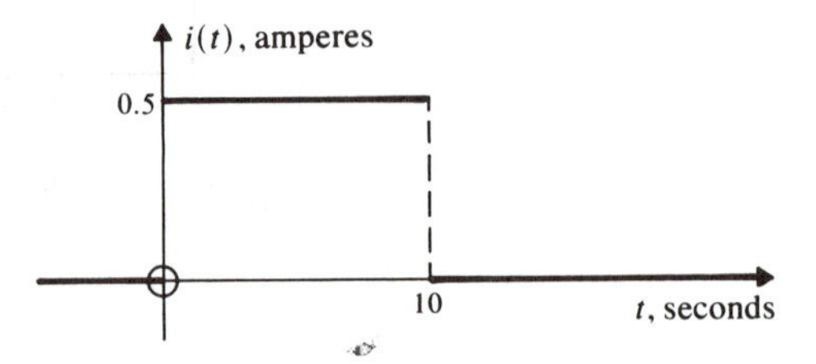


**Fig. 1-18**

From equation (*1.4*), $p(t) = dw(t)/dt$. This derivative is the *slope* of the $w(t)$ vs. $t$ curve, which is constant and equal to 10 watts from $t = 0$ to $t = 10$ seconds, and is zero elsewhere, as shown in Fig. 1-18. $i(t)$ follows at once from $p = vi$.

**1.6** Given the power curve $p(t)$ vs. $t$ in Fig. 1-19, sketch and label the curve for the work $w(t)$, if $w(0) = +100$ joules. Consider $t$ in the range $t \geq 0$.

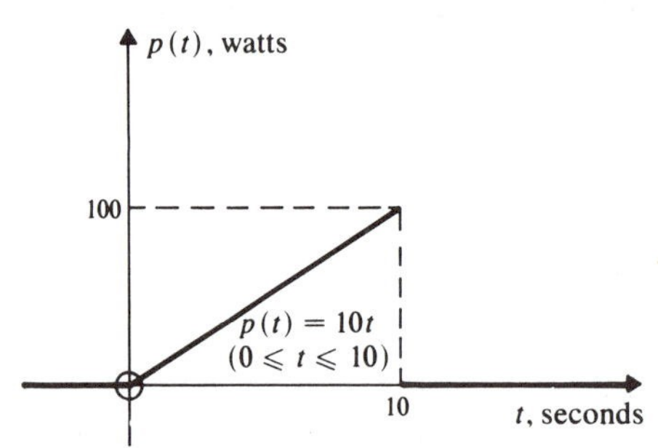


**Fig. 1-19**

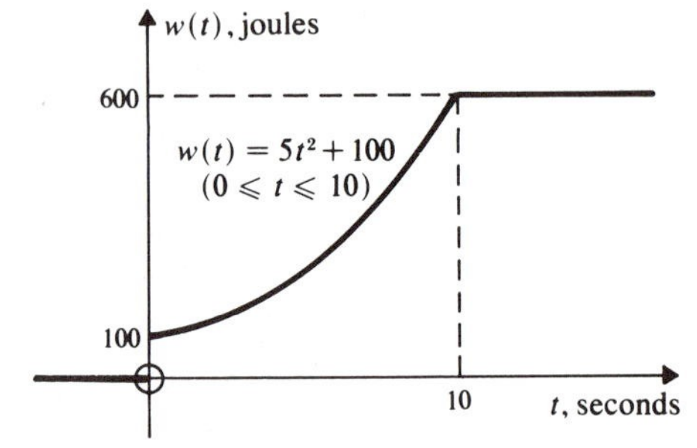


**Fig. 1-20**

From equation (*1.5*),

$$w(t) = \int_0^t p(\tau)\, d\tau + w(0) = \int_0^t 10\tau\, d\tau + 100 = 5t^2 + 100 \qquad \text{for } 0 \leq t \leq 10$$

and $w(t) = 600$ joules for $t \geq 10$ seconds, since $p(t) = 0$ for $t > 10$. The required result is given in Fig. 1-20.

### *Linear Passive Elements and Sources*

**1.7** (*a*) In Fig. 1-21*a* assume $i(t) = 4$ amperes. Determine $v(t)$.

(*b*) If for the same resistor the current changes to $i(t) = -7$ amperes, find $v(t)$. What is the polarity of the top end of the resistor with respect to the bottom end?

(*c*) If the current through the inductor of Fig. 1-21*b* is $i(t) = 2t$, find $v(t)$.

(*d*) If in the same inductor $i(t) = -2$ amperes, find $v(t)$.

(e) If the current in the inductor is $i(t) = 4e^{-t}$, determine the polarity of the *top* terminal with respect to the bottom.
(f) Given that $v(t) = 5e^t$ in Fig. 1-21c, what is the direction of the current $i(t)$ through the capacitor?
(g) If $v(t) = 5 \sin t$ in Fig. 1-21c, what is the direction of the current $i(t)$?

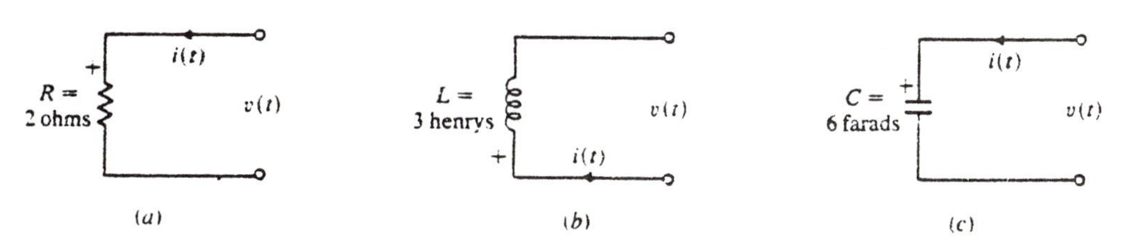


**Fig. 1-21**

Note that in all three parts of Fig. 1-21 the positive current direction enters the plus-marked end of the element, in agreement with the sign convention stated earlier. *It is important that we always assign the reference current direction and reference voltage polarity in this way.*

(a) $v_R = Ri_R$ and so $v(t) = (2)(4) = +8$ volts.
(b) Here $v(t) = -14$ volts, making the top end negative with respect to the bottom.
(c) $v_L = L\, di_L/dt$ and hence $v(t) = (3)(2) = +6$ volts. The bottom terminal is positive with respect to the top, since the polarity mark is at the bottom.
(d) $v(t) = 0$, since $di/dt = 0$.
(e) $v(t) = 3\frac{d}{dt}(4e^{-t}) = -12e^{-t}$. Since $e^{-t}$ is always positive, $v(t)$ will always be negative. That is, the bottom (plus-marked) terminal will always be negative with respect to the top. Therefore the *top* terminal must be *positive* with respect to the bottom!
(f) $i_C(t) = C\, dv_C/dt = +30e^t$. That is, the current will flow downward through the capacitor.
(g) $i(t) = 30 \cos t$ oscillates between positive and negative values. Thus the current will flow upward part of the time and downward the rest.

**1.8** Determine and sketch the current $i(t)$ in the capacitor of Fig. 1-22a when the voltage $v(t)$ is as shown in Fig. 1-22b.

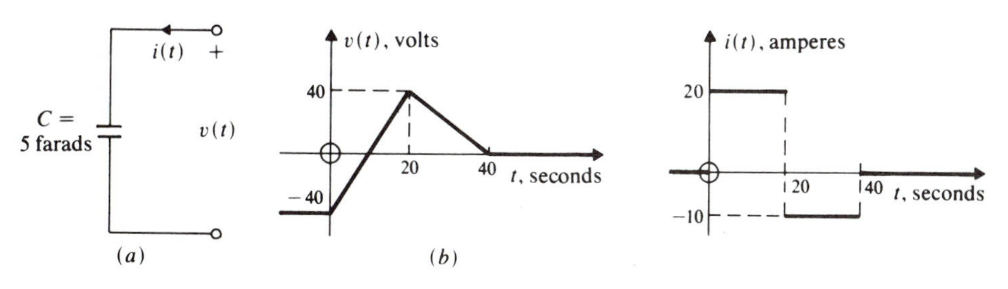


**Fig. 1-22** **Fig. 1-23**

The derivative of the voltage is the *slope* of the voltage vs. time curve, and $i = C\,dv/dt$. Thus for $0 < t < 20$, $dv/dt = 4$ and $i = (5)(4) = 20$ amperes. Similarly, for $20 < t < 40$, $dv/dt = -2$ and $i = (5)(-2) = -10$ amperes. The slope and therefore the current are zero elsewhere. The required function is graphed in Fig. 1-23.

**1.9** Given that the current in the capacitor of Fig. 1-21*c* is $i(t) = 18e^{-3t}$ for $t \geq 0$, and that $v(0) = -17$ volts, find the equation describing $v(t)$ for $t \geq 0$.

$$v(t) = \frac{1}{C}\int_0^t i(\tau)\,d\tau + v(0) = \frac{1}{6}\int_0^t 18e^{-3\tau}\,d\tau - 17$$

$$= \frac{1}{6}\,\frac{18e^{-3\tau}}{-3}\bigg|_0^t - 17 = -e^{-3t} - 16$$

*Alternative Solution*

Using the indefinite integral,

$$v(t) = \frac{1}{6}\int 18e^{-3t}\,dt = -e^{-3t} + K$$

But at time $t = 0$, $v(0) = -17$ volts. Substituting into the above, $-17 = -e^{-0} + K$ or $K = -16$ and so $v(t) = -e^{-3t} - 16$.

**1.10** Determine and sketch the current waveform $i(t)$ in the inductor of Fig. 1-24*a*, given that the applied voltage $v(t)$ is as shown in Fig. 1-24*b* and that $i(0) = -2.5$ amperes.

For $0 < t < 10$, $v(t) = 5$ volts, and using equation (*1.7*),

$$i(t) = \frac{1}{L}\int_0^t v(\tau)\,d\tau + i(0) = \frac{1}{10}\int_0^t 5\,d\tau - 2.5 = 0.5t - 2.5$$

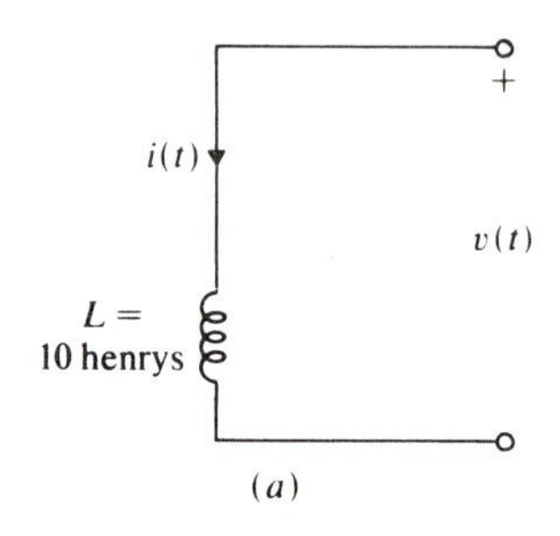


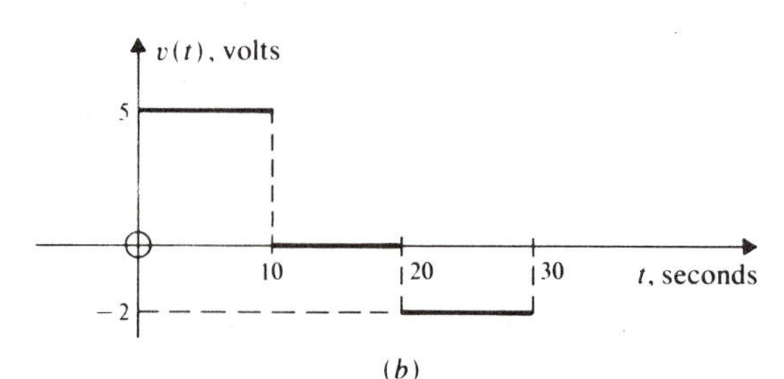


**Fig. 1-24**

Now the integrand of (*1.7*) is zero for $10 < t < 20$, hence $i(t)$ cannot change, and $i(10) = i(20) = 2.5$ amperes.

Finally, for $20 < t < 30$,

$$i(t) = \frac{1}{10}\int_{20}^t -2d\tau + 2.5 = -0.2t + 6.5$$

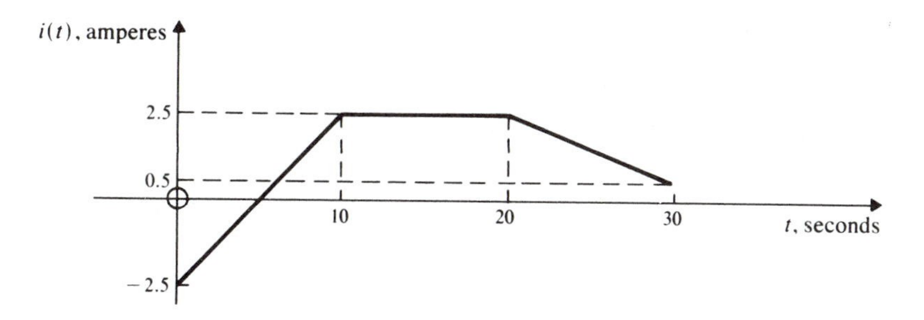


**Fig. 1-25**

The sketch of the waveform over the entire period $0 < t < 30$ is shown in Fig. 1-25.

*Note:* Although $v(t)$ is discontinuous, it can be seen from equation (*1.7*) that $i(t)$ *cannot* be discontinuous for *finite* $v(t)$. Thus although $v(t)$ is not defined at $t = 10$ seconds, for example, $i(t)$ is well-defined at that value of time.

### *Alternative Solution (Graphical)*

Equation (*1.7*) may be interpreted graphically: $\int_0^t v(\tau)\, d\tau$ is the area under the $v(\tau)$ vs. $\tau$ curve between the times 0 and $t$. Since this curve is identical to the $v(t)$ vs. $t$ curve, we find for example that

$$\int_0^{10} v(\tau)\, d\tau = \text{the area under the } v(t) \text{ curve between } t = 0 \text{ and } 10 \text{ seconds}$$
$$= (5)(10) = 50$$

Hence $i(10) = \frac{1}{10} \int_0^{10} v(\tau)\, d\tau - 2.5 = (\frac{1}{10})(50) - 2.5 = 2.5$. The $i(t)$ curve in Fig. 1-25 could be constructed by computing a number of points in this way.

**1.11** Which of the following statements are correct?

(*a*) The terminal voltage of an ideal, constant voltage source falls steadily as the current through the source increases.

(*b*) The voltage between the terminals of an ideal, constant current source is always zero.

(*c*) The voltage $v_R(t)$ in Fig. 1-26*a* is independent of the value of $R$.

(*d*) In Fig. 1-26*b*, $v_5(t) = 20$ volts.

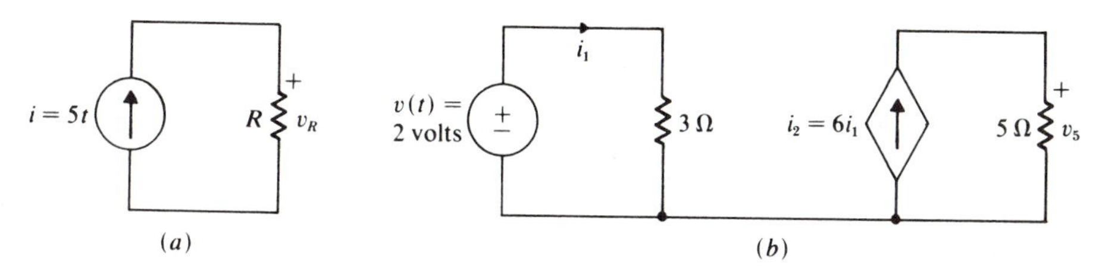


**Fig. 1-26**

(*a*) Incorrect. The terminal voltage of an *ideal* voltage source is *independent* of the current flowing through it.
(*b*) Incorrect. The terminal voltage of an ideal current source depends on the circuit into which it is connected; see, for example, the solution of part (*c*).
(*c*) Incorrect. The current through $R$ is $i(t) = 5t$, hence $v_R(t) = Ri = 5Rt$. That is, the voltage across the current source clearly depends on $R$.
(*d*) Correct. The voltage $v(t)$ is connected across the 3-ohm resistor making $i_1 = 2/3$ ampere. Thus $i_2 = 6i_1 = (6)(2/3) = 4$ amperes. Finally, $i_2$ flows through the 5-ohm resistor and $v_5 = 5i_2 = 20$ volts.

**1.12** Find the equation for $v_1(t)$ in the circuit of Fig. 1-27.

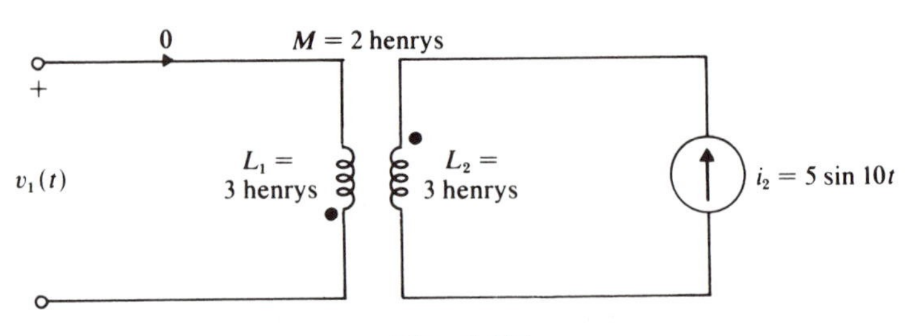


**Fig. 1-27**

The current $i_2$ enters a *dotted* terminal. Therefore the induced voltage relative to the *bottom* (dotted) terminal of the left-hand coil is

$$+M\frac{di_2}{dt} = 2\frac{d}{dt}(5 \sin 10t) = 100 \cos 10t$$

Hence relative to the polarity reference mark, $v_1(t) = -100 \cos 10t$.

**1.13** A current $i(t) = 2 \sin 3t$ flows in a 4-henry inductor. Determine the equations for the voltage $v(t)$ and the power delivered $p(t)$.

First, the current reference direction and the reference voltage polarity must be assigned (see Fig. 1-28*a*). Then

$$v(t) = L\frac{di}{dt} = 24 \cos 3t \qquad \text{(see Fig. 1-28}b\text{)}$$

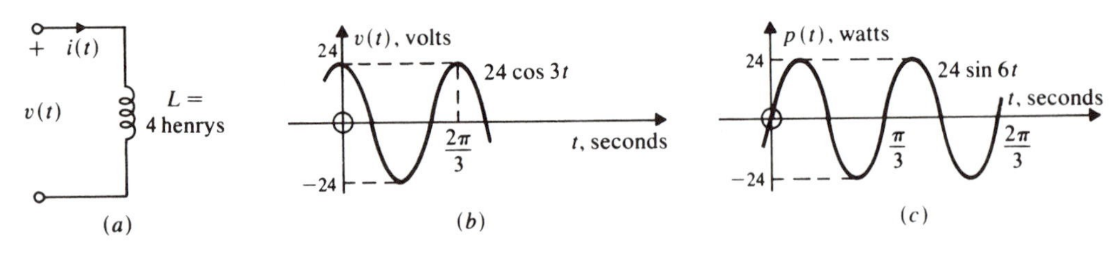


**Fig. 1-28**

and

$$p = vi = (24 \cos 3t)(2 \sin 3t) = \{24 \sin (3t + 90°)\}(2 \sin 3t)$$
$$= \tfrac{48}{2}\{\cos (+90°) - \cos (6t + 90°)\} = 24 \sin 6t \quad \text{(see Fig. 1-28}c\text{)}$$

Power is absorbed by the inductor between $t = 0$ and $\pi/6$ second, but the same power is given up from $t = \pi/6$ to $t = \pi/3$ seconds. Thus no net power is absorbed over one complete cycle: the power flows cyclically in and then out again.

### *Time-varying and Nonlinear Elements*

**1.14** Suppose that the capacitance of the motor-driven capacitor of Fig. 1-29*a* varies sinusoidally with time: $C(t) = C_0(1 + \sin \omega t)$. If the voltage impressed across this capacitor is $v(t) = V$, a constant, find the equation for the current $i(t)$.

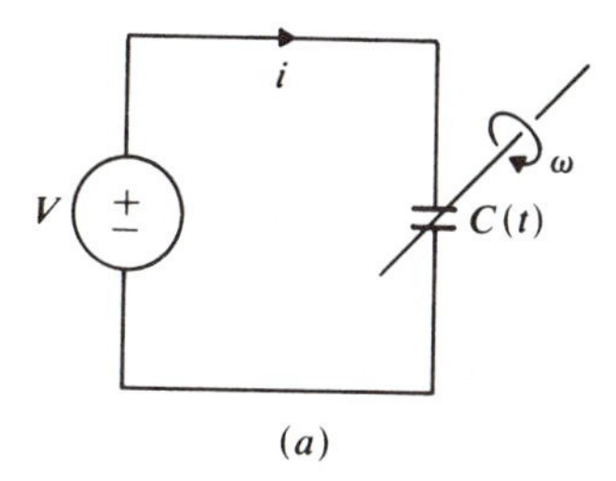


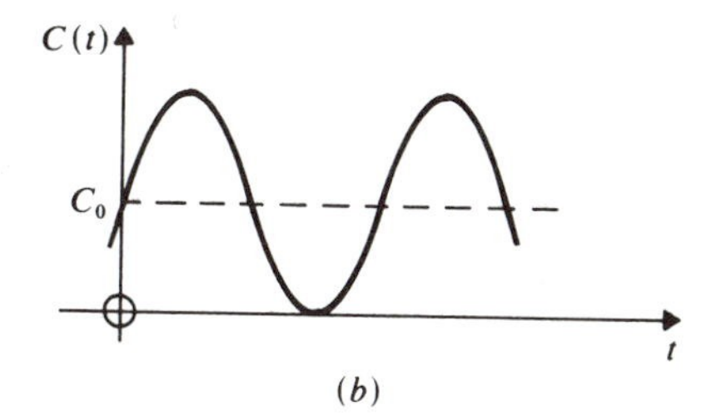


**Fig. 1-29**

Expanding equation (*1.11*), $i = \dfrac{dq}{dt} = \dfrac{d}{dt}(Cv)$, we obtain

$$i(t) = C(t)\frac{dv(t)}{dt} + v(t)\frac{dC(t)}{dt}$$
$$= C_0(1 + \sin \omega t)\frac{dV}{dt} + V\frac{d}{dt}\{C_0(1 + \sin \omega t)\}$$
$$= V\omega C_0 \cos \omega t$$

**1.15** To simplify analysis, the hysteresis loop for a power transformer (Fig. 1-30*a*) has been approximated by the $\lambda$ vs. $i$ curve of Fig. 1-30*b*. Find and sketch $v(t)$ vs. $t$ for $i(t) = 20 \sin 120\pi t$.

From equation (*1.14*), $v(t) = \dfrac{d\lambda(i)}{di}\dfrac{di}{dt}$. Here,

$$\frac{di}{dt} = 2400\pi \cos 120\pi t \qquad \text{and} \qquad \frac{d\lambda(i)}{di} = \begin{cases} 0 & \text{for } |i| > 10 \\ 0.10 & \text{for } |i| < 10 \end{cases}$$

Hence

$$v(t) = \begin{cases} 240\pi \cos 120\pi t & \text{for } |i| < 10 \\ 0 & \text{for } |i| > 10 \end{cases}$$

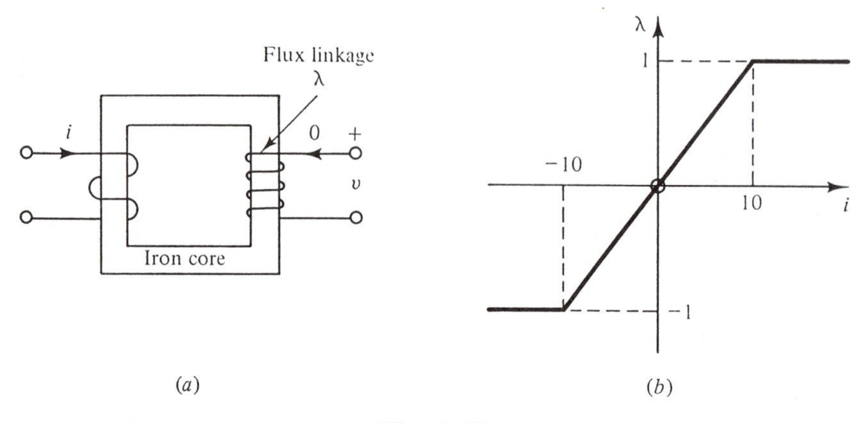


**Fig. 1-30**

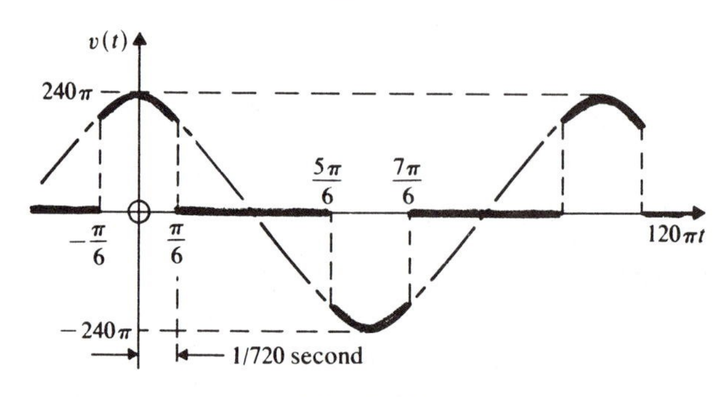


**Fig. 1-31**

But $|i| > 10$ when $\sin 120\pi t > 0.5$, i.e. $-\pi/6 < 120\pi t < \pi/6, 5\pi/6 < 120\pi t < 7\pi/6$, etc. The required sketch of $v(t)$ now follows as the solid curve in Fig. 1-31.

**1.16** A nonlinear resistor is described by

$$v = R(i)i = 2i + 3i^2$$

If the current flowing in this resistor is given by $i = 2 \cos 2\pi 60t$, determine the true frequencies (hertz) present in the impressed voltage $v$. (*Note:* $\cos^2 x = \frac{1}{2}(1 + \cos 2x)$.)

Substituting into the relation $v = 2i + 3i^2$,

$$v = 2(2 \cos 2\pi 60t) + 3(2 \cos 2\pi 60t)^2$$

$$= 4 \cos 2\pi 60t + 3(2)^2\left(\frac{1 + \cos 2\pi 120t}{2}\right)$$

$$= 6 + 4 \cos 2\pi 60t + 6 \cos 2\pi 120t$$

Thus there are two frequencies present in $v$, $f = 60$ hertz and $f = 120$ hertz. There is also a dc voltage present which is often thought of as a sinusoidal voltage of zero frequency!

**1.17** A nonlinear, time-invariant capacitor has the charge-voltage characteristic $q = \frac{1}{2}v^2$. If the voltage is $v(t) = e^{-t}$ for $t \geq 0$, find (*a*) the expression for $i(t)$, and then (*b*) the change in stored energy from $t = 0$ to $t = t_1$, where $t_1 > 0$.

(*a*) From equation (*1.13*),

$$i(t) = \frac{dq(v)}{dv}\frac{dv}{dt} = v(-e^{-t}) = -(e^{-t})(e^{-t}) = -e^{-2t}$$

(*b*) From equation (*1.5*),

$$w(t_1) - w(0) = \int_0^{t_1} p(\tau)\,d\tau = \int_0^{t_1} v(\tau)i(\tau)\,d\tau$$

$$= \int_0^{t_1} (e^{-\tau})(-e^{-2\tau})\,d\tau$$

$$= \frac{e^{-3\tau}}{3}\bigg|_0^{t_1} = \frac{1}{3}(e^{-3t_1} - 1)$$

The energy change is negative for $t_1 > 0$, so that energy is given up by the capacitor as the voltage falls.

### The Operational Amplifier

**1.18** The circuit of a *voltage follower* is shown in Fig. 1-32. Replace the op amp (assumed ideal) by its equivalent circuit (see Fig. 1-4*b*), and show that $v_o = v$.

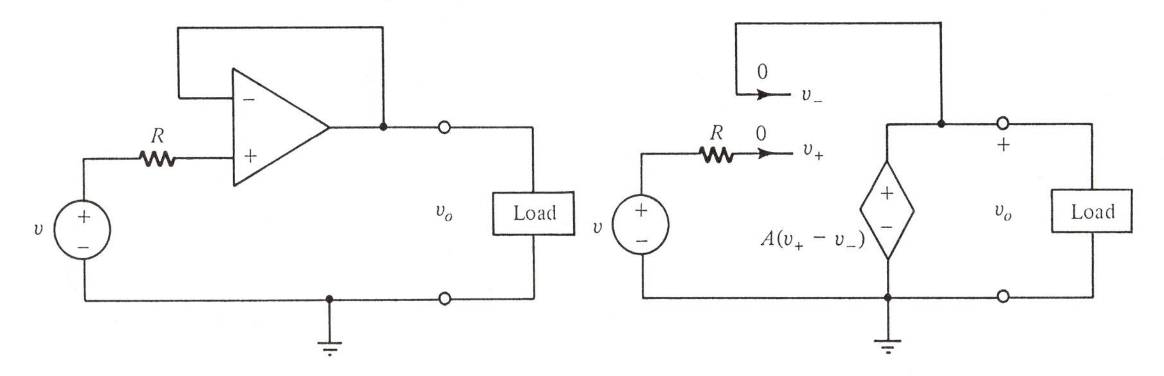


**Fig. 1-32**

Here $v_+$, the voltage at the op amp's + input, is equal to $v$, since no current flows through $R$ to cause a voltage drop. Also $v_-$, the voltage at the op amp's − terminal, is clearly equal to $v_o$. So we may write

$$v_o = A(v_+ - v_-) = A(v - v_o)$$

whence $v_o = Av/(A + 1) = v$ for $A \to \infty$.

Comment: The voltage follower circuit can accurately maintain $v_o = v$, while delivering current to a load and without drawing current from the voltage source $v$. Because of this *isolating* or *buffering* property, the voltage follower is much used.

### *Characteristics of Waveforms*

**1.19** A unit step function of voltage is impressed on a linear inductor of $L$ henrys. Calculate and sketch the current in the inductor, $i_L(t)$, given that $i_L(0) = 0$.

Recalling equation (*1.7*), and that $\mathbb{1}(t) = 1$ for $t > 0$,

$$i_L(t) = \frac{1}{L}\int_0^t v_L(\tau)\, d\tau + i_L(0) = \frac{1}{L}\int_0^t 1\, d\tau + 0 = \frac{1}{L}t, \qquad t \geq 0$$

For $t < 0$, $v_L(t) = 0$ and consequently equation (*1.7*) requires that $i_L(t)$ be constant; and the constant must be zero since $i_L(0) = 0$. Note also that although $v_L(t)$ is discontinuous, equation (*1.7*) requires that $i_L(t)$ be continuous, for finite $v_L(t)$.

The required curve of $i_L(t)$ vs. $t$ is plotted in Fig. 1-33. In the same figure it can be seen, as a graphical check, that $v_L(t) = L\, di_L/dt$, where $di_L(t)/dt$ is the slope of the $i_L(t)$ curve.

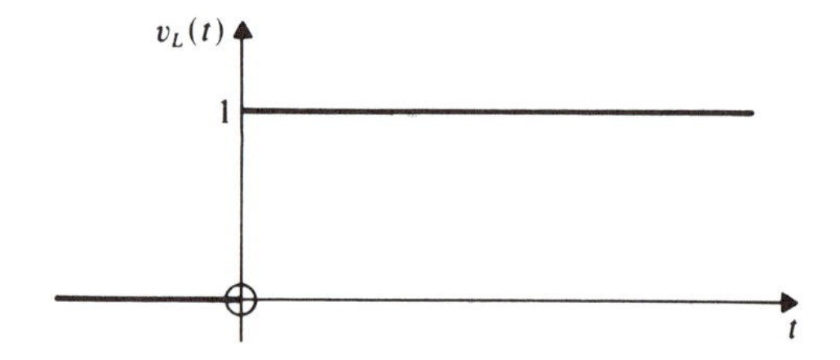


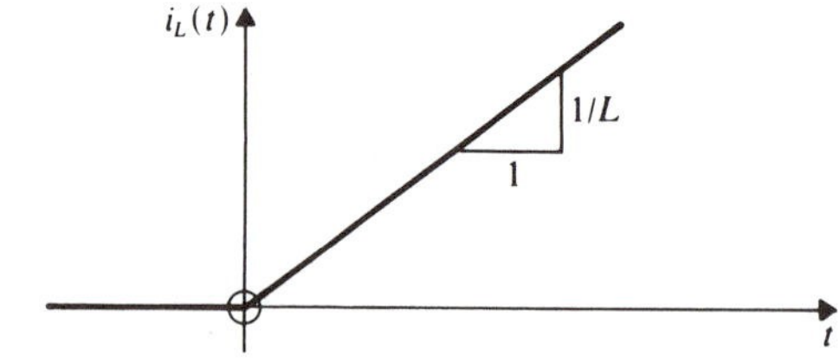


**Fig. 1-33**

**1.20** (*a*) Sketch the shifted (and negative) unit function, $-\mathbb{1}(t - a)$, where $a$ is positive.

(*b*) Add the functions $\mathbb{1}(t)$ and $-\mathbb{1}(t - a)$ graphically, to build a rectangular pulse, $p(t)$.

(*c*) Now synthesize the *unit* rectangular pulse $p_a(t)$, i.e. a pulse of unit *area*, following the same procedure as in part (*b*).

(*a*) To obtain $-\mathbb{1}(t - a)$, the unit step must be reversed in sign and moved $a$ units to the right, as in Fig. 1-34.

(*b*) If to $-\mathbb{1}(t - a)$ we now add $\mathbb{1}(t)$, the pulse $p(t)$ of Fig. 1-34 results.

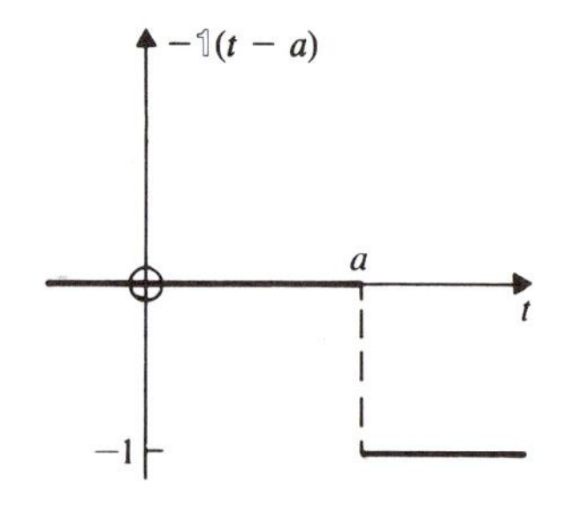


+

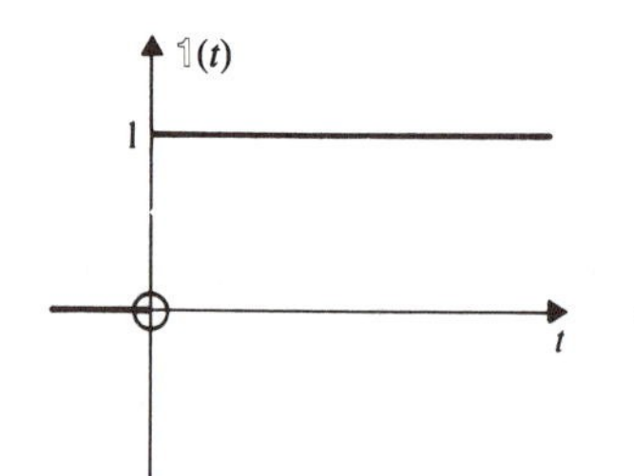


=

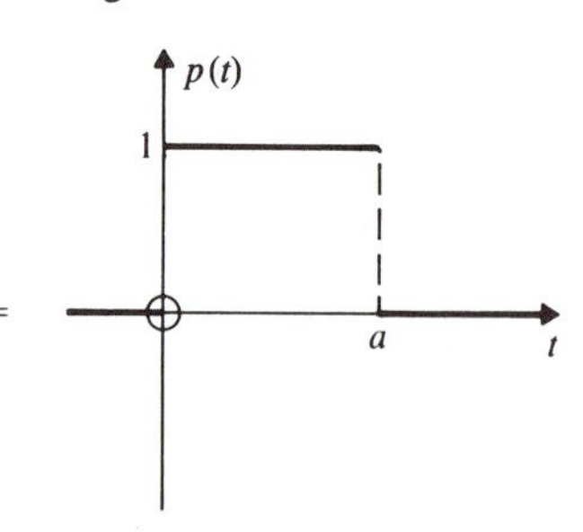


**Fig. 1-34**

(*c*) We have only to multiply the ordinates of each graph in Fig. 1-34 by the constant $1/a$ to obtain the *unit* rectangular pulse $p_a(t)$ of equation (*1.19*) and Fig. 1-10. Algebraically, $p_a(t) = \frac{1}{a}\{\mathbb{1}(t) - \mathbb{1}(t - a)\}$.

**1.21** In Problem 1.20 we constructed a unit rectangular pulse. Let us now consider the unit triangular pulse function. (*a*) First, show that the waveform in Fig. 1-35 could logically be described as a *unit* triangular pulse; and then (*b*) construct the unit triangular pulse from three appropriately shifted and signed ramp functions.

(*a*) The *area* under the graph of $p_t(t)$ in Fig. 1-35 is unity. Therefore, by analogy with the unit rectangular pulse, $p_t(t)$ could logically be called the *unit* triangular pulse.

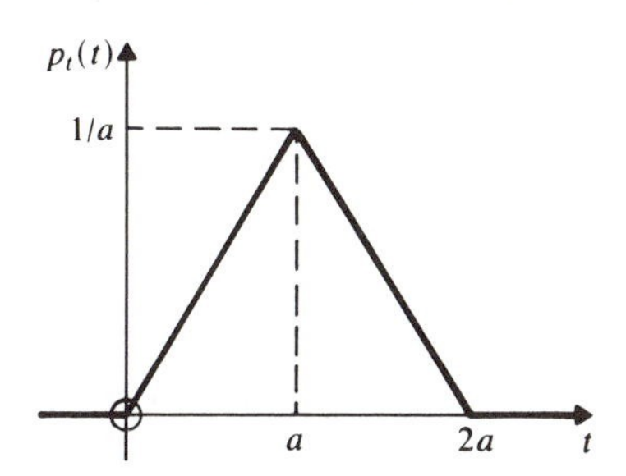


**Fig. 1-35**

(*b*) The graphical sum of the three functions $\frac{1}{a^2}r(t)$, $-\frac{2}{a^2}r(t - a)$, and $\frac{1}{a^2}r(t - 2a)$ of Fig. 1-36 can easily be seen to equal the unit triangular pulse $p(t)$ in Fig. 1-35.

Remark: As in the case of the unit rectangular pulse, we can obtain the unit impulse function $\delta(t)$ from $p_t(t)$ by allowing $a$ to approach the limit zero.

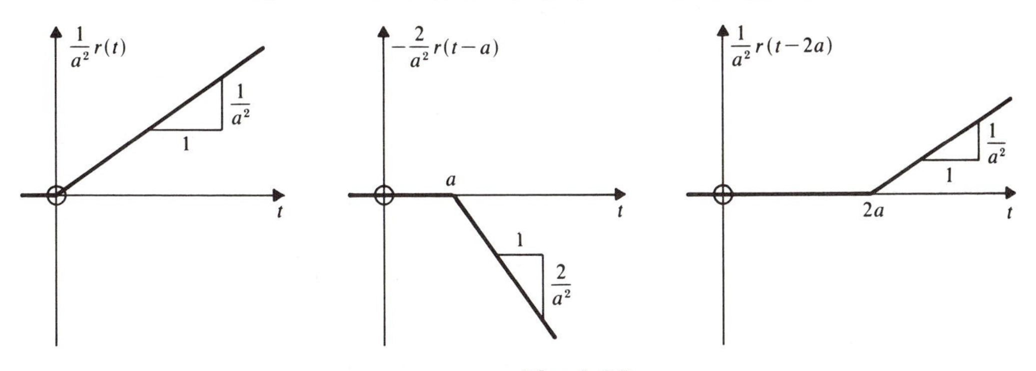


**Fig. 1-36**

**1.22** The waveform $v(t)$ shown in Fig. 1-37 is impressed on a capacitor of $C$ farads. Find and sketch the current waveform $i(t)$.

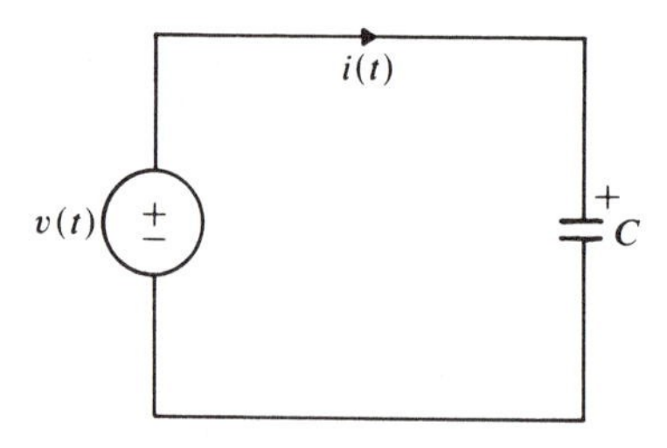


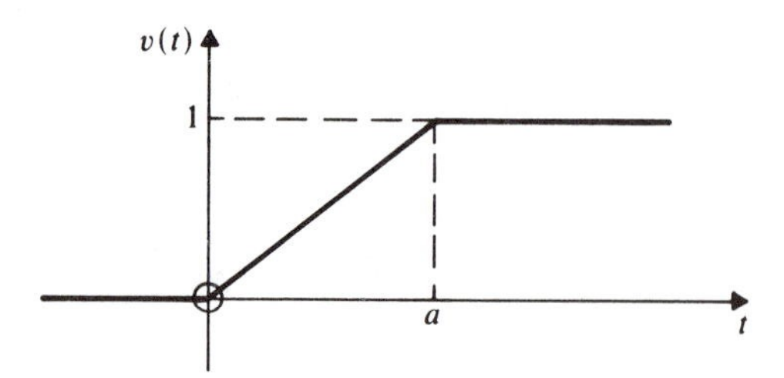


**Fig. 1-37**

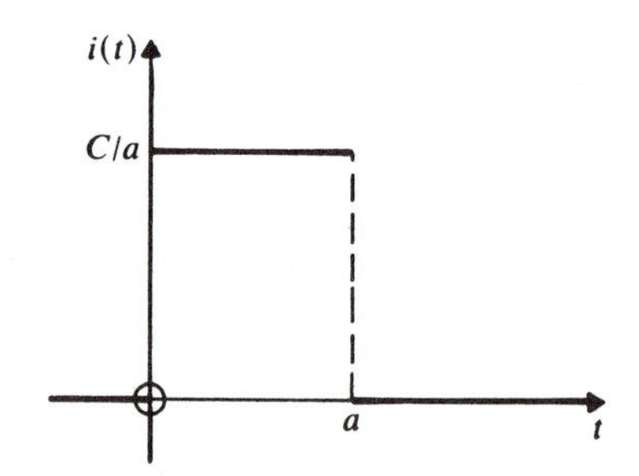


**Fig. 1-38**

Using equation (*1.8*) and differentiating the $v(t)$ waveform graphically,

$$i(t) = C\frac{dv}{dt} = \begin{cases} 0 & \text{for } t < 0 \\ C\left(\dfrac{1}{a}\right) & \text{for } 0 < t < a \\ 0 & \text{for } t > a \end{cases}$$

$i(t)$, which is undefined at $t = 0$ and $t = a$, is graphed in Fig. 1-38.

*Alternative Solution*

$v(t) = \dfrac{1}{a}\{r(t) - r(t - a)\}$ and since $dr(t)/dt = \mathbb{1}(t)$,

$$i(t) = C\frac{dv}{dt} = \frac{C}{a}\{\mathbb{1}(t) - \mathbb{1}(t - a)\},$$

which can be represented graphically by the rectangular pulse of Fig. 1-38.

**1.23** Consider the consequences of letting $a \to 0$ in problem 1.22. The voltage $v(t)$ becomes a unit step, $\mathbb{1}(t)$, but what happens to $i(t)$?

Graphically, the rectangular pulse in Fig. 1-38 has area $(C/a)(a) = C$. As $a \to 0$, therefore, it becomes an impulse function of magnitude (area) equal to $C$. That is,

$$\lim_{a \to 0} i(t) = C\delta(t)$$

Note that an impulse of current, corresponding to a sudden deposit of charge onto the capacitor, is needed to bring about an instantaneous change of capacitor voltage.

*Alternative Solution*

Since $v(t) = \mathbb{1}(t)$, and $d\mathbb{1}(t)/dt = \delta(t)$, it follows that $i(t) = C\dfrac{dv}{dt} = C\delta(t)$.

**1.24** Establish the relationships between adjacent members of the sequence $r(t)$, $\mathbb{1}(t)$, $\delta(t)$.

$r(t) = \begin{cases} 0 & t < 0 \\ t & t > 0 \end{cases}$ and $\dfrac{dr(t)}{dt} = \begin{cases} 0 & t < 0 \\ 1 & t > 0 \end{cases}$, which shows that $dr(t)/dt = \mathbb{1}(t)$.

Conversely, since $\mathbb{1}(t) = \begin{cases} 0 & t < 0 \\ 1 & t > 0 \end{cases}$, then $\displaystyle\int_{-\infty}^{t} \mathbb{1}(\tau)\, d\tau = \begin{cases} 0 & t < 0 \\ t & t > 0 \end{cases}$, i.e.

$$\int_{-\infty}^{t} \mathbb{1}(\tau)\, d\tau = r(t)$$

To differentiate the unit step function we will approximate the step as in Fig. 1-39*a*. Here $f(t) = \dfrac{1}{a}\{r(t) - r(t-a)\}$ and thus $df/dt = \dfrac{1}{a}\{\mathbb{1}(t) - \mathbb{1}(t-a)\}$. Since this derivative is a unit pulse (see Fig. 1-39*b*), we have in the limit as $a \to 0$,

$$\frac{d\mathbb{1}(t)}{dt} = \delta(t)$$

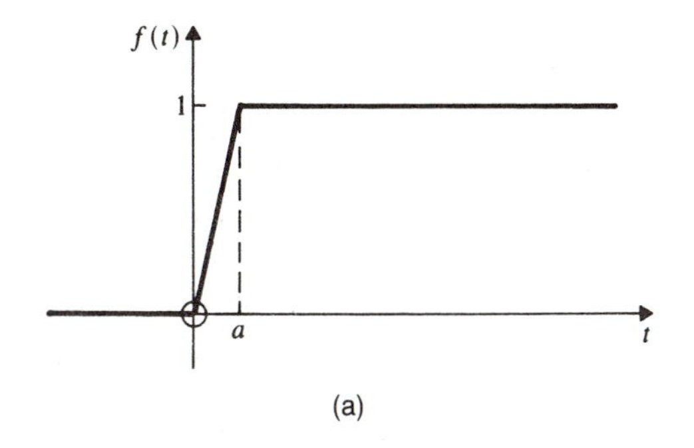


(a)

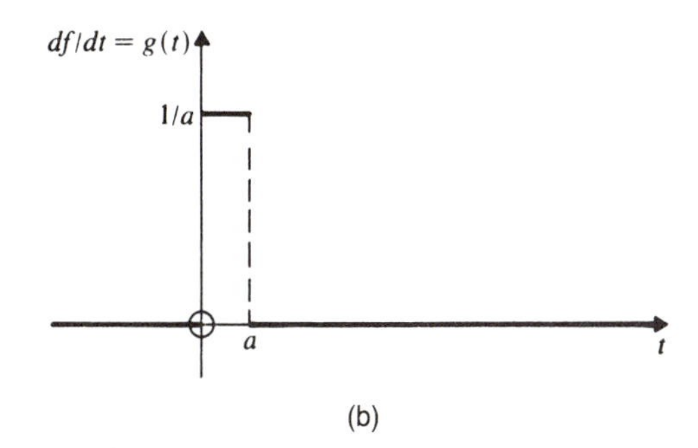


(b)

**Fig. 1-39**

Conversely, if we approximate the unit delta function by $g(t) = \dfrac{1}{a}\{\mathbb{1}(t) - \mathbb{1}(t-a)\}$, then

$$\int_{-\infty}^{t} g(\tau)\, d\tau = \begin{cases} 0 & t < 0 \\ \dfrac{1}{a}t & 0 < t < a \\ 1 & t > a \end{cases}$$

and so, in the limit as $a \to 0$, $\displaystyle\int_{-\infty}^{t} \delta(\tau)\, d\tau = \mathbb{1}(t)$.

**1.25** When a step function of voltage $v(t) = A\mathbb{1}(t)$ is applied to a linear circuit, the current response for $t > 0$ is found to be

$$i(t) = \tfrac{1}{2}e^{-t} - 2t + 3\sin 4t$$

Determine the current response if $v(t) = A\delta(t)$.

Consider a typical differential equation relating $v(t)$ to $i(t)$,

$$\frac{d^2 i}{dt^2} + 2\frac{di}{dt} + 3i = v(t)$$

Suppose we wanted the response due to an input $dv/dt$, instead of to $v(t)$ itself. If we differentiate the above equation,

$$\frac{d^2}{dt^2}\left(\frac{di}{dt}\right) + 2\frac{d}{dt}\left(\frac{di}{dt}\right) + 3\frac{di}{dt} = \frac{dv}{dt}$$

That is, differentiation of the input $v(t)$ results in differentiation of the response $i(t)$.
In this problem, $v(t) = A\mathbb{1}(t)$ and $dv/dt = A\delta(t)$. Therefore, the response due to $A\delta(t)$ must be the derivative of the response due to $A\mathbb{1}(t)$,

$$i(t) = \frac{d}{dt}\left(\tfrac{1}{2}e^{-t} - 2t + 3\sin 4t\right) = -\tfrac{1}{2}e^{-t} - 2 + 12\cos 4t \qquad t > 0$$

**1.26** The current response of a circuit element to a unit step of voltage $v(t) = \mathbb{1}(t)$ is $\frac{1}{L}r(t)$. Find the response due to (*a*) $v(t) = 4\delta(t)$ and (*b*) $v(t) = 7r(t)$.

(*a*) If $v(t) = 4\delta(t)$, the response $i(t)$ will be 4 times the *derivative* of the response due to a unit step. That is

$$i(t) = 4\frac{d}{dt}\left\{\frac{1}{L}r(t)\right\} = \frac{4}{L}\mathbb{1}(t)$$

Since the derivative of $r(t)$ is undefined at $t = 0$, $i(t)$ is also undefined. There is in fact a clear discontinuity in $i(t)$ at $t = 0$.

(*b*) If $v(t) = 7r(t)$, the response $i(t)$ will be 7 times the *integral* of the response due to a unit step.

As shown in Problem 1.24, the integral must run between the limits $-\infty$ and $t$. Thus

$$i(t) = 7\int_{-\infty}^{t} \frac{1}{L}r(\tau)\,d\tau = \begin{cases} 7\displaystyle\int_{-\infty}^{t} \frac{1}{L}(0)\,d\tau = 0 & t < 0 \\[2ex] 0 + 7\displaystyle\int_{0}^{t} \frac{1}{L}\tau\,d\tau = \frac{7}{2L}t^2 & t > 0 \end{cases}$$

In this case, the current $i(t)$ is continuous at $t = 0$.

Remarks: The circuit element in this problem can be identified as an inductor of $L$ henries. Further, in the presence of discontinuities we must be more careful about initial conditions. Thus in part (*a*), $i(0)$ when approached from the left (i.e. negative $t$) should be written $i(0^-) = 0$. When approached from the right, we must write $i(0^+) = 4/L$.

**1.27** A sinusoidal voltage $v_C(t) = V_m \cos(\omega t + \beta)$ is impressed on a linear capacitor of $C$ farads. Find the zero-to-peak amplitude of the capacitor current, and the phase angle of the capacitor *voltage* with respect to the *current*.

Using equation (*1.8*)

$$i_C(t) = C\frac{dv_C(t)}{dt} = C\frac{d}{dt}\{V_m \cos(\omega t + \beta)\}$$

$$= -V_m C\omega \sin(\omega t + \beta) = V_m C\omega \cos(\omega t + \beta + \pi/2)$$

since $-\sin x = \cos(x + \pi/2)$.

Thus we find the zero-to-peak amplitude of the capacitor current to be $V_m C\omega$, while the phase angle of the capacitor *voltage* with respect to the *current* is $+\beta - (\beta + \pi/2) = -\pi/2$ radians. We say that the sinusoidal voltage across a capacitor *lags* the current by 90°.

**1.28** A sinusoidal current $i(t) = 5\cos(2t - \pi/6)$ flows in a linear inductor of 0.10 henry.

(*a*) Find the value of $t$ at which $i(t)$ goes through its first maximum after $t = 0$. What is the value of $i(t)$ at this maximum?
(*b*) Determine the value of time where $i(t)$ first becomes zero, after $t = 0$.
(*c*) What is the equation for the voltage across the inductor?
(*d*) Sketch $i(t)$ vs. $t$.

(*a*) The current $i(t)$ will reach its first maximum when $2t - \pi/6 = 0$ or when $t = \pi/12$ second. At this point the value of $i(t)$ will be 5 amperes—the zero-to-peak amplitude.
(*b*) For the first zero value, we require $2t - \pi/6 = \pi/2$ or $t = \pi/3$ second.
(*c*) $v(t) = L\frac{di}{dt} = 0.1\frac{d}{dt}\{5\cos(2t - \pi/6)\} = -\sin(2t - \pi/6) = \cos(2t + \pi/3)$
(*d*) Using the above data, $i(t)$ can be graphed as in Fig. 1-40.

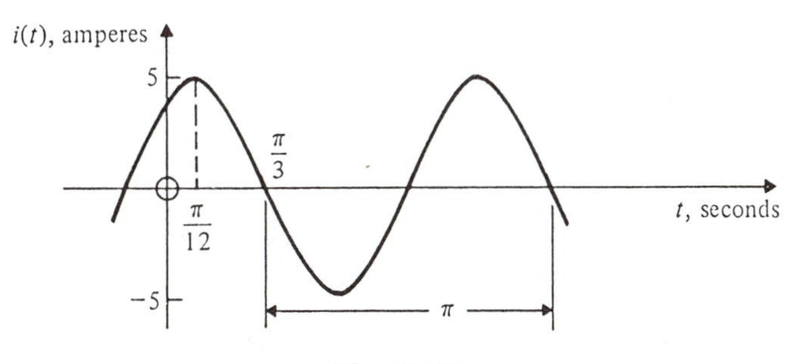


**Fig. 1-40**

### *Average and Effective Values of Repetitive Waveforms*

**1.29** Find the average value of the voltage waveforms shown in Fig. 1-41.

In Fig. 1-41*a*, $v(t) = A\sin\omega t = A\sin(2\pi/T)t$. Using equation (*1.22*), and choosing for convenience $t_0 = 0$,

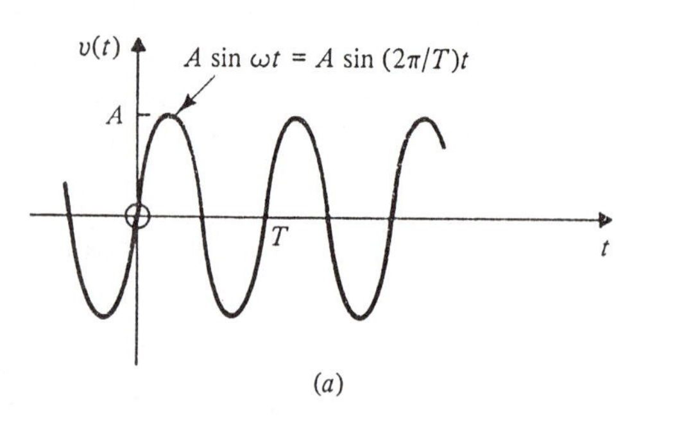

(a)

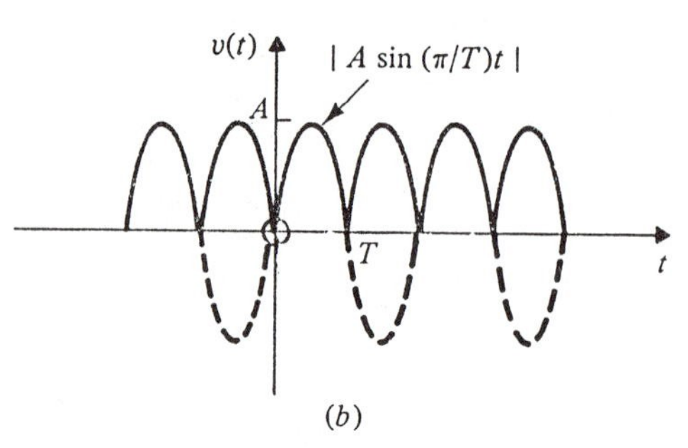

(b)

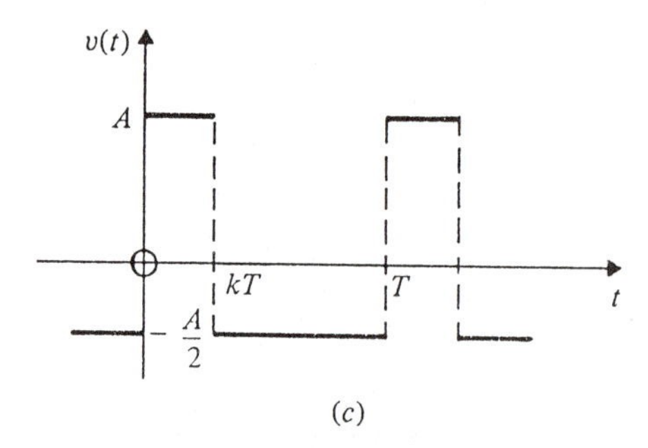

(c)

**Fig. 1-41**

$$V_{\text{av}} = \frac{1}{T}\int_{t_0}^{t_0+T} v(\tau)\,d\tau = \frac{1}{T}\int_0^T A\sin(2\pi/T)\tau\,d\tau$$

$$= \frac{1}{T}\left[-\frac{A\cos(2\pi/T)\tau}{2\pi/T}\right]_0^T = \frac{-A}{2\pi}\left[\cos(2\pi/T)\tau\right]_0^T = 0$$

Note: $\int_0^T A\sin\omega\tau\,d\tau$ represents the area under the $A\sin\omega t$ vs. $t$ curve over one period. This area is, by inspection, zero: graphical solutions are often quicker and less error-inducing than formal integrations.

In Fig. 1-41*b*, $v(t) = |A\sin(\pi/T)t|$. Thus

$$V_{\text{av}} = \frac{1}{T}\int_0^T A\sin(\pi/T)\tau\,d\tau = \frac{1}{T}\left[-\frac{A\cos(\pi/T)\tau}{\pi/T}\right]_0^T$$

$$= -\frac{A}{\pi}\left[\cos(\pi/T)\tau\right]_0^T = 2A/\pi \text{ volts}$$

Remark: This is the waveform produced by electronic *full-wave rectification* of a sinusoid.

For Fig. 1-41*c*,

$$V_{av} = \frac{1}{T}\left\{\int_0^{kT} A\,d\tau - \int_{kT}^T (A/2)\,d\tau\right\} = \frac{1}{T}\left\{A\tau\Big|_0^{kT} - (A/2)\tau\Big|_{kT}^{T}\right\}$$

$$= \frac{1}{T}\{AkT - (A/2)T + (Ak/2)T\} = \tfrac{1}{2}A\{3k - 1\}$$

### *Alternative Solution*

Using graphical integration,

$$V_{av} = \frac{1}{T}\int_0^T v(\tau)\,d\tau = \frac{1}{T}\{A(kT) - (A/2)(T - kT)\} = \tfrac{1}{2}A\{3k - 1\}$$

**1.30** Find the effective value of the voltage waveforms shown in Fig. 1-42.

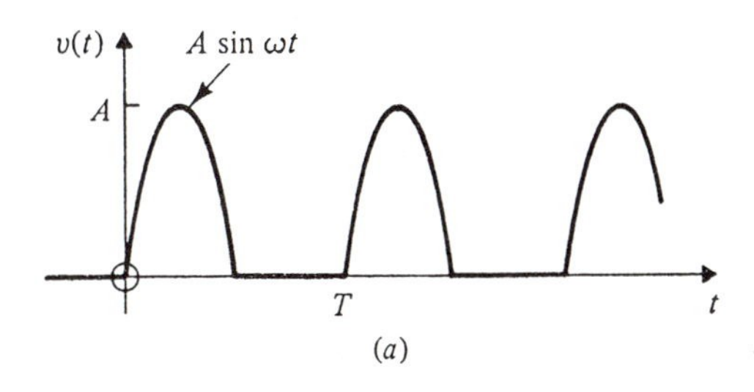


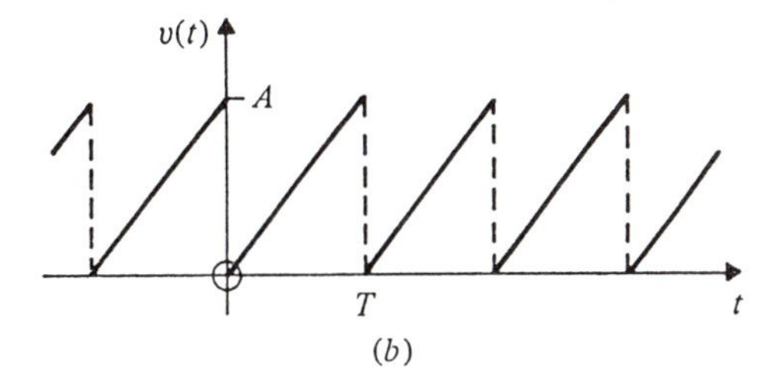


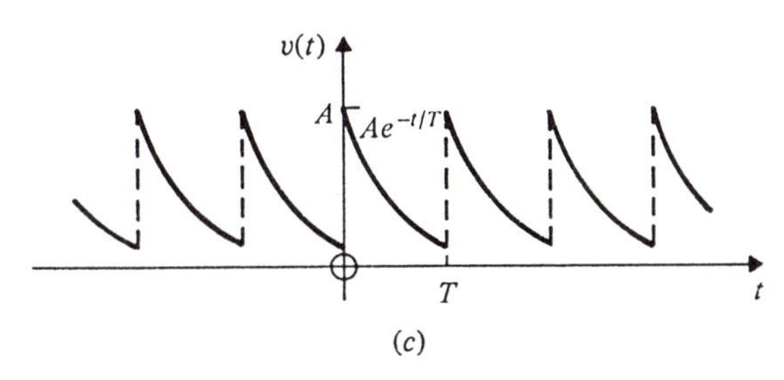


**Fig. 1-42**

Applying equation (*1.24*) to the waveform of Fig. 1-42*a*,

$$V_{eff}^2 = \frac{1}{T}\left\{\int_0^{T/2} (A \sin \omega\tau)^2\,d\tau + \int_{T/2}^{T} (0)^2\,d\tau\right\} = \frac{1}{T}\int_0^{T/2} \frac{A^2}{2}(1 - \cos 2\omega\tau)\,d\tau$$

$$= \frac{A^2}{2T}\left[\tau - \frac{\sin 2\omega\tau}{2\omega}\right]_0^{T/2} = \frac{A^2}{2T}\left[\tau - \frac{\sin (2\pi/T)\tau}{2\omega}\right]_0^{T/2} = \frac{A^2}{2T}\frac{T}{2}$$

or $V_{eff} = A/2$ volts.

For Fig. 1-42*b*, $v(t) = (A/T)t$ for $0 < t < T$, and so

$$V_{eff}^2 = \frac{1}{T}\int_0^T \frac{A^2\tau^2}{T^2}\,d\tau = \frac{A^2}{T^3}\left[\frac{\tau^3}{3}\right]_0^T$$

or $V_{eff} = A/\sqrt{3}$ volts.

For Fig. 1-42*c*,

$$V_{eff}^2 = \frac{1}{T}\int_0^T (Ae^{-\tau/T})^2\,d\tau = \frac{A^2}{T}\int_0^T e^{-2\tau/T}\,d\tau = \frac{A^2}{2}[-e^{-2} + 1]$$

or $V_{eff} = 0.656A$ volts.

**1.31** A current $i(t) = I_m \sin \omega t$ flows through a resistor of $R$ ohms. Calculate the average power dissipated.

Using equation (*1.22*),

$$P_{av} = \frac{1}{T}\int_0^T p(\tau)\,d\tau = \frac{1}{T}\int_0^T v(\tau)i(\tau)\,d\tau$$

But $v(\tau) = Ri(\tau) = RI_m \sin \omega\tau$. Thus

$$P_{av} = \frac{1}{T}\int_0^T (RI_m \sin \omega\tau)(I_m \sin \omega\tau)\,d\tau = \frac{RI_m^2}{2T}\int_0^T (1 - \cos 2\omega\tau)\,d\tau$$

$$= \frac{RI_m^2}{2T}\left[\tau - \frac{\sin 2\omega\tau}{2\omega}\right]_0^T = \frac{RI_m^2}{2T}[T - 0] = \frac{RI_m^2}{2} = RI^2{}_{eff}$$

$i(t)$ and $p(t)$ are graphed in Fig. 1-43$a$ and $b$ respectively.

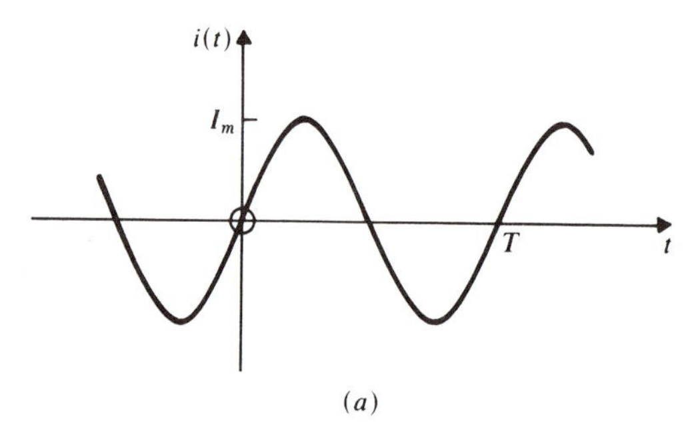


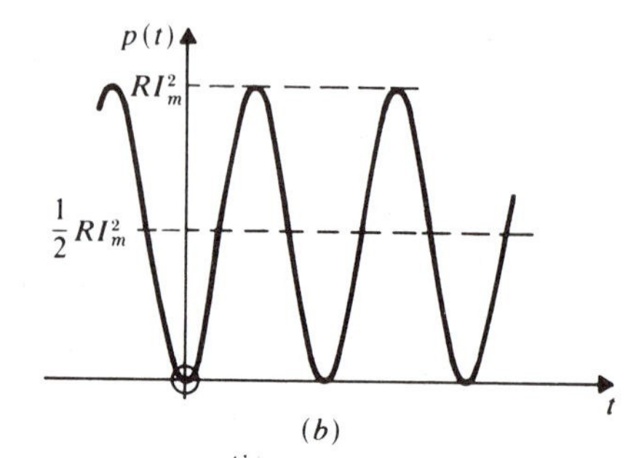


**Fig. 1-43**

**1.32** A current, having dc and ac components, is described by $i(t) = I + I_m \sin \omega t$. If this current flows through a resistor of $R$ ohms, find the average power dissipated per cycle.

From Ohm's law, $v(t) = Ri(t) = RI + RI_m \sin \omega t$. Thus from equation (*1.22*),

$$P_{av} = \frac{1}{T}\int_0^T p(\tau)\,d\tau = \frac{1}{T}\int_0^T v(\tau)i(\tau)\,d\tau$$

$$= \frac{1}{T}\int_0^T (RI + RI_m \sin \omega\tau)(I + I_m \sin \omega\tau)\,d\tau$$

$$= \frac{1}{T}\left\{\int_0^T RI^2\,d\tau + \int_0^T 2RII_m \sin \omega\tau\,d\tau + \int_0^T RI_m^2 \sin^2 \omega\tau\,d\tau\right\}$$

$$= \frac{1}{T}\{RI^2T + 0 + RI_m^2T/2\} = R(I^2 + I_m^2/2)$$

**1.33** A sinusoidal voltage $v(t) = A \sin \omega t$ is impressed across a capacitor of $C$

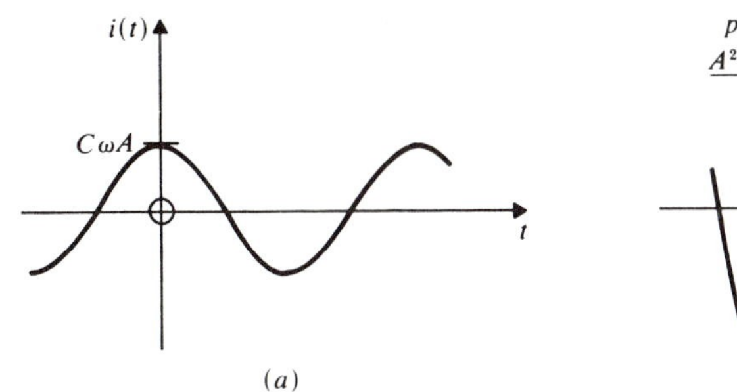


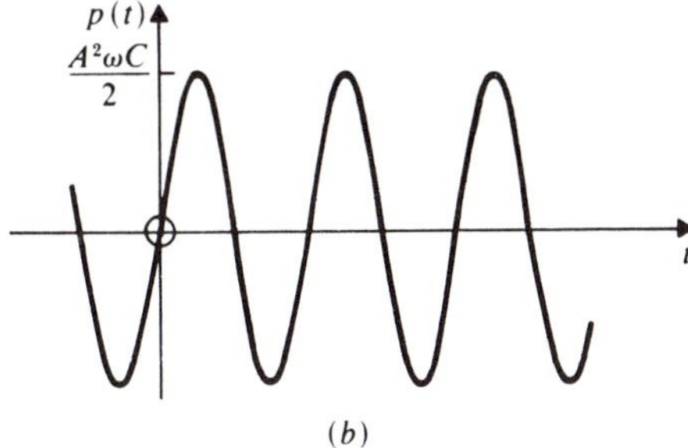


**Fig. 1-44**

farads. Determine the equations and sketch the waveforms of (*a*) the current $i(t)$, and (*b*) the power $p(t)$. Then compute $P_{av}$.

(a)
$$i(t) = C\frac{dv}{dt} = C\omega A \cos \omega t$$

(*b*)
$$\begin{aligned} p(t) &= v(t)i(t) = (A \sin \omega t)(C\omega A \cos \omega t) \\ &= A^2\omega C\{\sin \omega t\}\{\sin (\omega t + 90°)\} = \tfrac{1}{2}A^2\omega C \sin 2\omega t \end{aligned}$$

The two functions $i(t)$ and $p(t)$ are graphed in Fig. 1-44*a* and *b* respectively.

From the graph of $p(t)$ (Fig. 1-44*b*) it can be seen that $P_{av} = 0$ because the net area under the $p(t)$ vs. $t$ curve over one cycle is zero. Analytically,

$$\begin{aligned} P_{av} &= \frac{1}{T}\int_0^T p(\tau)\, d\tau = \frac{A^2\omega C}{2T}\int_0^T \sin 2\omega\tau\, d\tau = \frac{A^2\omega C}{2T}\left[-\frac{\cos 2\omega\tau}{2\omega}\right]_0^T \\ &= \frac{A^2\omega C}{2T}\left(\frac{-1+1}{2\omega}\right) = 0 \end{aligned}$$

# PROBLEMS

**1.34** The current flowing in a 2-microfarad capacitor is given by $i(t) = 10^{-3}e^{-100t}$ for $t \geq 0$, the capacitor being uncharged at $t = 0$. Show that for $t \geq 0$,
(*a*) $v(t) = 5(1 - e^{-100t})$
(*b*) $p(t) = 5 \times 10^{-3}(e^{-100t} - e^{-200t})$
(*c*) $w(t) = 10^{-5}(2.5 - 5e^{-100t} + 2.5e^{-200t})$

**1.35** A 3-millihenry inductor and its associated current are shown in Fig. 1-45. (*a*) Determine and sketch the waveforms of $v(t)$ and $p(t)$. (*b*) What is the

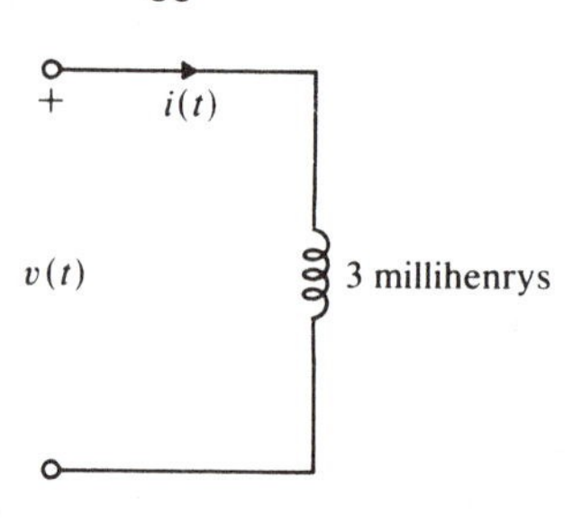


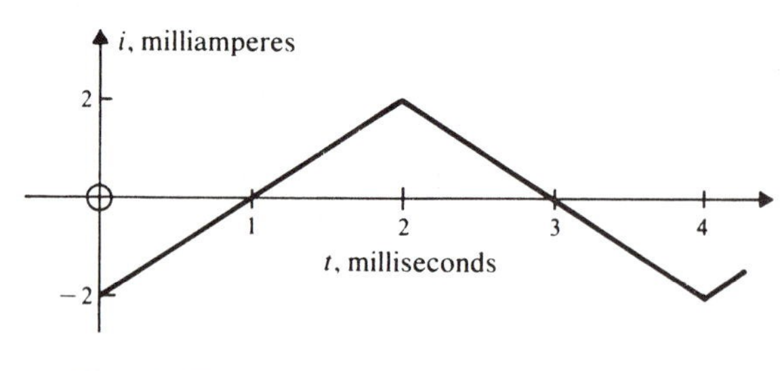


**Fig. 1-45**

frequency (in hertz) of the $i(t)$ and $v(t)$ waveforms? (*c*) What is the frequency of the power waveform?

**1.36** A 1-microfarad capacitor is initially charged to $-1.0$ volt and carries the current shown in Fig. 1-46. Calculate and graph $v_C(t)$ for $t \geq 0$.

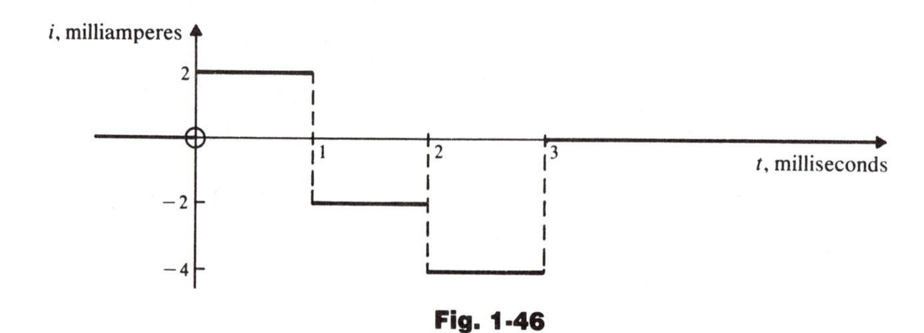


**Fig. 1-46**

**1.37** Show that the energy stored in a linear inductor is given by $w(t) = \frac{1}{2}Li^2(t)$. *Hint:* Assume an initial current (and stored energy) of zero, and calculate the work done in increasing the current to any finite value, $I$.

(We can also show that $w(t) = \frac{1}{2}Cv^2(t)$ for a capacitor.)

**1.38** (*a*) Given $i_1(t) = 10t$ in Fig. 1-47*a*, find $v_2(t)$. (*b*) In Fig. 1-47*b*, $i_1(t) = 10e^{-2t}$. Find the corresponding $v_2(t)$.

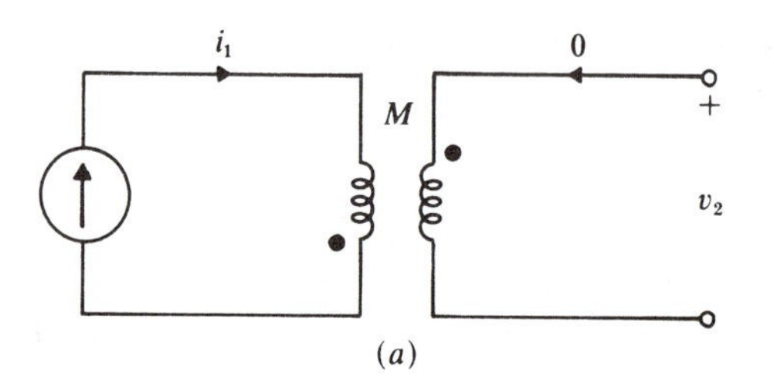


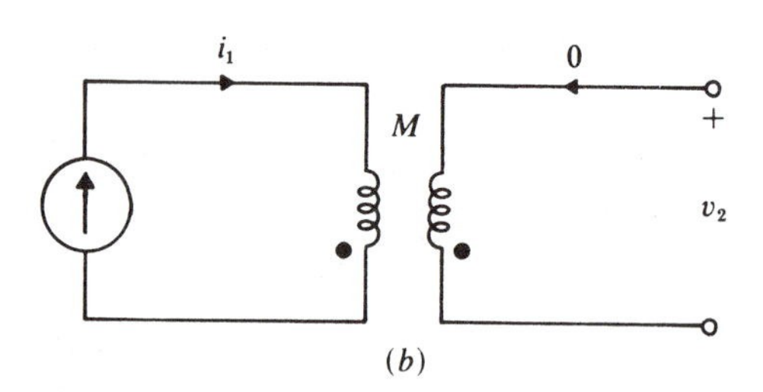


**Fig. 1-47**

**1.39** The charge and voltage in a nonlinear capacitor are related by $q = Av^{1/2}$. If $v(t) = Be^{\alpha t}$, find the power delivered $p(t)$.

**1.40** A nonlinear resistor is described by $R = v/i = 10 + |i|$. Compute the current when $v = 75$ volts.

**1.41** The two-terminal element in Fig. 1-48 is either an inductor or a capacitor. If $v(t) = 3 \cos 4t$ and $i(t) = 5 \sin 4t$, determine the type and value of the element.

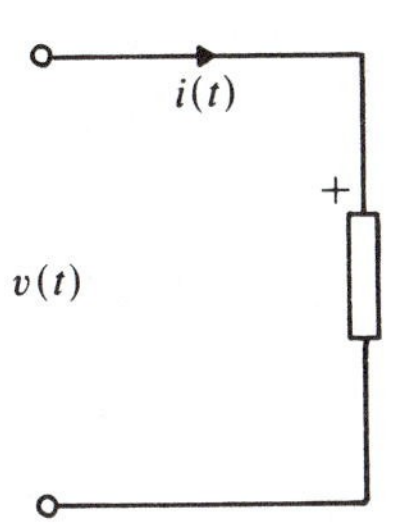


**Fig. 1-48**

**1.42** The voltage across a linear capacitor of $C$ farads is $v(t) = 1 - e^{-t}$. (*a*) Calculate the ensuing current. (*b*) Would the current be different if $v(t) = -e^{-t}$?

**1.43** Show why $p_g(t) = \mathbb{1}(t) - \mathbb{1}(t - a)$ is sometimes called a *gate function* by plotting, as an example, the graph of $v(t) = V_m \sin \omega t\{\mathbb{1}(t) - \mathbb{1}(t - \pi/\omega)\}$.

**1.44** A voltage impulse $v(t) = 7\delta(t)$ is impressed on a linear inductor of $L$ henries. Determine the resulting current, given that $i(t) = 0$ for $t < 0$.

**1.45** Find the average values of the three waveforms in Fig. 1-42.

**1.46** Find the effective values of the three waveforms in Fig. 1-41.

**1.47** An average power of 1000 watts is dissipated in a 10-ohm resistor carrying a periodic current. What is the effective value of this periodic current?

**1.48** Show, both analytically and by graphical means, that $P_{av} = 0$ for an inductor of $L$ henries subjected to a sinusoidal voltage.

**1.49** (*a*) The voltage $v(t) = A\{\mathbb{1}(t) - \mathbb{1}(t - a)\}$ is impressed across an inductor of $L$ henries. Find and graph the inductor current, given that $i(0) = 0$. (*b*) The same voltage is impressed on a capacitor of $C$ farads. Find and graph the capacitor current.

Chapter

# 2

# KIRCHHOFF'S LAWS AND CIRCUIT EQUATIONS

In Chapter 1 the characteristics of the *individual* circuit elements were established. Here we will investigate the relationships introduced by *connecting* elements into a circuit. This culminates in the writing of *circuit equations* which, when solved in the following chapters, yield the unknown currents and voltages.

## NETWORK TERMINOLOGY

A *circuit* or *network* is, for the present, simply a number of two-terminal elements (active or passive) connected together with resistanceless wire (or solder).

A *node* or *vertex* (or junction) is a point in a circuit where two or more elements are connected together. Two nodes may *not* be connected together with a resistanceless wire; such an arrangement constitutes *one* node only (see Fig. 2-1).

A *branch* is, for the moment, any circuit element joining two nodes. Each of the elements in Fig. 2-1 is a branch of the circuit.

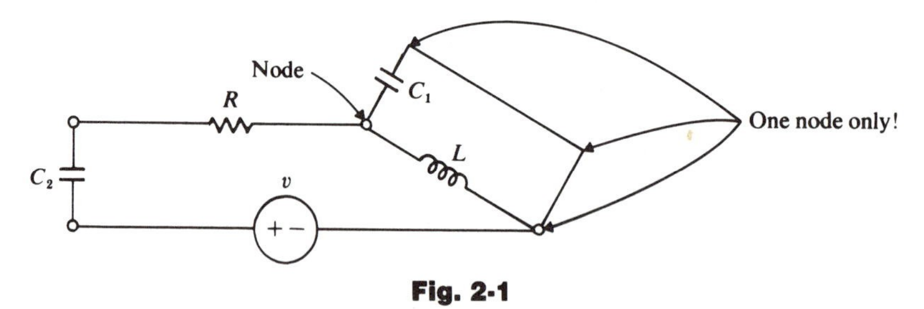


**Fig. 2-1**

A *loop* is any *closed* path in a circuit. For example, $v$, $C_2$, $R$, and $L$ form a loop in Fig. 2-1, as do $C_1$ and $L$, as well as $v$, $C_2$, $R$, and $C_1$. A loop around any "windowpane" of a circuit (for example, the loop through $C_1$ and $L$, or through $v$, $C_2$, $R$, and $L$ in Fig. 2-1) is called a *mesh*.

Two or more elements which always carry the same *current* are said to be connected in *series* (for example, $v$, $C_2$, and $R$ in Fig. 2-1).

Two or more elements which always have the same *voltage* across them are said to be connected *in parallel* (see $C_1$ and $L$ in Fig. 2-1).

Two terminals are *open circuit* if no current can flow between them. They are *short circuited* if no voltage can exist between them (see Fig. 2-2).

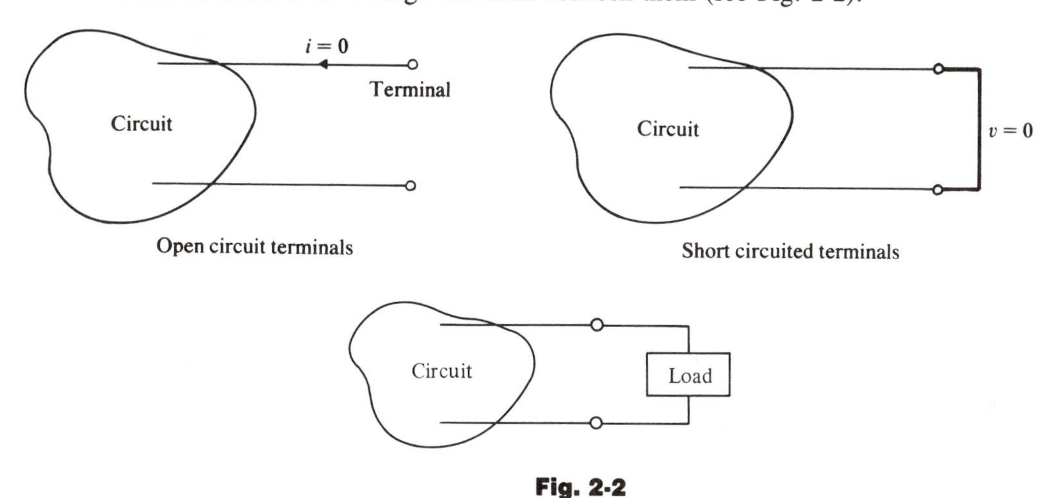


**Fig. 2-2**

An *input* is another name for an independent source. There may be more than one input in any circuit. We often say that a circuit is *driven* by its input(s). The source $v$ in Fig. 2-1 is an input.

An *output* is any unknown voltage or current in a network which is of interest to us. There may be more than one. For example, if we wanted to find the current $i_R$ and the voltage $v_L$ in Fig. 2-1, we would call $i_R$ and $v_L$ the outputs.

A *one-port network* or *one port* is any circuit with a single pair of accessible terminals as, for example, in Fig. 2-2.

A *load* is a one port, but not the one we are interested in (see Fig. 2-2). We are usually concerned with the *effect* of connecting a load to some circuit.

## KIRCHHOFF'S LAWS

1. Kirchhoff's current law (KCL): The *algebraic* sum of the currents incident at any node is zero.
2. Kirchhoff's voltage law (KVL): The *algebraic* sum of the voltages around any closed path is zero.

## LIKE ELEMENTS IN SERIES AND PARALLEL

Linear resistors $R_1$, $R_2$, . . . connected in series may be replaced by a single equivalent resistor,

$$R_{eq} = R_1 + R_2 + \cdots \tag{2.1}$$

If the resistors are in parallel,

$$\frac{1}{R_{eq}} = \frac{1}{R_1} + \frac{1}{R_2} + \cdots \tag{2.2}$$

Similarly, for linear capacitors and inductors (with *no* mutual coupling),

Series: $$\frac{1}{C_{eq}} = \frac{1}{C_1} + \frac{1}{C_2} + \cdots \tag{2.3}$$

Parallel: $$C_{eq} = C_1 + C_2 + \cdots \tag{2.4}$$

Series: $$L_{eq} = L_1 + L_2 + \cdots \tag{2.5}$$

Parallel: $$\frac{1}{L_{eq}} = \frac{1}{L_1} + \frac{1}{L_2} + \cdots \tag{2.6}$$

## NONLINEAR RESISTORS IN SERIES AND PARALLEL

If two nonlinear resistors, $\mathcal{R}_1$ and $\mathcal{R}_2$, are represented by their $v$-$i$ characteristics as shown in Fig. 2-3, then the $v$-$i$ curve of their series combination can be obtained graphically.

Since the elements are in series, they must carry the same current. And, applying KVL to the circuit in Fig. 2-3, $v = v_1 + v_2$. So at each current $i_1$ we can add the two voltages graphically to obtain the overall voltage, thus building up the $v$-$i$ characteristic of the series combination.

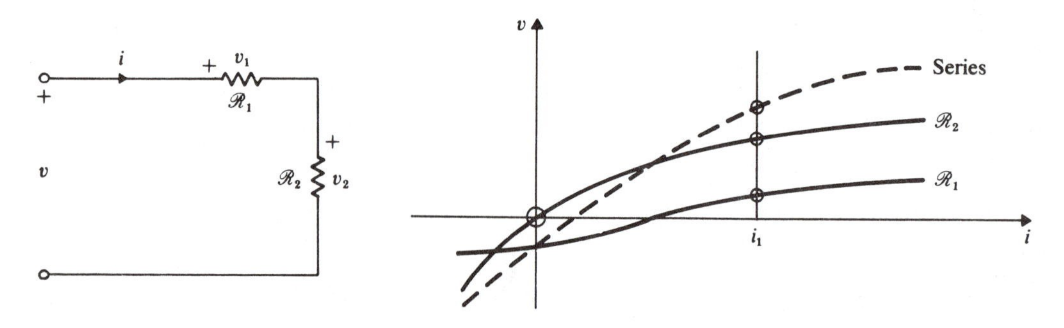


**Fig. 2-3**

A similar addition of the currents in $\mathscr{R}_1$ and $\mathscr{R}_2$ leads to the characteristic curve of their parallel combination.

## WRITING CIRCUIT EQUATIONS

There are several sets of equations we may write to describe the behavior of a given network. Today the loop, nodal, and state equations are the most common (each having advantages in specific situations) and we shall consider all three.

The *ultimate* objective is of course to solve for the voltages and currents in a given network, but the first step is necessarily to set up the equations whose solution will provide the information we seek.

Our intention is to develop clear procedures for obtaining a circuit's equations, procedures which if followed consistently will minimize the number of errors committed. The ability to write circuit equations easily and accurately is an essential prerequisite to circuit analysis.

## LOOP ANALYSIS

Let us for the sake of example obtain the *loop equations* for the linear, time-invariant circuit of Fig. 2-4, it being assumed that the circuit parameters, the independent source function $v(t)$, and the initial conditions are all given (known).

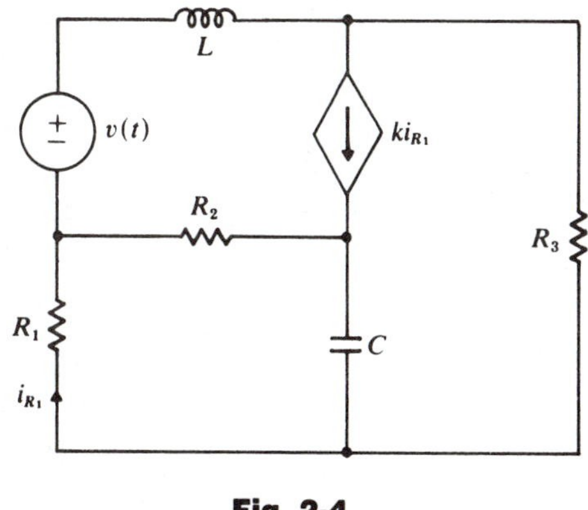

**Fig. 2-4** **Fig. 2-5**

**Step 1:** Indicate *reference* polarities on *all* elements which are not already so marked (e.g. $v(t)$ in Fig. 2-4). If we are to adhere to the associated sign convention, then the reference polarities for $R_1$ and for the dependent current source are predetermined, but the polarity markings on the remaining elements may be specified arbitrarily, as in Fig. 2-5.

**Step 2:** We now write a set of simultaneous equations, each of which corresponds to KVL around a loop. The *choice* of loops is not completely free. Thus:

1. The *number* of loops *must* equal the *number* of windowpanes. (This rule becomes more complicated if the circuit is *nonplanar*, that is, if it cannot be drawn without crossovers.)
2. As each new loop is chosen, at least one circuit element must be included which was *not* traversed by any of the earlier loops. (This is an over-safe rule to ensure that the final equations are independent and therefore soluble.) Each element must of course be traversed by at least one loop.

One appropriate choice would be the two left-hand meshes and the outer loop in Fig. 2-5. The corresponding KVL equations are

$$+v_{R_1} - v_{R_2} - v_C = 0 \qquad \text{(bottom-left mesh)}$$

$$-v(t) - v_L + v_i + v_{R_2} = 0 \qquad \text{(top-left mesh)}$$

$$+v_{R_1} - v(t) - v_L + v_{R_3} = 0 \qquad \text{(outer loop)}$$

where $v_i$ is the voltage across the current source.

**Step 3:** We "invent" a set of *loop currents* such as those in Fig. 2-6. Since a loop current both enters and leaves each node in its path, KCL is *automatically* satisfied. And if the loop currents are chosen properly (according to the same rules as for the choice of loops, above), any branch current can easily be related to the loop currents. For example, $i_{R_1} = -i_1$, $i_{R_2} = i_2$, and $i_L = i_1 - i_2$.

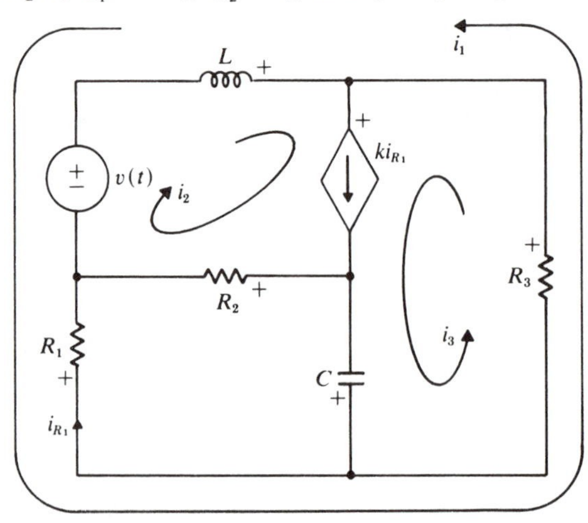


**Fig. 2-6**

Note that the loop currents may, but need not, follow the same paths as the loops chosen in step 2, and that their directions (clockwise or counterclockwise) may be chosen arbitrarily.

**Step 4:** Write the $v$-$i$ relations for each $L$, $R$, $C$, and dependent voltage source. Here, referring to Fig. 2-6,

$$v_{R_1} = R_1(-i_1) \qquad (-i_1 \text{ is the current through } R_1)$$

$$v_{R_2} = R_2 i_2$$

$$v_{R_3} = R_3(-i_1 - i_3) \qquad (-i_1 - i_3 \text{ is the current through } R_3)$$

$$v_L = L\frac{d}{dt}(i_1 - i_2)$$

$$v_C = \frac{1}{C}\int_0^t (-i_3)\,d\tau + v_C(0)$$

**Step 5:** There is always a relationship between each current source and one or more of the loop currents. Here $ki_{R_1} = i_2 + i_3$ where $i_{R_1} = -i_1$. For every current source we may therefore isolate one of the loop currents as a function of the others. Here, for example,

$$i_2 = -ki_1 - i_3$$

**Step 6:** Finally, we substitute the results of steps 4 and 5 back into the KVL equations of step 2, eliminating in the process the variable(s) isolated in step 5:

$$+R_1(-i_1) - R_2(-ki_1 - i_3) - \frac{1}{C}\int_0^t (-i_3)\,d\tau - v_C(0) = 0$$

$$-v(t) - L\frac{d}{dt}(i_1 + ki_1 + i_3) + v_i + R_2(-ki_1 - i_3) = 0$$

$$R_1(-i_1) - v(t) - L\frac{d}{dt}(i_1 + ki_1 + i_3) + R_3(-i_1 - i_3) = 0$$

or, rearranging,

$$(-R_1 + kR_2)\,i_1 + R_2 i_3 + \frac{1}{C}\int_0^t i_3\,d\tau = v_C(0)$$

$$kR_2 i_1 + (1 + k)L\frac{di_1}{dt} + R_2 i_3 + L\frac{di_3}{dt} - v_i = -v(t)$$

$$(R_1 + R_3)i_1 + (1 + k)L\frac{di_1}{dt} + R_3 i_3 + L\frac{di_3}{dt} = -v(t)$$

where $i_1$, $i_3$, and $v_i$ are the unknowns.

These are the required (integro-differential) loop equations which could be solved for the three unknowns. Any remaining loop current(s) would follow from the equation(s) in step 5, the branch currents as indicated in step 3, and the branch voltages from the $v$-$i$ relations in step 4.

## LOOP ANALYSIS WITH MUTUAL INDUCTANCE

In Fig. 2-7 the mutual inductance between $L_1$ and $L_2$ is indicated by the circular dots, that between $L_1$ and $L_3$ by the triangular dots, and that between $L_2$ and $L_3$ by the square dots. It is assumed the values of $M_{12}$, $M_{23}$, and $M_{31}$ are known.

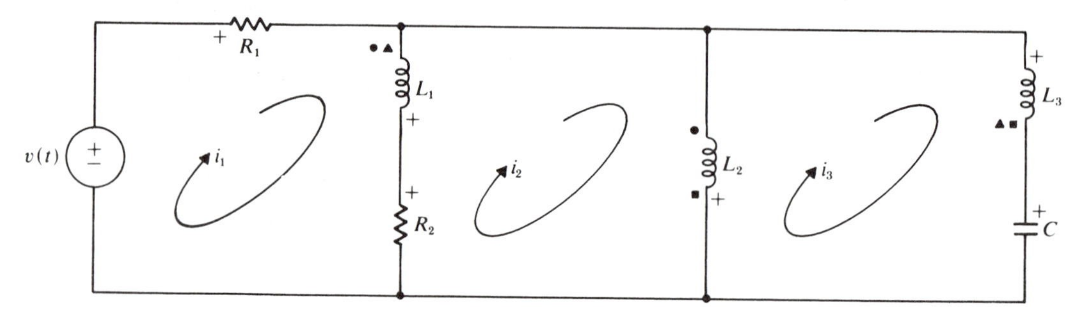


**Fig. 2-7**

The above procedure for loop analysis is unaltered, but the $v$-$i$ relations for the inductors must now include the voltages due to mutual coupling. If we assume that the polarities and loop currents have been chosen (steps 1 and 3) as shown in Fig. 2-7, then

$$v_{L_1} = \overset{\text{always positive}}{+}L_1 \frac{d}{dt} \underbrace{(i_2 - i_1)}_{\text{current entering + end of } L_1} \overset{\bullet \text{ and + are on } opposite \text{ ends of } L_1}{-} M_{12} \frac{d}{dt} \underbrace{(i_2 - i_3)}_{\text{current entering } \bullet \text{ end of } L_2} \overset{\blacktriangle \text{ and + are on } opposite \text{ ends of } L_1}{-} M_{31} \frac{d}{dt} \underbrace{(-i_3)}_{\text{current entering } \blacktriangle \text{ end of } L_3}$$

$$v_{L_2} = L_2 \frac{d}{dt}(i_3 - i_2) - M_{12} \frac{d}{dt}(i_1 - i_2) \overset{\blacksquare \text{ and + are on } same \text{ end of } L_2}{+} M_{23} \frac{d}{dt}(-i_3)$$

$$v_{L_3} = L_3 \frac{d}{dt}(i_3) - M_{31} \frac{d}{dt}(i_1 - i_2) - M_{23} \frac{d}{dt}(i_3 - i_2)$$

Note that both a current direction and a voltage polarity enter into the determination of sign for mutual terms. If these are kept quite separate, as indicated above, errors are less likely to be incurred. No significance attaches to the order of the subscripts, since it can be shown that $M_{ij} = M_{ji}$.

## NODAL ANALYSIS

Here we shall describe a procedure for obtaining a circuit's *nodal equations*, referring to the circuit of Fig. 2-4 as an example. As before, it will be assumed that the circuit parameters, the independent source function $v(t)$, and the initial conditions are given (known).

**Step 1:** Indicate current *reference* directions on *all* elements not already so marked (for example, $i_{R_1}$ in Fig. 2-4). If we adhere to the associated sign convention, then the reference direction for $v(t)$ is predetermined, but the direction may be specified arbitrarily on the other elements, as in Fig. 2-8.

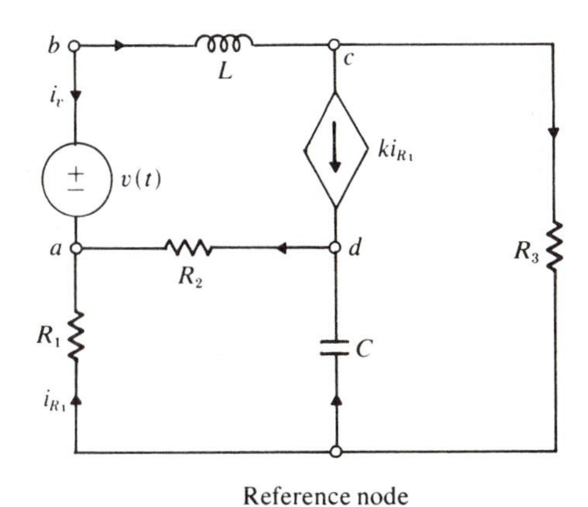


**Fig. 2-8**

**Step 2:** Identify and label *all* of the nodes, arbitrarily choosing one of them as the reference (see Fig. 2-8). As a check, there must be a node at both ends of each element, and no nodes may be connected by a (resistanceless) wire.

KCL equations are now written for all nodes except the reference:

Node $a$: $i_{R_1} + i_{R_2} + i_v = 0$
Node $b$: $i_v + i_L = 0$
Node $c$: $i_L - ki_{R_1} - i_{R_3} = 0$
Node $d$: $ki_{R_1} + i_C - i_{R_2} = 0$

where $i_v$ is the current through the voltage source $v(t)$.

**Step 3:** We work in terms of the *nodal voltages* $v_a, v_b, \ldots,$ corresponding to the labelled nodes, each voltage being measured *relative* to the reference node. The branch voltages are simply related to the nodal voltages; for example, $v_{R_3} = v_c$, $v_{R_1} = -v_a$, $v_{R_2} = v_d - v_a$.

**Step 4:** Write the $i$-$v$ relations for each $L$, $R$, $C$, and dependent current source. Here, referring to Fig. 2-8,

$$i_{R_1} = \frac{1}{R_1}(-v_a) \qquad (-v_a \text{ is the voltage across } R_1)$$

$$i_{R_2} = \frac{1}{R_2}(v_d - v_a) \qquad (v_d - v_a \text{ is the voltage across } R_2)$$

$$i_{R_3} = \frac{1}{R_3} v_c$$

$$i_L = \frac{1}{L}\int_0^t (v_b - v_c)\, d\tau + i_L(0)$$

$$i_C = C\frac{d}{dt}(-v_d)$$

$$ki_{R_1} = \frac{k}{R_1}(-v_a)$$ (substituting for $i_{R_1}$ in the equation for the dependent current source)

**Step 5:** There is always a relationship between each voltage source and one or more of the nodal voltages. Here $v(t) = v_b - v_a$. For every voltage source we may therefore isolate one of the nodal voltages as a function of the others. Here, for example,

$$v_b = v(t) + v_a$$

**Step 6:** Finally, we substitute the results of steps 4 and 5 back into the KCL equations of step 2, eliminating in the process the variable(s) isolated in step 5:

$$\frac{1}{R_1}(-v_a) + \frac{1}{R_2}(v_d - v_a) + i_v = 0$$

$$i_v + \frac{1}{L}\int_0^t \{v(\tau) + v_a - v_c\}\, d\tau + i_L(0) = 0$$

$$\frac{1}{L}\int_0^t \{v(\tau) + v_a - v_c\}\, d\tau + i_L(0) - \frac{k}{R_1}(-v_a) - \frac{1}{R_3}v_c = 0$$

$$\frac{k}{R_1}(-v_a) + C\frac{d}{dt}(-v_d) - \frac{1}{R_2}(v_d - v_a) = 0$$

or, rearranging,

$$-\left(\frac{1}{R_1} + \frac{1}{R_2}\right)v_a + \frac{1}{R_2}v_d + i_v = 0$$

$$\frac{1}{L}\int_0^t v_a\, d\tau - \frac{1}{L}\int_0^t v_c\, d\tau + i_v = -\frac{1}{L}\int_0^t v(\tau)\, d\tau - i_L(0)$$

$$\frac{k}{R_1}v_a + \frac{1}{L}\int_0^t v_a\, d\tau - \frac{1}{R_3}v_c - \frac{1}{L}\int_0^t v_c\, d\tau = -\frac{1}{L}\int_0^t v(\tau)\, d\tau - i_L(0)$$

$$\left(\frac{1}{R_2} - \frac{k}{R_1}\right)v_a - \frac{1}{R_2}v_d - C\frac{dv_d}{dt} = 0$$

where $v_a$, $v_c$, $v_d$, and $i_v$ are the unknowns.

These are the required (integro-differential) nodal equations which could be solved for the four unknowns. Any remaining nodal voltage(s) would follow from the equation(s) in step 5, the branch voltages as indicated in step 3, and the branch currents form the $i$-$v$ relationships in step 4.

## NOTES ON LOOP AND NODAL ANALYSIS

It is well worthwhile to cultivate the habit of following the above procedures. Attempts to combine steps mentally almost always lead to errors. And completely accurate equations are an obvious prerequisite to the successful analysis of a circuit.

It may help to go back and compare the two procedures. There is a strong relationship between them. If we are to analyze a circuit we must first choose between the two possible sets of equations, usually with the object of minimizing the number of simultaneous equations to be solved. This number is

Loop: The number of meshes (assuming a planar circuit)
Node: The number of nodes (*excluding* the reference)

However, it is often a simple matter to eliminate the voltage(s) across the current source(s) or the current(s) through the voltage source(s), thereby reducing the number of equations.

Sometimes, or even usually, one node of a circuit will be shown connected to ground (the zero reference of potential). This makes no difference whatsoever to either the loop or nodal analysis, since connecting *one* terminal to ground (or to any other potential) can have no effect on any of the currents or potential differences within the network.

Further, do *not* try to use the nodal method if there is mutual inductance. We can (and have) included mutual effects in the $v$-$i$ relations used for loop analysis, but there are no corresponding $i$-$v$ relationships for nodal analysis.

Finally, op amp circuits are normally best analyzed by the nodal method. However, *don't* write a KCL equation at the op amp's output node. Since we know the voltage at that point (in terms of $v_+$ and $v_-$), such a KCL equation is unnecessary and can cause confusion.

## THÉVENIN'S EQUIVALENT CIRCUIT

Any linear, resistive one-port network (see Fig. 2-9) has a linear algebraic relationship between its terminal current $i$ and its terminal voltage $v$. That is,

$$v = k_1 i + k_2 \quad \text{or} \quad i = k_3 v + k_4 \tag{2.7}$$

where the $k$'s are constants. This $v$-$i$ or $i$-$v$ relation can be found by loop or nodal analysis, even if we know nothing about the load. In other words, a one-port network's $v$-$i$ or $i$-$v$ relation is *independent of the load*.

It is easy to show (see Problem 2.31) that the Thévenin one port of Fig. 2-10 has the following $v$-$i$ and $i$-$v$ relations:

$$v = -R_{\text{Th}} i + v_{\text{Th}} \quad \text{or} \quad i = -(1/R_{\text{Th}})v + v_{\text{Th}}/R_{\text{Th}} \tag{2.8}$$

If we have found a one port's $v$-$i$ relation (i.e. evaluated $k_1$ and $k_2$), then a

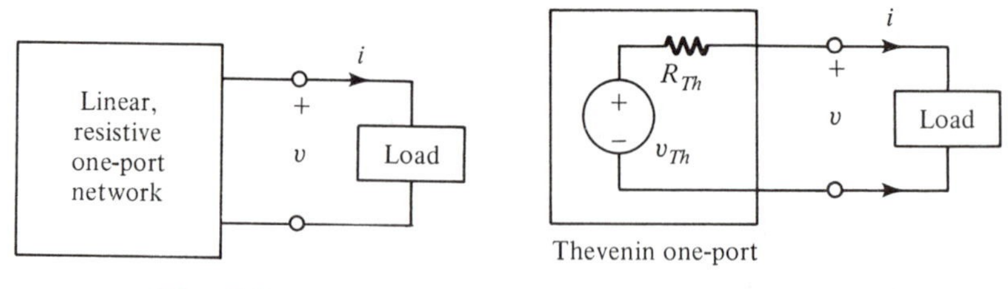


**Fig. 2-9** **Fig. 2-10**

comparison of equations (2.7) and (2.8) shows that the one port will have the same $v$-$i$ relation as a Thévenin one port for which

$$R_{\mathrm{Th}} = -k_1 \quad \text{and} \quad v_{\mathrm{Th}} = k_2$$

We say that we have found the *Thévenin equivalent* of the one port, *as seen by the load*. Put another way, any one port can be reduced to a single voltage source $v_{\mathrm{Th}}$ in series with a single resistor, $R_{\mathrm{Th}}$. *From the terminals* we cannot distinguish between a one port and its Thévenin equivalent.

## THE STATE EQUATIONS

In recent years, particularly since digital computers became generally available, modern systems analysis has leaned heavily on *state equation* formulations. The idea is to obtain a set of simultaneous *first-order* differential equations; in matrix notation,

$$\dot{\mathbf{x}} = \mathbf{f}(\mathbf{x}, \mathbf{u}, t) \tag{2.9}$$

where $\mathbf{x}$ is the vector of *state variables*, and $\mathbf{u}$ is the vector of *inputs* (independent sources). If the system (or circuit) contains only linear, time-invariant elements, then we will find that the state equations simplify to

$$\dot{\mathbf{x}} = \mathbf{A}\mathbf{x} + \mathbf{B}\mathbf{u} \tag{2.10}$$

where $\mathbf{A}$ and $\mathbf{B}$ are matrices of constants.

State equations lend themselves to computer solution, and can more easily take account of nonlinear elements than can the classical loop and nodal equations.

Here we will derive the state equations for the circuit considered earlier (see Fig. 2-4).

**Step 1:** Mark reference polarities and/or directions on all elements not already so marked, as in Fig. 2-11.

**Step 2:** Write the *derivative* relations for all inductors and capacitors. Here,

$$L\frac{di_L}{dt} = v_L \quad \text{and} \quad C\frac{dv_C}{dt} = i_C$$

The variables differentiated, in this case $i_L$ and $v_C$, are the state variables.

**Step 3:** Eliminate from the derivative equations all voltages and currents *except* the state variables and the inputs. Here we must substitute for $v_L$ and $i_C$ in terms of $i_L$, $v_C$, and $v(t)$.

There is a systematic procedure leading to this substitutiton, but it is out of place here. Instead we will regard the process as a test of our skill with Kirchhoff's laws and the $v$-$i$ (and $i$-$v$) relations for the remaining elements (resistors and dependent sources).

The first move might be to use KVL and KCL to write (from Fig. 2-11)

$$v_L = -v_{R_3} - v_{R_1} + v(t) \quad \text{and} \quad i_C = i_{R_2} + ki_{R_1}$$

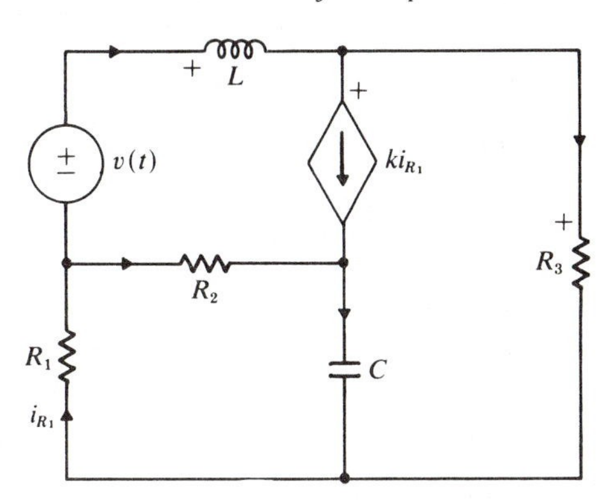


**Fig. 2-11**

*Now* we have only to find the voltages and/or currents in $R_1$, $R_2$, and $R_3$ in terms if $i_L$, $v_C$, and $v(t)$. From KVL, $v_{R_1} + v_{R_2} + v_C = 0$, or

$$R_1 i_{R_1} + R_2 i_{R_2} = -v_C$$

and from KCL,

$$i_{R_1} - i_{R_2} = i_L$$

These two equations can be solved by applying Cramer's rule,

$$i_{R_1} = \frac{\begin{vmatrix} -v_C & R_2 \\ i_L & -1 \end{vmatrix}}{\begin{vmatrix} R_1 & R_2 \\ 1 & -1 \end{vmatrix}} = \frac{R_2 i_L - v_C}{R_1 + R_2}, \qquad i_{R_2} = \frac{\begin{vmatrix} R_1 & -v_C \\ 1 & i_L \end{vmatrix}}{\begin{vmatrix} R_1 & R_2 \\ 1 & -1 \end{vmatrix}} = -\frac{R_1 i_L + v_C}{R_1 + R_2}$$

and from KCL again,

$$i_{R_3} = i_L - ki_{R_1} = \frac{(R_1 + R_2 - kR_2)i_L + kv_C}{R_1 + R_2}$$

Finally, substituting back into the equations of step 2,

$$\frac{di_L}{dt} = \frac{-R_3(R_1 + R_2 - kR_2)i_L - kR_3 v_C + R_1 v_C - R_1 R_2 i_L}{L(R_1 + R_2)} + \frac{1}{L} v(t)$$

$$\frac{dv_C}{dt} = \frac{-R_1 i_L - v_C + kR_2 i_L - kv_C}{C(R_1 + R_2)}$$

In matrix form, these two simultaneous first-order equations become

$$\begin{bmatrix} \dfrac{di_L}{dt} \\ \\ \dfrac{dv_C}{dt} \end{bmatrix} = \begin{bmatrix} \dfrac{-R_1R_2 - R_2R_3 - R_3R_1 + kR_2R_3}{L(R_1 + R_2)} & \dfrac{R_1 - kR_3}{L(R_1 + R_2)} \\ \\ \dfrac{kR_2 - R_1}{C(R_1 + R_2)} & \dfrac{-(1 + k)}{C(R_1 + R_2)} \end{bmatrix} \begin{bmatrix} i_L \\ \\ v_C \end{bmatrix} + \begin{bmatrix} \dfrac{1}{L} \\ \\ 0 \end{bmatrix} v(t)$$

Notice that these equations are in the form $\dot{\mathbf{x}} = \mathbf{Ax} + \mathbf{Bu}$ where, since there is only one input, $\mathbf{u} = v(t)$ is a scalar. Notice also that the variables appear in the same order ($v_C$ below $i_L$) on both sides of the state equation, and that **A** and **B** are indeed matrices of constants.

Forcing the circuit's equations into the desired form entailed an appreciable effort, but we will later find that the solution is much simplified. As we would expect, solution is possible only if we are given the circuit parameters, the input(s), and the initial conditions, in this case $i_L(0)$ and $v_C(0)$.

**Step 4:** We may, of course, want to find some of the other circuit variables. If the circuit is linear and time-invariant, these will be related to the state variables by a matrix equation of the form

$$\mathbf{y} = \mathbf{Cx} + \mathbf{Du}$$

where **y** is the vector of *outputs* (the variables which we want to find) and **C** and **D** are matrices of constants.

If, for example, we wish to regard $i_{R_2}$ and $v_{R_1}$ as outputs, then from the equations in step 3,

$$i_{R_2} = -\frac{R_1 i_L + v_C}{R_1 + R_2} \quad \text{and} \quad v_{R_1} = R_1 i_{R_1} = \frac{R_1R_2 i_L - R_1 v_C}{R_1 + R_2}$$

and in matrix form,

$$\mathbf{y} = \begin{bmatrix} i_{R_2} \\ \\ v_{R_1} \end{bmatrix} = \begin{bmatrix} \dfrac{-R_1}{R_1 + R_2} & \dfrac{-1}{R_1 + R_2} \\ \dfrac{R_1R_2}{R_1 + R_2} & \dfrac{-R_1}{R_1 + R_2} \end{bmatrix} \begin{bmatrix} i_L \\ \\ v_C \end{bmatrix} + \mathbf{0u}$$

where in this instance **D** is the null matrix **0**. Thus once the state equations have been solved, the desired outputs are easily calculated.

## THE OP AMP, AGAIN

In Chapter 1 the ideal op amp (Fig. 1-4) was defined to have an output voltage $v_o = A(v_+ - v_-)$, where $A \to \infty$. We noted then that this implies that $v_+ = v_-$ so long as $v_o$ is finite. It is often advantageous to use this approximation (i.e. $v_+ = v_-$) right from the start of a problem solution, since we can then use the op amp circuit diagram corresponding to Fig. 1-4*a* directly; there is no need for the dependent source model of Fig. 1-4*b*. We will also assume throughout that $i_+ = i_- = 0$.

# ILLUSTRATIVE PROBLEMS

## *Kirchhoff's Laws*

**2.1** Write Kirchhoff's current law for the node in Fig. 2-12, choosing the reference directions for $i_1$ and $i_3$ in accordance with the associated sign convention of Chapter 1. (The "boxes" represent two-terminal elements.)

Using the convention of Chapter 1, positive $i_1$ will be out of the node and positive $i_3$ will be into the node. Summing currents out of the node,

$$+i_1 + i_2 - i_3 - i_4 = 0$$

or, summing currents into the node,

$$-i_1 - i_2 + i_3 + i_4 = 0$$

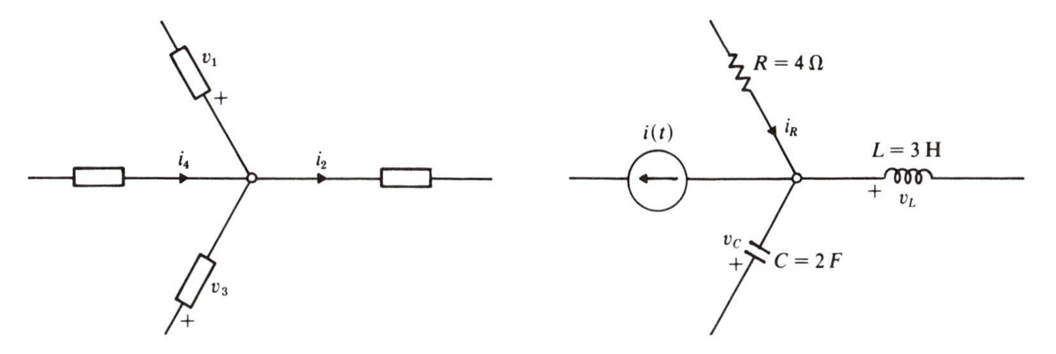


Fig. 2-12

Fig. 2-13

**2.2** With reference to Fig. 2-13, write KCL for the node. Solve this equation for $i_L$, and determine $v_L$ given $v_C = 5 \cos 6t$ V, $i_R = 7$ A, and $i(t) = 8t$ A.

By KCL, $i(t) - i_R + i_L - i_C = 0$ where $i_C = C\dfrac{dv_C}{dt} = 2\,\dfrac{d}{dt}\{5 \cos 6t\} = -60 \sin 6t$. Substituting the known currents,

$$8t - 7 + i_L - (-60 \sin 6t) = 0$$

from which $i_L = -8t + 7 - 60 \sin 6t$ and $v_L = 3\,\dfrac{di_L}{dt} = -24 - 1080 \cos 6t$.

**2.3** Write Kirchhoff's voltage law for the loop in Fig. 2-14.

According to the associated sign convention, the left end of element 4 and the top of element 5 should be marked with a plus sign. Then

$$+v_1 - v_2 + v_3 - v_4 - v_5 = 0$$

or, since we may sum voltage drops or rises, while traversing the loop clockwise or counterclockwise,

$$-v_1 + v_2 - v_3 + v_4 + v_5 = 0$$

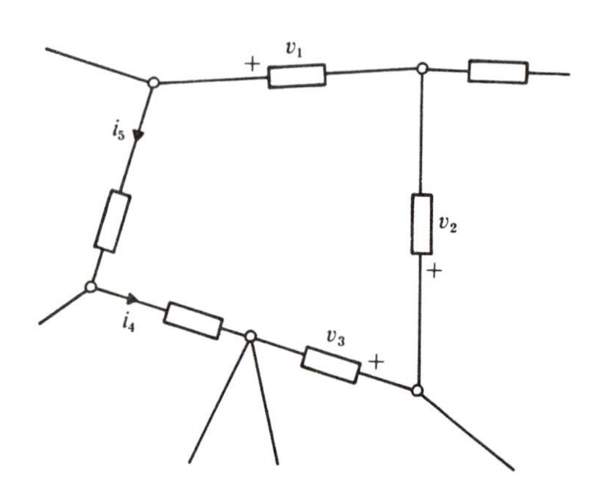


**Fig. 2-14**

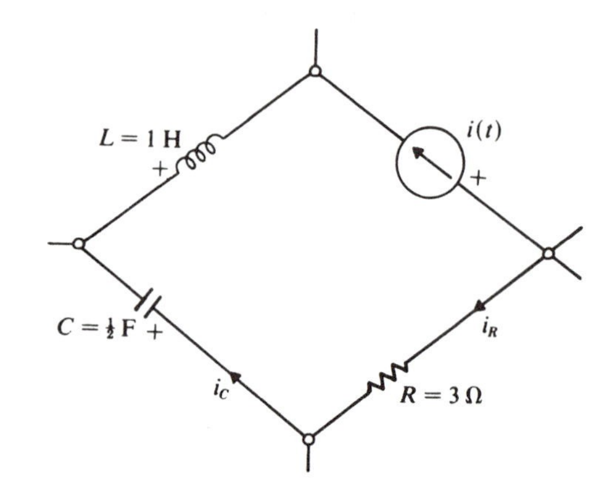


**Fig. 2-15**

**2.4** In Fig. 2-15, $i_R = 6e^{-t}$ A, $i_C = 5 \sin 10t$ A, $v_C(0) = -1$ V, $v_L = 10t$ V, and $i(t) = 17$ A. Write KVL for the loop and then solve for $v_i$, the voltage across the current source, and find $v_i$ at $t = 0$.

By KVL, $v_R + v_C + v_L - v_i = 0$ where $v_R = 3i_R = 18e^{-t}$ and $v_C = \dfrac{1}{0.5}\displaystyle\int_0^t 5 \sin 10\tau\, d\tau - 1 = -\cos 10t$. Then

$$18e^{-t} + (-\cos 10t) + 10t - v_i = 0$$

or $v_i(t) = 18e^{-t} - \cos 10t + 10t$, and so $v_i(0) = 18 - 1 + 0 = 17$ V.

### *Analysis of Some Simple Circuits*

**2.5** A *real* constant-current source may be "modeled" (represented) as shown within the dashed lines of Fig. 2-16. As the load resistor $R$ is varied, the terminal voltage $v$ and terminal current $i$ will change. Plot $i$ vs. $v$ for $0 \leq R \leq \infty$.

KCL at the upper terminal is $I - i_r - i = 0$ where $i_r = \dfrac{1}{r}v$. Thus we have

$$I - \frac{1}{r}v - i = 0 \quad \text{or} \quad i = -\frac{1}{r}v + I$$

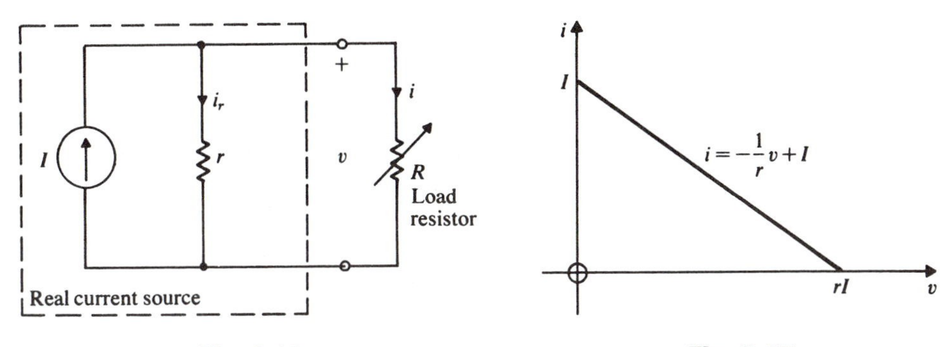


**Fig. 2-16** **Fig. 2-17**

which is plotted in Fig. 2-17. Note that this $i$-$v$ characteristic in no way depends on the nature of the *load*. It is a property of the source circuit.

**2.6** Answer the following questions about the circuit of the preceding problem.

(*a*) What is the open-circuit terminal current and voltage of the real constant-current source?

(*b*) What is the short-circuit terminal current and voltage?

(*c*) In the case of an ideal constant-current source, $r \to \infty$. Plot the corresponding $i$-$v$ characteristic and determine the open- and short-circuit terminal current and voltage.

(*a*) When $i = 0$ (output terminals open-circuit), $v = rI$. This follows from Fig. 2-17 or from the realization that the entire current $I$ will flow through $r$ when $i = 0$.

(*b*) When $v = 0$ (terminals shorted), $i = I$, which follows from Fig. 2-17 or from the fact that *no* current will now flow through $r$.

(*c*) In the case of the *ideal* source, $r \to \infty$ and *all* the current $I$ will flow through the output terminals. The $i$-$v$ characteristic will be a horizontal line $I$ units above the $v$ axis. When the terminals are open-circuit, the current $I$ will still flow (definition of an ideal current source), and the output voltage will be infinite. (Ideal sources obviously cannot exist in practice.) When the terminals are short-circuited, $v = 0$ and $i = I$.

**2.7** The known *output* voltage $v_2$ for a series *R-L* circuit (Fig. 2-18*a*) is shown in

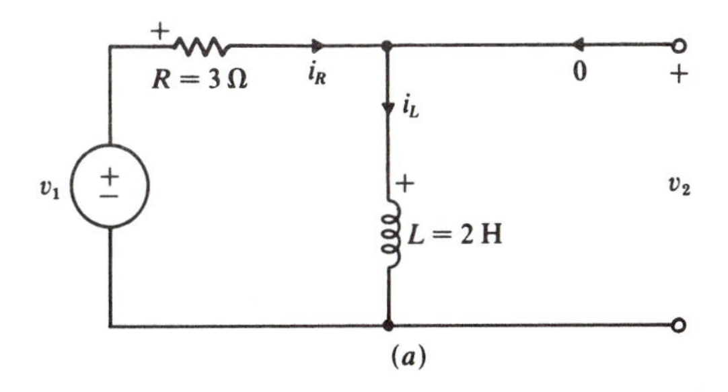


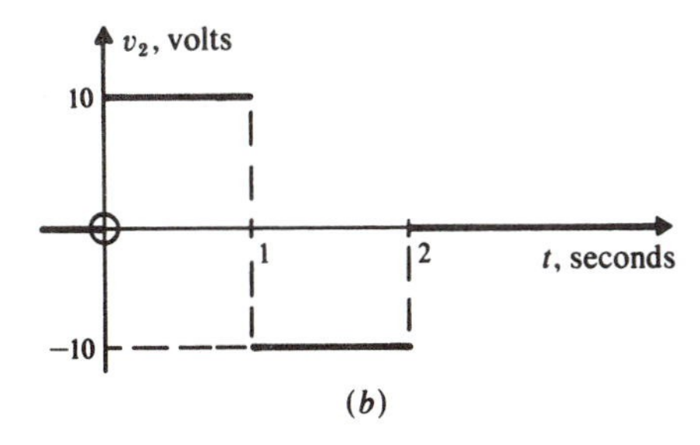


**Fig. 2-18**

Fig. 2-18*b*. It is also known that $i_L(0) = 0$. Determine and sketch the waveform of the *input* voltage $v_1$.

*Note*: Two terminals, as at the right of the circuit in Fig. 2-18*a*, through which no current (or negligible current) flows, can be thought of as *measurement terminals*. Quite different circuit behavior must be anticipated if current *is* drawn from these terminals.

Applying KCL at the top terminal, $i_R + 0 - i_L = 0$, or $i_R = i_L$. Therefore

$$v_R = Ri_R = Ri_L = \frac{R}{L}\int_0^t v_L\,d\tau + Ri_L(0) = \frac{3}{2}\int_0^t v_L\,d\tau$$

$v_L$ is easily integrated by the methods of Chapter 1, leading to $v_R$ as plotted against $t$ in Fig. 2-19*a*.

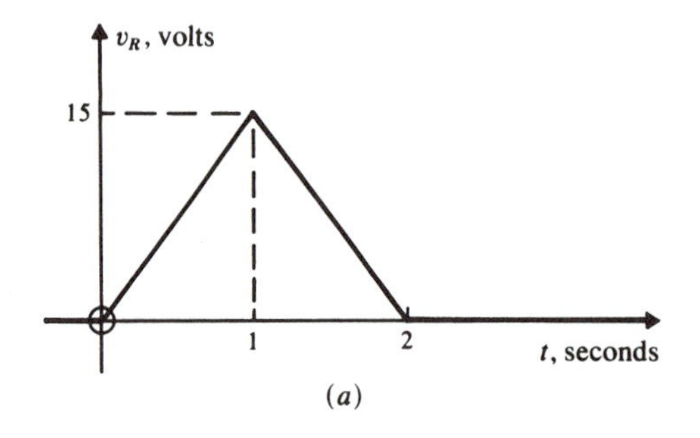


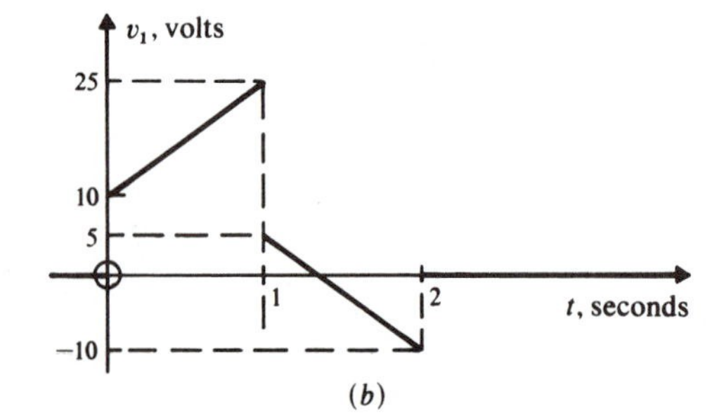


**Fig. 2-19**

If we apply KVL to the circuit of Fig. 2-18*a*, $-v_1 + v_R + v_L = 0$; or $v_1 = v_R + v_L = v_R + v_2$, since $v_L$ and $v_2$ describe the same voltage. We can now add $v_R$ and $v_2$ graphically to yield $v_1$ as shown in Fig. 2-19*b*.

**2.8** At $t = 0$ the following is known about the circuit of Fig. 2-20: $i_2(0) = +2$ A and $di_2(0)/dt = -10$ A/s. Determine the value of $L$.

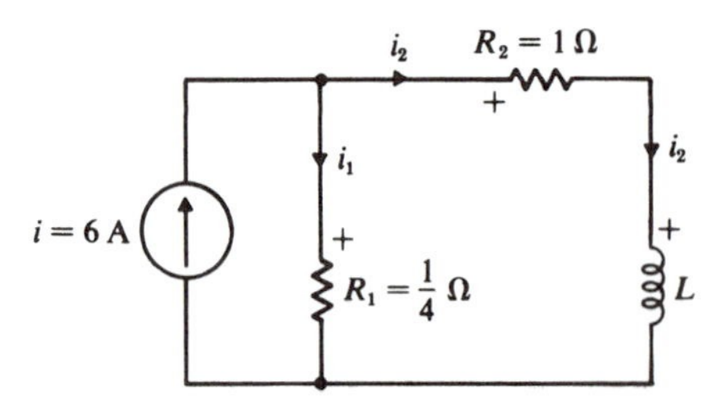


**Fig. 2-20**

From KCL, $i(0) - i_1(0) - i_2(0) = 0$, or

$$i_1(0) = 6 - 2 = 4 \text{ A}$$

and

$$v_{R_1}(0) = (\tfrac{1}{4})(4) = 1 \text{ V}$$

From KVL, $v_{R_1}(0) - v_L(0) - v_{R_2}(0) = 0$; and since $v_{R_2}(0) = (1)(2)$, $v_L(0) = 1 - 2 = -1$ V.

But $v_L(0) = L\, di_2(0)/dt$. That is, $-1 = L(-10)$ or $L = 0.1$ H.

**2.9** The two circuits of Fig. 2-21 have equivalent terminal characteristics. Determine the appropriate values of $R$ and $C$.

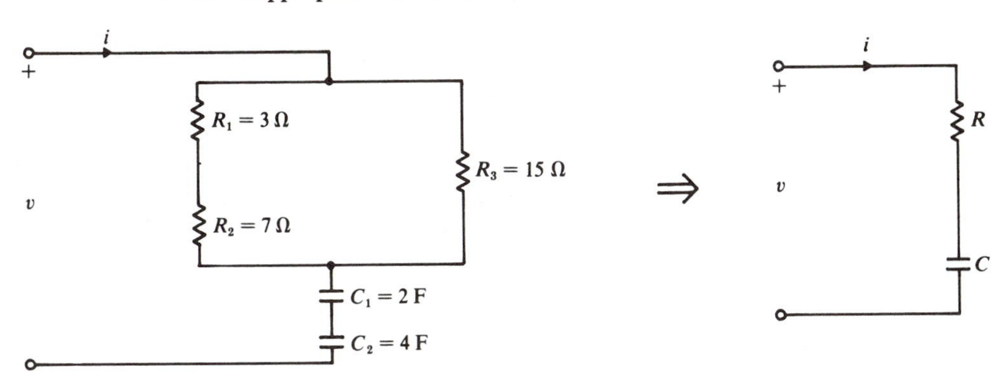


**Fig. 2-21**

The 3-Ω and 7-Ω resistors are in series and may be replaced by a 10-Ω resistor according to equation (*2.1*). Then from equation (*2.2*), $\frac{1}{R} = \frac{1}{10} + \frac{1}{15}$ or $R = 6$ Ω.

The two capacitors are in series and may be combined according to $\frac{1}{C} = \frac{1}{2} + \frac{1}{4}$, and so $C = 1.33$ F.

**2.10** (*a*) In Fig. 2-22*a*, are $R_1$ and $R_2$ in series, in parallel, or neither?
(*b*) What about $R_3$ and $R_4$ in Fig. 2-22*b*?

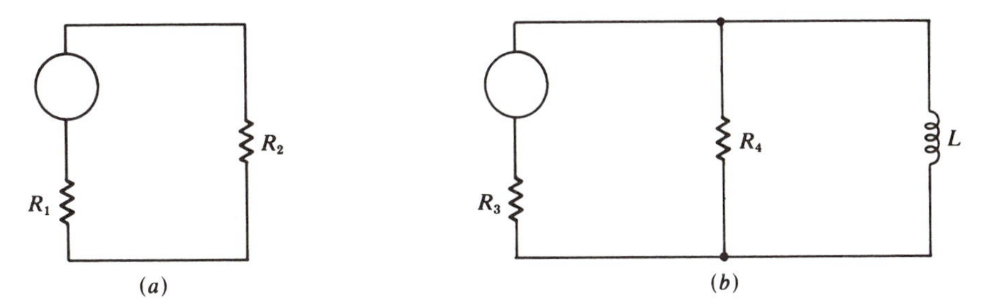


**Fig. 2-22**

(*a*) $R_1$ and $R_2$ are in series—they carry the same current.
(*b*) $R_3$ and $R_4$ are neither in series nor in parallel since, in general, they do not carry the same current, nor is the same voltage impressed upon them.

**2.11** Find the inductance $L_{eq}$ which will make the two circuits in Fig. 2-23 equivalent.

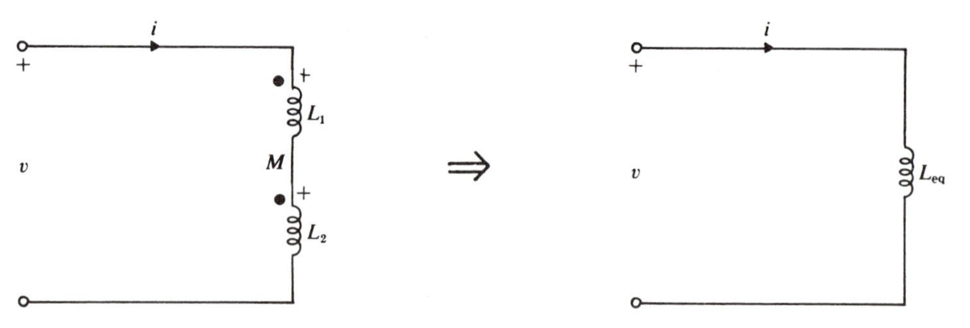


**Fig. 2-23**

The voltages $v_1$ and $v_2$ across $L_1$ and $L_2$ are given by

$$v_1 = L_1 \frac{di}{dt} + M \frac{di}{dt} \quad \text{and} \quad v_2 = L_2 \frac{di}{dt} + M \frac{di}{dt}$$

From KVL, $-v + v_1 + v_2 = 0$ and so

$$v = L_1 \frac{di}{dt} + L_2 \frac{di}{dt} + 2M \frac{di}{dt} = (L_1 + L_2 + 2M) \frac{di}{dt}$$

which shows that $L_{eq} = L_1 + L_2 + 2M$.

**2.12** The circuit of Fig. 2-24 models (i.e. represents) a transistor. The nodes $B$, $C$, and $E$ mark the transistor's *base*, *collector*, and *emitter* terminals. Find the quantities $I_B$, $I_C$, and $V_{CE}$ in the order given, treating $E$ as the reference node. (When $E$ is the reference node, $V_{CE}$, the voltage of node $C$ with respect to node $E$, is equivalent to $V_C$.)

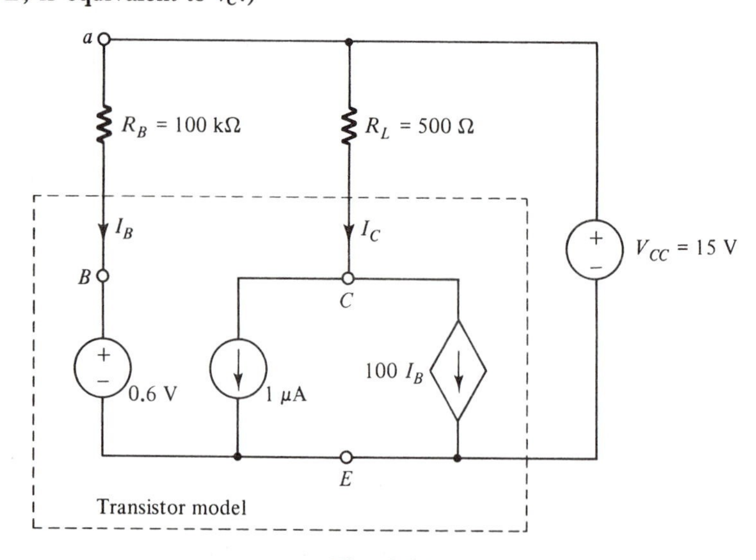


**Fig. 2-24**

By inspection of Fig. 2-24, $V_a = 15$ V, $I_B = (15 - 0.6)/10^5 = 144\ \mu$A, $I_C = 10^{-6} + (100)(144 \times 10^{-6}) = 14.4$ mA, and $V_{CE} = V_C = 15 - (500)(14.4 \times 10^{-3}) = 7.8$ V.

## *Some Graphical Methods for Resistive Circuits*

**2.13** The $v$-$i$ characteristics for two linear resistors, $R_a$ and $R_b$, are shown in Fig. 2-25. Sketch, using graphical methods, the combined $v$-$i$ characteristic if $R_a$ and $R_b$ are connected (*a*) in series and (*b*) in parallel.

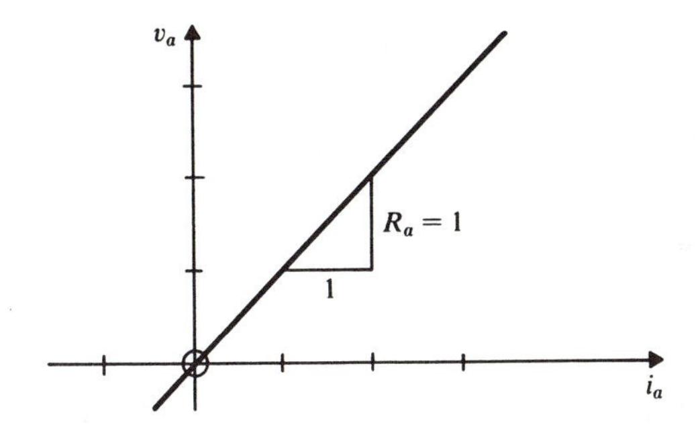


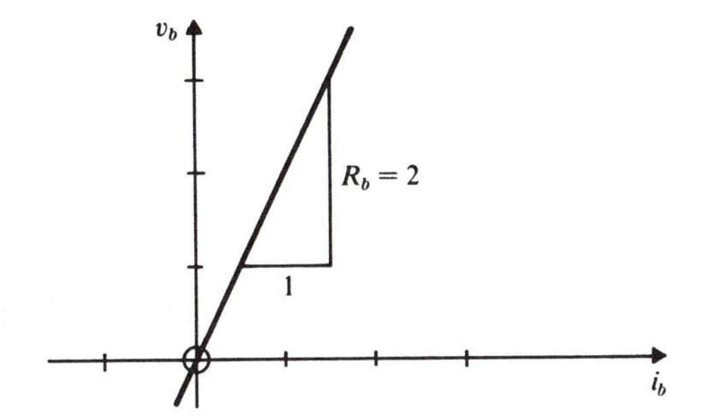


**Fig. 2-25**

(*a*) At each current $i_1$ for the series connection, KVL requires that we add the two voltages to build up the overall $v$-$i$ characteristic (see Fig. 2-26*a*).

(*b*) Here, by virtue of KCL we add the two currents $i_a$ and $i_b$ at each voltage $v_1$ to obtain the overall $v$-$i$ characteristic of Fig. 2-26*b*.

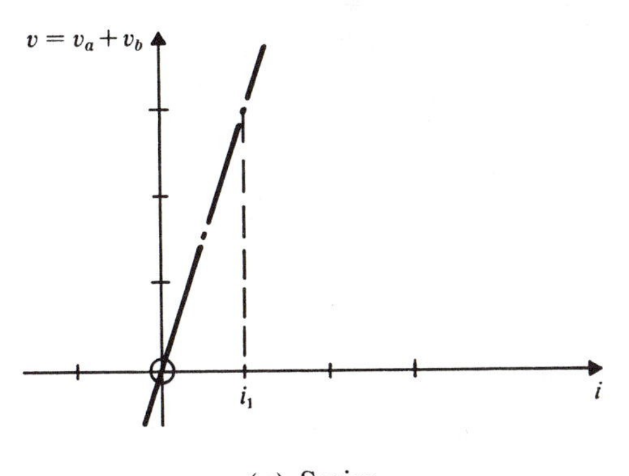


(*a*) Series

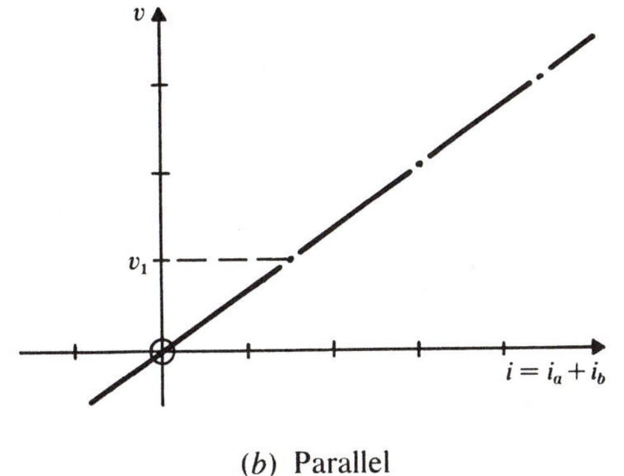


(*b*) Parallel

**Fig. 2-26**

**2.14** Sketch the $v$-$i$ characteristic of each of the following elements. (*a*) An ideal dc voltage source. (*b*) An ideal dc current source. (*c*) An open circuit. (*d*) A short circuit.

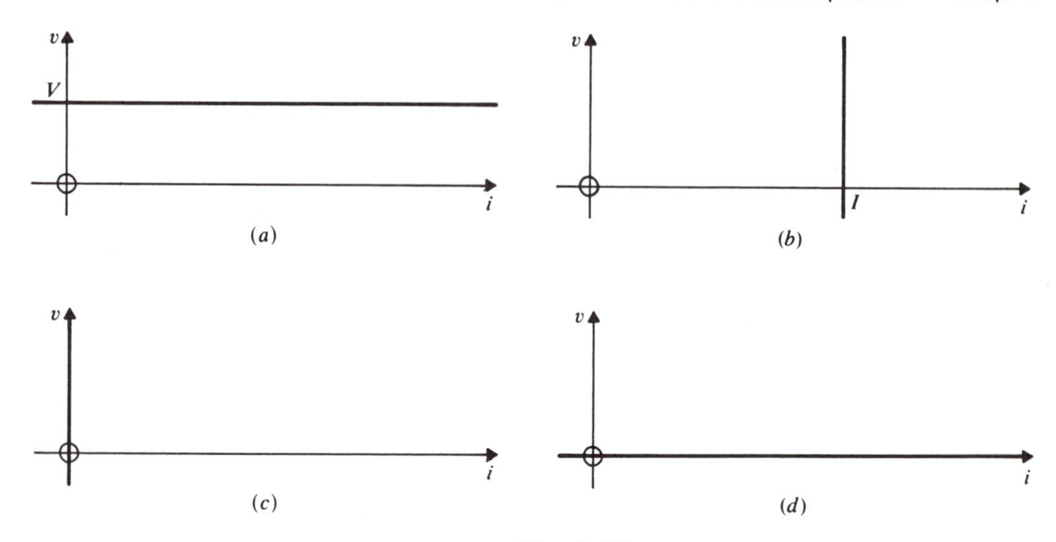


**Fig. 2-27**

(*a*) An ideal dc voltage source supplies a constant voltage regardless of the current flow (see Fig. 2-27*a*).
(*b*) A constant current for any terminal voltage, as in Fig. 2-27*b*.
(*c*) Zero current for any voltage, as in Fig. 2-27*c*.
(*d*) Zero voltage for any current, as in Fig. 2-27*d*.

**2.15** An ideal diode (Fig. 2-28*a*) may be described by the $v$-$i$ characteristic of Fig. 2-28*b*. Describe the diode's characteristics in words. (*Hint*: Refer to Problem 2.14.)

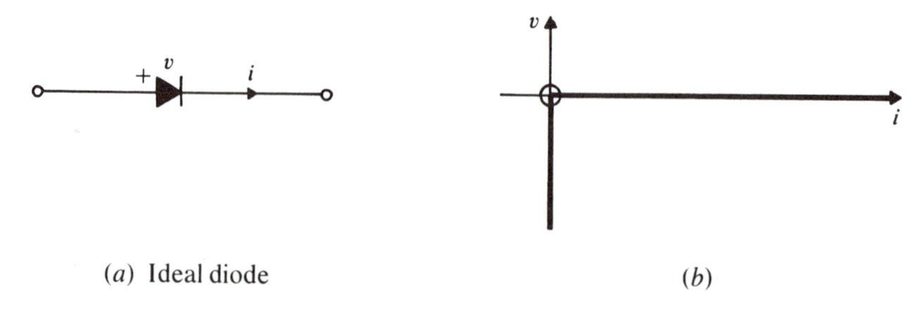


(*a*) Ideal diode (*b*)

**Fig. 2-28**

An ideal diode behaves as a short circuit to a *forward* voltage, and as an open circuit to a *reverse* one. (Forward is the word used to describe the "easy" current direction. This is the direction of the arrow in the diode's symbol.)

**2.16** A constant 10-V source, a 5-Ω linear resistor, and an ideal diode are connected in series as shown in Fig. 2-29. Determine the $v$-$i$ characteristic of this combination.

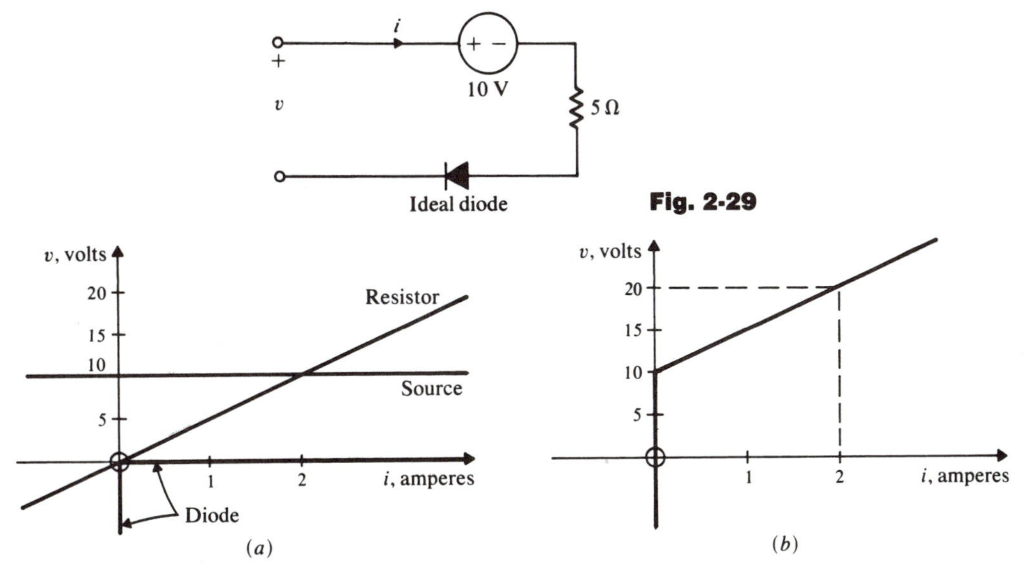


Fig. 2-29

Fig. 2-30

The $v$-$i$ characteristic for each component of the series combination is shown in Fig. 2-30*a*. Now by KVL the three voltages must be added at each current $i_1$ to obtain the overall characteristic. This of course amounts to adding the voltage source and resistor characteristics for positive (i.e. forward) currents, since the diode is then a short circuit. There can never be any negative (reverse) current, since the diode then acts as an open circuit. The required result is graphed in Fig. 2-30*b*.

**2.17** The 10-V ideal battery, the 5-Ω resistor, and the ideal diode of the previous problem are connected to a voltage source $v(t)$ as in Fig. 2-31. Use the results of the previous problem to find and sketch the waveform of the current $i(t)$ for the following inputs:
(*a*) $v(t) = 20$ V, (*b*) $v(t) = -20$ V, (*c*) $v(t) = 10t\{\mathbb{1}(t) - \mathbb{1}(t-3)\}$, (*d*) $v(t) = \{10 + 10 \sin 2\pi t\}\mathbb{1}(t)$.

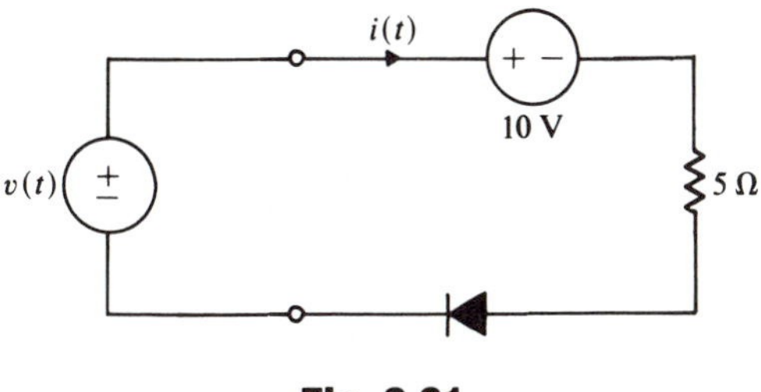


Fig. 2-31

(*a*) Referring to Fig. 2-30*b*, if $v(t) = 20$ V, then $i(t) = 2$ A. This result is plotted in Fig. 2-32*a*.

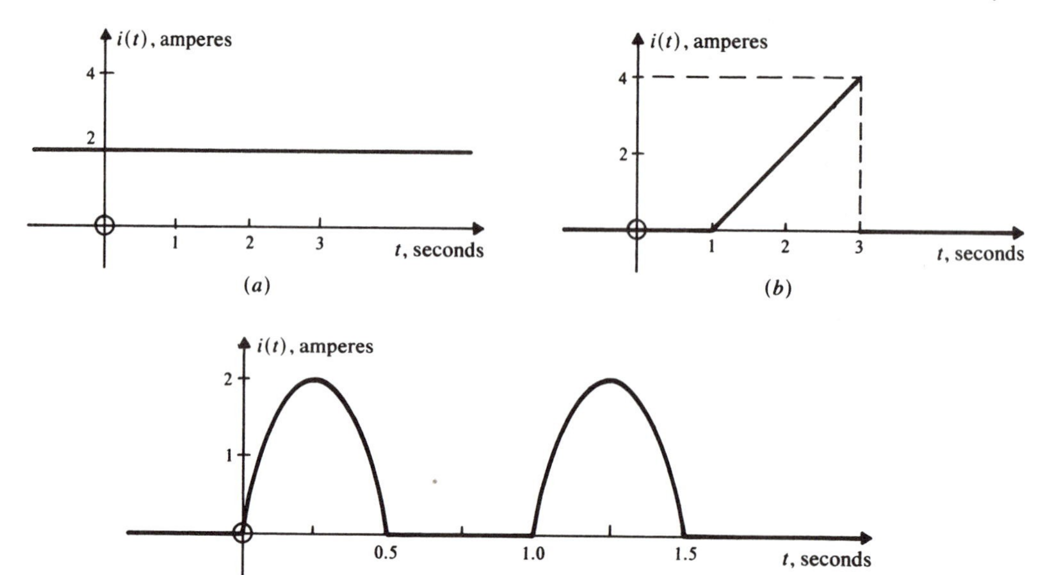


**Fig. 2-32**

(*b*) Referring again to Fig. 2-30*b*, $i(t) = 0$ for all $t$ if $v(t) = -20$ V. (This has not been graphed.)

(*c*) For $0 \le t < 1.0$, $v(t) < 10$ V and so $i(t) = 0$. For $1.0 \le t < 3.0$, $v(t) = 10t$, and using Fig. 2-30*b* we obtain the graph in Fig. 2-32*b*. (For example, when $t = 2$, $v(2) = 20$ V, and, from Fig. 2-32*b*, $i(2) = 2$ A.) For $t > 3$, $v(t)$ will be zero, as must $i(t)$.

(*d*) For $0 < t < 0.5$, $v(t) > 10$ and sinusoidal current will flow as in Fig. 2-32*c*; but for $0.5 \le t \le 1$, $v(t) \le 10$ and $i(t) = 0$. This waveform will repeat every half second as shown, for positive $t$.

**2.18** Two nonlinear resistors $\mathcal{R}_a$ and $\mathcal{R}_b$ are described in Fig. 2-33. Ascertain the $v$-$i$ characteristic of their parallel combination.

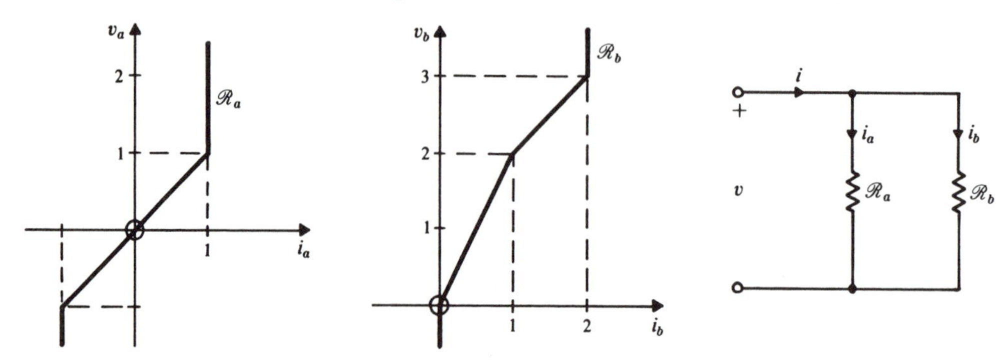


**Fig. 2-33**

| $v$ | $i_a$ | $i_b$ | $i = i_a + i_b$ |
|---|---|---|---|
| −1.0 | −1.0 | 0 | −1.0 |
| −0.5 | −0.5 | 0 | −0.5 |
| 0 | 0 | 0 | 0 |
| 0.5 | 0.5 | 0.25 | 0.75 |
| 1.0 | 1.0 | 0.50 | 1.50 |
| 1.5 | 1.0 | 0.75 | 1.75 |
| 2.0 | 1.0 | 1.0 | 2.0 |
| 2.5 | 1.0 | 1.5 | 2.50 |
| 3.0 | 1.0 | 2.0 | 3.0 |

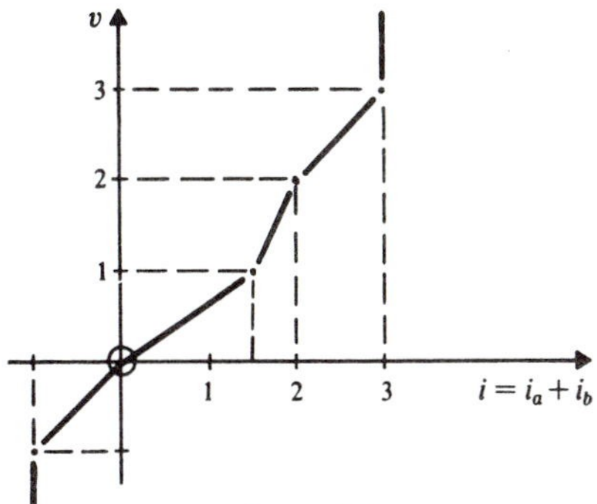


**Fig. 2-34**

For resistors in parallel, the overall $v$-$i$ characteristic is constructed by adding the parallel currents for each $v_1$, the consequence of KCL. This is shown in the tabulation above and the corresponding graph of $v$ vs. $i$ in Fig. 2-34. (Of course a purely graphical addition is possible.)

**2.19** A linear resistor, $R_c = 1\ \Omega$, is connected in series with the parallel combination of Problem 2.18 (see Fig. 2-35). Sketch the $v$-$i$ characteristic of the series-parallel combination.

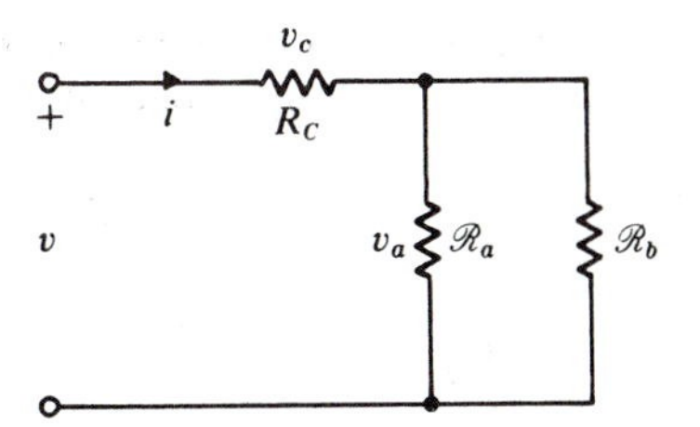


**Fig. 2-35**

Here we must add the series voltages for each current $i_1$ as in the following table, where the first two columns have been taken from Problem 2.18. The graph of $v$ vs. $i$ is plotted in Fig. 2-36.

| $i$ | $v_a$ for $\mathcal{R}_a$ and $\mathcal{R}_b$ in parallel | $v_c$ | $v = v_a + v_c$ |
|---|---|---|---|
| −1.0 | −1.0 | −1.0 | −2.0 |
| −0.5 | −0.5 | −0.5 | −1.0 |
| 0 | 0 | 0 | 0 |
| 0.75 | 0.5 | 0.75 | 1.25 |
| 1.5 | 1.0 | 1.5 | 2.5 |
| 1.75 | 1.5 | 1.75 | 3.25 |
| 2.0 | 2.0 | 2.0 | 4.0 |
| 2.5 | 2.5 | 2.5 | 5.0 |
| 3.0 | 3.0 | 3.0 | 6.0 |

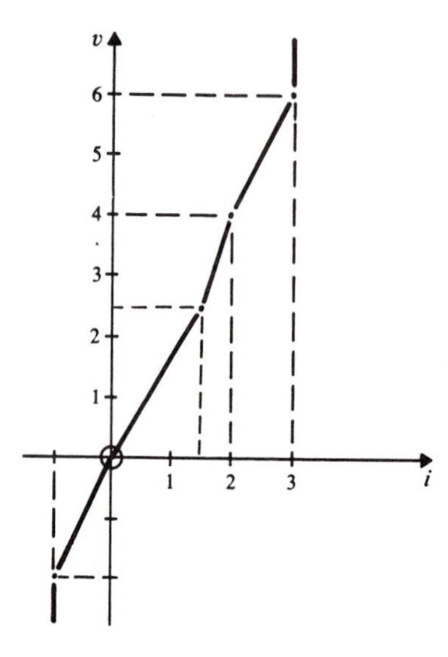


Fig. 2-36

### *Loop Analysis*

**2.20** Obtain a set of *loop* equations for the circuit of Fig. 2-37 following the procedure discussed under Loop Analysis (page 41).

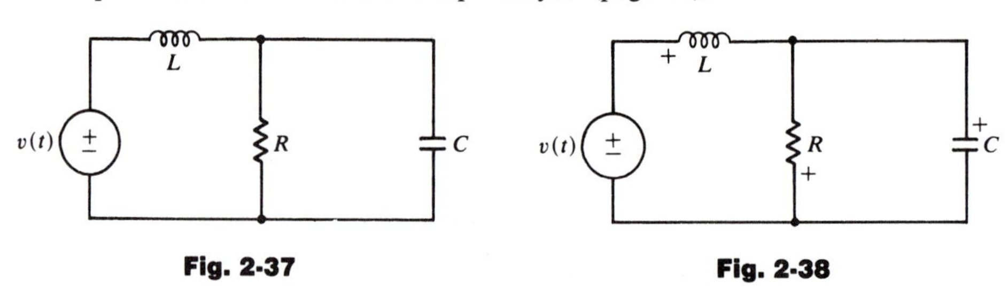


Fig. 2-37 Fig. 2-38

**Step 1:** Reference polarities are assigned arbitrarily for the $R$, $L$, and $C$ elements as in Fig. 2-38.

**Step 2:** One appropriate set of KVL equations is

$$-v(t) + v_L - v_R = 0 \qquad \text{(left windowpane)}$$

$$-v(t) + v_L + v_C = 0 \qquad \text{(outer loop)}$$

Note that we have satisfied all the requirements relating to the number and choice of loops.

**Step 3:** A legitimate set of loop currents is shown in Fig. 2-39. The currents need not follow the same paths as the loops chosen in step 2.

**Step 4:** Writing the $v$-$i$ relations,

$$v_L = L\frac{di_1}{dt}$$

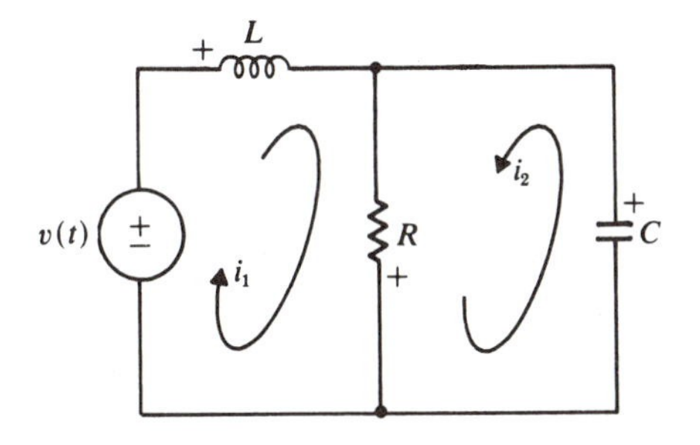


**Fig. 2-39**

$$v_R = R(-i_1 - i_2)$$

$$v_C = \frac{1}{C}\int_0^t -i_2\, d\tau + v_C(0)$$

**Step 5:** This step is not required since there are no current sources in this circuit.

**Step 6:** Substituting the $v$-$i$ relations of step 4 into the KVL equations of step 2,

$$-v(t) + L\frac{di_1}{dt} - R(-i_1 - i_2) = 0$$

$$-v(t) + L\frac{di_1}{dt} + \frac{1}{C}\int_0^t -i_2\, d\tau + v_C(0) = 0$$

### *Notes*

1. The final equations may look quite different if the KVL loops and/or the loop currents are chosen differently. However, if no errors are made, the *branch* currents and voltages will come out the same regardless of how the loops are chosen. If, for example, the KVL equations of step 2 had been

$$-v(t) + v_L - v_R = 0 \qquad \text{(left windowpane)}$$

$$v_R + v_C = 0 \qquad \text{(right windowpane)}$$

then the final equations would have been

$$-v(t) + L\frac{di_1}{dt} - R(-i_1 - i_2) = 0$$

$$R(-i_1 - i_2) + \frac{1}{C}\int_0^t -i_2\, d\tau + v_C(0) = 0$$

2. Don't try to take shortcuts. You should practice the application of the procedure until it becomes second nature. Only in this way you will avoid errors.

**2.21** Write a set of loop equations for the circuit of Fig. 2-40, given that the initial capacitor voltage is 2 V (bottom plate positive) and that the initial inductor current is 1 A (downward throught the inductor).

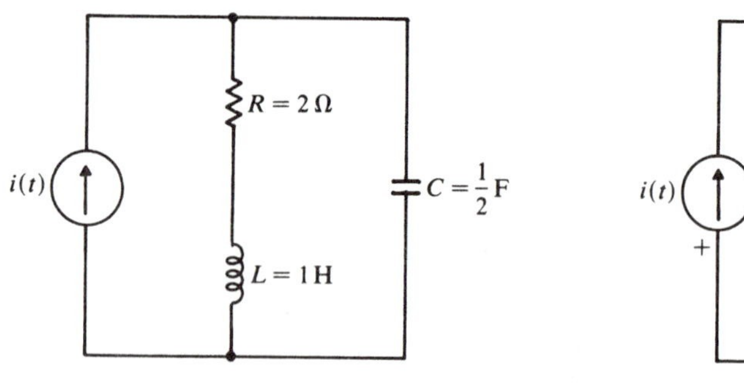


**Fig. 2-40** **Fig. 2-41**

**Step 1:** Reference polarities for $L$ and $C$ should, for convenience, correspond to the directions of the given initial conditions; the associated sign convention dictates the polarity marking for $i(t)$, while the marking of $R$ is arbitrary (see Fig. 2-41).

**Step 2:** The KVL equations around the two windowpanes are

$$v_1 + v_R + v_L = 0$$

$$-v_L - v_R - v_C = 0$$

**Step 3:** Here we shall choose the loop currents, $i_1$ and $i_2$, around the same paths as were used for the KVL equations (see Fig. 2-42).

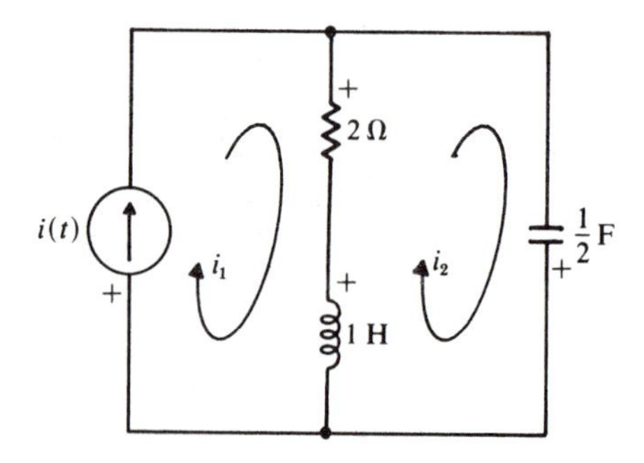


**Fig. 2-42**

**Step 4:** The $v$-$i$ relations are

$$v_R = 2(i_1 - i_2)$$

$$v_L = 1 \frac{d}{dt}(i_1 - i_2)$$

$$v_C = 2 \int_0^t -i_2 \, d\tau + 2$$

**Step 5:** Relating the loop currents to the current source, $i_1 = i(t)$.

**Step 6:** Substituting the results of steps 4 and 5 into the KVL equations of step 2 gives

$$v_i + 2\{i(t) - i_2\} + \frac{d}{dt}\{i(t) - i_2\} = 0$$

$$-\frac{d}{dt}\{i(t) - i_2\} - 2\{i(t) - i_2\} - 2\int_0^t -i_2\, d\tau - 2 = 0$$

We now have two loop equations in the two unknowns $v_i$ and $i_2$.

**2.22** Write an appropriate set of loop equations for the network of Fig. 2-43. The initial conditions $i_{L_1}(0)$, $i_{L_2}(0)$, $v_{C_1}(0)$, and $v_{C_2}(0)$ are given, as are $v(t)$, $i(t)$, and the parameter values.

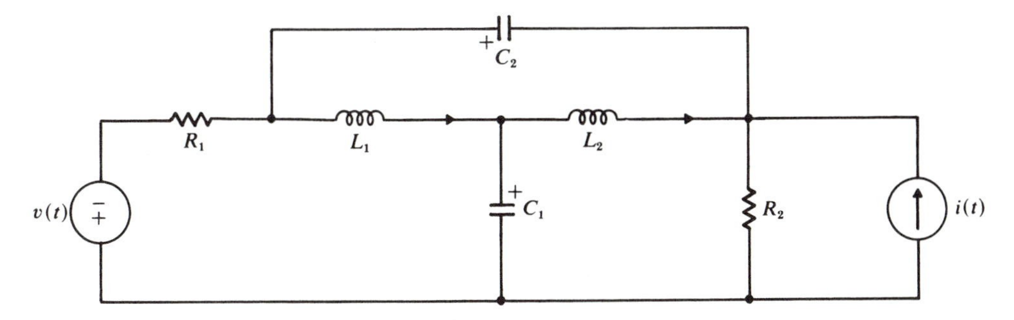


**Fig. 2-43**

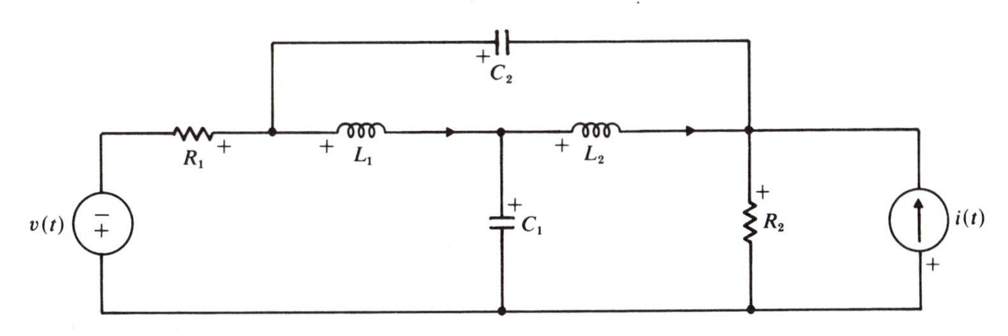


**Fig. 2-44**

**Step 1:** Reference polarities are assigned as in Fig. 2-44. For convenience, those for the $L$'s and $C$'s correspond to the given initial condition reference directions.

**Step 2:** This network has four windowpanes so we must write four KVL equations. For example,

$$v(t) - v_{R_1} + v_{L_1} + v_{C_1} = 0 \qquad \text{(left windowpane)}$$

$$v(t) - v_{R_1} + v_{C_2} - v_i = 0 \qquad \text{(outer loop)}$$

$$v_{L_1} + v_{L_2} - v_{C_2} = 0 \qquad \text{(top loop)}$$

$$v_{R_2} + v_i = 0 \qquad \text{(right windowpane)}$$

Notice that the rules for choosing loops have been obeyed.

**Step 3:** We will choose the loop currents around the windowpanes, as in Fig. 2-45.

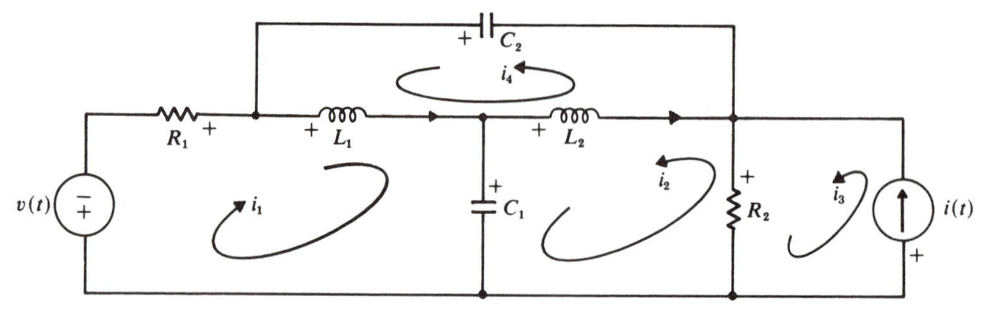


**Fig. 2-45**

**Step 4:** The $v$-$i$ relations are

$$v_{R_1} = R_1(-i_1)$$

$$v_{L_1} = L_1 \frac{d}{dt}(i_1 + i_4)$$

$$v_{C_1} = \frac{1}{C_1}\int_0^t (i_1 + i_2)\,d\tau + v_{C_1}(0)$$

$$v_{R_2} = R_2(-i_2 + i_3)$$

$$v_{L_2} = L_2 \frac{d}{dt}(-i_2 + i_4)$$

$$v_{C_2} = \frac{1}{C_2}\int_0^t -i_4\,d\tau + v_{C_2}(0)$$

**Step 5:**

$$i_3 = i(t)$$

**Step 6:** Substituting the equations of steps 4 and 5 into the KVL equations of step 2,

$$v(t) - R_1(-i_1) + L_1 \frac{d}{dt}(i_1 + i_4) + \frac{1}{C_1}\int_0^t (i_1 + i_2)\,d\tau + v_{C_1}(0) = 0$$

$$v(t) - R_1(-i_1) + \frac{1}{C_2}\int_0^t -i_4\,d\tau + v_{C_2}(0) - v_i = 0$$

$$L_1 \frac{d}{dt}(i_1 + i_4) + L_2 \frac{d}{dt}(-i_2 + i_4) - \frac{1}{C_2}\int_0^t -i_4\,d\tau - v_{C_2}(0) = 0$$

$$R_2\{-i_2 + i(t)\} + v_i = 0$$

Thus we have four loop equations in the four unknowns $i_1$, $i_2$, $i_4$, and $v_i$.

**2.23** The circuit of Fig. 2-46 contains a dependent voltage source and a crossover. First redraw the circuit without any crossovers, and then write an appropriate set of loop equations.

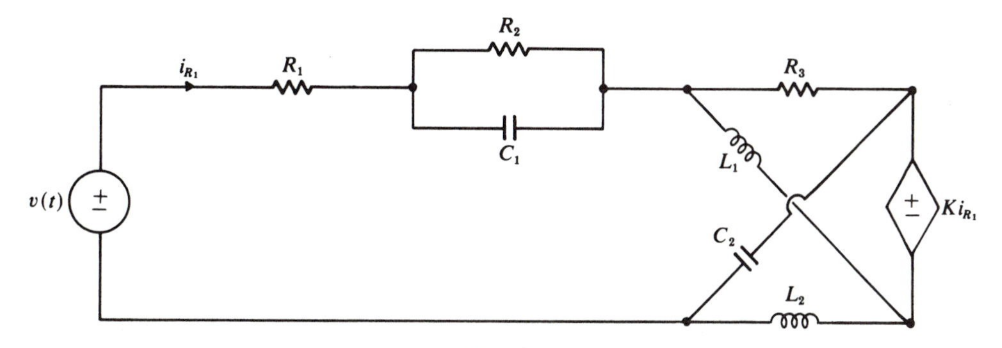


**Fig. 2-46**

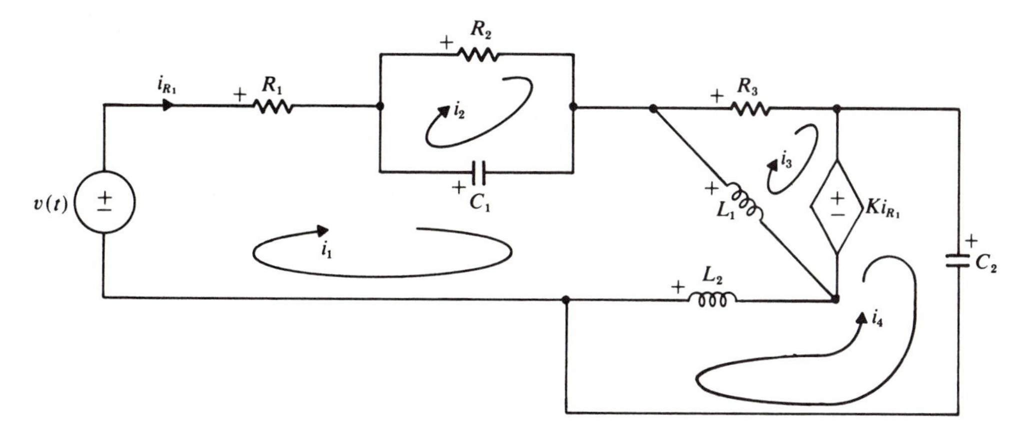


**Fig. 2-47**

**Step 1:** The redrawn circuit with assigned voltage polarities is shown in Fig. 2-47.

**Step 2:** Applying KVL around the windowpanes,

$$-v(t) + v_{R_1} + v_{C_1} + v_{L_1} - v_{L_2} = 0$$

$$v_{R_2} - v_{C_1} = 0$$

$$v_{R_3} + Ki_{R_1} - v_{L_1} = 0$$

$$-Ki_{R_1} + v_{C_2} + v_{L_2} = 0$$

**Step 3:** We will choose the loop currents around the windowpanes as in Fig. 2-47.

**Step 4:** The $v$-$i$ relations are

$$v_{R_1} = R_1 i_1, \qquad v_{L_1} = L_1 \frac{d}{dt}(i_1 - i_3), \qquad v_{C_1} = \frac{1}{C_1}\int_0^t (i_1 - i_2)\, d\tau + v_{C_1}(0)$$

$$v_{R_2} = R_2 i_2, \qquad v_{L_2} = L_2 \frac{d}{dt}(-i_1 + i_4), \qquad v_{C_2} = \frac{1}{C_2}\int_0^t i_4\, d\tau + v_{C_2}(0)$$

$$v_{R_3} = R_3 i_3$$

**Step 5:** Not required; but note that we can write $i_1$ for $i_{R_1}$ when describing the dependent source.

**Step 6:** After substituting the $v$-$i$ relationships and rearranging,

$$R_1 i_1 + \frac{1}{C_1}\int_0^t i_1\, d\tau + L_1 \frac{di_1}{dt} + L_2 \frac{di_1}{dt} + \frac{1}{C_1}\int_0^t -i_2\, d\tau - L_1 \frac{di_3}{dt} - L_2 \frac{di_4}{dt} = v(t) - v_{C_1}(0)$$

$$-\frac{1}{C_1}\int_0^t i_1\, d\tau + R_2 i_2 + \frac{1}{C_1}\int_0^t i_2\, d\tau = v_{C_1}(0)$$

$$K i_1 - L_1 \frac{di_1}{dt} + L_1 \frac{di_3}{dt} + R_3 i_3 = 0$$

$$-K i_1 - L_2 \frac{di_1}{dt} + L_2 \frac{di_4}{dt} + \frac{1}{C_2}\int_0^t i_4\, d\tau = -v_{C_2}(0)$$

These are the four loop equations in the four unknowns $i_1$, $i_2$, $i_3$, and $i_4$. The known function $v(t)$ and the known initial conditons have been moved to the other side of the equations from the unknowns.

**2.24** Write the loop equations for the circuit of Fig. 2-48, then with the help of Cramer's rule solve for the current $i_4$ flowing through the 3-Ω resistor. Use the prespecified reference polarities and loop currents shown in Fig. 2-48.

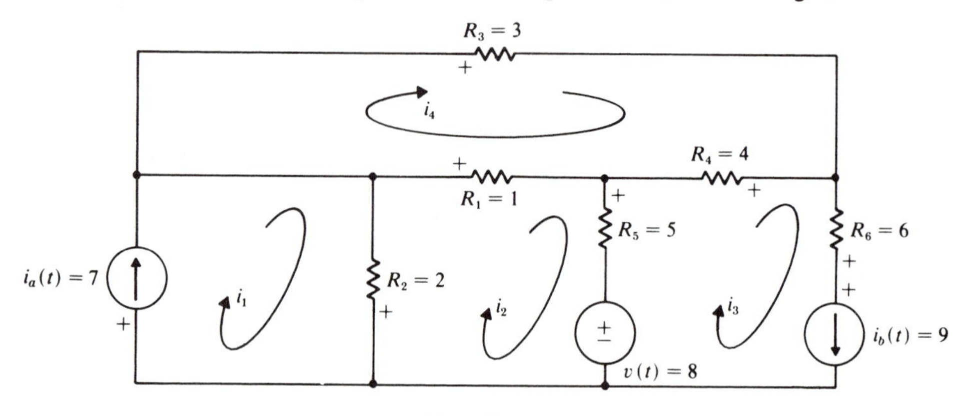


**Fig. 2-48**

Applying KVL around the windowpanes (step 2),

$$v_{i_a} - v_{R_2} = 0$$

$$v_{R_2} + v_{R_1} + v_{R_5} + v(t) = 0$$

$$-v_{R_5} - v_{R_4} - v_{R_6} + v_{i_b} - v(t) = 0$$

$$v_{R_3} + v_{R_4} - v_{R_1} = 0$$

Then substituting the $v$-$i$ relations (step 4) and the given numerical values,

$$v_{i_a} - 2(i_2 - i_1) = 0$$

$$2(i_2 - i_1) + 1(i_2 - i_4) + 5(i_2 - i_3) + 8 = 0$$

$$-5(i_2 - i_3) - 4(i_4 - i_3) - 6(-i_3) + v_{i_b} - 8 = 0$$

$$3(i_4) + 4(i_4 - i_3) - 1(i_2 - i_4) = 0$$

From step 5,

$$i_1 = i_a(t) = 7$$

$$i_3 = i_b(t) = 9$$

which when substituted into the preceding set of equations yields

$$v_{i_a} - 2i_2 = -14$$

$$8i_2 - i_4 = +51$$

$$v_{i_b} - 5i_2 - 4i_4 = -127$$

$$-i_2 + 8i_4 = +36$$

The required current (through the 3-Ω resistor) is $i_4$, which from Cramer's rule is

$$i_4 = \frac{\begin{vmatrix} 1 & 0 & -2 & -14 \\ 0 & 0 & 8 & 51 \\ 0 & 1 & -5 & -127 \\ 0 & 0 & -1 & 36 \end{vmatrix}}{\begin{vmatrix} 1 & 0 & -2 & 0 \\ 0 & 0 & 8 & -1 \\ 0 & 1 & -5 & -4 \\ 0 & 0 & -1 & 8 \end{vmatrix}}$$

$$= \frac{(1)\begin{vmatrix} 0 & 8 & 51 \\ 1 & -5 & -127 \\ 0 & -1 & 36 \end{vmatrix}}{(1)\begin{vmatrix} 0 & 8 & -1 \\ 1 & -5 & -4 \\ 0 & -1 & 8 \end{vmatrix}} \quad \text{(expanding with respect to the first column)}$$

$$= \frac{(-1)\begin{vmatrix} 8 & 51 \\ -1 & 36 \end{vmatrix}}{(-1)\begin{vmatrix} 8 & -1 \\ -1 & 8 \end{vmatrix}} = \frac{(-1)(288 + 51)}{(-1)(64 - 1)} = \frac{113}{21} \text{ A}$$

### *Notes*

1. The relationships of step 5 are simplest if the loop currents can be chosen so that only one passes through each current source.
2. The solution process is simplified if only one loop current passes through the element whose current or voltage is to be found.
3. Once the equations have been written, there is an uncomfortably high probability of making an error in the solution. One great advantage of the digital computer is that this probability can be reduced to almost zero. And there are, today, computer programs which can take over the task of equation *writing* as well as that of equation *solving*.

**2.25** Write a set of loop equations for the circuit of Fig. 2-49.

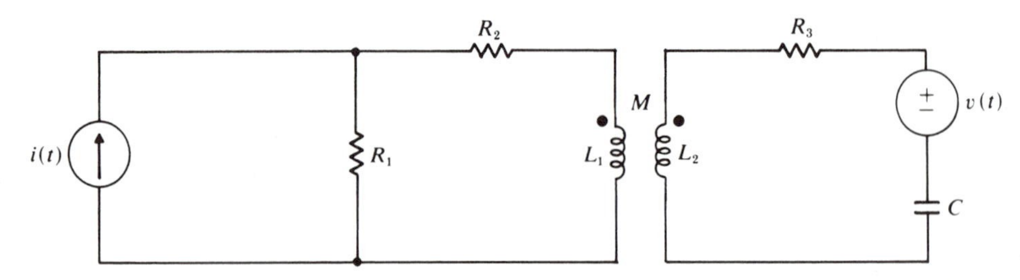


**Fig. 2-49**

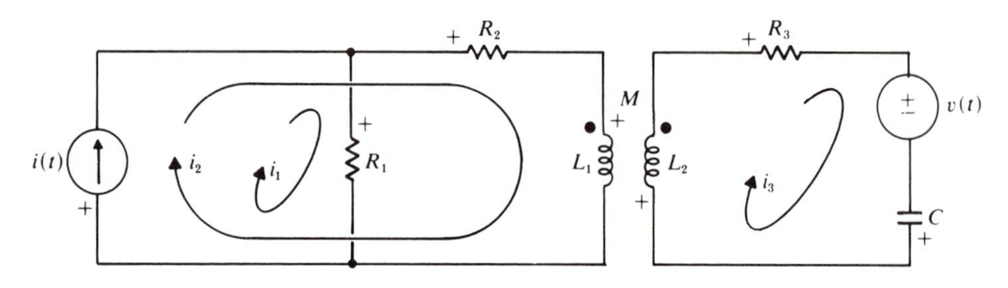


**Fig. 2-50**

Reference polarities have been assigned in Fig. 2-50, and an appropriate set of KVL equations is

$$v_i + v_{R_1} = 0$$

$$v_i + v_{R_2} + v_{L_1} = 0$$

$$v_{L_2} + v_{R_3} + v(t) - v_C = 0$$

The $v$-$i$ relations for the inductors (in terms of the loop currents of Fig. 2-50) are

$$v_{L_1} = +L_1 \frac{di_2}{dt} + M \frac{d}{dt}(-i_3)$$

$$v_{L_2} = +L_2 \frac{di_3}{dt} - M \frac{di_2}{dt}$$

Substituting all the $v$-$i$ relations and making use of the step 5 relation, $i_1 = i(t) - i_2$,

$$v_i + R_1\{i(t) - i_2\} = 0$$

$$v_i + R_2 i_2 + L_1 \frac{di_2}{dt} - M \frac{di_3}{dt} = 0$$

$$L_2 \frac{di_3}{dt} - M \frac{di_2}{dt} + R_3 i_3 + v(t) - \frac{1}{C}\int_0^t -i_3 \, d\tau - v_C(0) = 0$$

These are the network's three loop equations in the three unknowns $v_i$, $i_2$, and $i_3$.

Notice that this is the first circuit which has had more than one *separate part*. This has not had any effect on the procedure for writing loop equations.

### *Nodal Analysis*

**2.26** Use the procedure discussed above under Nodal Analysis (page 44) to obtain a set of *nodal* equations for the circuit of Fig. 2.51. (The *loop* equations for this circuit were written in Problem 2.20.)

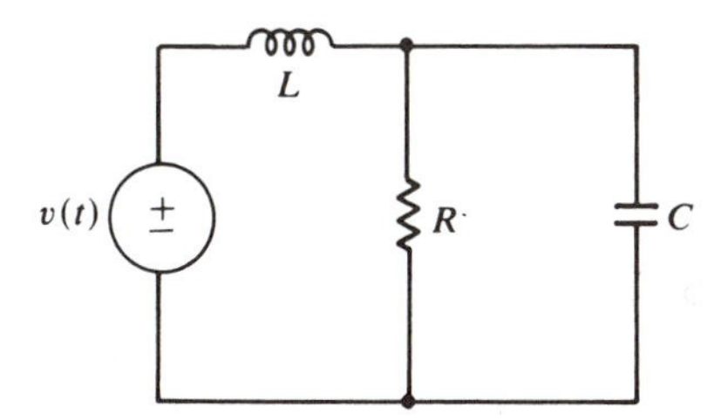


**Fig. 2-51**

**Step 1:** The current reference direction in the voltage source is determined by the associated sign convention. The others are arbitrarily assigned in Fig. 2-52.

**Step 2:** All the nodes have been labeled and a reference node has been selected in Fig. 2-53. We now write KCL equations for all the nodes except the reference node:

Node $a$: $i_v - i_L = 0$
Node $b$: $i_R + i_L - i_C = 0$

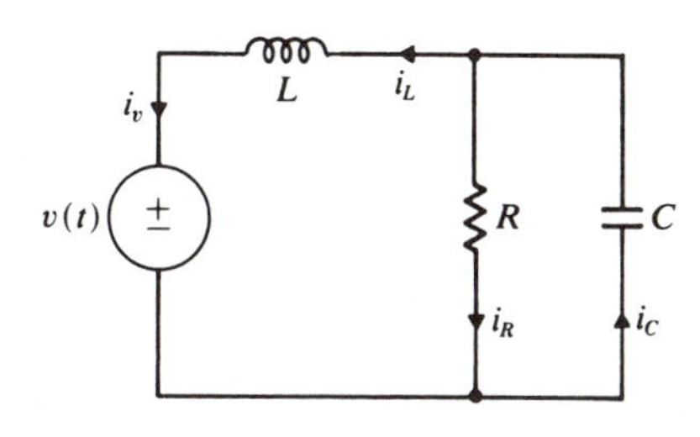


**Fig. 2-52**

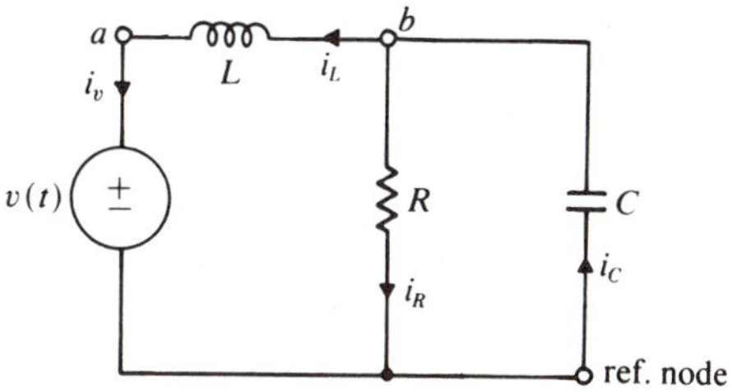

**Fig. 2-53**

**Step 3:** The branch voltages are related to the nodal voltages by

$$v(t) = v_a, \qquad v_L = v_b - v_a, \qquad v_C = -v_b, \qquad v_R = v_b$$

**Step 4:** The $i$-$v$ relations are

$$i_L = \frac{1}{L}\int_0^t (v_b - v_a)\, d\tau + i_L(0)$$

$$i_C = C\frac{d}{dt}(-v_b)$$

$$i_R = \frac{1}{R}v_b$$

**Step 5:** Here the nodal voltage $v_a$ is equal to the source voltage $v(t)$, i.e. $v_a = v(t)$.

**Step 6:** Substituting the results of steps 4 and 5 into the KCL equations of step 2,

$$i_v - \frac{1}{L}\int_0^t \{v_b - v(\tau)\}\, d\tau - i_L(0) = 0$$

$$\frac{1}{R} v_b + \frac{1}{L}\int_0^t \{v_b - v(\tau)\}\, d\tau + i_L(0) - C\frac{d}{dt}(-v_b) = 0$$

or, after rearrangement,

$$i_v - \frac{1}{L}\int_0^t v_b\, d\tau = -\frac{1}{L}\int_0^t v(\tau)\, d\tau + i_L(0)$$

$$\frac{1}{R} v_b + \frac{1}{L}\int_0^t v_b\, d\tau + C\frac{dv_b}{dt} = \frac{1}{L}\int_0^t v(\tau)\, d\tau - i_L(0)$$

Thus we have two nodal equations in the two unknowns $v_b$ and $i_v$. After solution we could find any branch voltage from step 3, and any branch current from step 4.

**2.27** Write a set of nodal equations for the circuit of Fig. 2-54, given that the initial capacitor voltage is 2 V (bottom plate positive) and that the initial inductor current is 1 A (downward). (See Problem 2.21 for the loop equations.)

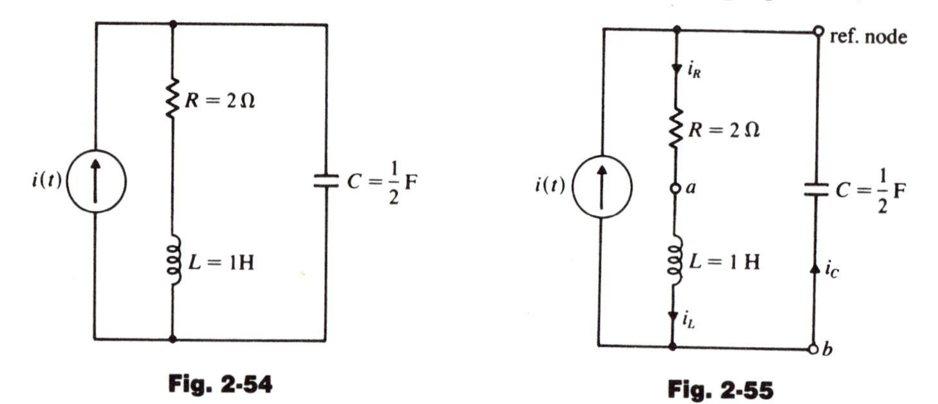


**Fig. 2-54** **Fig. 2-55**

**Steps 1 and 2:** See Fig. 2-55. Note that the reference directions of $i_L$ and $i_C$ have for convenience been chosen to coincide with the directions of the given initial conditions $v_C(0)$ and $i_L(0)$. The KCL equations are

Node $a$: $\quad i_L - i_R = 0$

Node $b$: $\quad i(t) - i_L + i_C = 0$

**Step 3:** The branch voltages in terms of the nodal voltages are

$$v_R = -v_a, \qquad v_L = v_a - v_b, \qquad v_C = v_b$$

**Step 4:** Writing the $i$-$v$ relations.

$$i_R = \frac{1}{2}(-v_a)$$

$$i_L = 1\int_0^t (v_a - v_b)\,d\tau + 1$$

$$i_C = \frac{1}{2}\frac{dv_b}{dt}$$

**Step 5:** Not required, since the circuit does not have any voltage sources.
**Step 6:** Substituting the $i$-$v$ relations of step 4 into the KCL equations of step 2 yields

$$\int_0^t (v_a - v_b)\,d\tau + 1 - \tfrac{1}{2}(-v_a) = 0$$

$$i(t) - \int_0^t (v_a - v_b)\,d\tau - 1 + \frac{1}{2}\frac{dv_b}{dt} = 0$$

Or, after rearranging into the preferred form,

$$\int_0^t v_a\,d\tau + \tfrac{1}{2}v_a - \int_0^t v_b\,d\tau = -1$$

$$-\int_0^t v_a\,d\tau + \int_0^t v_b\,d\tau + \frac{1}{2}\frac{dv_b}{dt} = 1 - i(t)$$

which are the required nodal equations in the unknowns $v_a$ and $v_b$.

**2.28** Write an appropriate set of nodal equations for the network of Fig. 2-56. (This network was described by loop equations in Problem 2.22.)

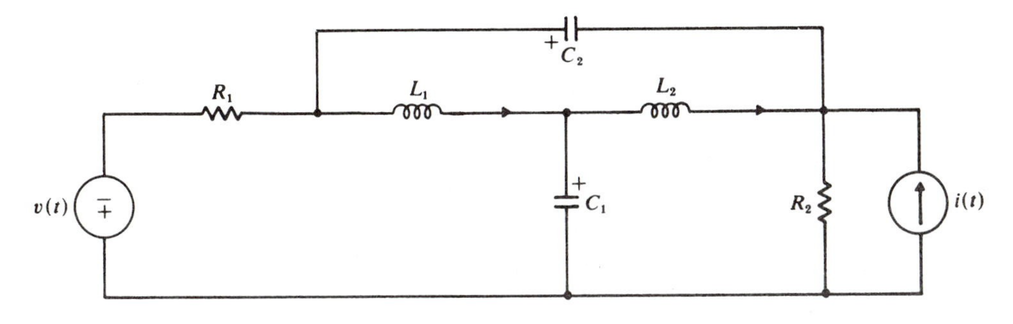


**Fig. 2-56**

**Steps 1 and 2:** See Fig. 2-57 for the chosen current reference directions and node labels. Then by KCL,

| | |
|---|---|
| Node 1: | $i_{R_1} - i_v = 0$ |
| Node 2: | $i_{L_1} + i_{C_2} - i_{R_1} = 0$ |
| Node 3: | $i_{L_2} + i_{C_1} - i_{L_1} = 0$ |
| Node 4: | $i_{R_2} - i_{L_2} - i_{C_2} - i(t) = 0$ |

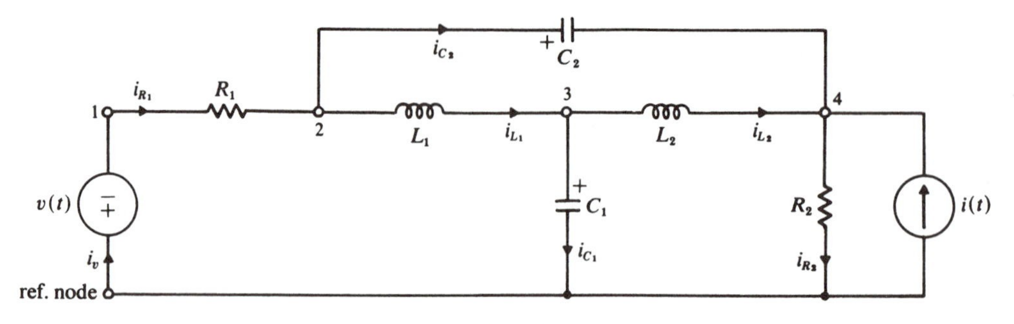


**Fig. 2-57**

**Step 3:** The branch voltages for the passive elements are

$$v_{R_1} = v_1 - v_2 \qquad v_{C_1} = v_3 \qquad v_{L_1} = v_2 - v_3$$

$$v_{R_2} = v_4 \qquad v_{C_2} = v_2 - v_4 \qquad v_{L_2} = v_3 - v_4$$

**Step 4:** Now, writing the $i$-$v$ relations,

$$i_{R_1} = \frac{1}{R_1}(v_1 - v_2) \qquad i_{C_1} = C_1 \frac{dv_3}{dt} \qquad i_{L_1} = \frac{1}{L_1}\int_0^t (v_2 - v_3)\, d\tau + i_{L_1}(0)$$

$$i_{R_2} = \frac{1}{R_2} v_4 \qquad i_{C_2} = C_2 \frac{d}{dt}(v_2 - v_4) \qquad i_{L_2} = \frac{1}{L_2}\int_0^t (v_3 - v_4)\, d\tau + i_{L_2}(0)$$

**Step 5:** $v_1 = -v(t)$

**Step 6:** Substituting steps 4 and 5 into the KCL equations yields

$$\frac{1}{R_1}\{-v(t) - v_2\} - i_v = 0$$

$$\frac{1}{L_1}\int_0^t (v_2 - v_3)\, d\tau + i_{L_1}(0) + C_2 \frac{d}{dt}(v_2 - v_4) - \frac{1}{R_1}\{-v(t) - v_2\} = 0$$

$$\frac{1}{L_2}\int_0^t (v_3 - v_4)\, d\tau + i_{L_2}(0) + C_1 \frac{dv_3}{dt} - \frac{1}{L_1}\int_0^t (v_2 - v_3)\, d\tau - i_{L_1}(0) = 0$$

$$\frac{1}{R_2} v_4 - \frac{1}{L_2}\int_0^t (v_3 - v_4)\, d\tau - i_{L_2}(0) - C_2 \frac{d}{dt}(v_2 - v_4) - i(t) = 0$$

Thus we have four nodal equations which could be solved for $v_2$, $v_3$, $v_4$, and $i_v$.

**2.29** Use nodal methods to find the current $i_3$ flowing through the 3-Ω resistor in the circuit of Fig. 2-58. (This problem was solved by loop analysis in Problem 2.24.)

Since only the current through the 3-Ω resistor is to be found, one end of this resistor should be chosen as the reference node so that we need only solve for one voltage, $v_a$. This choice, and that of the current reference directions and node designations, is indicated in Fig. 2-59.

Writing KCL equations at all the nodes except the reference node,

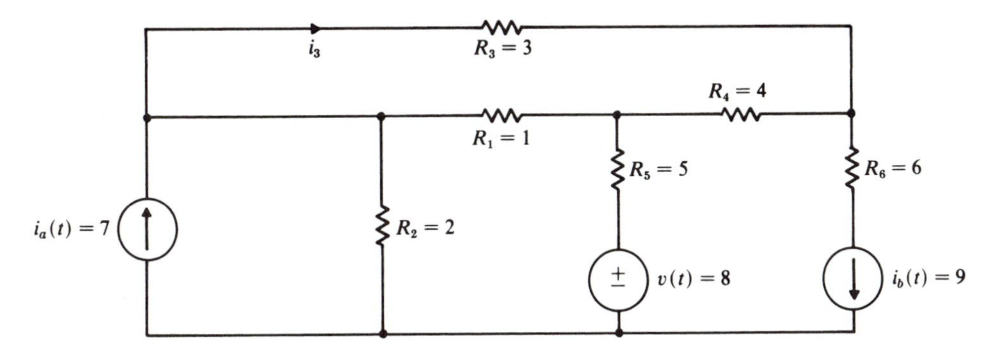


**Fig. 2-58**

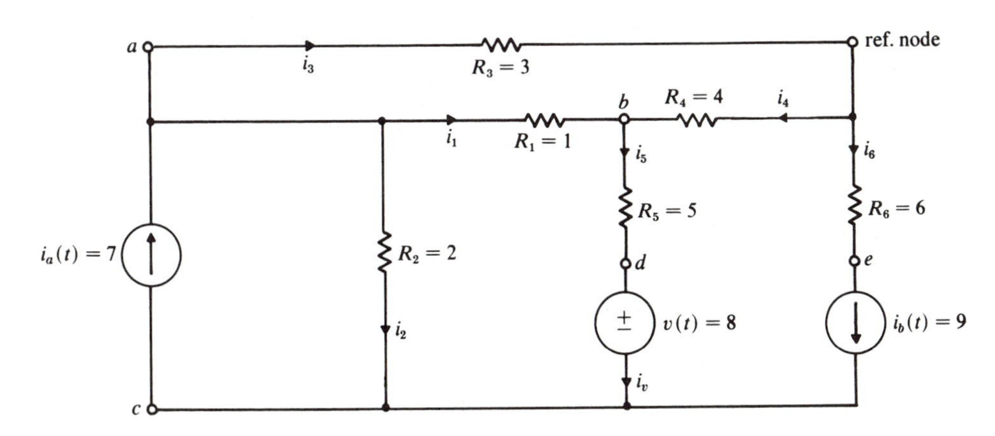


**Fig. 2-59**

$$i_a(t) - i_1 - i_2 - i_3 = 0$$
$$i_1 + i_4 - i_5 = 0$$
$$i_2 + i_v + i_b(t) - i_a(t) = 0$$
$$i_5 - i_v = 0$$
$$i_6 - i_b(t) = 0$$

Next, the $i$-$v$ relations are

$$i_1 = 1(v_a - v_b)$$
$$i_2 = \tfrac{1}{2}(v_a - v_c)$$
$$i_3 = \tfrac{1}{3}v_a$$
$$i_4 = \tfrac{1}{4}(-v_b)$$
$$i_5 = \tfrac{1}{5}(v_b - v_d)$$
$$i_6 = \tfrac{1}{6}(-v_e)$$

We also know (step 5) that

$$v_d = v_c + v(t)$$

Finally, if we substitute into the KCL equations, including the facts that $i_a(t) = 7$, $i_b(t) = 9$ and $v(t) = 8$, slight rearrangement yields

$$\begin{aligned}
-\tfrac{11}{6}v_a + v_b + \tfrac{1}{2}v_c &= -7\\
v_a - \tfrac{29}{20}v_b + \tfrac{1}{5}v_c &= -\tfrac{8}{5}\\
i_v + \tfrac{1}{2}v_a - \tfrac{1}{2}v_c &= -2\\
-i_v + \tfrac{1}{5}v_b - \tfrac{1}{5}v_c &= \tfrac{8}{5}\\
-\tfrac{1}{6}v_e &= 9
\end{aligned}$$

which is an appropriate set of five simultaneous nodal equations in the five unknowns $i_v$, $v_a$, $v_b$, $v_c$, and $v_e$.

The last equation can be omitted, since $v_e$ does not appear elsewhere and by adding the third and fourth equations we can eliminate $i_v$. The resulting three equations in the three unknowns $v_a$, $v_b$, and $v_c$ are

$$\begin{aligned}
-\tfrac{11}{6}v_a + v_b + \tfrac{1}{2}v_c &= -7\\
v_a - \tfrac{29}{20}v_b + \tfrac{1}{5}v_c &= -\tfrac{8}{5}\\
\tfrac{1}{2}v_a + \tfrac{1}{5}v_b - \tfrac{7}{10}v_c &= -\tfrac{2}{5}
\end{aligned}$$

Then, from Cramer's rule,

$$v_a = \frac{\begin{vmatrix} -7 & 1 & \frac{1}{2} \\ -\frac{8}{5} & -\frac{29}{20} & \frac{1}{5} \\ -\frac{2}{5} & \frac{1}{5} & -\frac{7}{10} \end{vmatrix}}{\begin{vmatrix} -\frac{11}{6} & 1 & \frac{1}{2} \\ 1 & -\frac{29}{20} & \frac{1}{5} \\ \frac{1}{2} & \frac{1}{5} & -\frac{7}{10} \end{vmatrix}} = \tfrac{113}{7}\text{ V}$$

Now the current through the 3-$\Omega$ resistor is

$$i_3 = \tfrac{1}{3}v_a = \tfrac{1}{3}\left(\tfrac{113}{7}\right) = \tfrac{113}{21}\text{ A}$$

which agrees with the solution to Problem 2.24. Note 3 in Problem 2.24 also applies here!

**2.30** A resistive circuit containing both an independent current source $i_1$ and a dependent current source $i_4$ is shown in Fig. 2-60. Find the voltage $v_3$ across the resistor $R_3$.

Proceeding by nodal analysis, the KCL equtions are

$$i_1 - i_{R_1} - i_{R_2} = 0$$

$$i_{R_2} - i_{R_3} - i_4 = 0$$

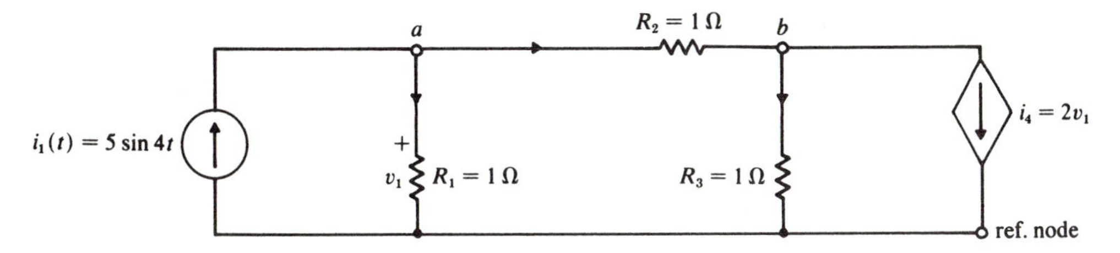


**Fig. 2-60**

and substituting the $i$-$v$ relations and $i_4 = 2v_1 = 2v_a$,

$$i_1 - v_a - (v_a - v_b) = 0$$
$$(v_a - v_b) - v_b - 2v_a = 0$$

Substituting $i_1 = 5 \sin 4t$ and rearranging,

$$2v_a - v_b = 5 \sin 4t$$
$$v_a + 2v_b = 0$$

Then, from Cramer's rule,

$$v_3 = v_b = \frac{\begin{vmatrix} 2 & 5 \sin 4t \\ 1 & 0 \end{vmatrix}}{\begin{vmatrix} 2 & -1 \\ 1 & 2 \end{vmatrix}} = \frac{-5 \sin 4t}{5} = -\sin 4t$$

### *Thévenin Equivalent Circuits*

**2.31** Show that the circuit of Fig. 2-10 is described by

$$v = -R_{\text{Th}}i + v_{\text{Th}}$$

With the + polarity reference mark at the left end of $R_{\text{Th}}$, $v_{R\text{Th}} = R_{\text{Th}}i$. Now the voltage across the load is $v$, so by KVL, $-v_{\text{Th}} + v_{R\text{Th}} + v = 0$, or $v = -v_{R\text{Th}} + v_{\text{Th}}$. Thus

$$v = -R_{\text{Th}}i + v_{\text{Th}}$$

Comment: By simple rearrangement we can write down the corresponding $i$-$v$ relation

$$i = -(1/R_{\text{Th}})v + v_{\text{Th}}/R_{\text{Th}}$$

**2.32** Use nodal analysis to obtain the $v$-$i$ relation for the one-port network of Fig. 2-61$a$, whose circuit diagram has already been marked with reference directions and node labels. Hence find the Thévenin equivalent for this one port.

**Step 2:** The KCL equations are

$$i_{R_1} + i_e = 0$$
$$i_{R_1} - i_{R_2} - i = 0$$

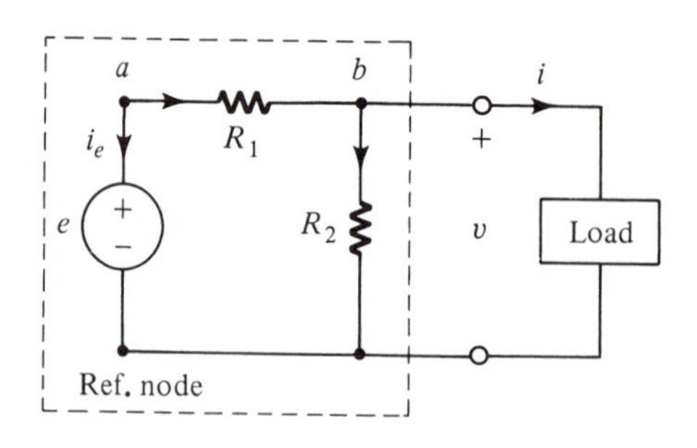


(*a*) A one-port network

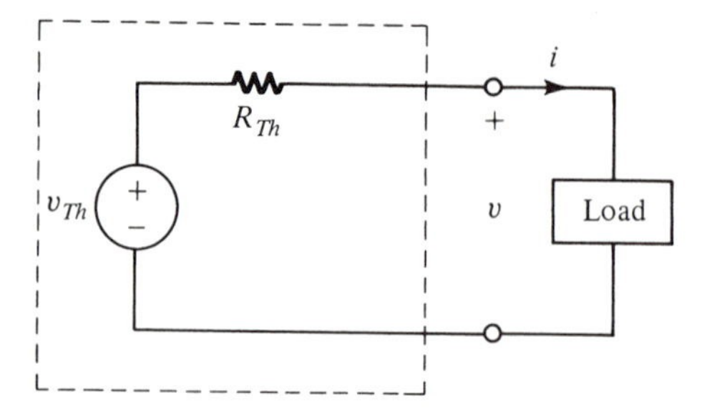


(*b*) Its Thevenin equivalent

**Fig. 2-61**

**Steps 3 and 4:** The $i$-$v$ relations are

$$i_{R_1} = \frac{1}{R_1}(v_a - v_b) \quad \text{and} \quad i_{R_2} = \frac{1}{R_2}v_b$$

**Step 5:** Here $v_a = e$. And by inspection we note that $v$ and $v_b$ describe the same voltage; that is, $v_b = v$.

**Step 6:** We make the usual substitutions, whence

$$\frac{1}{R_1}(e - v) + i_e = 0$$

$$\frac{1}{R_1}(e - v) - \frac{1}{R_2}v - i = 0$$

When rearranging we treat the terminal current $i$ as if it were known. Thus,

$$\frac{1}{R_1}v - i_e = \frac{1}{R_1}e$$

$$-\left(\frac{1}{R_1} + \frac{1}{R_2}\right)v = i - \frac{1}{R_1}e$$

We now solve for $v$, which in this case involves only the second equation, from which

$$v = -\frac{R_1R_2}{R_1 + R_2}i + \frac{R_2e}{R_1 + R_2}$$

Finally, by comparison with equation (*2.8*), which we derived in Problem 2.31,

$$R_{\text{Th}} = \frac{R_1R_2}{R_1 + R_2} \quad \text{and} \quad v_{\text{Th}} = \frac{R_2e}{R_1 + R_2}$$

which values apply to the Thévenin equivalent one port of Fig. 2-61*b*.

**2.33** Suppose that the load in the circuit of Fig. 2-61 is a resistor, $R_L$. Use the results of Problem 2.32 to find the terminal current $i$ and the terminal voltage $v$.

From Fig. 2-61*b* we see that $R_{\text{Th}}$ and $R_L$ are in series. Therefore,

$$i = \frac{v_{\text{Th}}}{R_{\text{Th}} + R_L} \quad \text{and} \quad v = R_L i = \frac{R_L v_{\text{Th}}}{R_{\text{Th}} + R_L}$$

where $v_{\text{Th}}$ and $R_{\text{Th}}$ can be substituted from the answer to Problem 2.32.

**2.34** Use loop analysis to find the Thévenin equivalent for the one port of Fig. 2-62, which has already been marked with reference polarities and appropriate loop currents. (It will usually simplify the analysis if the loop currents are chosen so that one and only one passes through the load.)

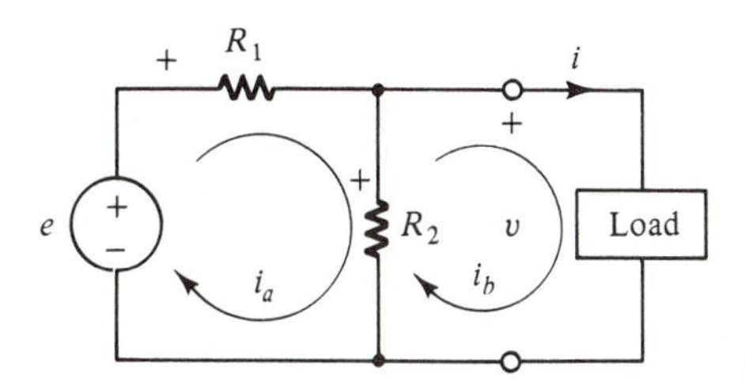


**Fig. 2-62**

**Step 2:** Applying KVL around the windowpanes,

$$-e + v_{R_1} + v_{R_2} = 0$$

$$-v_{R_2} + v = 0 \qquad (v \text{ is the voltage across the load})$$

**Step 4:** The $v$-$i$ relations are

$$v_{R_1} = R_1 i_a \quad \text{and} \quad v_{R_2} = R_2(i_a - i_b)$$

**Step 5:** Not applicable. But note that $i$ and $i_b$ describe the same current, so that $i_b = i$.

**Step 6:** We make the usual substitutions, and when rearranging we treat the terminal voltage $v$ as if it were known. Thus:

$$(R_1 + R_2)i_a - R_2 i = e$$

$$-R_2 i_a + R_2 i = -v$$

Finally, solving for $i$ by Cramer's rule,

$$i = \frac{\begin{vmatrix} R_1 + R_2 & e \\ -R_2 & -v \end{vmatrix}}{\begin{vmatrix} R_1 + R_2 & -R_2 \\ -R_2 & R_2 \end{vmatrix}} = -\frac{R_1 + R_2}{R_1 R_2} v + \frac{1}{R_1} e$$

When we compare this $i$-$v$ relation with equation *(2.8)*, it follows at once that

$$R_{\text{Th}} = \frac{R_1 R_2}{R_1 + R_2} \quad \text{and} \quad v_{\text{Th}} = \frac{R_2}{R_1 + R_2} e$$

### *Comments*

1. We can always find a one port's $v$-$i$ or $i$-$v$ relation by either nodal or loop

analysis. The Thévenin equivalent one port then follows by simple inspection.

2. There are other methods that can be used to find $v_{\text{Th}}$ and $R_{\text{Th}}$. These are often but not always quicker, and they are not always applicable. We will look at some of them in the next few problems.
3. Whenever we are interested in a one-port circuit as seen from its terminals, that is, as seen from the outside, the ultimate reduction of the Thévenin equivalent is often worthwhile, even if it takes a little effort to obtain it.

**2.35** By inspection of Fig. 2-10, show that

$$v_{\text{Th}} = v_{oc} \quad \text{and} \quad R_{\text{Th}} = v_{oc}/i_{sc}$$

where $v_{oc}$ is the terminal voltage of a one port when the load is an open circuit, and $i_{sc}$ is the terminal current when the load is a short circuit.

In Fig. 2-63*a* there is no voltage drop across $R_{\text{Th}}$, and therefore by KVL, $v = v_{oc} = v_{\text{Th}}$.

In Fig. 2-63*b* Ohm's law yields $i = i_{sc} = v_{Th}/R_{\text{Th}}$.

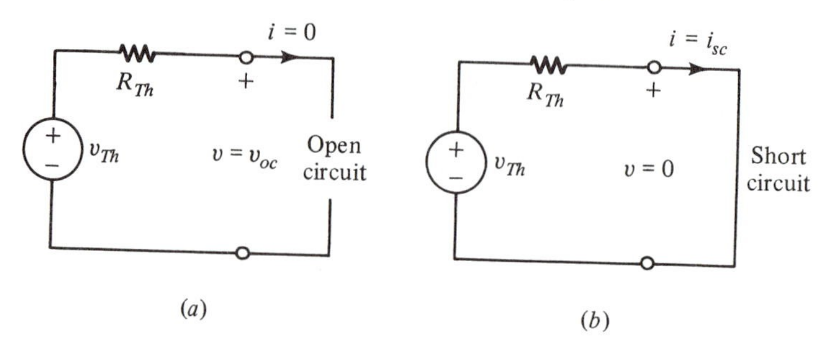


**Fig. 2-63**

**2.36** Find $v_{oc}$ and $i_{sc}$ for the one port of Fig. 2-64*a*. Hence find this one port's Thévenin equivalent.

In the circuit of Fig. 2-64*b*, $R_1$ and $R_2$ are in series, so that $i_{R_1} = i_{R_2} = e/(R_1 + R_2)$. Thus

$$v_{oc} = v_{R_2} = \frac{R_2 e}{R_1 + R_2}$$

In Fig. 2-64*c* there is no voltage across $R_2$, which may therefore be (mentally) removed from the circuit. Then by Ohm's law, $i_{sc} = e/R_1$.

It follows from the results of Problem 2.35 that

$$v_{\text{Th}} = v_{oc} = \frac{R_2 e}{R_1 + R_2} \quad \text{and} \quad R_{\text{Th}} = \frac{v_{oc}}{i_{sc}} = \frac{R_1 R_2}{R_1 + R_2}$$

This agrees with the results we obtained earlier (Problems 2.32 and 2.34) by nodal and then loop analysis. For this elementary circuit the present method is clearly simpler and quicker.

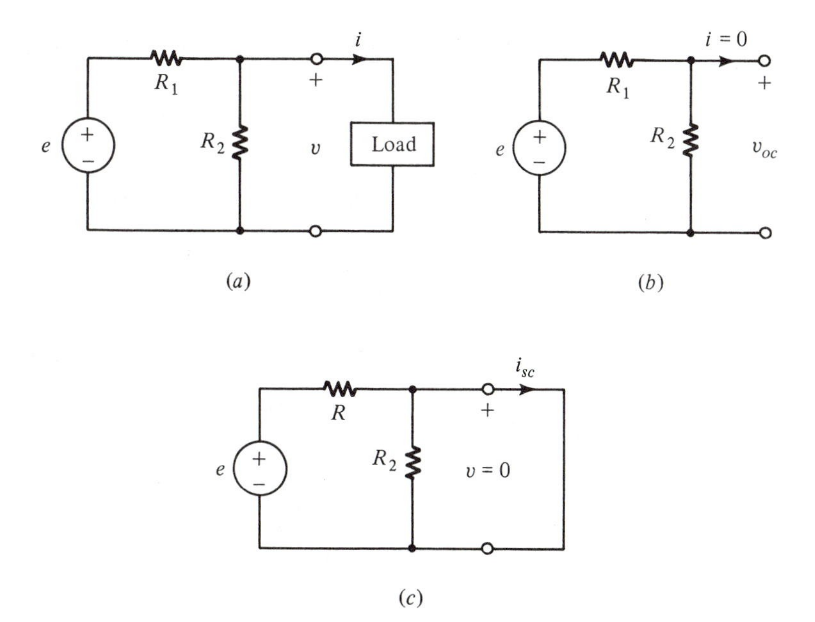


**Fig. 2-64**

**2.37** If we were to "kill" (i.e. set to zero) all of the *independent* sources in a one port, then $v_{Th}$ would be zero. Looking into the terminals of the (dead) Thévenin one port (see Fig. 2-10) we would see only the resistance $R_{Th}$. (Note: A dead voltage source is equivalent to a short circuit; while a dead current source is equivalent to an open circuit.) We can often use this concept to find $R_{Th}$. Do so for the one port of Fig. 2-65*a*.

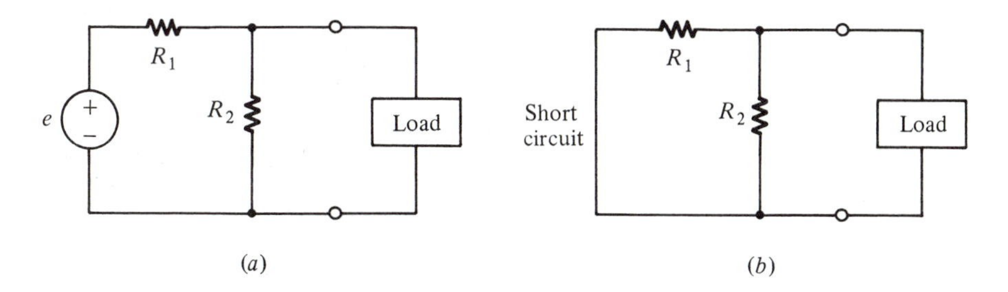


**Fig. 2-65**

The one port's source has been killed in Fig. 2-65*b*, leaving $R_1$ and $R_2$ in parallel. Hence $R_{Th} = R_{eq} = R_1R_2/(R_1 + R_2)$.

This method is obviously simpler and quicker than loop or nodal analysis—for this Spartan circuit. But note that we have only found $R_{Th}$. We would still have to find $v_{Th}$.

**2.38** Find the Thévenin equivalent of the circuit in Fig. 2-66 by as many methods as you can.

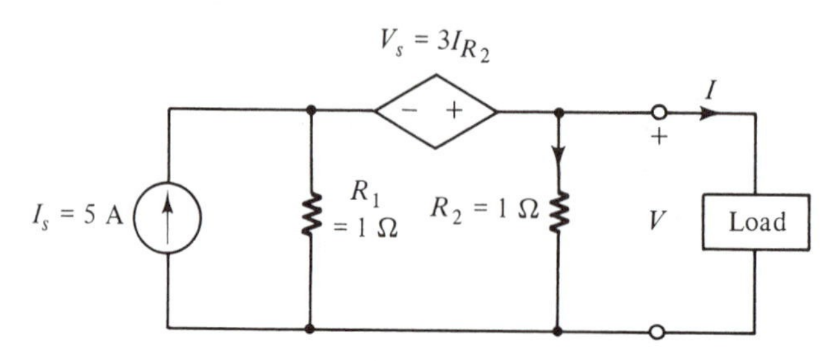


**Fig. 2-66**

1. *Nodal analysis:* If you follow the usual steps, you should obtain two simultaneous equations—with $I$ on the side of the knowns. Solving for $V$ yields

$$V = I - 5$$

and so $R_{\text{Th}} = -1\ \Omega$ and $V_{\text{Th}} = -5$ V.

A negative resistance is always a possibility when there is a dependent source in the circuit.

2. *Loop analysis:* The normal procedure leads to three simultaneous equations—with $V$ treated as a known. Solving for $I$ yields

$$I = V + 5$$

which means that $R_{\text{Th}} = -1\ \Omega$ and $V_{\text{Th}} = -5$ V.

3. *To find $V_{oc}$:* We make the load an open circuit, so that $I = 0$. It is not immediately obvious how we can "easily" find $V_{oc}$ here. (There *is* a relatively simple maneuver.) The alternative is to use loop or nodal analysis, but open circuiting the output does not simplify matters much; it represents a lot of work just to find $V_{oc}$.
4. *To find $I_{sc}$:* We make the load a short circuit, so that $V = 0$. For this circuit there *is* an obvious simple route. $V = V_{R_2} = 0$, and so $I_{R_2}$ and $V_s$ are also zero. It follows that $V_{R_1} = 0$ and $I_{R_1} = 0$. Therefore, $I_{sc} = I_s = 5$ A.
5. *To find $R_{Th}$:* We can kill the *independent* source, but in this case we *cannot* combine resistances because of the presence of the dependent source. Alternatively, if you found $V_{oc} = -5$ V and $I_{sc} = 5$ A in (3) and (4), above, then $R_{\text{Th}} = V_{oc}/I_{sc} = -1\ \Omega$.

Comment: The short-cut methods are not usually helpful if a circuit contains one or more dependent sources.

### *State Equations*

**2.39** Derive the state equations for the network of Fig. 2-67. (The loop equations were written in Problem 2.20, the nodal equations in Problem 2.26.)

**Step 1:** Reference polarities or current reference directions have already been assigned.

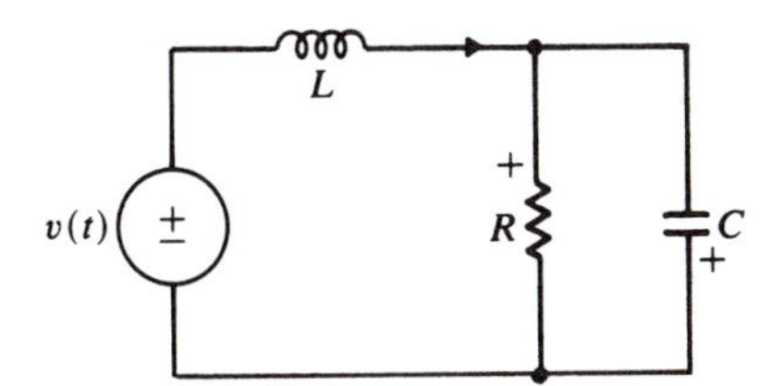


**Fig. 2-67**

**Step 2:** The derivative relations for the inductor and capacitor are

$$C\frac{dv_C}{dt} = i_C \quad \text{and} \quad L\frac{di_L}{dt} = v_L$$

where $v_C$ and $i_L$ are the state variables.

**Step 3:** We must now eliminate $i_C$ and $v_L$ in terms of $v_C$, $i_L$ and $v(t)$.

$$-i_C - i_L + i_R = 0 \qquad \text{(KCL at top-right node)}$$

$$-v(t) + v_L + v_R = 0 \qquad \text{(KVL for left windowpane)}$$

Rearranging,

$$i_C = i_R - i_L \quad \text{and} \quad v_L = v(t) - v_R$$

But $v_R = -v_c$ (KVL for right windowpane), and $i_R = \frac{1}{R}v_R = -\frac{1}{R}v_C$. Therefore,

$$C\frac{dv_C}{dt} = -\frac{1}{R}v_C - i_L \quad \text{and} \quad L\frac{di_L}{dt} = v(t) + v_C$$

Thus the state equations are

$$\frac{dv_C}{dt} = -\frac{1}{RC}v_C - \frac{1}{C}i_L$$

$$\frac{di_L}{dt} = \frac{1}{L}v_C + \frac{1}{L}v(t)$$

In the matrix form, $\dot{\mathbf{x}} = \mathbf{Ax} + \mathbf{Bu}$,

$$\begin{bmatrix} \frac{dv_C}{dt} \\ \frac{di_L}{dt} \end{bmatrix} = \begin{bmatrix} -\frac{1}{RC} & -\frac{1}{C} \\ \frac{1}{L} & 0 \end{bmatrix} \begin{bmatrix} v_C \\ i_L \end{bmatrix} + \begin{bmatrix} 0 \\ \frac{1}{L} \end{bmatrix} v(t)$$

In order to solve these equations for $v_C(t)$ and $i_L(t)$ we would need to be given the expression for $v(t)$, the values of $L$, $R$, and $C$, and the initial states $v_C(0)$ and $i_L(0)$.

**2.40** Obtain a set of state equations for the network of Fig. 2-68. Use the given reference current directions and polarities.

There will be three state variables: $v_{C_1}$, $i_L$, and $v_{C_2}$. For the first,

$$C_1\frac{dv_{C_1}}{dt} = i_{C_1}$$

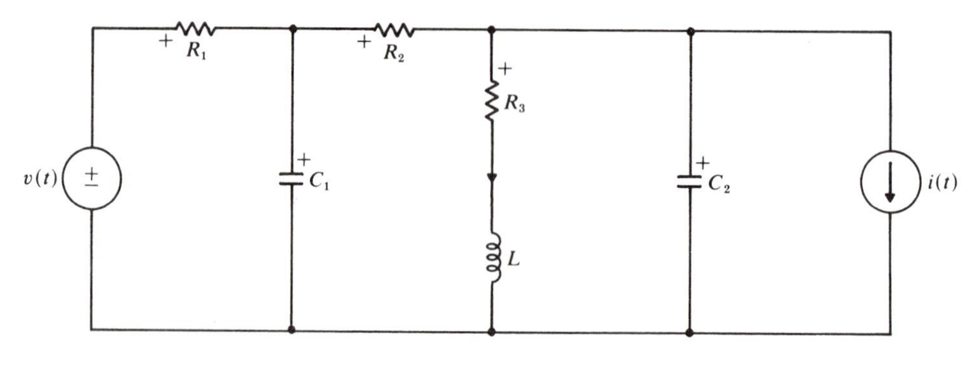


**Fig. 2-68**

But by KCL, $i_{C_1} = i_{R_1} - i_{R_2}$ where $i_{R_1} = \dfrac{1}{R_1}\{v(t) - v_{C_1}\}$ and $i_{R_2} = \dfrac{1}{R_2}(v_{C_1} - v_{C_2})$. Thus the first state equation is

$$\frac{dv_{C_1}}{dt} = \frac{1}{C_1} i_{C_1} = \frac{1}{C_1}(i_{R_1} - i_{R_2}) = \frac{1}{R_1 C_1}\{v(t) - v_{C_1}\} - \frac{1}{R_2 C_1}(v_{C_1} - v_{C_2})$$

Next, for the inductor,

$$\frac{di_L}{dt} = \frac{1}{L} v_L = \frac{1}{L}(v_{C_2} - v_{R_3}) = \frac{1}{L}(v_{C_2} - R_3 i_L)$$

Finally, for the second capacitor,

$$\frac{dv_{C_2}}{dt} = \frac{1}{C_2} i_{C_2} = \frac{1}{C_2}\{i_{R_2} - i_L - i(t)\} = \frac{1}{C_2}\left\{\frac{1}{R_2}(v_{C_1} - v_{C_2}) - i_L - i(t)\right\}$$

Now, putting these three state equations together in the form $\dot{\mathbf{x}} = \mathbf{Ax} + \mathbf{Bu}$, we have

$$\begin{bmatrix} \dfrac{dv_{C_1}}{dt} \\ \dfrac{di_L}{dt} \\ \dfrac{dv_{C_2}}{dt} \end{bmatrix} = \begin{bmatrix} -\left(\dfrac{1}{R_1 C_1} + \dfrac{1}{R_2 C_1}\right) & 0 & \dfrac{1}{R_2 C_1} \\ 0 & -\dfrac{R_3}{L} & \dfrac{1}{L} \\ \dfrac{1}{R_2 C_2} & -\dfrac{1}{C_2} & -\dfrac{1}{R_2 C_2} \end{bmatrix} \begin{bmatrix} v_{C_1} \\ i_L \\ v_{C_2} \end{bmatrix} + \begin{bmatrix} \dfrac{1}{R_1 C_1} & 0 \\ 0 & 0 \\ 0 & -\dfrac{1}{C_2} \end{bmatrix} \begin{bmatrix} v(t) \\ i(t) \end{bmatrix}$$

**2.41** Obtain the state and output equations ($\dot{\mathbf{x}} = \mathbf{Ax} + \mathbf{Bu}$ and $\mathbf{y} = \mathbf{Cx} + \mathbf{Du}$) for the network of Fig. 2-69, given that the desired output is

$$\mathbf{y} = \begin{bmatrix} v_{C_1} \\ i_{R_2} \end{bmatrix}$$

The appropriate derivative relations are

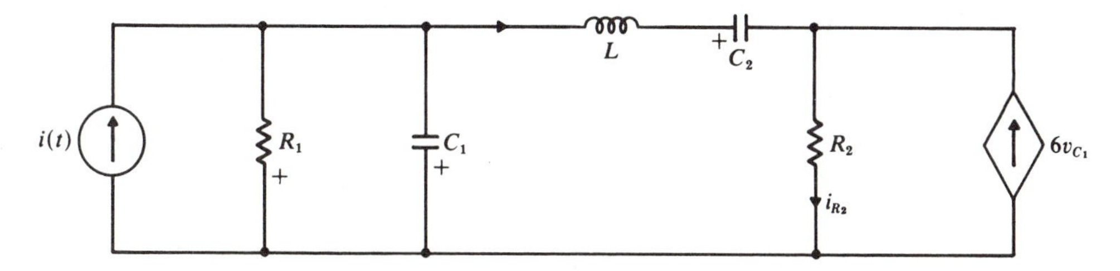


**Fig. 2-69**

$$C_1\frac{dv_{C1}}{dt} = i_{C1}, \qquad L\frac{di_L}{dt} = v_L, \qquad C_2\frac{dv_{C2}}{dt} = i_{C2}$$

where we must now eliminate $i_{C1}$, $i_{C2}$, and $v_L$ in terms of the states $v_{C1}$, $v_{C2}$, and $i_L$, and the input $i(t)$. Starting with the first equation,

$$\frac{dv_{C1}}{dt} = \frac{1}{C_1}i_{C1} = \frac{1}{C_1}\{-i_{R_1} - i(t) + i_L\} = \frac{1}{C_1}\left\{-\frac{1}{R_1}v_{C1} - i(t) + i_L\right\}$$

Next,

$$\frac{di_L}{dt} = \frac{1}{L}v_L = \frac{1}{L}\{-v_{C2} - v_{R2} - v_{C1}\} = \frac{1}{L}\{-v_{C2} - R_2(i_L + 6v_{C1}) - v_{C1}\}$$

$$= \frac{1}{L}\{-v_{C2} - R_2 i_L - (6R_2 + 1)v_{C1}\}$$

And next,

$$\frac{dv_{C2}}{dt} = \frac{1}{C_2}i_{C2} = \frac{1}{C_2}i_L$$

In matrix form,

$$\begin{bmatrix} \dfrac{dv_{C1}}{dt} \\ \dfrac{di_L}{dt} \\ \dfrac{dv_{C2}}{dt} \end{bmatrix} = \begin{bmatrix} -\dfrac{1}{R_1C_1} & \dfrac{1}{C_1} & 0 \\ -\dfrac{6R_2+1}{L} & -\dfrac{R_2}{L} & -\dfrac{1}{L} \\ 0 & \dfrac{1}{C_2} & 0 \end{bmatrix} \begin{bmatrix} v_{C1} \\ i_L \\ v_{C2} \end{bmatrix} + \begin{bmatrix} -\dfrac{1}{C_1} \\ 0 \\ 0 \end{bmatrix} i(t)$$

Turning now to the outputs, $v_{C_1}$ is a state and will be known as soon as the state equation has been solved, while the second output is given by

$$i_{R2} = i_L + 6v_{C1}$$

That is, $i_{R_2}$ is a function of two of the states. In standard matrix form,

$$\mathbf{y} = \begin{bmatrix} v_{C1} \\ i_{R2} \end{bmatrix} = \begin{bmatrix} 1 & 0 & 0 \\ 6 & 1 & 0 \end{bmatrix} \begin{bmatrix} v_{C1} \\ i_L \\ v_{C2} \end{bmatrix} + \begin{bmatrix} 0 \\ 0 \end{bmatrix} i(t)$$

***The Op Amp, Again***

**2.42** Show that the op amp circuit of Fig. 2-70 will integrate the input voltage $v$.

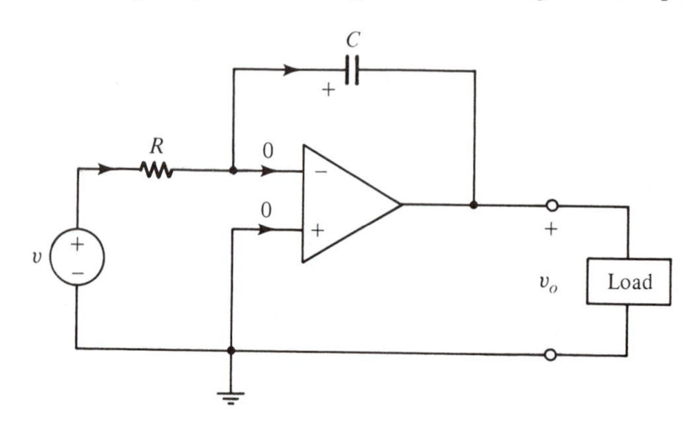


**Fig. 2-70**

The approximation $i_+ = i_- = 0$ is already shown in Fig. 2-70. Now, since the op amp's + input terminal is connected to ground, $v_+$ and therefore $v_-$ are both zero. Thus

$$i_R = \frac{1}{R}v$$

$$i_C = i_R = \frac{1}{R}v \qquad \text{(KCL)}$$

$$v_C = -v_o = \frac{1}{C}\int_0^t i_C\, d\tau + v_C(0)$$

or

$$v_o = -\frac{1}{RC}\int_0^t v\, d\tau - v_C(0)$$

This is the equation of an integrator—with a sign change. That is, the output is (minus) the integral of the input, together with an initial value equal to (minus) the initial capacitor voltage.

In this op amp circuit as well as in the voltage follower of Problem 1.18 and the op amp circuits that we will discuss later, the load current does not enter into the circuit equations. Thus a load does not affect the behavior of an (ideal) op amp circuit.

**2.43** Show that the op amp circuit of Fig. 2-71 can be used to add the three input voltages.

As in the preceding problem, $v_+ = v_- = 0$. Then,

$$i_{R_1} = \frac{1}{R_1}v_1 \qquad i_{R_2} = \frac{1}{R_2}v_2 \qquad i_{R_3} = \frac{1}{R_3}v_3$$

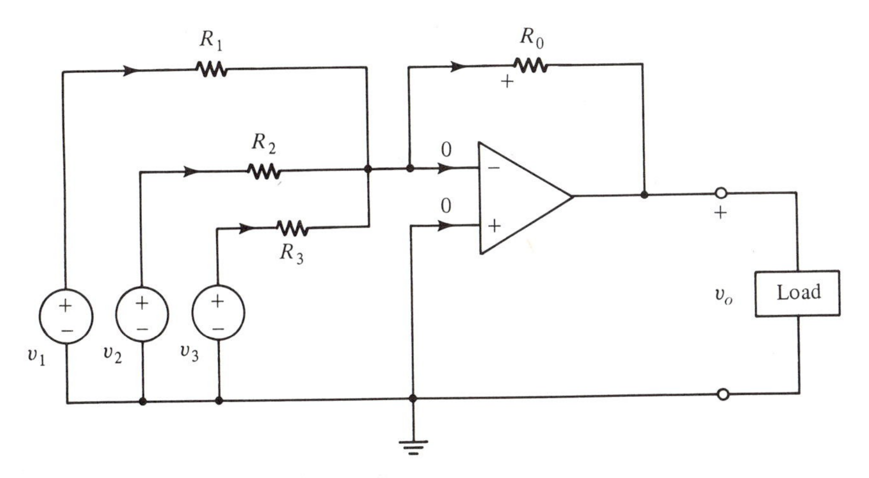


**Fig. 2-71**

and from KCL,

$$i_{R_o} = i_{R_1} + i_{R_2} + i_{R_3}$$

and so

$$v_{R_o} = -v_o = R_o \sum_{n=1}^{3} i_{R_n}$$

or

$$v_o = -R_o \sum_{n=1}^{3} (v_n/R_n)$$

Clearly, if all the resistances are equal, the output $v_o$ will equal the (negative) sum of the three input voltages.

Comment: If $R_o$ in Fig. 2-71 is replaced by a capacitor $C$, the result is a *summing integrator* for which, upon combining the solutions of this and the previous problem,

$$v_o = -\frac{1}{C}\int_0^t \left\{\frac{1}{R_1}v_1 + \frac{1}{R_2}v_2 + \frac{1}{R_3}v_3\right\} d\tau - v_C(0)$$

**2.44** Show that $v_o = v_1 - v_2$ for the circuit of Fig. 2-72 when $R_1 = R_2 = R_3 = R_4$.

The op amp does not affect the current $i_{R_1} = i_{R_2} = v_1/(R_1 + R_2)$. Thus $v_+ = v_- = R_2 i_{R_2} = R_2 v_1/(R_1 + R_2)$. Now

$$i_{R_3} = i_{R_4} = \frac{1}{R_3}(v_2 - v_-)$$

and

$$v_{R_4} = R_4 i_{R_4} = \frac{R_4}{R_3}v_2 - \frac{R_4}{R_3}\frac{R_2}{R_1 + R_2}v_1$$

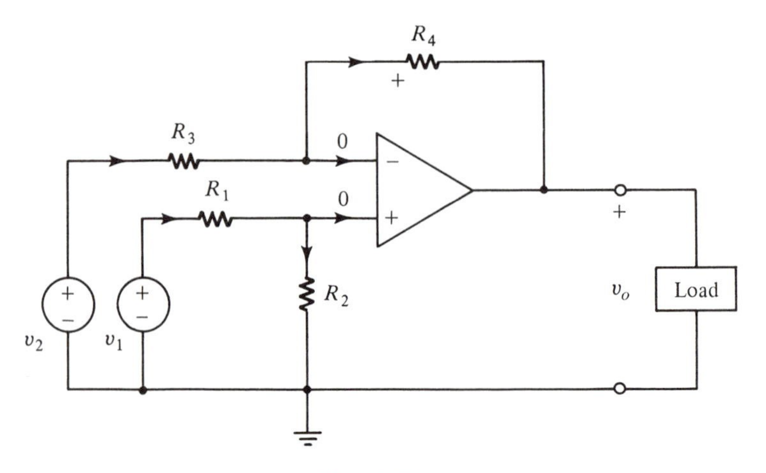


**Fig. 2-72**

Finally,

$$v_o = v_- - v_{R4} = \frac{R_2}{R_1 + R_2} v_1 - \frac{R_4}{R_3} v_2 + \frac{R_4}{R_3} \frac{R_2}{R_1 + R_2} v_1$$

If we now set $R_1 = R_2 = R_3 = R_4 = R$, then

$$v_o = v_1 - v_2$$

# PROBLEMS

**2.45** In the circuit of Fig. 2-73*a* it is known that $i_1(t) = 10 \sin t$, $v_2(t) = 2t$, and that $i_3(t)$ is as shown in Fig. 2-73*b*. Find $v_4(t)$ for $0 < t < 2\pi$.

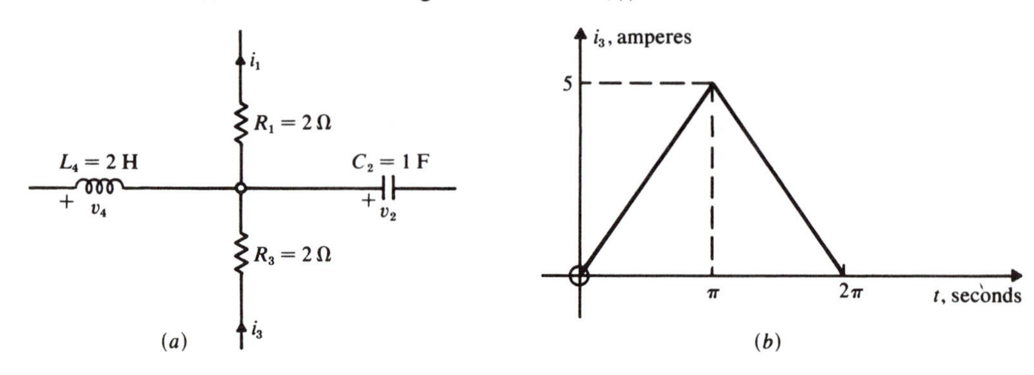


**Fig. 2-73**

**2.46** For the circuit and the current waveforms of Fig. 2-74, show that

$$v_C(0) = -3(1 + \pi) \qquad i_C\left(\frac{1}{2}\right) = \frac{3\pi^2}{4\sqrt{2}} - 2$$

$$v_C(1) = -7$$

$$v_C\left(\frac{3}{2}\right) = -5 + \frac{3\pi}{\sqrt{2}} \qquad i_C\left(\frac{3}{2}\right) = \frac{3\pi^2}{4\sqrt{2}} + 2$$

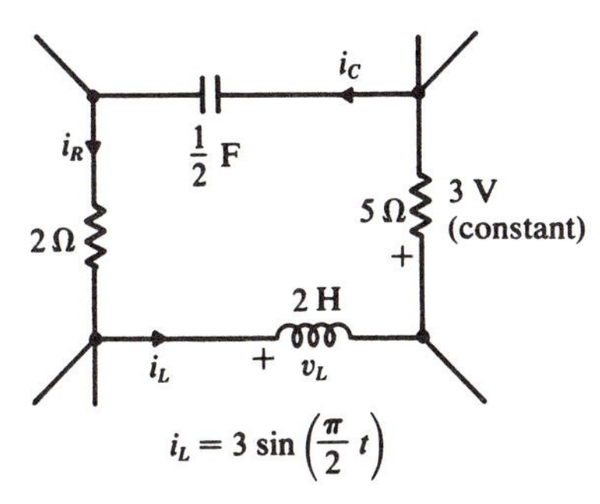


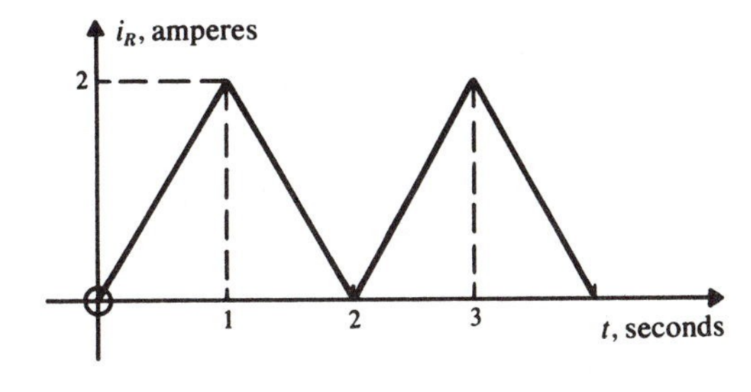


**Fig. 2-74**

**2.47** A simple *L-R* circuit is shown in Fig. 2-75, together with its inductor current. Verify that $v_1(\frac{1}{2}) = 5$, $v_1(\frac{3}{2}) = 6$, and $v_1(\frac{5}{2}) = 1$.

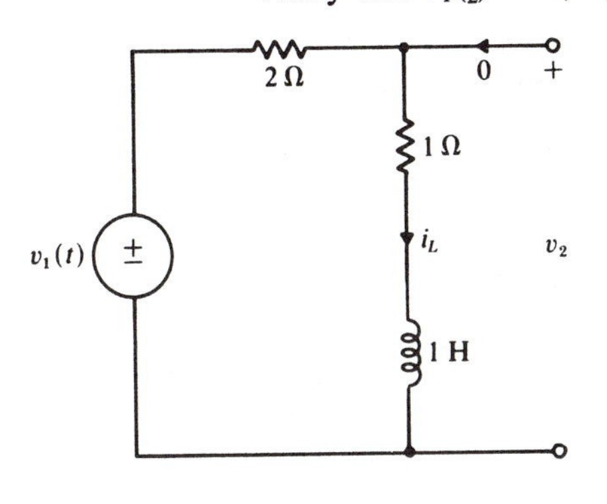


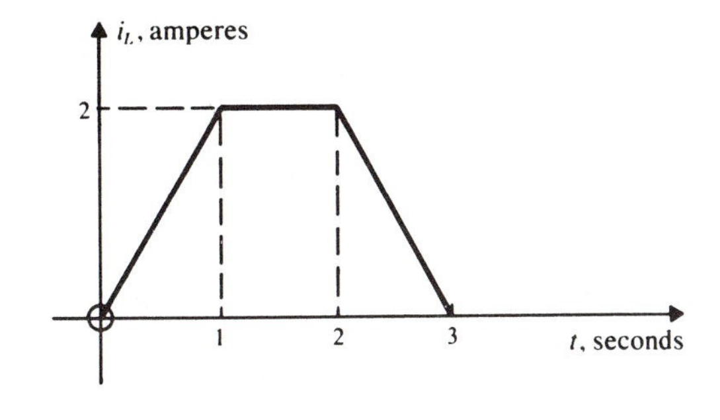


**Fig. 2-75**

**2.48** Show that the following simultaneous nodal equations are correct for the circuit of Fig. 2-76.

$$\frac{1}{R_1}v_b + C\frac{dv_b}{dt} + \frac{1}{L_1}\int_0^t v_b\,d\tau - \frac{1}{L_1}\int_0^t v_c\,d\tau = i(t) - i_{L_1}(0) + \frac{1}{R_1}v(t)$$

$$-\frac{1}{L_1}\int_0^t v_b\,d\tau + \frac{1}{R_2}v_c + \frac{1}{L_1}\int_0^t v_c\,d\tau - \frac{1}{R_2}v_d = -i(t) + i_{L_1}(0)$$

$$-\frac{1}{R_2}v_c + \frac{1}{R_2}v_d + \frac{1}{L_2}\int_0^t v_d\,d\tau = -i_{L_2}(0)$$

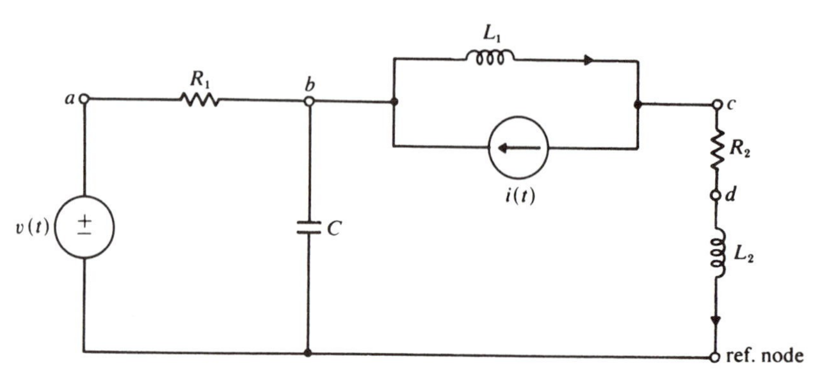


Fig. 2-76

**2.49** Given the circuit and the designated loop currents of Fig. 2-77, show that the following loop equations are correct.

$$R_1 i_2 + (L_1 + L_2)\frac{di_2}{dt} + 2M_{12}\frac{di_2}{dt} - M_{13}\frac{di_3}{dt} - M_{23}\frac{di_3}{dt} = -R_1 i(t)$$

$$-M_{13}\frac{di_2}{dt} - M_{23}\frac{di_2}{dt} + L_3\frac{di_3}{dt} + R_2 i_3 + \frac{1}{C}\int_0^t i_3\,d\tau = -v_C(0)$$

*Note:* The current $i_2$ passes through $L_2$, but *not* through $L_3$.

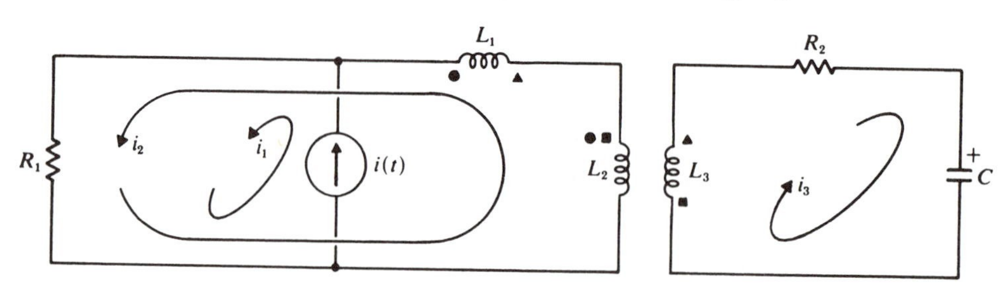


Fig. 2-77

**2.50** In the circuit of Fig. 2-78, it is known that at some instant of time $t_0$,

$$v_1(t_0) = 2\text{ V}, \qquad v_2(t_0) = 5\text{ V}, \qquad \left.\frac{dv_2}{dt}\right|_{t_0} = -10\text{ V/s}$$

Find the value of the resistor $R$.

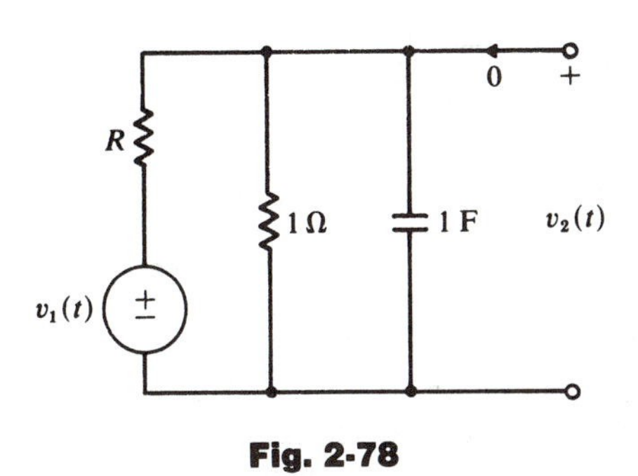


Fig. 2-78

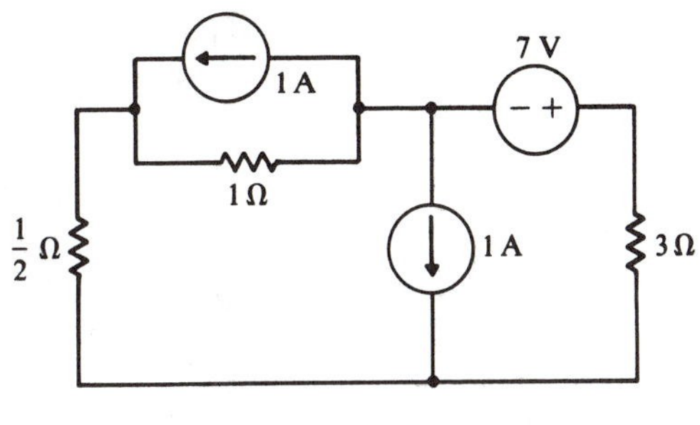


Fig. 2-79

**2.51** Use the loop method of analysis to show that the current through the 3-Ω resistor in Fig. 2-79 is 1 A in a downward direction.

**2.52** Solve for the same current as in Problem 2.51, using the nodal method of analysis.

**2.53** Verify the following loop equations for the network of Fig. 2-80.

$$v_i + Ri_1 = -L\frac{di(t)}{dt}$$

$$-Ri_1 + \frac{1}{C}\int_0^t -i_1\, d\tau = \frac{1}{C}\int_0^t -\, i(\tau)\, d\tau + v(t) - v_C(0)$$

where $v_i$ is the voltage across the current source.

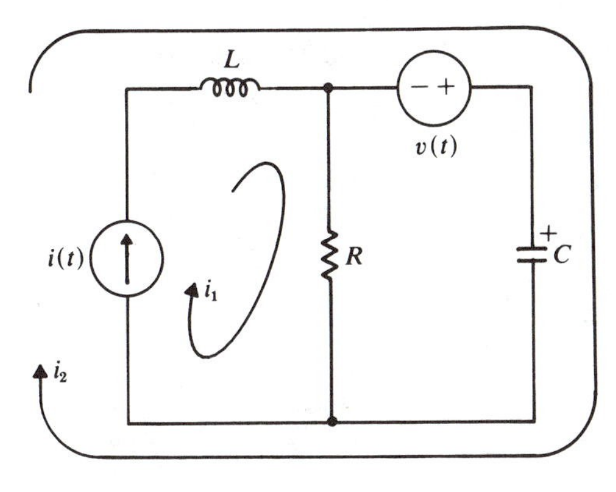


Fig. 2-80

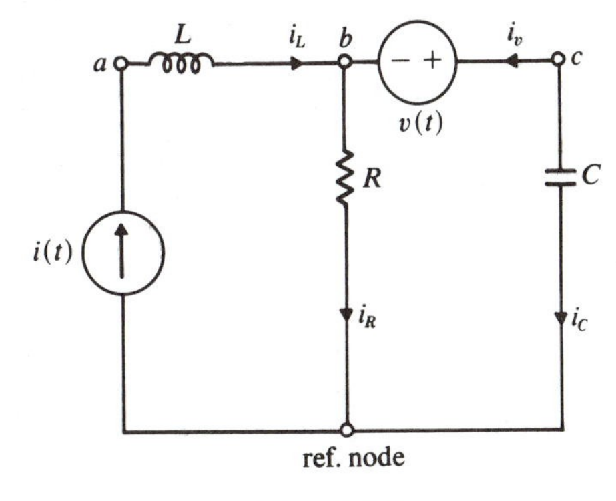


Fig. 2-81

**2.54** The following nodal equations have been written for the circuit in Fig. 2-81. Are they correct?

$$\frac{1}{L}\int_0^t v_a\,d\tau - \frac{1}{L}\int_0^t v_b\,d\tau = i(t) - i_L(0)$$

$$\frac{1}{L}\int_0^t v_a\,d\tau - \frac{1}{L}\int_0^t v_b\,d\tau - \frac{1}{R}v_b + i_v = -i_L(0)$$

$$C\frac{dv_b}{dt} + i_v = -C\frac{dv(t)}{dt}$$

**2.55** A *real* constant-voltage source may be "modeled" by an ideal constant voltage source $V$ in series with a (small) resistor $r$. Plot the terminal characteristics, $v$ vs. $i$, of the real source. Discuss the short circuit and open circuit characteristics for both the real source and the ideal source ($r \to 0$).

**2.56** The inductance as measured between the terminals of the circuit in Fig. 2-82*a* is 6 mH. When the connections to the second coil are reversed, as in Fig. 2-82*b*, the measured inductance falls to 2 mH. Find the value of $M$. (This is a common method for measuring mutual inductance.)

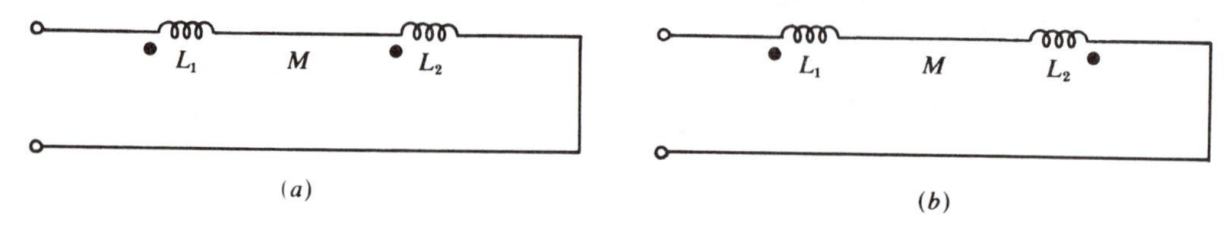


**Fig. 2-82**

**2.57** Verify the correctness of the equations below, which were written for the circuit of Fig. 2-83. (Note that one is a loop equation and the other a nodal equation.)

$$(R + R_1)i_1 - \mu v_2 = e(t)$$

$$\alpha i_1 + \frac{R_2 + R_L}{R_2 R_L} v_2 = 0$$

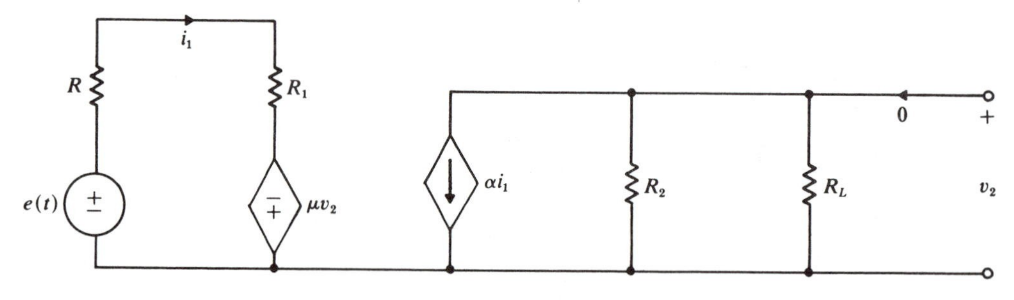


**Fig. 2-83**

**2.58** Use nodal analysis to find the $V$-$I$ relation for the one-port network of Fig. 2-84. Hence find the Thévenin equivalent as seen by the load. Use the Thévenin equivalent to find the load current $I$ when the load is

(*a*) A 1-Ω resistor.
(*b*) A 1-V voltage source with its + terminal up.
(*c*) A 1-A current source pointing upward.

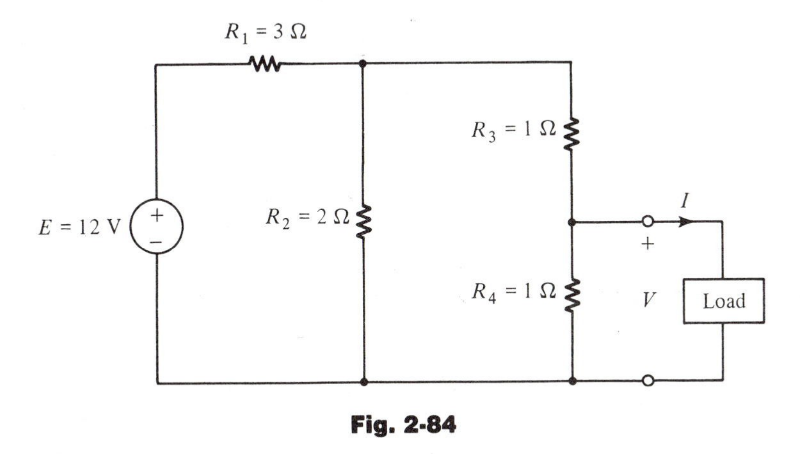


**Fig. 2-84**

**2.59** Use loop analysis to find the Thévenin equivalent of the one port of Fig. 2-84.

**2.60** Find $V_{oc}$ and $I_{sc}$ for the one port of Fig. 2-84. Hence find the Thévenin equivalent of the one port.

**2.61** Use the source killing method to find $R_{\mathrm{Th}}$ for the circuit of Fig. 2-84.

**2.62** Use loop and/or nodal analysis to show that $R_{\mathrm{Th}} = 2.46\ \Omega$ and $V_{\mathrm{Th}} = 0$ for the one port network of Fig. 2-85. (Note: Although this circuit appears to be dead, so that you might expect to find $v$ and $i$ both zero, this need not be the case, since the *load* may contain one or more independent sources.)

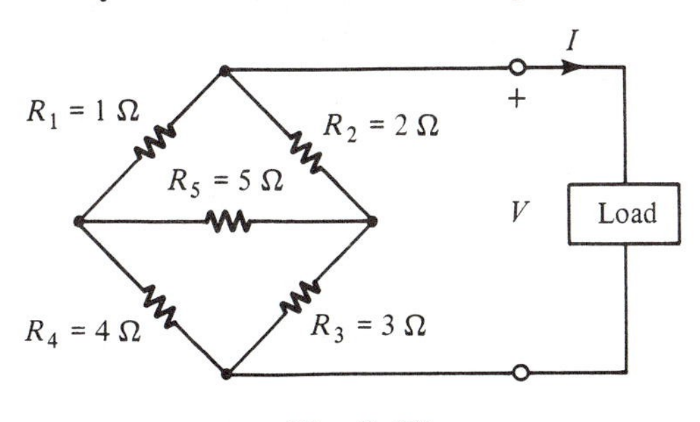


**Fig. 2-85**

**2.63** Explain why you cannot find $R_{\mathrm{Th}} = V_{oc}/I_{sc}$ for the one port of Fig. 2-85.

**2.64** Explain why the source killing method fails you when applied to the circuit of Fig. 2-85. (There *are* no independent sources to kill, but that is not the problem. You simply omit the killing.)

**2.65** Write the state equations for the two networks of Fig. 2-86.

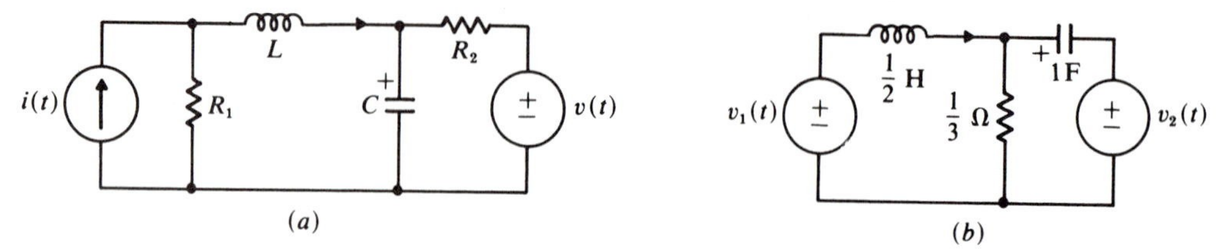


**Fig. 2-86**

**2.66** An ideal 10-A constant current source, a linear 2-Ω resistor, and an ideal diode are connected in parallel as in Fig. 2-87. Sketch the $v$-$i$ characteristic of the combination.

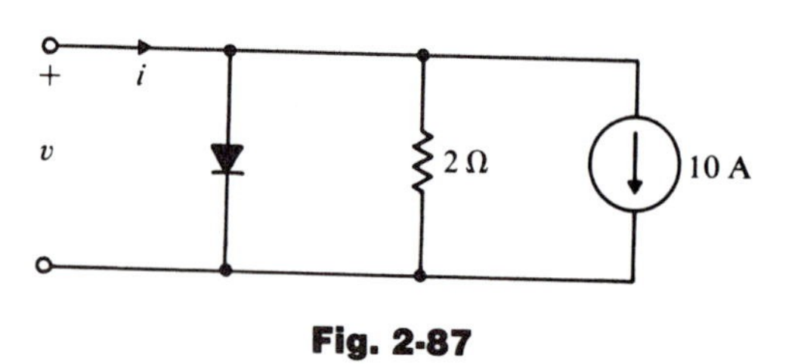


**Fig. 2-87**

**2.67** Show that the op amp circuit of Fig. 2-88 can be used to change the sign of a voltage. That is, prove that $v_o = -v$.

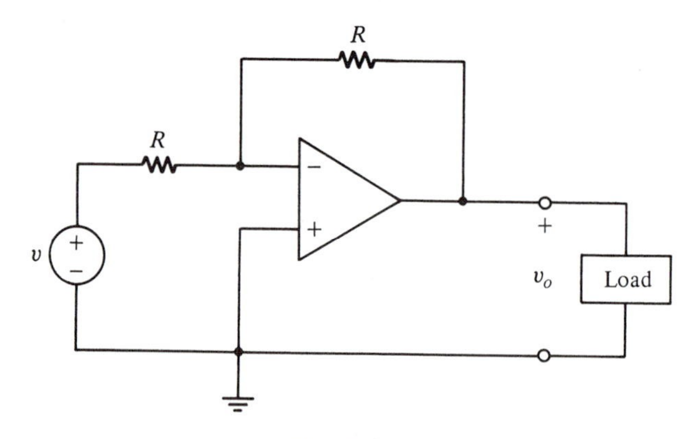


**Fig. 2-88**

Chapter

# 3

# RESPONSE OF SIMPLE SOURCE-FREE CIRCUITS

You are now well versed in the *writing* of equations to describe an electric circuit. And if the circuit is resistive, you can *solve* the *algebraic* equations for the unknown currents and voltages. But if the circuit contains inductance and/or capacitance, we are faced with the solution of a *differential* equation (or simultaneous differential equations).

In this chapter the solution of single first- or second-order differential equations will be taken up—for the special case of circuits without sources. That is, the circuits will respond to the initial conditions only. The effects of sources will be considered in Chapter 4.

## FIRST-ORDER CIRCUITS

The simple $R$-$C$ and $R$-$L$ networks of Fig. 3-1 may be described by the equations

$$\frac{1}{R}v + C\frac{dv}{dt} = 0 \quad \text{and} \quad Ri + L\frac{di}{dt} = 0$$

These are both first-order differential equations of the form

$$\frac{dx}{dt} + \alpha x = 0 \tag{3.1}$$

where $x$ and $\alpha$ are as tabulated in Table 3-1.

**Table 3-1**

| Circuit | $x(t)$ | $\alpha$ | Initial condition, $x(0)$ |
|---|---|---|---|
| $R$-$C$ | $v(t)$ | $\dfrac{1}{RC}$ | $v(0)$ |
| $R$-$L$ | $i(t)$ | $\dfrac{R}{L}$ | $i(0)$ |

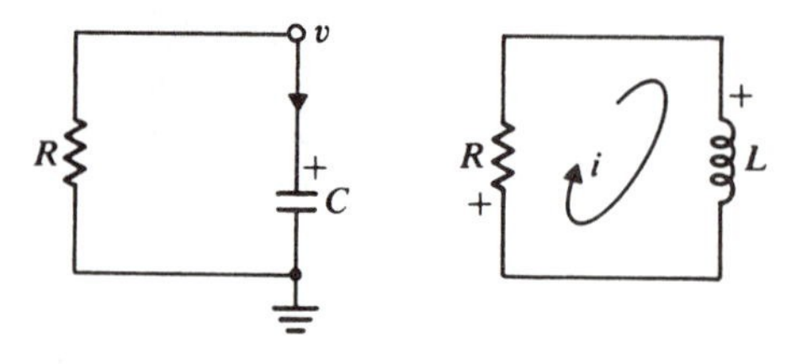


**Fig. 3-1**

To solve equation (*3.1*), the simplest procedure is to assume or "guess" a *trial* solution,

$$x(t) = ke^{st} \tag{3.2}$$

This, of course, is not really a guess; it is based on long experience. We must now show that it satisfies the differential equation and the given initial condition $x(0)$. In the process we will evaluate the constants $k$ and $s$. Thus substituting equation (*3.2*) into (*3.1*),

$$ske^{st} + \alpha ke^{st} = 0$$

which is true if $s = -\alpha$. Finally, setting $t = 0$ in equation (*3.2*) to check the initial condition, we find $k = x(0)$. The solution of the original differential equation (*3.1*) is therefore

$$x(t) = x(0)e^{-\alpha t} \tag{3.3}$$

More specifically, for the $R$-$C$ and $R$-$L$ circuits of Fig. 3-1,

$$v(t) = v(0)e^{-(1/RC)t} \quad \text{and} \quad i(t) = i(0)e^{-(R/L)t}$$

where $v(0)$ is the initial capacitor voltage in the $R$-$C$ circuit, and $i(0)$ is the initial inductor current in the $R$-$L$ circuit.

The solution can be plotted, as shown in Fig. 3-2. Note that this is a standard

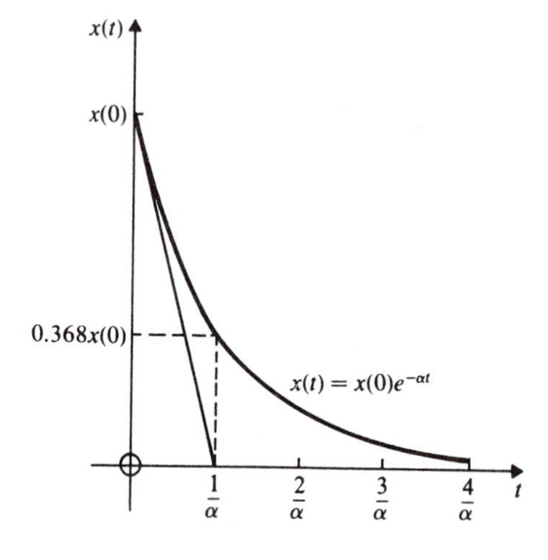


**Fig. 3-2**

(negative) exponential curve, and that the initial tangent intersects the $t$ axis at $t = 1/\alpha$, where $x = x(0)e^{-1} = 0.368x(0)$.

The period of time $1/\alpha$ seconds is called the *time constant* $\tau$ of the first-order circuit. It is a measure of the rapidity with which the variable decays from its initial value toward zero. Thus after *four* time constants, the response is $e^{-4}x(0)$ or about 2% of its initial value.

For simple *R-L* and *R-C* circuits the time constant $1/\alpha$ is $L/R$ and $RC$ respectively. The time constant is strictly a property of *first-order* circuits, although the concept is sometimes extended to higher-order networks.

## SECOND-ORDER CIRCUITS

For circuits which can be described by a second-order differential equation we may consider the general equation

$$\frac{d^2x}{dt^2} + \alpha \frac{dx}{dt} + \beta x = 0 \tag{3.4}$$

As was the case for the first-order equation, we assume a trial solution

$$x(t) = ke^{st} \tag{3.5}$$

Then substituting equation (*3.5*) into (*3.4*), we find

$$s^2ke^{st} + \alpha ske^{st} + \beta ke^{st} = 0$$

or

$$s^2 + \alpha s + \beta = 0 \tag{3.6}$$

This is called the *characteristic* or *auxiliary* equation and it may be solved, using the quadratic formula, for the *two* characteristic roots $s_1$ and $s_2$.

### Case I

If the two roots $s_1$ and $s_2$ are real and unequal, then both

$$x_1 = k_1e^{s_1t} \quad \text{and} \quad x_2 = k_2e^{s_2t}$$

will satisfy the original differential equation and it is easily shown by direct substitution that the sum

$$x(t) = k_1e^{s_1t} + k_2e^{s_2t} \tag{3.7}$$

must also satisfy equation (*3.4*). That is, equation (*3.7*) is the solution of equation (*3.4*). However, $k_1$ and $k_2$ are not yet known and must be found from the initial conditions. Thus if $x(0)$ and $\dot{x}(0)$ are given,* we set $t = 0$ in equation (*3.7*) and its

*The derivative, evaluated at $t = 0$, should strictly be written $(dx/dt)|_{t=0}$, but the notations $dx(0)/dt$ and $\dot{x}(0)$ are in common use.

derivative to obtain

$$x(0) = k_1 + k_2 \quad \text{and} \quad \dot{x}(0) = k_1 s_1 + k_2 s_2 \tag{3.8}$$

These two simultaneous equations may be solved for $k_1$ and $k_2$ to complete the solution.

### *Case II*

If the characteristic roots are complex conjugates, say $s_{1,2} = \sigma \pm j\omega$, then

$$\begin{aligned} x(t) &= k_1 e^{s_1 t} + k_2 e^{s_2 t} \\ &= e^{\sigma t}\{k_1 e^{j\omega t} + k_2 e^{-j\omega t}\} \\ &= e^{\sigma t}\{k_1(\cos \omega t + j \sin \omega t) + k_2(\cos \omega t - j \sin \omega t)\} \end{aligned}$$

or

$$x(t) = e^{\sigma t}\{k_3 \cos \omega t + k_4 \sin \omega t\} \tag{3.9}$$

where $j = \sqrt{-1}$, $k_3 = k_1 + k_2$, and $k_4 = j(k_1 - k_2)$. It is more convenient to evaluate $k_3$ and $k_4$ directly, rather than from $k_1$ and $k_2$. Thus from equation (*3.9*) and its derivative, when $t = 0$:

$$x(0) = k_3 \quad \text{and} \quad \dot{x}(0) = \sigma k_3 + \omega k_4 \tag{3.10}$$

Solving (*3.10*) for $k_3$ and $k_4$ and substituting back into (*3.9*) completes the solution.

Alternatively, equation (*3.9*) may be put in the form

$$x(t) = k_5 e^{\sigma t} \cos(\omega t + \phi)$$

Then equation (*3.10*) must be appropriately modified in order to evaluate $k_5$ and $\phi$.

### *Case III*

Finally, we will consider the case of real *and equal* characteristic roots. If $s_1$ is the repeated root, the solution cannot simply be

$$x(t) = k_1 e^{s_1 t} + k_2 e^{s_1 t} = k_3 e^{s_1 t}$$

since there must be *two* "free" constants corresponding to the two initial conditions of a second-order equation. Suppose, now, that the two roots are not quite equal; that is,

$$\begin{aligned} x(t) &= k_1 e^{s_1 t} + k_2 e^{(s_1 + \Delta s_1)t} \\ &= (k_1 + k_2)e^{s_1 t} + k_2\{e^{(s_1+\Delta s_1)t} - e^{s_1 t}\} \\ &= k_3 e^{s_1 t} + k_4 \frac{e^{(s_1+\Delta s_1)t} - e^{s_1 t}}{\Delta s_1} \end{aligned}$$

(Here $k_2 e^{s_1 t}$ has been both added and subtracted on the right-hand side, and two new constants $k_3 = k_1 + k_2$ and $k_4 = k_2 \Delta s_1$ have been defined.) Now, proceeding to the limit $\Delta s_1 \to 0$, $s_1$ and $s_2$ become equal, as specified, and

$$x(t) = k_3 e^{s_1 t} + k_4 \frac{d}{ds_1}(e^{s_1 t})$$

or

$$x(t) = k_3 e^{s_1 t} + k_4 t e^{s_1 t} \qquad (3.11)$$

which is the solution sought.

To find the constants $k_3$ and $k_4$, we solve the simultaneous equations

$$x(0) = k_3 \quad \text{and} \quad \dot{x}(0) = k_3 s_1 + k_4 \qquad (3.12)$$

(The case of equal roots is really of academic interest only, since there is zero probability of the circuit elements having *exactly* the values necessary to yield equal roots.)

## ALTERNATIVE VIEW OF SECOND-ORDER CIRCUITS

It is informative to rewrite the original differential equation (*3.4*) in the form

$$\frac{d^2x}{dt^2} + 2\zeta\omega_n \frac{dx}{dt} + \omega_n^2 x = 0 \qquad (3.13)$$

This in no way alters equation (*3.4*), but merely expresses the constants in terms of two new parameters, $\zeta$ and $\omega_n$. Both of these, like $\alpha$ and $\beta$, must be real numbers if we are dealing with a real circuit.

The characteristic equation is now

$$s^2 + 2\zeta\omega_n s + \omega_n^2 = 0 \qquad (3.14)$$

and the characteristic roots are

$$s_{1,2} = \omega_n\{-\zeta \pm \sqrt{\zeta^2 - 1}\} \qquad (3.15)$$

These equations in $\zeta$ and $\omega_n$ are not particularly helpful when we are calculating the solution of a *particular* second-order equation. However, they do allow us to categorize the *type* of solution to be expected for any value of $\zeta$, as shown in Table 3-2 and Fig. 3-3.

**Table 3-2**

| $\zeta$ | Characteristic roots | Description | Figure |
|---|---|---|---|
| $\zeta > 1$ | real, negative, unequal | overdamped | 3-3*a* |
| $\zeta = 1$ | real, negative, equal | critically damped | |
| $0 < \zeta < 1$ | complex conjugates; real parts negative | underdamped | 3-3*b* |
| $\zeta = 0$ | imaginary conjugates | undamped | 3-3*c* |
| $-1 < \zeta < 0$ | complex conjugates; real parts positive | oscillatory instability | 3-3*d* |
| $\zeta = -1$ | real, positive, equal | divergent instability | 3-3*e* |
| $\zeta < -1$ | real, positive, unequal | | |

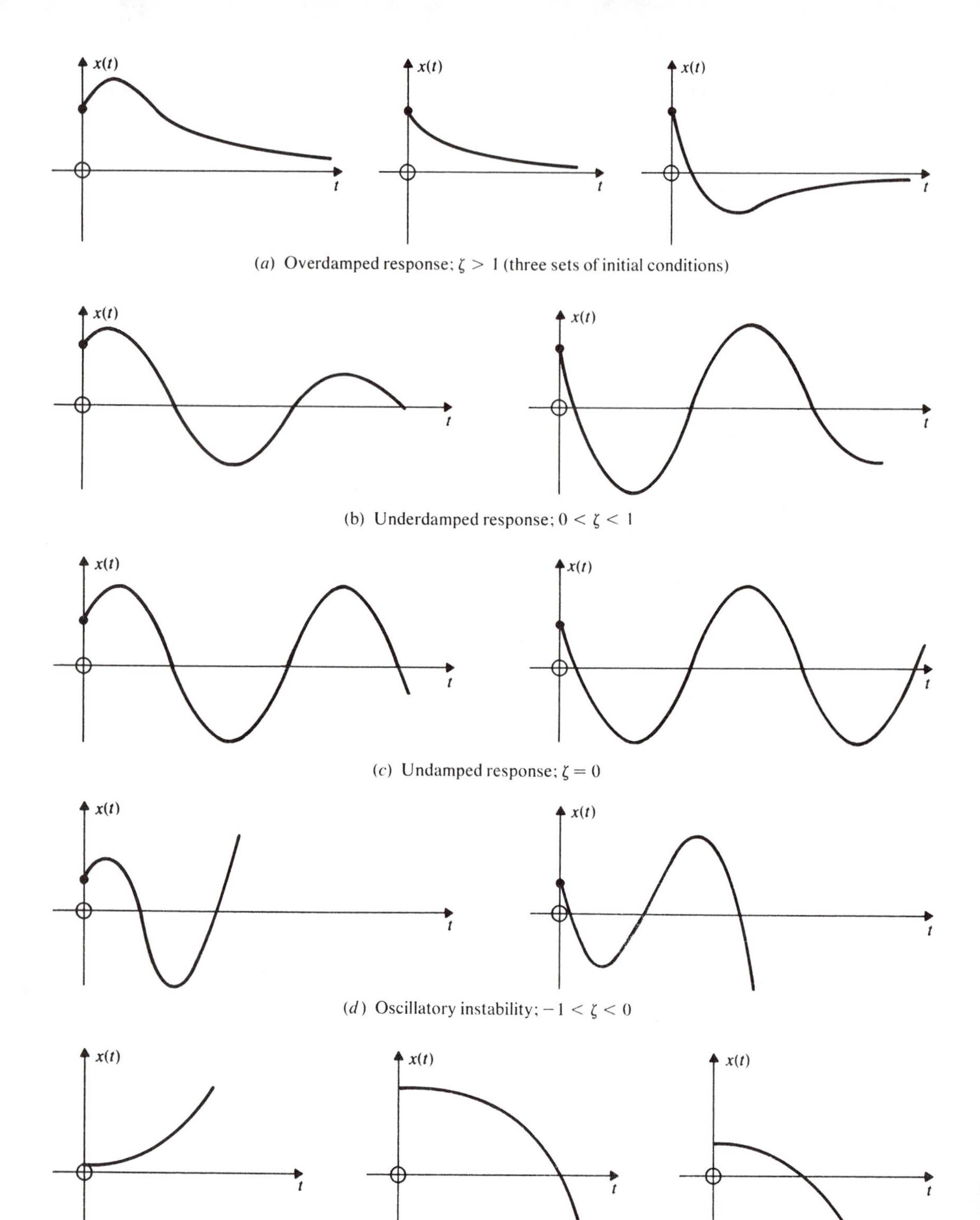
x(t)
t
(a) Overdamped response; ζ > 1 (three sets of initial conditions)
(b) Underdamped response; 0 < ζ < 1
(c) Undamped response; ζ = 0
(d) Oscillatory instability; −1 < ζ < 0
(e) Divergent instability; ζ < −1

**Fig. 3-3**

Note in Fig. 3-3 that the value of $\dot{x}(0)$ relative to $x(0)$ changes the shape of the curves to some extent. However, we can say, for example, that an overdamped solution will cut the $t$ axis no more than once, while an underdamped solution will continue to oscillate (with decreasing amplitude) indefinitely. The case of $\zeta = 1$ separates the latter two regimes and is given the name *critical*. When $\zeta = 0$, an oscillation will continue indefinitely at constant amplitude, corresponding to the behavior of an oscillator. If $\zeta < 0$, the solution will *grow* indefinitely with time, which means that energy must be fed into the circuit in some way.

As can be seen above, $\zeta$ is a measure of the damping of a second-order circuit: it is called the *damping ratio*. On the other hand, $\omega_n$ is a measure of the speed with which a circuit responds to its initial conditions. Thus from equations (*3.7*) and (*3.15*),

$$x(t) = k_1 e^{\omega_n\{-\zeta+\sqrt{\zeta^2-1}\}t} + k_2 e^{\omega_n\{-\zeta-\sqrt{\zeta^2-1}\}t} \qquad (3.16)$$

which shows that $\omega_n$ acts as a time-scale factor—the larger $\omega_n$, the faster will be the response. Looked at in another way, if $\zeta = 0$ in equation (*3.15*), then $s_{1,2} = \pm j\omega_n$ and therefore from equation (*3.9*) the (undamped) solution will be

$$x(t) = k_3 \cos \omega_n t + k_4 \sin \omega_n t$$

That is, the *undamped* or *natural frequency* of the circuit's response is $\omega_n$ rad/s. But, be careful! If $0 < \zeta < 1$, then the *actual* frequency of the oscillation is $\omega = \omega_n\sqrt{1 - \zeta^2}$, and if $\zeta \geq 1$, the circuit will not oscillate at all!

Finally, just as the time-constant applies only to first-order circuits, the undamped natural frequency $\omega_n$ and the damping ratio $\zeta$ apply, strictly speaking, to second-order circuits only.

## HIGHER-ORDER CIRCUITS

If a circuit can be described by the differential equation

$$\frac{d^n x}{dt^n} + a_{n-1}\frac{d^{n-1}x}{dt^{n-1}} + \cdots + a_1\frac{dx}{dt} + a_0 x = 0$$

then the corresponding characteristic equation will be

$$s^n + a_{n-1}s^{n-1} + \cdots + a_1 s + a_0 = 0$$

and there will be $n$ characteristic roots, $s_1, s_2, \ldots, s_n$. If, further, none of these roots is repeated, the solution will be

$$x(t) = k_1 e^{s_1 t} + k_2 e^{s_2 t} + \cdots + k_{n-1}e^{s_{n-1}t} + k_n e^{s_n t}$$

and $k_1 e^{s_1 t}, k_2 e^{s_2 t}, \ldots, k_{n-1}e^{s_{n-1}t}, k_n e^{s_n t}$ are called the *natural modes* of the response. The procedure for evaluating the constants $k_1, k_2, \ldots, k_n$ from the $n$ (given) initial conditions follows as for the second-order case.

The solution of *simultaneous* differential equations will be deferred to Chapter 5.

# ILLUSTRATIVE PROBLEMS

## First-Order Circuits

**3.1** The *R-C* circuit of Fig. 3-4 may be described by

$$v_R + v_C = 0 \quad \text{or} \quad Ri + \frac{1}{C}\int_0^t i\,d\tau + v_C(0) = 0$$

Solve for $i(t)$ using the method discussed in the section on first-order circuits. Assume that $v_C(0)$ is the given initial condition.

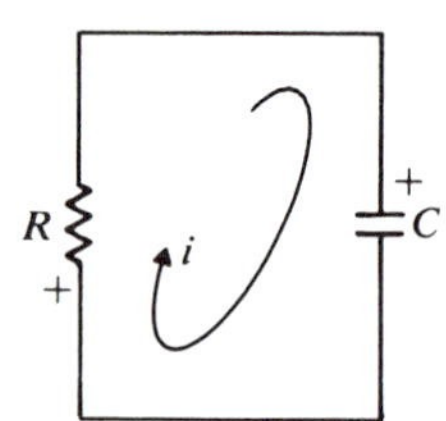


**Fig. 3-4**

The first move is to obtain a true differential equation in the form of equation (*3.1*) by differentiating the given integro-differential equation. This results in

$$R\frac{di}{dt} + \frac{1}{C}i = 0 \quad \text{or} \quad \frac{di}{dt} + \frac{1}{RC}i = 0$$

Substituting the trial solution $i(t) = ke^{st}$ into the differential equation yields

$$\frac{d}{dt}(ke^{st}) + \frac{1}{RC}(ke^{st}) = 0 \quad \text{or} \quad ske^{st} + \frac{1}{RC}ke^{st} = 0$$

We therefore find $s = -1/RC$, and $i(t) = ke^{-(1/RC)t}$.

If we set $t = 0$, then $k = i(0)$. But we were given $v_C(0)$ as the initial condition, from which we must now deduce $i(0)$. Applying KVL at $t = 0$,

$$v_R(0) + v_C(0) = 0 \quad \text{or} \quad Ri(0) + v_C(0) = 0$$

and

$$i(0) = -\frac{1}{R}v_C(0)$$

Thus the required solution is

$$i(t) = -\frac{1}{R}v_C(0)e^{-(1/RC)t}$$

Note: This (negative) solution indicates that the current will flow in a counter-clockwise direction while discharging the capacitor.

**3.2** Find the equation for the open-circuit voltage $v_2(t)$ in the circuit of Fig. 3-5, given that $i_1(0) = -2$ A.

The two loop equations are

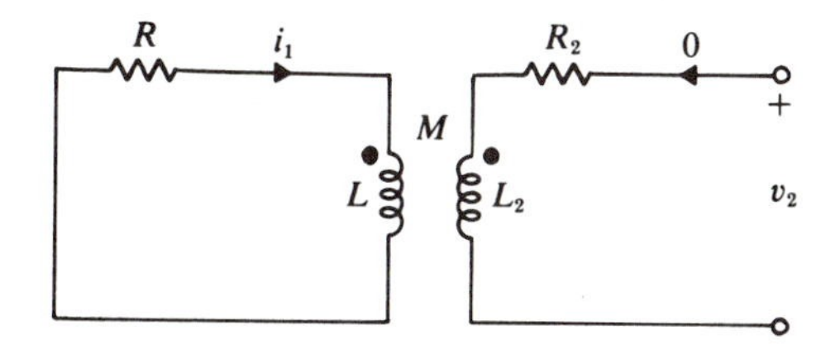


Fig. 3-5

$$L\frac{di_1}{dt} + Ri_1 = 0 \quad \text{and} \quad M\frac{di_1}{dt} = v_2$$

Solving the first equation for $i_1$ by assuming $i_1 = ke^{st}$ yields $s = -R/L$ and

$$i_1(t) = ke^{-(R/L)t} = -2e^{-(R/L)t}$$

since $k = i_1(0) = -2$. Then from the second equation,

$$v_2(t) = M\frac{di_1}{dt} = M\frac{d}{dt}\{-2e^{-(R/L)t}\} = \frac{2MR}{L}\,e^{-(R/L)t}$$

**3.3** The current in a resistive-inductive circuit is given by the equation $i(t) = -5e^{-10t}$.

(*a*) What is the time constant?

(*b*) Determine the initial rate of change of current.

(*c*) Find the approximate value of the current at $t = 0.40$ s.

(*d*) What is the initial value of the inductor current?

Comparing the current equation with the general first-order response $x(t) = x(0)e^{-\alpha t}$ of equation (*3.3*), we find:

(*a*) The time constant $= 1/\alpha = 1/10$ s.

(*b*) $di/dt = 50e^{-10t}$, therefore $(di/dt)|_{t=0} = 50$ A/s.

(*c*) $i(0.4) = -5e^{-4} = -(5)(0.02) = -0.10$ A; or nearly zero, compared with $i(0) = -5$ A.

(*d*) $i(0) = -5$ A.

$i(t)$ is sketched in Fig. 3-6.

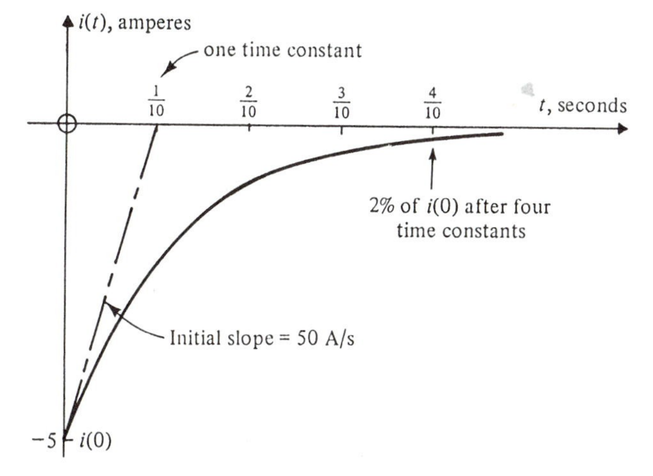


Fig. 3-6

**3.4** Write the nodal equation and hence find the time constant of the circuit in Fig. 3-7.

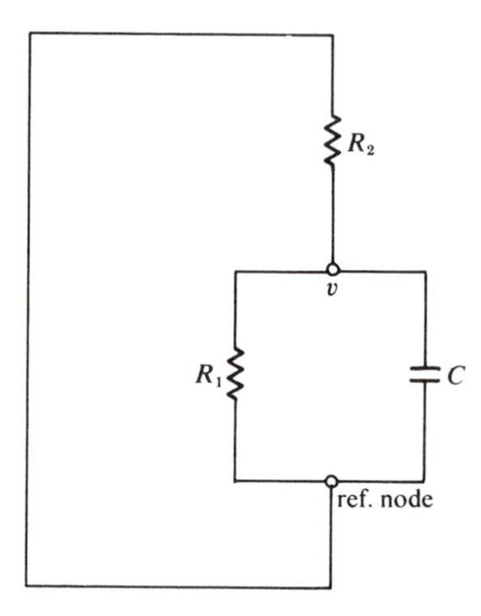


**Fig. 3-7**

Assuming all three current reference directions to be away from (or toward) the node $v$,

$$i_C + i_{R1} + i_{R2} = 0$$

and

$$C\frac{dv}{dt} + \frac{v}{R_1} + \frac{v}{R_2} = 0 \quad \text{or} \quad \frac{dv}{dt} + \frac{R_1 + R_2}{R_1 R_2 C} v = 0$$

Comparing this result with equation *(3.1)*, $(dx/dt) + \alpha x = 0$, we see that $\alpha = (R_1 + R_2)/R_1 R_2 C$, and hence the time constant is

$$\tau = \frac{1}{\alpha} = \frac{R_1 R_2 C}{R_1 + R_2}$$

Alternatively, the two resistors $R_1$ and $R_2$ are in parallel and may be combined as

$$\frac{1}{R_{\text{eq}}} = \frac{1}{R_1} + \frac{1}{R_2} \quad \text{or} \quad R_{\text{eq}} = \frac{R_1 R_2}{R_1 + R_2}$$

Thus the time constant, which equals $RC$ for an $R$-$C$ network, is

$$\tau = R_{\text{eq}} C = \frac{R_1 R_2 C}{R_1 + R_2}$$

**3.5** Redraw the circuit of Fig. 3-8*a* as a one port with the capacitor $C$ as the load. Then find $R_{\text{Th}}$ for the one port and hence the circuit's time constant. (Each resistor is 10 k$\Omega$ and $C = 1\ \mu$F.)

The circuit can be redrawn in many equivalent "shapes." The arrangement of Fig. 3-8*b* emphasizes the symmetry, which leads to a short-cut solution for $R_{\text{Th}}$.

Because the resistors are all equal, symmetry tells us that $v_a = v_b$ (see Fig. 3-8*b*).

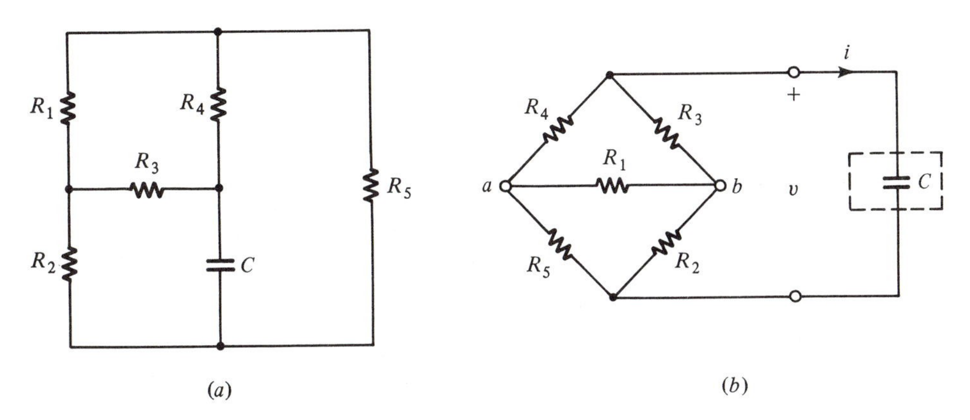


**Fig. 3-8**

Thus no current flows through $R_1$, which can therefore be (mentally) removed from the circuit. *Now* we see that $R_4$ and $R_5$ are in series, as are $R_3$ and $R_2$. These series pairs are in parallel and it follows at once that $R_{eq} = R_{Th} = 10\ \text{k}\Omega$.

Alternatively, if you did not see the symmetry, you would have to write three nodal (or three loop) equations and then solve for the terminal voltage $v$ (or the terminal current $i$). You would obtain $v = -10^4 i$ (or $i = -10^{-4} v$), which means of course that $R_{Th} = 10\ \text{k}\Omega$.

The time constant is simply $R_{Th}C = 10^4 \times 10^{-6} = 10$ ms, since the capacitor "sees" the 10-kΩ Thévenin resistance of the one port.

### *Second-Order Circuits and Their Characteristics*

**3.6** Write the KVL equation for the circuit of Fig. 3-9 and put it into the form of equation (*3.4*). Then find

(*a*) the characteristic equation,
(*b*) the characteristic roots, and
(*c*) the solution $i(t)$, given that $i(0) = 5$ A and $(di/dt)|_0 = -10$ A/s.

Using KVL,

$$v_L + v_R + v_C = 0 \quad \text{or} \quad \frac{1}{20}\frac{di}{dt} + \frac{5}{4}i + 5\int_0^t i\,d\tau + v_C(0) = 0$$

If we differentiate and multiply by 20, we obtain the desired form,

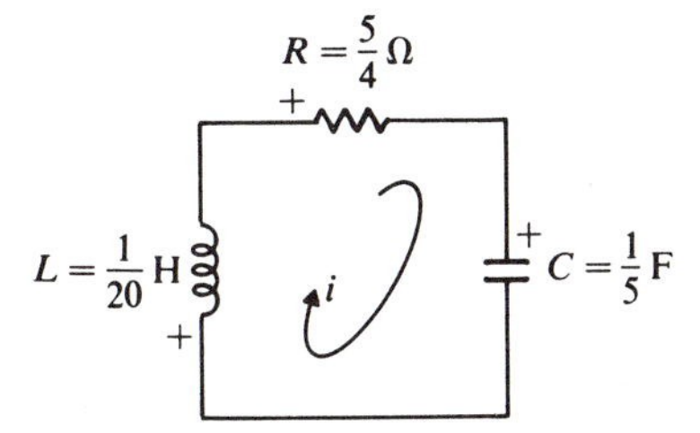


**Fig. 3-9**

$$\frac{d^2i}{dt^2} + 25\frac{di}{dt} + 100i = 0$$

(*a*) Substituting $i(t) = ke^{st}$ yields

$$s^2ke^{st} + 25ske^{st} + 100\,ke^{st} = 0 \quad \text{or} \quad s^2 + 25s + 100 = 0$$

which is the characteristic equation.

In practice, the characteristic equation may be obtained by direct inspection of the differential equation (compare equations (*3.4*) and (*3.6*)).

(*b*) Using the quadratic formula, we find that the characteristic roots are $s_1 = -5$ and $s_2 = -20$ (the order is unimportant).

(*c*) From equation (*3.7*), the solution is of the form $i(t) = k_1e^{-5t} + k_2e^{-20t}$ where $k_1$ and $k_2$ must be evaluated from the initial conditions, as in equation (*3.8*). That is,

$$i(0) = 5 = k_1 + k_2 \quad \text{and} \quad \frac{di(0)}{dt} = -10 = -5k_1 - 20k_2$$

from which $k_1 = 6$ and $k_2 = -1$, giving us the solution

$$i(t) = 6e^{-5t} - e^{-20t}$$

As a check, we set $t = 0$ in the solution and its derivative,

$$i(0) = 6e^{-0} - e^{-0} = 5 \quad \text{and} \quad \frac{di(0)}{dt} = -30e^{-0} + 20e^{-0} = -10$$

which agrees with the given initial conditions.

**3.7** Find the damping ratio $\zeta$ and the undamped natural frequency $\omega_n$ for the circuit of Problem 3.6 (Fig. 3-9).

If we compare the two forms for the characteristic equation,

$$s^2 + 25s + 100 = 0 \quad \text{and} \quad s^2 + 2\zeta\omega_n s + \omega_n^2 = 0$$

then by inspection,

$$\omega_n^2 = 100 \quad \text{or} \quad \omega_n = 10$$

and

$$2\zeta\omega_n = 25 \quad \text{or} \quad \zeta = 1.25$$

Since $\zeta > 1$, the current in Problem 3.6 will look "something like" either the middle or the right-hand graph of Fig. 3-3*a*. (We would have to plot a few points of the curve $i(t) = 6e^{-5t} - e^{-20t}$ vs. $t$ to see exactly what the response is like.)

**3.8** (*a*) Write two state equations (two first-order differential equations) for the circuit of Fig. 3-10.

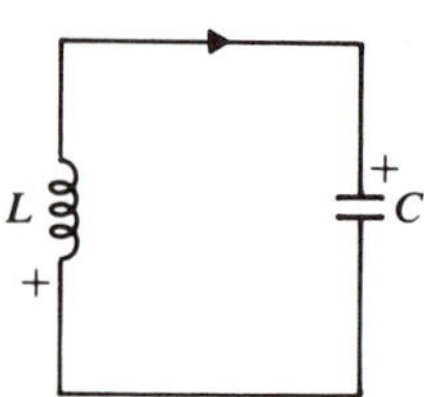


**Fig. 3-10**

(*b*) Eliminate one of the state variables to obtain a second-order differential equation for this circuit.
(*c*) Determine the natural frequency $\omega_n$ and the damping ratio $\zeta$.

(*a*) We first write the derivative relations,

$$L\frac{di_L}{dt} = v_L \quad \text{and} \quad C\frac{dv_C}{dt} = i_C$$

To eliminate the (nonstate) variables $v_L$ and $i_C$ we note that $v_L + v_C = 0$ and $i_L = i_C$, so that the state equations are

$$\frac{di_L}{dt} = -\frac{1}{L}v_C \quad \text{and} \quad \frac{dv_C}{dt} = \frac{1}{C}i_L$$

(*b*) To eliminate $i_L$ (alternatively we could eliminate $v_C$) we write

$$\frac{d}{dt}\left\{C\frac{dv_C}{dt}\right\} = -\frac{1}{L}v_C \quad \text{or} \quad \frac{d^2v_C}{dt^2} + \frac{1}{LC}v_C = 0$$

(*c*) Here $\omega_n = \sqrt{1/LC}$ and $\zeta = 0$. That is, the response will be oscillatory and undamped (see Fig. 3-3*c*).

**3.9** Write the nodal equation for the *L-R-C* circuit of Fig. 3-11, and then find

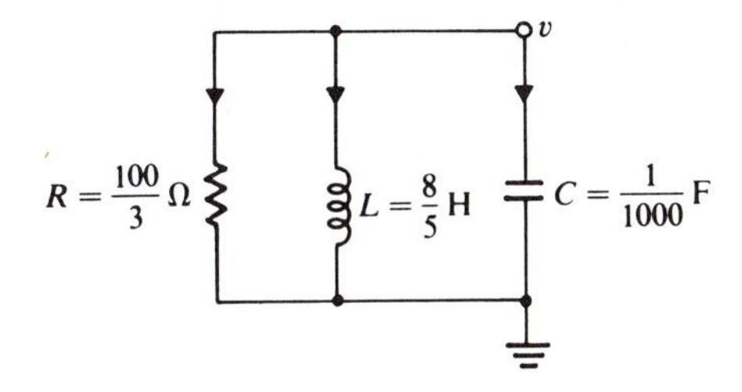


**Fig. 3-11**

(*a*) the characteristic equation,
(*b*) the characteristic roots,
(*c*) $v(t)$, given that $v(0) = 10$ V and $\dot{v}(0) = -50$ V/s,
(*d*) the damping ratio $\zeta$ and the natural frequency $\omega_n$,
(*e*) the graph of Fig. 3-3 that most nearly "looks like" the solution $v(t)$.

From KCL,

$$i_R + i_L + i_C = 0 \quad \text{or} \quad \frac{1}{R}v + \frac{1}{L}\int_0^t v\,d\tau + i_L(0) + C\frac{dv}{dt} = 0$$

Substituting the data, differentiating and rearranging,

$$\frac{d^2v}{dt^2} + 30\frac{dv}{dt} + 625v = 0$$

(*a*) By inspection, the characteristic equation is $s^2 + 30s + 625 = 0$.
(*b*) The quadratic formula yields the characteristic roots $s_{1,2} = -15 \pm j20$.
(*c*) The roots are complex conjugates with $\sigma = -15$ and $\omega = 20$. Therefore from equation (*3.9*), the solution will be of the form

$$v(t) = e^{-15t}\{k_3 \cos 20t + k_4 \sin 20t\}$$

Now with the help of equation (*3.10*),

$$v(0) = 10 = k_3 \quad \text{and} \quad \dot{v}(0) = -50 = -15k_3 + 20k_4$$

from which $k_3 = 10$ and $k_4 = 5$, and so

$$v(t) = e^{-15t}\{10 \cos 20t + 5 \sin 20t\}$$

(*d*) Comparing $s^2 + 30s + 625 = 0$ with $s^2 + 2\zeta\omega_n s + \omega_n^2 = 0$, we find $\omega_n = 25$ and $\zeta = 0.6$.

(*e*) With $\zeta = 0.6$, the response is underdamped; and since $\dot{v}(0)$ is negative, the right-hand graph of Fig. 3-3*b* should be most like the response $v(t)$.

**3.10** In Problem 3.9 (Fig. 3-11), the initial conditions were given as $v(0) = 10$ V and $\dot{v}(0) = -50$ V/s. Calculate the initial current $i_L(0)$.

From KCL at $t = 0$, $i_R(0) + i_L(0) + i_C(0) = 0$ and thus

$$\frac{1}{R}v(0) + i_L(0) + C\frac{dv(0)}{dt} = 0$$

Upon substituting the data and solving for $i_L(0)$, we find $i_L(0) = -0.25$ A.

Thus an initial inductor current of $-0.25$ A and an initial capacitor voltage of $+10$ V are the "true or physical" initial conditions for this circuit (more physical, at least, then $\dot{v}(0)$ which we would regard as a "mathematical" initial condition). In state variable terminology, $i_L(0)$ and $v(0)$ are the initial states of the network.

**3.11** The series *L*-*R*-*C* circuit of Fig. 3-12 is described by the equation

$$\frac{d^2i}{dt^2} + \frac{R}{L}\frac{di}{dt} + \frac{1}{LC}i = 0$$

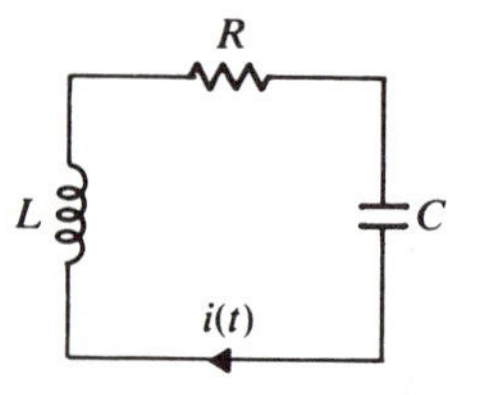


**Fig. 3-12**

(*a*) For $R = 1\,\Omega$, $L = 1$ H, $C = 1$ F, $i(0) = 0$, and $di(0)/dt = 0.866$ A/s, find $i(t)$, $\zeta$, $\omega_n$, and sketch $i(t)$ vs. $t$.

(*b*) Repeat part (*a*) for $R = 100\,\Omega$, $L = 0.1$ mH, $C = 10$ nF, $i(0) = 0$, and $di(0)/dt = 0.866 \times 10^6$ A/s.

(*a*) The characteristic equation is $s^2 + s + 1 = 0$ and the characteristic roots are $s_{1,2} = -0.5 \pm j0.866$. Also $\omega_n = 1$ and $\zeta = 0.5$.

From equations (*3.9*) and (*3.10*) the waveform, which is sketched in Fig. 3-13*a*, is

$$i(t) = e^{-0.5t} \sin 0.866t$$

(*b*) The characteristic equation corresponding to the new data is $s^2 + 10^6 s + 10^{12} = 0$ and the characteristic roots are $s_{1,2} = 10^6(-0.5 \pm j0.866)$. Also,

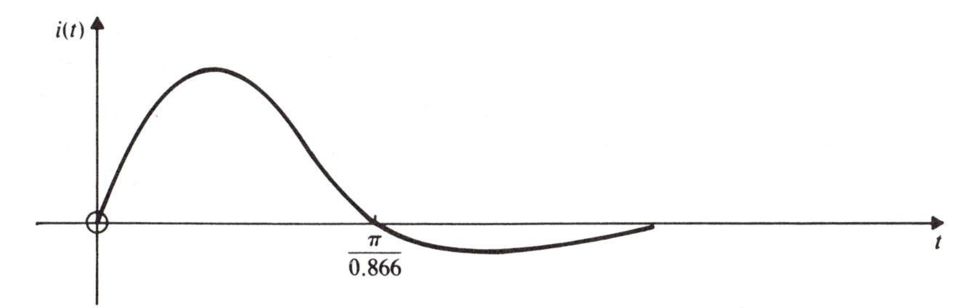


**Fig. 3-13*a*** ***R*** **= 1 Ω,** ***L*** **= 1** ***H,*** ***C*** **= 1 F**

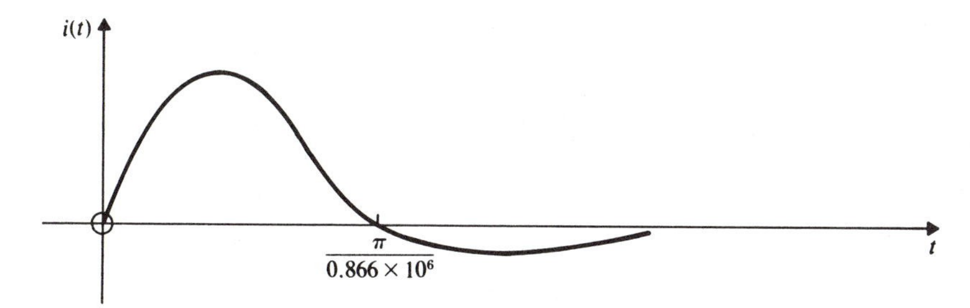


**Fig. 3-13*b*** ***R*** **= 100 Ω,** ***L*** **= 0.1 mH,** ***C*** **= 10 nF**

$\omega_n = 10^6$ and $\zeta = 0.5$ (as before). Here

$$i(t) = e^{-0.5\times 10^6 t} \sin(0.866 \times 10^6 t)$$

which is sketched in Fig. 3-13*b*.

*Comment:* Most of the problems in this book use "nonphysical" values of the circuit elements (*R, L,* and *C* of the order of unity). This avoids the necessity for carrying awkward powers of 10 through the calculations. Notice that "real" data does not change the general character of the results. This is brought out in the nondimensional graph of Fig. 3-13*c*, which is *identical* to the graph of Fig. 3-13*a*. Thus the second set of data in this problem is such that only the *speed* of response is changed.

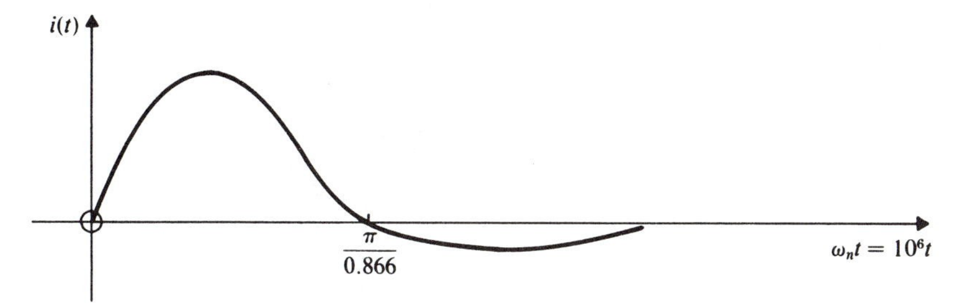


**Fig. 3-13*c*** **Normalized time scale (*R*** **= 100 Ω,** ***L*** **= 0.1 mH,** ***C*** **= 10 nF**

**3.12** For the circuit of Problem 3.11 (Fig. 3-12), let $R = 2000\ \Omega$, $L = 1$ H, $C = 1\ \mu$F, $i(0) = -2$ A, and $di(0)/dt = 4200$ A/s. Find the current $i(t)$.

The characteristic equation is $s^2 + 2000s + 10^6 = 0$ and the characteristic roots are $s_1 = s_2 = -1000$. Thus the solution will take the form of equation (*3.11*), $i(t) = k_3e^{-1000t} + k_4te^{-1000t}$. Using equation (*3.12*) to calculate $k_3$ and $k_4$, we have

$$i(0) = -2 = k_3 \quad \text{and} \quad \frac{di(0)}{dt} = 4200 = -1000k_3 + k_4$$

Solving these equations yields $k_3 = -2$ and $k_4 = 2200$, so that

$$i(t) = -2e^{-1000t} + 2200t\,e^{-1000t}$$

**3.13** In the *L-C* circuit of Fig. 3-14 both switches are open for $t < 0$ and are closed at $t = 0$. At this time, $i_L(0) = 0$, $v_{C_1}(0) = 1$ V, and $v_{C_2}(0) = 2$ V, with the top plates of both capacitors positive. Solve for the clockwise current flowing in the circuit for $t \geq 0$.

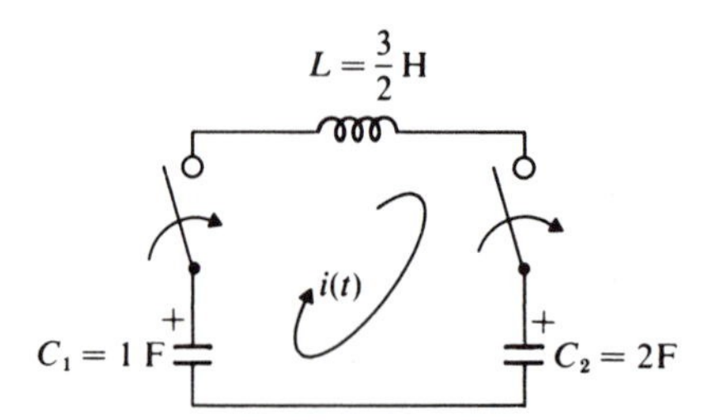


**Fig. 3-14**

From KVL, $-v_{C_1} + v_L + v_{C_2} = 0$ or

$$-\int_0^t (-i)d\tau - 1 + \frac{3}{2}\frac{di}{dt} + \frac{1}{2}\int_0^t i\,d\tau + 2 = 0$$

which, after differentiation and rearrangement, becomes

$$\frac{d^2i}{dt^2} + i = 0$$

The characteristic roots are $s_{1,2} = \pm j1$ and the solution will be

$$i(t) = k_3 \cos t + k_4 \sin t$$

Now $i_L(0) = 0$, and applying KVL at $t = 0$, $-v_{C_1}(0) + v_L(0) + v_{C_2}(0) = 0$ or

$$-1 + \frac{3}{2}\frac{di(0)}{dt} + 2 = 0$$

and thus

$$\frac{di(0)}{dt} = -\frac{2}{3}$$

Finally, setting $t = 0$ in the above solution for $i(t)$ and its derivative,

$$i(0) = 0 = k_3 \quad \text{and} \quad \frac{di(0)}{dt} = -\frac{2}{3} = k_4$$

Therefore the required current is

$$i(t) = -\frac{2}{3} \sin t$$

### Higher-Order Circuits

**3.14** The characteristic equation of a linear circuit is $s^3 + 6s^2 + 11s + 6 = 0$. Find the natural modes of response given that $s = -1$ is one of the characteristic roots.

Since $s = -1$ is a root, $(s + 1)$ will be a factor which we must divide into the characteristic polynomial to find the other roots,

$$\begin{array}{r|l} & s^2 + 5s + 6 \\ s+1 & s^3 + 6s^2 + 11s + 6 \\ & \underline{s^3 + s^2} \\ & 5s^2 + 11s \\ & \underline{5s^2 + 5s} \\ & 6s + 6 \\ & \underline{6s + 6} \end{array}$$

Since $s^2 + 5s + 6 = 0$ has roots of $s = -2$ and $s = -3$, it follows that the three natural modes of response are

$$K_1e^{-t}, \quad K_2e^{-2t}, \quad \text{and} \quad K_3e^{-3t}$$

In the absence of the known root we might use Newton's algorithm to search for one of the roots, or we could go to a standard root-finding program in a calculator or computer.

**3.15** The natural modes of response of a linear circuit are known to be

$$K_1 \quad \text{and} \quad e^{-t}\{K_2 \cos 2t + K_3 \sin 2t\}$$

(*a*) What are the characteristic roots of this circuit?
(*b*) Find the characteristic equation.

(*a*) The natural mode, $K_1$ or $K_1e^{0t}$, is characteristic of the root $s = 0$. Now $e^{-t}\{K_2 \cos 2t + K_3 \sin 2t\}$ is of the same form as equation (*3.9*) with $\sigma = -1$ and $\omega = 2$, thus making the other roots $s = -1 \pm j2$.
(*b*) The characteristic equation is

$$(s)(s + 1 - j2)(s + 1 + j2) = 0 \quad \text{or} \quad s^3 + 2s^2 + 5s = 0$$

# PROBLEMS

**3.16** Determine the equation for the current in the circuit of Fig. 3-15 given $v_{C_1}(0) = 1$ V and $v_{C_2}(0) = 2$ V.

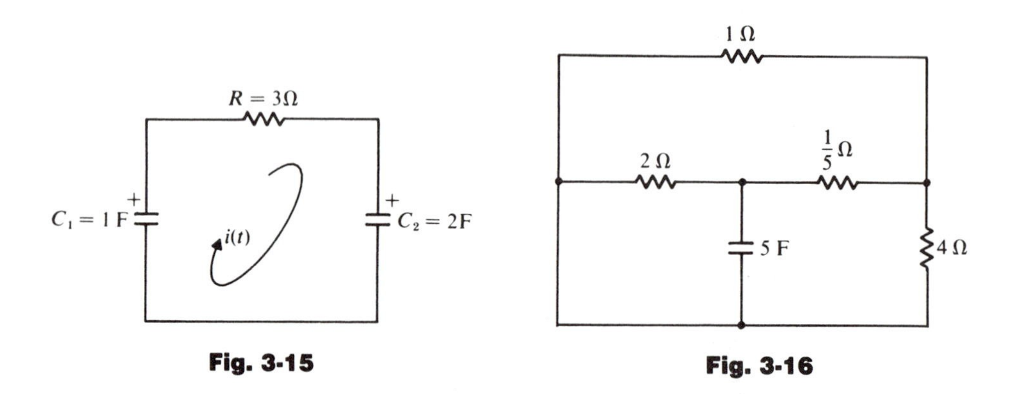


**Fig. 3-15**

**Fig. 3-16**

**3.17** Find the time constant of the circuit in Fig. 3-16.

**3.18** For the circuit of Fig. 3-17 it is known that $i(0) = 5$ A and $v_C(0) = -6$ V. Find $di(0)/dt$ and $dv_C(0)/dt$.

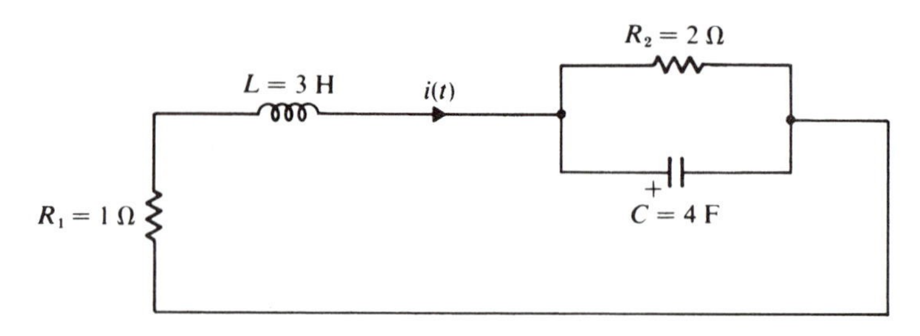


**Fig. 3-17**

**3.19** The current in an *L-R-C* circuit is described by the differential equation

$$\frac{d^2i}{dt^2} + k\frac{di}{dt} + 100\,i = 0$$

It is also known that $di(0)/dt = 76$ and $i(0) = 17$. Given the values of $k$ in the following table, show that all the other entries are correct.

| $k$ | $\zeta$ | Characteristic roots | Form of current response |
|---|---|---|---|
| 0 | 0 | $\pm j10$ | Fig. 3-3$c$, left graph |
| 12 | 0.6 | $-6 \pm j8$ | Fig. 3-3$b$, left graph |
| 52 | 2.6 | $-2; -50$ | Fig. 3-3$a$, left graph |
| $-25$ | $-1.25$ | $+5; +20$ | Fig. 3-3$e$, left graph |

**3.20** Find the open-circuit voltage $v(t)$ for the circuit of Fig. 3-18, given $v(0) = (4/3)$ V and $i_L(0) = 2$ A.

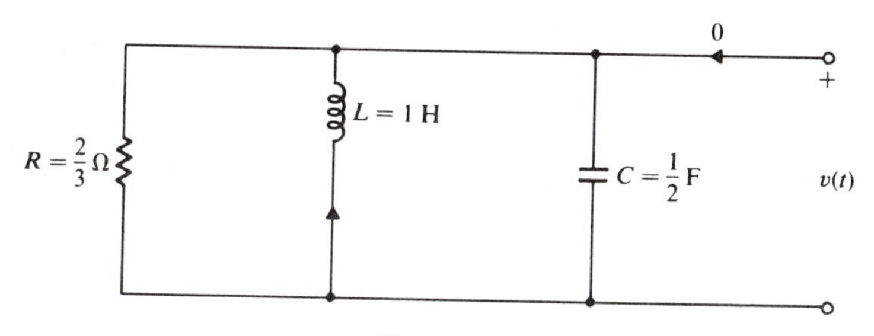


**Fig. 3-18**

**3.21** Ascertain the natural frequency $\omega_n$ and the damping ratio $\zeta$ for the circuit in Fig. 3-19.

**3.22** Given $i_L(0) = -3$ A and $v_C(0) = 2$ V in the circuit of Fig. 3-19, find $v_C(t)$ for $t \geq 0$.

**3.23** Find $i_L(t)$ corresponding to the circuit and data of Problem 3.22.

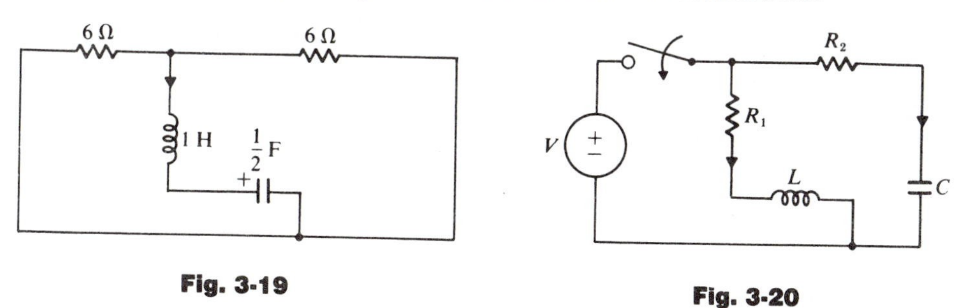


**Fig. 3-19**

**Fig. 3-20**

**3.24** The switch in the circuit of Fig. 3-20 has been open for a long time prior to its closing at $t = 0$. Determine ($a$) $i_{R_1}(0)$, ($b$) $i_{R_2}(0)$, and ($c$) $di_L(0)/dt$.

# Chapter 4

# RESPONSE OF SIMPLE DRIVEN CIRCUITS

This chapter complements the preceding one by extending the process of solution to cases where the circuit contains one or more sources. This is, after all, more the rule than the exception. Of course, the initial conditions are still important and continue to play their role in the determination of the free constants of integration.

## FIRST-ORDER DRIVEN CIRCUITS

The simple *R-C* and *R-L* networks of Fig. 4-1 are *driven* by the sources $i(t)$ and $e(t)$ respectively. The appropriate circuit equations are

$$\frac{1}{R}v + C\frac{dv}{dt} = i(t) \quad \text{and} \quad Ri + L\frac{di}{dt} = e(t)$$

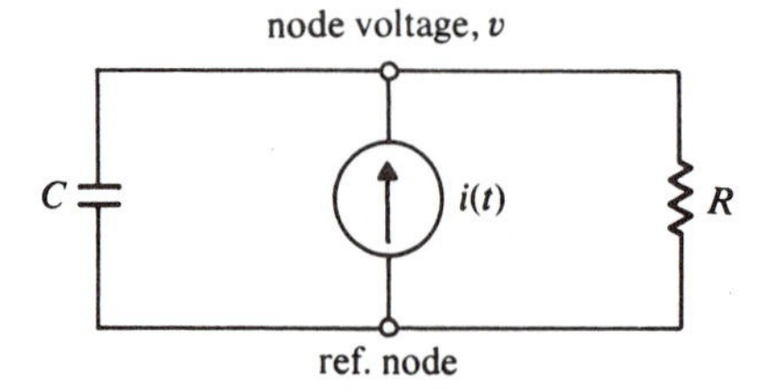


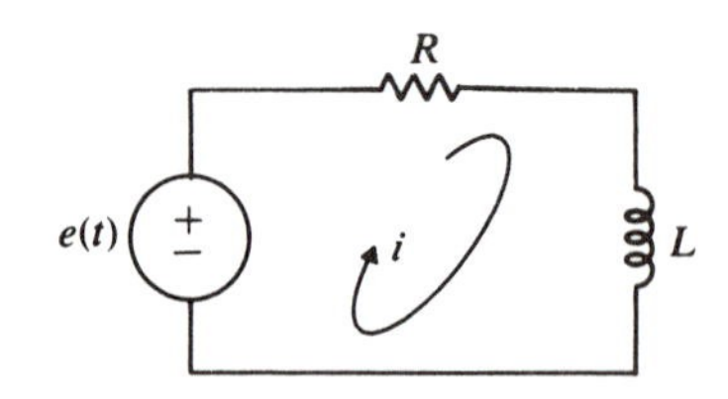


**Fig. 4-1**

or, in general,

$$\frac{dx}{dt} + \alpha x = \frac{1}{\beta} f(t) \tag{4.1}$$

where $x(t)$ is the variable and $f(t)$ is the *driving source* or *input*.

There will in general be two (additive) parts of the solution: the *complementary solution* $x_c(t)$ which satisfies the source-free (homogeneous) equation

$$\frac{dx}{dt} + \alpha x = 0 \tag{4.2}$$

and the *particular solution* $x_p(t)$ which satisfies the forced (nonhomogeneous) equation (*4.1*).

Now when $x_p(t)$ is substituted into the left-hand side of equation (*4.1*), the result must be $f(t)/\beta$. And when $x_c(t)$ is similarly substituted, the result must be zero. Therefore, if $x_c(t) + x_p(t)$ is substituted, the result will be $0 + f(t)/\beta = f(t)/\beta$. That is, $x_c + x_p$ satisfies the original equation (*4.1*), and this sum is called the *total solution*,

$$x(t) = x_c(t) + x_p(t) \tag{4.3}$$

The search for the complementary solution follows the procedure of Chapter 3. Thus the characteristic equation corresponding to equation (*4.2*) is

$$s + \alpha = 0 \tag{4.4}$$

and the characteristic root is $s = -\alpha$. Accordingly, the complementary solution (the solution of the homogeneous equation) is

$$x_c(t) = k_1 e^{-\alpha t} \tag{4.5}$$

Next we must obtain the particular solution—most easily by "trial." If, for example, $f(t) = A$, a constant, then our first guess, or trial, would probably be

$$x_p(t) = k_2 \tag{4.6}$$

The validity of the trial must of course be tested by substitution into the original nonhomogeneous equation. At the same time (assuming a valid trial) the constant $k_2$ should be evaluated. Thus substituting equation (*4.6*) into (*4.1*),

$$\alpha k_2 = \frac{1}{\beta} A \quad \text{or} \quad k_2 = \frac{A}{\alpha\beta} \tag{4.7}$$

The *total* solution now follows,

$$x(t) = x_c(t) + x_p(t) = k_1 e^{-\alpha t} + \frac{A}{\alpha\beta} \tag{4.8}$$

To evaluate the remaining constant $k_1$, we substitute the given initial condition $x(0)$ into equation (*4.8*),

$$x(0) = k_1 + \frac{A}{\alpha\beta} \quad \text{or} \quad k_1 = x(0) - \frac{A}{\alpha\beta} \tag{4.9}$$

In summary, the solution we seek is

$$x(t) = \left\{x(0) - \frac{A}{\alpha\beta}\right\} e^{-\alpha t} + \frac{A}{\alpha\beta} \tag{4.10}$$

(Recall that we assumed a constant input $A$. A different input would lead to a different solution.)

## TRIAL SOLUTION FOR THE PARTICULAR SOLUTION

There is no need to guess for a trial solution, since appropriate lists, such as Table 4-1, were compiled long ago.

If the forcing function $f(t)$ is the sum of two or more terms, then the particular solution is the corresponding sum from the right-hand column of the table. And if one of the terms in $f(t)$ corresponds to a term in the complementary solution, then we must follow the procedure for repeated roots, as in Chapter 3.

**Table 4-1**

| Term in $f(t)$ | Choice of particular solution |
|---|---|
| $A$ | $K$ |
| $At$ | $K_1t + K_0$ |
| $At^n$ | $K_nt^n + k_{n-1}t^{n-1} + \cdots + K_1t + K_0$ |
| $Ae^{\alpha t}$ | $Ke^{\alpha t}$ |
| $A \cos \omega t$<br>or<br>$A \sin \omega t$ | $K_1 \cos \omega t + K_2 \sin \omega t$<br>or<br>$K_3 \cos (\omega t + \phi)$ |
| $Ae^{\alpha t} \cos \omega t$<br>or<br>$Ae^{\alpha t} \sin \omega t$ | $e^{\alpha t}(K_1 \cos \omega t + K_2 \sin \omega t)$<br>or<br>$K_3e^{\alpha t} \cos (\omega t + \phi)$ |

## HIGHER-ORDER CIRCUITS

The procedure for obtaining the total solution for circuit equations of higher than first order follows precisely the same lines, but simultaneous differential equations pose a somewhat different problem, which we will take up in Chapter 5.

## DC STEADY-STATE RESPONSE

If a circuit's input is a constant $A$, then the solution of the circuit equation will be of the form

$$x(t) = k_1e^{s_1t} + k_2e^{s_2t} + \cdots + k_ne^{s_nt} + K$$

*If* all the terms except the last go to zero as $t \to \infty$, that is, *if* the real part of each $s_i$ is negative, then

$$x_{ss}(t) = K$$

which is called the *dc steady state solution,* since it is the steady solution remaining after all the *transients* $k_ie^{s_it}$ have died away.

The dc steady-state solution can be found directly from the circuit (or its equations). We recall that

$$i_C = C\frac{dv_C}{dt} \quad \text{and} \quad v_L = L\frac{di_L}{dt}$$

but in the dc steady-state (provided that one exists) all rates of change must be zero, as must all capacitor currents and inductor voltages. That is, as $t \to \infty$, all capacitors act as if they were open circuited, and all inductors act as if they were short circuited.

### Example 4.1

Let us find the dc steady-state inductor current and capacitor voltage in the circuit of Fig. 4-2*a*.

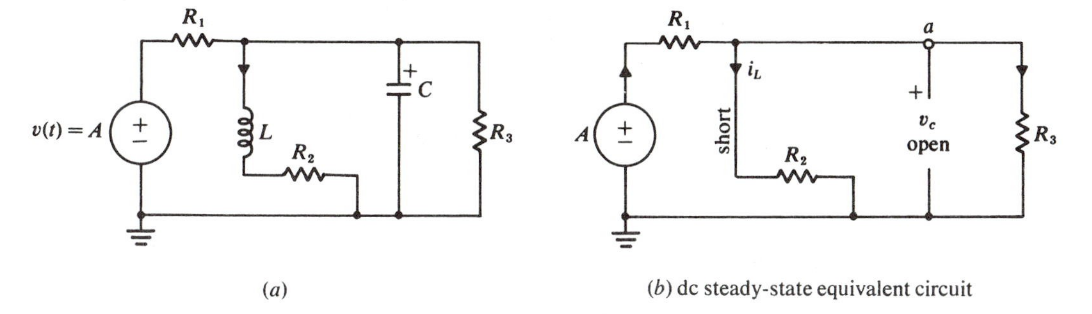


(*a*) (*b*) dc steady-state equivalent circuit

**Fig. 4-2**

The first step is to replace the inductor by a short circuit and the capacitor by an open circuit, as in Fig. 4-2*b*. Then applying KCL at node *a*,

$$i_{R1} - i_L - i_{R3} = 0 \quad \text{or} \quad \frac{A - v_a}{R_1} - \frac{v_a}{R_2} - \frac{v_a}{R_3} = 0$$

Substituting $v_a = v_C$ and rearranging,

$$\left(\frac{1}{R_1} + \frac{1}{R_2} + \frac{1}{R_3}\right) v_C = \frac{A}{R_1} \quad \text{or} \quad v_C = \frac{AR_2R_3}{R_1R_2 + R_2R_3 + R_3R_1}$$

and since $i_L = v_C/R_2$ in the steady-state,

$$i_L = \frac{AR_3}{R_1R_2 + R_2R_3 + R_3R_1}$$

These values of $v_C$ and $i_L$, being derived from the dc equivalent circuit, are the required dc steady-state values. ■

It is evident from the above example that the solution of dc steady-state problems leads to algebraic rather than differential equations.

## AC STEADY-STATE

If a circuit's input is a sinusoid $A \cos \omega t$, then the solution of the circuit equation will be of the form

$$x(t) = k_1 e^{s_1 t} + k_2 e^{s_2 t} + \cdots + k_n e^{s_n t} + K \cos(\omega t + \phi)$$

*If* all the terms except the last go to zero as $t \to \infty$, then

$$x_{sss}(t) = K \cos(\omega t + \phi) \tag{4.11}$$

Equation (*4.11*) is called the *ac* or *sinusoidal steady-state solution* since it represents the unchanging sinusoid remaining after all the transients have decayed to zero.

The dc steady-state problem reduced to the solution of algebraic equations. The ac or sinusoidal steady-state problem reduces to the solution of complex algebraic equations—but this must wait until Chapter 7.

## EVALUATION OF INITIAL CONDITIONS

It has been assumed throughout the last two chapters that if we have to solve an $n$th order differential equation in $x$, then the initial conditions $x(0)$, $\dot{x}(0)$, $\ddot{x}(0)$, . . . , $x^{(n-1)}(0)$ would be given as data. In fact, the *initial condition of a network* is usually specified in quite a different way, and from this specification $x(0)$, $\dot{x}(0)$, $\ddot{x}(0)$, . . . must be found. There is no standard gambit, so it is not surprising that the evaluation has been compared with a challenging game of chess.

### Example 4.2

Suppose that the switch in the circuit of Fig. 4-3 has been in position $a$ for a long time and that at some instant, conveniently designated $t = 0$, the switch is moved to position $b$. What are the appropriate values of $i(0)$ and $di(0)/dt$? ($V_1$ and $V_2$ are constants.)

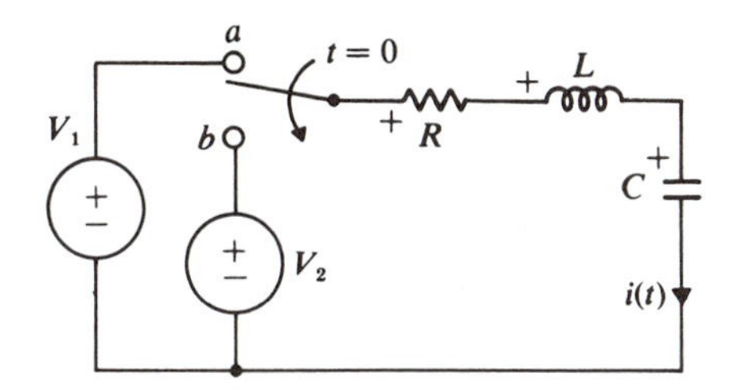


Fig. 4-3

If the switch has been at $a$ long enough for the dc steady-state to be established, the capacitor acts as an open circuit and $i(0) = 0$. It follows that both $v_R$ and $v_L$ are initially zero, and therefore from KVL, $v_C(0) = V_1$.

Now neither $i_L$ nor $v_C$ can change at the *instant* of switching, since infinite $v_L$ or $i_C$ would result. Therefore $i(0) = 0$ and $v_C(0) = V_1$ still apply at the instant following the switching action.

Immediately *after* switching, KVL yields

$$Ri(0) + L\frac{di(0)}{dt} + v_C(0) - V_2 = 0$$

or

$$\frac{di(0)}{dt} = \frac{1}{L}\{V_2 - v_C(0) - Ri(0)\} = \frac{1}{L}(V_2 - V_1)$$

The initial conditions $i(0) = 0$ and $di(0)/dt = (V_2 - V_1)/L$ are those we would need if we wished to solve the (second-order) loop equation for the response following the act of switching. ■

# ILLUSTRATIVE PROBLEMS

### *First-Order Driven Circuits*

**4.1** Given the circuit of Fig. 4-4 with $v(t) = A\cos\omega t$ and $i(0) = I_0$, find the *total* solution $i(t)$.

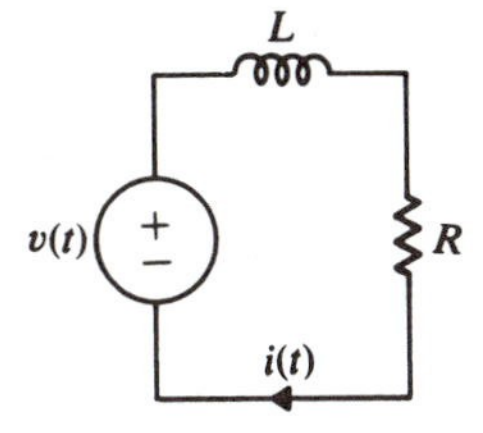


Fig. 4-4

Here the appropriate differential equation is

$$L\frac{di}{dt} + Ri = A\cos\omega t \quad \text{or} \quad \frac{di}{dt} + \frac{R}{L}i = \frac{A}{L}\cos\omega t$$

The complementary (source-free) solution is

$$i_c(t) = k_1 e^{-(R/L)t}$$

and from Table 4-1, the particular solution is of the form

$$i_p(t) = k_2\cos\omega t + k_3\sin\omega t$$

Upon substituting $i_p$ into the original differential equation, we obtain

$$L\{-k_2\omega\sin\omega t + k_3\omega\cos\omega t\} + R\{k_2\cos\omega t + k_3\sin\omega t\} = A\cos\omega t$$

This equation must be satisfied for all $t$, and so $k_2$ and $k_3$ may be found by equating the coefficients of like functions on the two sides of the equation. That is,

$$-Lk_2\omega + Rk_3 = 0 \quad \text{and} \quad Lk_3\omega + Rk_2 = A$$

from which, solving simultaneously,

$$k_2 = \frac{AR}{R^2 + \omega^2L^2} \quad \text{and} \quad k_3 = \frac{A\omega L}{R^2 + \omega^2L^2}$$

Thus the total solution is

$$i(t) = i_c(t) + i_p(t) = k_1e^{-(R/L)t} + \frac{A}{R^2 + \omega^2L^2}\{R\cos\omega t + \omega L\sin\omega t\}$$

To evaluate $k_1$, we set $t = 0$, obtaining

$$i(0) = I_0 = k_1 + \frac{AR}{R^2 + \omega^2L^2} \quad \text{or} \quad k_1 = I_0 - \frac{AR}{R^2 + \omega^2L^2}$$

Finally,

$$i(t) = \left\{I_0 - \frac{AR}{R^2 + \omega^2L^2}\right\}e^{-(R/L)t} + \frac{A}{R^2 + \omega^2L^2}\{R\cos\omega t + \omega L\sin\omega t\}$$

### *Comments*

1. A long time ($t \gg L/R$) after the circuit is "started" with its initial condition and sinusoidal input, the first term $i_c$ will have decayed to zero, leaving the sinusoidal steady-state solution

$$i_{sss}(t) = i_p(t) = \frac{A}{R^2 + \omega^2L^2}\{R\cos\omega t + \omega L\sin\omega t\}$$

2. The solution of a differential equation entails much algebra, with a correspondingly high probability of error. It is always possible (if tedious) to *check* a solution—the given initial conditions can be recalculated from the solution; the particular integral should satisfy the original nonhomogeneous equation; and the complementary function should satisfy the homogeneous equation.

**4.2** If the source voltage in Fig. 4-4 is changed to $v(t) = At$, with $i(0) = I_0$ as before, what is $i(t)$?

The complementary solution is still $i_c(t) = k_1 e^{-(R/L)t}$ and from Table 4-1, $i_p(t) = k_2 t + k_3$.

The substitution of $i_p$ into the circuit's differential equation yields

$$Lk_2 + Rk_2 t + Rk_3 = At$$

from which, upon equating coefficients of like terms,

$$Rk_2 = A \quad \text{and} \quad Lk_2 + Rk_3 = 0$$

or $k_2 = A/R$ and $k_3 = -AL/R^2$.

The total solution is therefore

$$i(t) = k_1 e^{-(R/L)t} + \frac{A}{R}t - \frac{AL}{R^2}$$

To evaluate $k_1$ we set $t = 0$,

$$i(0) = I_0 = k_1 + 0 - \frac{AL}{R^2} \quad \text{or} \quad k_1 = I_0 + \frac{AL}{R^2}$$

and the desired solution for $i(t)$ becomes

$$i(t) = \left(I_0 + \frac{AL}{R^2}\right) e^{-(R/L)t} + \frac{A}{R}t - \frac{AL}{R^2}$$

**4.3** In the $R$-$C$ circuit of Fig. 4-5, $i(t) = 5e^{-t}$ and $v_C(0) = 2$ V. Find $v(t)$ for $t \geq 0$.

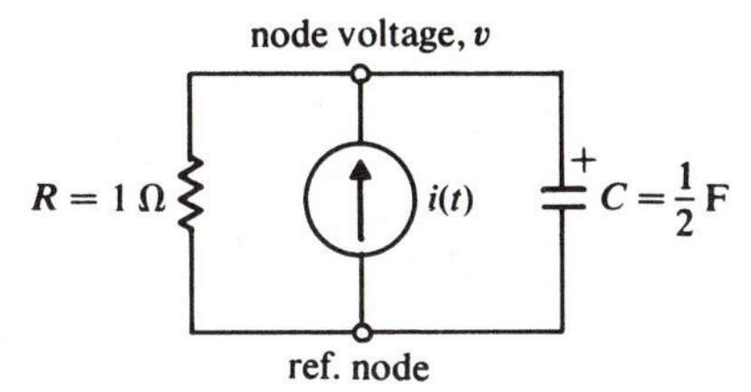


**Fig. 4-5**

Writing the KCL equation and substituting the $i$-$v$ relationships and the given data, we obtain

$$\frac{1}{2}\frac{dv}{dt} + v = 5e^{-t}$$

The complementary solution is $v_c = k_1 e^{-2t}$; and from Table 4-1 (or common sense), $v_p = k_2 e^{-t}$.

To evaluate $k_2$, we substitute $v_p$ into the circuit's differential equation, yielding

$$-\frac{k_2}{2}e^{-t} + k_2 e^{-t} = 5e^{-t} \quad \text{or} \quad k_2 = 10$$

It follows that the total solution is

$$v(t) = k_1e^{-2t} + 10e^{-t}$$

and setting $t = 0$ to evaluate $k_1$, $v(0) = 2 = k_1 + 10$, or $k_1 = -8$.

Finally,

$$v(t) = -8e^{-2t} + 10e^{-t}$$

### *Second-Order Driven Circuits*

**4.4** Confirm that the current $i(t)$ in the circuit of Fig. 4-6 is of the algebraic *form*

$$i(t) = k_1e^{-5t} + k_2e^{-20t} + k_3 \cos (15t + \phi)$$

The circuit's loop equation is

$$\frac{1}{20}\frac{di}{dt} + \frac{5}{4}i + 5\int_0^t i\, d\tau + v_C(0) = 10 \sin 15t$$

and if we differentiate and multiply by 20, we obtain the differential equation

$$\frac{d^2i}{dt^2} + 25\frac{di}{dt} + 100i = 3000 \cos 15t$$

The characteristic roots (the roots of $s^2 + 25s + 100 = 0$) are $s_1 = -5$ and $s_2 = -20$. Therefore,

$$i_c = k_1e^{-5t} + k_2e^{-20t}$$

and from Table 4-1, $i_p = k_3 \cos (15t + \phi)$.

Thus the total solution is of the algebraic form

$$i(t) = k_1e^{-5t} + k_2e^{-20t} + k_3 \cos (15t + \phi)$$

**4.5** For the circuit of Fig. 4-6, find the algebraic *form* of (*a*) the inductor voltage $v_L(t)$ and (*b*) the capacitor voltage $v_C(t)$. (Use the solution for $i(t)$ in Problem 4.4.)

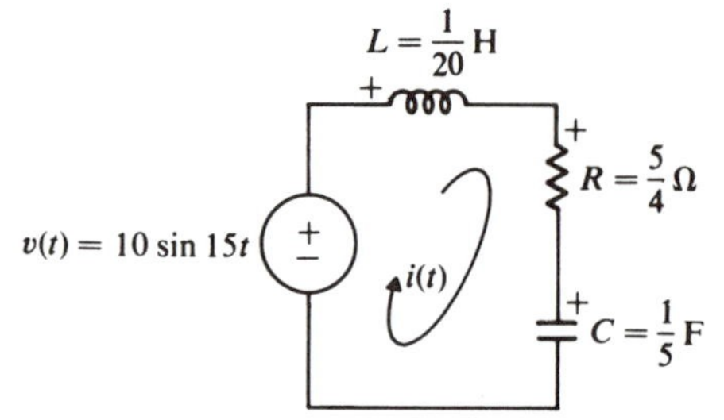


**Fig. 4-6**

(*a*)
$$v_L(t) = \frac{1}{20}\frac{di}{dt} = \frac{1}{20}\frac{d}{dt}\{k_1e^{-5t} + k_2e^{-20t} + k_3 \cos (15t + \phi)\}$$

$$= \frac{1}{20}\{-5k_1e^{-5t} - 20k_2e^{-20t} - 15k_3 \sin (15t + \phi)\}$$

$$= k_4e^{-5t} + k_5e^{-20t} + k_6 \sin (15t + \phi)$$

(*b*)
$$v_C(t) = 5\int_0^t i\,d\tau + v_C(0)$$
$$= 5\int_0^t \{k_1e^{-5\tau} + k_2e^{-20\tau} + k_3\cos(15\tau+\phi)\}\,d\tau + v_C(0)$$
$$= 5\left\{\frac{k_1e^{-5\tau}}{-5}\Bigg|_0^t + \frac{k_2e^{-20\tau}}{-20}\Bigg|_0^t + \frac{k_3}{15}\sin(15\tau+\phi)\Bigg|_0^t\right\} + v_C(0)$$
$$= k_7e^{-5t} + k_8e^{-20t} + k_9\sin(15t+\phi) + k_{10}$$

**4.6** In the circuit of Fig. 4-6, $R$ is changed to $\frac{3}{5}\,\Omega$, all else remaining the same. Find the algebraic *form* of $i(t)$.

The new loop equation is

$$\frac{1}{20}\frac{di}{dt} + \frac{3}{5}i + 5\int_0^t i\,d\tau + v_C(0) = 10\sin 15t$$

or

$$\frac{d^2i}{dt^2} + 12\frac{di}{dt} + 100i = 3000\cos 15t$$

The characteristic roots are now $s_{1,2} = -6 \pm j8$, and so

$$i_c = e^{-6t}(k_1\cos 8t + k_2\sin 8t) = k_3e^{-6t}\cos(8t+\beta)$$

Since from Table 4-1 $i_p = k_4\cos(15t+\phi)$, the total solution is

$$i(t) = k_3e^{-6t}\cos(8t+\beta) + k_4\cos(15t+\phi)$$

The first term, $i_c(t)$, is a *damped* oscillation of angular frequency 8 rad/s which will decay to zero as $t \to \infty$, leaving the particular solution. The latter, which is also the sinusoidal steady-state solution, is an *undamped* sinusoid with an angular frequency of 15 rad/s (equal to that of the input sinusoid).

**4.7** Solve for the output voltage $v(t)$ in the circuit of Fig. 4-7, given that $v(0) =$ 4 V and $i_L(0) = 6$ A.

Assuming that all the current reference directions are away from node $n$,

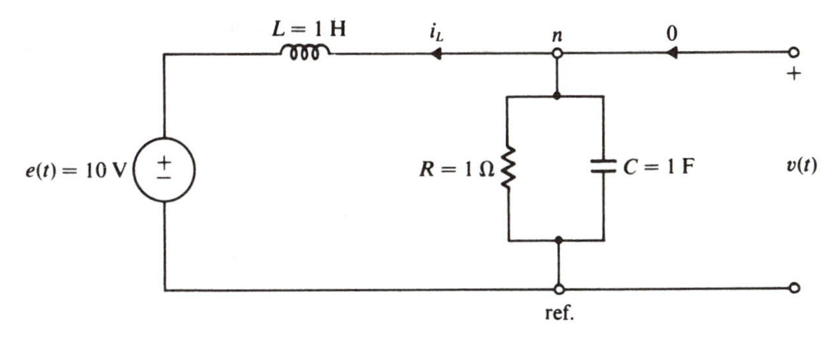


**Fig. 4-7**

$i_L + i_R + i_C = 0$, or

$$\frac{1}{L}\int_0^t (v - e)\,d\tau + i_L(0) + \frac{1}{R}v + C\frac{dv}{dt} = 0$$

If we now differentiate, substitute the given data, and rearrange,

$$\frac{d^2v}{dt^2} + \frac{dv}{dt} + v = 10$$

The characteristic roots are $s_{1,2} = -0.5 \pm j0.866$ and the complementary solution will therefore be

$$v_c = e^{-0.5t}(k_1 \cos 0.866t + k_2 \sin 0.866t)$$

Next, from Table 4-1, $v_p = k_3$, which when substituted into the circuit's differential equation yields $k_3 = 10$. Thus the total solution is of the form

$$v(t) = e^{-0.5t}(k_1 \cos 0.866t + k_2 \sin 0.866t) + 10$$

In order to evaluate $k_1$ and $k_2$ we must know $v(0)$ and $\dot{v}(0)$. The former is given, and to find the latter we write the KCL equation (at $t = 0$) at node $n$:

$$C\frac{dv(0)}{dt} + \frac{1}{R}v(0) + i_L(0) = 0$$

Then, substituting the given data, we obtain $\dot{v}(0) = -i_L(0) - v(0) = -10$.

Finally, relating $v(0)$ and $\dot{v}(0)$ to the total solution $v(t)$,

$$v(0) = 4 = k_1 + 10$$

$$\dot{v}(0) = -10 = -0.5k_1 + 0.866k_2$$

The solution of these two equations is $k_1 = -6$ and $k_2 = -15$ so the total solution is

$$v(t) = e^{-0.5t}(-6 \cos 0.866t - 15 \sin 0.866t) + 10$$

### Steady-State Solutions

**4.8** Draw the dc steady-state equivalent of the circuit in Fig. 4-7, and hence find $v(t)$ as $t \to \infty$.

See Fig. 4-8: the inductor becomes a short, the capacitor an open circuit. Then from KVL, $v_{ss}(t) = e(t) = 10$ V.

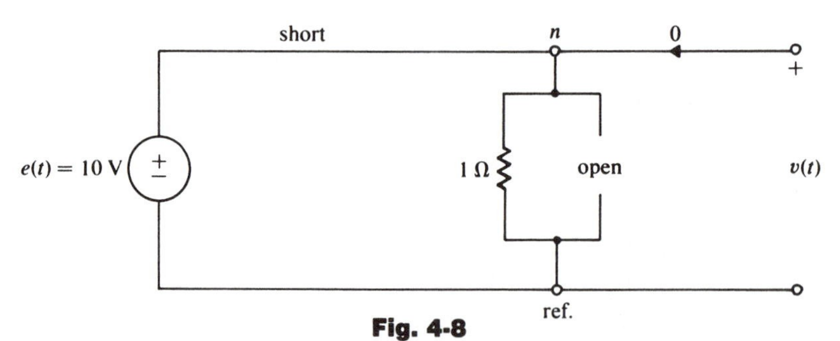


**Fig. 4-8**

**4.9** Draw the dc steady-state equivalent of Fig. 4-9 and then compute the voltage across the current source as $t \to \infty$.

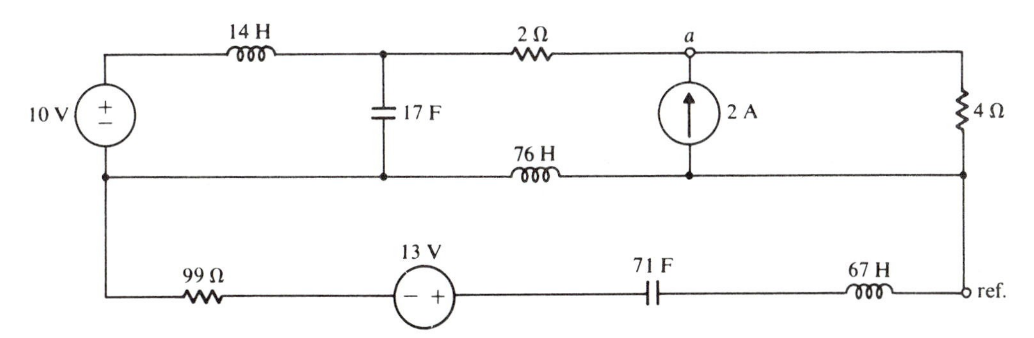


**Fig. 4-9**

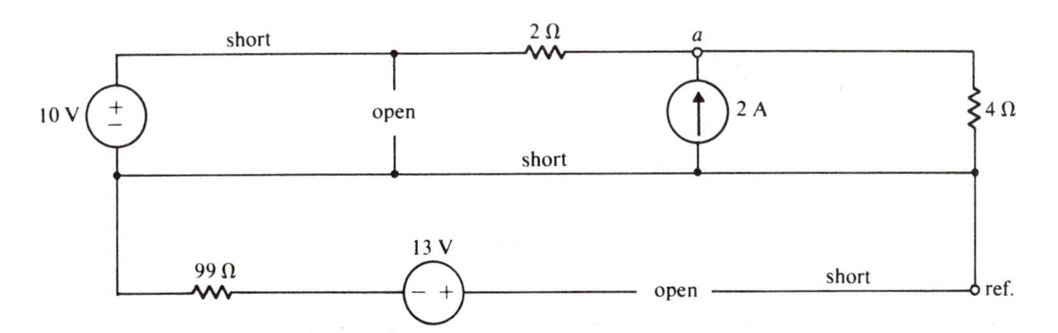


**Fig. 4-10**

See Fig. 4-10 for the dc steady-state circuit. Applying KCL at node $a$ of Fig. 4-10,

$$\frac{1}{4}v_a + \frac{1}{2}(v_a - 10) = 2 \quad \text{or} \quad v_a = \frac{28}{3}\text{ V}$$

This is the required voltage across the current source.

**4.10** The circuit of Problem 4.4 (Fig. 4-6) is described by the differential equation

$$\frac{d^2i}{dt^2} + 25\frac{di}{dt} + 100i = 3000 \cos 15t$$

Determine the sinusoidal steady-state current $i_{sss}(t)$.

Here we need only calculate $i_p$ because $i_c \to 0$ as $t \to \infty$ (the characteristic roots are $s_1 = -5$ and $s_2 = -20$). As usual, we substitute the trial soltuion

$$i_p = k_1 \cos 15t + k_2 \sin 15t$$

into the nonhomogeneous differential equation. The result is

$$-225k_1 \cos 15t - 225k_2 \sin 15t - 375k_1 \sin 15t + 375k_2 \cos 15t$$
$$+ 100k_1 \cos 15t + 100k_2 \sin 15t = 3000 \cos 15t$$

Equating coefficients of like terms,

$$-125k_1 + 375k_2 = 3000$$
$$-375k_1 - 125k_2 = 0$$

from which $k_1 = -2.4$ and $k_2 = 7.2$, and so

$$i_p(t) = i_{sss}(t) = -2.4 \cos 15t + 7.2 \sin 15t$$

### Initial Conditions

**4.11** In the circuit of Fig. 4-11, after the switch has been in the open position for a long time, it is closed at $t = 0$. Find $v(t)$ for $t \geq 0$.

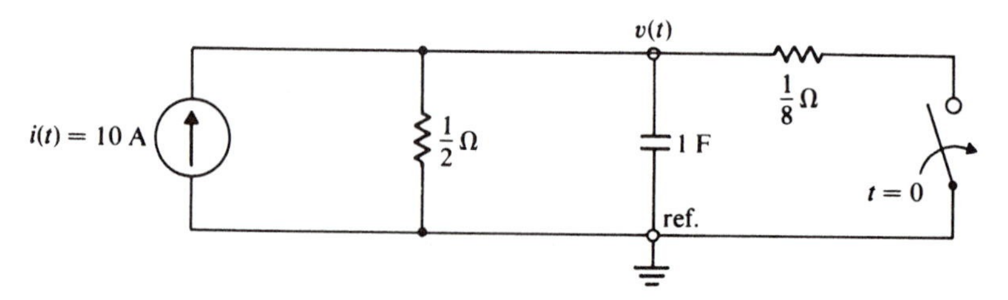


**Fig. 4-11**

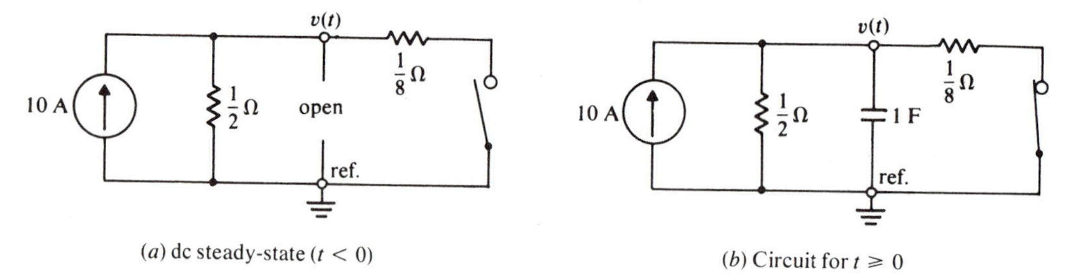


($a$) dc steady-state ($t < 0$)

($b$) Circuit for $t \geq 0$

**Fig. 4-12**

From Fig. 4-12$a$, $v_{ss}(t) = (10)(\frac{1}{2}) = 5$ V for $t < 0$. Since the capacitor voltage cannot change instantaneously, $v_{ss}(t)$ will be the capacitor voltage immediately before *and* immediately after the switch closes. That is, $v(0) = v_{ss}(t) = 5$ V.

Then from Fig. 4-12$b$, which applies for $t \geq 0$,

$$-10 + 2v + \frac{dv}{dt} + 8v = 0 \quad \text{or} \quad \frac{dv}{dt} + 10v = 10$$

The complementary and particular solutions are $v_c = k_1e^{-10t}$ and $v_p = k_2$. And if we substitute $v_p$ into the differential equation, we find $k_2 = 1$. Hence

$$v(t) = k_1 e^{-10t} + 1$$

Setting $t = 0$, and knowing that $v(0) = 5$ V, it follows that $k_1 = 4$. Thus the required solution is

$$v(t) = 4e^{-10t} + 1, \qquad t \geq 0$$

**4.12** The circuit of Fig. 4-13 has been turned on, with the swith *open*, for a long time prior to $t = 0$. First, draw the corresponding dc steady-state equivalent circuit, and then compute the current that will flow through the switch at the instant of closing.

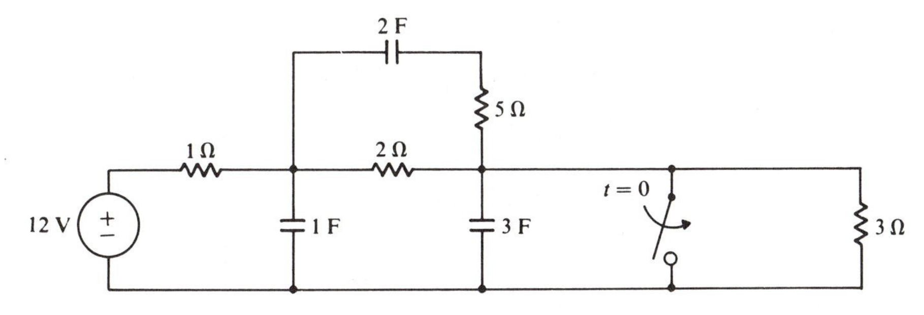


**Fig. 4-13**

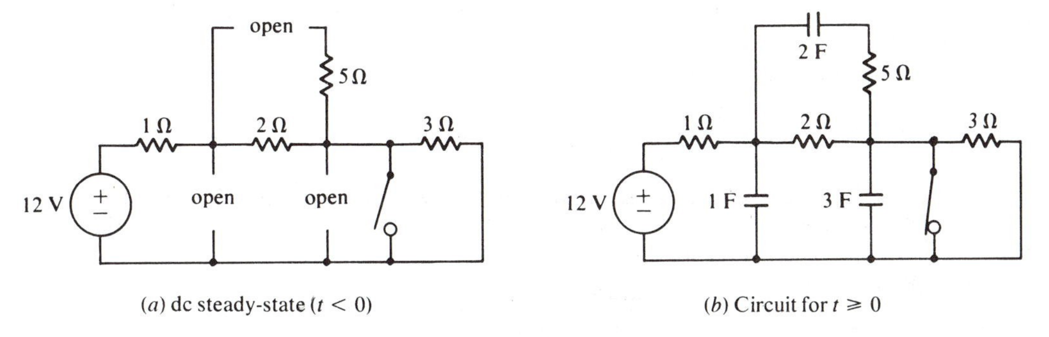


($a$) dc steady-state ($t < 0$) ($b$) Circuit for $t \geqslant 0$

**Fig. 4-14**

Figure 4-14$a$ is the required equivalent circuit. A current of $12/(1 + 2 + 3) = 2$ A will flow around the resistive loop when the switch is open. Therefore the voltage across the switch (and the 3-F capacitor) will be $(3)(2) = 6$ V at this time. When the switch closes, the voltage across the 3-F capacitor will drop instantaneously to zero (the voltage across a short circuit). And since $i_C = C\, dv_C/dt$, the capacitor will discharge an *infinite* current through the switch at the instant of closing.

More realistically, if the wiring resistance plus the contact resistance in the closed switch were $r$, then an initial current of $6/r$ A would flow.

**4.13** Given the circuit of Fig. 4-15 with $e(t) = 10$ V, $i(t) = 5e^{-3t}$ A, $i_L(0) = -3$ A, and $v_C(0) = 2$ V, find $i_e(t)$, the current flowing through the voltage source.

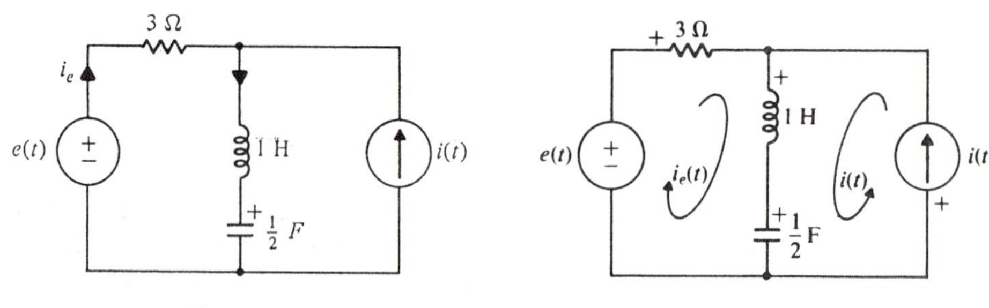


**Fig. 4-15** **Fig. 4-16**

Writing a KVL equation around the left-hand mesh of Fig. 4-16,

$$3i_e + 1\frac{d}{dt}\{i_e + i(t)\} + 2\int_0^t \{i_e + i(\tau)\}\,d\tau + 2 = e(t)$$

If we now differentiate and rearrange,

$$\frac{d^2i_e}{dt^2} + 3\frac{di_e}{dt} + 2i_e = -\frac{d^2i(t)}{dt^2} - 2i(t) + \frac{de(t)}{dt} = -55e^{-3t}$$

Then in the usual way, $i_{e_c} = k_1e^{-t} + k_2e^{-2t}$ and $i_{e_p} = k_3e^{-3t}$. Substitution of the particular solution yields

$$9k_3e^{-3t} - 9k_3e^{-3t} + 2k_3e^{-3t} = -55e^{-3t} \quad \text{or} \quad k_3 = -\frac{55}{2}$$

and the total solution is

$$i_e(t) = k_1e^{-t} + k_2e^{-2t} - \frac{55}{2}e^{-3t}$$

To evaluate $k_1$ and $k_2$ we must know $i_e(0)$ and $di_e(0)/dt$. From Fig. 4-16, $i_e(0) = i_L(0) - i(0) = -3 - 5 = -8$ A. If we differentiate the relationship $i_e = i_L - i(t)$, and set $t = 0$,

$$\frac{di_e(0)}{dt} = \frac{di_L(0)}{dt} - \frac{di(0)}{dt} = v_L(0) + 15$$

But from KVL, $v_L(0) = e(0) - 3i_e(0) - v_C(0) = 10 + 24 - 2 = 32$ V. Therefore $di_e(0)/dt = 32 + 15 = 47$ A/s. The two free constants now follow in the normal way from

$$i_e(0) = -8 = k_1 + k_2 - \frac{55}{2}$$

$$di_e(0)/dt = 47 = -k_1 - 2k_2 + \frac{165}{2}$$

Thus $k_1 = 7/2$, $k_2 = 16$, and

$$i_e(t) = \frac{7}{2}e^{-t} - 16e^{-2t} - \frac{55}{2}e^{-3t}$$

**4.14** Your instructor has solved the differential equation

$$\frac{d^2v}{dt^2} + 3\frac{dv}{dt} + 2v = 20$$

with the initial conditions $v(0) = 5$ V and $\dot{v}(0) = -30$ V/s. He claims the solution to be

$$v(t) = 10 - 40e^{-t} + 35e^{-2t}$$

Is he correct?

Your instructor is obviously claiming that $v_c = -40e^{-t} + 35e^{-2t}$ is the complementary solution, which must satisfy the homogeneous equation

$$\frac{d^2v}{dt^2} + 3\frac{dv}{dt} + 2v = 0$$

By substitution,

$$\frac{d^2}{dt^2}(-40e^{-t} + 35e^{-2t}) + 3\frac{d}{dt}(-40e^{-t} + 35e^{-2t}) + 2(-40e^{-t} + 35e^{-2t}) \stackrel{?}{=} 0$$

or

$$-40e^{-t} + 140e^{-2t} + 120e^{-t} - 210e^{-2t} - 80e^{-t} + 70e^{-2t} \stackrel{?}{=} 0$$

The left side does indeed equal zero, so the complementary solution is at least a possible one. (The constants $-40$ and $+35$ are *not* checked by this substitution.)

Your instructor is also claiming that $v_p = 10$ is the particular solution. In this he is correct, since $v_p = 10$ satisfies the given differential equation

$$\frac{d^2v}{dt^2} + 3\frac{dv}{dt} + 2v = 20$$

Finally, verify that the initial conditions are satisfied:

$$v(0) = 5 \stackrel{?}{=} \{10 - 40e^{-t} + 35e^{-2t}\}|_0$$

$$\dot{v}(0) = -30 \stackrel{?}{=} \frac{d}{dt}\{10 - 40e^{-t} + 35e^{-2t}\}|_0$$

These also check, vindicating your instructor's assertion.

**4.15** The input to the op amp circuit of Fig. 4-17 is $e(t) = 10$ V and the capacitor's initial voltage is $v_C(0) = 2$ V. Find the circuit's output $v_o(t)$.

Recall that for an ideal op amp $v_+ = v_-$. Thus node $a$ in Fig. 4-17 is at ground potential. It follows that

$$i_{R_1} = \frac{1}{R_1}e \qquad i_{R_o} = -\frac{1}{R_o}v_o \qquad i_C = -C\frac{dv_o}{dt}$$

$$i_{R_1} - i_{R_o} - i_C = 0 \qquad \text{(KCL)}$$

$$\frac{1}{R_1}e + \frac{1}{R_o}v_o + C\frac{dv_o}{dt} = 0$$

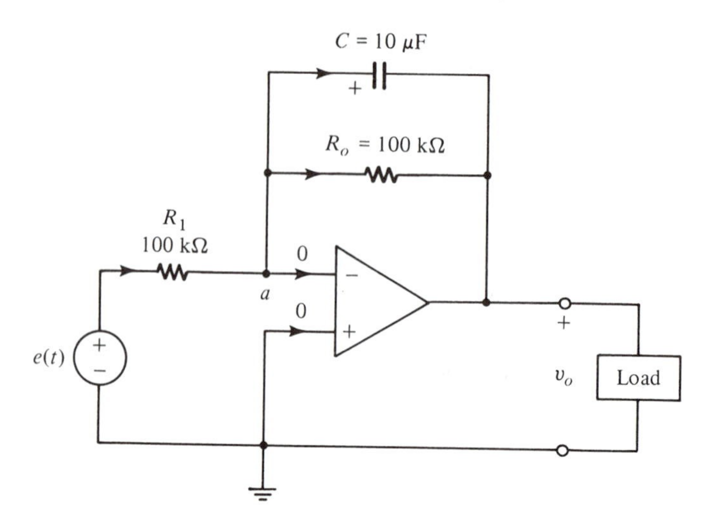


**Fig. 4-17**

or

$$\frac{dv_o}{dt} + \frac{1}{R_oC}v_o = -\frac{1}{R_1C}e$$

In the normal way, $v_{o_c} = k_1e^{-(1/R_oC)t}$ and $v_{o_p} = k_2$. When we substitute the data and solve for $k_1$ and $k_2$, noting that $v_o(0) = -v_C(0)$,

$$v_o = v_{o_c} + v_{o_p} = 8e^{-t} - 10$$

# PROBLEMS

**4.16** In the *R-C* circuit of Fig. 4-18, it is known that $v_C(0) = 5$ V and that $e(t) = 20e^{-5t}$. Find the output voltage $v(t)$. *Hint*: Use nodal analysis to obtain the circuit equation.

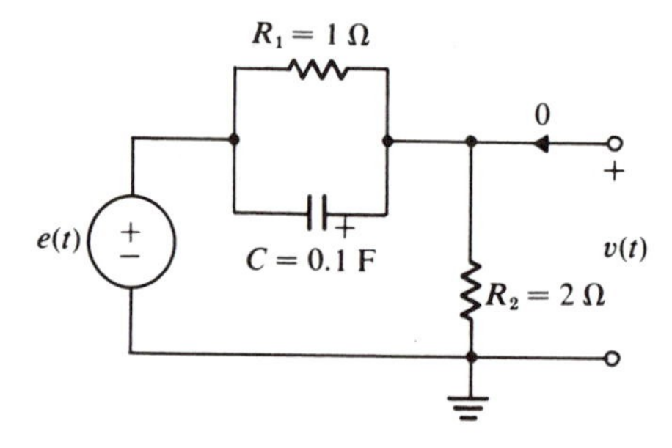


**Fig. 4-18**

**4.17** The input current $i(t)$ and the output voltage $v(t)$ of Fig. 4-19 are related by the differential equation

$$\frac{d^2v}{dt^2} + 10\frac{dv}{dt} + 100v = 10\frac{di(t)}{dt} + 100i(t)$$

For $i(t) = 16t$, $v(0) = A$, and $i_L(0) = B$, confirm that the algebraic *form* of the total solution is

$$v(t) = e^{-5t}(k_1 \cos 8.66t + k_2 \sin 8.66t) + k_3 t + k_4$$

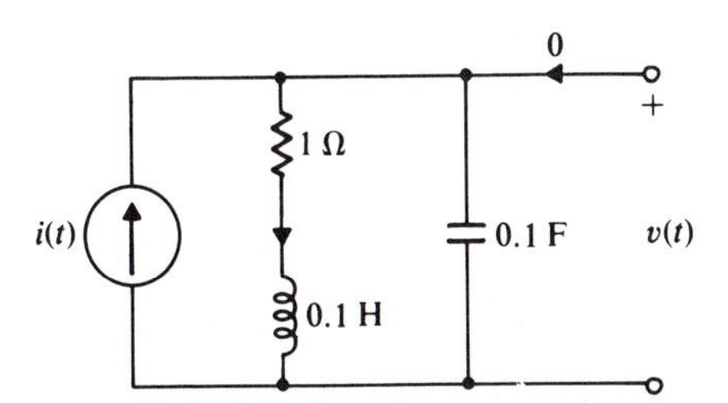


**Fig. 4-19**

**4.18** If, in the circuit of Fig. 4-19, the input current is changed to $i(t) = 16$, calculate the dc steady-state output voltage $v_{ss}(t)$.

**4.19** If $i(t) = 20 \sin 10t$ in the circuit of Fig. 4-19 (its differential equation is given in Problem 4.17), determine the sinusoidal steady-state output voltage $v_{sss}(t)$.

**4.20** Given $e(t) = 4 \sin 3t$ and $i(0) = 2$ A in the circuit of Fig. 4-20, find $v(t)$.

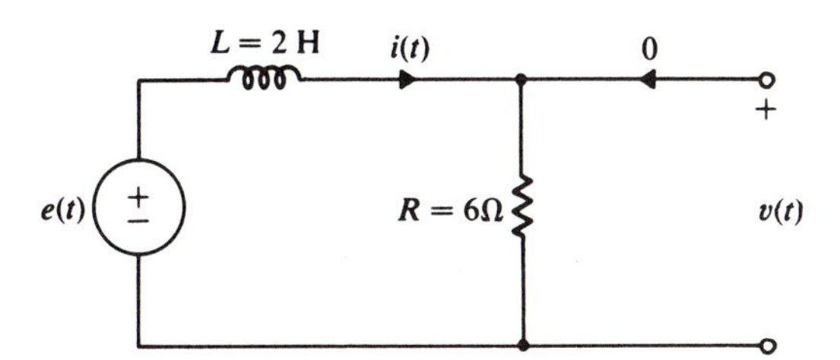


**Fig. 4-20**

**4.21** Both switches in the circuit of Fig. 4-21 are opened at $t = 0$. If $v(0)$ and $i_L(0)$ are both zero, find $v(t)$.

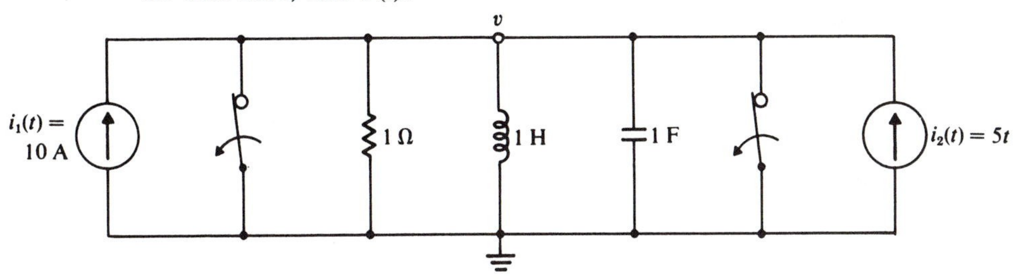


**Fig. 4-21**

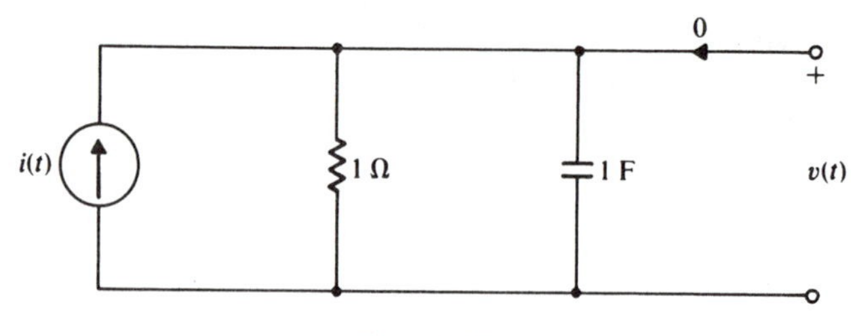


**Fig. 4-22**

**4.22** Given $i(t) = 5t^2$ and $v(0) = 2$ V in the circuit of Fig. 4-22, find $v(t)$.

**4.23** If $i(t) = 10e^{-t}$ and $v(0) = 2$ V in the circuit of Fig. 4-22, find $v(t)$.

Chapter
5

# THE LAPLACE TRANSFORM AND STATE VARIABLE METHODS OF SOLUTION

The classical method for solving differential equations discussed in Chapters 3 and 4 has certain disadvantages—primarily the difficulty posed by *simultaneous* differential equations, and the need to solve a set of simultaneous algebraic equations in order to evaluate the free constants of integration. Here we will develop the method of the Laplace transformation, which overcomes both of these difficulties.

It is worth mentioning that the Laplace transform leads to techniques of circuit (and systems) analysis which are valuable in their own right, quite apart from the question of differential equation solving. These techniques will be introduced briefly here and then developed further in the following chapters (especially Chapter 6).

In the second part of the chapter we will take up the question of solving a circuit's or system's state equations.

## THE LAPLACE TRANSFORM METHOD

This approach to the solution of differential equations consists of *transforming* differential equations into algebraic equations which can be solved for the algebraic (transformed) variables. An *inverse transformation* then yields the final solution.

The Laplace transform of a function $f(t)$ is defined by

$$\mathcal{L}[f(t)] = \int_0^\infty f(t)e^{-st}\,dt \tag{5.1}$$

where $s$ is a new algebraic variable which is independent of $t$. We will show that this transformation does indeed convert the processes of calculus into those of algebra.

As you will verify in Problem 5.1, we can use this definition to transform a function of time (e.g. $A\cos\omega t$) into an algebraic expression in the new variable $s$ (e.g. $As/(s^2 + \omega^2)$ for the above example). A number of useful time functions are listed with their $s$-transforms in Table 5-1. *Now* we can obtain our transforms by table look-up, without further reference to equation (*5.1*). Note that the transform of a function of time $f(t)$ is written as $\mathcal{L}[f(t)]$ or more succinctly $F(s)$, which latter emphasizes the transform's dependence on the new variable $s$. Notice, too, the notation $t \geq 0$ at the top of the table. The Laplace transform, as defined here, applies *only* to behavior *after* time zero. It could not be otherwise, since the lower limit of the defining integral is zero.

**Table 5-1 Laplace Transform Pairs**

| $f(t),\quad t \geqslant 0$ | $\mathcal{L}[f(t)] = F(s)$ |
|---|---|
| 1. $A$ | $A/s$ |
| 2. $At^n$ | $An!/s^{n+1}$ $\quad(n = \text{positive integer})$ |
| 3. $Ae^{at}$ | $A/(s-a)$ |
| 4. $A\cos\omega t$ | $As/(s^2+\omega^2)$ |
| 5. $A\sin\omega t$ | $A\omega/(s^2+\omega^2)$ |
| 6. $2Me^{at}\cos(\omega t+\theta)$ | $\dfrac{Me^{j\theta}}{s-a-j\omega}+\dfrac{Me^{-j\theta}}{s-a+j\omega}=\dfrac{2M\cos\theta\,(s-a-\omega\tan\theta)}{(s-a)^2+\omega^2}$ |
| 7. $A\delta(t)$ $\quad[\delta(t) = \text{unit impulse function}]$ | $A$ |

We can also use equation (*5.1*) to derive a set of theorems which will allow us to transform the time-domain *operations* of summation, differentiation, integration, and so on into algebraic operations involving $s$. Such a set is listed below. (You will verify one of these theorems in Problem 5.2.) Notice that the operations of time-domain calculus *are* transformed into algebraic operations—integration with respect to $t$ becomes division by $s$, for example.

**Table 5-2 Laplace Transform Theorems**

| | |
|---|---|
| ***Theorem 5.1:*** | $\mathcal{L}[af(t) + bg(t)] = aF(s) + bG(s)$ $\quad$ ($a$ and $b$ are constants) |
| ***Theorem 5.1a:*** | $\mathcal{L}[af(t)] = aF(s)$ |
| ***Theorem 5.2:*** | $\mathcal{L}[df(t)/dt] = sF(s) - f(0)$ |
| ***Theorem 5.3:*** | $\mathcal{L}[d^nf/dt^n] = s^nF(s) - s^{n-1}f(0) - s^{n-2}df(0)/dt - \cdots - sd^{n-2}f(0)/dt^{n-2} - d^{n-1}f(0)/dt^{n-1}$ |
| ***Theorem 5.4:*** | $\mathcal{L}[\int_0^t f(\tau)\,d\tau] = F(s)/s$ |
| ***Theorem 5.5:*** | $\mathcal{L}[e^{at}f(t)] = F(s-a)$ |
| ***Theorem 5.6:*** | $\mathcal{L}[f(t-a)\mathbb{1}(t-a)] = F(s)e^{-as}$ |

Tables 5-1 and 5-2 will be adequate for all our needs in this book, and so we will have no further recourse to equation (*5.1*). Much more extensive tables of transform pairs and theorems are available to you in your library.

## Example 5.1

We will use the Laplace transform method to solve the equation

$$Ri + \frac{1}{C}\int_0^t i(\tau)\, d\tau + v_C(0) = 0$$

The first step is to Laplace transform each term in the given equation. Thus from Theorem 5.1*a*,

$$\mathcal{L}[Ri(t)] = RI(s)$$

(Since $i(t)$ is the unknown, we cannot look up its transform in Table 5-1. $I(s)$ is the best we can do.) Next, from Theorem 5.4,

$$\int_0^t i(\tau)\, d\tau = I(s)/s$$

or, using Theorem 5.1*a* again,

$$\mathcal{L}\left[\frac{1}{C}\int_0^t i(\tau)\, d\tau\right] = \frac{1}{Cs} I(s)$$

And, since $v_C(0)$ is a constant, entry 1 in Table 5-1 yields

$$\mathcal{L}[v_C(0)] = v_C(0)/s$$

Theorem 5.1 shows that it is legitimate to add the three $s$-domain terms corresponding to the three additive terms in the original equation, whose transform is therefore

$$RI(s) + \frac{1}{Cs} I(s) + \frac{v_C(0)}{s} = 0$$

This transformed equation is algebraic, and so is easily solved for the transformed variable $I(s)$:

$$I(s) = \frac{-v_C(0)/s}{R + 1/Cs} = \frac{-v_C(0)/R}{s + 1/RC}$$

The last step was a rearrangement to make the expression look like entry 3 in Table 5-1, which it does, with $A = -v_C(0)/R$ and $a = -1/RC$.

It follows that we can now write down the inverse transform of $I(s)$, which is the solution we seek:

$$i(t) = -\{v_C(0)/R\}e^{-(1/RC)t}, \qquad t \geq 0$$

■

## Example 5.2

Suppose we wish to solve for $i_2(t)$ in the circuit of Fig. 5-1. The loop equations are

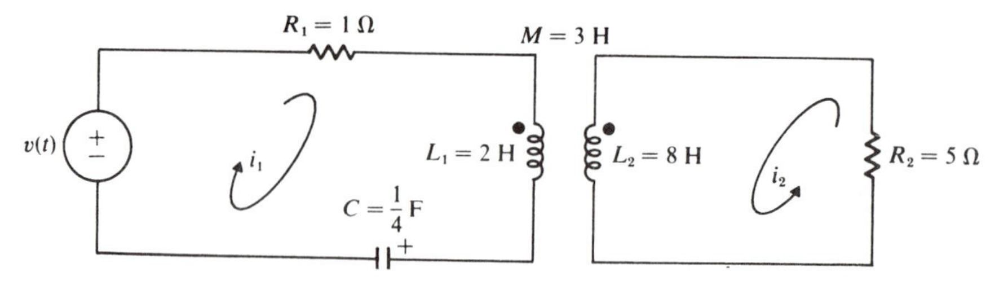


**Fig. 5-1**

$$R_1 i_1 + L_1 \frac{di_1}{dt} + \frac{1}{C}\int_0^t i_1\, d\tau + M\frac{di_2}{dt} = v(t) - v_C(0)$$

$$M\frac{di_1}{dt} + R_2 i_2 + L_2\frac{di_2}{dt} = 0$$

If we transform these two equations with the help of Theorems 5.1, 5.2, and 5.4, and the transform pair 1 of Table 5-1,

$$R_1 I_1(s) + L_1\{sI_1(s) - i_1(0)\} + \frac{I_1(s)}{Cs} + M\{sI_2(s) - i_2(0)\} = V(s) - \frac{v_C(0)}{s}$$

$$M\{sI_1(s) - i_1(0)\} + R_2 I_2(s) + L_2\{sI_2(s) - i_2(0)\} = 0$$

Notice that all three "physical" initial conditions, $v_C(0)$, $i_1(0)$, and $i_2(0)$, are already part of the calculation.

The original *differential* equations have now been transformed into *algebraic* equations which may be solved for $I_2(s)$. Substituting the values of the circuit elements in Fig. 5-1, and given that $v(t) = 5$ V, $v_C(0) = 1$ V, $i_1(0) = 2$ A, and $i_2(0) = 3$ A, the circuit equations become

$$\left(2s + 1 + \frac{4}{s}\right)I_1(s) + 3sI_2(s) = \frac{5}{s} - \frac{1}{s} + 4 + 9$$

$$3sI_1(s) + (8s + 5)I_2(s) = 6 + 24$$

and from Cramer's rule,

$$I_2(s) = \frac{\begin{vmatrix} 2s + 1 + \frac{4}{s} & \frac{4}{s} + 13 \\ 3s & 30 \end{vmatrix}}{\begin{vmatrix} 2s + 1 + \frac{4}{s} & 3s \\ 3s & 8s + 5 \end{vmatrix}} = \frac{21s^2 + 18s + 120}{7s^3 + 18s^2 + 37s + 20} = \frac{3(s^2 + \frac{6}{7}s + \frac{40}{7})}{s^3 + \frac{18}{7}s^2 + \frac{37}{7}s + \frac{20}{7}}$$

Now that the transformed variable $I_2(s)$ has been found, we have only to perform the inverse transformation to obtain $i_2(t)$.

The first (and most difficult) step of the inverse transformation is to factor the denominator of the expression for $I_2(s)$. If we equate this denominator to zero, we obtain the circuit's characteristic equation

$$s^3 + \frac{18}{7}s^2 + \frac{37}{7}s + \frac{20}{7} = 0$$

and then use a computer or a calculator to find the characteristic roots, and hence the required factors. In the interests of clarity we will complete this solution in literal form.

If the characteristic roots are found to be $-\alpha$, $-\beta$, and $-\gamma$, then $I_2(s)$ can be expanded in partial fractions,

$$I_2(s) = \frac{3(s^2 + \frac{6}{7}s + \frac{40}{7})}{(s + \alpha)(s + \beta)(s + \gamma)} \equiv \frac{K_1}{s + \alpha} + \frac{K_2}{s + \beta} + \frac{K_3}{s + \gamma}$$

The simplest method for finding the partial fraction constant $K_1$ is to multiply the last equation through by $(s + \alpha)$, and then set $s = -\alpha$. This will eliminate the $K_2$ and $K_3$ terms, leaving

$$K_1 = \left.\frac{3(s^2 + \frac{6}{7}s + \frac{40}{7})\cancel{(s + \alpha)}}{\cancel{(s + \alpha)}(s + \beta)(s + \gamma)}\right|_{s=-\alpha}$$

which is easily evaluated, as are the similar expressions for $K_2$ and $K_3$.

Finally, the inverse transform is obtained from Table 5-1. In this instance, using entry 3,

$$i_2(t) = K_1e^{-\alpha t} + K_2e^{-\beta t} + K_3e^{-\gamma t}, \qquad t \geq 0$$

Note that the inverse transformation yields *no* information about $i_2(t)$ for $t < 0$. ■

These two examples illustrate the basic technique of the Laplace transform method of solution. Some special methods which are needed to handle some of the algebra will be developed in the problem set.

## TRANSFORM IMPEDANCE

When transformed, the $v$-$i$ relations of equations (*1.6*), (*1.7*), and (*1.8*) become

$$V_R(s) = RI_R(s) \qquad I_R(s) = \frac{1}{R}V_R(s) \tag{5.2}$$

$$V_L(s) = L\{sI_L(s) - i_L(0)\} \qquad I_L(s) = \frac{1}{Ls}V_L(s) + \frac{i_L(0)}{s} \tag{5.3}$$

$$V_C(s) = \frac{1}{Cs}I_C(s) + \frac{v_C(0)}{s} \qquad I_C(s) = C\{sV_C(s) - v_C(0)\} \tag{5.4}$$

The quantities $R$, $Ls$, and $1/Cs$ are called the *transform impedance* of $R$, $L$, and $C$ respectively, while $1/R$, $1/Ls$, and $Cs$ are called their *transform admittance*. The symbol $Z(s)$ is used to represent transform impedance, and $Y(s)$ transform admittance. Thus

$$Z_R(s) = R \qquad Y_R(s) = \frac{1}{R} \tag{5.5}$$

$$Z_L(s) = Ls \qquad Y_L(s) = \frac{1}{Ls} \tag{5.6}$$

$$Z_C(s) = \frac{1}{Cs} \qquad Y_C(s) = Cs \tag{5.7}$$

In Fig. 5-2 we can see how an initial condition may be replaced by an equivalent *initial condition source*. Then the $L$ or $C$ element, *freed of its initial condition*, may be described by

$$V(s) = Z(s)I(s) \quad \text{or} \quad I(s) = Y(s)V(s) \tag{5.8}$$

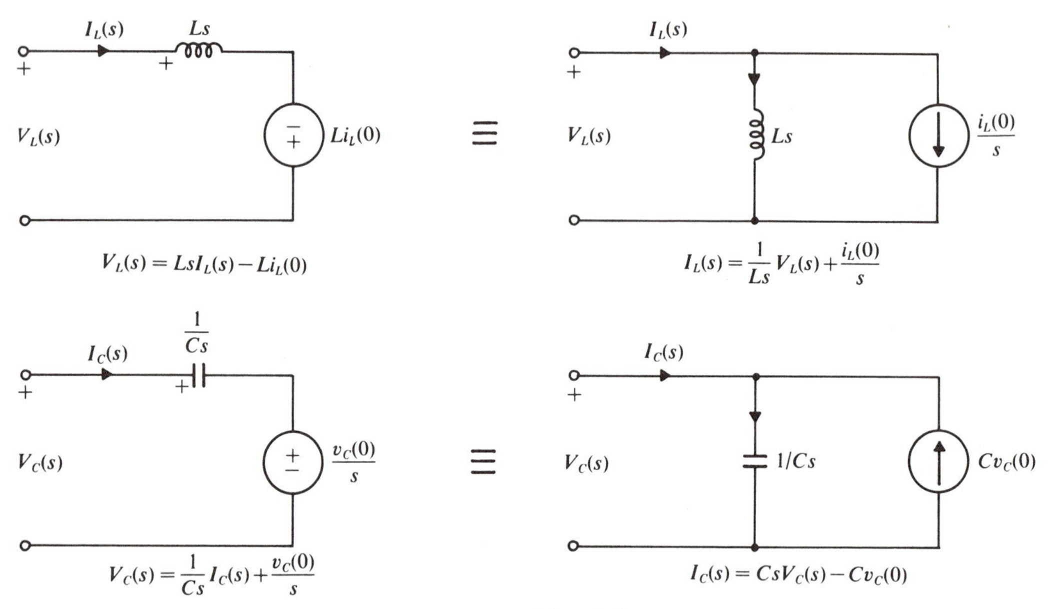


**Fig. 5-2**

Some comments regarding Fig. 5-2 are in order. First, it will be observed that there are *two* equivalent transform networks for both inductors and capacitors. This follows from the fact that we can write both *v-i* and *i-v* relations. Notice also that $Li_L(0)$ and $Cv_C(0)$ correspond to impulse functions in the time-domain (see entry 7 in Table 5-1), while $i_L(0)/s$ and $v_C(0)/s$ represent constants in time (entry 1). Watch their polarity (or direction)!

It is important to realize that the circuit pairs of Fig. 5-2 are equivalent *only* as far as the world *outside* their terminals is concerned. That is, they have the same *terminal characteristics*. *Inside* the equivalent networks the voltages and currents are quite different, as must be obvious from the fact that one equivalent in each pair is a series circuit, the other a parallel circuit.

The concept of impedance and admittance has some important applications. Most obvious and useful is the idea of drawing *transform circuit diagrams*.

Before we investigate transform circuit diagrams, however, we should give a moment's thought to the Kirchhoff laws in the transform domain. By transformation,

$$\sum_{\text{node}} i(t) = 0 \;\Rightarrow\; \sum_{\text{node}} I(s) = 0 \tag{5.9}$$

and

$$\sum_{\text{loop}} v(t) = 0 \;\Rightarrow\; \sum_{\text{loop}} V(s) = 0 \tag{5.10}$$

That is, the Kirchhoff laws also apply in the transform domain.

### Example 5.3

The transform representation in the center of Fig. 5-3 follows directly from the ordinary circuit diagram on the left as a consequence of equations (*5.2*), (*5.3*), and (*5.4*).

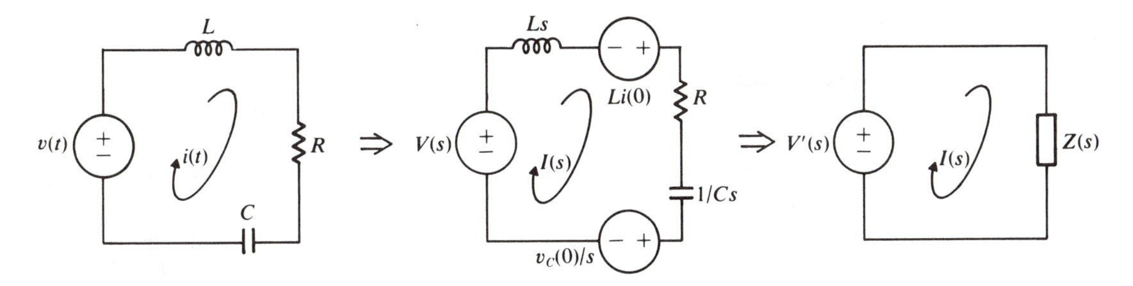


**Fig. 5-3**

Then if we apply KVL (transformed) to the center diagram,

$$LsI(s) + RI(s) + \frac{1}{Cs}I(s) = V(s) + Li(0) - \frac{v_C(0)}{s}$$

or

$$\{Z_L(s) + Z_R(s) + Z_C(s)\}I(s) = V'(s)$$

or

$$Z(s)I(s) = V'(s)$$

where $Z(s) = Z_L(s) + Z_R(s) + Z_C(s)$ and $V'(s) = V(s) + Li(0) - v_C(0)/s$.

The last equation justifies the asserted equivalence of the right-hand transform circuit diagram in Fig. 5-3. ■

The result of Example 5.3 may be generalized. If $Z_1(s)$, $Z_2(s)$, . . . are in series, then they may be replaced by a single equivalent impedance given by

$$Z_{eq}(s) = Z_1(s) + Z_2(s) + \cdots \tag{5.11}$$

Similarly it can be shown that if $Z_1(s)$, $Z_2(s)$, . . . are in parallel, then they may be replaced by a single equivalent impedance given by

$$\frac{1}{Z_{eq}(s)} = \frac{1}{Z_1(s)} + \frac{1}{Z_2(s)} + \cdots \tag{5.12}$$

or more neatly, in terms of admittance,

$$Y_{eq}(s) = Y_1(s) + Y_2(s) + \cdots \tag{5.13}$$

These results assume that any nonzero initial conditions have been replaced by equivalent sources. An impedance or admittance cannot, *by itself,* represent an inductor or capacitor with a nonzero initial condition.

The impedance-admittance concept, which has been briefly introduced here, will be extended in the following chapters.

## SOLUTION OF STATE EQUATIONS

In many ways the simplest and best procedure for the solution of state equations is to seek the aid of your Computer Center personnel. They should be able to offer several programs for the solution of ordinary differential equations, and references to texts which discuss the numerical methods used. Although the program titles will probably not mention the words *state variable,* the programs will almost certainly require that the equations be written in what *you* know to be state variable form.

As an alternative, you might see if analog computers are available to you. They, too, normally require the equations to be in state variable form, even though the manuals and textbooks may not use this term.

Typical computer solutions list the calculated values of the variables at successive instants of time and/or provide equivalent graphs. This should be contrasted with closed-form analytical solutions. Two comments are relevant: first, analytical solutions are not often possible if the system is nonlinear; and, second, it is generally *much* easier to extract information about engineering trends and dependencies from an analytical solution than it is from pages of raw data and graphs from a computer. In general the two approaches, analytical and computational, should be regarded as complementary.

## ANALYTICAL SOLUTION IN THE TIME-DOMAIN

First let us consider the solution of the single linear, first-order differential equation

$$\dot{x} = ax + bu$$

where $x(t)$ is the (unknown) variable, $u(t)$ is the input, and $a$ and $b$ are constants.

To solve this equation analytically we first multiply by the *integrating factor* $e^{-at}$. Then after rearrangement,

$$\frac{dx}{dt} e^{-at} - axe^{-at} = e^{-at}bu$$

or

$$\frac{d}{dt}\{xe^{-at}\} = e^{-at}bu$$

Now it is possible to integrate,

$$\int_0^t \frac{d}{d\tau}\{x(\tau)e^{-a\tau}\}\, d\tau = \int_0^t e^{-a\tau}bu(\tau)\, d\tau$$

$$x(t)e^{-at} - x(0) = \int_0^t e^{-a\tau}bu(\tau)\, d\tau$$

or finally,

$$x(t) = e^{at}x(0) + e^{at}\int_0^t e^{-a\tau}bu(\tau)\, d\tau$$

This result can be generalized to apply to the linear state equation

$$\dot{\mathbf{x}} = \mathbf{Ax} + \mathbf{Bu}$$

That is, the *matrix* solution can be shown to be*

$$\mathbf{x}(t) = e^{\mathbf{A}t}\mathbf{x}(0) + e^{\mathbf{A}t}\int_0^t e^{-\mathbf{A}\tau}\mathbf{Bu}(\tau)\, d\tau \qquad (5.14)$$

where the *matrix exponential* is defined to be

$$e^{\mathbf{A}t} = \mathbf{I} + \mathbf{A}t + \mathbf{A}^2\frac{t^2}{2!} + \cdots + \mathbf{A}^n\frac{t^n}{n!} + \cdots \qquad (5.15)$$

and $\mathbf{I}$ is the identity matrix.

Although it is possible to obtain analytic solutions in this way, the evaluation of equation (*5.14*) is *very* tedious, due to the matrix series form of each exponential. It

*See, for example, Richard Bellman, *Introduction to Matrix Analysis,* (New York: McGraw-Hill, 1960), pp. 165–169.

is better to think of equation (*5.14*) as the basis for a numerical (computer) algorithm for the solution of *linear* state equations. However, this will not be pursued further here.

### ANALYTICAL SOLUTION IN THE TRANSFORM DOMAIN

If we take the Laplace transform of the linear state equation $\dot{\mathbf{x}} = \mathbf{Ax} + \mathbf{Bu}$, we find

$$s\mathbf{X}(s) - \mathbf{x}(0) = \mathbf{AX}(s) + \mathbf{BU}(s)$$

Now $s\mathbf{X}(s) - \mathbf{AX}(s) \neq (s - \mathbf{A})\mathbf{X}(s)$, since $(s - \mathbf{A})$ does not even exist. Instead, $s\mathbf{X}(s) - \mathbf{AX}(s) = (s\mathbf{I} - \mathbf{A})\mathbf{X}(s)$. Therefore

$$(s\mathbf{I} - \mathbf{A})\mathbf{X}(s) = \mathbf{x}(0) + \mathbf{BU}(s)$$

or

$$\begin{aligned} \mathbf{X}(s) &= (s\mathbf{I} - \mathbf{A})^{-1}\mathbf{x}(0) + (s\mathbf{I} - \mathbf{A})^{-1}\mathbf{BU}(s) \qquad (5.16) \\ &= \mathbf{\Phi}(s)\mathbf{x}(0) + \mathbf{\Phi}(s)\mathbf{BU}(s) \end{aligned}$$

Once the matrix inverse $(s\mathbf{I} - \mathbf{A})^{-1} = \mathbf{\Phi}(s)$ has been evaluated—a tedious chore—the inverse Laplace transform of each element of $\mathbf{X}(s)$ follows in the usual—tedious—way!

Both of these analytical methods for solving state equations have important theoretical ramifications in systems analysis, but for a quick, more or less painless solution, the computer has much to recommend it.

# ILLUSTRATIVE PROBLEMS

### The Laplace Transform Method

**5.1** Verify the validity of entry 3 in Table 5-1, page 134. That is, derive the result $\mathcal{L}[Ae^{at}] = A/(s - a)$.

From equation (*5.1*), with $f(t) = Ae^{at}$, and noting that $s$ is independent of $t$,

$$\begin{aligned} \mathcal{L}[Ae^{at}] &= \int_0^\infty Ae^{at}e^{-st}\,dt \\ &= \int_0^\infty Ae^{-(s-a)t}\,dt \end{aligned}$$

$$= \frac{-A}{s-a}[e^{-(s-a)t}]_0^\infty$$

$$= \frac{A}{s-a}$$

provided that $s$ is such that $e^{-(s-a)t} \to 0$ as $t \to \infty$.

### *Comments*

1. If you set $a = 0$ in the above result, you can check entry 1 in Table 5-1.
2. The sine and cosine functions can be written in terms of exponentials with the help of the Euler relation. Hence you can verify entries 4 and 5 in the table.

**5.2** Prove Theorem 5.2. That is, show that $\mathcal{L}[df(t)/dt] = sF(s) - f(0)$.

From equation (*5.1*), integrating by parts,

$$\mathcal{L}[df(t)/dt] = \int_0^\infty \frac{df(t)}{dt} e^{-st}\, dt$$

$$= \int_0^\infty \left[ -f(t)\frac{de^{-st}}{dt} + \frac{d}{dt}\{f(t)e^{-st}\} \right] dt$$

$$= s\int_0^\infty f(t)e^{-st}\, dt + \int_0^\infty \frac{d}{dt}\{f(t)e^{-st}\}\, dt$$

$$= sF(s) + [f(t)e^{-st}]_0^\infty$$

$$= sF(s) - f(0)$$

provided that $f(t)e^{-st} \to 0$ as $t \to \infty$.

**5.3** Turning now to the method of the Laplace transform, find $V(s)$ for the circuit of Fig. 5-19, given $e(t) = 10$ V, $v(0) = 5$ V, and $i_L(0) = 15$ A.

Writing a single nodal equation at $a$ is the most direct approach. No equation is needed at node $b$ since $v_b$ is known [equal to $e(t)$]. Thus

$$2\int_0^t (v - 10)\, d\tau + 15 + 3v + \frac{dv}{dt} = 0$$

or after rearrangement,

$$2\int_0^t v\, d\tau + 3v + \frac{dv}{dt} = 2\int_0^t 10\, d\tau - 15 = 20t - 15$$

We can now transform the differential equation with the aid of Table 5-1, page 134, and the transform theorems,

$$\frac{2}{s}V(s) + 3V(s) + sV(s) - 5 = \frac{20}{s^2} - \frac{15}{s}$$

or

$$\left(\frac{2}{s} + 3 + s\right) V(s) = \frac{20}{s^2} - \frac{15}{s} + 5 = \frac{5}{s^2}(s^2 - 3s + 4)$$

and

$$V(s) = \frac{5(s^2 - 3s + 4)}{s(s^2 + 3s + 2)}$$

The solution for $v(t)$ will be completed in the following problem.

**5.4** Given $V(s) = 5(s^2 - 3s + 4)/s(s^2 + 3s + 2)$ from Problem 5.3, expand $V(s)$ in partial fractions and then, using Table 5-1, page 134, find $v(t)$.

First we must find the factors of the denominator of $V(s)$, which in this case is quite easy,

$$s(s^2 + 3s + 2) = s(s + 1)(s + 2)$$

Then $V(s)$ may be written as the sum of the three partial fractions,

$$V(s) = \frac{5(s^2 - 3s + 4)}{s(s + 1)(s + 2)} \equiv \frac{K_1}{s} + \frac{K_2}{s + 1} + \frac{K_3}{s + 2}$$

To evaluate $K_1$ we first multiply by $s$,

$$sV(s) = \frac{5(s^2 - 3s + 4)\cancel{s}}{\cancel{s}(s + 1)(s + 2)} = \frac{K_1\cancel{s}}{\cancel{s}} + \frac{K_2 s}{s + 1} + \frac{K_3 s}{s + 2}$$

If $s$ is now set to zero, the last two terms on the right will vanish, leaving

$$K_1 = \left.\frac{5(s^2 - 3s + 4)\cancel{s}}{\cancel{s}(s + 1)(s + 2)}\right|_{s=0} = \frac{20}{2} = 10$$

A similar procedure is used to find $K_2$: first multiply $V(s)$ by $(s + 1)$,

$$(s + 1)V(s) = \frac{5(s^2 - 3s + 4)\cancel{(s + 1)}}{s\cancel{(s + 1)}(s + 2)} \equiv \frac{K_1(s + 1)}{s} + \frac{K_2\cancel{(s + 1)}}{\cancel{s + 1}} + \frac{K_3(s + 1)}{s + 2}$$

and then set $s = -1$ to remove the first and third terms on the right, leaving

$$K_2 = \left.\frac{5(s^2 - 3s + 4)\cancel{(s + 1)}}{s\cancel{(s + 1)}(s + 2)}\right|_{s=-1} = -40$$

In a similar manner,

$$K_3 = \left.\frac{5(s^2 - 3s + 4)\cancel{(s + 2)}}{s(s + 1)\cancel{(s + 2)}}\right|_{s=-2} = 35$$

It is always possible and worthwhile to check the algebra of the partial fraction expansion. That is,

$$\frac{5(s^2 - 3s + 4)}{s(s + 1)(s + 2)} \stackrel{?}{=} \frac{10}{s} - \frac{40}{s + 1} + \frac{35}{s + 2}$$

$$= \frac{10(s + 1)(s + 2) - 40s(s + 2) + 35s(s + 1)}{s(s + 1)(s + 2)} \checkmark$$

Finally, having found that

$$V(s) = \frac{10}{s} - \frac{40}{s+1} + \frac{35}{s+2}$$

we can use Table 5-1 to write the inverse transform,

$$v(t) = 10 - 40e^{-t} + 35e^{-2t}, \qquad t \geq 0$$

As a final check on our work we can set $t = 0$, $v(0) = 10 - 40 + 35 = 5$ V, which agrees with the initial condition given in Problem 5.3.

**5.5** The circuit of Fig. 5-4 is described by

$$2\int_0^t v(\tau)\,d\tau + i_L(0) + 3v + \frac{dv}{dt} = 2\int_0^t e(\tau)\,d\tau$$

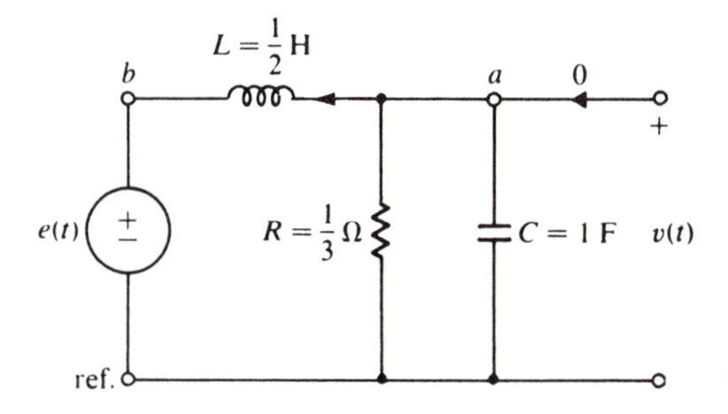


**Fig. 5-4**

Use the Laplace transform method to find $v(t)$ when $e(t) = 4e^{-3t}$, $v(0) = 5$ V, and $i_L(0) = 10$ A.

Transforming the circuit equation yields

$$\frac{2}{s}V(s) + \frac{i_L(0)}{s} + 3V(s) + sV(s) - v(0) = \frac{2}{s}E(s)$$

or

$$(s^2 + 3s + 2)V(s) = 2\left(\frac{4}{s+3}\right) - 10 + 5s = \frac{5(s^2 + s - 4.4)}{s+3}$$

so

$$V(s) = \frac{5(s^2 + s - 4.4)}{(s+1)(s+2)(s+3)} \equiv \frac{K_1}{s+1} + \frac{K_2}{s+2} + \frac{K_3}{s+3}$$

where

$$K_1 = \left.\frac{5(s^2 + s - 4.4)\cancel{(s+1)}}{\cancel{(s+1)}(s+2)(s+3)}\right|_{s=-1} = -11$$

$$K_2 = \left.\frac{5(s^2 + s - 4.4)\cancel{(s+2)}}{(s+1)\cancel{(s+2)}(s+3)}\right|_{s=-2} = 12$$

$$K_3 = \left.\frac{5(s^2 + s - 4.4)\cancel{(s+3)}}{(s+1)(s+2)\cancel{(s+3)}}\right|_{s=-3} = 4$$

Thus

$$V(s) = \frac{-11}{s+1} + \frac{12}{s+2} + \frac{4}{s+3}$$

which can easily be shown to be the correct partial fraction expansion.

Now we take the inverse transform using Table 5-1, page 134,

$$v(t) = -11e^{-t} + 12e^{-2t} + 4e^{-3t}, \qquad t \geq 0$$

with checks with the given value of $v(0)$ when we set $t = 0$.

**5.6** Given the circuit equations

$$\frac{di_1}{dt} + i_1 - i_2 = v_1(t) = 10$$

$$-i_1 + i_2 + \int_0^t i_2\, d\tau + v_C(0) = -v_2(t) = -5t$$

with $i_1(0) = 0$ and $v_C(0) = 0$, solve for $i_1(t)$.

Transforming the given equations,

$$sI_1(s) - \cancel{i_1(0)} + I_1(s) - I_2(s) = V_1(s) = \frac{10}{s}$$

$$-I_1(s) + I_2(s) + \frac{1}{s}I_2(s) + \frac{1}{s}\cancel{v_C(0)} = -V_2(s) = \frac{-5}{s^2}$$

and by Cramer's rule,

$$I_1(s) = \frac{\begin{vmatrix} \dfrac{10}{s} & -1 \\ -\dfrac{5}{s^2} & 1 + \dfrac{1}{s} \end{vmatrix}}{\begin{vmatrix} s+1 & -1 \\ -1 & 1 + \dfrac{1}{s} \end{vmatrix}} = \frac{10/s + 10/s^2 - 5/s^2}{s + 1 + 1/s} = \frac{10(s+0.5)}{s(s^2+s+1)}$$

$$= \frac{10(s+0.5)}{s(s+0.5-j0.866)(s+0.5+j0.866)}$$

$$\equiv \frac{K_1}{s} + \frac{K_2}{s+0.5-j0.866} + \frac{K_3}{s+0.5+j0.866}$$

Notice *carefully* that since the objective is to obtain the inverse transformation with the help of Table 5-1, page 134, the *order* of the last two terms *must* match that of entry 6. That is, the $s - a - j\omega$ denominator *must* precede the $s - a + j\omega$ denominator.

We find $K_1$, $K_2$, and $K_3$ in the usual way,

$$K_1 = \left.\frac{10(s+0.5)\cancel{s}}{\cancel{s}(s^2+s+1)}\right|_{s=0} = 5$$

$$K_2 = \left. \frac{10(s + 0.5)\cancel{(s + 0.5 - j0.866)}}{s\cancel{(s + 0.5 - j0.866)}(s + 0.5 + j0.866)} \right|_{s=-0.5+j0.866} = \frac{5}{-0.5 + j0.866}$$

$$= \frac{5}{De^{j\phi}}$$

where $D = \sqrt{0.5^2 + 0.866^2} = 1$ and $\phi = 90° + \tan^{-1}(0.5/0.866) = 120°$ (as shown in Fig. 5-5). That is,

$$K_2 = 5e^{-j120°}$$

(Your calculator probably has built-in routines for rectangular-to-polar and polar-to-rectangular conversions. It is still a good idea, however, to habitually make a sketch equivalent to that of Fig. 5-5 as a rough check on your complex-number arithmetic.)

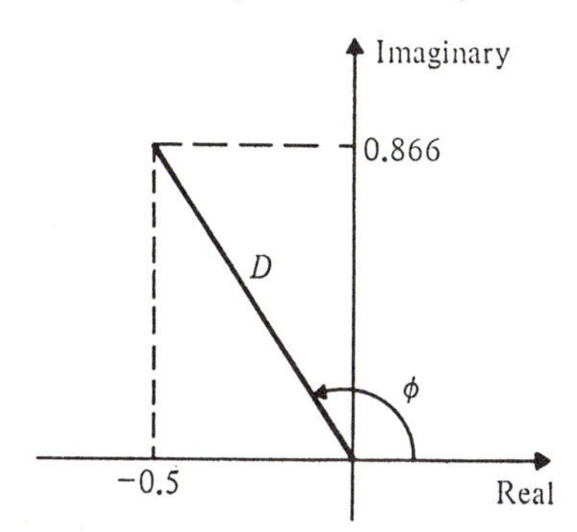


**Fig. 5-5**

Similarly we could show that $K_3 = 5e^{j120°}$. Then, in summary,

$$I_1(s) = \frac{5}{s} + \frac{5e^{-j120°}}{s + 0.5 - j0.866} + \frac{5e^{j120°}}{s + 0.5 + j0.866}$$

and from entries 1 and 6 in Table 5-1,

$$i_1(t) = 5 + 10e^{-0.5t} \cos (0.866t - 120°), \qquad t \geq 0$$

### *Comments*

1. If the two complex conjugate terms are *not* kept in the proper order to match entry 6 in Table 5-1, then the inverse transform will be in error to the extent of a reversed sign in the phase angle of the cosine.
2. The second partial fraction constant of such a pair will always be the complex conjugate of the first (here $K_3 = K_2^*$). We may therefore omit the formal evaluation of the second constant with some saving of time and effort.

**5.7** Use the method of the Laplace transform to find the *impulse response* of the network in Fig. 5-6. That is, find $v(t)$ for $i(t) = \delta(t)$ given that the initial value of $v(t)$ just before the impulse is applied is zero.

From KCL the differential equation is

$$C\frac{dv}{dt} + \frac{1}{R}v = \delta(t)$$

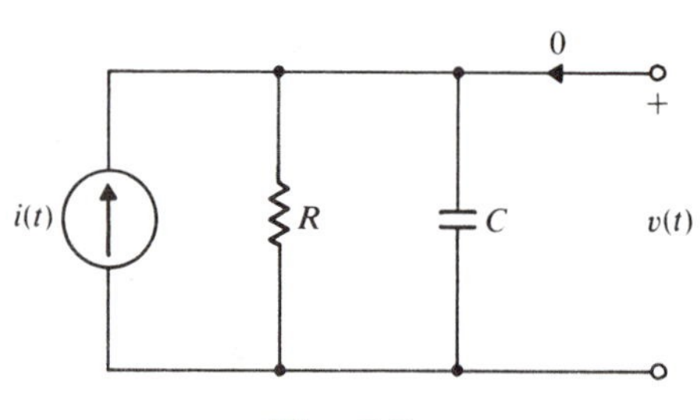


**Fig. 5-6**

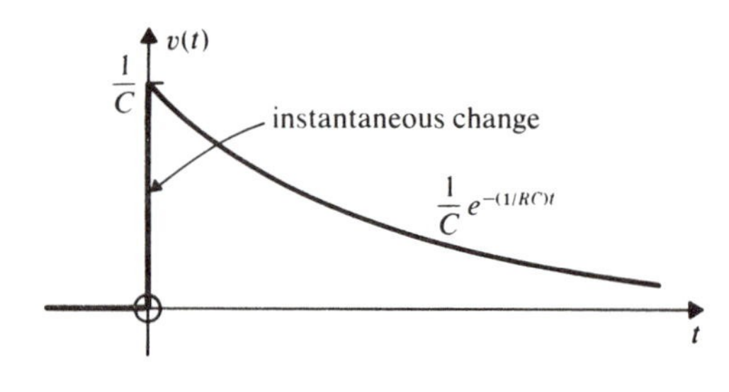


**Fig. 5-7**

Transforming, with $\mathscr{L}[\delta(t)] = 1$ from Table 5-1,

$$C\{sV(s) - \cancel{v(0)}\} + \frac{1}{R}V(s) = 1$$

Thus

$$V(s) = \frac{1/C}{s + 1/RC} \quad \text{and} \quad v(t) = \frac{1}{C}e^{-(1/RC)t}, \qquad t \geq 0$$

This function is graphed in Fig. 5-7.

*Notes on the impulse function:* The unit impulse function $\delta(t)$ is an infinitely high pulse of infinitely small width whose *area* is unity. This pulse occurs at $t = 0$.

In the present problem an impulse of current was applied. This is, of course, quite nonphysical. However, since the integral of current is charge, we may think of one unit of charge being supplied at the instant $t = 0$, thus raising the capacitor's charge instantaneously from 0 to 1, and its voltage from 0 to $1/C$ as shown in Fig. 5-7.

It is clear that we must be careful with initial conditions in the presence of impulse functions. Here, as illustrated in Fig. 5-7, $v(0^-) = 0$ and $v(0^+) = 1/C$. A capacitor's voltage *can* change instantaneously if its current is infinite!

**5.8** Find the *step response* for the network of Fig. 5-6. That is, find $v(t)$ for $i(t) = \mathbb{1}(t)$ with $v(0) = 0$.

Here the circuit's differential equation is

$$C\frac{dv}{dt} + \frac{1}{R}v = \mathbb{1}(t)$$

from which

$$C\{sV(s) - \cancel{v(0)}\} + \frac{1}{R}V(s) = \frac{1}{s}$$

and

$$V(s) = \frac{1/C}{s(s + 1/RC)} \equiv \frac{K_1}{s} + \frac{K_2}{s + 1/RC}$$

where

$$K_1 = \left.\frac{(1/C)\cancel{s}}{\cancel{s}(s + 1/RC)}\right|_{s=0} = R$$

$$K_2 = \left.\frac{(1/C)\cancel{(s + 1/RC)}}{s\cancel{(s + 1/RC)}}\right|_{s=-1/RC} = -R$$

Therefore

$$v(t) = R\{1 - e^{-(1/RC)t}\}, \qquad t \geq 0$$

*Note:* The unit impulse function is the derivative of the unit step. Thus the impulse response should be the derivative of the step response, which is confirmed if we compare the above result with the answer to Problem 5.7. That is,

$$\frac{d}{dt}[R\{1 - e^{-(1/RC)t}\}] = \frac{1}{C}e^{-(1/RC)t}$$

**5.9** In the circuit of Figure 5-6 let $R = 1\,\Omega$, $C = 1\,F$, $v(0) = 2$ V, and $i(t) = 5t^2$ A. Use the Laplace transform method to find $v(t)$.

The nodal equation is now

$$\frac{dv}{dt} + v = 5t^2$$

which after transformation becomes

$$sV(s) - 2 + V(s) = 10/s^3 \quad \text{or} \quad V(s) = \frac{2(s^3 + 5)}{s^3(s + 1)}$$

Here the denominator of $V(s)$ is said to have *repeated roots* and the most general form of the partial fraction expansion is

$$V(s) = \frac{2(s^3 + 5)}{s^3(s + 1)} \equiv \frac{K_1}{s + 1} + \frac{K_2}{s^3} + \frac{K_3}{s^2} + \frac{K_4}{s}$$

In the usual way,

$$K_1 = \left.\frac{2(s^3 + 5)\cancel{(s + 1)}}{s^3\cancel{(s + 1)}}\right|_{s=-1} = \frac{2(4)}{-1} = -8$$

and

$$K_2 = \left.\frac{2(s^3 + 5)\cancel{s^3}}{\cancel{s^3}(s + 1)}\right|_{s=0} = 10$$

If we attempt to find $K_3$ in the same way, we would first multiply $V(s)$ by $s^2$,

$$\frac{2(s^3 + 5)\cancel{s^2}}{s^{\cancel{3}}(s + 1)} = \frac{K_1 s^2}{s + 1} + \frac{k_2\cancel{s^2}}{s^{\cancel{3}}} + \frac{K_3\cancel{s^2}}{\cancel{s^2}} + \frac{K_4 s^{\cancel{2}}}{\cancel{s}}$$

and then set $s$ equal to zero. But the term on the left and the second term on the right would then go to infinity. However, if we first *combine* the two misbehaving terms and substitute $K_2 = 10$,

$$\frac{K_1 s^2}{s + 1} + K_3 + K_4 s = \frac{2(s^3 + 5)}{s(s + 1)} - \frac{10}{s} = \frac{2s^3 + 10 - 10s - 10}{s(s + 1)} = \frac{2\cancel{s}(s^2 - 5)}{\cancel{s}(s + 1)}$$

*Now* we may safely set $s = 0$ and obtain

$$K_3 = \left.\frac{2(s^2 - 5)}{s + 1}\right|_{s=0} = -10$$

The evaluation of $K_4$ follows a similar pattern: multiply $V(s)$ by $s$, rearrange, and *then* set $s = 0$,

$$\frac{2(s^3 + 5)s}{s^3(s + 1)} = \frac{K_1 s}{s + 1} + \frac{K_2 s}{s^3} + \frac{K_3 s}{s^2} + \frac{K_4 s}{s}$$

But $K_2 = 10$ and $K_3 = -10$, so

$$K_4 = \left.\left\{\frac{2(s^3 + 5)}{s^2(s + 1)} - \frac{10}{s^2} + \frac{10}{s}\right\}\right|_{s=0}$$

$$= \left.\frac{2s^3 + 10 - 10s - 10 + 10s^2 + 10s}{s^2(s + 1)}\right|_{s=0}$$

$$= \left.\frac{2s^2(s + 5)}{s^2(s + 1)}\right|_{s=0} = 10$$

Recapitulating,

$$V(s) = \frac{-8}{s + 1} + \frac{10}{s^3} - \frac{10}{s^2} + \frac{10}{s}$$

so

$$v(t) = -8e^{-t} + 5t^2 - 10t + 10, \qquad t \geq 0$$

*Note on repeated roots:* There *is* a "trick"! We *must* evaluate the partial fraction constants in a particular order. Here, with

$$V(s) = \frac{K_1}{s + 1} + \frac{K_2}{s^3} + \frac{K_3}{s^2} + \frac{K_4}{s}$$

we *must* find $K_2$, $K_3$, and $K_4$ *in that order*. That is, we find the numerator associated with the *highest* power of the repeated root *first*, and so on down the list.

**5.10** Referring again to Fig. 5-6 with $R = 1\,\Omega$, $C = 1\,\text{F}$, $v(0) = 2$ V, and $i(t) = 10e^{-t}$ A, find $v(t)$.

The nodal equation is

$$\frac{dv}{dt} + v = 10e^{-t}$$

from which

$$sV(s) - 2 + V(s) = \frac{10}{s + 1} \quad \text{and} \quad V(s) = \frac{2(s + 6)}{(s + 1)^2}$$

(This is an academic problem as there is zero probability of the input time constant *exactly* equaling the circuit's time constant.)

Expanding in partial fractions,

$$V(s) = \frac{2(s + 6)}{(s + 1)^2} \equiv \frac{K_1}{(s + 1)^2} + \frac{K_2}{s + 1}$$

where as usual,

$$K_1 = \left. \frac{2(s+6)\cancel{(s+1)^2}}{\cancel{(s+1)^2}} \right|_{s=-1} = 10$$

$$K_2 = \left\{ \frac{2(s+6)}{s+1} - \frac{10}{s+1} \right\} \Bigg|_{s=-1} = \left. \frac{2s+12-10}{s+1} \right|_{s=-1} = \left. \frac{2\cancel{(s+1)}}{\cancel{s+1}} \right|_{s=-1} = 2$$

and

$$V(s) = \frac{10}{(s+1)^2} + \frac{2}{s+1}$$

To find the inverse transform of $10/(s+1)^2$ we use entry 2 in Table 5-1 (with $n = 1$), and Theorem 5.5, page 134. That is,

$$\mathcal{L}^{-1}\left[ \frac{10}{(s+1)^2} \right] = 10te^{-t}$$

and

$$v(t) = 10te^{-t} + 2e^{-t}, \qquad t \geq 0$$

**5.11** Find the unit ramp response for the network of Fig. 5-6, (*a*) using Laplace transform methods, and (*b*) by integrating the step response of Problem 5.8. (The unit ramp function $t\mathbb{1}(t)$ is the integral of the unit step function $\mathbb{1}(t)$.)

(*a*) The differential equation and its transform are

$$C\frac{dv}{dt} + \frac{1}{R}v = t\mathbb{1}(t) \quad \text{and} \quad C\{sV(s) - \cancel{v(0)}\} + \frac{1}{R}V(s) = \frac{1}{s^2}$$

Hence

$$V(s) = \frac{1/C}{s^2(s + 1/RC)} \equiv \frac{K_1}{s^2} + \frac{K_2}{s} + \frac{K_3}{s + 1/RC}$$

where

$$K_1 = \left. \frac{(1/C)\cancel{s^2}}{\cancel{s^2}(s + 1/RC)} \right|_{s=0} = R$$

$$K_2 = \left\{ \frac{1/C}{s(s + 1/RC)} - \frac{R}{s} \right\} \Bigg|_{s=0} = \left. \frac{1/C - Rs - 1/C}{s(s + 1/RC)} \right|_{s=0} = \left. \frac{-R\cancel{s}}{\cancel{s}(s + 1/RC)} \right|_{s=0}$$

$$= -R^2C$$

$$K_3 = \left. \frac{(1/C)\cancel{(s + 1/RC)}}{s^2\cancel{(s + 1/RC)}} \right|_{s=-1/RC} = R^2C$$

It then follows that

$$v(t) = Rt - R^2C + R^2Ce^{-(1/RC)t}, \qquad t \geq 0$$

(*b*) In Problem 5.8 the step response was found to be

$$v(t) = R\{1 - e^{-(1/RC)t}\}$$

Integrating the right-hand side yields the ramp response,

$$R\tau\Big|_0^t + \frac{Re^{-(1/RC)\tau}}{1/RC}\Big|_0^t = Rt + R^2Ce^{-(1/RC)t} - R^2C$$

**5.12** Given zero initial conditions in the circuit of Fig. 5-8, find the output voltage $v(t)$ for an input voltage $e(t) = 5\delta(t)$.

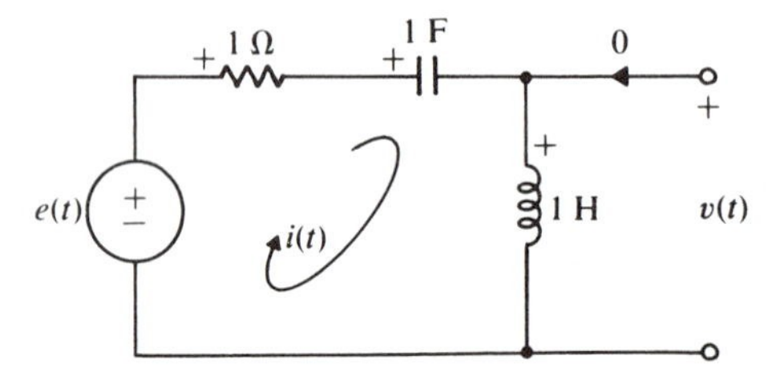


**Fig. 5-8**

Since the objective is to find $v(t)$, it is convenient to describe the circuit by the simultaneous loop equations

$$i + \int_0^t i\,d\tau + v_C(0) + v(t) = 5\delta(t)$$

$$-\frac{di}{dt} + v(t) = 0$$

It is perfectly legitimate to apply KVL around the right-hand "mesh," even though it is not a closed *circuit*. That is, in writing the second equation we have, as required, applied KVL around a closed *path*.

Transforming, setting $i(0) = 0$, and arranging in matrix form gives us

$$\begin{bmatrix} 1 + \dfrac{1}{s} & 1 \\ -s & 1 \end{bmatrix} \begin{bmatrix} I(s) \\ V(s) \end{bmatrix} = \begin{bmatrix} 5 \\ 0 \end{bmatrix}$$

Then by Cramer's rule,

$$V(s) = \frac{\begin{vmatrix} 1 + \dfrac{1}{s} & 5 \\ -s & 0 \end{vmatrix}}{\begin{vmatrix} 1 + \dfrac{1}{s} & 1 \\ -s & 1 \end{vmatrix}} = \frac{5s}{1 + 1/s + s} = \frac{5s^2}{s^2 + s + 1}$$

Since the numerator and denominator of $V(s)$ are of the same order, we cannot proceed directly to the partial fraction expansion. First we must divide the numerator by the denominator as follows,

$$\begin{array}{r} 5 \\ s^2 + s + 1 \overline{)\,5s^2 \qquad\qquad} \\ \underline{5s^2 + 5s + 5} \\ -\,5s - 5 \end{array}$$

so that

$$V(s) = 5 - \frac{5(s+1)}{s^2+s+1} \qquad (1)$$

The second term on the right of (1) can be expanded in the usual way, remembering that the two terms must be in the correct order, to match entry 6 in Table 5-1,

$$\frac{5(s+1)}{(s+0.5-j0.866)(s+0.5+j0.866)} \equiv \frac{K_1}{s+0.5-j0.866} + \frac{K_2}{s+0.5+j0.866}$$

where

$$K_1 = \left.\frac{5\cancel{(s+0.5-j0.866)}(s+1)}{\cancel{(s+0.5-j0.866)}(s+0.5+j0.866)}\right|_{s=-0.5+j0.866}$$

$$= \frac{5(0.5+j0.866)}{j1.732} = \frac{4.33-j2.5}{1.732} = 2.885e^{-j30^\circ}$$

and $K_2 = K_1^*$, the complex conjugate of $K_1$. Recapitulating,

$$V(s) = 5 - \frac{2.885e^{-j30^\circ}}{s+0.5-j0.866} - \frac{2.885e^{j30^\circ}}{s+0.5+j0.866}$$

from which

$$v(t) = 5\delta(t) - 5.77e^{-0.5t}\cos(0.866t - 30^\circ)$$

### *Alternative Solution*

Somewhat more directly, the loop equation for the circuit of Fig. 5-8 is

$$i + \int_0^t i\,d\tau + \cancel{v_C(0)} + \frac{di}{dt} = 5\delta(t)$$

which yields

$$I(s) = \frac{5s}{s^2+s+1}$$

Then since $v(t) = di/dt$, $V(s) = sI(s) - \cancel{i(0)} = 5s^2/(s^2+s+1)$ from which $v(t)$ follows as before.

**5.13** The voltage pulse $v(t) = 10\{\mathbb{1}(t) - \mathbb{1}(t-1)\}$ of Fig. 5-9*a* is applied to the

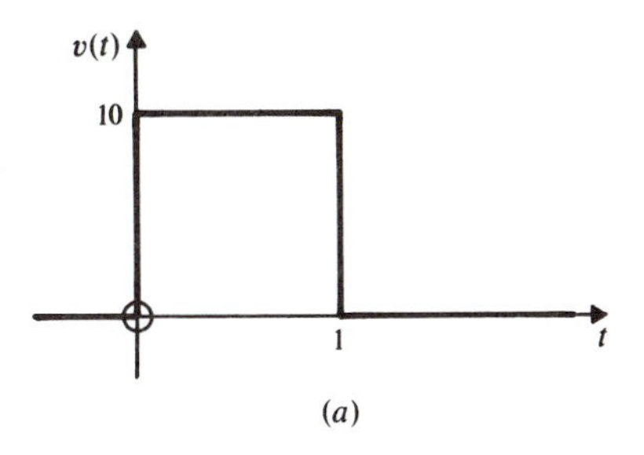

(*a*)

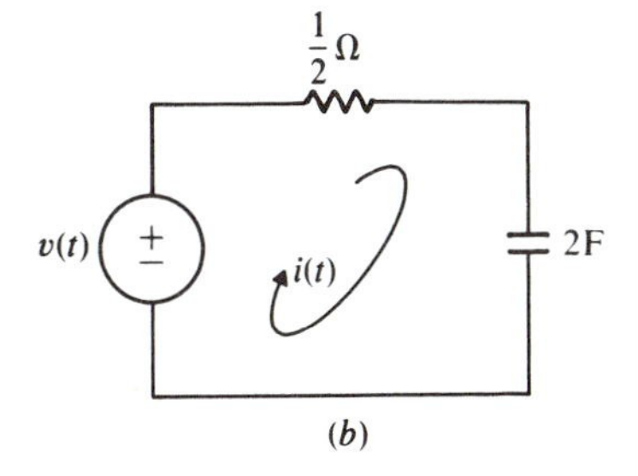

(*b*)

**Fig. 5-9**

circuit of Fig. 5-9*b* where the initial condition is $v_C(0^-) = 0$. Find $i(t)$ and sketch the waveform $i(t)$ vs. $t$.

The appropriate loop equation is

$$\frac{1}{2}i + \frac{1}{2}\int_0^t i\,d\tau + \cancel{v_C(0)} = 10\{\mathbb{1}(t) - \mathbb{1}(t-1)\}$$

Transforming, with the help of Theorem 5.6,

$$\frac{I(s)}{2} + \frac{1}{2s}I(s) = \frac{10}{s}(1 - e^{-s})$$

or

$$I(s) = \frac{20(1 - e^{-s})}{s + 1}$$

To find the inverse transform of $I(s)$ we invoke Theorem 5.6 again to obtain

$$i(t) = 20e^{-t} - 20e^{-(t-1)}\mathbb{1}(t-1), \qquad t \geq 0$$

This waveform is sketched in Fig. 5-10. (The function $20e^{-(t-1)}\mathbb{1}(t-1)$ is simply $20e^{-t}\mathbb{1}(t)$ moved one unit of time towards the right.)

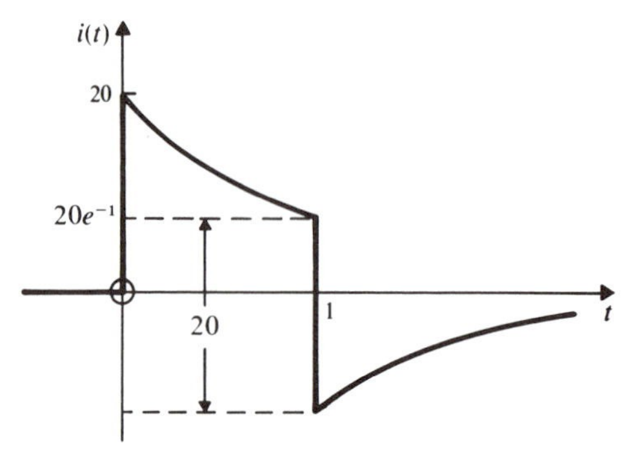


**Fig. 5-10**

### *Transformed Circuit Diagrams*

**5.14** Draw the transform circuit diagram for the circuit of Fig. 5-11 and from this diagram obtain the appropriate transformed nodal equation(s). Then solve for $V(s)$.

Since nodal equations are called for, the initial conditions $v(0)$ and $i_L(0)$ are best represented as equivalent transformed current sources in *parallel* with $Z_C$ and $Z_L$ (see Fig. 5-2, page 138). This has the advantage of creating no additional nodes (and therefore no additional nodal equations). The corresponding transform circuit diagram is shown in Fig. 5-12.

There is no need to write an equation at node $b$, since $V_b(s)$ is already known [equal to $E(s)$]. Therefore applying KCL at node $a$,

$$I_R(s) + I_L(s) + I_C(s) = 0$$

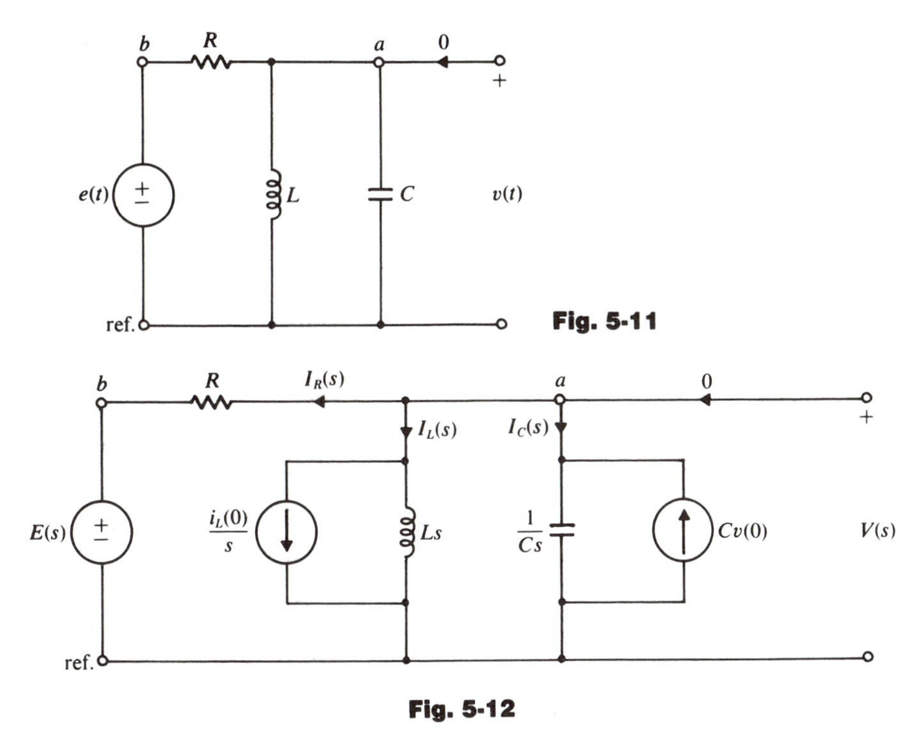


Fig. 5-11

Fig. 5-12

From the transformed circuit diagram we can write the $i$-$v$ relations

$$I_R(s) = \frac{1}{R}\{V(s) - E(s)\}, \quad I_L(s) = \frac{1}{Ls}V(s) + \frac{i_L(0)}{s}, \quad I_C(s) = CsV(s) - Cv(0)$$

which when substituted into the KCL equation yields

$$\left(Cs^2 + \frac{1}{R}s + \frac{1}{L}\right)V(s) = \frac{s}{R}E(s) - i_L(0) + Csv(0)$$

or

$$V(s) = \frac{(s/RC)E(s) - (1/C)i_L(0) + sv(0)}{s^2 + (1/RC)s + 1/LC}$$

**5.15** Draw the transform circuit diagram for the circuit of Fig. 5-13 and obtain the appropriate transformed loop equation(s) given $i_L(0) = 1$ A and $v_C(0) = 3$ V. Then solve for $I_R(s)$.

With loop analysis as the objective, the initial conditions are best represented by transformed voltage sources in *series* with $Z_C$ and $Z_L$ (see Fig. 5-2), thus minimizing the number of meshes.

Choosing loop currents in the transform circuit diagram as shown in Fig. 5-14, $I_b(s)$ is known [equal to $I(s)$] and only one KVL equation is needed to find $I_R(s)$,

$$-V_C(s) + V_R(s) + V_L(s) = 0$$

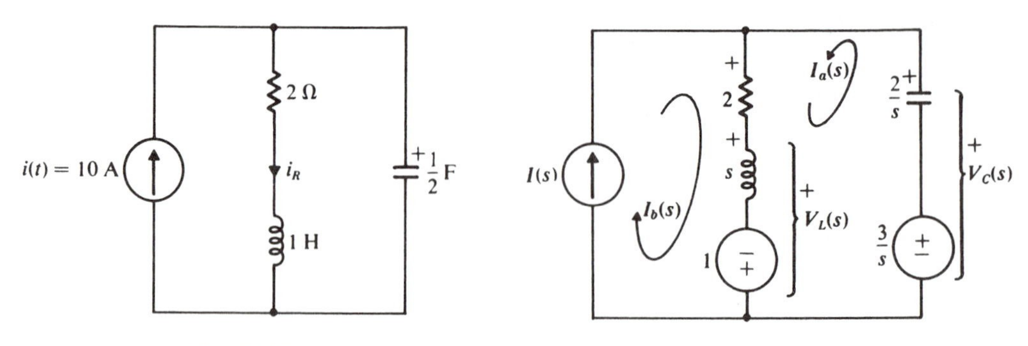


Fig. 5-13 Fig. 5-14

The $v$-$i$ relations are

$$V_C(s) = -\frac{2}{s}I_a(s) + \frac{3}{s}, \qquad V_R(s) = 2\{I_a(s) + I(s)\}$$

$$V_L(s) = s\{I_a(s) + I(s)\} - 1, \qquad I(s) = \frac{10}{s}$$

which when substituted into the KVL equation yields

$$-\left\{-\frac{2}{s}I_a(s) + \frac{3}{s}\right\} + 2\left\{I_a(s) + \frac{10}{s}\right\} + s\left\{I_a(s) + \frac{10}{s}\right\} - 1 = 0$$

or

$$I_a(s) = \frac{-9(s + 17/9)}{s^2 + 2s + 2}$$

But $I_R(s) = I_a(s) + I_b(s) = I_a(s) + I(s)$ so that

$$I_R(s) = \frac{-9(s + 17/9)}{s^2 + 2s + 2} + \frac{10}{s} = \frac{s^2 + 3s + 20}{s(s^2 + 2s + 2)}$$

**5.16** Draw the transform circuit diagram for the network of Fig. 5-15, write the appropriate loop equations, and then complete their matrix representation

$$\begin{bmatrix} R_1 + R_2 + L_1 s & & \\ & & \\ & & \end{bmatrix} \begin{bmatrix} I_1(s) \\ I_2(s) \\ I_3(s) \end{bmatrix} = \begin{bmatrix} V_1(s) + L_1 i_{L_1}(0) \\ \\ \end{bmatrix}$$

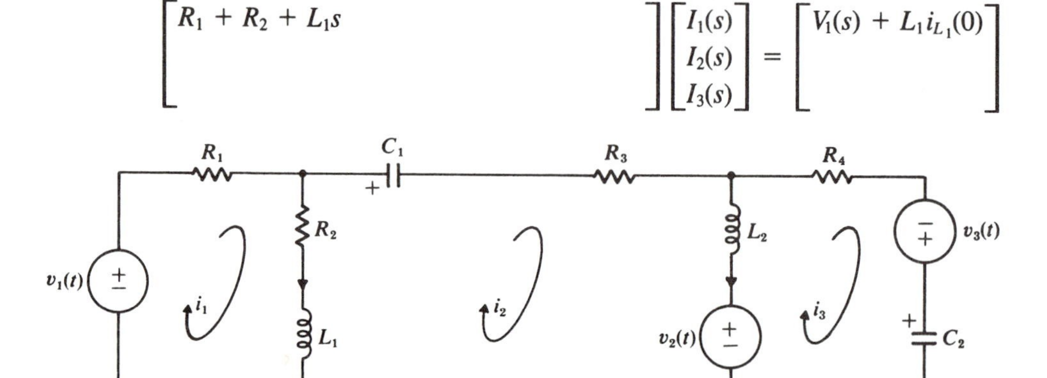


Fig. 5-15

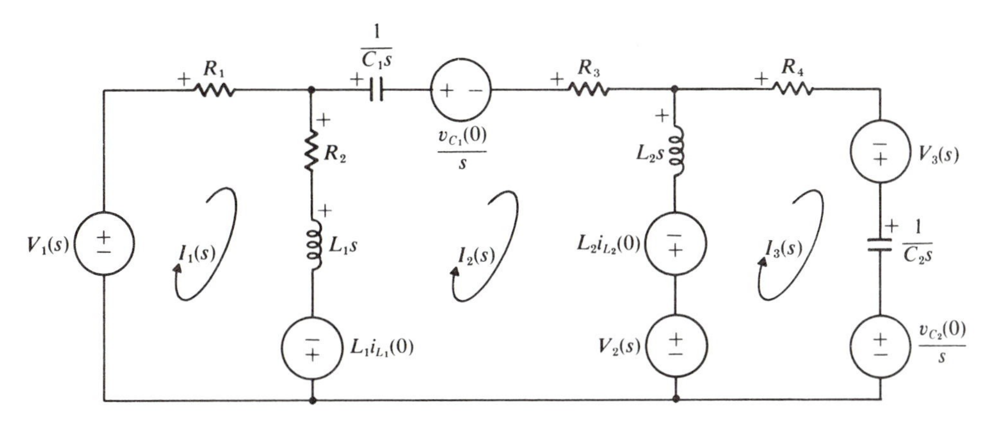


**Fig. 5-16**

The transform circuit diagram is shown in Fig. 5-16 for which the loop equations are (with some steps omitted),

$$-V_1(s) + R_1 I_1(s) + (L_1 s + R_2)\{I_1(s) - I_2(s)\} - L_1 i_{L_1}(0) = 0$$

$$-(L_1 s + R_2)\{I_1(s) - I_2(s)\} + \left(\frac{1}{C_1 s} + R_3\right) I_2(s) + \frac{v_{C_1}(0)}{s}$$
$$+ L_2 s\{I_2(s) - I_3(s)\} - L_2 i_{L_2}(0) + V_2(s) + L_1 i_{L_1}(0) = 0$$

$$-V_2(s) + L_2 i_{L_2}(0) - L_2 s\{I_2(s) - I_3(s)\} + \left(R_4 + \frac{1}{C_2 s}\right) I_3(s) - V_3(s) + \frac{v_{C_2}(0)}{s} = 0$$

or in the matrix form,

$$\begin{bmatrix} R_1 + R_2 + L_1 s & -R_2 - L_1 s & 0 \\ -R_2 - L_1 s & R_2 + R_3 + L_1 s + L_2 s + \dfrac{1}{C_1 s} & -L_2 s \\ 0 & -L_2 s & L_2 s + R_4 + \dfrac{1}{C_2 s} \end{bmatrix} \begin{bmatrix} I_1(s) \\ I_2(s) \\ I_3(s) \end{bmatrix}$$

$$= \begin{bmatrix} V_1(s) + L_1 i_{L_1}(0) \\ L_2 i_{L_2}(0) - \dfrac{v_{C_1}(0)}{s} - L_1 i_{L_1}(0) - V_2(s) \\ -L_2 i_{L_2}(0) + V_2(s) + V_3(s) - \dfrac{v_{C_2}(0)}{s} \end{bmatrix}$$

### Comments

1. The three diagonal terms in the square *impedance matrix* correspond to the transform impedances in the first, second, and third meshes, respectively. These terms are all positive.
2. The off-diagonal terms are all negative (or zero) and symmetric, and correspond to the impedances which are *common* to each pair of meshes. That is,

the element in row 1, column 2 is minus the impedance common to the first and second meshes.

3. In the next chapter we will find that we can write the loop (or nodal) equations for many circuits by inspection, without going through the formal steps of setting up the equations.

**5.17** Draw the transform circuit diagram for the network of Fig. 5-17 given zero initial conditions. Then write transformed nodal equations at $a$ and $b$ and solve for $V_a(s)$.

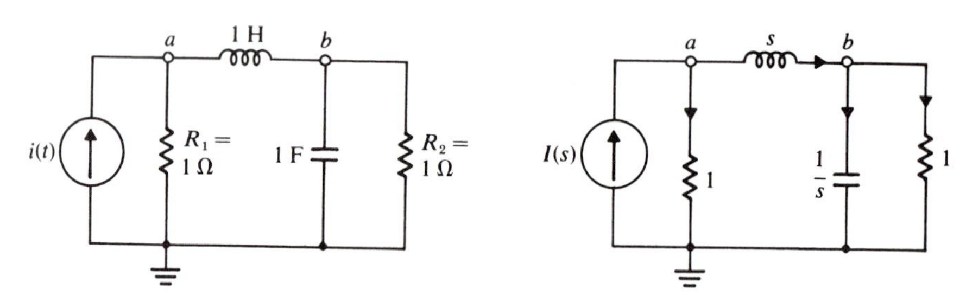


**Fig. 5-17** **Fig. 5-18**

The transform circuit is shown in Fig. 5-18, and applying KCL at nodes $a$ and $b$,

$$I(s) - I_{R1}(s) - I_L(s) = 0$$

$$I_L(s) - I_C(s) - I_{R2}(s) = 0$$

The transformed $i$-$v$ relations are (for zero initial conditions),

$$I_{R1}(s) = \frac{V_a(s)}{1}, \qquad I_L(s) = \frac{V_a(s) - V_b(s)}{s}, \qquad I_C(s) = sV_b(s), \qquad I_{R2}(s) = \frac{V_b(s)}{1}$$

which when substituted into the KCL equations yield the required nodal equations

$$\begin{bmatrix} \frac{1}{s} + 1 & -\frac{1}{s} \\ -\frac{1}{s} & \frac{1}{s} + s + 1 \end{bmatrix} \begin{bmatrix} V_a(s) \\ V_b(s) \end{bmatrix} = \begin{bmatrix} I(s) \\ 0 \end{bmatrix}$$

Then from Cramer's rule

$$V_a(s) = \frac{\begin{vmatrix} I(s) & -\frac{1}{s} \\ 0 & \frac{1}{s} + s + 1 \end{vmatrix}}{\begin{vmatrix} \frac{1}{s} + 1 & -\frac{1}{s} \\ -\frac{1}{s} & \frac{1}{s} + s + 1 \end{vmatrix}} = \frac{(1/s + s + 1)I(s)}{1/s^2 + 1 + 1/s + 1/s + s + 1 - 1/s^2}$$

$$= \frac{s^2 + s + 1}{s^2 + 2s + 2} I(s)$$

**5.18** Derive the relationship between $V_a(s)$ and $I(s)$ in Problem 5.17 by combining transform impedances in series and in parallel.

The "trick" is to start at the far end of the circuit, and use reduction techniques to work back toward the source. (No impedances are in series or parallel if we start at the source end of the circuit.)

The first move is to combine the parallel impedances $1/s$ and 1 in Fig 5-18 into the single impedance $Z_{bg}$ shown in Fig. 5-19,

$$\frac{1}{Z_{bg}(s)} = 1 + \frac{1}{1/s} = s + 1 \quad \text{or} \quad Z_{bg}(s) = \frac{1}{s+1}$$

(The subscript $bg$ draws attention to the path from node $b$ to the ground node $g$.)

Next we add the series impedances $s$ and $1/(s + 1)$ in Fig. 5-19, yielding

$$Z_{abg}(s) = s + \frac{1}{s+1} = \frac{s^2 + s + 1}{s+1}$$

which is shown in Fig. 5-20.

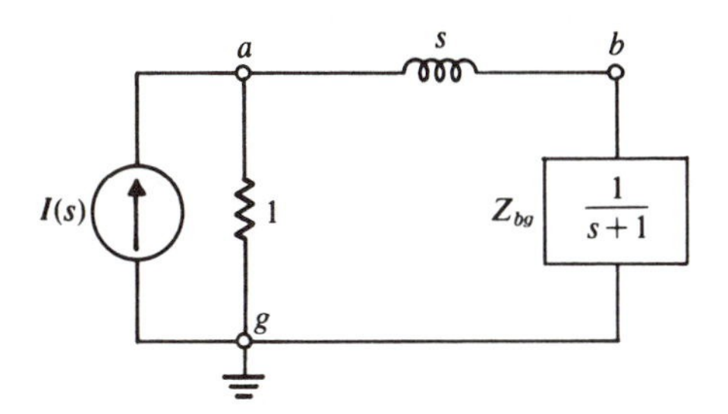


**Fig. 5-19**

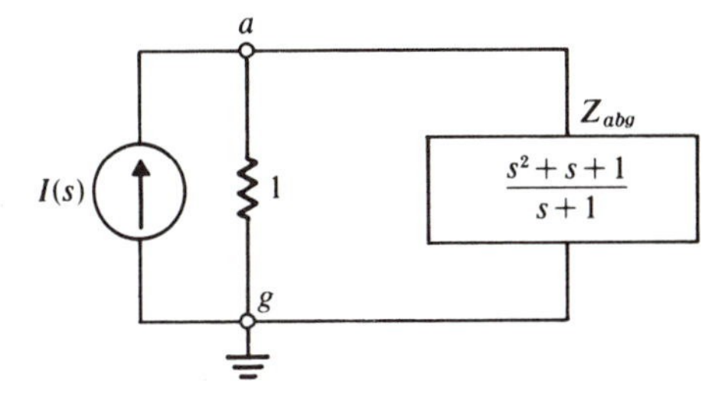


**Fig. 5-20**

Finally, we combine the two impedances 1 and $(s^2 + s + 1)/(s + 1)$ in Fig. 5-20,

$$\frac{1}{Z_{ag}(s)} = \frac{1}{1} + \frac{s+1}{s^2 + s + 1} = \frac{s^2 + 2s + 2}{s^2 + s + 1}$$

or

$$Z_{ag}(s) = \frac{s^2 + s + 1}{s^2 + 2s + 2}$$

as shown in Fig. 5-21.

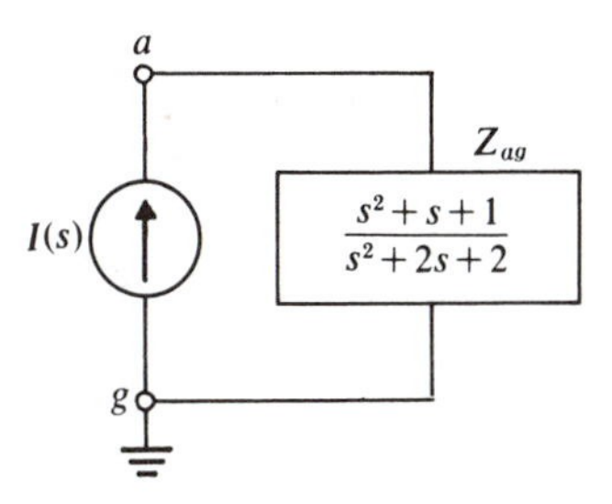


**Fig. 5-21**

It now follows from equation (5.8), page 138, that

$$V_a(s) = Z_{ag}(s)I(s) = \frac{s^2 + s + 1}{s^2 + 2s + 2}I(s)$$

**5.19** Draw the transform circuit diagram for Fig. 5-22 and use impedance concepts to relate $I(s)$ to $E(s)$, $V(s)$ to $I(s)$, and finally $V(s)$ to $E(s)$. All initial conditions are zero.

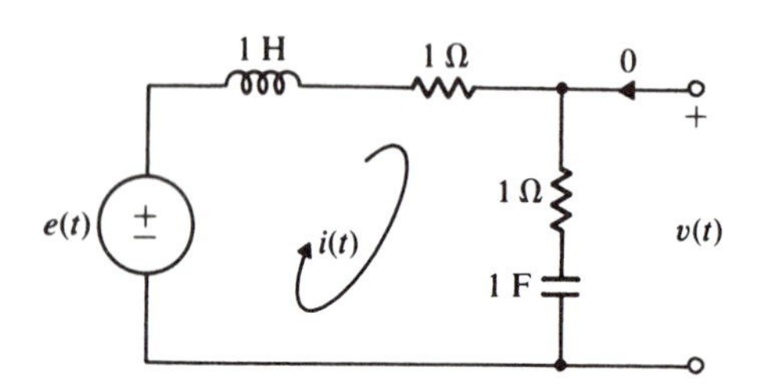


**Fig. 5-22**

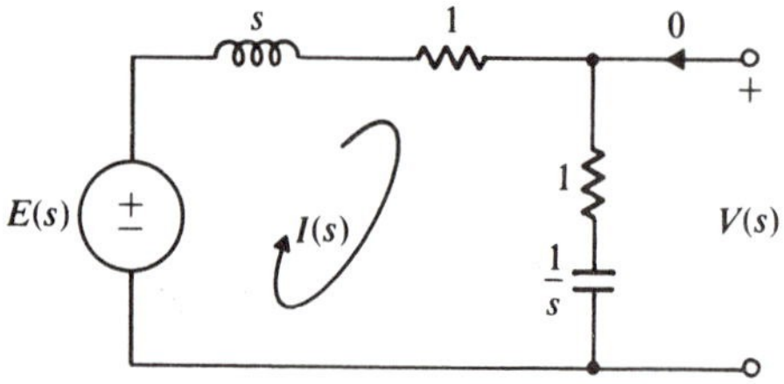

**Fig. 5-23**

The transform circuit diagram is shown in Fig. 5-23. First, because impedances in series add,

$$\left(s + 1 + 1 + \frac{1}{s}\right)I(s) = E(s)$$

which appropriately relates $I(s)$ to $E(s)$. Next, the impedance between the output terminals is $(1 + 1/s)$ through which "flows" the transformed current $I(s)$. Therefore

$$V(s) = \left(1 + \frac{1}{s}\right)I(s) = \left(1 + \frac{1}{s}\right)\frac{E(s)}{s + 2 + 1/s} = \frac{s + 1}{s^2 + 2s + 1}E(s)$$

*Note:* Alternatively we can relate $V(s)$ and $E(s)$ by establishing simultaneous equations in $I(s)$ and $V(s)$. From KVL around the left-hand and right-hand meshes of the circuit in Fig. 5-23,

$$(s + 1)I(s) + V(s) = E(s)$$

$$-\left(1 + \frac{1}{s}\right)I(s) + V(s) = 0$$

Then from Cramer's rule,

$$V(s) = \frac{\begin{vmatrix} 1 + s & E(s) \\ -\left(1 + \frac{1}{s}\right) & 0 \end{vmatrix}}{\begin{vmatrix} 1 + s & 1 \\ -\left(1 + \frac{1}{s}\right) & 1 \end{vmatrix}} = \frac{(1 + 1/s)E(s)}{1 + s + 1 + 1/s} = \frac{s + 1}{s^2 + 2s + 1}E(s)$$

**5.20** Use impedance concepts to find $V(s)$ in Fig. 5-24 when $e(t) = A \cos \omega t$ and the initial conditions are zero.

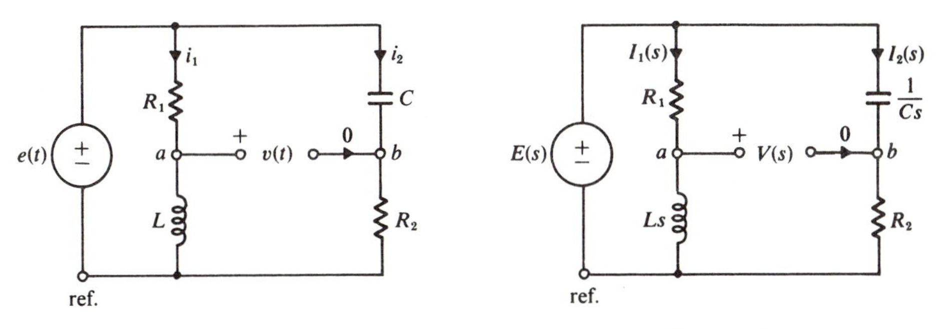


**Fig. 5-24** **Fig. 5-25**

From the transform circuit diagram of Fig. 5-25,

$$I_1(s) = \frac{E(s)}{R_1 + Ls} \quad \text{and} \quad I_2(s) = \frac{E(s)}{1/Cs + R_2}$$

But by definition,

$$V(s) = V_a(s) - V_b(s) = LsI_1(s) - R_2I_2(s) = \frac{LsE(s)}{R_1 + Ls} - \frac{R_2E(s)}{1/Cs + R_2}$$

which after a little algebra and the substitution of $E(s) = As/(s^2 + \omega^2)$ becomes

$$V(s) = \frac{(1/R_2C - R_1/L)s}{s^2 + (R_1/L + 1/R_2C)s + R_1/R_2LC} \frac{As}{s^2 + \omega^2}$$

### *Solution of State Equations*

**5.21** A linear time-invariant circuit is represented by the state equation

$$\dot{\mathbf{x}} = \begin{bmatrix} 0 & 1 \\ -1 & -1 \end{bmatrix} \mathbf{x} + \begin{bmatrix} 1 & 0 \\ 0 & 1 \end{bmatrix} \mathbf{u}$$

The objective here is to use the time-domain method to solve for $\mathbf{x}(t)$ given $\mathbf{x}(0)$ and $\mathbf{u}(t)$. As the first step, evaluate the matrix exponential $e^{\mathbf{A}t}$ for $t = 0.1$ s. Then find the initial condition response at $t = 0.1$ s, given that the *transpose* of $\mathbf{x}(0)$ is $\mathbf{x}'(0) = [2 \ -3]$.

From the definition of $e^{\mathbf{A}t}$ in equation (*5.15*), page 141,

$$e^{\mathbf{A}t} = \mathbf{I} + \mathbf{A}t + \frac{\mathbf{A}^2}{2!}t^2 + \cdots$$

$$= \begin{bmatrix} 1 & 0 \\ 0 & 1 \end{bmatrix} + \begin{bmatrix} 0 & 1 \\ -1 & -1 \end{bmatrix} t + \begin{bmatrix} 0 & 1 \\ -1 & -1 \end{bmatrix}\begin{bmatrix} 0 & 1 \\ -1 & -1 \end{bmatrix}\frac{t^2}{2!} + \cdots$$

$$= \begin{bmatrix} 1 & 0 \\ 0 & 1 \end{bmatrix} + \begin{bmatrix} 0 & t \\ -t & -t \end{bmatrix} + \begin{bmatrix} -\frac{t^2}{2} & -\frac{t^2}{2} \\ \frac{t^2}{2} & 0 \end{bmatrix} + \cdots$$

$$= \begin{bmatrix} 1 - \frac{t^2}{2} + \cdots & t - \frac{t^2}{2} + \cdots \\ -t + \frac{t^2}{2} - \cdots & 1 - t + \cdots \end{bmatrix}$$

Setting $t = 0.1$ s,

$$e^{0.1\mathbf{A}} = \begin{bmatrix} 1 - 0.005 + \cdots & 0.1 - 0.005 + \cdots \\ -0.1 + 0.005 - \cdots & 1 - 0.1 + \cdots \end{bmatrix} \doteq \begin{bmatrix} 0.995 & 0.095 \\ -0.095 & 0.900 \end{bmatrix}$$

The initial condition solution is given by the first term on the right of equation (*5.14*), namely

$$\mathbf{x}_{IC}(t) = e^{\mathbf{A}t}\mathbf{x}(0)$$

Therefore for $t = 0.1$ s,

$$\mathbf{x}_{IC}(0.1) \doteq \begin{bmatrix} 0.995 & 0.095 \\ -0.095 & 0.900 \end{bmatrix} \begin{bmatrix} 2 \\ -3 \end{bmatrix} = \begin{bmatrix} 1.705 \\ -2.890 \end{bmatrix}$$

### *Comments*

1. If a digital computer is handy, $e^{0.1\mathbf{A}}$ could easily be evaluated to greater accuracy by taking more terms in the series.
2. To find $\mathbf{x}_{IC}(0.2)$ we can use $\mathbf{x}'(0.1) = [1.705 \; -2.890]$ as the transpose of a *new* set of initial conditions and the *same* $e^{0.1\mathbf{A}}$ matrix to move forward another 0.1 s in time. And so on, 0.1 s at a time.
3. This method has the advantages that (*a*) only one value of $e^{\mathbf{A}t}$ need be evaluated, (*b*) $e^{\mathbf{A}t}$ converges rapidly for small values of $t$, and (*c*) the step-by-step analysis is easily implemented on a computer.

**5.22** Find the forced response at $t = 0.1$ s caused by the input $\mathbf{u}(t)$ applied to the circuit of Problem 5.21, given $\mathbf{u}'(t) = [1 \; -t]$.

The forced response is given by the second term on the right of equation (*5.14*), namely

$$\mathbf{x}_F(t) = e^{\mathbf{A}t} \int_0^t e^{-\mathbf{A}\tau}\mathbf{B}\mathbf{u}(\tau)\, d\tau$$

We calculated $e^{\mathbf{A}t}$ in Problem 5.21, and since $e^{-\mathbf{A}\tau} = e^{\mathbf{A}t}|_{t=-\tau}$ we can write the integrand as

$$e^{-\mathbf{A}\tau}\mathbf{Bu}(\tau) = \begin{bmatrix} 1 + \frac{\tau^2}{2} + \cdots & -\tau - \frac{\tau^2}{2} - \cdots \\ \tau + \frac{\tau^2}{2} + \cdots & 1 + \tau + \cdots \end{bmatrix} \begin{bmatrix} 1 & 0 \\ 0 & 1 \end{bmatrix} \begin{bmatrix} 1 \\ -\tau \end{bmatrix}$$

$$= \begin{bmatrix} 1 + \frac{\tau^2}{2} + \cdots \\ -\frac{\tau^2}{2} + \cdots \end{bmatrix}$$

Therefore setting $t = 0.1$ s,

$$\int_0^{0.1} e^{-\mathbf{A}\tau}\mathbf{Bu}(\tau)\, d\tau = \int_0^{0.1} \begin{bmatrix} 1 + \frac{\tau^2}{2} + \cdots \\ -\frac{\tau^2}{2} + \cdots \end{bmatrix} d\tau = \begin{bmatrix} \tau + \frac{\tau^3}{6} + \cdots \\ -\frac{\tau^3}{6} + \cdots \end{bmatrix}_0^{0.1}$$

$$\doteq \begin{bmatrix} 0.10017 \\ -0.00017 \end{bmatrix}$$

Finally, substituting $e^{0.1\mathbf{A}}$ from Problem 5.21,

$$\mathbf{x}_F(0.1) = e^{0.1\mathbf{A}} \int_0^{0.1} e^{-\mathbf{A}\tau}\mathbf{Bu}(\tau)\, d\tau \doteq \begin{bmatrix} 0.995 & 0.095 \\ -0.095 & 0.900 \end{bmatrix} \begin{bmatrix} 0.10017 \\ -0.00017 \end{bmatrix}$$

$$= \begin{bmatrix} 0.0997 \\ -0.0097 \end{bmatrix}$$

### *Comments*

1. It is somewhat more difficult than was the case with $\mathbf{x}_{IC}$ to extend the forced solution to $t = 0.2, 0.3, \ldots$ while avoiding unnecessary computation. However, quite efficient computer algorithms are available.
2. The total solution can, of course, be found from $\mathbf{x} = \mathbf{x}_{IC} + \mathbf{x}_F$. That is, we can add the solutions in this and the previous problem to find $\mathbf{x}(0.1)$.

**5.23** Now let us solve the state equation of Problems 5.21 and 5.22 in the transform domain.

First obtain the matrix $(s\mathbf{I} - \mathbf{A})^{-1}$. Then find the transform of the initial condition solution $\mathbf{X}_{IC}(s)$, due to $\mathbf{x}(0)$, given $\mathbf{x}'(0) = [2\ -3]$. Finally, find $\mathbf{X}_F(s)$ due to $\mathbf{u}(t)$ when $\mathbf{u}'(t) = [1\ -t]$.

From Problem 5.21,

$$\mathbf{A} = \begin{bmatrix} 0 & 1 \\ -1 & -1 \end{bmatrix}$$

and so

$$(s\mathbf{I} - \mathbf{A}) = \begin{bmatrix} s & 0 \\ 0 & s \end{bmatrix} - \begin{bmatrix} 0 & 1 \\ -1 & -1 \end{bmatrix} = \begin{bmatrix} s & -1 \\ 1 & s+1 \end{bmatrix}$$

Since there are many ways to take the inverse of a matrix, we will simply show the result,

$$(s\mathbf{I} - \mathbf{A})^{-1} = \begin{bmatrix} \dfrac{s+1}{s^2+s+1} & \dfrac{1}{s^2+s+1} \\ \dfrac{-1}{s^2+s+1} & \dfrac{s}{s^2+s+1} \end{bmatrix}$$

The first term on the right-hand side of equation (*5.16*), page 142, is the transform of the initial condition solution, namely

$$\mathbf{X}_{IC}(s) = (s\mathbf{I} - \mathbf{A})^{-1}\mathbf{x}(0) = \begin{bmatrix} \dfrac{s+1}{s^2+s+1} & \dfrac{1}{s^2+s+1} \\ \dfrac{-1}{s^2+s+1} & \dfrac{s}{s^2+s+1} \end{bmatrix} \begin{bmatrix} 2 \\ -3 \end{bmatrix}$$

$$= \begin{bmatrix} \dfrac{2(s-0.5)}{s^2+s+1} \\ \dfrac{-3(s+2/3)}{s^2+s+1} \end{bmatrix}$$

The second term on the right-hand side of equation (*5.16*) is the transform of the forced solution,

$$\mathbf{X}_F(s) = (s\mathbf{I} - \mathbf{A})^{-1}\mathbf{B}\mathbf{U}(s)$$
$$= \mathbf{\Phi}(s)\,\mathbf{B}\mathbf{U}(s)$$

Since from Problem 5.21,

$$\mathbf{B} = \begin{bmatrix} 1 & 0 \\ 0 & 1 \end{bmatrix}$$

we have

$$\mathbf{X}_F(s) = \begin{bmatrix} \dfrac{s+1}{s^2+s+1} & \dfrac{1}{s^2+s+1} \\ \dfrac{-1}{s^2+s+1} & \dfrac{s}{s^2+s+1} \end{bmatrix} \begin{bmatrix} 1 & 0 \\ 0 & 1 \end{bmatrix} \begin{bmatrix} \dfrac{1}{s} \\ -\dfrac{1}{s^2} \end{bmatrix} = \begin{bmatrix} \dfrac{s^2+s-1}{s^2(s^2+s+1)} \\ \dfrac{-2}{s(s^2+s+1)} \end{bmatrix}$$

### Comments

1. If we wished to compute $\mathbf{x}_{IC}(t)$, $\mathbf{x}_F(t)$, or $\mathbf{x}(t) = \mathbf{x}_{IC}(t) + \mathbf{x}_F(t)$, we would have to expand in partial fractions to evaluate the inverse transform. For example,

$$x_{1_{IC}}(t) = \mathcal{L}^{-1}\left[\frac{2(s - 0.5)}{s^2 + s + 1}\right]$$

$$= \mathcal{L}^{-1}\left[\frac{Me^{j\theta}}{s + 0.5 - j0.866} + \frac{Me^{-j\theta}}{s + 0.5 + j0.866}\right]$$

from which, after computing $M$ and $\theta$ in the usual way,

$$x_{1_{IC}}(t) = 2Me^{-0.5t} \cos (0.866t + \theta), \qquad t \geq 0$$

2. This method clearly leads to a closed-form solution which is valid for any $t \geq 0$. This contrasts with the numerical time-domain solution, which is computed at discrete instants of time.

# PROBLEMS

**5.24** An electric circuit is described by the differential equation

$$\frac{d^3v}{dt^3} + 5\frac{d^2v}{dt^2} + 7v = 2\frac{di(t)}{dt} + 4i(t)$$

with initial conditions $d^2v(0)/dt^2 = A$, $dv(0)/dt = B$, $v(0) = C$, and $i(0) = D$. Use the Laplace transform method to verify that

$$V(s) = \frac{Cs^2 + (B + 5C)s + A + 5B - 2D + \{2s + 4\}I(s)}{s^3 + 5s^2 + 7}$$

**5.25** The Laplace transform of a current $i(t)$ is $I(s) = 2(s^2 - 2s + 2)/s^2(s^2 + 4)$. Find $i(t)$ for $t \geq 0$.

**5.26** In the circuit of Fig. 5-26, $v(t) = 4 \cos (2t - 45°)$ and $i_L(0) = 0$. Use the method of Laplace tranformation to find $v_L(t)$ for $t \rightarrow \infty$.

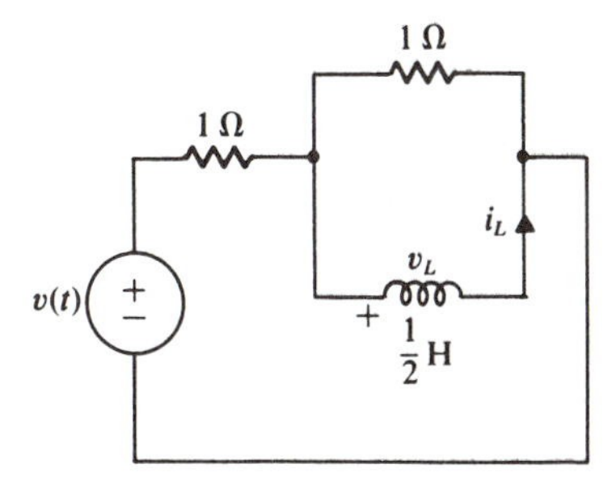


**Fig. 5-26**

**5.27** An electric circuit is described by the simultaneous differential equations

$$\frac{1}{2}\frac{dv}{dt} + i = f(t), \qquad -v + \frac{di}{dt} + 2i = 0$$

with $v(0) = -2$ V and $i(0) = 1$ A. Show that

$$V(s) = \frac{2\{s + 2\}F(s) - 2(s + 3)}{s^2 + 2s + 2}$$

**5.28** For the circuit of Problem 5.27, find $v(t)$ if $f(t) = e^{-2t}$.

**5.29** In the circuit of Fig. 5-27, $e(t) = 4$ V, $i(t) = 6$ A, and all the initial conditions are zero. Use the Laplace transform method to find $v(t)$ for $t \geq 0$.

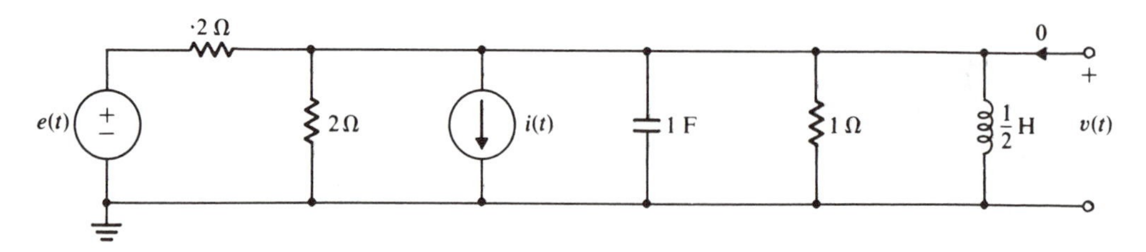


**Fig. 5-27**

**5.30** Given the transform circuit diagram of Fig. 5-28, use impedance concepts to show that

$$\frac{I(s)}{V(s)} = \frac{s + 1}{s^2 + 2s + 2}$$

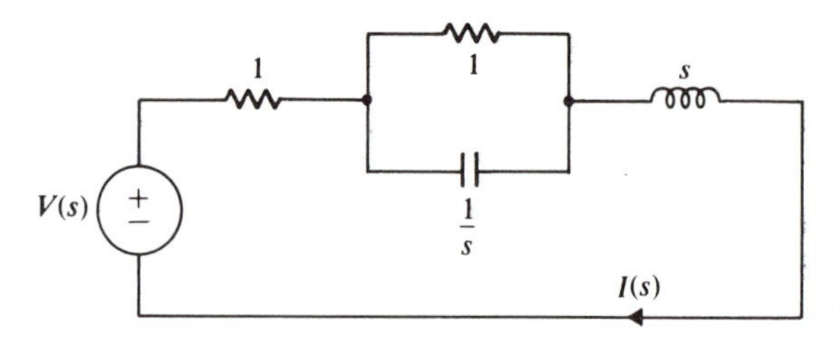


**Fig. 5-28**

**5.31** Confirm that Fig. 5-29*b* is a correct transformed equivalent of Fig. 5-29*a*.

**5.32** Using the transform circuit diagram of Fig. 5-29*b*, find $V(s)$.

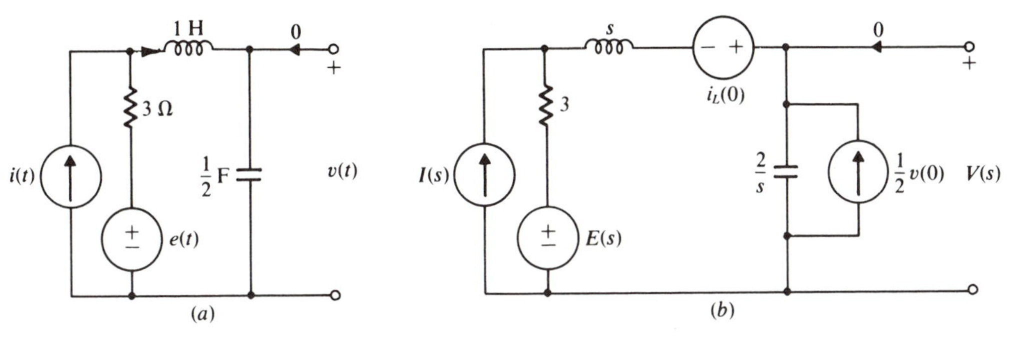


**Fig. 5-29**

**5.33** Draw a transform circuit diagram suitable for loop analysis of the circuit of Fig. 5-30. Then write the equations. (The same initial curent $i_L(0)$ flows in both inductors.)

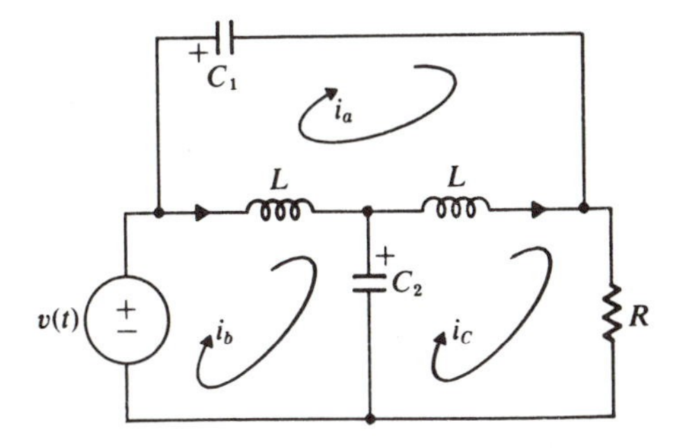


**Fig. 5-30**

**5.34** Given the state equation

$$\dot{\mathbf{x}} = \begin{bmatrix} 0 & 1 & 0 \\ 0 & 0 & 1 \\ 0 & 0 & -2 \end{bmatrix} \mathbf{x} + \begin{bmatrix} 0 \\ 0 \\ 3 \end{bmatrix} \mathbf{u}$$

find the forced response at $t = 0.1$ s when $\mathbf{u} = u_1 = \delta(t)$.

**5.35** For the state equation of Problem 5.34, show that

$$\mathbf{X}_F(s) = \begin{bmatrix} \dfrac{3}{s^2(s+2)} \\ \dfrac{3}{s(s+2)} \\ \dfrac{3}{s+2} \end{bmatrix}$$

Chapter 6

# NETWORK FUNCTIONS, POLES AND ZEROS, AND STABILITY

In the first five chapters we have seen how to go about the quite direct processes of describing an electric circuit mathematically and then solving the equations for the unknown circuit variables. (The emphasis has been on linear circuits, not because nonlinearity is either unusual or unimportant, but because there is little we can do to solve such problems in a general way.) We have also seen that the above processes—equation writing and solving—can be quite tedious. Some of the tedium can be avoided by turning the labor over to a computer, but sometimes it is better to search for "tricks of the trade": the shortcuts, special cases, and partial solutions. It is this second approach which will occupy us for the next six chapters.

We should be quick to point out that these two approaches are complementary. Although the most obvious application of a computer—the solution of a particular set of equations—is usually a rather barren exercise, the computer can back up the "tricks of the trade" to provide information about the behavior of *classes* of circuits in response to various inputs and initial conditions. This brings us a step nearer to circuit synthesis (i.e. design).

## INITIAL CONDITIONS

Unless otherwise specified, all initial conditions will hereafter be assumed zero. This does not constitute a restriction. If the transform method is adopted, any nonzero initial conditions may be replaced by equivalent transform sources.

## IMPEDANCE AND ADMITTANCE

Impedance and admittance were introduced in Chapter 5 in the equations

$$V(s) = Z(s)I(s) \quad \text{and} \quad I(s) = Y(s)V(s) \tag{6.1}$$

Clearly $Z(s)$ is a generalization of resistance $R$, and $Y(s)$ is a generalization of conductance $G = 1/R$. These are the first two members of a more general class of *network functions*.

## SERIES AND PARALLEL IMPEDANCES

If $Z_1(s), Z_2(s), \ldots, Z_n(s)$ are in series, then they are equivalent to

$$Z_{eq}(s) = Z_1(s) + Z_2(s) + \cdots + Z_n(s) \tag{6.2}$$

If the elements are in parallel, then

$$Y_{eq}(s) = Y_1(s) + Y_2(s) + \cdots + Y_n(s) \tag{6.3}$$

## A NOTE ON EQUIVALENCY

Two "black boxes" are said to be equivalent, or more precisely, to have *equivalent terminal characteristics*, if one cannot be distinguished from the other as a result of electrical measurements made at the terminals of the boxes. (The term *black box* implies that the network inside the box is not open to direct inspection.) Quite different *internal* behavior is possible in two black boxes having identical *terminal* characteristics.

To overlook this interpretation of equivalency is to leave ourselves open to error, misunderstanding, and confusion.

As an example, the two one-port black boxes of Fig. 6-1 have identical *terminal* characteristics (that is, they have identical graphs of $v$ vs. $i$), but under open-circuit conditions, for example, the left-hand circuit is quiescent, while the one on the right is generating and dissipating a steady 1 W *inside the box*.

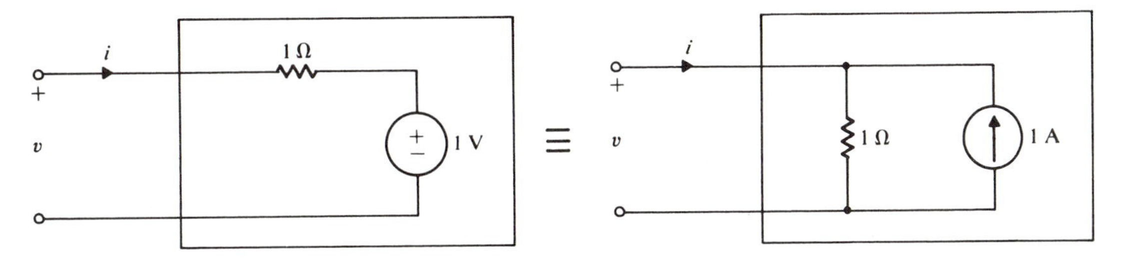


**Fig. 6-1**

## SOURCE TRANSFORMATIONS

It is sometimes convenient to replace a voltage source by an equivalent current source, or vice versa. This may be achieved by the *transformation* of Fig. 6-2. (Figure 6-1 is a special case.) One application is suggested in the following section.

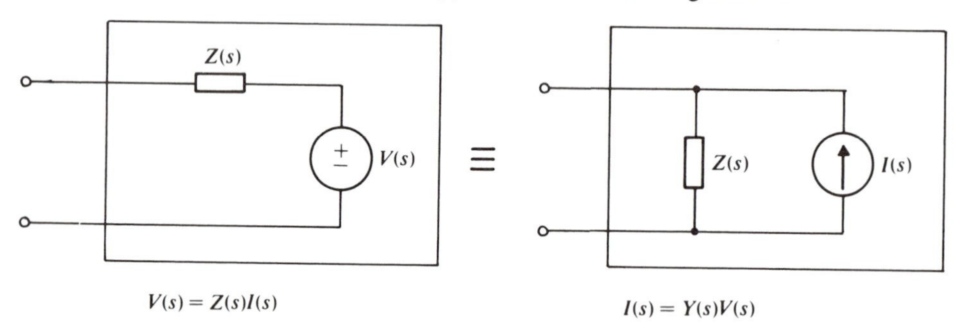


**Fig. 6-2**

## THE LOOP AND NODAL EQUATIONS

The transform loop equations for any $n$-mesh, connected, planar network containing no current sources, dependent sources, or mutual inductance can always be written by inspection.

If we agree to write the KVL equations around the $n$ window panes (meshes), and further, if we choose the $n$ loop currents to be around the $n$ meshes, either all clockwise or all counterclockwise, then

$$[Z][I(s)] = [V(s)] \tag{6.4}$$

or

$$\begin{bmatrix} Z_{11} & Z_{12} & \cdots & Z_{1n} \\ Z_{21} & Z_{22} & \cdots & Z_{2n} \\ \cdots & \cdots & \cdots & \cdots \\ Z_{n1} & Z_{n2} & & Z_{nn} \end{bmatrix} \begin{bmatrix} I_1(s) \\ I_2(s) \\ \cdots \\ I_n(s) \end{bmatrix} = \begin{bmatrix} V_1(s) \\ V_2(s) \\ \cdots \\ V_n(s) \end{bmatrix}$$

where

$Z_{ii}$ = total impedance in mesh $i$ $(i = 1, 2, \ldots, n)$,

$-Z_{ij} = -Z_{ji}$ = total impedance common to meshes $i$ and $j$ $(i, j = 1, 2, \ldots, n,\ i \neq j)$,

$I_j(s)$ = transform of the (unknown) current around the $j$th mesh,

$V_j(s)$ = total transform source voltage *rise* around the $j$th mesh (traveling in the direction of $I_j$).

*Notes*

1. If the given circuit contains independent current sources, we may use source transformations to convert them into voltage sources.
2. Otherwise, or if any of the other conditions are violated, we must revert to the formal equation-writing procedure of Chapter 2, a procedure which may be applied in the $s$-domain, as shown in Chapter 5.

Similarly we can write by inspection the nodal equations for any $n$-node, connected network containing no voltage sources, dependent sources, or mutual inductance. If we choose to call the reference node $\#n$, then

$$[Y][V(s)] = [I(s)] \tag{6.5}$$

or

$$\begin{bmatrix} Y_{11} & Y_{12} & \cdots & Y_{1,n-1} \\ Y_{21} & Y_{22} & \cdots & Y_{2,n-1} \\ \cdots & \cdots & \cdots & \cdots \\ Y_{n-1,1} & Y_{n-1,2} & \cdots & Y_{n-1,n-1} \end{bmatrix} \begin{bmatrix} V_1(s) \\ V_2(s) \\ \cdots \\ V_{n-1}(s) \end{bmatrix} = \begin{bmatrix} I_1(s) \\ I_2(s) \\ \cdots \\ I_{n-1}(s) \end{bmatrix}$$

where

$Y_{ii}$ = total admittance connected to node $i$ ($i = 1, 2, \ldots, n-1$),

$-Y_{ij} = -Y_{ji}$ = total admittance connected between nodes $i$ and $j$ ($i, j = 1, 2, \ldots, n-1, i \neq j$),

$V_i(s)$ = transform of the (unknown) voltage of the $i$th node, relative to the voltage of the reference node ($\#n$),

$I_i(s)$ = total transform source current *entering* node $i$.

*Notes*

1. If the given circuit contains independent voltage sources, we may use source transformations to convert them into current sources.
2. Otherwise, or if any of the other conditions are violated, we must revert to the formal equation-writing procedure of Chapter 2 (or Chapter 5).

Although these algorithms make the *writing* of circuit equations very easy, the latter have still to be *solved*! We should therefore consider the possibility of eliminating meshes (for loop analysis) or nodes (for nodal analysis) by appropriate reduction techniques (series and parallel combinations, for example) *before* writing down any circuit equations.

## *n*-PORT NETWORKS

Our interest in a circuit often centers on its terminal characteristics rather than on its internal behavior. We then speak of *one-port*, *two-port*, or in general, *n-port* networks, where $n$ is the number of "ports" of access. Laboratory power supplies and oscillators are examples of one ports, while an electronic amplifier with its input and output terminal pairs typifies a two port (see Fig. 6-3).

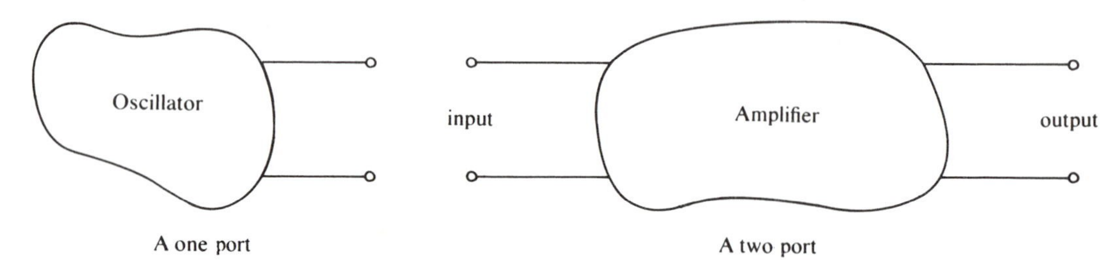


**Fig. 6-3**

## THE ONE PORT AND ITS THÉVENIN EQUIVALENT

In Chapter 2 we saw that any linear, resistive one port can be reduced to a single voltage source in series with a single resistor—as far as the world outside the one port's terminals is concerned. The reduced network was called the Thévenin equivalent of the original one port. Now we will find that any (linear) $s$-domain one port can be reduced to an $s$-domain Thévenin equivalent consisting of a single impedance $Z_{\text{Th}}(s)$ in series with a single transform voltage source $V_{\text{Th}}(s)$, as is shown in Fig. 6-4.

We can find the $s$-domain Thévenin equivalent by the same methods that we used before—now transformed into the $s$-domain. That is, we can choose from:

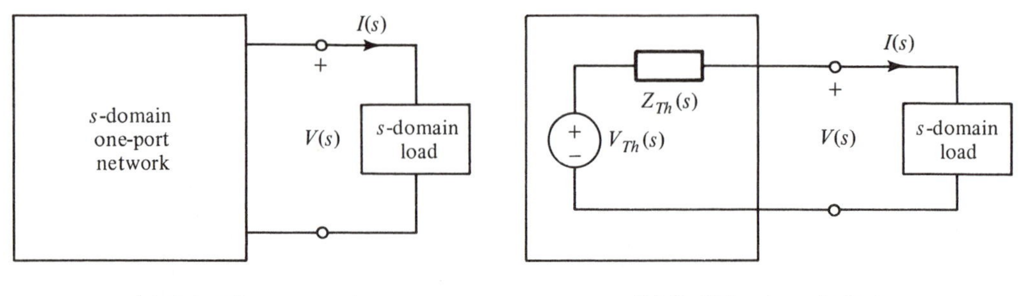


**Fig. 6-4**

1. Obtaining a transform one port's $V(s)$-$I(s)$ relation by $s$-domain nodal analysis. The result can then be compared with the Thévenin one port's $V(s)$-$I(s)$ relation,

$$V(s) = Z_{\text{Th}}(s)I(s) + V_{\text{Th}}(s) \tag{6.6}$$

to yield $Z_{\text{Th}}(s)$ and $V_{\text{Th}}(s)$.

2. Obtaining a one port's $I(s)$-$V(s)$ relation by loop analysis. This can then be compared with

$$I(s) = -Y_{\text{Th}}(s)V(s) + Y_{\text{Th}}(s)V_{\text{Th}}(s) \tag{6.7}$$

to yield $Y_{\text{Th}}(s)$ and $V_{\text{Th}}(s)$.

3. Finding $V_{oc}(s)$ and $I_{sc}(s)$, from which

$$V_{\text{Th}}(s) = V_{oc}(s) \quad \text{and} \quad Z_{\text{Th}}(s) = V_{oc}(s)/I_{sc}(s)$$

4. Killing all the independent transform sources (including any initial condition sources) in the one port, and then reducing the remaining circuit to a single impedance $Z_{\text{Th}}(s)$.

You will recall that the first two methods always work, while the second two, while sometimes quicker and easier, do not always yield the results we seek.

## A ONE PORT'S INPUT OR OUTPUT IMPEDANCE

As a matter of nomenclature, $Z_{\text{Th}}(s)$ is often called the one port's *output impedance* if the one port is driving the load. If, on the other hand, the load is driving the one port, we call $Z_{\text{Th}}(s)$ the one port's *input* or *driving point impedance*. There is no substantive difference between input and output impedance; just a different point of view.

To find a one port's output (or input) impedance, we simply find $Z_{\text{Th}}(s)$ by any appropriate method. As long as we do not also want to find $V_{\text{Th}}(s)$, it is perfectly legitimate to set all the one port's independent sources (including initial condition sources) to zero.

### Example 6.1

We will calculate the input (or output) impedance of the one port of Fig. 6-5 in three different ways.

#### *Solution 1*

Since we want only the one port's impedance, we may set all the initial conditions to zero, as in the transform diagram of Fig. 6-6*a*. We now know how to write the nodal equations for that circuit by inspection (see equation (*6.5*) and following):

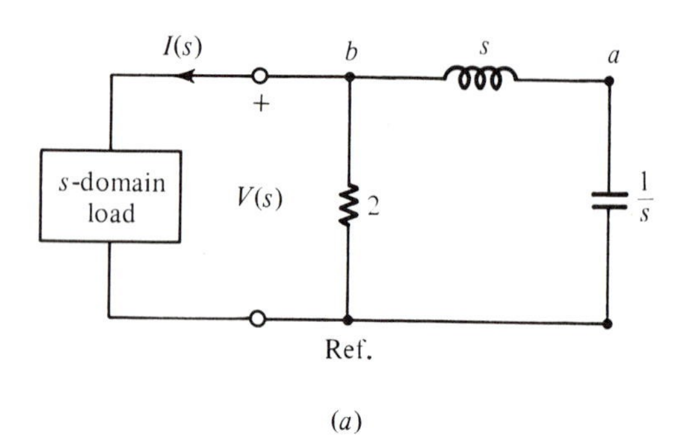


(a)

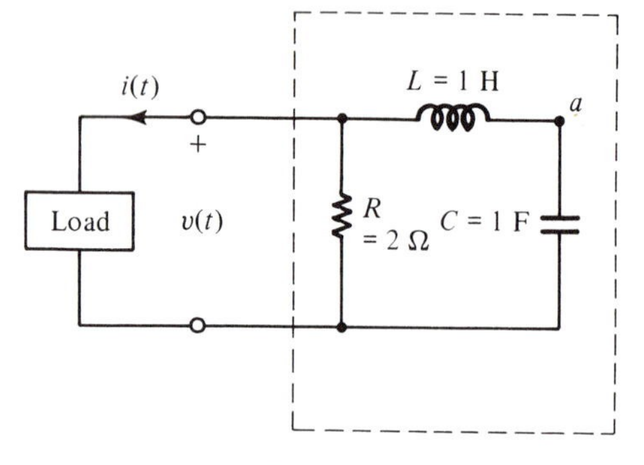


**Fig. 6-5**

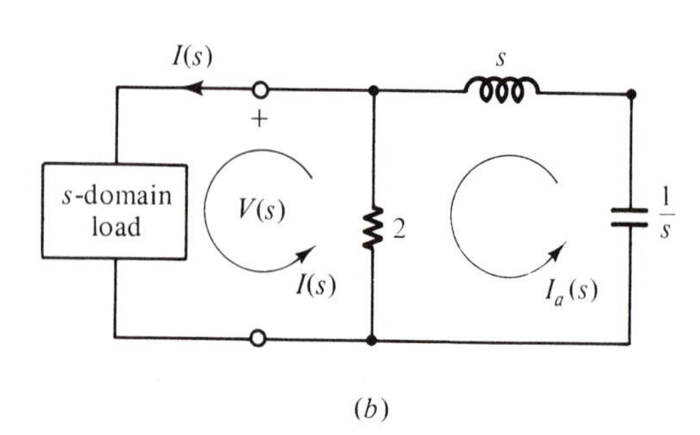


(b)

**Fig. 6-6**

$$\begin{bmatrix} s + \frac{1}{s} & -\frac{1}{s} \\ -\frac{1}{s} & \frac{1}{s} + \frac{1}{2} \end{bmatrix} \begin{bmatrix} V_a(s) \\ V(s) \end{bmatrix} = \begin{bmatrix} 0 \\ -I(s) \end{bmatrix}$$

where, as usual, the terminal current $I(s)$ has been treated as a known. By Cramer's rule,

$$V(s) = -\frac{2(s^2 + 1)}{s^2 + 2s + 1} I(s)$$

Finally, when we compare this $V(s) - I(s)$ relation with equation (*6.6*) for the transform Thévenin one port, it follows at once that $Z_{\text{Th}}(s) = 2(s^2 + 1)/(s^2 + 2s + 1)$.

### *Solution 2*

We can now write the loop equations for the circuit of Fig. 6-6*b* by inspection (see equation (*6.4*) and following):

$$\begin{bmatrix} s + 2 + \frac{1}{s} & -2 \\ -2 & 2 \end{bmatrix} \begin{bmatrix} I_a(s) \\ I(s) \end{bmatrix} = \begin{bmatrix} 0 \\ -V(s) \end{bmatrix}$$

where, as usual, the terminal voltage $V(s)$ has been treated as a known. Cramer's rule yields

$$I(s) = -\frac{s^2 + 2s + 1}{2(s^2 + 1)} V(s)$$

By comparison with equation (*6.7*) for the Thévenin one port, we see that $Y_{Th}(s) = 0.5(s^2 + 2s + 1)/(s^2 + 1)$, which agrees with the result from *Solution 1*.

### *Solution 3*

This time we will find the one port's Thévenin impedance by combining the circuit elements as in Fig. 6-7.

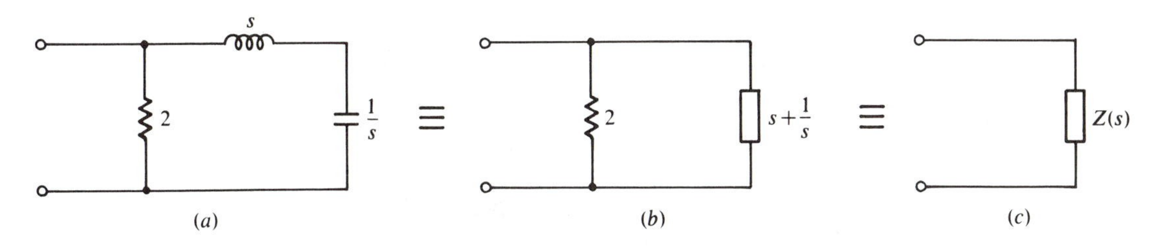


**Fig. 6-7**

In moving from Fig. 6-7*a* to Fig. 6-7*b* we have combined $Z_L(s) = s$ and $Z_C(s) = 1/s$ in series; the impedances add. To obtain Fig. 6-7*c* we must combine $Z_R(s) = 2$ in parallel with the impedance $s + 1/s = (s^2 + 1)/s$. Thus from equation (*6.3*),

$$\frac{1}{Z(s)} = Y(s) = \frac{1}{2} + \frac{s}{s^2 + 1} = \frac{0.5(s^2 + 2s + 1)}{s^2 + 1}$$

### *Comment*

A one port's Thévenin impedance, its input (or driving point) impedance, and its output impedance are functionally synonymous. They all refer to the impedance seen when looking into the one port's terminals with all internal independent sources (including initial condition sources) zero. ■

## THE TWO-PORT NETWORK

There are now two ports of access and two loads, as shown in Fig. 6-8. One point to note is that the conventional choice of current direction is *into* each plus-marked terminal. This is the *opposite* of the choice usually made when working with one ports and their Thévenin equivalents—in Fig. 6-4, for example. Given a little extra alertness, this should not cause any difficulty.

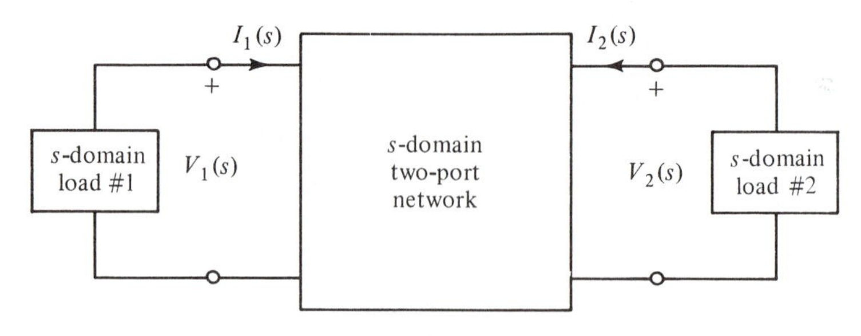


**Fig. 6-8**

## THE TWO PORT'S NETWORK FUNCTIONS

If we treat the two port together with its load #2 as a one port (see Fig. 6-9*a*), then we can find the input (or output) impedance seen by load #1, just as we did for one ports, above. When doing so we will *always* kill *all* of the independent sources (including initial condition sources) in the two port *and* in load #2. *Then* it follows that we can write, from the defining equations (*6.1*),

$$Z_1(s) = \frac{V_1(s)}{I_1(s)} \quad \text{or} \quad Y_1(s) = \frac{I_1(s)}{V_1(s)}$$

Still with reference to Fig. 6-9*a*, we can also define the ratios

$$H_1(s) = \frac{V_2(s)}{V_1(s)}, \qquad H_2(s) = \frac{V_2(s)}{I_1(s)},$$

$$H_3(s) = \frac{I_2(s)}{I_1(s)}, \quad \text{and} \quad H_4(s) = \frac{I_2(s)}{V_1(s)}$$

where by convention the numerator quantity in each ratio is regarded as the output, while the corresponding denominator quantity is assumed to be the *sole* input causing the numerator output. It is for this reason that we must kill all other independent sources and initial conditions (see Fig. 6-9*a* again). Each of the six ratios above will usually depend upon load #2, which must therefore be specified.

The $H(s)$ ratios are called *transfer functions*, since each describes a transfer of information from port #1 to port #2. $Z_1(s)$, $Y_1(s)$, and the $H(s)$ ratios are collectively called *network functions*.

Similarly, with reference now to Fig. 6-9*b*, we can define the impedance (or admittance) seen by load #2, namely

$$Z_2(s) = \frac{V_2(s)}{I_2(s)} \quad \text{or} \quad Y_2(s) = \frac{I_2(s)}{V_2(s)}$$

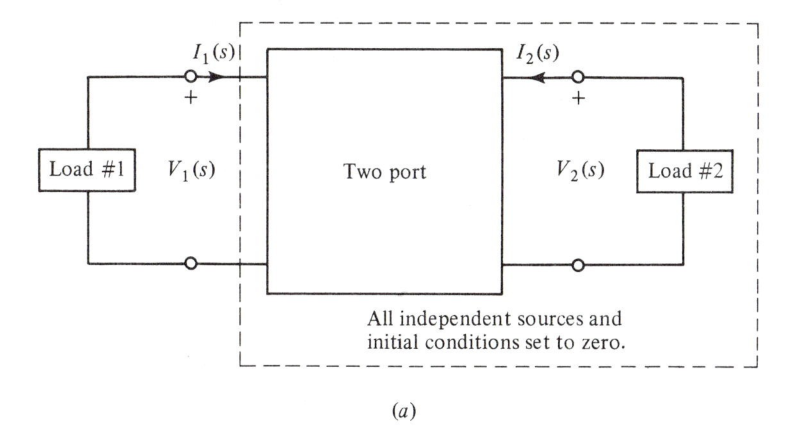


(a)

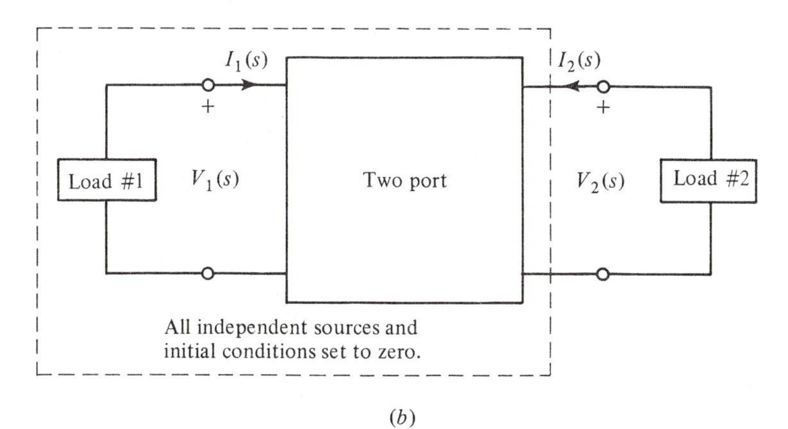


(b)

**Fig. 6-9**

and the transfer functions

$$H_5(s) = \frac{V_1(s)}{V_2(s)}, \qquad H_6(s) = \frac{V_1(s)}{I_2(s)},$$

$$H_7(s) = \frac{I_1(s)}{I_2(s)}, \quad \text{and} \quad H_8(s) = \frac{I_1(s)}{V_2(s)}$$

Each of *these* network functions will usually depend on load #1, which must be specified. And whenever we evaluate one of these network functions we must (by definition) kill all of the independent sources (including initial conditions) in the two port *and* in load #1. (The *numbering* of the transfer functions, $H_1(s)$, $H_2(s)$, . . . , $H_8(s)$ is without significance.)

### Example 6.2

Calculate $H_1(s) = V_2(s)/V_1(s)$ and $H_2(s) = V_1(s)/V_2(s)$ for the two port of Fig. 6-10, given in both cases that the output terminals are open circuit.

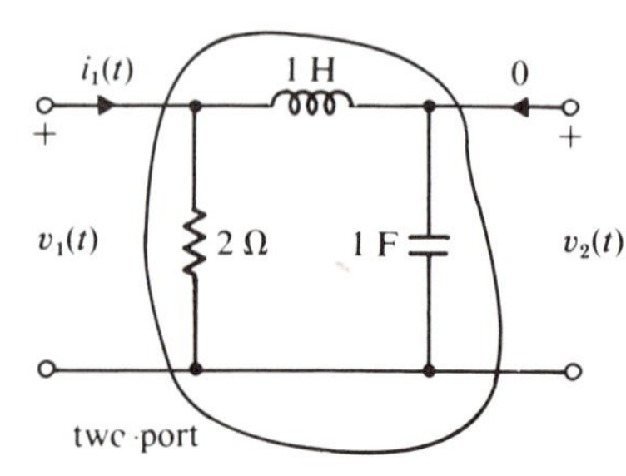


**Fig. 6-10**

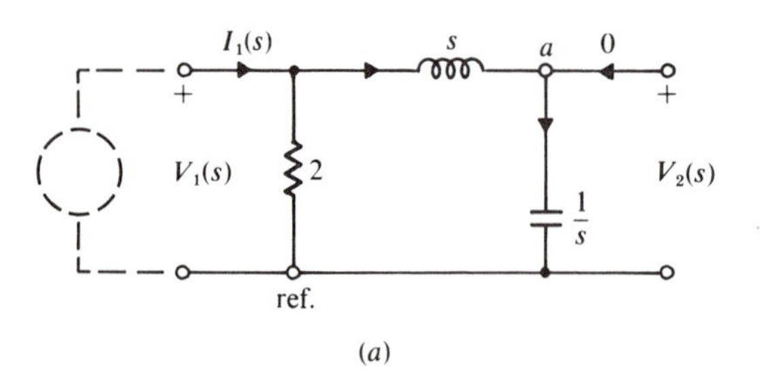


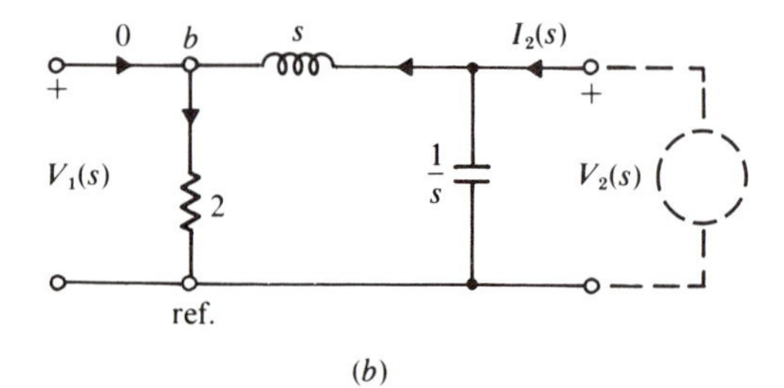


**Fig. 6-11**

To calculate $H_1(s)$, we want the voltage $V_2(s)$ due to $V_1(s)$, as shown in Fig. 6-11*a*. Note that all initial conditions have been set to zero. Applying KCL at node $a$, $I_L(s) - I_C(s) = 0$, or

$$\frac{1}{s}\{V_1(s) - V_2(s)\} - sV_2(s) = 0$$

from which $V_2(s) = V_1(s)/(s^2 + 1)$ and so $H_1(s) = V_2(s)/V_1(s) = 1/(s^2 + 1)$.

To calculate $H_2(s)$ we want $V_1(s)$ due to $V_2(s)$, as indicated in Fig. 6-11*b*. If we apply KCL at node $b$,

$$\frac{1}{s}\{V_2(s) - V_1(s)\} - \frac{1}{2}V_1(s) = 0$$

from which $V_1(s) = 2V_2(s)/(s + 2)$ and so $H_2(s) = V_1(s)/V_2(s) = 2/(s + 2)$. ■

Note that in Example 6.2, $H_1(s) = V_2(s)/V_1(s)$ is *not* the reciprocal of $H_2(s) = V_1(s)/V_2(s)$. The reason is not hard to find. The two ratios are calculated for different circuits, as can be seen by comparing Figs. 6-11*a* and *b*. Only in the case of ratios calculated at one terminal pair, for example $Z_1(s) = V_1(s)/I_1(s)$ and $Y_1(s) = I_1(s)/V_1(s)$, is there a reciprocal relationship, here $Y_1(s) = 1/Z_1(s)$.

These concepts are easily extended to $n$-port networks, and for the general transfer function we write

$$H(s) = \frac{Y(s)}{U(s)} = \frac{\text{transform of the output, } y(t)}{\text{transform of the input, } u(t)} \tag{6.8}$$

(Do not confuse $Y(s)$, the transform of $y(t)$, with the same notation used for admittance.)

## APPLICATION OF NETWORK FUNCTIONS

Network functions are useful when we are primarily concerned with one input and one resulting output: when we do *not* want to solve for all the currents and voltages in a circuit. The transform of the desired output is given by equation (6.8),

$$Y(s) = H(s)U(s)$$

from which the response due to $u(t)$ follows by inverse transformation,

$$y(t) = \mathscr{L}^{-1}[Y(s)]$$

### *An Aside . . .*

The network functions, which we have seen to be independent of the particular input and output functions, are of the form $N(s)/D(s)$, where $N(s)$ and $D(s)$ are polynomials in $s$ with real coefficients. That is,

$$H(s) = \frac{Y(s)}{U(s)} = \frac{N(s)}{D(s)} \quad \text{or} \quad D(s)Y(s) = N(s)U(s)$$

This corresponds to the differential equation

$$a_n \frac{d^n y}{dt^n} + a_{n-1} \frac{d^{n-1} y}{dt^{n-1}} + \cdots + a_1 \frac{dy}{dt} + a_0 y$$

$$= b_m \frac{d^m u}{dt^m} + b_{m-1} \frac{d^{m-1} u}{dt^{m-1}} + \cdots + b_1 \frac{du}{dt} + b_0 u$$

where the $a$'s and $b$'s are the coefficients of the polynomials $D(s)$ and $N(s)$. It follows that if we are given a circuit's transfer function and the initial conditions $y(0)$, $\dot{y}(0)$, . . . , $y^{(n-1)}(0)$, then we can calculate the initial condition response.

### *. . . and a Caution*

Except in some special cases we *cannot* simply connect some known driving source to a network, measure the output, and take the ratio of these *time functions* to find the network function. The latter is a function of $s$ which does not lead a real-world existence. The measurement of a transfer function is quite difficult, as we might expect from the fact that all the coefficients of the polynomials $N(s)$ and $D(s)$ would have to be found.

## VOLTAGE AND CURRENT DIVIDERS

We commonly meet the *divider networks* of Fig. 6-12, and it is therefore worth filing away the following easily-proved results:

$$\frac{V_2(s)}{V(s)} = \frac{Z_2(s)}{Z_1(s) + Z_2(s)} \tag{6.9}$$

$$\frac{I_2(s)}{I(s)} = \frac{Y_2(s)}{Y_1(s) + Y_2(s)} \tag{6.10}$$

Note that equation (*6.9*) assumes that no current is drawn from the output terminals.

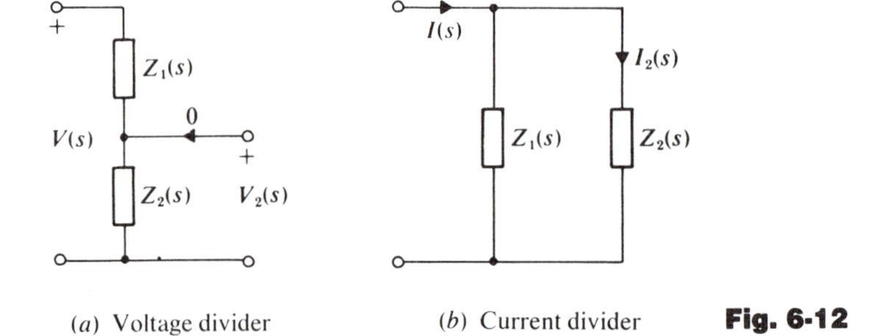


(*a*) Voltage divider (*b*) Current divider **Fig. 6-12**

## POLES AND ZEROS

Suppose $X(s)$ is either a network function or the transform of a voltage or current $x(t)$. Then

$$X(s) = \frac{N(s)}{D(s)} = \frac{K(s - z_1)(s - z_2) \cdots (s - z_n)}{(s - p_1)(s - p_2) \cdots (s - p_m)}$$

where $z_1, z_2, \ldots, z_n$ are called the *zeros* of $X(s)$ and $p_1, p_2, \ldots, p_m$ are called the *poles*. Since poles and zeros are, in general, complex numbers, they may be plotted on a *pole-zero diagram* as shown in Fig. 6-13. The *scale factor K* should also be noted on this diagram.

Note that if $s$ is set equal to one of the *zeros*, then $X(s) = 0$: and if $s$ is set equal to one of the *poles*, then $X(s) \rightarrow \infty$.

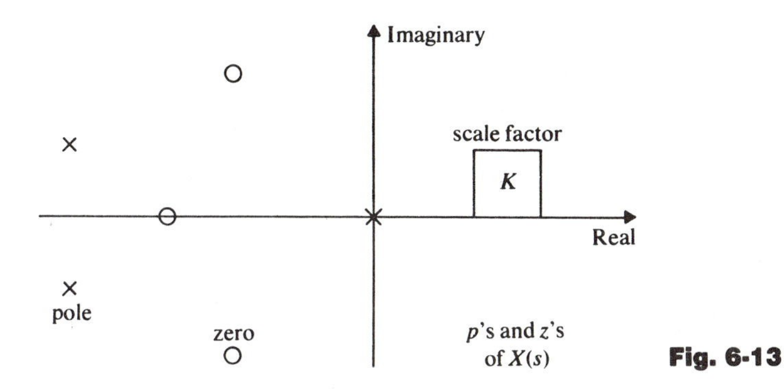


Fig. 6-13

If $m > n$, then $X(s) \rightarrow 0$ as $s \rightarrow \infty$ and we say that $X(s)$ has $m - n$ zeros at infinity. If $m < n$, then $X(s) \rightarrow \infty$ as $s \rightarrow \infty$ and we say that $X(s)$ has $n - m$ poles at infinity. This is a mathematical dodge to ensure that the total number of poles always equals the total number of zeros.

Suppose now that $X(s)$ is the transform of a time function $x(t)$. Then, assuming that $m > n$ and that there are no repeated poles,

$$X(s) = \frac{N(s)}{D(s)} = \frac{K(s - z_1)(s - z_2) \cdots (s - z_n)}{(s - p_1)(s - p_2) \cdots (s - p_m)}$$

$$\equiv \frac{K_1}{s - p_1} + \frac{K_2}{s - p_2} + \cdots + \frac{K_m}{s - p_m}$$

or, taking the inverse transform,

$$x(t) = K_1 e^{p_1 t} + K_2 e^{p_2 t} + \cdots + K_m e^{p_m t}$$

It follows that the *poles* determine the form or the *natural modes* of the function $x(t)$. If $p_i$ is real, the corresponding natural mode will be $K_i e^{p_i t}$, and the closer this real pole approaches the origin of the pole-zero diagram, the slower will be the exponential decay (or growth) of the natural mode (see Fig. 6-14).

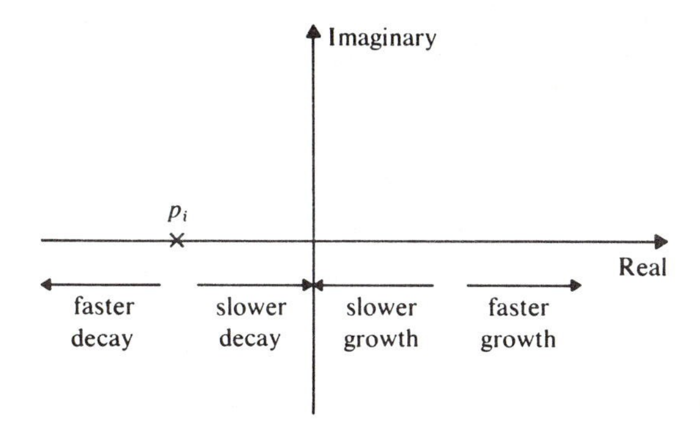


**Fig. 6-14 Natural mode $K_i e^{p_i t}$**

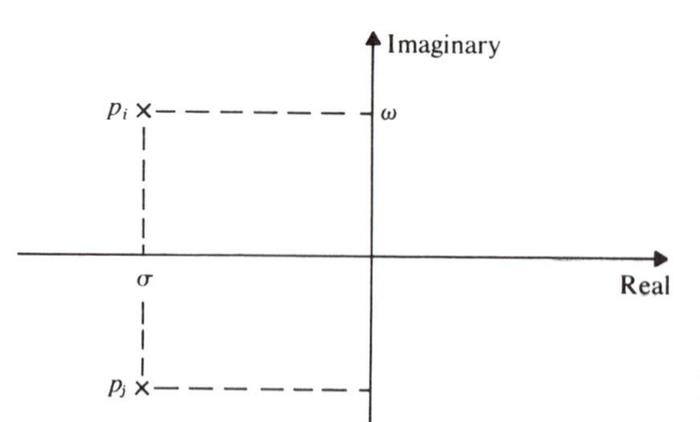


**Fig. 6-15 Natural Mode $2|K_i|e^{\sigma t}\cos(\omega t + \underline{/K_i})$**

In the case of a pair of complex conjugate poles, $p_{i,j} = \sigma \pm j\omega$, the corresponding natural mode will be $2|K_i|e^{\sigma t}\cos(\omega t + \underline{/K_i})$, for which the pole-zero geometry is shown in Fig. 6-15. As for a real pole, the speed of response will decrease as the distance of $p_i$ from the origin decreases.

## GRAPHICAL EVALUATION OF PARTIAL FRACTION CONSTANTS

The amplitude of each natural mode (i.e. the value of each partial fraction constant $K_i$) depends upon the poles, the zeros, and the scale factor $K$.

From Chapter 5 we know that the partial-fraction constants are given by

$$K_i = \left.\frac{K(s - z_1)(s - z_2)\cdots(s - z_n)\cancel{(s - p_i)}}{(s - p_1)(s - p_2)\cdots\cancel{(s - p_i)}\cdots(s - p_m)}\right|_{s=p_i}$$

$$= \frac{K(p_i - z_1)(p_i - z_2)\cdots(p_i - z_n)}{(p_i - p_1)(p_i - p_2)\cdots(p_i - p_m)} = \frac{KN_1e^{j\psi_1}N_2e^{j\psi_2}\cdots N_ne^{j\psi_n}}{D_1e^{j\phi_1}D_2e^{j\phi_2}\cdots D_me^{j\phi_m}}$$

$$= \frac{KN_1N_2\cdots N_n}{D_1D_2\cdots D_m}e^{j(\psi_1+\psi_2+\cdots+\psi_n-\phi_1-\phi_2-\cdots-\phi_m)}$$

Now the poles and zeros of $X(s)$ may be plotted as in Fig. 6-16*a*, where we can identify the typical quantities $p_i - p_j$ and $p_i - z_k$. And, as in Fig. 6-16*b*, we can also identify the corresponding magnitudes and arguments $D_j$, $N_k$, $\phi_j$, and $\psi_k$.

The following procedure may thus be established:

> To find $K_i$, join all the finite poles and zeros *to* $p_i$. Measure (or calculate, trigonometrically) all of the $N$'s, $D$'s, $\psi$'s, and $\phi$'s. Then compute $K_i = |K_i|e^{j\underline{/K_i}}$ from $|K_i| = (KN_1N_2\cdots N_n)/(D_1D_2\cdots D_m)$ and $\underline{/K_i} = \psi_1 + \psi_2 + \cdots + \psi_n - (\phi_1 + \phi_2 + \cdots + \phi_m)$.

This method for evaluating $K_i$ is a useful alternative to the ordinary algebraic method which was used in Chapter 5. It is not possible to obtain all of the $K_i$ graphically if there are repeated poles.

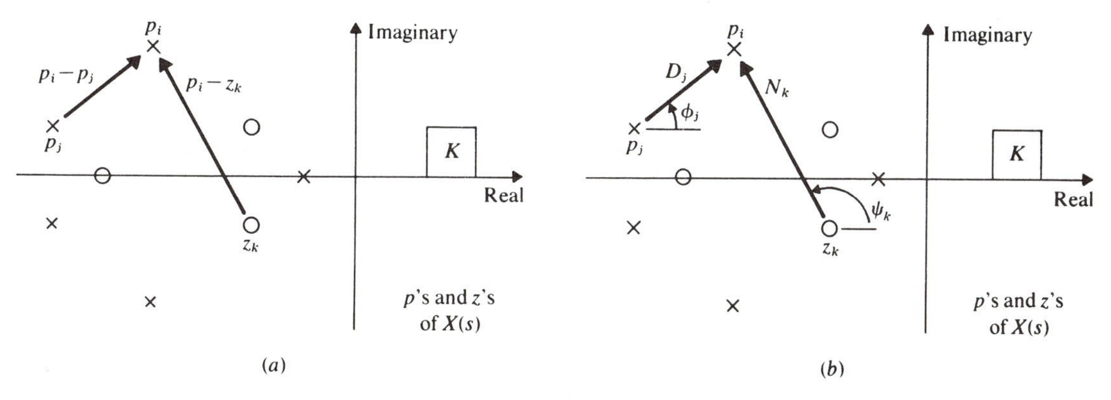


**Fig. 6-16**

## Example 6.3

Given a circuit whose input and output are related by $H(s) = Y(s)/U(s) = 5s/(s + 2)(s^2 + 2s + 2)$, we will find the *impulse response* (the output due to a unit impulse) with the aid of a pole-zero diagram.

The transform of the output $y(t)$ is

$$Y(s) = H(s)U(s) = \frac{5s}{(s+2)(s^2+2s+2)} \equiv \frac{K_1}{s+2} + \frac{K_2}{s+1-j1} + \frac{K_3}{s+1+j1}$$

since $U(s) = \mathscr{L}[\delta(t)] = 1$ and $s^2 + 2s + 2 = (s + 1 - j1)(s + 1 + j1)$.

The poles and zeros of $Y(s)$ are plotted in Fig. 6-17*a*, together with the vectors needed to find $K_1$. With reference to this figure and the procedure just established,

$$|K_1| = \frac{(5)(2)}{(\sqrt{2})(\sqrt{2})} = 5 \quad \text{and} \quad \underline{/K_1} = 180° - (-135° + 135°) = 180°$$

and so $K_1 = 5e^{j180°} = -5$. Similarly, with reference to Fig. 6-17*b*,

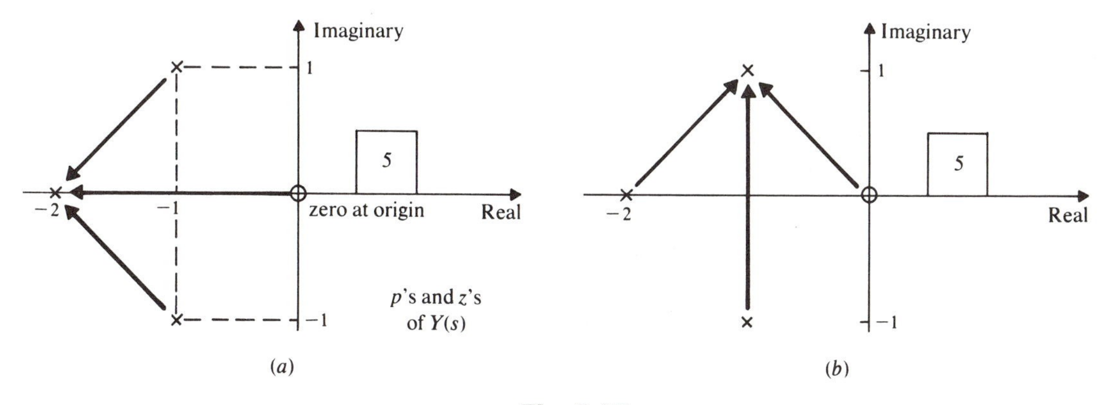


**Fig. 6-17**

$$|K_2| = \frac{(5)(\sqrt{2})}{(\sqrt{2})(2)} = \frac{5}{2} \quad \text{and} \quad \underline{/K_2} = 135° - (45° + 90°) = 0°$$

resulting in $K_2 = 2.5$ and $K_3 = K_2^* = 2.5$.

Completing the normal procedure of Chapter 5, the inverse transform is

$$y(t) = -5e^{-2t} + 5e^{-t}\cos t, \quad t \geq 0$$

■

## STABILITY

A linear circuit is said to be *stable* if all of its characteristic roots are either real and negative or complex with negative real parts. That is, all the poles of the network function must lie to the *left* of the imaginary axis in the pole-zero diagram. If any poles lie to the right of the imaginary axis, the corresponding transient terms in the solution will grow with time and the network would be called *unstable*. If nonrepeated poles lie *on* the imaginary axis, the network is *marginally stable*. (*Repeated* poles *on* the imaginary axis give rise to unstable behavior.)

A polynomial $D(s) = 0$ with all roots in the left-half plane is called a Hurwitz polynomial. A necessary condition is that all of the coefficients of $D(s)$ be greater than zero. However, this is not a sufficient condition, and a number of methods have been devised to test a polynomial without the labor of finding the roots of $D(s) = 0$. One of these is the Routh test, which now follows.

We write $D(s) = a_0 s^n + b_0 s^{n-1} + a_1 s^{n-2} + b_1 s^{n-3} + \cdots$ and set up the array

$$\begin{array}{cccc} a_0 & a_1 & a_2 & \cdots \\ b_0 & b_1 & b_2 & \cdots \\ c_0 & c_1 & c_2 & \cdots \\ d_0 & d_1 & d_2 & \cdots \\ \cdots & \cdots & \cdots & \cdots \end{array}$$

where

$$c_0 = \frac{b_0 a_1 - a_0 b_1}{b_0}, \qquad c_1 = \frac{b_0 a_2 - a_0 b_2}{b_0}, \qquad c_2 = \frac{b_0 a_3 - a_0 b_3}{b_0}, \qquad \cdots$$

and where each succeeding row is obtained (like the $c$'s) from the two rows preceding it. Zeros are used to complete any row shorter than the row of $a$'s. This procedure must be continued to and including the $(n + 1)$th row.

$D(s)$ will be a Hurwitz polynomial provided that all elements in the first column are greater than zero. And if the denominator of a network function is a Hurwitz polynomial, the network will be stable.

The Routh test can give additional information. For example, if the elements change sign $r$ times as we read down the first column, then there will be $r$ roots of $D(s) = 0$ in the right-half plane.

# ILLUSTRATIVE PROBLEMS

### *Network Functions*

**6.1** In a certain electric circuit it is known that

$$\frac{I(s)}{V(s)} = Y(s) = \frac{s^2 + 2s + 3}{4s^3 + 5s^2 + 6s + 7}$$

Find the differential equation relating $v(t)$ and $i(t)$.

First, $Y(s)V(s) = I(s)$ or

$$\{4s^3 + 5s^2 + 6s + 7\}\, I(s) = \{s^2 + 2s + 3\}\, V(s)$$

Recalling that $\mathscr{L}[d^n x(t)/dt^n] = s^n X(s)$ + initial condition terms, the differential equation must be

$$4\frac{d^3 i}{dt^3} + 5\frac{d^2 i}{dt^2} + 6\frac{di}{dt} + 7i = \frac{d^2 v}{dt^2} + 2\frac{dv}{dt} + 3v$$

**6.2** Use series and parallel combinations to determine the impedance $Z(s) = V(s)/I(s)$ in the circuit of Fig. 6-18.

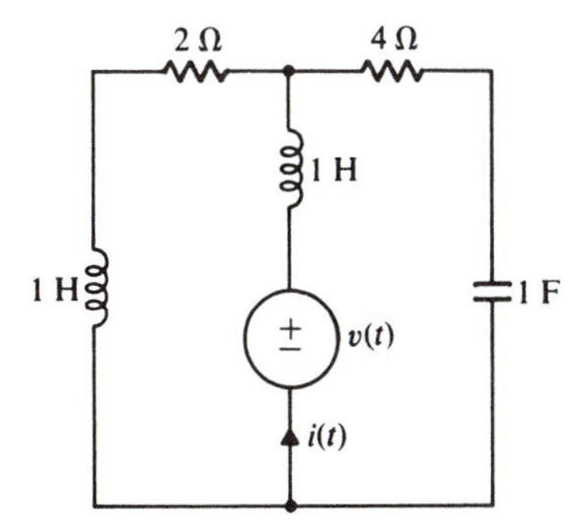


**Fig. 6-18**

In the transform circuit diagram of Fig. 6-19*a*, in which all initial conditions have been set to zero, the elements in series have been replaced by equivalent impedances. Next it can be seen that the impedances $s + 2$ and $4 + 1/s$ are in parallel. They may therefore be replaced by the equivalent impedance

$$\frac{1}{Z_{eq}(s)} = \frac{1}{s+2} + \frac{1}{4 + 1/s} = \frac{4s + 1 + s^2 + 2s}{(s+2)(4s+1)}$$

or, as in Fig. 6-19*b*,

$$Z_{eq}(s) = \frac{4s^2 + 9s + 2}{s^2 + 6s + 1}$$

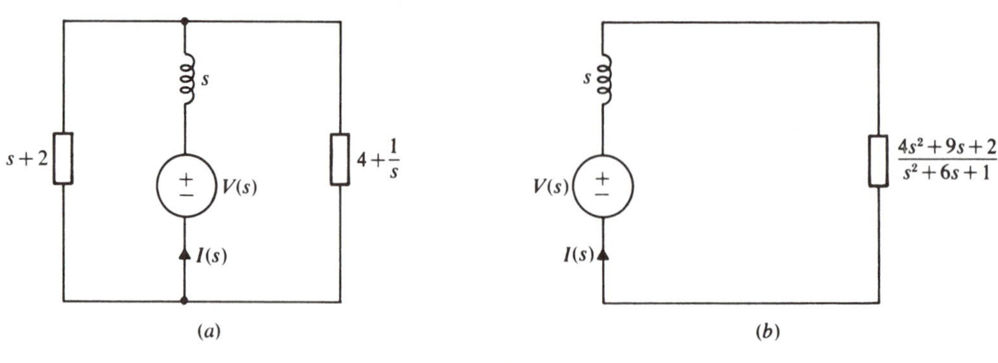


**Fig. 6-19**

Finally, combining $(4s^2 + 9s + 2)/(s^2 + 6s + 1)$ and $s$ in series,

$$\frac{V(s)}{I(s)} = Z(s) = s + \frac{4s^2 + 9s + 2}{s^2 + 6s + 1} = \frac{s^3 + 10s^2 + 10s + 2}{s^2 + 6s + 1}$$

*Comment*

We can think of this result in one-port terms. $Z(s)$ is the Thévenin impedance seen by the transform source $V(s)$.

**6.3** Show that the two one-port networks of Fig. 6-20 are equivalent as far as their input terminals are concerned if $R = \sqrt{L/C}$.

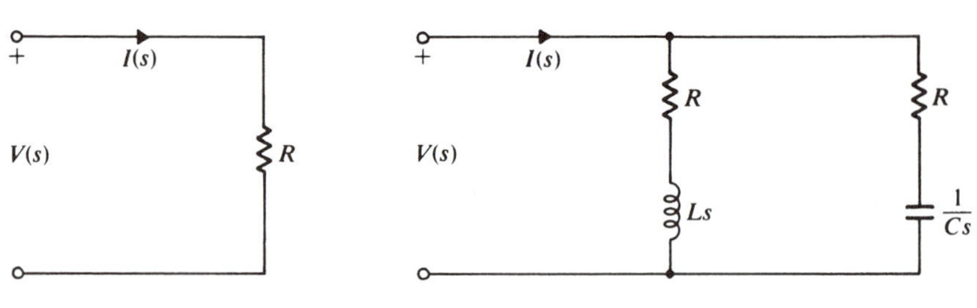


**Fig. 6-20**

In Fig. 6-20*a*,

$$\frac{V(s)}{I(s)} = Z(s) = R$$

while in Fig. 6-20*b*,

$$\frac{V(s)}{I(s)} = \frac{1}{1/Z_1(s) + 1/Z_2(s)} = \frac{1}{1/(R + Ls) + 1/(R + 1/Cs)}$$

$$= \frac{(RCs + 1)(R + Ls)}{RCs + 1 + Cs(R + Ls)} = \frac{RLCs^2 + (L + R^2C)s + R}{LCs^2 + 2RCs + 1}$$

Substituting $\sqrt{L/C}$ for $R$,

$$\frac{V(s)}{I(s)} = \frac{\sqrt{L/C}(LCs^2 + 2\sqrt{LC}s + 1)}{LCs^2 + 2\sqrt{LC}s + 1} = \sqrt{L/C} = R$$

Thus the two networks have the same $V(s)$-$I(s)$ relationship and are therefore equivalent as seen from outside their terminals.

**6.4** Show that the two circuits in Fig. 6-21 have identical terminal characteristics.

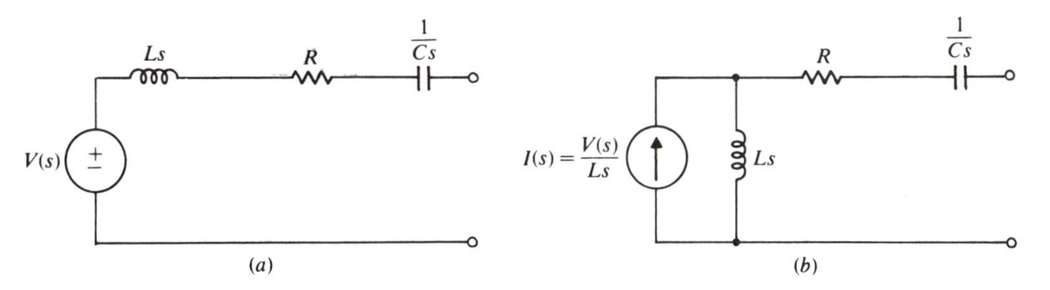


**Fig. 6-21**

We are to show that the two one ports have the same terminal $V(s)$-$I(s)$ relations. Therefore we connect an arbitrary load to each of the circuits as shown in Fig. 6-22.

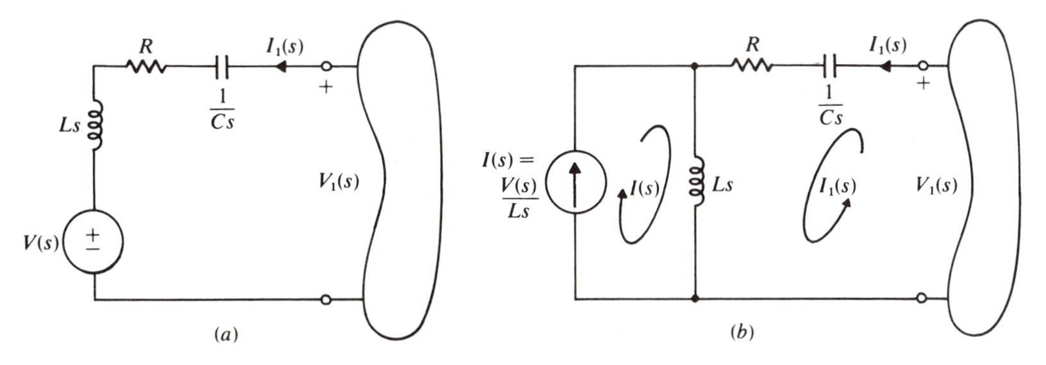


**Fig. 6-22**

In Fig. 6-22$a$, KVL yields

$$V_1(s) = V(s) + \left(Ls + R + \frac{1}{Cs}\right)I_1(s)$$

while in Fig. 6-22$b$,

$$V_1(s) = Ls\{I(s) + I_1(s)\} + \left(R + \frac{1}{Cs}\right) I_1(s)$$

$$= Ls\frac{V(s)}{Ls} + \left(Ls + R + \frac{1}{Cs}\right) I_1(s)$$

$$= V(s) + \left(Ls + R + \frac{1}{Cs}\right) I_1(s)$$

Thus the two circuits have identical terminal characterisitics.

*Alternative Solution*

Apply the source transformation of Fig. 6-2 to the series combination of $V(s)$ and $Ls$ in Fig. 6-21*a*. This results in the parallel combination of $I(s) = V(s)/Ls$ and $Ls$ as in Fig. 6-21*b*.

**6.5** An automobile battery has been modeled by the circuit of Fig. 6-23*a*. Show that the circuit of Fig. 6-23*b* has equivalent terminal characterisitics.

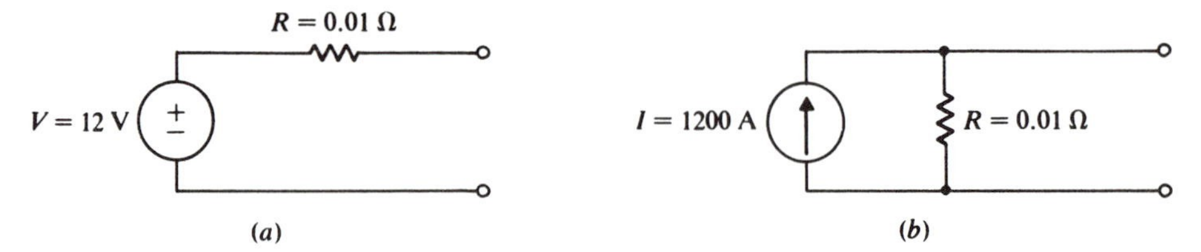


**Fig. 6-23**

The result follows at once from the source transformation of Fig. 6-2. Rather obviously these two circuits do *not* have similar *internal* characteristics!

Alternatively, we could use the first method of Problem 6.4.

**6.6** Show that $I_3(s) = Y_3(s)I(s)/\{Y_1(s) + Y_2(s) + Y_3(s)\}$ in Fig. 6-24.

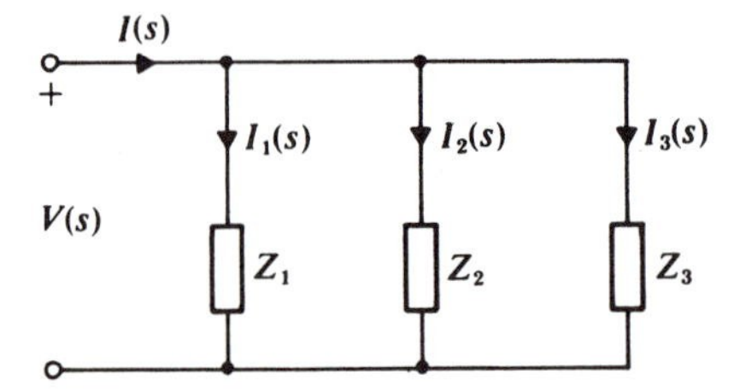


**Fig. 6-24**

Applying KCL, $I_1(s) + I_2(s) + I_3(s) = I(s)$ or

$$Y_1(s)V(s) + Y_2(s)V(s) + Y_3(s)V(s) = I(s)$$

But $I_3(s) = Y_3(s)V(s)$, and dividing $I_3(s)$ by $I(s)$,

$$\frac{I_3(s)}{I(s)} = \frac{Y_3(s)V(s)}{\{Y_1(s) + Y_2(s) + Y_3(s)\}V(s)}$$

which proves the proposition.

This is the current divider relation for the case of three parallel paths.

**6.7** Find the transfer function $H(s) = V(s)/E(s)$ for the network of Fig. 6-25.

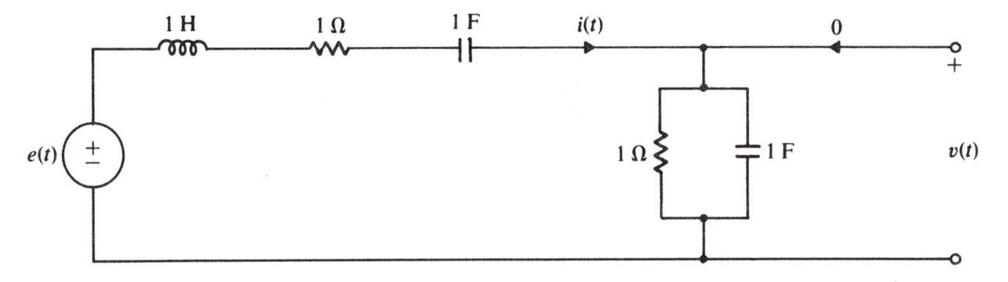


**Fig. 6-25**

The transformed diagram, which is shown in Fig. 6.26*a* with zero initial conditions, can be simplified as in Fig. 6-26*b*. From the latter figure, $V(s) = Z_2(s)I(s)$ and $I(s) = E(s)/\{Z_1(s) + Z_2(s)\}$. Therefore

$$\frac{V(s)}{E(s)} = \frac{Z_2(s)}{Z_1(s) + Z_2(s)}$$

which proves the general voltage divider relationship of equation (*6.9*)! Substituting for $Z_1(s)$ and $Z_2(s)$,

$$\frac{V(s)}{E(s)} = \frac{1/(s+1)}{s + 1 + 1/s + 1/(s+1)} = \frac{s}{s^3 + 2s^2 + 3s + 1}$$

a result which is valid only when the output terminals are open circuit.

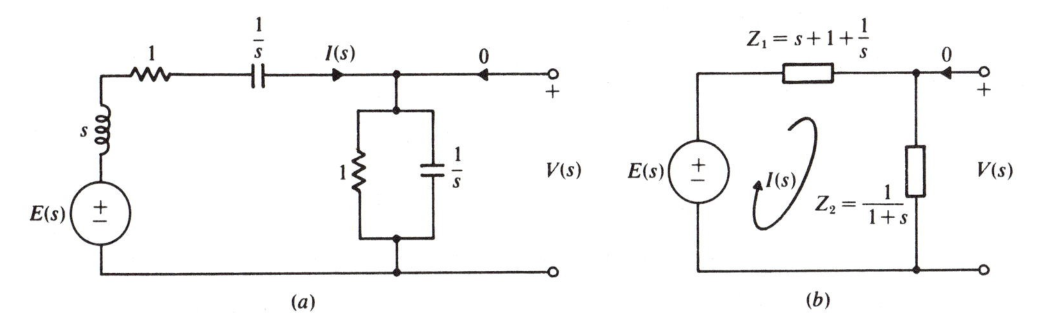


**Fig. 6-26**

**6.8** For the network of Problem 6.7 (Fig. 6-25), find $V(s)$ when: $e(t) = 5 \cos 10t$; $e(t) = 5t$; and $e(t) = 5$.

From Problem 6.7,

$$\frac{V(s)}{E(s)} = \frac{s}{s^3 + 2s^2 + 3s + 1}$$

and so

$$V(s) = \frac{s}{s^3 + 2s^2 + 3s + 1} E(s)$$

If we now look up $E(s)$ in Table 5-1, for each of the three cases,

$$V(s) = \left(\frac{s}{s^3 + 2s^2 + 3s + 1}\right)\left(\frac{5s}{s^2 + 10^2}\right), \quad \left(\frac{s}{s^3 + 2s^2 + 3s + 1}\right)\left(\frac{5}{s^2}\right),$$

$$\text{and} \quad \left(\frac{s}{s^3 + 2s^2 + 3s + 1}\right)\left(\frac{5}{s}\right)$$

**6.9** Find the differential equation relating $v(t)$ to $e(t)$ in the circuit of Problem 6.7 (see Fig. 6-25).

From Problem 6.7,

$$\frac{V(s)}{E(s)} = \frac{s}{s^3 + 2s^2 + 3s + 1}$$

whence, proceeding as in Problem 6.1,

$$\dddot{v} + 2\ddot{v} + 3\dot{v} + v = \dot{e}$$

**6.10** Find the transfer function $H(s) = V(s)/I(s)$ for the network of Fig. 6-27. Hint: First transform the current source to a voltage source and then use the voltage divider idea.

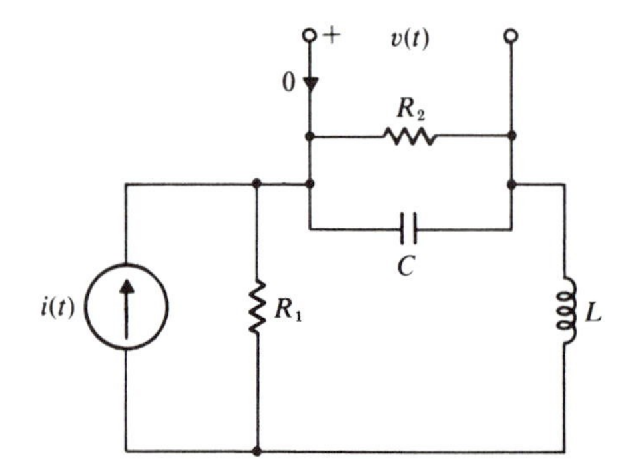


**Fig. 6-27**

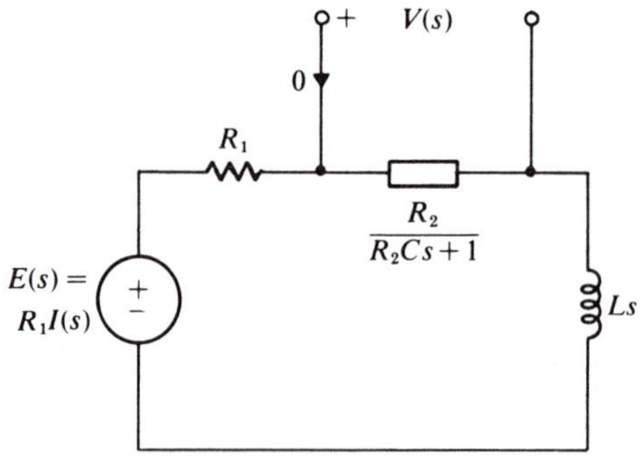

**Fig. 6-28**

In the transform diagram of Fig. 6-28, the series combination of $E(s) = R_1I(s)$ and $R_1$ replaces the parallel combination of $I(s)$ and $R_1$. And $R_2/(R_2Cs + 1)$ is equivalent to $R_2$ and $1/Cs$ in parallel. Then, using the voltage divider concept of equation (*6.9*),

$$\frac{V(s)}{E(s)} = \frac{V(s)}{R_1I(s)} = \frac{R_2/(R_2Cs + 1)}{R_2/(R_2Cs + 1) + R_1 + Ls}$$

or

$$\frac{V(s)}{I(s)} = \frac{R_1R_2}{R_2LCs^2 + (R_1R_2C + L)s + R_1 + R_2}$$

Remember that when we obtain a network function we are relating the output *to the input that caused it*. Consequently we must automatically set all initial conditions and all other independent sources to zero. This is implicit in Fig. 6-28, for example.

**6.11** If $i_1(t)$ is the component of $i(t)$ due to $f(t)$ in the circuit of Fig. 6-29, find the transfer function $H_1(s) = I_1(s)/F(s)$.

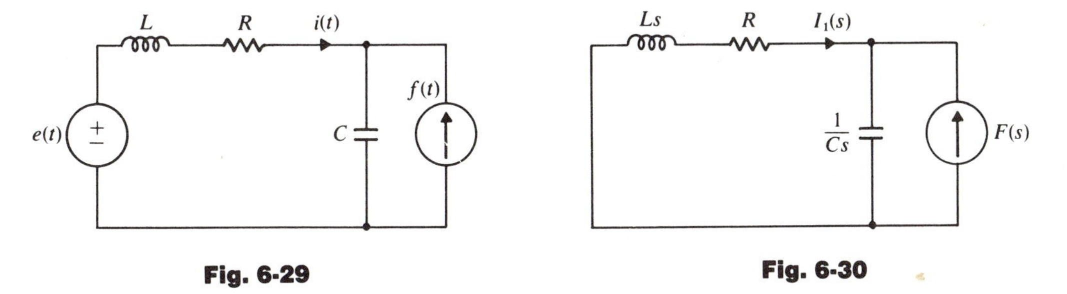


**Fig. 6-29** **Fig. 6-30**

After setting $e(t)$ and the two initial conditions to zero, since we want the current due to $f(t)$, the $s$-domain diagram of Fig. 6-30 results. Then using the current divider rationale of equation (*6.10*),

$$\frac{I_1(s)}{F(s)} = -\frac{1/(Ls + R)}{1/(Ls + R) + Cs} = -\frac{1}{LCs^2 + RCs + 1}$$

where the minus sign is due to the reference direction of $i(t)$.

**6.12** Synthesize a simple series network having a driving point impedance $Z(s) = 2s + 3 + 4/s$.

See Fig. 6-31.

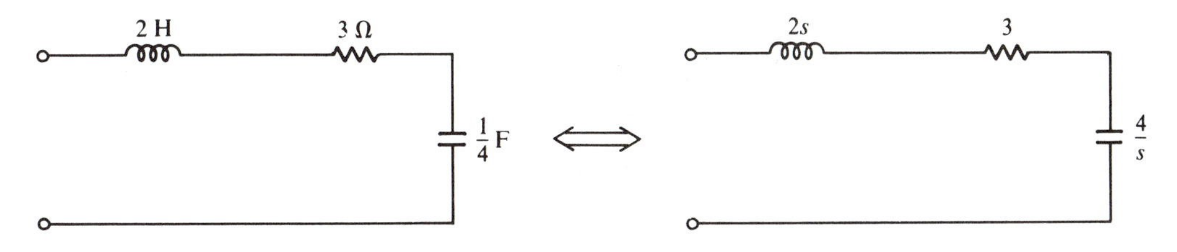


**Fig. 6-31**

**6.13** Determine the driving point impedance at the left-hand terminals of the network in Fig. 6-32 if

(*a*) the right-hand terminals are open circuit,

(*b*) the right-hand terminals are short circuit, and

(*c*) the right-hand terminals are terminated in a 2-Ω resistor.

(*a*) The transform diagram, with the output open circuit, is shown in Fig. 6-33*a*. Here

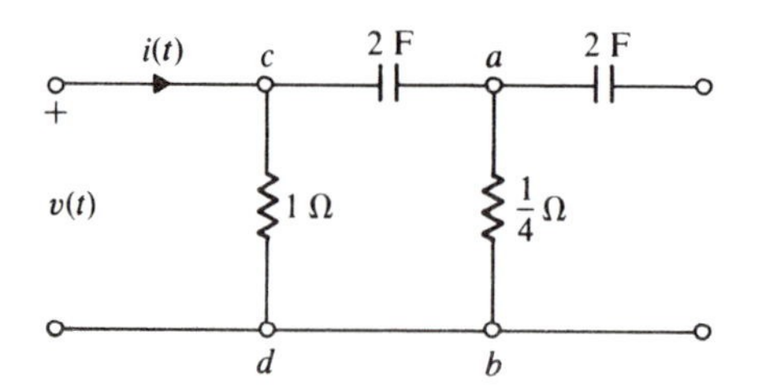


Fig. 6-32

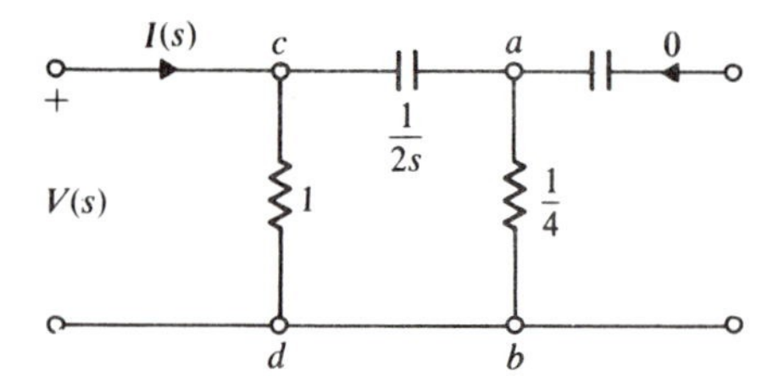


Fig. 6-33a

$$Z_{cab}(s) = \frac{1}{4} + \frac{1}{2s} = \frac{s+2}{4s}$$

Therefore

$$\frac{1}{Z(s)} = \frac{1}{Z_{cab}(s)} + \frac{1}{1} = \frac{4s}{s+2} + 1 = \frac{5s+2}{s+2}$$

or

$$Z(s) = \frac{s+2}{5s+2}$$

(*b*) See Fig. 6-33*b* for the short-circuit condition. In this case

$$\frac{1}{Z_{ab}(s)} = 4 + 2s, \quad Z_{cab}(s) = \frac{1}{2s} + \frac{1}{2s+4} = \frac{s+1}{s(s+2)}$$

and

$$\frac{1}{Z(s)} = 1 + \frac{s(s+2)}{s+1} = \frac{s^2+3s+1}{s+1} \quad \text{or} \quad Z(s) = \frac{s+1}{s^2+3s+1}$$

(*c*) For the terminating resistance of 2Ω, see Fig. 6-33*c*. Here

$$\frac{1}{Z_{ab}(s)} = 4 + \frac{1}{2+1/2s} = \frac{18s+4}{4s+1}, \quad Z_{cab}(s) = \frac{1}{2s} + \frac{4s+1}{18s+4} = \frac{8s^2+20s+4}{36s^2+8s},$$

and

$$\frac{1}{Z(s)} = 1 + \frac{36s^2+8s}{8s^2+20s+4} \quad \text{or} \quad Z(s) = \frac{2s^2+5s+1}{11s^2+7s+1}$$

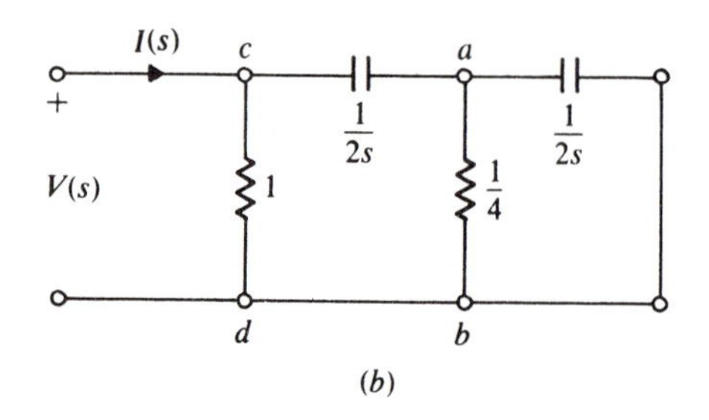


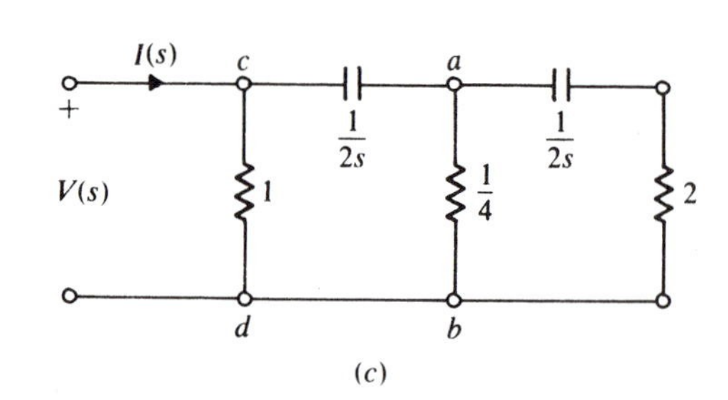


Fig. 6-33b,c

## Loop and Nodal Equations

**6.14** The circuit of Fig. 6-34 is to be analyzed by the loop method from a transform circuit diagram in which sources are to represent the initial conditions. As the first step, draw the $s$-domain diagram most suitable for loop analysis.

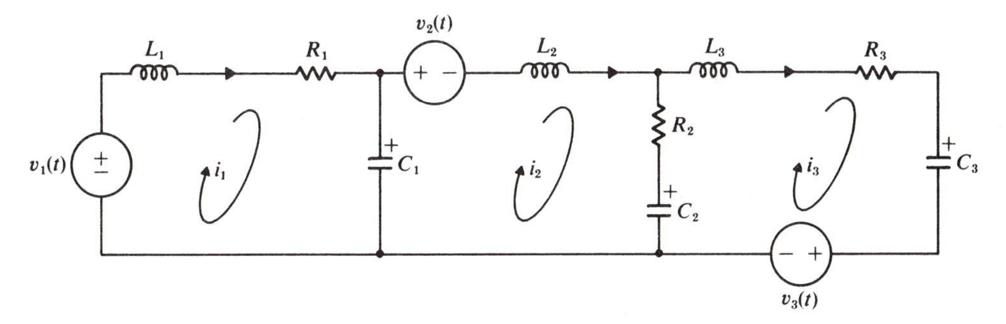


**Fig. 6-34**

See Fig. 6-35. Note that the initial conditions have been replaced by equivalent *series voltage* sources (see Fig. 5-3), thereby minimizing the number of meshes in the transform diagram and ensuring that there are no *current* sources.

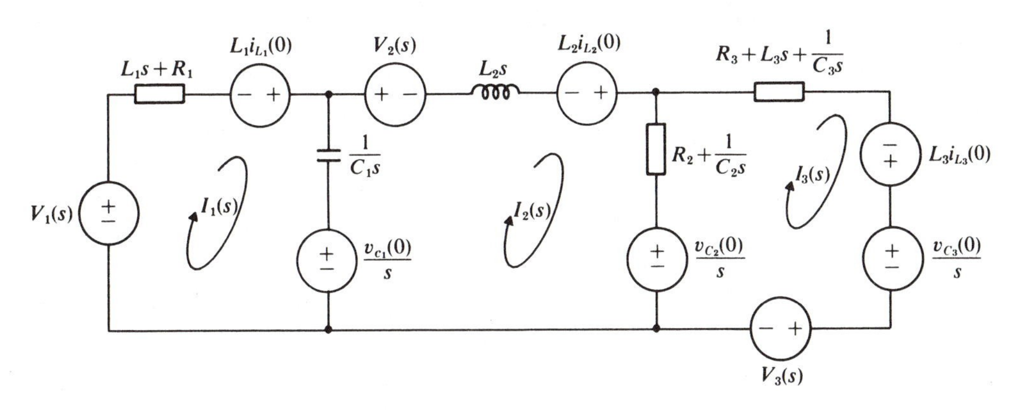


**Fig. 6-35**

**6.15** A matrix equation of the form $[Z][I(s)] = [V(s)]$ is to be written for the circuit of Problem 6.14. Write the *diagonal* elements of $[Z]$ by inspection of Fig. 6-35.

The network of Fig. 6-35 is connected, planar, and without current sources, dependent sources, or mutual inductance. And the loop currents are around the meshes, all in a clockwise direction. We may therefore use the technique described for equation (*6.4*) to write down

$$Z_{11} = L_1 s + R_1 + \frac{1}{C_1 s} \quad \text{(the sum of the impedances around mesh 1 in Fig. 6-35)}$$

$$Z_{22} = L_2 s + R_2 + \frac{1}{C_1 s} + \frac{1}{C_2 s}$$

$$Z_{33} = L_3 s + R_2 + R_3 + \frac{1}{C_2 s} + \frac{1}{C_3 s}$$

**6.16** Continue Problem 6.15 by finding the remaining elements in [Z]. (See Fig. 6-35.)

Following once more the procedure specified for equation (*6.4*), we can write down by inspection,

$$Z_{12} = Z_{21} = -\frac{1}{C_1 s} \quad \text{(minus the impedance common to meshes 1 and 2 in Fig. 6-35)}$$

$$Z_{23} = Z_{32} = -\left(R_2 + \frac{1}{C_2 s}\right)$$

$$Z_{31} = Z_{13} = 0 \quad \text{(no common impedance)}$$

**6.17** Using the results of Problems 6.15 and 6.16, write the complete matrix loop equation corresponding to Fig. 6-35.

$$\begin{bmatrix} L_1 s + R_1 + \frac{1}{C_1 s} & -\frac{1}{C_1 s} & 0 \\ -\frac{1}{C_1 s} & L_2 s + R_2 + \frac{1}{C_1 s} + \frac{1}{C_2 s} & -R_2 - \frac{1}{C_2 s} \\ 0 & -R_2 - \frac{1}{C_2 s} & L_3 s + R_2 + R_3 + \frac{1}{C_2 s} + \frac{1}{C_3 s} \end{bmatrix} \begin{bmatrix} I_1(s) \\ I_2(s) \\ I_3(s) \end{bmatrix}$$

$$= \begin{bmatrix} V_1(s) + L_1 i_{L_1}(0) - \frac{v_{C_1}(0)}{s} \\ -V_2(s) + L_2 i_{L_2}(0) - \frac{v_{C_2}(0)}{s} + \frac{v_{C_1}(0)}{s} \\ L_3 i_{L_3}(0) - \frac{v_{C_3}(0)}{s} - V_3(s) + \frac{v_{C_2}(0)}{s} \end{bmatrix}$$

$[V(s)]$ was obtained by summing the transformed source voltage *rises* around each mesh, traveling in the direction of the mesh currents.

**6.18** A set of matrix nodal equations $[Y][V(s)] = [I(s)]$ is to be written for the

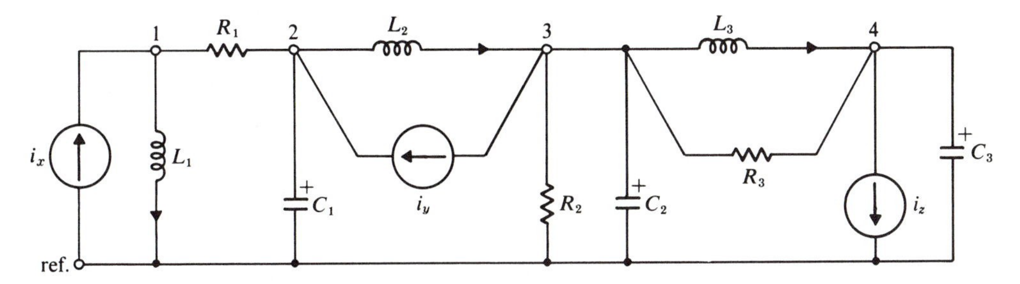


**Fig. 6-36**

circuit of Fig. 6-36. First draw the appropriate $s$-domain diagram, including all initial conditions as sources, and then write the matrix equation.

This time we will replace the initial conditions by *parallel current* sources in order to minimize the number of nodes and to ensure that there are no *voltage* sources. This has been done, with reference to Fig. 5-3, in Fig. 6.37.

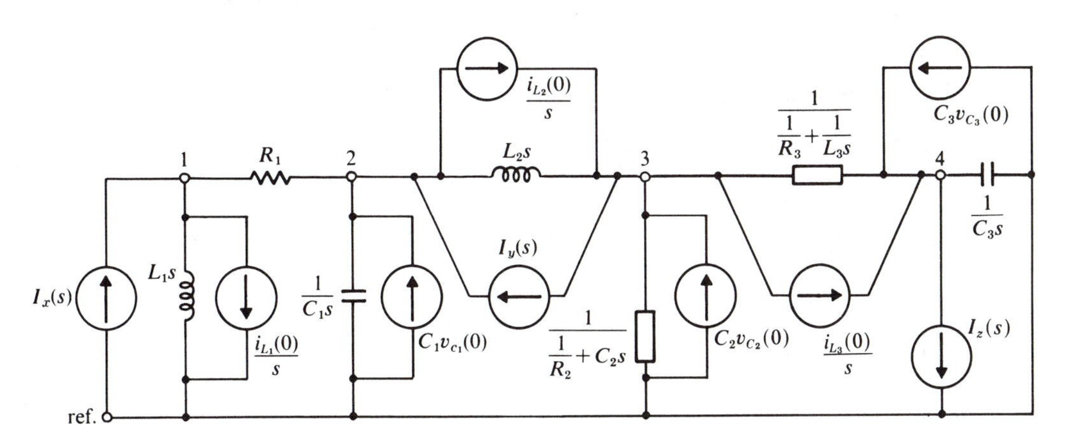


**Fig. 6-37**

Now, following the procedure specified for equation (*6.5*),

$$\begin{bmatrix} \frac{1}{L_1 s}+\frac{1}{R_1} & -\frac{1}{R_1} & 0 & 0 \\ -\frac{1}{R_1} & \frac{1}{R_1}+C_1 s+\frac{1}{L_2 s} & -\frac{1}{L_2 s} & 0 \\ 0 & -\frac{1}{L_2 s} & \frac{1}{R_2}+\frac{1}{R_3}+C_2 s+\frac{1}{L_2 s}+\frac{1}{L_3 s} & -\frac{1}{R_3}-\frac{1}{L_3 s} \\ 0 & 0 & -\frac{1}{R_3}-\frac{1}{L_3 s} & C_3 s+\frac{1}{R_3}+\frac{1}{L_3 s} \end{bmatrix} \begin{bmatrix} V_1(s) \\ V_2(s) \\ V_3(s) \\ V_4(s) \end{bmatrix}$$

$$= \begin{bmatrix} I_x(s) - \dfrac{i_{L_1}(0)}{s} \\ C_1 v_{C_1}(0) + I_y(s) - \dfrac{i_{L_2}(0)}{s} \\ \dfrac{i_{L_2}(0)}{s} - I_y(s) + C_2 v_{C_2}(0) - \dfrac{i_{L_3}(0)}{s} \\ \dfrac{i_{L_3}(0)}{s} + C_3 v_{C_3}(0) - I_z(s) \end{bmatrix}$$

**6.19** The network of Fig. 6-38 is to be analyzed using *loop* methods in the transform domain. Before we can write a set of equations by inspection we must eliminate the current source. To this end, convert everything to the right of the terminals *d*-*g* into a voltage source and series impedance in the *s*-domain.

Moving into the transform domain, we first obtain the circuit of Fig. 6-39*a*. Two source transformations in succession take us through Fig. 6-39*b* to Fig. 6-39c.

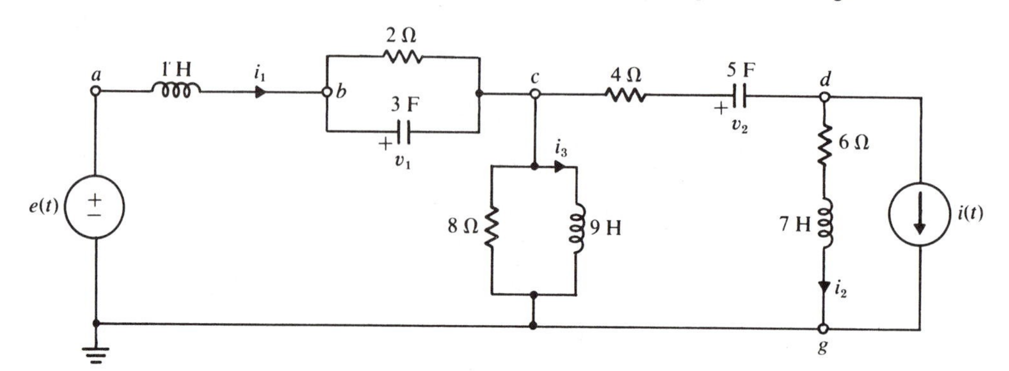


**Fig. 6-38**

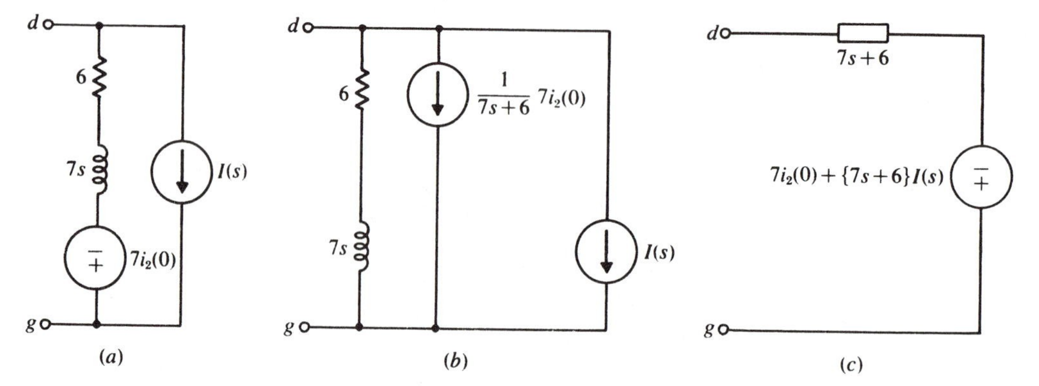


**Fig. 6-39**

**6.20** Now draw separate transform circuit diagrams for the networks between the terminals $b$-$c$ and $c$-$g$ in the circuit of Fig. 6-38. Manipulate these diagrams into the form most suited for loop analysis.

The network $b$-$c$ takes the form of Fig. 6-40$a$ in the $s$-domain. A source transformation then leads to Fig. 6-40$b$.

Similarly the network $c$-$g$ can be represented first by Fig. 6-40$c$, then by Fig. 6-40$d$.

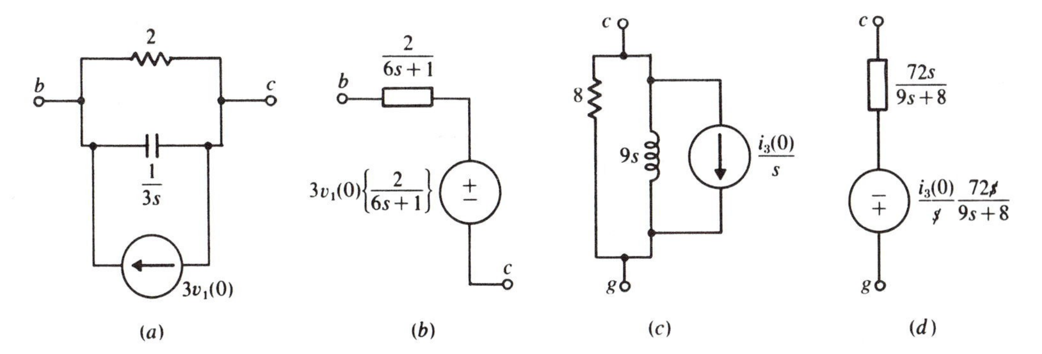


**Fig. 6-40**

**6.21** Finally, draw the transform circuit diagram for the entire circuit of Fig. 6-38, making use of the equivalent circuits produced in Problems 6.19 and 6.20. Then write the matrix equation $[Z][I(s)] = [V(s)]$.

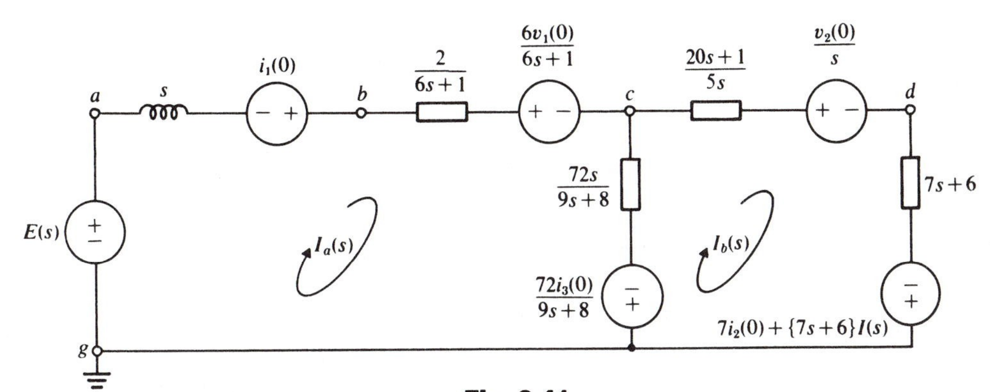


**Fig. 6-41**

The complete $s$-domain circuit is shown in Fig. 6-41, together with the clockwise mesh currents $I_a(s)$ and $I_b(s)$. Then following the procedure associated with equation (6.4), we can write, by inspection,

$$\begin{bmatrix} s + \dfrac{2}{6s+1} + \dfrac{72s}{9s+8} & \dfrac{-72s}{9s+8} \\ \dfrac{-72s}{9s+8} & 10 + \dfrac{1}{5s} + 7s + \dfrac{72s}{9s+8} \end{bmatrix} \begin{bmatrix} I_a(s) \\ I_b(s) \end{bmatrix}$$

$$= \begin{bmatrix} E(s) + i_1(0) - \dfrac{6v_1(0)}{6s+1} + \dfrac{72i_3(0)}{9s+8} \\ -\dfrac{72i_3(0)}{9s+8} - \dfrac{v_2(0)}{s} + 7i_2(0) + \{7s+6\}I(s) \end{bmatrix}$$

**6.22** Circuit equations in the nodal form $[Y][V(s)] = [I(s)]$ are to be written for the network of Fig. 6-38. Draw the most suitable transform circuit diagram and write the appropriate matrix equation.

The transformed circuit diagram, set up for nodal analysis, is shown in Fig. 6-42. The procedure for obtaining this circuit follows that of Problems 6.19–6.21. In fact, much of the work done in Problems 6.19 and 6.20 (Figs. 6-39 to 6-41) can be reapplied here.

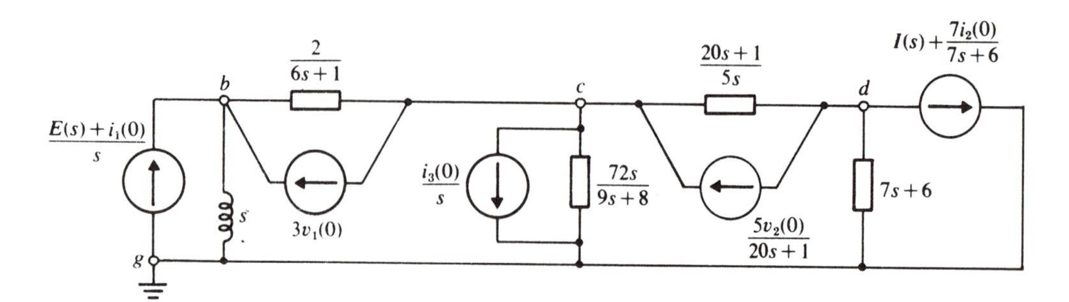


Fig. 6-42

Then from Fig. 6-42, following the procedure associated with equation (6.5),

$$\begin{bmatrix} \dfrac{1}{s} + \dfrac{6s+1}{2} & -\dfrac{6s+1}{2} & 0 \\ -\dfrac{6s+1}{2} & \dfrac{6s+1}{2} + \dfrac{9s+8}{72s} + \dfrac{5s}{20s+1} & \dfrac{-5s}{20s+1} \\ 0 & \dfrac{-5s}{20s+1} & \dfrac{5s}{20s+1} + \dfrac{1}{7s+6} \end{bmatrix} \begin{bmatrix} V_b(s) \\ V_c(s) \\ V_d(s) \end{bmatrix}$$

$$= \begin{bmatrix} \dfrac{E(s) + i_1(0)}{s} + 3v_1(0) \\ -3v_1(0) - \dfrac{i_3(0)}{s} + \dfrac{5v_2(0)}{20s+1} \\ -\dfrac{5v_2(0)}{20s+1} - I(s) - \dfrac{7i_2(0)}{7s+6} \end{bmatrix}$$

### *A Miscellany*

**6.23** Find the network function $H(s) = V(s)/E(s)$ for the *ladder network* of Fig. 6-43.

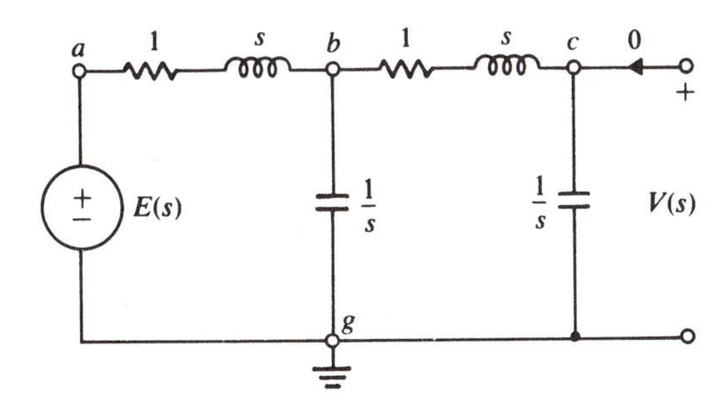


**Fig. 6-43**

A "trick" that we may use with a ladder network is to assume an output quite arbitrarily, and to then trace the consequences of this assumption from the output back to the input.

Let us assume that the transformed output is known and equal to $V(s)$. Then at node $c$ in Fig. 6-43,

$$I_{cg}(s) = \frac{V(s)}{1/s} = sV(s)$$

where $i_{cg}$ is the current *from c to g*.

At node $b$,

$$V_b(s) = V(s) + (1 + s)I_{bc}(s) = V(s) + (1 + s)I_{cg}(s)$$
$$= V(s) + (1 + s)sV(s) = (s^2 + s + 1)V(s)$$

and

$$I_{bg}(s) = \frac{V_b(s)}{1/s} = (s^3 + s^2 + s)V(s)$$

so that

$$I_{ab}(s) = I_{bg}(s) + I_{bc}(s) = (s^3 + s^2 + 2s)V(s)$$

Moving next to node $a$,

$$E(s) = V_b(s) + (1 + s)I_{ab}(s) = (s^2 + s + 1)V(s) + (1 + s)(s^3 + s^2 + 2s)V(s)$$
$$= (s^4 + 2s^3 + 4s^2 + 3s + 1)V(s)$$

Thus the desired network function is $V(s)/E(s) = 1/(s^4 + 2s^3 + 4s^2 + 3s + 1)$.

*Note:* We chose to assume a transformed output $V(s)$. The algebra can be kept a little neater by assuming that $V(s) = 1$. Then $V(s)$ will be replaced everywhere by 1 in the above solution, leading finally to

$$E(s) = s^4 + 2s^3 + 4s^2 + 3s + 1$$

Since $V(s)$ was assumed to be 1, the desired transfer function must be

$$\frac{V(s)}{E(s)} = \frac{1}{E(s)} = \frac{1}{s^4 + 2s^3 + 4s^2 + 3s + 1}$$

### Alternative Solution

After a source transformation we can write the nodal equations by inspection,

$$\begin{bmatrix} \dfrac{2}{s+1} + s & \dfrac{-1}{s+1} \\ \dfrac{-1}{s+1} & s + \dfrac{1}{s+1} \end{bmatrix} \begin{bmatrix} V_b(s) \\ V_c(s) \end{bmatrix} = \begin{bmatrix} \dfrac{E(s)}{s+1} \\ 0 \end{bmatrix}$$

Therefore

$$V_c(s) = V(s) = \frac{\begin{vmatrix} \dfrac{2}{s+1} + s & \dfrac{E(s)}{s+1} \\ \dfrac{-1}{s+1} & 0 \end{vmatrix}}{\begin{vmatrix} \dfrac{2}{s+1} + s & \dfrac{-1}{s+1} \\ \dfrac{-1}{s+1} & s + \dfrac{1}{s+1} \end{vmatrix}}$$

$$= \frac{E(s)/(s+1)^2}{(s^4 + 2s^3 + 4s^2 + 3s + 1)/(s+1)^2} = \frac{E(s)}{s^4 + 2s^3 + 4s^2 + 3s + 1}$$

and so $V(s)/E(s) = 1/(s^4 + 2s^3 + 4s^2 + 3s + 1)$.

*Note:* If the ladder is long, the determinant evaluation would become intolerable. But the first (step-by-step) procedure can be continued quite easily along a longer ladder; there are no simultaneous equations to be solved.

**6.24** Use the step-by-step method of the preceding problem to verify that

$$H_2(s) = \frac{I_v(s)}{I(s)} = \frac{3}{8s^5 + 12s^4 + 16s^3 + 18s^2 + 6s + 3}$$

for the circuit of Fig. 6-44.

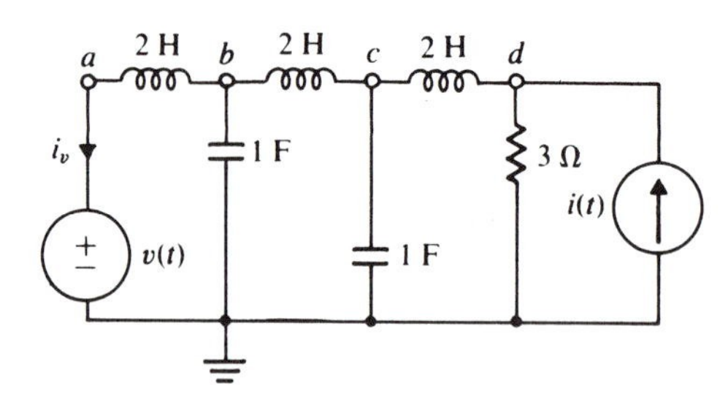


**Fig. 6-44**

Since we want the response due to $i(t)$, we must set $v(t)$ and all the initial conditions to zero, as in Fig. 6-45. And we will assume, arbitrarily, that the transformed output is $I_v(s) = 1$.

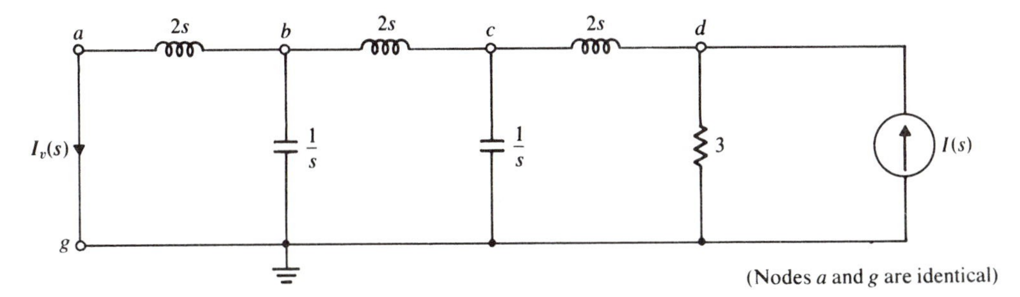


**Fig. 6-45**

At node $b$,

$$V_b(s) = 2sI_v(s) = 2s$$

$$I_{bg}(s) = \frac{V_b(s)}{1/s} = 2s^2$$

and

$$I_{bc}(s) = I_{ab}(s) - I_{bg}(s) = -1 - 2s^2$$

At node $c$,

$$V_c(s) = V_b(s) - 2sI_{bc}(s) = 4s^3 + 4s$$

$$I_{cg}(s) = \frac{V_c(s)}{1/s} = 4s^4 + 4s^2$$

and

$$I_{cd}(s) = I_{bc}(s) - I_{cg}(s) = -4s^4 - 6s^2 - 1$$

At node $d$,

$$V_d(s) = V_c(s) - 2sI_{cd}(s) = 8s^5 + 16s^3 + 6s$$

$$I_{dg}(s) = \frac{V_d(s)}{3} = \frac{8s^5 + 16s^3 + 6s}{3}$$

and

$$I(s) = I_{dg}(s) - I_{cd}(s) = \frac{8s^5 + 16s^3 + 6s}{3} + 4s^4 + 6s^2 + 1$$

$$= \frac{8s^5 + 12s^4 + 16s^3 + 18s^2 + 6s + 3}{3}$$

Finally, since we assumed that $I_v(s) = 1$,

$$H_2(s) = \frac{I_v(s)}{I(s)} = \frac{1}{I(s)} = \frac{3}{8s^5 + 12s^4 + 16s^3 + 18s^2 + 6s + 3}$$

**6.25** The circuit of Fig. 6-46 models a field effect transistor (FET) amplifier. The nodes $G$, $D$, and $S$ mark the transistor's gate, drain, and source terminals. Use nodal analysis to find the $s$-domain Thévenin equivalent of the one port seen by the load. Hence find $V_{R_L}(s)$ as a function of $R_L$. Both capacitor voltages are initially zero.

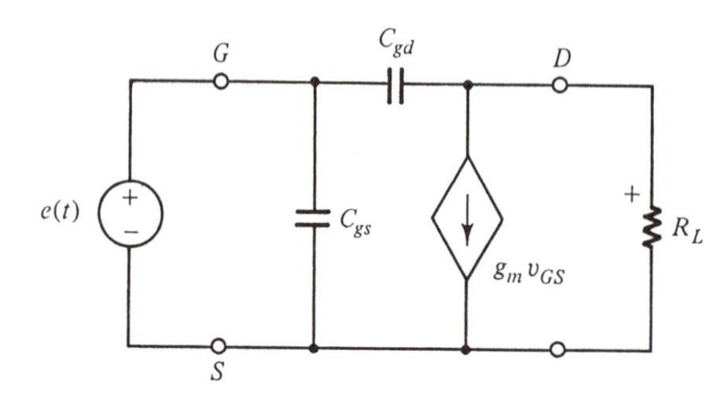


**Fig. 6-46**

Treating $S$ as the reference node in the $s$-domain circuit of Fig. 6-47$a$, KCL at node $D$ yields

$$C_{gd}sV(s) = -I(s) + \{C_{gd}s - g_m\}E(s)$$

or

$$V(s) = -\frac{1}{C_{gd}s}I(s) + \frac{s - g_m/C_{gd}}{s}E(s)$$

It therefore follows from equation (6.6) that

$$V_{\text{Th}}(s) = \frac{s - g_m/C_{gd}}{s}E(s) \quad \text{and} \quad Z_{\text{Th}}(s) = 1/C_{gd}s$$

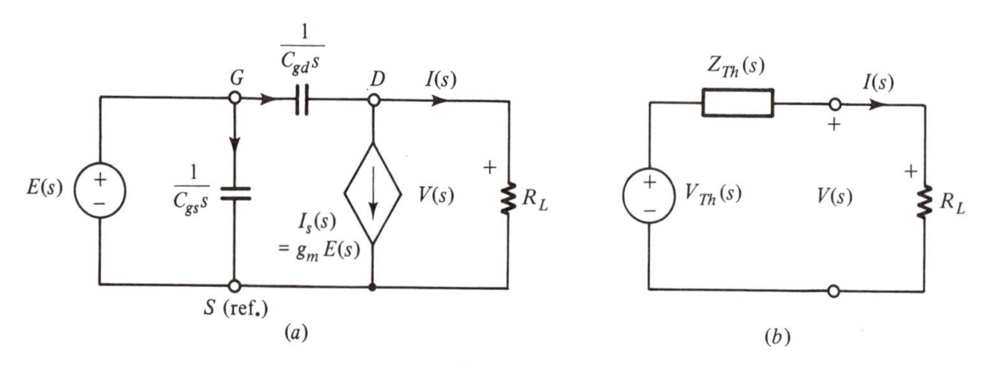


**Fig. 6-47**

Finally, with reference to Fig. 6-47$b$, the voltage divider relation yields

$$V_{R_L}(s) = \frac{R_L}{R_L + Z_{\text{Th}}(s)}V_{\text{Th}}(s) = \frac{s - g_m/C_{gd}}{s + 1/R_L C_{gd}}E(s)$$

**6.26** Find the transform voltage gain $A_v(s) = V_{R_L}(s)/E(s)$ and the $s$-domain transconductance $G_m(s) = I_{R_L}(s)/E(s)$ for the amplifier circuit of Fig. 6-46.

Using the final result from the previous problem, it follows at once that

$$A_v(s) = \frac{s - g_m/C_{gd}}{s + 1/R_L C_{gd}}$$

Next,

$$G_m(s) = \frac{I_{R_L}(s)}{E(s)} = \frac{V_{R_L}(s)/R_L}{E(s)} = \frac{A_v(s)}{R_L}$$
$$= \frac{(1/R_L)(s - g_m/C_{gd})}{s + 1/R_L C_{gd}}$$

These two quantities, $A_v(s)$ and $G_m(s)$, are of course transfer functions relating conditions at the amplifier's output to those at its input.

**6.27** Draw the transform circuit corresponding to the op amp circuit of Fig. 6-48. Both capacitor voltages are initially zero. Hence find the transfer function $H(s) = V(s)/E(s)$. (*Note:* An ideal op amp is described by algebraic relations which transform into corresponding algebraic relations in the $s$-domain—see Theorem 5.1.)

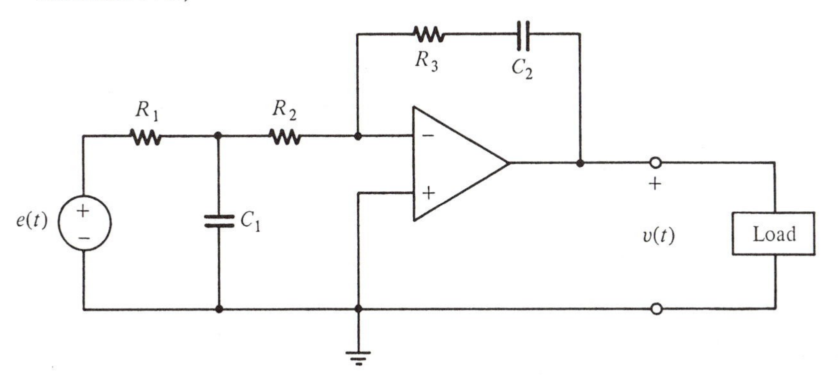


**Fig. 6-48**

The transformed diagram is shown in Fig. 6-49. As in the time domain, the ideal op amp is assumed to have zero input currents and $V_+(s) = V_-(s)$.

In this circuit the op amp's + input terminal is grounded and so $V_-(s)$ is also zero. Then from KCL at node $b$ in Fig. 6-49,

$$\frac{1}{R_1}\{V_b(s) - E(s)\} + C_1 s V_b(s) + \frac{1}{R_2} V_b(s) = 0$$

or

$$V_b(s) = \frac{R_2 E(s)}{R_1 R_2 C_1 s + R_1 + R_2}$$

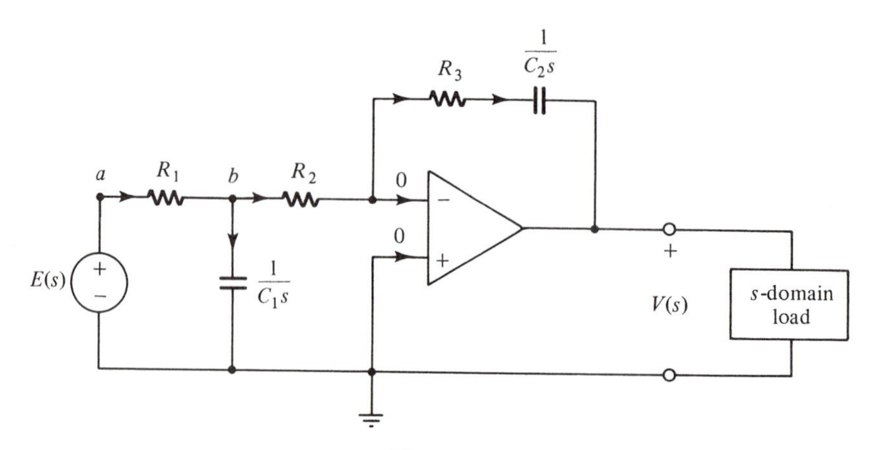


**Fig. 6-49**

Therefore, with reference to Fig. 6-49,

$$I_{R_2}(s) = I_{R_3}(s) = I_{C_2}(s) = \frac{E(s)}{R_1 R_2 C_1 s + R_1 + R_2}$$

and $V(s) = -\{R_3 + 1/C_2 s\} I_{R_3}$. Thus

$$H(s) = \frac{V(s)}{E(s)} = -\frac{R_3 C_2 s + 1}{s\{R_1 R_2 C_1 C_2 s + C_2(R_1 + R_2)\}}$$

**6.28** It is sometimes advantageous to reduce a transform one port to its Norton equivalent—a current source $I_N(s)$ in parallel with an impedance $Z_N(s)$, as shown with the Thévenin one port in Fig. 6-50. Derive the relationships between these two one ports.

Applying KVL to the circuit of Fig. 6-50*a*,

$$V(s) = -Z_{\text{Th}}(s) I(s) + V_{\text{Th}}(s)$$

which was given earlier without proof as equation (6.6).

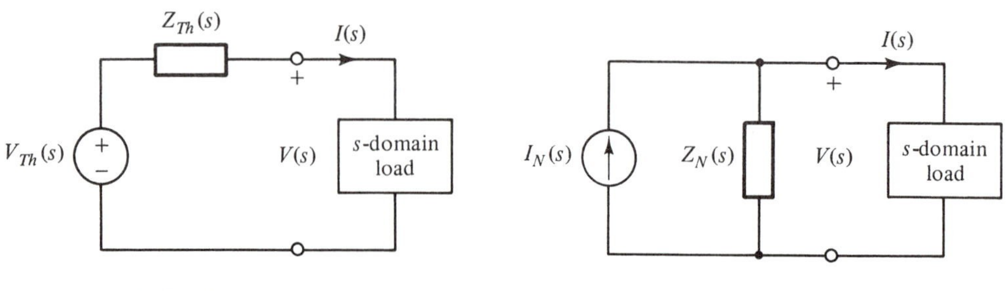


(*a*) The Thévenin one port

(*b*) The Norton one port

**Fig. 6-50**

Similarly, application of KCL to the circuit of Fig. 6-50$b$ yields

$$I(s) = -Y_N(s)V(s) + I_N(s)$$

This equation is easily put in the form

$$V(s) = -Z_N(s)I(s) + Z_N(s)I_N(s)$$

Comparison of the two $V(s)$-$I(s)$ relations above requires that for equivalence

$$Z_N(s) = Z_{\text{Th}}(s) \quad \text{and} \quad I_N(s) = Y_{\text{Th}}(s)V_{\text{Th}}(s)$$

Alternatively, the same result follows from the source transformation of Fig. 6-2. Thus, once a one port's Thévenin equivalent has been found, it is easy to obtain the corresponding Norton equivalent.

**6.29** Obtain the Norton equivalent of the FET amplifier of Problem 6.25 and Fig. 6-46.

It follows at once from the results of Problems 6.25 and 6.28 that the Norton equivalent consists of an impedance

$$Z_N(s) = Z_{\text{Th}}(s) = 1/C_{gd}s$$

in parallel with a current source

$$I_N(s) = Y_{\text{Th}}(s)V_{\text{Th}}(s) = \{C_{gd}s - g_m\}E(s)$$

Thus the FET amplifier—as seen by its load—can be reduced to the Thévenin equivalent obtained in Problem 6.25 *or* to the Norton equivalent found here.

### *Poles and Zeros*

**6.30** Find the poles and zeros of the transfer function $V(s)/E(s)$ in the network of Fig. 6-51.

Applying the voltage divider concept of equation (*6.9*) to the transform circuit of Fig. 6-52, where all initial conditions have been set to zero,

$$\frac{V(s)}{E(s)} = \frac{1 + 2/s}{s + 1 + 1 + 2/s} = \frac{s + 2}{s^2 + 2s + 2} = \frac{s + 2}{(s + 1 - j1)(s + 1 + j1)}$$

Thus there are two poles, $s_{1,2} = -1 \pm j1$, and one finite zero, $s = -2$. (There is also one zero at infinity.)

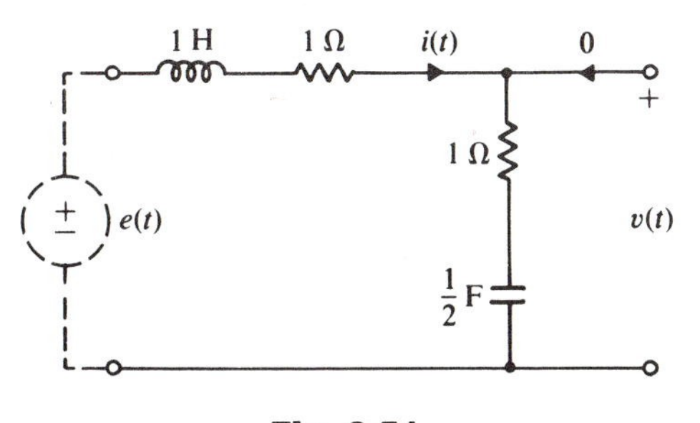


**Fig. 6-51**

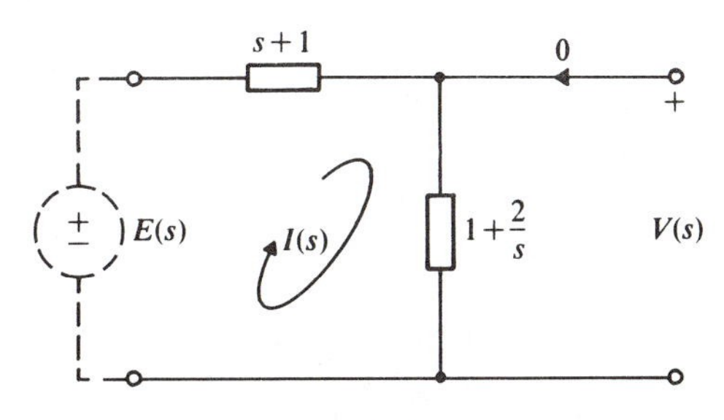


**Fig. 6-52**

**6.31** Find the poles and zeros of the driving point impedance of the network of Fig. 6-51.

From Fig. 6-52,

$$Z(s) = \frac{E(s)}{I(s)} = s + 1 + 1 + \frac{2}{s} = \frac{s^2 + 2s + 2}{s}$$

The zeros are $s_{1,2} = -1 \pm j1$, and there is a single finite pole at $s = 0$. (There is also one pole at infinity.)

**6.32** What are the poles and zeros of the network function $I_0(s)/I(s)$ in the network of Fig. 6-53?

Using current divider tactics,

$$\frac{I_0(s)}{I(s)} = \frac{1/2}{1/2 + 1/(1 + 0.1s)} = \frac{1 + 0.1s}{1 + 0.1s + 2} = \frac{s + 10}{s + 30}$$

The pole is $s = -30$, and the zero is $s = -10$.

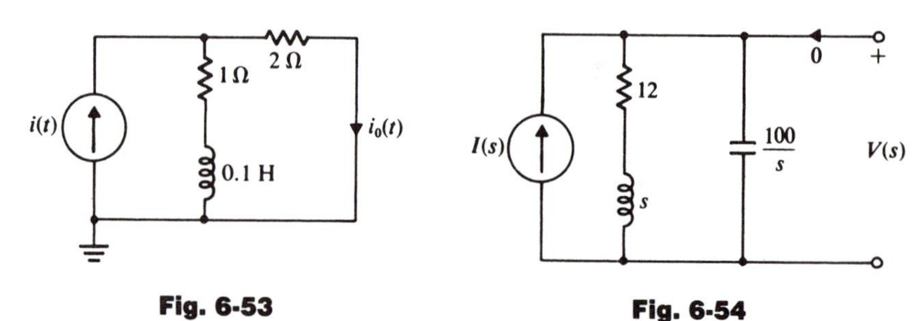


Fig. 6-53

Fig. 6-54

**6.33** Given the transform network in Fig. 6-54, find

(*a*) the poles and zeros of the output impedance $Z_o(s)$,
(*b*) the poles and zeros of the transfer function $H(s) = V(s)/I(s)$,
(*c*) the poles and zeros of the output $V(s)$ when $i(t) = 10$,
(*d*) the poles and zeros of $V(s)$ when $i(t) = 10t$,
(*e*) the poles and zeros of $V(s)$ when $i(t) = 11e^{-12t}$.

(*a*) To find the output impedance, the internal independent source $i(t)$ must be set to zero—an open circuit. Then combining impedances in parallel,

$$\frac{1}{Z_o(s)} = \frac{s}{100} + \frac{1}{12 + s} = \frac{0.01(s^2 + 12s + 100)}{s + 12} = \frac{0.01(s + 6 - j8)(s + 6 + j8)}{s + 12}$$

and

$$Z_o(s) = \frac{100(s + 12)}{(s + 6 - j8)(s + 6 + j8)}$$

Therefore the poles are $s_{1,2} = -6 \pm j8$ and the finite zero is $s = -12$. (There is one zero at infinity.)

(*b*) Here by KCL,

$$\frac{s}{100}V(s) + \frac{1}{s+12}V(s) = I(s)$$

which yields

$$H(s) = \frac{V(s)}{I(s)} = \frac{100(s+12)}{s^2+12s+100} = \frac{100(s+12)}{(s+6-j8)(s+6+j8)}$$

Thus the poles and zeros are the same as for $Z_o(s)$. This is no coincidence!

(*c*) For $\dfrac{V(s)}{I(s)} = \dfrac{100(s+12)}{(s+6-j8)(s+6+j8)}$ and $I(s) = \dfrac{10}{s}$,

$$V(s) = \frac{100(s+12)}{(s+6-j8)(s+6+j8)}\left(\frac{10}{s}\right)$$

which has one finite zero, $s = -12$, and three poles: $s = -6 \pm j8$ and $s = 0$.

(*d*)
$$V(s) = \frac{100(s+12)}{(s+6-j8)(s+6+j8)}\left(\frac{10}{s^2}\right)$$

which has one finite zero, $s = -12$, and four poles: $s = -6 \pm j8$, and *two* at $s = 0$.

(*e*)
$$V(s) = \frac{100\cancel{(s+12)}}{(s+6-j8)(s+6+j8)}\left(\frac{11}{\cancel{s+12}}\right)$$

which has *no* finite zeros and two poles, $s = -6 \pm j8$. (The pole at $s = -12$ cancels the zero at the same point—an academic situation.)

*Note:* It is always advisable to put network functions into a standard form. Either

$$\frac{K(s^n + a_{n-1}s^{n-1} + \cdots + a_0)}{s^m + b_{m-1}s^{m-1} + \cdots + b_0} \quad \text{or} \quad \frac{K(s-z_1)(s-z_2)\cdots(s-z_n)}{(s-p_1)(s-p_2)\cdots(s-p_m)}$$

**6.34** The pole-zero plots of $X(s)$, $Y(s)$, and $Z(s)$ are shown in Fig. 6-55.
(*a*) Write algebraic expressions for $X(s)$, $Y(s)$, and $Z(s)$.
(*b*) What is the algebraic form of $x(t)$, $y(t)$, and $z(t)$?

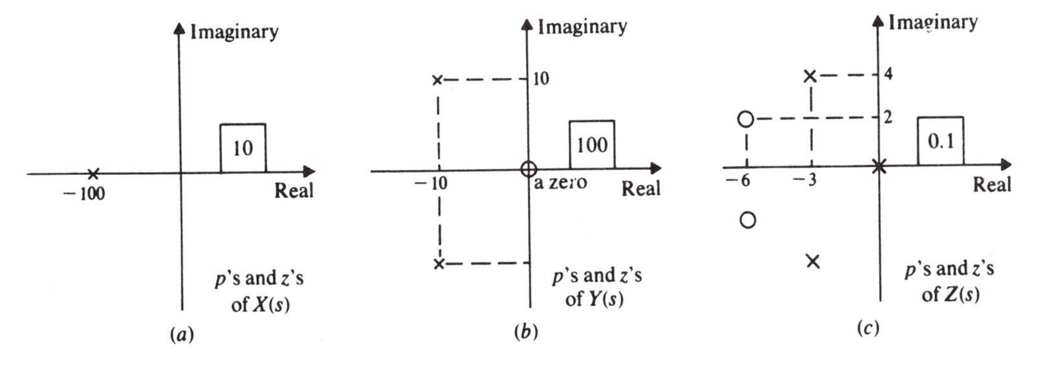


**Fig. 6-55**

(c) Find the time constant $\tau$ of $x(t)$.
(d) Find the damping ratio $\zeta$ and the undamped natural frequency $\omega_n$ for $y(t)$.

(a)
$$X(s) = \frac{10}{s + 100}, \qquad Y(s) = \frac{100s}{(s + 10 - j10)(s + 10 + j10)}$$

and

$$Z(s) = \frac{0.1(s + 6 - j2)(s + 6 + j2)}{s(s + 3 - j4)(s + 3 + j4)}$$

(b)
$$x(t) = K_1 e^{-100t}, \qquad y(t) = K_2 e^{-10t} \cos (10t + \phi_1)$$

and

$$z(t) = K_3 + K_4 e^{-3t} \cos (4t + \phi_2)$$

(c)
$$\tau = \frac{1}{100} = 0.01 \text{ s}$$

(d) From Part (a),

$$Y(s) = \frac{100s}{(s + 10 - j10)(s + 10 + j10)} = \frac{100s}{s^2 + 20s + 200}$$

Comparing $s^2 + 20s + 200$ with $s^2 + 2\zeta\omega_n s + \omega_n^2$, we find $\omega_n = \sqrt{200} = 14.1$ rad/s and $\zeta = 20/(2)(14.1) = 0.707$.

**6.35** The transform of a voltage in an electric circuit is known to be

$$V(s) = \frac{100(s + 6)}{s(s + 3 - j4)(s + 3 + j4)}$$

Evaluate the partial fraction constants graphically, and then determine $v(t)$.

Expressing $V(s)$ in partial fractions,

$$V(s) = \frac{100(s + 6)}{s(s + 3 - j4)(s + 3 + j4)} \equiv \frac{K_1}{s} + \frac{K_2}{s + 3 - j4} + \frac{K_3}{s + 3 + j4}$$

From the corresponding pole-zero diagram, shown in Fig. 6-56a,

$$K_1 = \frac{100(6e^{j0^\circ})}{(5e^{-j53.2^\circ})(5e^{j53.2^\circ})} = 24e^{j0^\circ} = 24$$

From Fig. 6-56b,

$$K_2 = \frac{100(5e^{j53.2^\circ})}{(5e^{j126.8^\circ})(8e^{j90^\circ})} = 12.5e^{-j163.6^\circ} \quad \text{and} \quad K_3 = K_2^* = 12.5e^{j163.6^\circ}$$

Therefore using Table 5-1,

$$v(t) = 24 + 25e^{-3t} \cos (4t - 163.6^\circ), \qquad t \geq 0$$

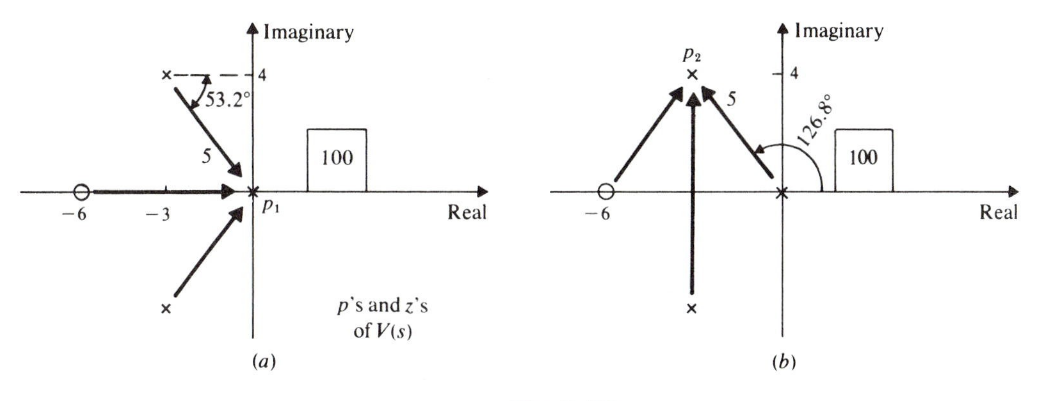


**Fig. 6-56**

### Stability

**6.36** The transform of the impulse response of an electronic feedback amplifier is known to be

$$V(s) = \frac{10(s - 4)}{s^2 - 8s + 25}$$

(*a*) Using the graphical procedure to evaluate the partial fraction constants, find $v(t)$.

(*b*) Is this a stable amplifier?

(*a*) Using the pole-zero diagram in Fig. 6-57,

$$V(s) = \frac{10(s - 4)}{(s - 4 - j3)(s - 4 + j3)} \equiv \frac{K_1}{s - 4 - j3} + \frac{K_1^*}{s - 4 + j3}$$

where

$$K_1 = \frac{10(3e^{j90°})}{6e^{j90°}} = 5e^{j0°} = 5 \qquad \text{and} \qquad K_2 = K_1^* = 5$$

Thus $v(t) = 10e^{4t} \cos 3t$, $t \geq 0$ which is an exponentially growing sinusoid.

(*b*) $V(s)$ must be the amplifier's network function, since the input was $\delta(t)$, which

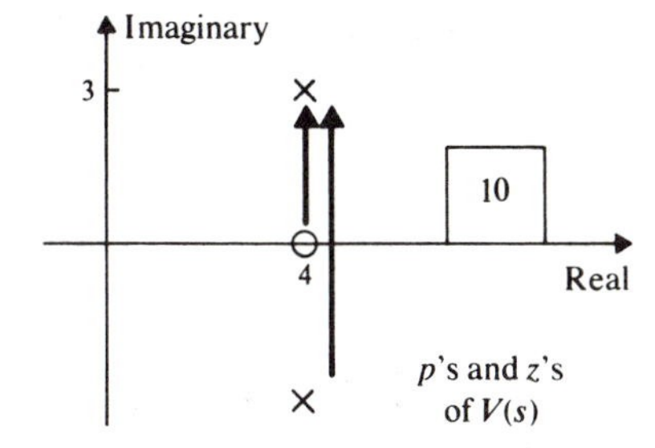


**Fig. 6-57**

becomes 1 after transformation. Therefore the amplifier is *unstable;* the poles of its network function are in the right-half plane. The same conclusion follows from the fact that the impulse response grows indefinitely as $t \to \infty$.

Although unnecessary, since the location of the poles is known, we could use the Routh test on the denominator of $V(s)$. The Routh array is

$$\begin{matrix} 1 & 25 \\ -8 & 0 \\ 25 & 0 \end{matrix}$$

Since there is a negative element in the first column, the amplifier is unstable. And since there are two sign changes down the first column, there are exactly two poles in the right-half plane.

**6.37** Use the Routh test to ascertain which of the following initial condition responses were generated by stable circuits.

(*a*) $$V(s) = \frac{2s - 7}{s^3 + 6s^2 + 11s + 6}$$

(*b*) $$I(s) = \frac{6}{s^4 + 2s^3 + 3s^2 + 4s + 5}$$

(*c*) $$I(s) = \frac{76s + 17}{s^4 + 2s^3 + 3s^2 + 2s + 2}$$

(*a*) The circuit's characteristic equation is $s^3 + 6s^2 + 11s + 6 = 0$ from which we can generate the Routh array.

$$\begin{matrix} 1 & 11 \\ 6 & 6 \\ 10 & 0 \\ 6 & 0 \end{matrix}$$

Since all the entries in the first column are greater than zero, the circuit is stable.

(*b*) The Routh array is

$$\begin{matrix} 1 & 3 & 5 \\ 2 & 4 & 0 \\ 1 & 5 & 0 \\ -6 & 0 & 0 \\ 5 & 0 & 0 \end{matrix}$$

This circuit is unstable, and since there are two sign changes down the first column, there are two poles in the right-half plane.

(*c*) The Routh array is

$$\begin{matrix} 1 & 3 & 2 \\ 2 & 2 & 0 \\ 2 & 2 & 0 \\ 0 & 0 & 0 \end{matrix}$$

Here the array terminates with a row of zeros before the $(n + 1)$th (the 5th) row has been obtained. The circuit is not stable. (The Routh test can be extended to determine whether this amounts to marginal stability or complete instability.)

# PROBLEMS

**6.38** The terminal characteristics of the passive network in Fig. 6-58 are described by the integro-differential equation

$$\frac{d^2 i}{dt^2} + 6\frac{di}{dt} + 11i + 6\int_0^t i\,d\tau = 2\frac{dv}{dt} + 4v + 4\int_0^t v\,d\tau$$

Determine the driving point impedance $Z(s) = V(s)/I(s)$.

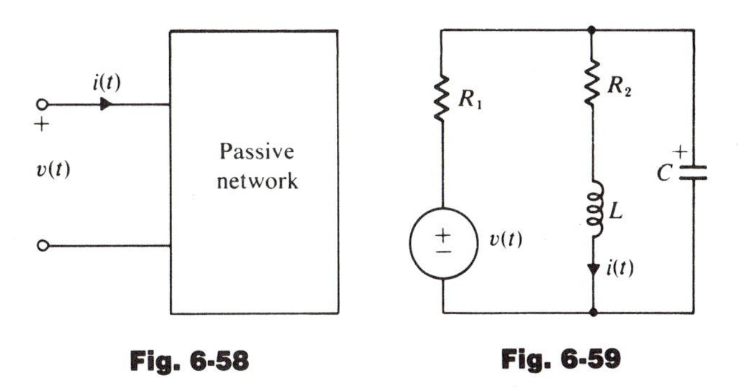


**Fig. 6-58** **Fig. 6-59**

**6.39** Find the transfer function $H(s) = I(s)/V(s)$ for the network in Fig. 6-59.

**6.40** Verify that the following matrix equation correctly represents the network of Fig. 6-59, where $i_1(t)$ and $i_2(t)$ are the clockwise mesh currents.

$$\begin{bmatrix} R_1 + R_2 + Ls & -R_2 - Ls \\ -R_2 - Ls & Ls + R_2 + \dfrac{1}{Cs} \end{bmatrix} \begin{bmatrix} I_1(s) \\ I_2(s) \end{bmatrix} = \begin{bmatrix} V(s) + Li(0) \\ -Li(0) - \dfrac{v_C(0)}{s} \end{bmatrix}$$

**6.41** The circuit in Fig. 6-60 contains two dependent voltage sources, $v_x(t)$ and $v_y(t)$

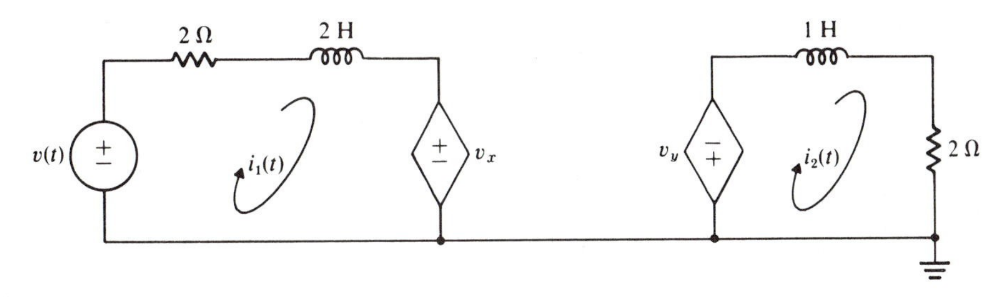


**Fig. 6-60**

where $v_x = di_2/dt$ and $v_y = di_1/dt$. Verify the following matrix equation for this network.

$$\begin{bmatrix} 2s+2 & s \\ s & s+2 \end{bmatrix} \begin{bmatrix} I_1(s) \\ I_2(s) \end{bmatrix} = \begin{bmatrix} V(s) + 2i_1(0) + i_2(0) \\ i_1(0) + i_2(0) \end{bmatrix}$$

**6.42** Show that the network of Fig. 6-61 is correctly represented by the matrix equation

$$\begin{bmatrix} \frac{1}{R_1} + Cs & -Cs \\ -Cs & Cs + \frac{1}{Ls} + \frac{1}{R_2} \end{bmatrix} \begin{bmatrix} V_a(s) \\ V_b(s) \end{bmatrix} = \begin{bmatrix} I(s) + Cv_C(0) \\ -Cv_C(0) - i_L(0)/s \end{bmatrix}$$

**6.43** Find the transfer function $H(s) = V_b(s)/I(s)$ for the network of Fig. 6-61.

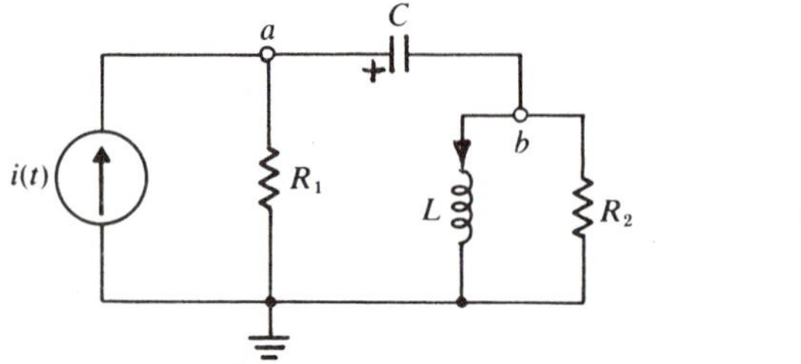


**Fig. 6-61**

**Fig. 6-62**

**6.44** If the initial condition is zero in the network of Fig. 6-62, find $V(s)$ for (*a*) $e(t) = 10$, (*b*) $e(t) = 10e^{-5t}$, (*c*) $e(t) = 10e^{-10t}$, and (*d*) $e(t) = 10\delta(t)$.

**6.45** Find the output impedance for the network of Fig. 6-63.

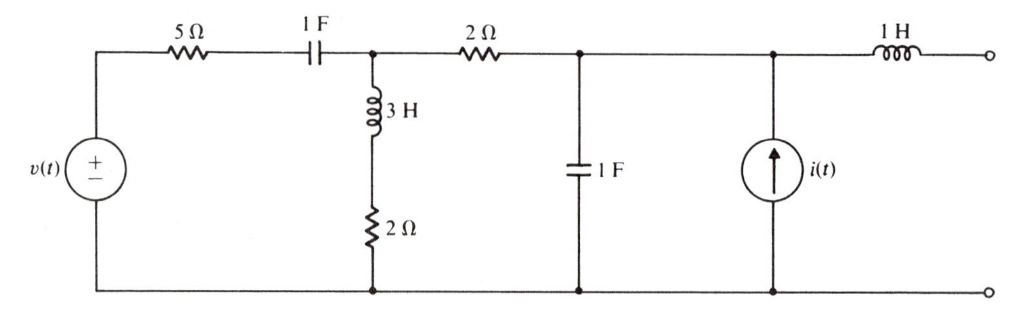


**Fig. 6-63**

**6.46** Show that Fig. 6-64*b* is a valid transform-domain representation of Fig. 6-64*a*.

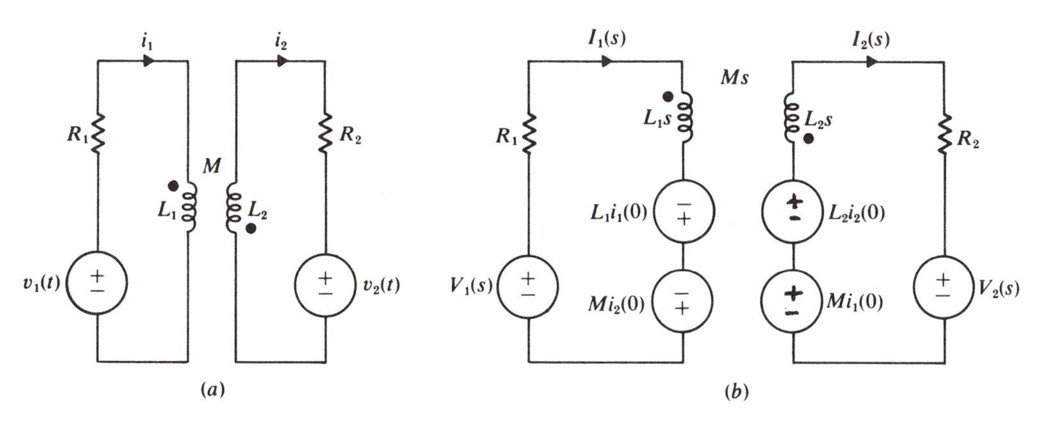


Fig. 6-64

**6.47** Use the step-by-step reduction technique for the ladder network of Fig. 6-65 to determine the transfer function $H(s) = V(s)/E(s)$, given $i(t) = 0$.

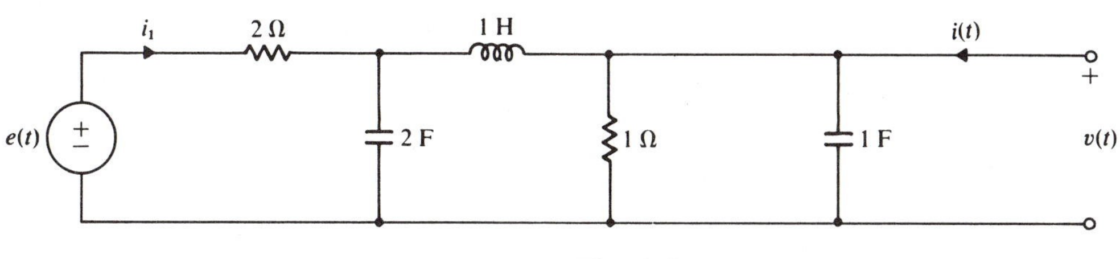


Fig. 6-65

**6.48** Find the output impedance of the ladder network in Fig. 6-65.

**6.49** For $v(0) = 0$ and $i(t) = 10 \cos 2t$ in Fig. 6-66,

$$V(s) = \frac{10s}{(s+2)(s^2+4)} \equiv \frac{K_1}{s+2} + \frac{K_2}{s-j2} + \frac{K_2^*}{s+j2}$$

(*a*) List the poles and zeros of $V(s)$.

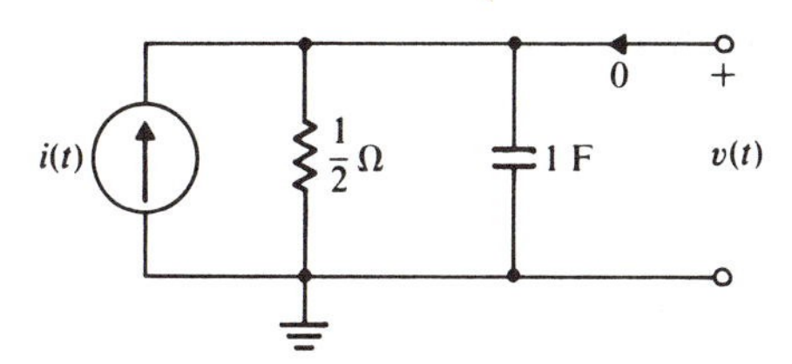


Fig. 6-66

(*b*) Use the graphical method to evaluate $K_1$ and $K_2$ in the partial fraction expansion of $V(s)$.
(*c*) Find $v(t)$.

**6.50** The voltage transfer function for the active network in Fig. 6-67 is given by

$$H(s) = \frac{V(s)}{E(s)} = \frac{b_2 s^3 + b_1 s + b_0}{a_3 s^3 + a_2 s^2 + a_1 s + a_0}$$

State the conditions necessary and sufficient for stability.

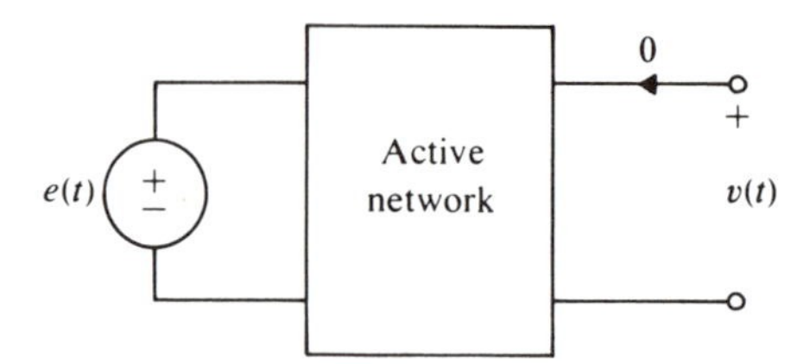


Fig. 6-67

**6.51** Find $H(s) = V_2(s)/I_1(s)$ for the *symmetrical lattice* network of Fig. 6-68.

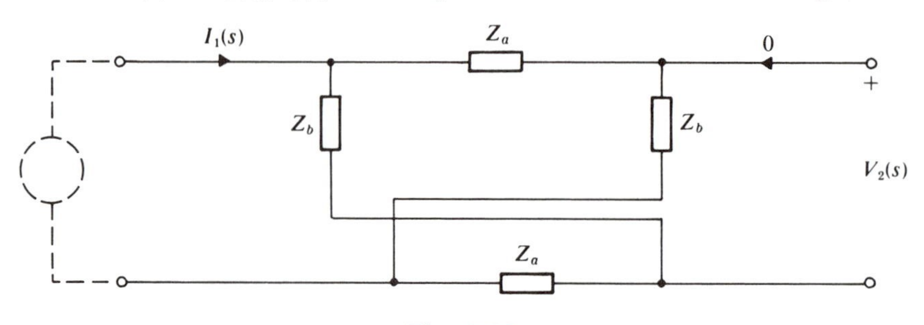


Fig. 6-68

**6.52** List the poles and zeros of $H(s)$ for the network of Problem 6.51 if $Z_a(s) = s/(s^2 + 1)$ and $Z_b(s) = (s^2 + 1)/s$. Is this network stable?

**6.53** Establish the validity of the source transformation of Fig. 6-2.

**6.54** Show that $I_N(s) = I_{sc}(s)$.

**6.55** Find the Thévenin and Norton equivalents of the circuit of Fig. 6-61 as seen by $R_2$. You are given that $v_C(0)$ is zero. (*Hint:* Re-label the source current to avoid confusion with the terminal current.)

# Chapter 7

# SINUSOIDAL STEADY-STATE RESPONSE

The concept of sinusoidal steady-state or ac response was introduced in Chapter 4. It is simply the continuing sinusoidal response of a stable network to a sinusoidal input after all the transients have decayed to zero.

Although sinusoidal response problems can and have been solved by the general methods of Chapters 4 and 5, the sinusoidal driving function is so common in electrical engineering that we are well justified in searching out special techniques to simplify the solution of this particular type of problem. To emphasize the ubiquity of electrical sinusoids, consider the 60-Hz sinusoid of normal utility power; the sinusoids of communications; ac control and instrumentation systems; ac motors and generators; and sinusoidal test oscillators. And more complicated waveforms are often treated sinusoidally with the aid of Fourier series.

It has already been demonstrated that the solution of dc steady-state problems reduces to the solution of simultaneous *algebraic* equations. Here we will show that the same is true of ac steady-state problems, but the algebraic equations will now be complex.

## THE SINUSOIDAL WAVEFORM

We discussed the properties of sinusoids in Chapter 1; a brief review will be adequate here. We will usually work with the cosine function

$$x(t) = X_m \cos (\omega t + \theta) \tag{7.1}$$

where $X_m$ is the *zero-to-peak amplitude*, $\omega$ is the *angular frequency*, and $\theta$ is the *phase angle*. Positive phase is said to be *leading*, and negative phase to be *lagging*.

In Fig. 7-1 a given network's input $u(t) = U_m \cos \omega t$ and its ac output $y(t) = Y_m \cos(\omega t + \theta)$ is sketched on the assumption that $\theta > 0$. We see that a *positive* value of $\theta$ moves $y(t)$ to the *left*. (A given event, such as a maximum, occurs *earlier* in time as $\theta$ becomes more positive.) Usually we do not have, or need, a fixed time reference, such as the instant when the input sinusoid was switched on. Instead we *compare* one sinusoid, such as $y(t)$, with a *reference* sinusoid, such as $u(t)$, and speak of the *phase shift* of $y(t)$ *relative* to $u(t)$. Then the location of "time zero" is irrelevant, and usually far in the past.

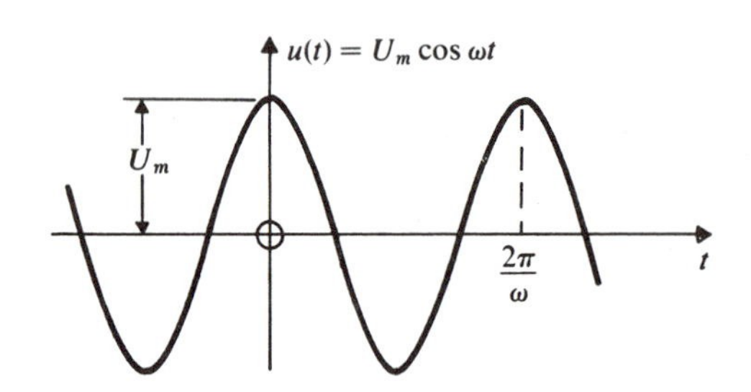


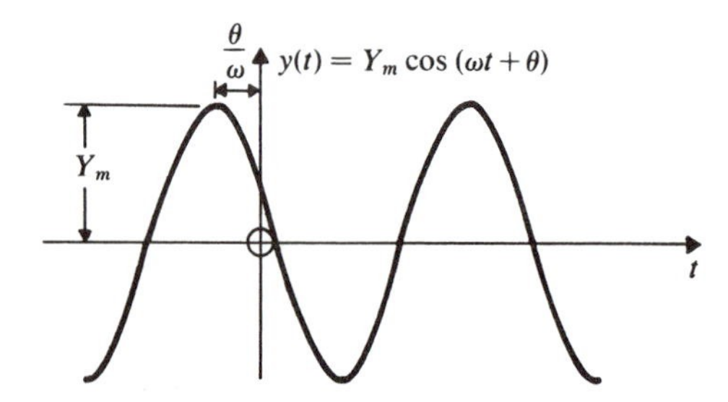


**Fig. 7-1**

In Chapter 1 we also defined

$$T = \frac{2\pi}{\omega} \qquad \text{and} \qquad f = \frac{1}{T} = \frac{\omega}{2\pi}$$

the *period* and true *frequency* of the sinusoid. Finally, it is always a simple matter to convert a sine function into a cosine, or vice versa:

$$\sin(\omega t + \theta) = \cos(\omega t + \theta - \tfrac{1}{2}\pi)$$

$$\cos(\omega t + \theta) = \sin(\omega t + \theta + \tfrac{1}{2}\pi)$$

## DIRECT CALCULATION OF SINUSOIDAL STEADY-STATE RESPONSE

If a circuit's input is $u(t) = U_m \cos(\omega t + \theta)$ and if its input and output are related by the network function $H(s) = Y(s)/U(s)$ whose poles are all in the left-half plane, then the sinusoidal steady-state output is given by

$$y_{sss}(t) = U_m |H(j\omega)| \cos(\omega t + \theta + \underline{/H(j\omega)}) \tag{7.2}$$

where $H(s)|_{s=j\omega} = H(j\omega) = |H(j\omega)| e^{j\underline{/H(j\omega)}}$.

*Proof:* The transform of the output caused by $u(t) = U_m \cos(\omega t + \theta)$ is

$$Y(s) = H(s)U(s) = H(s)\left\{\frac{U_m(\cos\theta)(s - \omega\tan\theta)}{s^2 + \omega^2}\right\}$$
$$= \frac{K_1}{s - p_1} + \frac{K_2}{s - p_2} + \cdots + \frac{K_\omega}{s - j\omega} + \frac{K_\omega^*}{s + j\omega}$$

where we have used entry 6 from Table 5-1, page 134, to transform $u(t)$. Provided all the poles $p_i$ of $H(s)$ are in the left-half plane, the inverse transform of all terms except the last two will be transients which decay to zero as $t \to \infty$. Therefore we need consider only the final pair, which corresponds to the sinusoidal steady-state response.

We evaluate the partial fraction constant $K_\omega$ in the usual way,

$$K_\omega = H(s)\left\{\frac{U_m(\cos\theta)(s - \omega\tan\theta)\cancel{(s - j\omega)}}{\cancel{(s - j\omega)}(s + j\omega)}\right\}\Bigg|_{s=j\omega}$$
$$= H(j\omega)\left\{\frac{U_m(\cos\theta)(j\omega - \omega\tan\theta)}{j2\omega}\right\}$$
$$= \tfrac{1}{2}U_m H(j\omega)(\cos\theta + j\sin\theta) = \tfrac{1}{2}U_m H(j\omega)e^{j\theta}$$
$$= \tfrac{1}{2}U_m|H(j\omega)|e^{j(\theta + \underline{/H(j\omega)})}$$

And the inverse transform of the final pair of terms in the partial fraction expansion is

$$y_{sss}(t) = U_m|H(j\omega)|\cos(\omega t + \theta + \underline{/H(j\omega)})$$

which is the result we set out to derive.

This equation makes the evaluation of a circuit's ac response so simple that we are tempted to refer to it as *the magic formula*! In words, it tells us that when a sinusoidal signal "passes through" a network, then

1. its amplitude is multiplied by $|H(j\omega)|$, and
2. its phase is shifted by an angle $\underline{/H(j\omega)}$. If the angle is positive, the output sinusoid *leads* the input. A negative angle corresponds to a phase *lag*.

### Example 7.1

In Example 6.2, page 178, we found that $H_2(s) = V_1(s)/V_2(s) = 2/(s + 2)$ for the two port of Fig. 6-9. Now we will determine the ac output corresponding to the input $v_2(t) = 3\cos 2t$.

Here $\omega = 2$ rad/s and thus

$$H_2(j2) = \frac{2}{s + 2}\Bigg|_{s=j2} = \frac{2}{2(1 + j)} = \frac{1}{\sqrt{2}}e^{-j45°}$$

From equation (7.2)

$$v_{1,sss}(t) = \frac{3}{\sqrt{2}}\cos(2t - 45°)$$

which indicates that the sinusoidal steady-state output is smaller in amplitude than the input by a factor of $1/\sqrt{2}$, and that the output phase lags the input by 45°.

We *cannot* use equation (7.2) to calculate the ac output corresponding to $H_1(s) = V_2(s)/V_1(s) = 1/(s^2 + 1)$, since *this is not a stable situation*. The "transient" sinusoid corresponding to the poles of $H_1(s)$ at $\pm j1$ will *not* decay to zero with increasing time. (If one or more poles of $H(s)$ were in the right-half plane, *there would be no steady-state*. The transient part of the solution would *grow* exponentially with time.) ■

The availability of equation (7.2) means that *for sinusoidal steady-state problems* we need not go through the full procedure of solving the circuit's differential equations. It has been done *once,* in the proof of equation (7.2); it need not be repeated.

Note that initial conditions can have *no* effect on the sinusoidal steady-state solution. They *do* affect the transient terms, but these decay to zero as $t \to \infty$.

## THE PHASOR CONCEPT

The direct application of equation (7.2) is an effective time-saver as it stands, but it has long been common practice to employ an *operational* procedure for solving sinusoidal problems.

From the Euler relation,

$$X_m \cos(\omega t + \theta) + jX_m \sin(\omega t + \theta) = X_m e^{j(\omega t + \theta)}$$

or

$$X_m \cos(\omega t + \theta) = \mathrm{Re}\,[X_m e^{j(\omega t + \theta)}] = \mathrm{Re}\,[\mathbf{X}] \tag{7.3}$$

where $\mathbf{X}$ is called a *rotating phasor*. Graphically, as shown in Fig. 7-2, $\mathbf{X}$ may be regarded as a rotating vector which we can "freeze" at some convenient time, say $t = 0$, so that it can be seen and drawn. The horizontal component of the rotating vector $\mathbf{X}$ is the original sinusoid $X_m \cos(\omega t + \theta)$.

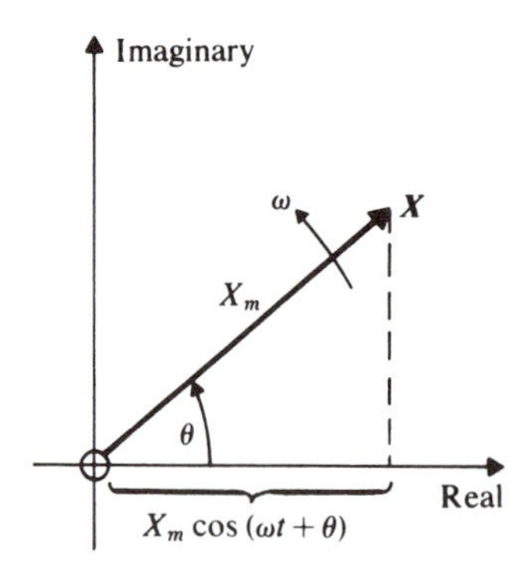


*X* is frozen at $t = 0$ **Fig. 7-2**

We say that $\mathbf{X}$ *represents* the sinusoid operationally, and using this representation we may rewrite equation (7.2) in phasor form. Thus if $u(t) = U_m \cos(\omega t + \theta)$,

$$\mathbf{Y} = H(j\omega)\mathbf{U} = U_m |H(j\omega)| e^{j(\omega t + \theta + \angle H(j\omega))}$$

since, if we take the real part of both sides, we obtain

$$y_{sss}(t) = U_m |H(j\omega)| \cos(\omega t + \theta + \angle H(j\omega))$$

To restate this relationship,

$$H(j\omega) = \frac{\mathbf{Y}}{\mathbf{U}} \tag{7.4}$$

which is the $j\omega$- or phasor-domain analog of $H(s) = Y(s)/U(s)$ in the $s$- or transform-domain. $H(j\omega)$ is called the *sinusoidal network function* relating the input and output phasors $\mathbf{U}$ and $\mathbf{Y}$.

Often the effective or RMS amplitude is used instead of the zero-to-peak amplitude, i.e. $\mathbf{X}' = X_{\text{eff}} e^{j(\omega t + \theta)}$. This does not invalidate equation (7.4) so long as *both* $\mathbf{Y}$ and $\mathbf{U}$ are RMS phasors. But note that if $\mathbf{X}'$ is an RMS phasor,

$$x(t) = \sqrt{2}\,\text{Re}\,[\mathbf{X}']$$

since the effective or RMS amplitude of the sinusoid $X_m \cos(\omega t + \theta)$ is $X_{\text{eff}} = X_m/\sqrt{2}$ (see Problem 1.46).

A *stationary phasor* $\mathbf{X}'' = X_m e^{j\theta}$ is often used to represent the sinusoid $X_m \cos(\omega t + \theta)$. We suggest that this can be confusing, since both $\mathbf{X}''$ and a sinusoidal network function $H(j\omega)$ will appear to have similar properties—they are both complex numbers. On the other hand the *rotating* phasor $\mathbf{X} = X_m e^{j(\omega t + \theta)}$ has a time varying or rotary property not possessed by $H(j\omega) = |H(j\omega)| e^{j\angle H(j\omega)}$, thus emphasizing their essentially different natures.

## APPLICATION OF PHASOR ANALYSIS

If equation (7.4) is applied to simple $R$, $L$, and $C$ elements, we obtain

$$\mathbf{V}_R = R\mathbf{I}_R \qquad \mathbf{I}_R = \frac{1}{R}\mathbf{V}_R \tag{7.5}$$

$$\mathbf{V}_L = jL\omega\mathbf{I}_L \qquad \mathbf{I}_L = \frac{1}{jL\omega}\mathbf{V}_L \tag{7.6}$$

$$\mathbf{V}_C = \frac{1}{jC\omega}\mathbf{I}_C \qquad \mathbf{I}_C = jC\omega\mathbf{V}_C \tag{7.7}$$

or, in general,

$$\mathbf{V} = Z(j\omega)\mathbf{I} \qquad \mathbf{I} = Y(j\omega)\mathbf{V} \tag{7.8}$$

where $Z(j\omega)$ is called the *sinusoidal impedance* and $Y(j\omega)$ the *sinusoidal admittance*.

Further, the phasor form of the two Kirchhoff laws is

$$\sum_{\text{node}} \mathbf{I} = 0 \quad \text{and} \quad \sum_{\text{loop}} \mathbf{V} = 0 \tag{7.9}$$

Now we can draw phasor- or $j\omega$-domain circuit diagrams, just as we did earlier in the $s$-domain.

### Example 7.2

The phasor representation in the center of Fig. 7-3 follows directly from the ordinary circuit diagram on the left as a consequence of equations (7.5), (7.6), and (7.7).

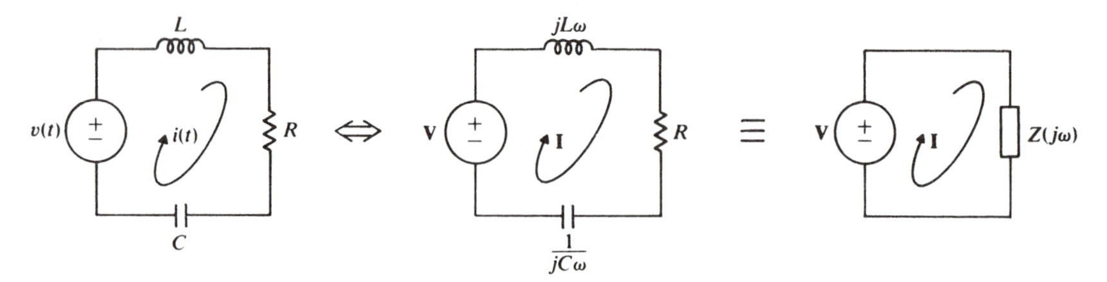


**Fig. 7-3**

Then if we apply the phasor form of KVL to the center diagram,

$$jL\omega\mathbf{I} + R\mathbf{I} + \frac{1}{jC\omega}\mathbf{I} = \mathbf{V}$$

or

$$\{Z_L(j\omega) + Z_R(j\omega) + Z_C(j\omega)\}\mathbf{I} = \mathbf{V}$$

or

$$Z(j\omega)\mathbf{I} = \mathbf{V}$$

where $Z(j\omega) = Z_L(j\omega) + Z_R(j\omega) + Z_C(j\omega)$.

The last equation justifies the asserted equivalence of the right-hand diagram in Fig. 7-3.

It must again be emphasized that since the initial conditions cannot effect the ac response, they never enter into phasor calculations. Bearing in mind this comment on initial conditions, it is instructive to compare the phasor procedure of this example with the $s$-domain procedure of Example 5-3 on page 139.

### Example 7.3

The circuit on the left in Fig. 7-4 is shown with its phasor equivalent on the right.

The $j\omega$-domain KVL equations around the two meshes are

$$\mathbf{V}_{L_1} + \mathbf{V}_R + \mathbf{V}_C = \mathbf{V}$$

$$-\mathbf{V}_R + \mathbf{V}_{L_2} = 0$$

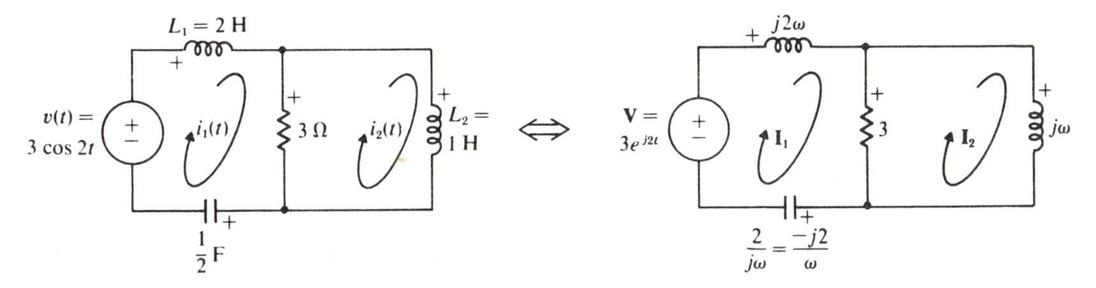


**Fig. 7-4**

and the phasor-domain $v$-$i$ relations are

$$\mathbf{V}_{L_1} = j2\omega\mathbf{I}_1 \qquad \mathbf{V}_R = 3(\mathbf{I}_1 - \mathbf{I}_2)$$

$$\mathbf{V}_{L_2} = j\omega\mathbf{I}_2 \qquad \mathbf{V}_C = -(j2/\omega)\mathbf{I}_1$$

On substituting the $v$-$i$ relations into the KVL equations,

$$(j2\omega + 3 - j2/\omega)\mathbf{I}_1 - 3\mathbf{I}_2 = \mathbf{V}$$

$$-3\mathbf{I}_1 + (3 + j\omega)\mathbf{I}_2 = 0$$

These are *complex algebraic equations,* and if we wish to solve for $\mathbf{I}_1$, with $\omega = 2$ rad/s as specified,

$$\mathbf{I}_1 = \frac{\begin{vmatrix} \mathbf{V} & -3 \\ 0 & 3 + j2 \end{vmatrix}}{\begin{vmatrix} 3 + j3 & -3 \\ -3 & 3 + j2 \end{vmatrix}} = \frac{(3 + j2)\mathbf{V}}{-6 + j15} = \frac{(3.60e^{j33.7^\circ})(3e^{j2t})}{16.16e^{j111.8^\circ}} = 0.668e^{j(2t - 78.1^\circ)}$$

Therefore,

$$i_{1,\text{sss}}(t) = 0.668 \cos(2t - 78.1^\circ) \qquad \blacksquare$$

Note that the circuit differential equations in the time-domain have been "transformed" into complex algebraic equations in the $j\omega$-domain.

Finally, observe that any sinusoidal network function can be reduced to a complex number. For example, we often write

$$Z(j\omega) = R(\omega) + jX(\omega) \quad \text{and} \quad Y(j\omega) = G(\omega) + jB(\omega)$$

where $R(\omega)$ is the *resistive part* of $Z(j\omega)$, $X(\omega)$ is the *reactive part*, or *reactance*, $G(\omega)$ is the *conductance*, and $B(\omega)$ the *susceptance*.

## SERIES AND PARALLEL IMPEDANCE

If $Z_1(j\omega), Z_2(j\omega), \ldots, Z_n(j\omega)$ are in series, then they are equivalent to

$$Z_{\text{eq}}(j\omega) = Z_1(j\omega) + Z_2(j\omega) + \cdots + Z_n(j\omega) \tag{7.10}$$

If the elements are in parallel, then

$$Y_{eq}(j\omega) = Y_1(j\omega) + Y_2(j\omega) + \cdots + Y_n(j\omega) \tag{7.11}$$

## SOURCE TRANSFORMATIONS

It is sometimes convenient to replace a phasor voltage source by an equivalent phasor current source, or vice versa, as shown in Fig. 7-5. One application is suggested in the following section.

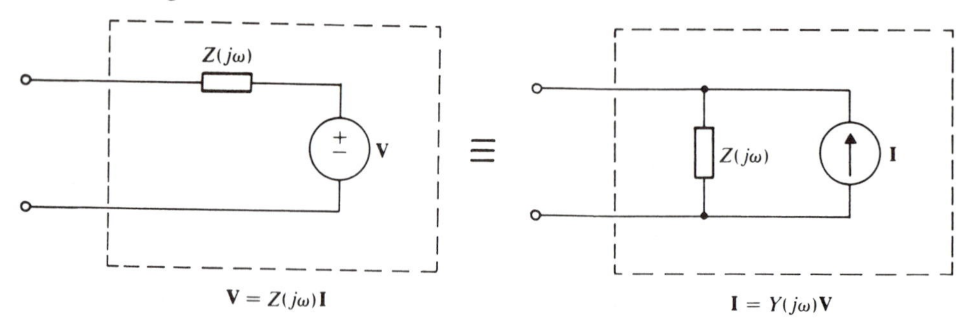


**Fig. 7-5**

## LOOP AND NODAL EQUATIONS

The phasor loop equations for any $n$-mesh, connected, planar network containing no current sources, dependent sources, or mutual inductance can always be written by inspection.

If we agree to write the KVL equations around the $n$ windowpanes (meshes), and, further, if we choose the $n$ loop currents to be around the $n$ meshes, either all clockwise or all counterclockwise, then

$$[Z(j\omega)][\mathbf{I}] = [\mathbf{V}] \tag{7.12}$$

or

$$\begin{bmatrix} Z_{11} & Z_{12} & \cdots & Z_{1n} \\ Z_{21} & Z_{22} & \cdots & Z_{2n} \\ \cdots & \cdots & \cdots & \cdots \\ Z_{n1} & Z_{n2} & \cdots & Z_{nn} \end{bmatrix} \begin{bmatrix} \mathbf{I}_1 \\ \mathbf{I}_2 \\ \cdots \\ \mathbf{I}_n \end{bmatrix} = \begin{bmatrix} \mathbf{V}_1 \\ \mathbf{V}_2 \\ \cdots \\ \mathbf{V}_n \end{bmatrix}$$

where

$Z_{ii}$ = total sinusoidal impedance in mesh $i$ ($i = 1, 2, \ldots, n$),

$-Z_{ij} = -Z_{ji}$ = total sinusoidal impedance common to meshes $i$ and $j$ ($i, j = 1,$

$2, \ldots, n, i \neq j$),

$\mathbf{I}_j$ = (unknown) phasor current around the $j$th mesh,

$\mathbf{V}_j$ = total phasor source voltage *rise* around the $j$th mesh (traveling in the direction of $\mathbf{I}_j$).

*Notes*

1. If the given circuit contains independent current sources, we may use source transformations to convert them into voltage sources.
2. Otherwise, or if any of the other conditions are violated, we must revert to the formal procedure for writing loop equations.

Similarly, we can write by inspection the nodal equations for any $n$-node, connected network containing no voltage sources, dependent sources, or mutual inductance. If we choose to call the reference node #$n$, then

$$[Y(j\omega)][\mathbf{V}] = [\mathbf{I}] \qquad (7.13)$$

or

$$\begin{bmatrix} Y_{11} & Y_{12} & \cdots & Y_{1,n-1} \\ Y_{21} & Y_{22} & \cdots & Y_{2,n-1} \\ \cdots & \cdots & \cdots & \cdots \\ Y_{n-1,1} & Y_{n-1,2} & \cdots & Y_{n-1,n-1} \end{bmatrix} \begin{bmatrix} \mathbf{V}_1 \\ \mathbf{V}_2 \\ \cdots \\ \mathbf{V}_{n-1} \end{bmatrix} = \begin{bmatrix} \mathbf{I}_1 \\ \mathbf{I}_2 \\ \cdots \\ \mathbf{I}_{n-1} \end{bmatrix}$$

where

$Y_{ii}$ = total sinusoidal admittance connected to node $i$ ($i = 1, 2, \ldots, n-1$),

$-Y_{ij} = -Y_{ji}$ = total sinusoidal admittance connected between nodes $i$ and $j$ ($i, j = 1, 2, \ldots, n-1, i \neq j$),

$\mathbf{V}_i$ = (unknown) phasor voltage of the $i$th node, relative to the phasor voltage of the reference node (#$n$),

$\mathbf{I}_i$ = total phasor source current *entering* node $i$.

*Notes*

1. If the given circuit contains independent voltage sources, we may use source transformations to convert them into current sources.
2. Otherwise, or if any of the other conditions are violated, we must revert to the formal procedure for writing nodal equations.

Although these algorithms make the *writing* of circuit equations very easy, the latter still have to be *solved*! We should therefore consider the possibility of eliminating meshes (for loop analysis) or nodes (for nodal analysis) by appropriate reduction techniques *before* writing any circuit equations.

### THE ONE PORT IN THE PHASOR-DOMAIN

The phasor-domain one port of Fig. 7-6*a* can be reduced to a Thévenin one port whose sinusoidal $v$-$i$ relation is

$$\mathbf{V} = -Z_{\text{Th}}(j\omega)\mathbf{I} + \mathbf{V}_{\text{Th}} \tag{7.14}$$

where, as in the $s$-domain, the Thévenin impedance is often called the input (or driving point) impedance, or the output impedance—as appropriate.

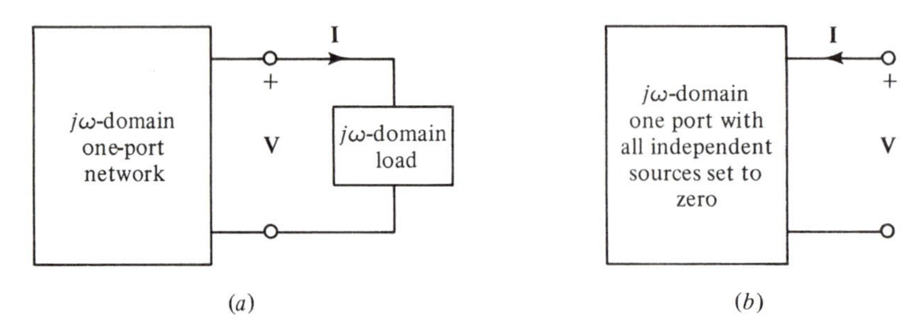


**Fig. 7-6**

Alternatively, if we want only the one port's sinusoidal impedance (or admittance), we can use the defining equation (*7.8*),

$$Z(j\omega) = \frac{\mathbf{V}}{\mathbf{I}} \quad \text{or} \quad Y(j\omega) = \frac{\mathbf{I}}{\mathbf{V}}$$

where $\mathbf{V}$ is the phasor terminal voltage due *entirely* to an applied phasor current $\mathbf{I}$, or vice versa. Therefore, all other independent sources must be set to zero as in Fig. 7-6*b*. Note, too, the chosen current direction in that figure. As in the $s$-domain, we speak of the impedance or admittance seen by the external circuit looking into the one port's terminals.

### SINUSOIDAL TRANSFER FUNCTIONS AND NETWORK FUNCTIONS

If a network has more than one port of access, we may define the sinusoidal impedance and admittance seen at each port. Thus with reference to the two port of Fig. 7-7,

$$Z_1(j\omega) = \frac{\mathbf{V}_1}{\mathbf{I}_1} \qquad Y_1(j\omega) = \frac{\mathbf{I}_1}{\mathbf{V}_1}$$

$$Z_2(j\omega) = \frac{\mathbf{V}_2}{\mathbf{I}_2} \qquad Y_2(j\omega) = \frac{\mathbf{I}_2}{\mathbf{V}_2}$$

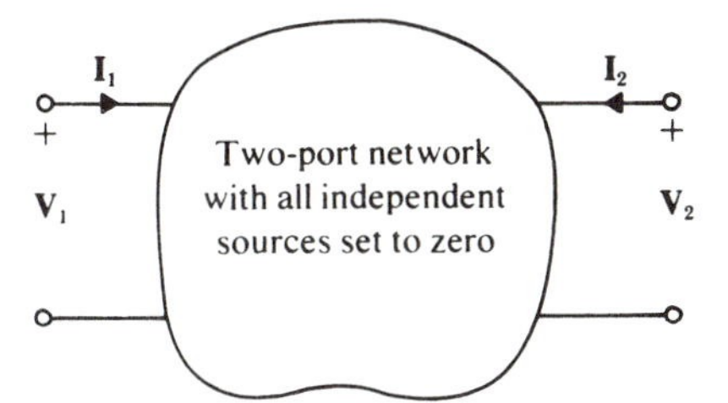


**Fig. 7-7**

*However*, $Z_1(j\omega)$, for example, will depend on the circuit (if any) connected to the other port. And if a circuit *is* connected to port #2, then all independent sources therein must be set to zero (since we want $\mathbf{V}_1$ *due to* $\mathbf{I}_1$, or vice versa).

Moving a step further, still in relation to Fig. 7-7, we may define other ratios, such as

$$H_1(j\omega) = \frac{\mathbf{V}_2}{\mathbf{V}_1} \qquad H_2(j\omega) = \frac{\mathbf{V}_1}{\mathbf{V}_2}$$

$$H_3(j\omega) = \frac{\mathbf{I}_2}{\mathbf{I}_1} \qquad H_4(j\omega) = \frac{\mathbf{I}_2}{\mathbf{V}_1}$$

These are called *sinusoidal transfer functions* since they all involve some kind of ac signal transfer from one pair of terminals to another.

It is conventional for the numerator to be the output phasor, and for the denominator to be the input phasor, *which causes the output*. Thus $H_4(j\omega) = \mathbf{I}_2/\mathbf{V}_1$ relates the output current at port #2 to the voltage at port #1, which is causing the current. Note that $\mathbf{I}_2$ will depend upon the circuit connected at port #2, and that if a circuit *is* connected to port #2, all independent sources therein must be set to zero.

## Example 7.4

Calculate $H_1(j\omega) = \mathbf{V}_2/\mathbf{V}_1$ and $H_2(j\omega) = \mathbf{V}_1/\mathbf{V}_2$ for the two port of Fig. 7-8, given in both cases that the output terminals are open circuit.

To calculate $H_1(j\omega)$ we want $\mathbf{V}_2$ due to $\mathbf{V}_1$ as shown in Fig. 7-9$a$. Applying KCL at node $a$, $\mathbf{I}_L - \mathbf{I}_C = 0$, or

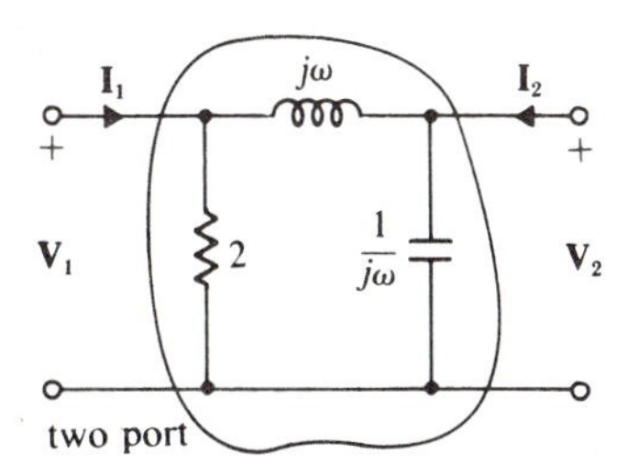


**Fig. 7-8**

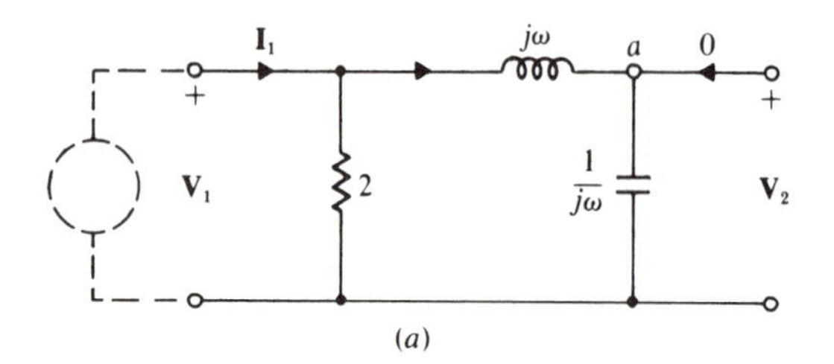


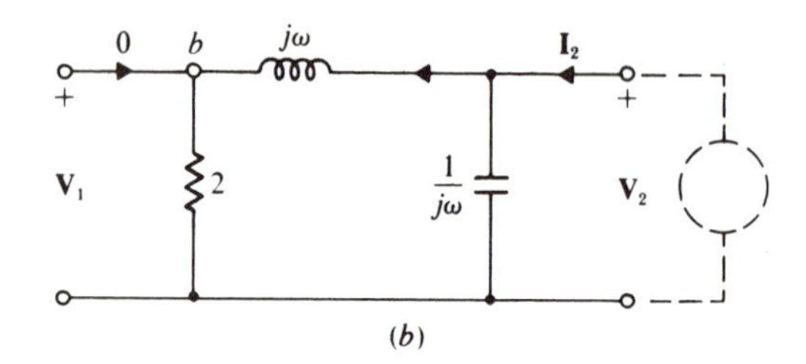


**Fig. 7-9**

$$\frac{1}{j\omega}(\mathbf{V}_1 - \mathbf{V}_2) - j\omega\mathbf{V}_2 = 0$$

from which $\mathbf{V}_2 = \mathbf{V}_1/(-\omega^2 + 1)$ and $H_1(j\omega) = \mathbf{V}_2/\mathbf{V}_1 = 1/(-\omega^2 + 1)$.

To calculate $H_2(j\omega)$ we want $\mathbf{V}_1$ due to $\mathbf{V}_2$ as indicated in Fig. 7-9*b*. If we apply KCL at node $b$,

$$\frac{1}{j\omega}(\mathbf{V}_2 - \mathbf{V}_1) - \tfrac{1}{2}\mathbf{V}_1 = 0$$

and so $\mathbf{V}_1 = 2\mathbf{V}_2/(j\omega + 2)$ and $H_2(j\omega) = \mathbf{V}_1/\mathbf{V}_2 = 2/(j\omega + 2)$. ■

It is instructive to compare the $j\omega$-domain Example 7.4 with the $s$-domain Example 6.2, page 178.

Also note that we could have taken advantage of the solution in the $s$-domain to find

$$H_1(j\omega) = H_1(s)\big|_{s=j\omega} \quad \text{and} \quad H_2(j\omega) = H_2(s)\big|_{s=j\omega}$$

These concepts are readily extended to $n$-ports, and for the general sinusoidal transfer function we write

$$H(j\omega) = \frac{\mathbf{Y}}{\mathbf{U}} = \frac{\text{output phasor}}{\text{input phasor}} \tag{7.15}$$

Sinusoidal transfer functions, admittances, and impedances are collectively referred to as *sinusoidal network functions*.

## APPLICATION OF SINUSOIDAL NETWORK FUNCTIONS

If our primary objective is to relate one specific sinusoidal steady-state output to one given input, the network function approach is appropriate and the phasor output is given by

$$\mathbf{Y} = H(j\omega)\mathbf{U}$$

## MEASUREMENT OF SINUSOIDAL TRANSFER FUNCTIONS

A *sinusoidal* transfer function may be evaluated experimentally (at any given frequency) by applying a sinusoidal input $u(t) = U_m \cos \omega t$, measuring the magnitude and phase of the steady-state output $y_{sss}(t) = Y_m \cos(\omega t + \phi)$, and calculating $|H(j\omega)| = Y_m/U_m$ and $\underline{/H(j\omega)} = \phi$. Then $H(j\omega) = |H(j\omega)|e^{j\underline{/H(j\omega)}}$.

## SINUSOIDAL VOLTAGE AND CURRENT DIVIDERS

We commonly meet the *divider networks* of Fig. 7-10, and it is therefore worth filing away the following easily proved results:

$$\frac{\mathbf{V}_2}{\mathbf{V}} = \frac{Z_2(j\omega)}{Z_1(j\omega) + Z_2(j\omega)} \tag{7.16}$$

$$\frac{\mathbf{I}_2}{\mathbf{I}} = \frac{Y_2(j\omega)}{Y_1(j\omega) + Y_2(j\omega)} \tag{7.17}$$

Note that equation (*7.16*) assumes that no current is drawn from the output terminals.

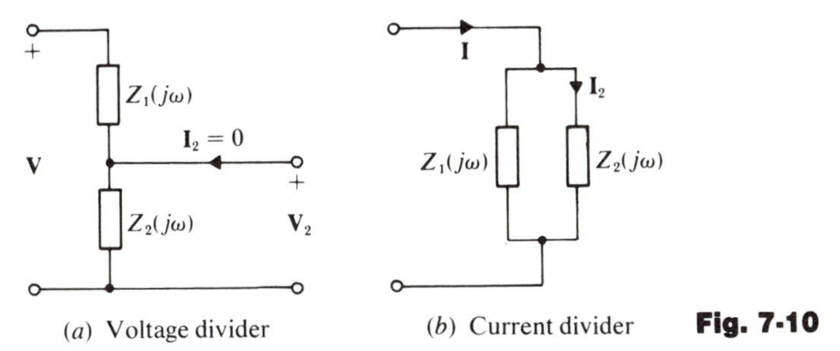


(*a*) Voltage divider (*b*) Current divider **Fig. 7-10**

## GRAPHICAL POLE–ZERO EVALUATION OF H(jω)

Poles and zeros were defined in Chapter 6. Now, if

$$H(s) = \frac{K(s - z_1)(s - z_2)\cdots(s - z_n)}{(s - p_1)(s - p_2)\cdots(s - p_m)}$$

then

$$H(j\omega) = \frac{K(j\omega - z_1)(j\omega - z_2)\cdots(j\omega - z_n)}{(j\omega - p_1)(j\omega - p_2)\cdots(j\omega - p_m)}$$

$$= \frac{K(N_1 e^{j\psi_1})(N_2 e^{j\psi_2}) \cdots (N_n e^{j\psi_n})}{(D_1 e^{j\phi_1})(D_2 e^{j\phi_2}) \cdots (D_m e^{j\phi_m})}$$

$$= \frac{KN_1 N_2 \cdots N_n}{D_1 D_2 \cdots D_m} e^{j(\psi_1 + \psi_2 + \cdots + \psi_n - \phi_1 - \phi_2 - \cdots - \phi_m)}$$

The $N$'s, $D$'s, $\psi$'s, and $\phi$'s can all be identified in the pole-zero plot of $H(s)$, as indicated in Fig. 7-11.

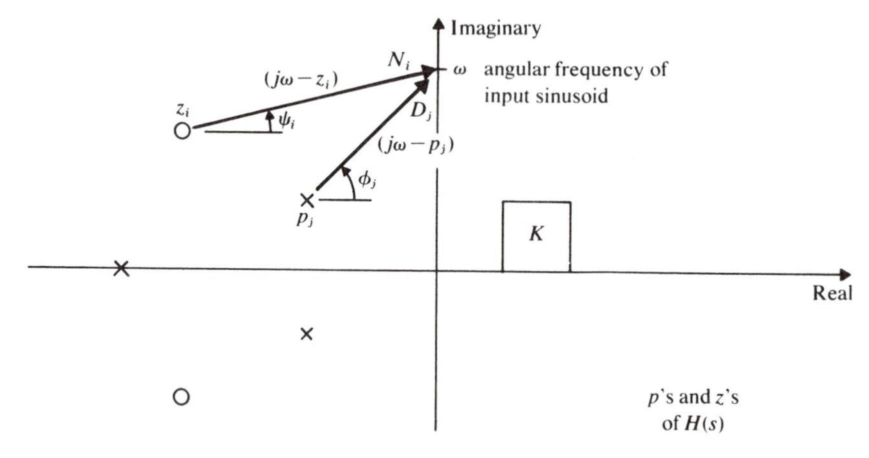


**Fig. 7-11**

Thus we can define the procedure:

To find $H(j\omega)$, join all the finite poles and zeros *to* the point $j\omega$. Measure (or calculate, trigonometrically) all of the $N$'s, $D$'s, $\psi$'s, and $\phi$'s. Compute $H(j\omega) = |H(j\omega)|\, e^{j\underline{/H(j\omega)}}$ from

$$|H(j\omega)| = \frac{KN_1 N_2 \cdots N_n}{D_1 D_2 \cdots D_m}$$

and

$$\underline{/H(j\omega)} = \psi_1 + \psi_2 + \cdots + \psi_n - (\phi_1 + \phi_2 + \cdots + \phi_m)$$

Once $H(j\omega)$ has been evaluated, the ac output follows immediately from equation (*7.2*) or (*7.4*).

## PHASOR DIAGRAMS

Since sinusoidal circuit equations are algebraic and complex, they may often be solved with advantage graphically in the complex plane. But be careful not to confuse phasor diagram analysis with pole-zero calculations. They are quite different.

Consider, for example, the simple series $L$-$R$-$C$ circuit of Fig. 7-3. The corresponding phasor loop equation is

$$jL\omega\mathbf{I} + R\mathbf{I} + \frac{1}{jC\omega}\mathbf{I} = \mathbf{V}$$

We first choose the common phasor **I** as the *reference*, even though it is the unknown variable. Then, as in Fig. 7-12, we draw **I** with zero phase and convenient length.

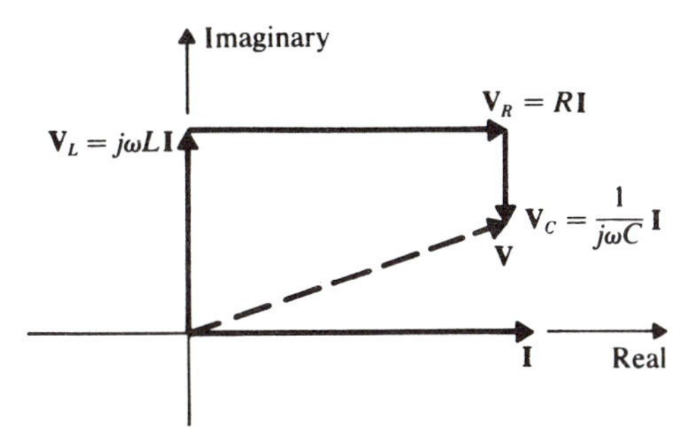


**Fig. 7-12**

Now multiplication by $j$ corresponds to a counterclockwise graphical rotation by 90°. Therefore $jL\omega\mathbf{I}$ is a phasor 90° counterclockwise from **I** (see Fig. 7-12). Similarly $(1/jC\omega)\mathbf{I}$ is 90° clockwise from **I**. And, of course, $R\mathbf{I}$ is parallel to **I**.

Once **I** has been drawn, therefore, the other phasors on the left of the phasor equation may be drawn tail-to-head as shown in Fig. 7-12. Their sum must equal **V**, on the right of the equation.

Now **V** is presumably known (given). And from the phasor diagram we may read off the *relative* magnitudes of **V** and **I**, and their *relative* angle. Hence **I** has been found.

### Example 7.5

An ac voltmeter has been used to measure $V_R = 40$ V and $V_C = 30$ V in the circuit of Fig. 7-13 under sinusoidal steady-state conditions. We will use a phasor diagram to find $v_{sss}(t)$ and the phase of $i_L(t)$ relative to $i(t)$.

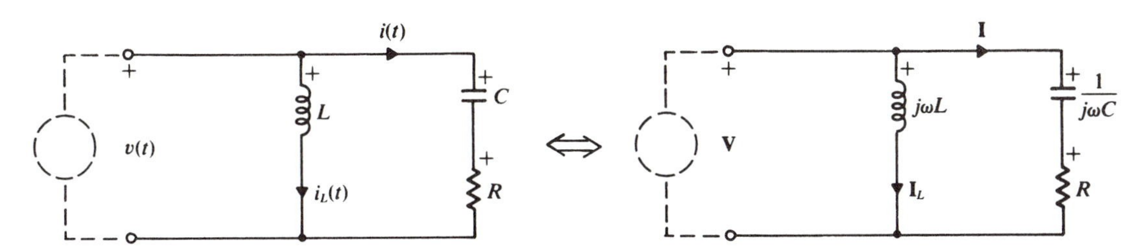


**Fig. 7-13**

From KVL, $\mathbf{V}_C + \mathbf{V}_R = \mathbf{V}$. And we know that $\mathbf{V}_C = (1/jC\omega)\mathbf{I}$ must lag $\mathbf{I}$ by 90°. We can therefore find $\mathbf{V}$ by adding $\mathbf{V}_C$ and $\mathbf{V}_R$ graphically, as shown in Fig. 7-14. Thus

$$\mathbf{V} = \sqrt{30^2 + 40^2}\, e^{j\{\omega t - \tan^{-1}(30/40)\}} = 50 e^{j(\omega t - 36.8^\circ)}$$

or $v_{sss}(t) = 50 \cos(\omega t - 36.8^\circ)$ relative to $i(t)$ as the reference of phase.

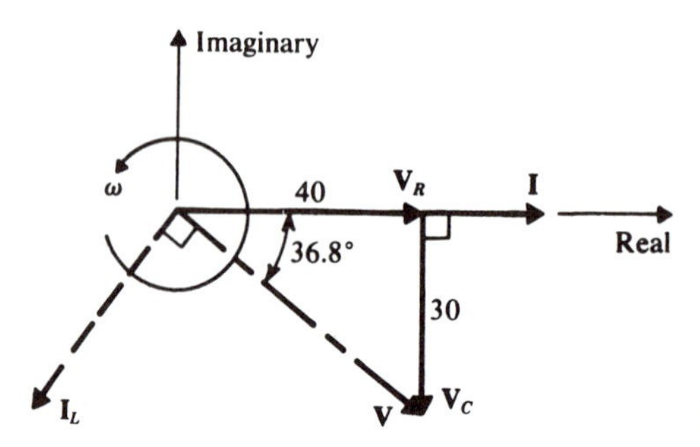


Fig. 7-14

Then, since $\mathbf{I}_L = (1/jL\omega)\mathbf{V}$, $\mathbf{I}_L$ must lag $\mathbf{V}$ by 90°, and so we find from Fig. 7-14 that

$$\angle \mathbf{I}_L - \angle \mathbf{I} = -(90^\circ + 36.8^\circ) = -126.8^\circ$$

which is the angle of $i_L(t)$ relative to $i(t)$.

The phasor diagram can always be employed as a graphical alternative to algebraic methods of solution when there is only one unknown, or when the equations have been reduced to leave one unknown. It does not lend itself to the solution of simultaneous equations. ■

# ILLUSTRATIVE PROBLEMS

## *Sinusoidal Steady-State*

**7.1** The input voltage to a linear circuit is $v_i(t) = 100 \cos(10^6 t + 76^\circ)$ while the steady-state output is $v_{0,sss}(t) = 50 \sin(10^6 t + 67^\circ)$.

(*a*) What is the angular frequency of the input and output sinusoids?
(*b*) What is the true frequency of these signals?
(*c*) Find the gain and the phase shift produced by this circuit.
(*d*) Determine the period of the input and output voltages.

(*a*) $\omega = 10^6$ rad/s.
(*b*) $f = 10^6/2\pi$ Hz.
(*c*) The circuit has a gain of $50/100 = 0.5$ (or an attenuation of $1/0.5 = 2.0$). For phase shift determination, both signals should be expressed in terms of the same trigonometric function. So,

$$v_{0,sss}(t) = 50 \sin(10^6 t + 67^\circ) = 50 \cos(10^6 t - 23^\circ)$$

Comparing $v_{0,\text{sss}}(t)$ with $v_i(t)$, it is easy to see that the phase of the output is $(23° + 76°) = 99°$ behind, or lagging, the input. In a graph or on an oscilloscope $v_{0,\text{sss}}(t)$ would be displaced to the *right* relative to $v_i(t)$.
(*d*) $T = 1/f = 2\pi/10^6$ s.

**7.2** In the circuit of Fig. 7-15 we want to find the sinusoidal steady-state voltage $v_{\text{sss}}(t)$ for several different sinusoidal inputs $e(t)$.

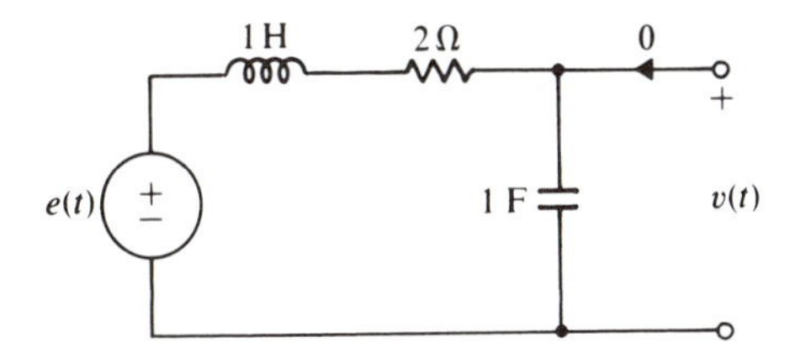


**Fig. 7-15**

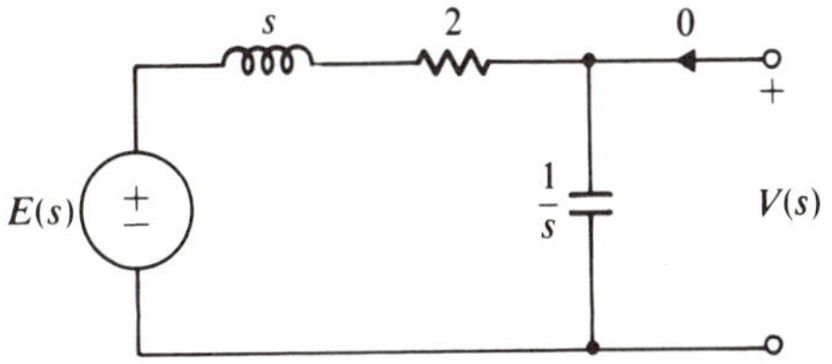

**Fig. 7-16**

(*a*) As the first step in applying equation (7.2), page 216, to this problem, find the transfer function $H(s) = V(s)/E(s)$.
(*b*) Then find $v_{\text{sss}}(t)$ for $e(t) = 20 \cos 3t$.
(*c*) What is $v_{\text{sss}}(t)$ if $e(t) = 20 \cos (t + 30°)$?
(*d*) Finally, determine $v_{\text{sss}}(t)$ for $e(t) = 20 \sin (t + 30°)$.

(*a*) First we draw an *s*-domain circuit diagram as in Fig. 7-16. Then by inspection,

$$H(s) = \frac{V(s)}{E(s)} = \frac{1/s}{s + 2 + 1/s} = \frac{1}{s^2 + 2s + 1}$$

(*b*) When $e(t) = 20 \cos 3t$, $\omega = 3$ and

$$H(j3) = \left.\frac{1}{s^2 + 2s + 1}\right|_{s=j3} = \frac{1}{(j3)^2 + 2(j3) + 1} = \frac{1}{-8 + j6} = 0.1e^{-j143.2°}$$

Thus $|H(j3)| = 0.1$ and $\angle H(j3) = -143.2°$. And by direct substitution into equation (7.2),

$$v_{\text{sss}}(t) = (20)(0.1) \cos (3t + 0° - 143.2°) = 2 \cos (3t - 143.2°)$$

(*c*) Here $\omega = 1$ and $H(j1) = 1/\{(j1)^2 + 2j + 1\} = 0.5e^{-j90°}$. Therefore

$$v_{\text{sss}}(t) = (20)(0.5) \cos (t + 30° - 90°) = 10 \cos (t - 60°)$$

(*d*) Converting the input into a cosine function, we have

$$e(t) = 20 \cos (t + 30° - 90°) = 20 \cos (t - 60°)$$

Then with $H(j1) = 0.5e^{-j90°}$ from part (*c*),

$$v_{\text{sss}}(t) = (20)(0.5) \cos (t - 60° - 90°) = 10 \cos (t - 150°) \quad \text{or} \quad 10 \sin (t - 60°)$$

*Comment:* The "magic formula" may be expressed in terms of the sine function. Thus if the input is $u(t) = U_m \sin (\omega t + \theta)$, the sinusoidal steady-state output will be

$$y_{\text{sss}}(t) = U_m |H(j\omega)| \sin (\omega t + \theta + \angle H(j\omega))$$

The solution to part $(d)$ of this problem then follows directly as

$$v_{sss}(t) = (20)(0.5)\sin(t + 30° - 90°) = 10\sin(t - 60°)$$

**7.3** It is known that the network function for an electric circuit is $H(s) = V(s)/E(s) = 10(s + 1)/(s^2 + 2s + 3)$. Find the sinusoidal steady-state solution $v_{sss}(t)$ when $e(t) = 4\cos 2t$.

Here the angular frequency $\omega$ is 2 rad/s, so

$$H(j2) = H(s)\big|_{s=j2} = \left.\frac{10(s+1)}{s^2 + 2s + 3}\right|_{s=j2} = \frac{10(j2+1)}{-4 + j4 + 3} = \frac{10\sqrt{5}\,e^{j63.4°}}{\sqrt{17}\,e^{j104°}} = 5.44e^{-j40.6°}$$

Substituting into equation (7.2), page 216, $v_{sss}(t) = (4)(5.44)\cos(2t - 40.6°) = 21.76\cos(2t - 40.6°)$.

**7.4** Use loop equations to determine the transfer function $H(s) = I_1(s)/E(s)$ for the network of Fig. 7-17. Then for $e(t) = 7\cos(3t + 4°)$, write $i_{1,sss}(t)$ with the aid of equation (7.2).

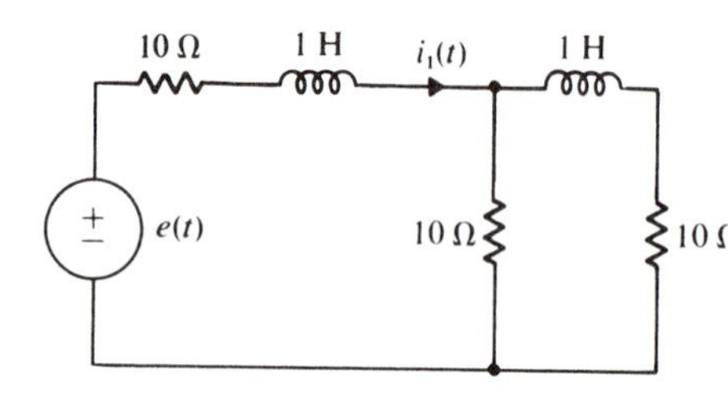


**Fig. 7-17**

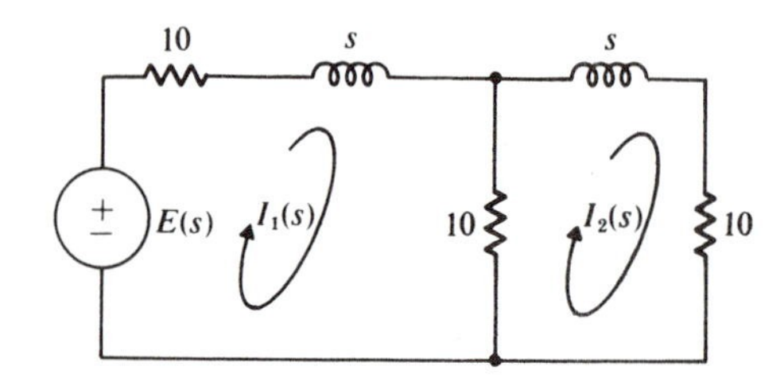


**Fig. 7-18**

The $s$-domain diagram is shown in Fig. 7-18, and an appropriate set of loop equations is

$$\{20 + s\}I_1(s) - 10I_2(s) = E(s)$$

$$-10I_1(s) + \{20 + s\}I_2(s) = 0$$

Using Cramer's rule, we find

$$I_1(s) = \frac{\begin{vmatrix} E(s) & -10 \\ 0 & 20 + s \end{vmatrix}}{\begin{vmatrix} 20 + s & -10 \\ -10 & 20 + s \end{vmatrix}} = \frac{\{20 + s\}E(s)}{s^2 + 40s + 300}$$

and the required transfer function is

$$H(s) = \frac{I_1(s)}{E(s)} = \frac{s + 20}{s^2 + 40s + 300}$$

Now with $\omega = 3$,

$$H(j3) = \frac{20 + j3}{(j3)^2 + 40(j3) + 300} = \frac{20 + j3}{291 + j120} = 0.064e^{-j13.9°}$$

Finally, from equation (7.2), $i_{1,\text{sss}}(t) = (7)(0.064)\cos(3t + 4° - 13.9°) = 0.448 \cos(3t - 9.9°)$.

**7.5** Apply nodal analysis to the circuit of Fig. 7-19 to find $H(s) = V_a(s)/E(s)$ and then use equation (7.2) to determine $v_{a,\text{sss}}(t)$ for $e(t) = 5\cos(2t + 117°)$.

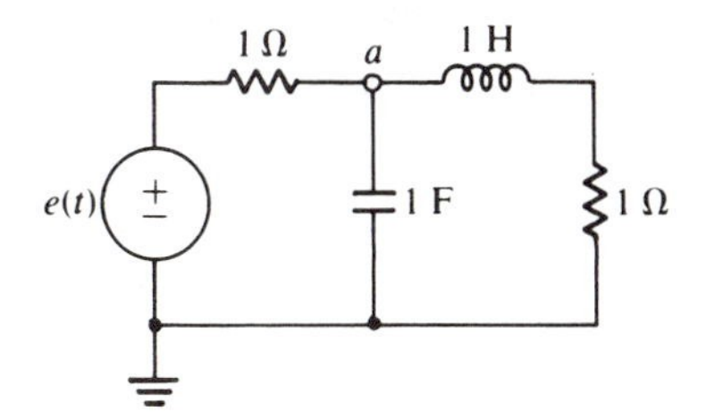


Fig. 7-19

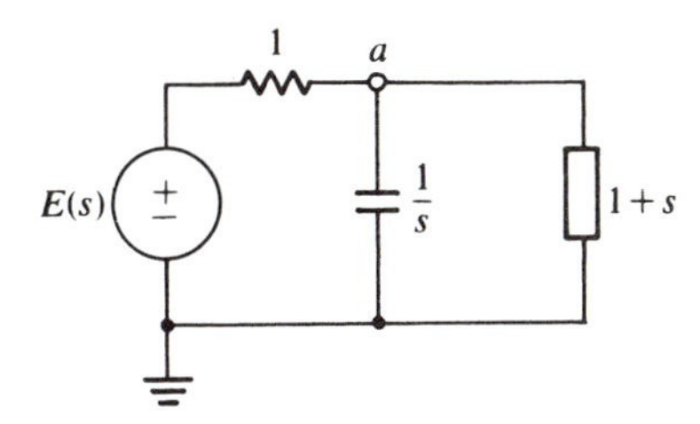


Fig. 7-20

From the $s$-domain circuit of Fig. 7-20 we can write

$$\frac{V_a(s)}{1+s} + \frac{V_a(s)}{1/s} + \frac{V_a(s) - E(s)}{1} = 0$$

which when simplified yields

$$\frac{s^2 + 2s + 2}{s+1} V_a(s) = E(s) \quad \text{or} \quad H(s) = \frac{V_a(s)}{E(s)} = \frac{s+1}{s^2 + 2s + 2}$$

Then with $\omega = 2$,

$$H(j2) = \frac{j2 + 1}{(j2)^2 + 2(j2) + 2} = \frac{1 + j2}{-2 + j4} = \frac{2.23e^{j63.4°}}{4.48e^{j116.6°}} = 0.5e^{-j53.2°}$$

And using equation (7.2), $v_{a,\text{sss}}(t) = (5)(0.5)\cos(2t + 117° - 53.2°) = 2.5\cos(2t + 63.8°)$.

## The Phasor Concept and Phasor Analysis

**7.6** (*a*) Write $\mathbf{U}$, the rotating phasor that represents $u(t) = 10\cos(20t + 30°)$.
(*b*) Find the sinusoidal signal $v(t)$ which is represented by the rotating phasor $\mathbf{V} = 3e^{j(5t+4°)}$.

(*a*) $\mathbf{U} = 10e^{j(20t+30°)}$.
(*b*) From equation (7.3), page 218,

$$v(t) = \text{Re}[\mathbf{V}] = \text{Re}[3e^{j(5t+4°)}]$$
$$= \text{Re}[3\cos(5t + 4°) + 3j\sin(5t + 4°)] = 3\cos(5t + 4°)$$

**7.7** A sinusoidal input $u(t) = 3\cos(4t + 5°)$ causes the sinusoidal steady-state output $y_{\text{sss}}(t) = 6\cos(4t + 58.2°)$. (*a*) Represent $u(t)$ and $y_{\text{sss}}(t)$ by rotating phasors. (*b*) Find the sinusoidal network function $H(j4) = \mathbf{Y}/\mathbf{U}$.

(*a*) $\mathbf{U} = 3e^{j(4t+5°)}$ and $\mathbf{Y} = 6e^{j(4t+58.2°)}$.
(*b*) From equation (*7.4*), page 219,

$$H(j4) = \frac{\mathbf{Y}}{\mathbf{U}} = \frac{6e^{j(4t+58.2°)}}{3e^{j(4t+5°)}} = 2e^{j53.2°}$$

**7.8** In Problem 7.3 a network function for a linear circuit was given as $H(s) = V(s)/E(s) = 10(s+1)/(s^2+2s+3)$. Use equation (*7.4*) to find $v_{sss}(t)$ given $e(t) = 4\cos 2t$.

Here $\omega = 2$ and

$$H(j2) = H(s)\big|_{s=j2} = \frac{10(j2+1)}{(j2)^2 + 2(j2) + 3} = 5.44e^{-j40.6°}$$

And from equation (*7.4*),

$$\mathbf{V} = H(j\omega)\mathbf{E} = (5.44e^{-j40.6°})(4e^{j2t})$$

Finally,

$$v_{sss}(t) = \text{Re}\,[\mathbf{V}] = 21.76\cos(2t - 40.6°)$$

**7.9** Use the classical method of Chapter 4 to find $i_{sss}(t)$ in the circuit of Fig. 7-21 when $v(t) = V_m\cos(\omega t + \phi)$.

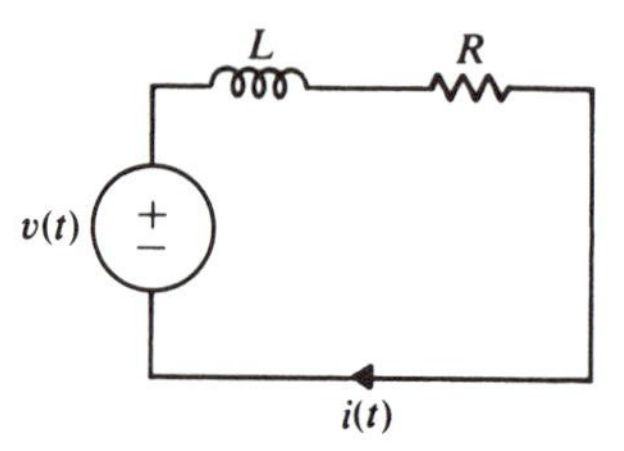


**Fig. 7-21**

The appropriate differential equation is $L(di/dt) + Ri = V_m\cos(\omega t + \phi)$. Following the classical procedure of Chapter 4 we assume a particular solution of the form $i_p(t) = I_m\cos(\omega t + \beta)$. This will also be the sinusoidal steady-state solution, since the complementary function $i_c(t)$ dies out in time.

Further, using the Euler relationship,

$$i_p(t) = i_{sss}(t) = I_m\cos(\omega t + \beta) = \text{Re}\,[I_m e^{j(\omega t+\beta)}]$$

and

$$v(t) = \text{Re}\,[V_m e^{j(\omega t+\phi)}]$$

Substituting into the differential equation yields

$$L\frac{d}{dt}\{\text{Re}\,[I_m e^{j(\omega t+\beta)}]\} + R\{\text{Re}\,[I_m e^{j(\omega t+\beta)}]\} = \text{Re}\,[V_m e^{j(\omega t+\phi)}]$$

Now

$$\frac{d}{dt}\{\text{Re}\,[I_m e^{j(\omega t+\beta)}]\} = \frac{d}{dt}\{I_m\cos(\omega t+\beta)\} = -\omega I_m\sin(\omega t+\beta)$$
$$= \text{Re}\,[j\omega I_m e^{j(\omega t+\beta)}]$$

Thus we can write

$$\operatorname{Re}\left[j\omega L I_m e^{j(\omega t+\beta)}\right] + \operatorname{Re}\left[R I_m e^{j(\omega t+\beta)}\right] = \operatorname{Re}\left[V_m e^{j(\omega t+\phi)}\right]$$

As usual, it is convenient to make use of phasor notation while completing the solution. Here

$$(j\omega L + R)\mathbf{I} = \mathbf{V} \quad \text{or} \quad \mathbf{I} = \frac{1}{j\omega L + R}\mathbf{V}$$

Therefore

$$\mathbf{I} = \frac{V_m}{\sqrt{\omega^2 L^2 + R^2}} e^{j\{\omega t + \phi - \tan^{-1}(\omega L/R)\}}$$

and

$$i_{sss}(t) = \frac{V_m}{\sqrt{\omega^2 L^2 + R^2}} \cos\{\omega t + \phi - \tan^{-1}(\omega L/R)\}$$

*Comment:* At no time have we used equation (*7.2*) or (*7.4*). Nor was it *necessary* to use phasors. We can, if we do not mind the extra work, solve ac problems by the classical method.

**7.10** The network of Fig. 7-22 is assumed to be operating in the sinusoidal steady-state with $v(t) = v_{sss}(t) = 2\cos(0.5t - 30°)$. Use the phasor concept to find $e(t)$. *Hint:* First draw a phasor equivalent of Fig. 7-22 and then apply equations (*7.5*)–(*7.9*) to find the phasor $\mathbf{E}$.

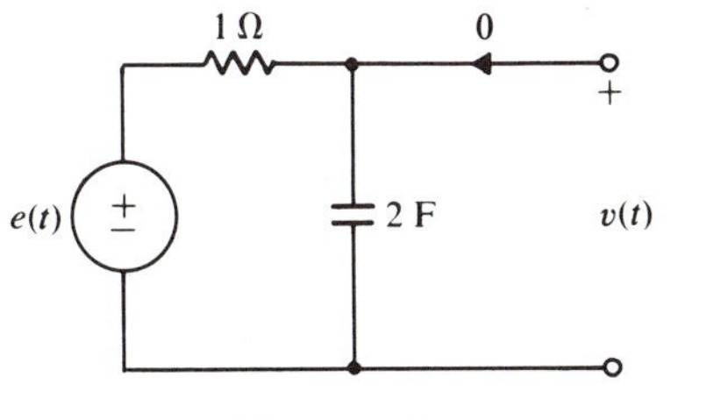


**Fig. 7-22**

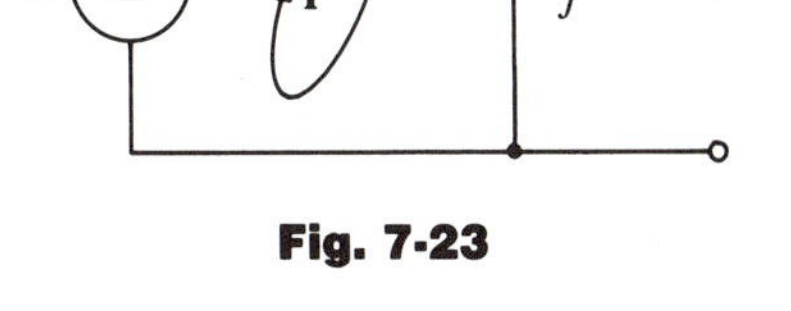


**Fig. 7-23**

In Fig. 7-23 we have drawn the phasor equivalent of the circuit in Fig. 7-22, marking the impedances $Z_R(j0.5) = 1$ and $Z_C(j0.5) = 1/j(2)(0.5) = 1/j$. Then applying KVL, $-\mathbf{E} + \mathbf{V}_R + \mathbf{V} = 0$, or

$$\mathbf{E} = (1)\mathbf{I} + \frac{1}{j}\mathbf{I} = (1 - j)\mathbf{I}$$

But $\mathbf{I} = \mathbf{V}/(1/j) = j\mathbf{V} = j2e^{j(0.5t-30°)} = 2e^{j(0.5t+60°)}$.

Thus

$$\mathbf{E} = (1 - j)(2e^{j(0.5t+60°)}) = (1.41e^{-j45°})(2e^{j(0.5t+60°)}) = 2.82e^{j(0.5t+15°)}$$

Finally,

$$e(t) = \operatorname{Re}\left[2.82e^{j(0.5t+15°)}\right] = 2.82\cos(0.5t + 15°)$$

**7.11** Draw the phasor domain equivalent of Fig. 7-24, write the phasor loop equations, and then solve for $i_{R,\text{sss}}(t)$ if $i(t) = 10 \cos t$.

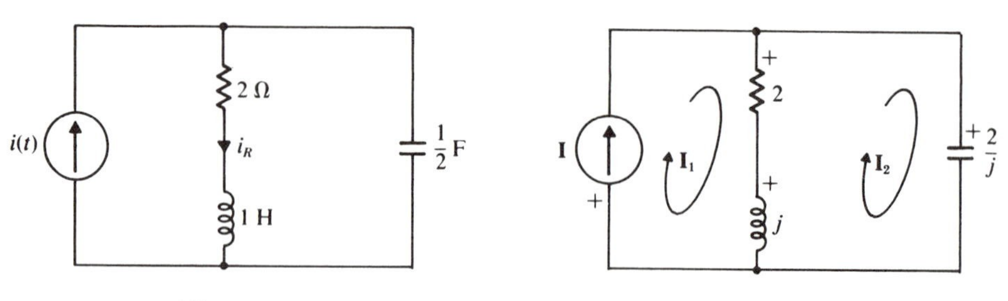


Fig. 7-24 Fig. 7-25

The phasor equivalent of Fig. 7-24 with $\omega = 1$ is shown in Fig. 7-25. Since the left-hand loop current is known, we need only one KVL equation $-\mathbf{V}_L - \mathbf{V}_R + \mathbf{V}_C = 0$. Substituting the sinusoidal $v$-$i$ relations yields

$$-j(\mathbf{I} - \mathbf{I}_2) - 2(\mathbf{I} - \mathbf{I}_2) + (-2j)\mathbf{I}_2 = 0$$

from which

$$\mathbf{I}_2 = \frac{2 + j}{2 - j}\mathbf{I} = e^{j53.1^\circ}\mathbf{I} = (0.6 + j0.8)\mathbf{I}$$

Therefore

$$\mathbf{I}_R = \mathbf{I} - \mathbf{I}_2 = \mathbf{I} - (0.6 + j0.8)\mathbf{I} = (0.4 - j0.8)\mathbf{I}$$
$$= (0.894e^{-j63.4})(10e^{jt}) = 8.94e^{j(t-63.4^\circ)}$$

and

$$i_{R,\text{sss}}(t) = \text{Re}\,[\mathbf{I}_R] = 8.94 \cos (t - 63.4^\circ)$$

**7.12** Use the phasor concept to find $v_{\text{sss}}(t)$ given $i(t) = A \cos \omega t$ in the network of Fig. 7-26.

Figure 7-27 is the $j\omega$-domain equivalent of Fig. 7-26. If we apply KCL, $-\mathbf{I} + \mathbf{I}_R + \mathbf{I}_L = 0$. That is,

$$-\mathbf{I} + \frac{\mathbf{V}}{R} + \frac{\mathbf{V}}{j\omega L} = 0 \quad \text{or} \quad \left(\frac{1}{R} + \frac{1}{j\omega L}\right)\mathbf{V} = \mathbf{I}$$

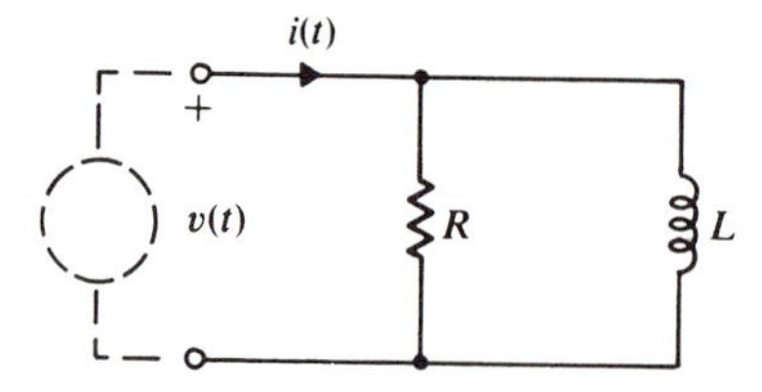


Fig. 7-26

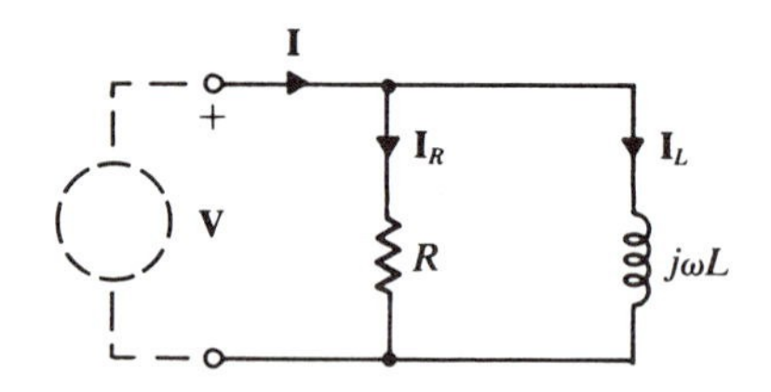


Fig. 7-27

Solving for **V**,

$$\mathbf{V} = \frac{j\omega LR}{R + j\omega L}\mathbf{I} = \frac{(\omega LRe^{j90^\circ})(Ae^{j\omega t})}{\sqrt{R^2 + \omega^2L^2}\,e^{j\tan^{-1}(\omega L/R)}} = \frac{A\omega LR}{\sqrt{R^2 + \omega^2L^2}}e^{j\{\omega t + 90^\circ - \tan^{-1}(\omega L/R)\}}$$

Finally,

$$v_{sss}(t) = \text{Re}\,[\mathbf{V}] = \frac{A\omega LR}{\sqrt{R^2 + \omega^2L^2}}\cos\{\omega t + 90^\circ - \tan^{-1}(\omega L/R)\}$$

*Comment:* From the above sinusoidal $v$-$i$ relation we see that this network's Thévenin or input impedance is $j\omega LR/(R + j\omega L)$.

### *Sinusoidal Impedance*

**7.13** Find the resistive and reactive parts of the sinusoidal input impedance $Z(j\omega) = \mathbf{V}/\mathbf{I}$ for the $j\omega$-domain circuit of Fig. 7-27.

Since the resistor and inductor are in parallel, we can write

$$\frac{1}{Z(j\omega)} = 1/R + 1/j\omega L = \frac{R + j\omega L}{j\omega LR}$$

and rationalizing,

$$Z(j\omega) = \frac{j\omega LR}{R + j\omega L} = \left(\frac{j\omega LR}{R + j\omega L}\right)\left(\frac{R - j\omega L}{R - j\omega L}\right) = \frac{\omega^2L^2R}{R^2 + \omega^2L^2} + j\frac{R^2\omega L}{R^2 + \omega^2L^2}$$

Thus the resistive and reactive parts of $Z(j\omega)$ are

$$R(\omega) = \frac{\omega^2L^2R}{R^2 + \omega^2L^2} \quad\text{and}\quad X(\omega) = \frac{R^2\omega L}{R^2 + \omega^2L^2}$$

*Note:* The conductance $G(\omega)$ and susceptance $B(\omega)$ are defined by $Y(j\omega) = G(\omega) + jB(\omega)$. But although $Y(j\omega) = 1/Z(j\omega)$, $G(\omega) \neq 1/R(\omega)$ and $B(\omega) \neq 1/X(\omega)$. Here, for example,

$$Y(j\omega) = \frac{1}{Z(j\omega)} = \frac{R + j\omega L}{j\omega LR} = \frac{\omega L - jR}{\omega LR}$$

and so

$$G(\omega) = \frac{1}{R} \quad\text{and}\quad B(\omega) = -\frac{1}{\omega L}$$

**7.14** Given the circuit of Fig. 7-28, draw the phasor-domain diagram, find $Z(j\omega) = \mathbf{V}/\mathbf{I}$ by sinusoidal impedance combinations, and determine the resistive and reactive parts of $Z(j\omega)$.

See Fig. 7-29 for the $j\omega$-domain circuit diagram. Then

$$\frac{1}{Z_{ab}(j\omega)} = \frac{1}{1/4} + \frac{1}{1/j2\omega} = 4 + j2\omega$$

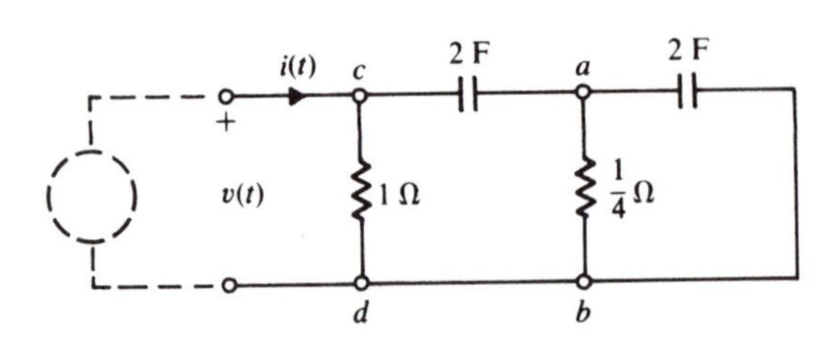


Fig. 7-28

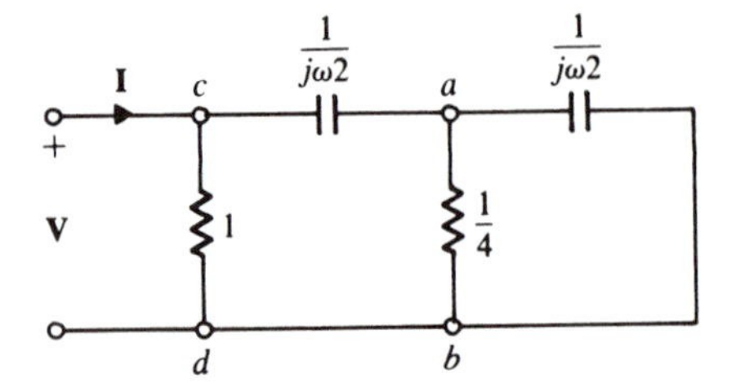


Fig. 7-29

Next,

$$Z_{cab}(j\omega) = \frac{1}{j2\omega} + \frac{1}{4 + j2\omega} = \frac{4 + j2\omega + j2\omega}{j2\omega(4 + j2\omega)}$$

And finally,

$$\frac{1}{Z_{cd}(j\omega)} = \frac{1}{Z_{cab}(j\omega)} + 1 = \frac{j2\omega(4 + j2\omega)}{4 + j4\omega} + 1 = \frac{4 - 4\omega^2 + 12j\omega}{4 + 4j\omega}$$

or

$$Z_{cd}(j\omega) = Z(j\omega) = \frac{1 + j\omega}{1 - \omega^2 + 3j\omega}$$

To find $R(\omega)$ and $X(\omega)$ we must rationalize $Z(j\omega)$, i.e.

$$Z(j\omega) = \left\{\frac{1 + j\omega}{(1 - \omega^2) + 3j\omega}\right\}\left\{\frac{(1 - \omega^2) - 3j\omega}{(1 - \omega^2) - 3j\omega}\right\}$$

$$= \frac{1 - \omega^2 + 3\omega^2 + j(\omega - \omega^3 - 3\omega)}{(1 - \omega^2)^2 + 9\omega^2}$$

so that

$$R(\omega) = \frac{1 + 2\omega^2}{(1 - \omega^2)^2 + 9\omega^2} \quad \text{and} \quad X(\omega) = \frac{-\omega^3 - 2\omega}{(1 - \omega^2)^2 + 9\omega^2}$$

**7.15** Use impedance reductions and the phasor concept to find $i_{sss}(t)$ for the circuit in Fig. 7-30.

From the $j\omega$-domain diagram of Fig. 7-31, in which $\omega = 1$,

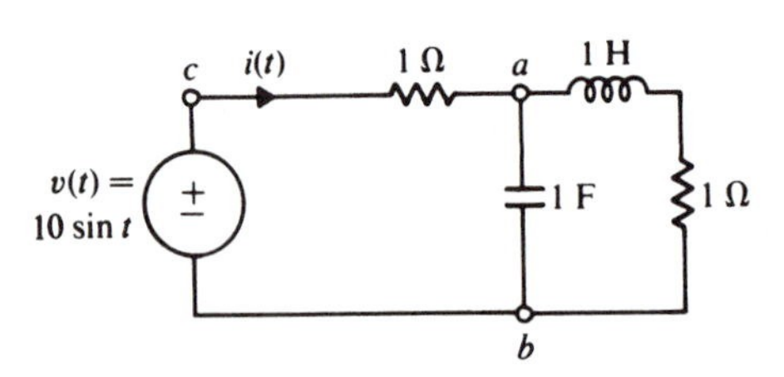

Fig. 7-30

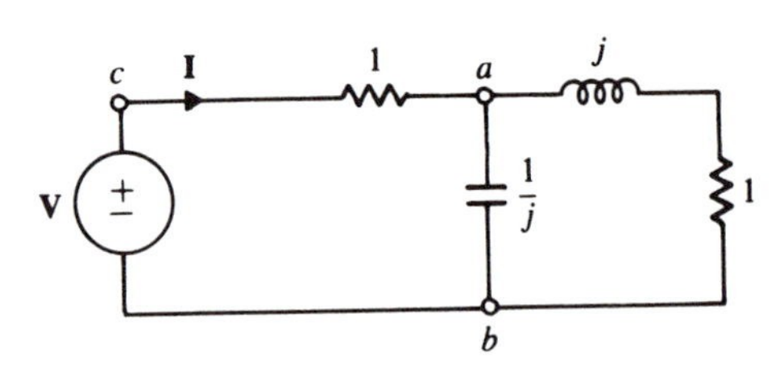


Fig. 7-31

$$\frac{1}{Z_{ab}(j1)} = j + \frac{1}{1+j} = \frac{j}{1+j} \quad \text{or} \quad Z_{ab}(j1) = 1 - j$$

and

$$Z_{cab}(j1) = 1 + 1 - j = 2 - j = \sqrt{5}\, e^{-j26.6^\circ}$$

Finally, *if* we represent the *sine* function $v(t) = 10 \sin t$ by the phasor $\mathbf{V} = 10e^{jt}$, then

$$\mathbf{I} = \frac{\mathbf{V}}{Z_{cab}(j1)} = \frac{10e^{jt}}{\sqrt{5}\, e^{-j26.6^\circ}} = 2\sqrt{5}\, e^{j(t+26.6^\circ)}$$

This phasor represents the *sine* function

$$i_{sss}(t) = 2\sqrt{5} \sin (t + 26.6^\circ)$$

*Note:* If the input is a sine function, it is *not* necessary to convert to a cosine. Only the *relative* phase is relevant, so it does not matter which function is used, provided that the *same* function, sine or cosine, is used for both input and output.

**7.16** Find the sinusoidal network function $H(j\omega) = \mathbf{V}/\mathbf{E}$ for the network of Fig. 7-32. Work from a $j\omega$-domain diagram.

Writing a nodal equation at $a$ in Fig. 7-33$a$, $\mathbf{I}_{R_1} + \mathbf{I}_{R_2} + \mathbf{I}_C = 0$ or

$$\frac{\mathbf{V} - \mathbf{E}}{10} + \frac{\mathbf{V}}{20} + \frac{\mathbf{V}}{20/j\omega} = 0$$

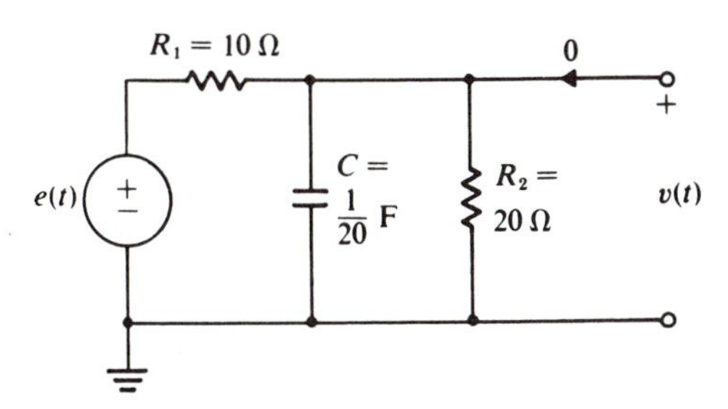


Fig. 7-32

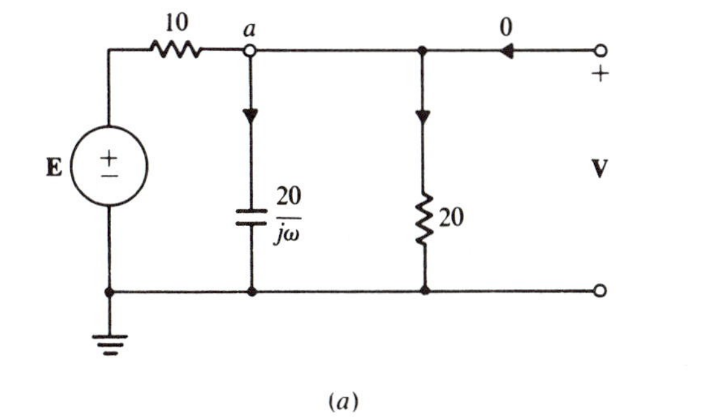


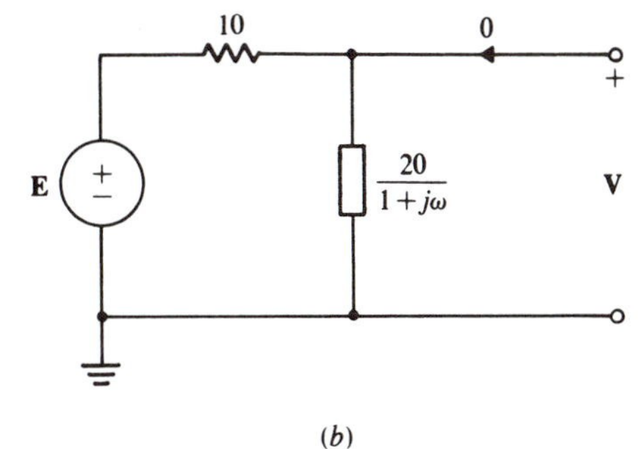


Fig. 7-33

Thus

$$\left(\frac{1}{10} + \frac{1}{20} + \frac{j\omega}{20}\right)\mathbf{V} = \frac{1}{10}\mathbf{E}$$

and

$$H(j\omega) = \frac{\mathbf{V}}{\mathbf{E}} = \frac{2}{3 + j\omega}$$

*Alternative Solution*

After combining the two parallel elements, we can apply the voltage divider concept to the circuit of Fig. 7-33*b*. That is,

$$H(j\omega) = \frac{\mathbf{V}}{\mathbf{E}} = \frac{20/(j\omega + 1)}{10 + 20/(j\omega + 1)} = \frac{20}{10j\omega + 10 + 20} = \frac{2}{3 + j\omega}$$

## Loop and Nodal Equations

**7.17** Draw the $j\omega$-domain diagram for the circuit of Fig. 7-34, given $v(t) = 7 \cos(3t + 4°)$. Then write the loop equations in the matrix form $[Z(j\omega)][\mathbf{I}] = [\mathbf{V}]$ and solve for $\mathbf{I}_1$ and $i_{1,\text{sss}}(t)$.

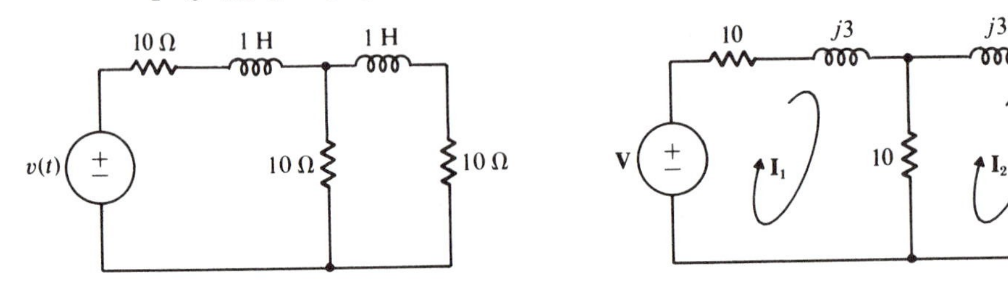


**Fig. 7-34** **Fig. 7-35**

The $j\omega$-domain diagram with properly assigned mesh currents is shown in Fig. 7-35. Then by inspection,

$$\begin{bmatrix} 20 + j3 & -10 \\ -10 & 20 + j3 \end{bmatrix}\begin{bmatrix} \mathbf{I}_1 \\ \mathbf{I}_2 \end{bmatrix} = \begin{bmatrix} \mathbf{V} \\ 0 \end{bmatrix}$$

Using Cramer's rule,

$$\mathbf{I}_1 = \frac{\begin{vmatrix} \mathbf{V} & -10 \\ 0 & 20 + j3 \end{vmatrix}}{\begin{vmatrix} 20 + j3 & -10 \\ -10 & 20 + j3 \end{vmatrix}} = \frac{(20 + j3)\mathbf{V}}{291 + j120} = \frac{(20e^{j8.5°})(7e^{j(3t+4°)})}{315e^{j22.4°}} = 0.448e^{j(3t-9.9°)}$$

Finally,

$$i_{1,\text{sss}}(t) = 0.448 \cos(3t - 9.9°)$$

This problem was solved using the "magic formula" in Problem 7.4.

**7.18** The sinusoidal steady-state capacitor voltage $v_{n,\text{sss}}(t)$ is to be determined for the circuit in Fig. 7-36, where $v(t) = 5 \cos 2t$. Find $\mathbf{V}_n$ by writing a *single* nodal equation and verify the results using the voltage divider concept.

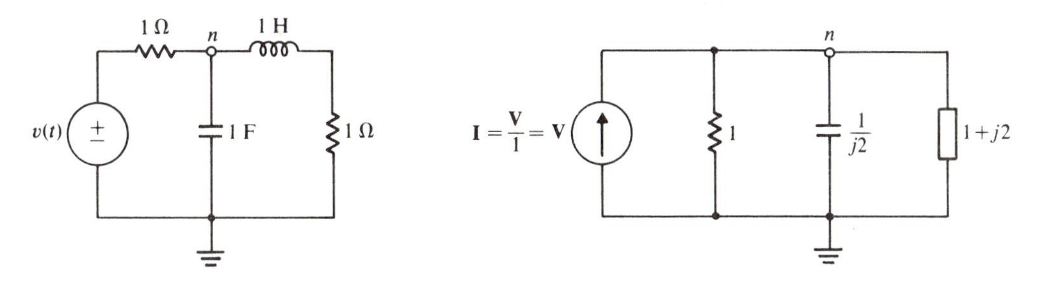


**Fig. 7-36** **Fig. 7-37**

The phasor circuit diagram after a source transformation is shown in Fig. 7-37. Then at node $n$,

$$\left(1 + j2 + \frac{1}{1 + j2}\right)\mathbf{V}_n = \left(1 + j2 + \frac{1 - j2}{5}\right)\mathbf{V}_n = \mathbf{I}$$

or

$$\mathbf{V}_n = \frac{5\mathbf{I}}{6 + j8} = 0.5e^{-j53.2^\circ}\mathbf{V}$$

$$= (0.5e^{-j53.2^\circ})(5e^{j2t}) = 2.5e^{j(2t-53.2^\circ)}$$

To apply the voltage divider concept, the circuit of Fig. 7-36 must be put in the form of Fig. 7-38, where $Z_1(j2) = 1$ and

$$Z_2(j2) = \frac{1}{j2 + 1/(1 + j2)} = \frac{1 + j2}{-3 + j2}$$

It now follows that

$$\frac{\mathbf{V}_n}{\mathbf{V}} = \frac{Z_2(j2)}{Z_1(j2) + Z_2(j2)} = \left(\frac{1 + j2}{-3 + j2}\right)\Big/\left(1 + \frac{1 + j2}{-3 + j2}\right) = \frac{1 + j2}{-2 + j4}$$

$$= \frac{\sqrt{5}\, e^{j63.4^\circ}}{\sqrt{20}\, e^{j116.6^\circ}} = 0.5e^{-j53.2^\circ}$$

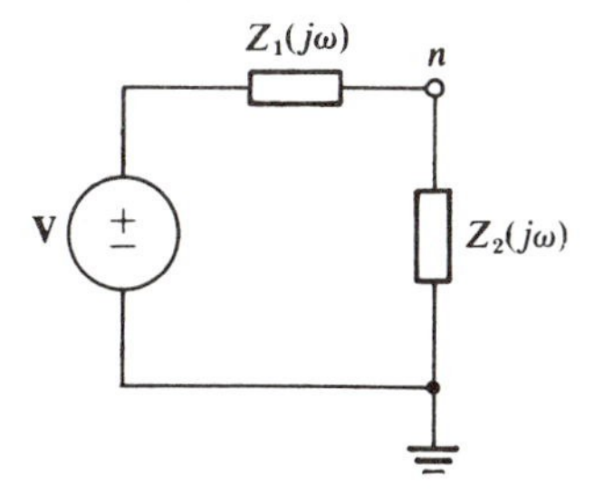


**Fig. 7-38**

and finally,

$$\mathbf{V}_n = (0.5e^{-j53.2°})(5e^{j2t}) = 2.5e^{j(2t-53.2°)}$$

which checks with the previous solution. Returning to the time-domain,

$$v_{n,\text{sss}}(t) = 2.5 \cos (2t - 53.2°)$$

**7.19** Phasor loop equations $[Z(j\omega)][\mathbf{I}] = [\mathbf{V}]$ are to be written for the network of Fig. 7-39. Prepare a $j\omega$-domain diagram suitable for this method of analysis and then write the phasor loop equations.

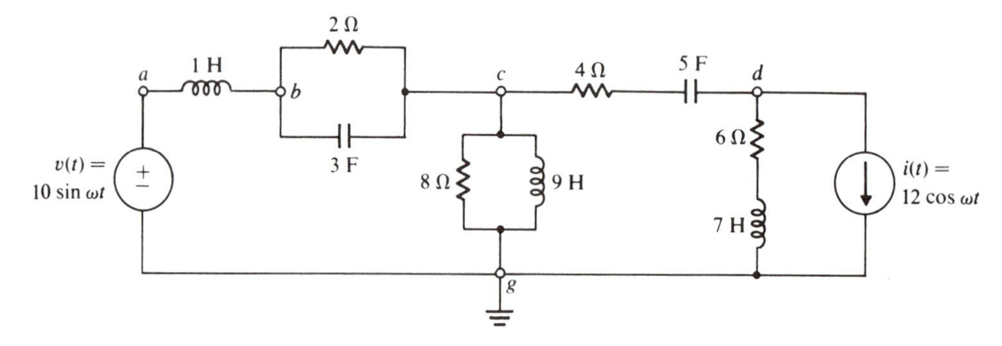


**Fig. 7-39**

Before we can write loop equations by inspection, we must convert the current source into a voltage source. We must also rewrite the sine function as a cosine (or vice versa). That is, $v(t) = 10 \sin \omega t = 10 \cos (\omega t - 90°)$. The $j\omega$-domain diagram of Fig. 7-40 now follows.

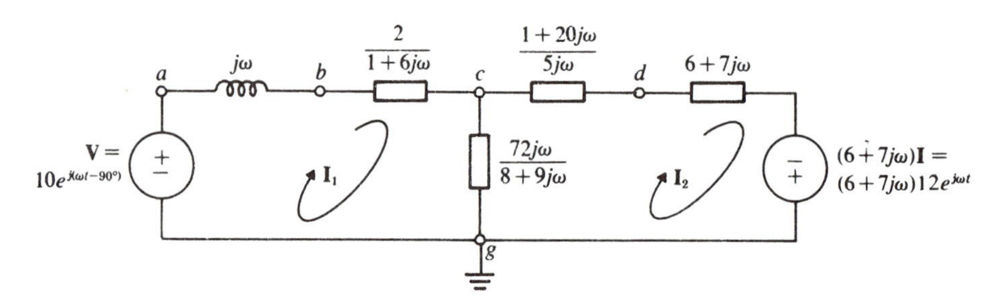


**Fig. 7-40**

Then by inspection, since $\mathbf{I}_1$ and $\mathbf{I}_2$ have both been chosen clockwise,

$$\begin{bmatrix} j\omega + \dfrac{2}{1+6j\omega} + \dfrac{72j\omega}{8+9j\omega} & \dfrac{-72j\omega}{8+9j\omega} \\ \dfrac{-72j\omega}{8+9j\omega} & 10 + 7j\omega + \dfrac{1}{5j\omega} + \dfrac{72j\omega}{8+9j\omega} \end{bmatrix} \begin{bmatrix} \mathbf{I}_1 \\ \mathbf{I}_2 \end{bmatrix} = \begin{bmatrix} 10e^{j(\omega t-90°)} \\ 12(6+7j\omega)e^{j\omega t} \end{bmatrix}$$

*Comments*

1. The first term in $[Z(j\omega)]$ is the impedance in loop #1; the last is that in loop #2. The off-diagonal terms are the negative of the impedance *common* to the two loops.
2. The terms on the right of the equation are the voltage *rises* in loops 1 and 2 as we travel in the direction of the loop currents.
3. The two source frequencies *must* be the same. Otherwise $[Z(j\omega)]$ would be two different matrices at once! We will discover how to solve such problems in Chapter 10.

**7.20** Write phasor equations of the form $[Y(j\omega)][\mathbf{V}] = [\mathbf{I}]$ for the network of Fig. 7-39.

This time we must replace the voltage source by an equivalent current source. This leads to the $j\omega$-domain diagram of Fig. 7-41 from which we can write the matrix nodal equation by inspection,

$$\begin{bmatrix} \frac{1}{j\omega} + \frac{1+6j\omega}{2} & -\frac{1+6j\omega}{2} & 0 \\ -\frac{1+6j\omega}{2} & \frac{1+6j\omega}{2} + \frac{8+9j\omega}{72j\omega} + \frac{5j\omega}{1+20j\omega} & \frac{-5j\omega}{1+20j\omega} \\ 0 & \frac{-5j\omega}{1+20j\omega} & \frac{1}{6+7j\omega} + \frac{5j\omega}{1+20j\omega} \end{bmatrix} \begin{bmatrix} \mathbf{V}_b \\ \mathbf{V}_c \\ \mathbf{V}_d \end{bmatrix}$$

$$= \begin{bmatrix} \frac{10}{j\omega} e^{j(\omega t - 90^\circ)} \\ 0 \\ -12e^{j\omega t} \end{bmatrix} = \begin{bmatrix} -\frac{10}{\omega} e^{j\omega t} \\ 0 \\ -12e^{j\omega t} \end{bmatrix}$$

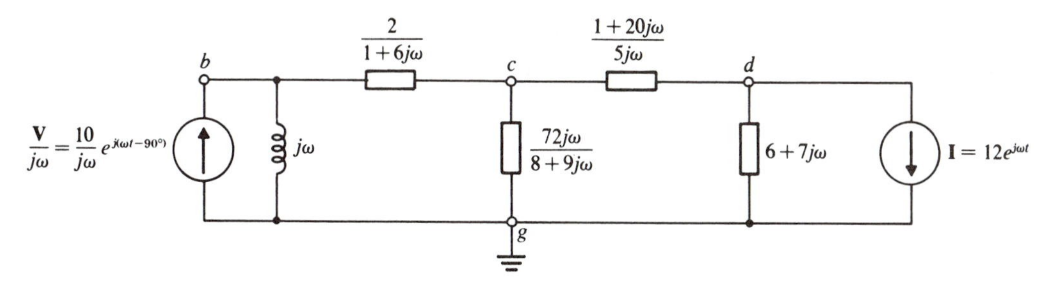


**Fig. 7-41**

*Comments*

1. The three diagonal terms in $[Y(j\omega)]$ are the admittances connected to the three nodes $b$, $c$, and $d$. The off-diagonal terms are the negatives of the admittances connected *between* each appropriate pair of nodes.

2. The matrix on the right of the equation contains the source currents *entering* the three nodes.
3. Problems 7.19 and 7.20 are *much* simpler than the corresponding $s$-domain Problems 6.19–6.22, since we do not have to cope with the initial condition sources in the $j\omega$-domain.

**7.21** Explain how we could use the results of Problems 7.19 and 7.20 to find the ac current flowing from left to right in the 2-Ω resistor of Fig. 7-39.

We could apply Cramer's rule to the loop equations of Problem 7.19 to compute $\mathbf{I}_1$. Then $\mathbf{V}_{bc} = 2\mathbf{I}_1/(1 + 6j\omega)$ and $\mathbf{I}_{R_2} = \mathbf{V}_{bc}/2 = \mathbf{I}_1/(1 + 6j\omega)$.

Or, having found $\mathbf{I}_1$ as above, $\mathbf{I}_{R_2}$ follows from the current divider concept,

$$\mathbf{I}_{R_2} = \frac{1/2}{1/2 + 3j\omega}\mathbf{I}_1 = \frac{\mathbf{I}_1}{1 + 6j\omega}$$

Alternatively, we could compute $\mathbf{V}_b$ and $\mathbf{V}_c$ by applying Cramer's rule to the nodal equations of Problem 7.20. Then $\mathbf{V}_{bc} = \mathbf{V}_b - \mathbf{V}_c$ and $\mathbf{I}_{R_2} = \mathbf{V}_{bc}/2$.

**7.22** The network of Fig. 7-42 contains a dependent voltage source $v_x(t) = 5i_L$, and $v(t) = 10 \sin (3t + 9°)$. Find the effective value of $i_L$ in the sinusoidal steady-state.

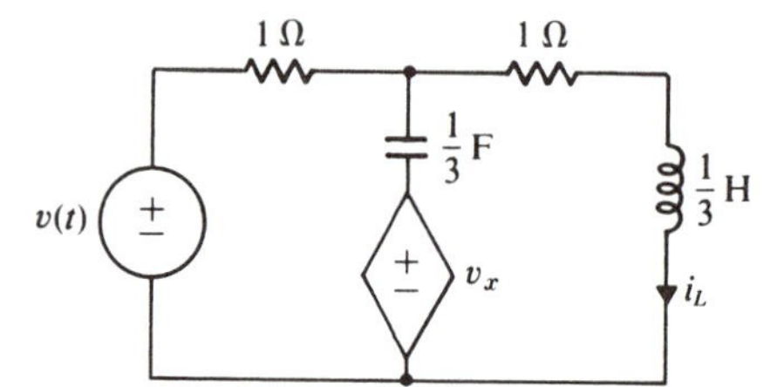


**Fig. 7-42**

Whenever a circuit contains one or more dependent sources, instability is a possibility. A stability test should therefore precede any sinusoidal steady-state calculations.

From Fig. 7-43, the $s$-domain circuit diagram with zero initial conditions, we can obtain the transform loop equations

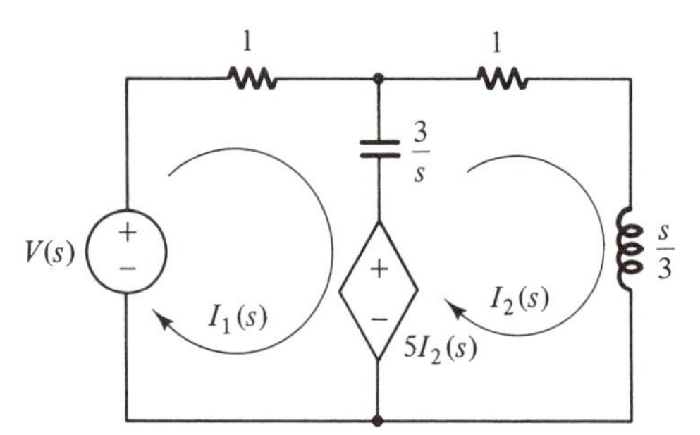


**Fig. 7-43**

$$\begin{bmatrix} 1 + \frac{3}{s} & 5 - \frac{3}{s} \\ -\frac{3}{s} & \frac{s}{3} - 4 + \frac{3}{s} \end{bmatrix} \begin{bmatrix} I_1(s) \\ I_2(s) \end{bmatrix} = \begin{bmatrix} V(s) \\ 0 \end{bmatrix}$$

To test these equations for stability we can solve for $I_1(s)$ or $I_2(s)$. Thus from Cramer's rule,

$$I_2(s) = \frac{9V(s)}{s^2 - 9s + 18}$$

Since the denominator polynomial $D(s)$ has one negative coefficient, it follows at once that the circuit is unstable. A steady state does not exist, and a calculation of $i_{L,\,sss}(t)$ would yield a meaningless result.

**7.23** Find the output impedance of the circuit in Fig. 7-44.

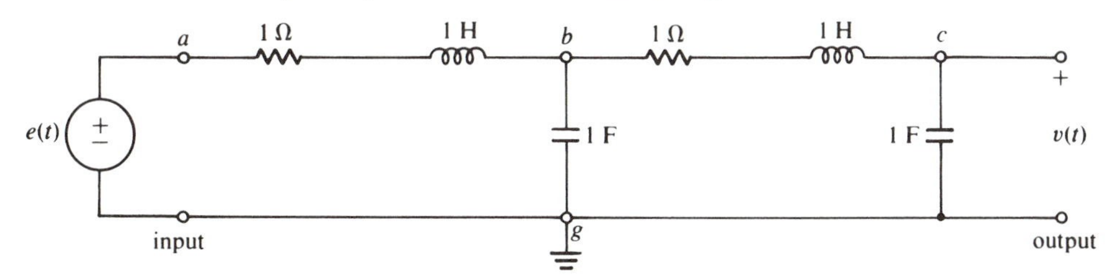


**Fig. 7-44**

Setting the source $e(t)$ to zero, we obtain the $j\omega$-domain diagram of Fig. 7-45. To find the output impedance $Z_0(j\omega) = V/I$ we must start combining elements from the *left-hand* end of Fig. 7-45. If we were to start at the right we would find no $L$, $R$, or $C$ elements in series or parallel because of the presence of the external circuit. Now, therefore,

$$Y_{bg}(j\omega) = j\omega + \frac{1}{1 + j\omega} = \frac{-\omega^2 + j\omega + 1}{1 + j\omega}$$

Next,

$$Z_{cbg}(j\omega) = 1 + j\omega + \frac{1 + j\omega}{-\omega^2 + 1 + j\omega} = \frac{-2\omega^2 + 2 + j(-\omega^3 + 3\omega)}{-\omega^2 + 1 + j\omega}$$

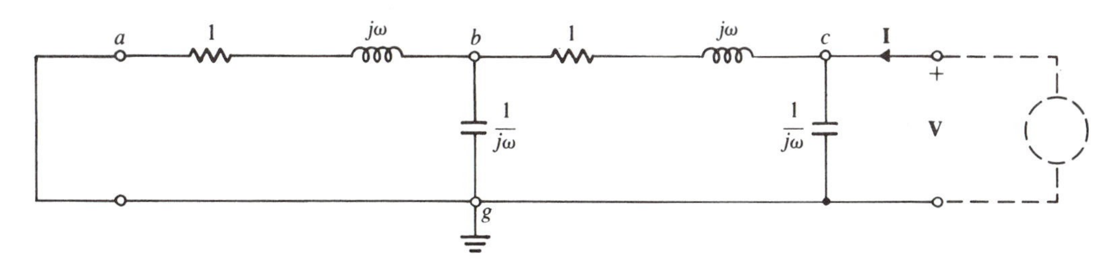


**Fig. 7-45**

Finally,

$$Y_0(j\omega) = Y_{cg}(j\omega) = j\omega + \frac{-\omega^2 + 1 + j\omega}{-2\omega^2 + 2 + j(-\omega^3 + 3\omega)}$$

$$= \frac{\omega^4 - 4\omega^2 + 1 + j(-2\omega^3 + 3\omega)}{-2\omega^2 + 2 + j(-\omega^3 + 3\omega)}$$

or

$$Z_0(j\omega) = \frac{1}{Y_0(j\omega)} = \frac{-2\omega^2 + 2 + j(-\omega^3 + 3\omega)}{\omega^4 - 4\omega^2 + 1 + j(-2\omega^3 + 3\omega)}$$

Note that we have found the output impedance *with the input source connected.* That is, we set $e(t) = 0$; we did *not* disconnect $e(t)$.

### *Alternative Solution*

By inspection of Fig. 7-45 we can write the circuit's nodal equation,

$$\begin{bmatrix} j\omega + \dfrac{1}{1 + j\omega} & \dfrac{-1}{1 + j\omega} \\ \dfrac{-1}{1 + j\omega} & \dfrac{2}{1 + j\omega} + j\omega \end{bmatrix} \begin{bmatrix} \mathbf{V}_c \\ \mathbf{V}_b \end{bmatrix} = \begin{bmatrix} \mathbf{I} \\ 0 \end{bmatrix}$$

Thus

$$\mathbf{V}_c = \mathbf{V} = \frac{\begin{vmatrix} \mathbf{I} & \dfrac{-1}{1 + j\omega} \\ 0 & \dfrac{2}{1 + j\omega} + j\omega \end{vmatrix}}{\begin{vmatrix} j\omega + \dfrac{1}{1 + j\omega} & \dfrac{-1}{1 + j\omega} \\ \dfrac{-1}{1 + j\omega} & \dfrac{2}{1 + j\omega} + j\omega \end{vmatrix}} = \frac{(-\omega^2 + 2 + j\omega)\,(1 + j\omega)}{\omega^4 - 4\omega^2 + 1 + j(-2\omega^3 + 3\omega)}\,\mathbf{I}$$

and

$$Z_0(j\omega) = \frac{\mathbf{V}}{\mathbf{I}} = \frac{-2\omega^2 + 2 + j(-\omega^3 + 3\omega)}{\omega^4 - 4\omega^2 + 1 + j(-2\omega^3 + 3\omega)}$$

**7.24** Find the driving point impedance for the two port of Fig. 7-44 when the output terminals are short circuited. Then determine $i_{ab,\,\mathrm{sss}}(t)$ given that $e(t) = 10 \cos (2t + 13^\circ)$.

The appropriate $j\omega$-domain diagram, with $\omega = 2$, is shown in Fig. 7-46. This time we must start from the right-hand end of the circuit. Thus

$$Y_{bg}(j2) = j2 + \frac{1}{1 + j2} = \frac{-3 + j2}{1 + j2}$$

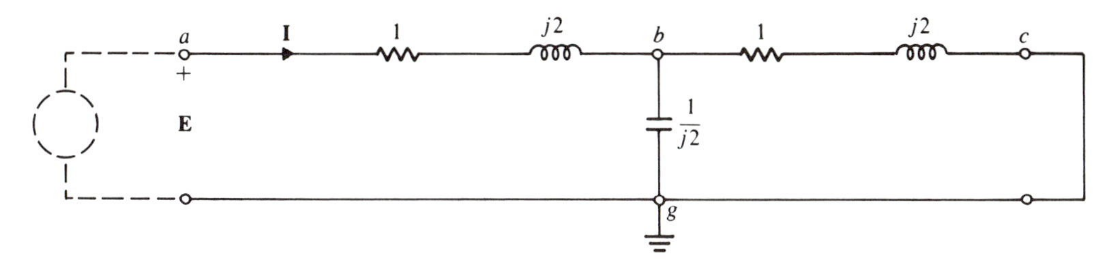


Fig. 7-46

and

$$Z_{dp}(j2) = 1 + j2 + \frac{1 + j2}{-3 + j2} = \frac{-6 - j2}{-3 + j2} = \frac{6.33e^{-j161.6°}}{3.60e^{j146.3°}} = 1.76e^{j52.1°}$$

Therefore

$$\mathbf{I}_{ab} = \frac{\mathbf{E}}{Z_{dp}(j2)} = \frac{10e^{j(2t+13°)}}{1.76e^{j52.1°}} = 5.69e^{j(2t-39.1°)}$$

and

$$i_{ab,\,\text{sss}}(t) = 5.69 \cos (2t - 39.1°)$$

**7.25** Rework Problem 7.24 with the output terminated by a 1-Ω resistor.

Using the circuit diagram in Fig. 7-47,

$$Z_{bcg}(j2) = 1 + j2 + \frac{1}{1 + j2} = \frac{-2 + j4}{1 + j2}$$

$$Y_{bg}(j2) = j2 + \frac{1 + j2}{-2 + j4} = \frac{-7 - j2}{-2 + j4}$$

and

$$Z_{dp}(j2) = 1 + j2 + \frac{-2 + j4}{-7 - j2} = \frac{5 + j12}{7 + j2} = 1.78e^{j51.4°}$$

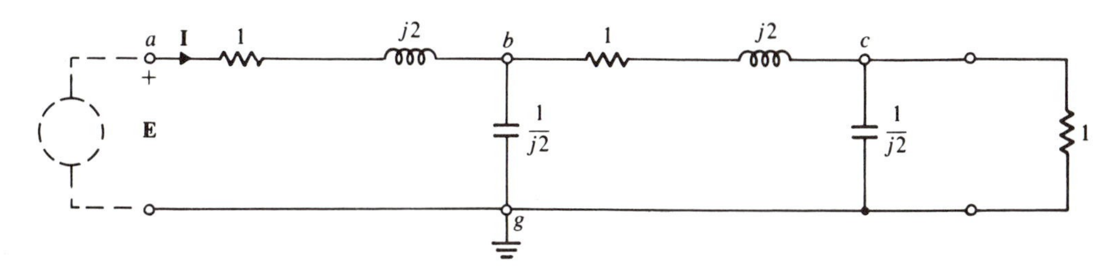


Fig. 7-47

Thus

$$\mathbf{I}_{ab} = \frac{10e^{j(2t+13°)}}{1.78e^{j51.4°}} = 5.61e^{j(2t-38.4°)}$$

and

$$i_{ab,\,\text{sss}}(t) = 5.61 \cos (2t - 38.4°)$$

The 1-Ω termination changes the magnitude and phase of the input current as compared with the previous case when the output termination was a short circuit.

**7.26** If in Fig. 7-48, $v(t) = 120 \cos (20t + 43°)$ and $i(t) = 12 \sin (20t + 73°)$ in the sinusoidal steady-state, find the driving-point impedance of the network. Then obtain a series combination of elements having this driving-point impedance (when $\omega = 20$ rad/s).

Putting both time functions into the same trigonometric form,

$$v(t) = 120 \cos (20t + 43°) \quad \text{and} \quad i(t) = 12 \cos (20t - 17°)$$

Then

$$Z_{dp}(j20) = \frac{\mathbf{V}}{\mathbf{I}} = \frac{120e^{j(20t+43°)}}{12e^{j(20t-17°)}} = 10e^{j60°} = 10 (\cos 60° + j \sin 60°) = 5 + j8.66$$

This suggests the circuit of Fig. 7-49 since $j\omega L = j(20)(0.433) = j8.66$.

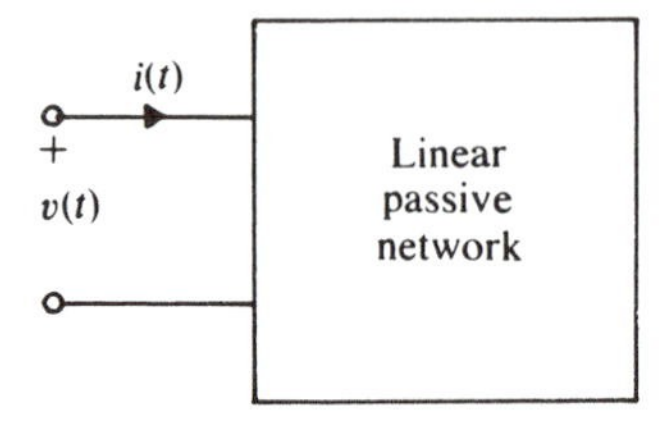


Fig. 7-48

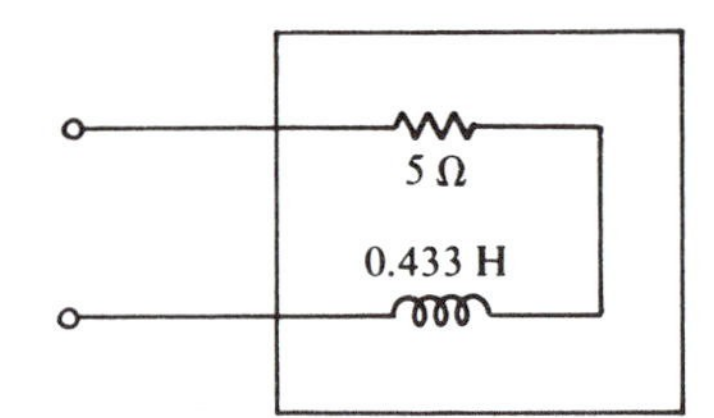


Fig. 7-49

**7.27** Find the frequency for which the real part of the sinusoidal transfer function $H(j\omega) = \mathbf{V}/\mathbf{E}$ is zero in the $j\omega$-domain circuit of Fig. 7-50.

With reference to the voltage divider concept,

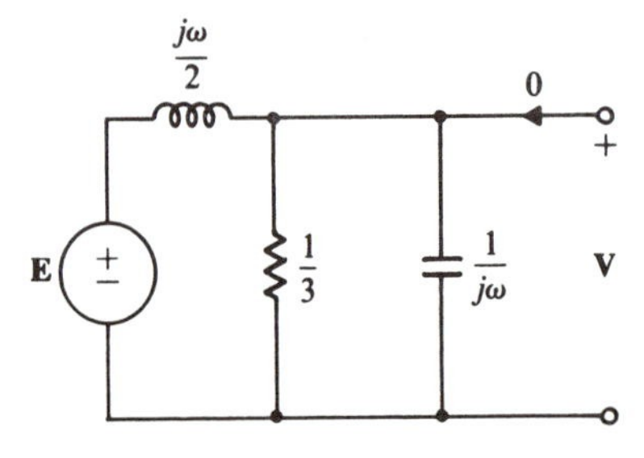


Fig. 7-50

$$H(j\omega) = \frac{\mathbf{V}}{\mathbf{E}}$$
$$= \left(\frac{1}{j\omega + 3}\right) \bigg/ \left(\frac{j\omega}{2} + \frac{1}{j\omega + 3}\right) = \frac{2}{-\omega^2 + 3j\omega + 2}$$
$$= \left(\frac{2}{-\omega^2 + 2 + 3j\omega}\right)\left(\frac{-\omega^2 + 2 - 3j\omega}{-\omega^2 + 2 - 3j\omega}\right)$$
$$= \frac{-2\omega^2 + 4 - 6j\omega}{(-\omega^2 + 2)^2 + 9\omega^2}$$

and

$$\text{Re}[H(j\omega)] = \frac{-2\omega^2 + 4}{(-\omega^2 + 2)^2 + 9\omega^2}$$

This will be zero when $-2\omega^2 + 4 = 0$ or when $\omega = \sqrt{2}$ rad/s and $f = \sqrt{2}/2\pi$ Hz.

**7.28** If $v_{sss}(t) = 2 \sin 2t$ in the circuit of Fig. 7-51, find $e(t)$.

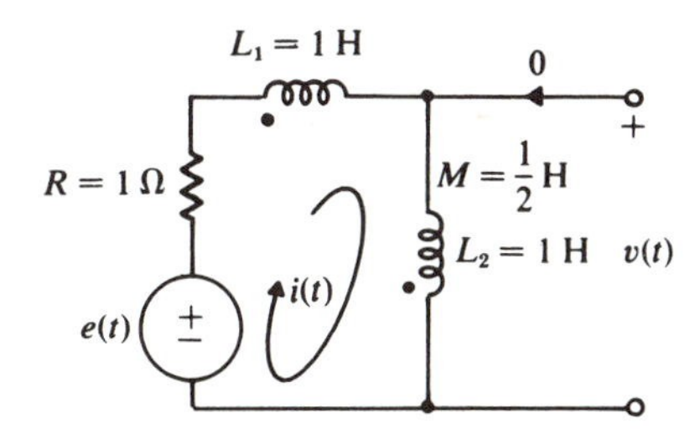


**Fig. 7-51**

Here the appropriate differential equations are

$$Ri + L_1\frac{di}{dt} - M\frac{di}{dt} + L_2\frac{di}{dt} - M\frac{di}{dt} = e(t)$$
$$L_2\frac{di}{dt} - M\frac{di}{dt} = v(t)$$

Substituting the numerical data and writing these equations in phasor form gives us

$$\mathbf{I} + j2\mathbf{I} - j\frac{2}{2}\mathbf{I} + j2\mathbf{I} - j\frac{2}{2}\mathbf{I} = \mathbf{E}$$
$$j2\mathbf{I} - j\frac{2}{2}\mathbf{I} = \mathbf{V}$$

From the second equation,

$$\mathbf{I} = \frac{\mathbf{V}}{j} = \frac{2e^{j2t}}{j} = 2e^{j(2t - 90°)}$$

And from the first,

$$\mathbf{E} = (1 + j2)\mathbf{I} = (\sqrt{5}e^{j63.4^\circ})\,(2e^{j(2t-90^\circ)})$$

Therefore $e(t) = 2\sqrt{5}\sin(2t - 26.6^\circ)$.

Notice that we have not bothered to convert to the cosine function, but have let the phasors represent sines.

**7.29** Find **V** in the $j\omega$-domain diagram of Fig. 7-52 given $\mathbf{I} = 10e^{j\omega t}$.

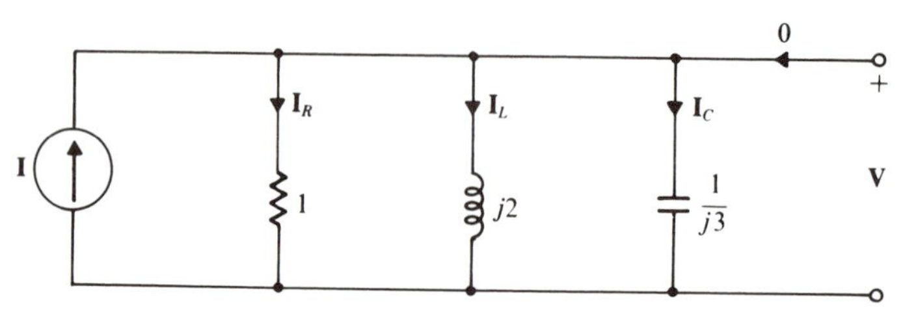


**Fig. 7-52**

The easiest approach to this problem is through the relation $Y(j\omega)\mathbf{V} = \mathbf{I}$, where

$$Y(j\omega) = 1 + \frac{1}{j2} + j3 = 1 + j2.5 = 2.7e^{j68.3^\circ}$$

Thus

$$V = \frac{10e^{j\omega t}}{2.7e^{j68.3^\circ}} = 3.72e^{j(\omega t - 68.3^\circ)}$$

### *A Miscellany*

**7.30** Determine the sinusoidal transfer function $H(j\omega) = \mathbf{V}/\mathbf{E}$ for the *ladder network* of Fig. 7-53.

The "trick" here is to assume a convenient value for the output (in this case, **V**), and then to work back to calculate the input (in this case, **E**).

If we choose to work with the *stationary* phasor $\mathbf{V} = 1e^{j0^\circ} = 1$, then we have at node $c$,

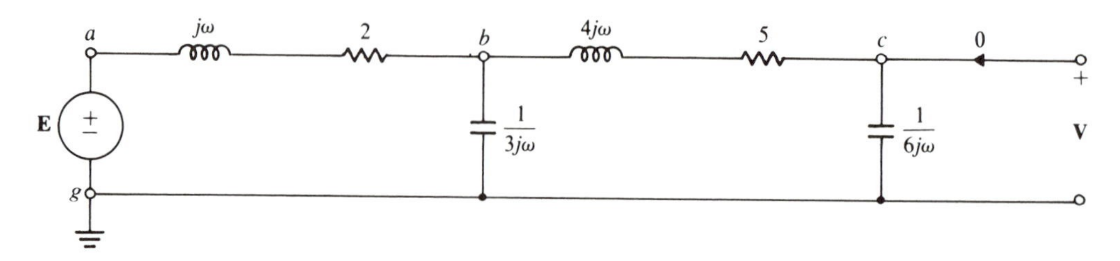


**Fig. 7-53**

$$\mathbf{I}_{cg} = \frac{1}{1/6j\omega} = 6j\omega$$

At node $b$,

$$\mathbf{V}_b = \mathbf{V} + (5 + 4j\omega)\mathbf{I}_{cg} = 1 + (5 + 4j\omega)\,(6j\omega) = 1 - 24\omega^2 + 30j\omega$$

and

$$\mathbf{I}_{bg} = \frac{\mathbf{V}_b}{1/3j\omega} = \frac{1 - 24\omega^2 + 30j\omega}{1/3j\omega} = -90\omega^2 + 3j\omega - 72j\omega^3$$

At node $a$,

$$\begin{aligned}\mathbf{V}_a &= \mathbf{V}_b + (2 + j\omega)\,(\mathbf{I}_{bc} + \mathbf{I}_{bg}) \\ &= 1 - 24\omega^2 + 30j\omega + (2 + j\omega)\,(9j\omega - 90\omega^2 - 72j\omega^3) \\ &= 72\omega^4 - 213\omega^2 + 1 + j(48\omega - 234\omega^3)\end{aligned}$$

But $\mathbf{V}_a = \mathbf{E}$ and we chose $\mathbf{V} = 1$, so

$$\frac{\mathbf{V}}{\mathbf{E}} = \frac{1}{\mathbf{V}_a} = \frac{1}{72\omega^4 - 213\omega^2 + 1 + j(48\omega - 234\omega^3)}$$

We used stationary phasors in this problem simply to avoid the burden of carrying $e^{j\omega t}$ terms through the calculation.

### *Alternative Solution*

After mentally transforming the voltage source into a current source, we can write by inspection,

$$\begin{bmatrix} 3j\omega + \dfrac{1}{2 + j\omega} + \dfrac{1}{5 + 4j\omega} & \dfrac{-1}{5 + 4j\omega} \\ \dfrac{-1}{5 + 4j\omega} & 6j\omega + \dfrac{1}{5 + 4j\omega} \end{bmatrix} \begin{bmatrix} \mathbf{V}_b \\ \mathbf{V} \end{bmatrix} = \begin{bmatrix} \dfrac{\mathbf{E}}{2 + j\omega} \\ 0 \end{bmatrix}$$

Thus

$$\mathbf{V} = \frac{\begin{vmatrix} 3j\omega + \dfrac{1}{2 + j\omega} + \dfrac{1}{5 + 4j\omega} & \dfrac{\mathbf{E}}{2 + j\omega} \\ \dfrac{-1}{5 + 4j\omega} & 0 \end{vmatrix}}{\begin{vmatrix} 3j\omega + \dfrac{1}{2 + j\omega} + \dfrac{1}{5 + 4j\omega} & \dfrac{-1}{5 + 4j\omega} \\ \dfrac{-1}{5 + 4j\omega} & 6j\omega + \dfrac{1}{5 + 4j\omega} \end{vmatrix}}$$

We could persevere, but this is torture! The systematic step-by-step solution is far simpler, especially for longer ladders.

**7.31** Find $H(j\omega) = \mathbf{I}_v/\mathbf{I}$ in the network of Fig. 7-54.

We can proceed as in the previous problem; that is, we assume an output and then

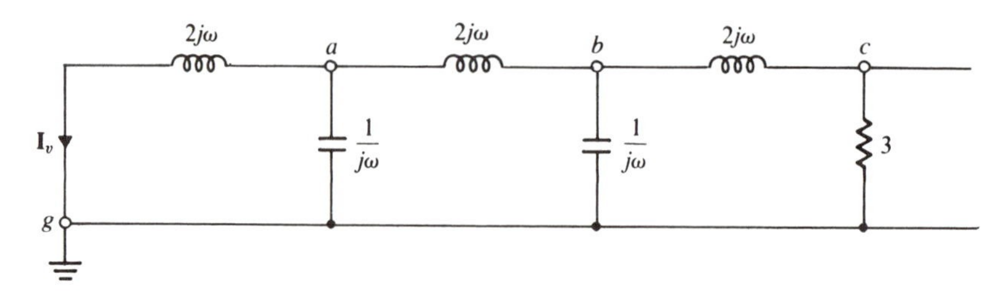


**Fig. 7-54**

move step by step to the input. If we choose $\mathbf{I}_v = 1$,

$$\mathbf{V}_a = 2j\omega, \qquad \mathbf{I}_{ag} = \frac{2j\omega}{1/j\omega} = -2\omega^2, \qquad \mathbf{I}_{ab} = -\mathbf{I}_v - \mathbf{I}_{ag} = -1 + 2\omega^2$$

At node $b$,

$$\mathbf{V}_b = \mathbf{V}_a - 2j\omega\mathbf{I}_{ab} = 2j\omega - 2j\omega(-1 + 2\omega^2) = 4j\omega - 4j\omega^3$$

$$\mathbf{I}_{bg} = \frac{\mathbf{V}_b}{1/j\omega} = j\omega(4j\omega - 4j\omega^3) = -4\omega^2 + 4\omega^4$$

$$\mathbf{I}_{bc} = \mathbf{I}_{ab} - \mathbf{I}_{bg} = -1 + 2\omega^2 - (-4\omega^2 + 4\omega^4) = -1 + 6\omega^2 - 4\omega^4$$

At node $c$,

$$\mathbf{V}_c = \mathbf{V}_b - 2j\omega\mathbf{I}_{bc} = 4j\omega - 4j\omega^3 - 2j\omega(-1 + 6\omega^2 - 4\omega^4)$$
$$= 6j\omega - 16j\omega^3 + 8j\omega^5$$

and

$$\mathbf{I} = \mathbf{I}_{cg} - \mathbf{I}_{bc} = \frac{\mathbf{V}_c}{3} - \mathbf{I}_{bc}$$
$$= \frac{8j\omega^5 - 16j\omega^3 + 6j\omega}{3} - (-1 + 6\omega^2 - 4\omega^4)$$
$$= \frac{8j\omega^5 - 16j\omega^3 + 6j\omega + 3 - 18\omega^2 + 12\omega^4}{3}$$

Therefore, since $\mathbf{I}_v = 1$,

$$\frac{\mathbf{I}_v}{\mathbf{I}} = \frac{3}{12\omega^4 - 18\omega^2 + 3 + j(8\omega^5 - 16\omega^3 + 6\omega)}$$

*Comment:* In Problem 6.24, page 200, we found

$$H(s) = \frac{I_v(s)}{I(s)} = \frac{3}{8s^5 + 12s^4 + 16s^3 + 18s^2 + 6s + 3}$$

Evaluating $H(j\omega) = H(s)|_{s=j\omega}$ clearly confirms our answer to this problem.

**7.32** Draw the $j\omega$-domain circuit corresponding to the FET amplifier of Fig. 7-55, and find its Thévenin equivalent as seen by the load resistor. Hence find the circuit's phasor voltage gain, $A_v(j\omega) = \mathbf{V}/\mathbf{E}$.

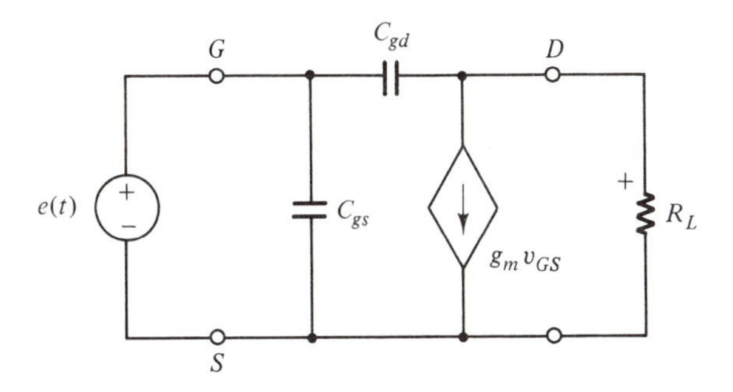


**Fig. 7-55**

The phasor-domain diagram follows as in Fig. 7-56*a*. Then, applying KCL at node *D* in that figure,

$$j\omega C_{gd}(\mathbf{E} - \mathbf{V}) - g_m\mathbf{E} - \mathbf{I} = 0$$

or

$$\mathbf{V} = \frac{-1}{j\omega C_{gd}}\mathbf{I} + \left(1 + j\frac{g_m}{\omega C_{gd}}\right)\mathbf{E}$$

whence by comparison with equation (*7.14*),

$$Z_{\text{Th}}(j\omega) = 1/j\omega C_{gd} \quad \text{and} \quad \mathbf{V}_{\text{Th}} = (1 + jg_m/\omega C_{gd})\mathbf{E}$$

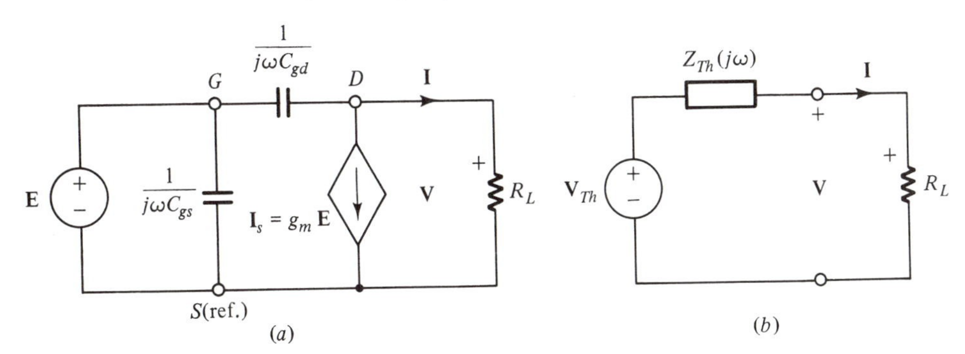


**Fig. 7-56**

It now follows from Fig. 7-56*b* that

$$\mathbf{V} = \frac{R_L}{R_L + Z_{\text{Th}}(j\omega)}\mathbf{V}_{\text{Th}} = \frac{R_L + jR_L g_m/\omega C_{gd}}{R_L - j/\omega C_{gd}}\mathbf{E}$$

Thus

$$A_v(j\omega) = (j\omega - g_m/C_{gd})/(j\omega + 1/R_L C_{gd})$$

This is a result which can easily be checked by setting $s = j\omega$ in $A_v(s)$ as obtained in Problem 6.26.

**7.33** Draw the phasor-domain circuit corresponding to the op amp circuit of Fig. 7-57. Hence find the sinusoidal transfer function $H(j\omega) = \mathbf{V}/\mathbf{E}$. Given that all component values are unity and that $e(t) = 10 \cos t$, find $v_{\text{sss}}(t)$.

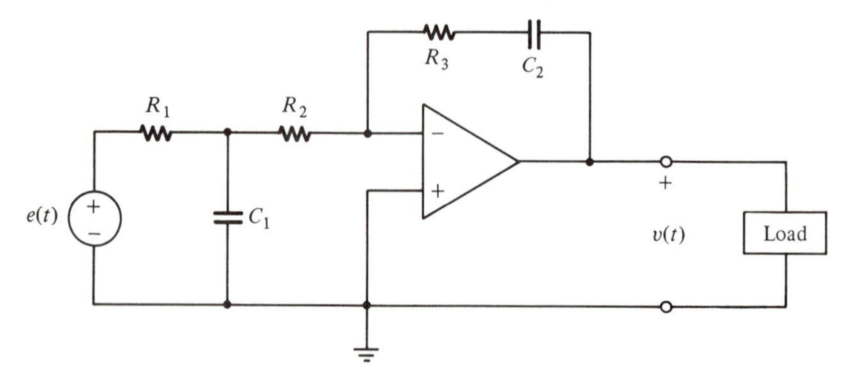


Fig. 7-57

The phasor-domain circuit follows as in Fig. 7-58. The ideal op amp, being described by algebraic relations, transfers directly into the phasor-domain. Thus with the op amp's + terminal connected to ground we have $\mathbf{V}_- = \mathbf{V}_+ = 0$. And the amplifier's input currents have, as usual, been assumed to be zero.

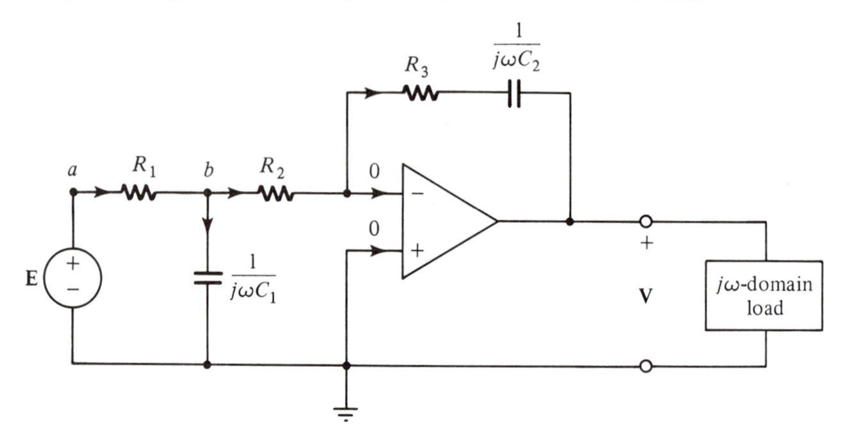


Fig. 7-58

Applying KCL at node $b$ in Fig. 7-58,

$$\frac{1}{R_1}\{\mathbf{V}_b - \mathbf{E}\} + j\omega C_1 \mathbf{V}_b + \frac{1}{R_2}\mathbf{V}_b = 0$$

or

$$\mathbf{V}_b = \frac{R_2 \mathbf{E}}{R_1 + R_2 + j\omega R_1 R_2 C_1}$$

Therefore, with reference again to Fig. 7-58,

$$\boldsymbol{I}_{R_2} = \boldsymbol{I}_{R_3} = \boldsymbol{I}_{C_2} = \frac{\mathbf{E}}{R_1 + R_2 + j\omega R_1 R_2 C_1}$$

and $\mathbf{V} = -\{R_3 + 1/j\omega C_2\}\mathbf{I}_{R_3}$. Thus,

$$H(j\omega) = \frac{\mathbf{V}}{\mathbf{E}} = \frac{1 + j\omega R_3 C_2}{R_1 R_2 C_1 C_2 \omega^2 - j\omega C_2(R_1 + R_2)}$$

This result is easily checked by setting $s = j\omega$ in the expression obtained for $H(s)$ in Problem 6.27.

Now, substitution of the data yields

$$\mathbf{V} = H(j1)\mathbf{E} = \frac{1 + j}{1 - j2} 10e^{jt} = 6.32e^{j(t + 108.4°)}$$

whence $v_{sss}(t) = 6.32 \cos (t + 108.4°)$.

**7.34** For the network of Fig. 7-59 it is known that under sinusoidal steady-state conditions,

$$v_2(t) = 100 \cos \omega t, \qquad v_3(t) = 200 \cos (\omega t + 60°)$$
$$v_4(t) = 400 \cos (\omega t + 120°)$$

With the help of a phasor diagram, find $v_1(t)$.

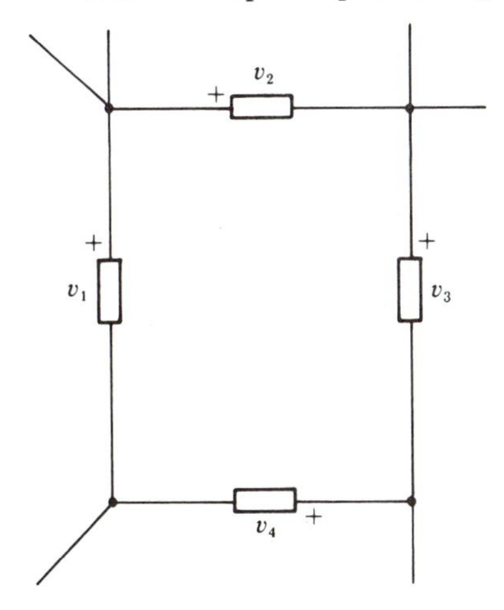


**Fig. 7-59**

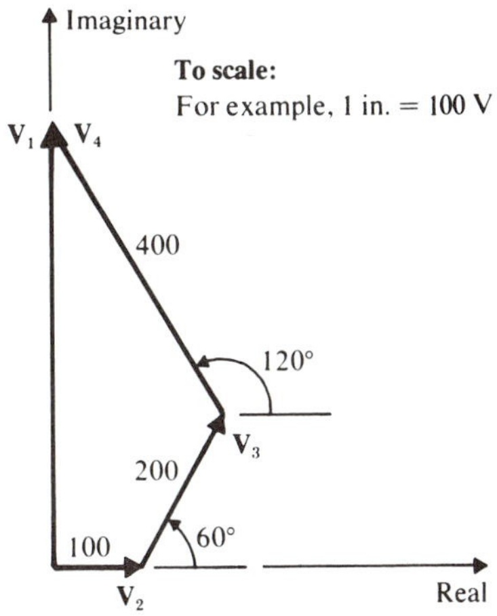

**Fig. 7-60**

From KVL, $-v_1 + v_2 + v_3 + v_4 = 0$. Therefore under sinusoidal conditions as postulated,

$$-\mathbf{V}_1 + \mathbf{V}_2 + \mathbf{V}_3 + \mathbf{V}_4 = 0 \quad \text{or} \quad \mathbf{V}_1 = \mathbf{V}_2 + \mathbf{V}_3 + \mathbf{V}_4$$

Thus

$$\mathbf{V}_1 = 100e^{j\omega t} + 200e^{j(\omega t + 60°)} + 400e^{j(\omega t + 120°)}$$

In order to draw these rotating phasors we must "freeze" them; at $t = 0$, say. This has been done in Fig. 7-60, where the frozen phasors $\mathbf{V}_2$, $\mathbf{V}_3$, and $\mathbf{V}_4$ have been added to yield $\mathbf{V}_1$. By measurement,

$$|\mathbf{V}_1| = 520 \quad \text{and} \quad \underline{/\mathbf{V}_1} = \omega t + 90°$$

Therefore

$$\mathbf{V}_1 = 520e^{j(\omega t + 90°)} \quad \text{and} \quad v_1(t) = 520\cos(\omega t + 90°)$$

*Alternative Solution*

We can, of course, work this problem in a strictly analytical manner. Using *stationary* phasors for convenience,

$$\mathbf{V}_1 = \mathbf{V}_2 + \mathbf{V}_3 + \mathbf{V}_4 = 100e^{j0°} + 200e^{j60°} + 400e^{j120°}$$

Then, using your calculator to convert these phasors into rectangular form,

$$\mathbf{V}_1 = (100 + j0) + (100 + j173) + (-200 + j346)$$
$$= j519 = 519e^{j90°}$$

and $v_1(t) = 519 \cos(\omega t + 90°)$.

**7.35** The following sinusoidal steady-state voltmeter measurements have been made in the circuit of Fig. 7-61: $E_{\text{eff}} = 200$ V, $V_{R,\text{eff}} = 60$ V, $V_{\text{eff}} = 163$ V. The frequency was 60 Hz or $\omega = 2\pi(60) = 377$ rad/s. Use a phasor diagram to determine the phase angle $\theta$ of $v(t)$ relative to $i(t)$, and then find $L$ and $R$.

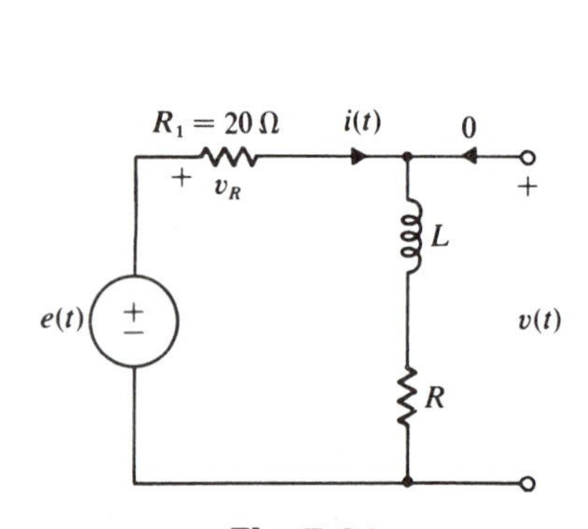


**Fig. 7-61**

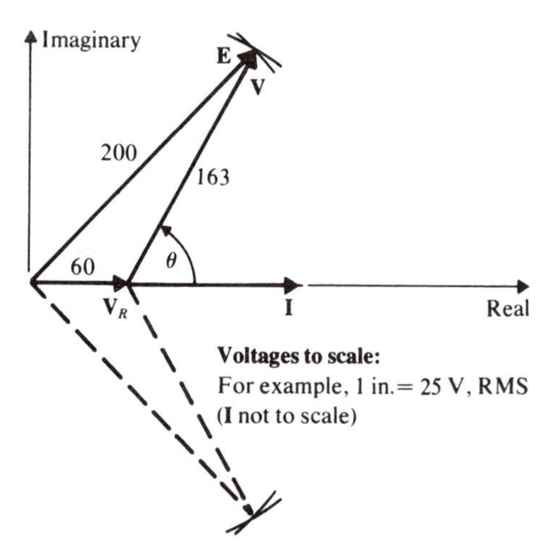


**Fig. 7-62**

We will work with RMS phasors. The appropriate phasor equations for this circuit are

$$\mathbf{V}_R + \mathbf{V} = \mathbf{E}, \qquad \mathbf{I} = \frac{1}{R_1}\mathbf{V}_R = \frac{1}{20}\mathbf{V}_R$$

and if we think of these phasors as "frozen" (or as stationary phasors), we can start drawing the phasor diagram of Fig. 7-62. First we note that $v_R(t)$ and $i(t)$ are in phase. Therefore if we choose the phasor $\mathbf{I}$ as reference, we can draw $\mathbf{V}_R$ to scale, say 1 inch = 25 V, RMS.

Next we must add $\mathbf{V}_R$ and $\mathbf{V}$ graphically to obtain $\mathbf{E}$. We do not know any of the angles, but we can swing arcs of appropriate radii to obtain the phasor triangle of Fig. 7-62. Both the solid and dashed parts of Fig. 7-62 satisfy KVL, but the bottom half is discarded because $\mathbf{V}$ is known to lead $\mathbf{I}$ in this inductive circuit.

Then by measurement we find $\theta = 60°$. Further, $\mathbf{I} = \mathbf{V}_R/20 = 3e^{j\omega t}$, so

$$Z_{RL}(j\omega) = R + j\omega L = \frac{\mathbf{V}}{\mathbf{I}} = \frac{163e^{j(\omega t+60°)}}{3e^{j\omega t}} = 54.3e^{j60°}$$

$$= 54.3(\cos 60° + j \sin 60°) = 27.2 + j47$$

Therefore $R = 27.2\ \Omega$ and $L = 47/\omega = 47/377$ H.

*Comment:* This situation arises when only voltmeters are available for measurement, which is a fairly common occurrence. Also, $R$ will often be the resistance of an inductor. Then $L$ and $R$ cannot be separated physically; that is, we could not measure the voltage across either $L$ or $R$ individually.

**7.36** At a source frequency of 400 Hz the following RMS amplitudes have been measured in the circuit of Fig. 7-63: $I_{\text{eff}} = 0.25$ A, $I_{1,\text{eff}} = 0.10$ A, and $I_{2,\text{eff}} = 0.20$ A. With the aid of a phasor diagram, find $R$ and $C$.

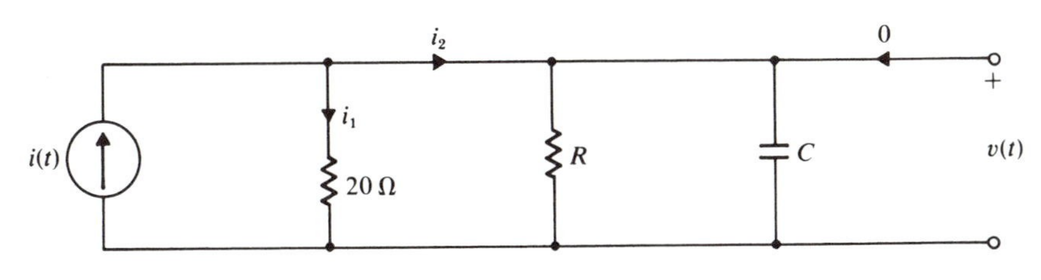


**Fig. 7-63**

The appropriate phasor equations are $\mathbf{I}_1 + \mathbf{I}_2 = \mathbf{I}$ and $\mathbf{V} = 20\mathbf{I}_1$; and, choosing $\mathbf{I}_1$ as a reference, the phasor diagram of Fig. 7-64 may be constructed. By measurement, $\theta = 72°$. Now $\mathbf{V} = 20\mathbf{I}_1 = 2e^{j\omega t}$ and so

$$Y_{RC}(j\omega) = \frac{1}{R} + j\omega C = \frac{\mathbf{I}_2}{\mathbf{V}} = \frac{0.20e^{j(\omega t+72°)}}{2e^{j\omega t}} = 0.10e^{j72°} = 0.0309 + j0.0951$$

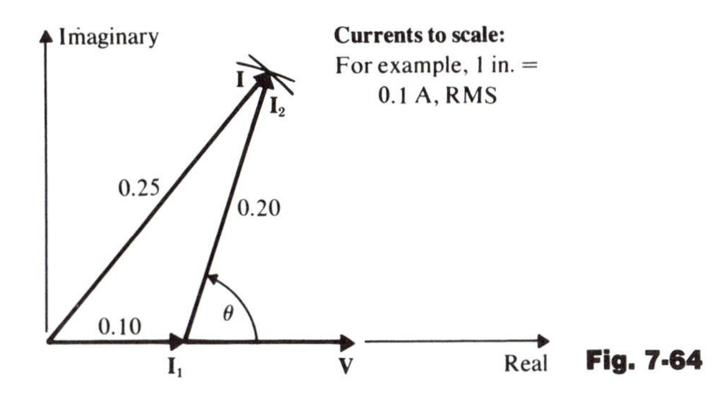


**Fig. 7-64**

Hence

$$R = \frac{1}{0.0309} = 32.4\ \Omega, \qquad C = \frac{0.0951}{2\pi(400)} = 38\ \mu F$$

As in the previous problem, we have worked with RMS phasors which we have "frozen" in order to draw Fig. 7-64.

**7.37** The following RMS amplitudes have been measured in the circuit of Fig. 7-65: $E_{\text{eff}} = 100$ V, $V_{\text{eff}} = 100$ V, $V_{C,\text{eff}} = 50$ V, and $I_{R,\text{eff}} = 20$ A. Find $R$, $\omega L$, and $1/\omega C$.

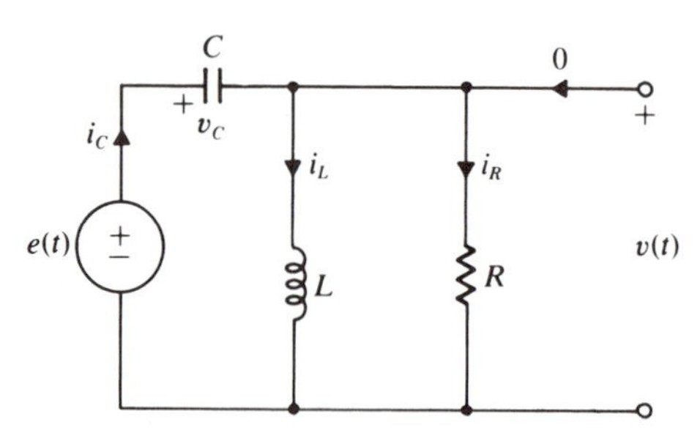


**Fig. 7-65**

If we choose the stationary phasor $\mathbf{V} = 100e^{j0°}$ as reference, then since $\mathbf{V}_C + \mathbf{V} = \mathbf{E}$, we can draw the *stationary* phasor diagram of Fig. 7-66*a*. By measurement, $\theta = 105°$.

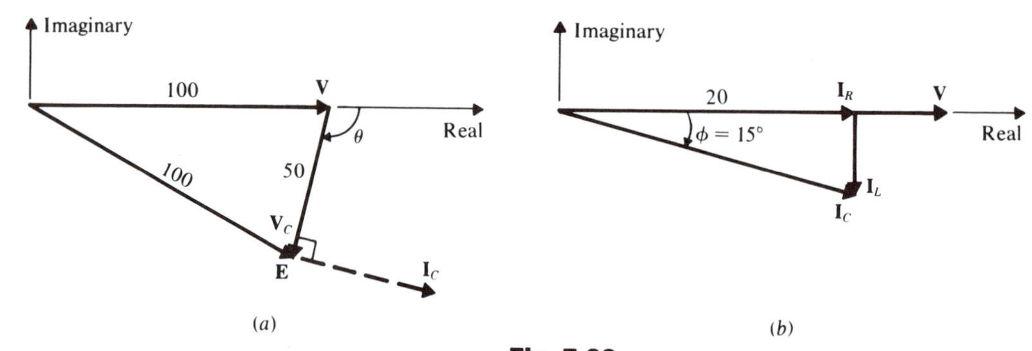


**Fig. 7-66**

Now $\mathbf{I}_C = j\omega C\,\mathbf{V}_C$, which means that $\mathbf{I}_C$ leads $\mathbf{V}_C$ by 90°, as shown in Fig. 7-66*a*. That is, $\angle\mathbf{I}_C = -(105° - 90°) = -15°$.

At this point we know that $\mathbf{I}_C = \mathbf{I}_R + \mathbf{I}_L$, $\mathbf{I}_R = 20e^{j0°}$, $\angle\mathbf{I}_C = -15°$, and $\angle\mathbf{I}_L = -90°$; hence we can draw the current phasor diagram of Fig. 7-66*b*. By measurement, $|\mathbf{I}_L| = 5.36$ A and $|\mathbf{I}_C| = 20.7$ A. Finally we compute

$$R = \frac{\mathbf{V}}{\mathbf{I}_R} = \frac{100e^{j0°}}{20e^{j0°}} = 5\ \Omega, \quad j\omega L = \frac{\mathbf{V}}{\mathbf{I}_L} = \frac{100e^{j0°}}{5.36e^{-j90°}} = 18.7e^{j90°} \quad \text{or} \quad \omega L = 18.7\ \Omega$$

and

$$\frac{1}{j\omega C} = \frac{\mathbf{V}_C}{\mathbf{I}_C} = \frac{50e^{-j105°}}{20.7e^{-j15°}} = 2.42e^{-j90°} \quad \text{or} \quad \frac{1}{\omega C} = 2.42\ \Omega$$

### *Graphical Evaluation of $H(j\omega)$*

**7.38** Given the circuit in Fig. 7-67, (*a*) Find the poles and zeros of the transfer function $H(s) = I(s)/F(s)$, (*b*) Plot the poles and zeros of $H(s)$ and determine $H(j10)$ graphically, (*c*) Determine $i_{sss}(t)$ if $f(t) = 9 \cos(10t + 30°)$.

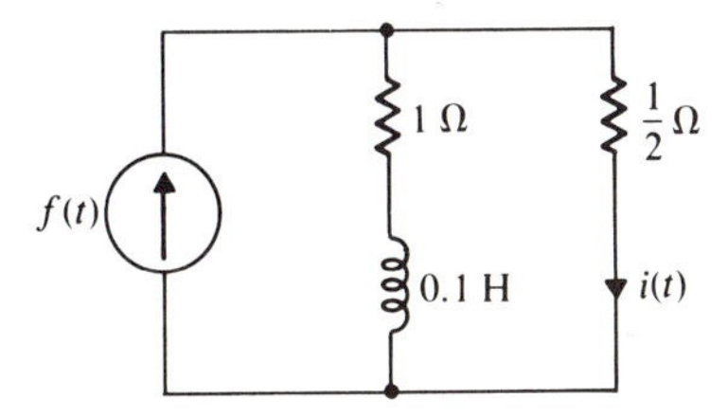


**Fig. 7-67**

(*a*) From Fig. 7-68*a*, and using the current divider notion,

$$\frac{I(s)}{F(s)} = \frac{2}{2 + 1/(1 + 0.1s)} = \frac{2 + 0.2s}{0.2s + 3} = \frac{s + 10}{s + 15}$$

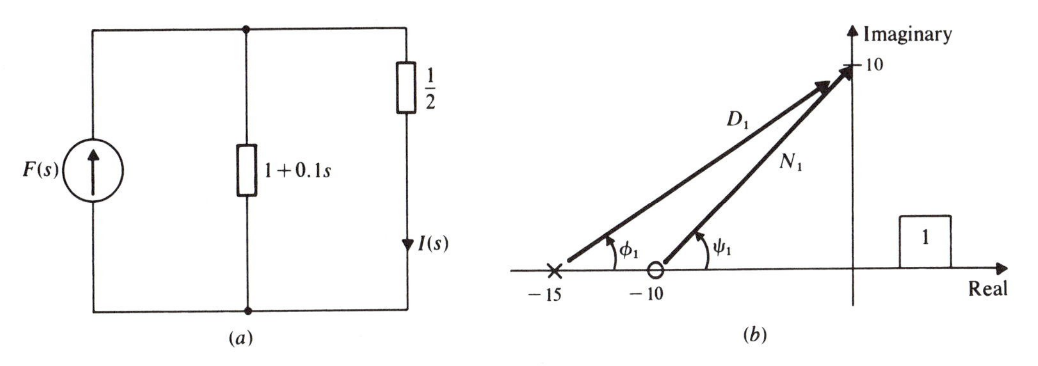


**Fig. 7-68**

(*b*) There is a pole at $s = -15$ and a zero at $s = -10$. From Fig. 7-68*b*,

$$H(j10) = \frac{K(N_1 e^{j\psi_1})}{D_1 e^{j\phi_1}} = \frac{1(10\sqrt{2}e^{j45°})}{18e^{j33.7°}} = 0.79e^{j11.3°}$$

(*c*) Using the phasor concept, with $\mathbf{F} = 9e^{j(10t+30°)}$,

$$\mathbf{I} = H(j10)\mathbf{F} = (0.79e^{j11.3°})(9e^{j(10t+30°)}) = 7.11e^{j(10t+41.3°)}$$

and $i_{sss}(t) = 7.11 \cos(10t + 41.3°)$.

**7.39** A *symmetrical lattice* network is shown in Fig. 7-69. Find the $s$-domain transfer function $H(s) = V(s)/I(s)$.

The lattice network is more easily visualized if it is redrawn as in Fig. 7-70. Now

$$V(s) = V_{ba}(s) + V_{ad}(s)$$

where $V_{ba}(s) = V_b - V_a = -Z_1(s)I_1(s)$ and $V_{ad}(s) = V_a - V_d = Z_2(s)I_2(s)$. But

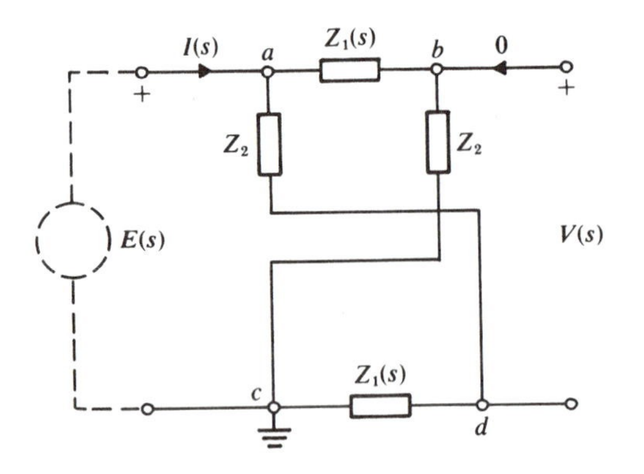


Fig. 7-69

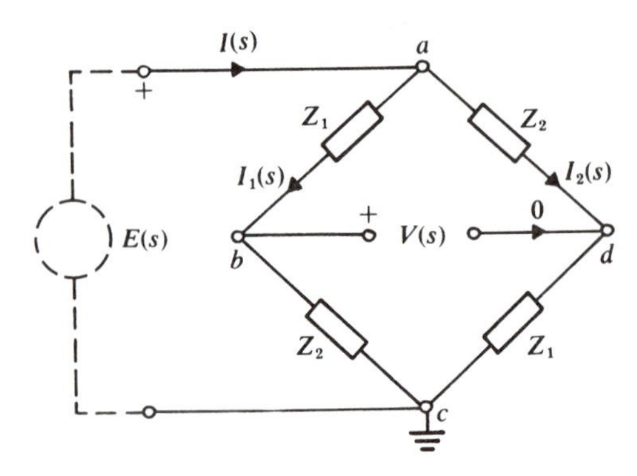


Fig. 7-70

$$I_1(s) = \frac{E(s)}{Z_1(s) + Z_2(s)} \quad \text{and} \quad I_2(s) = \frac{E(s)}{Z_2(s) + Z_1(s)}$$

Therefore

$$I_1(s) = I_2(s), \qquad I(s) = I_1(s) + I_2(s) = \frac{2E(s)}{Z_1(s) + Z_2(s)}$$

and

$$V(s) = \frac{-Z_1(s)E(s)}{Z_1(s) + Z_2(s)} + \frac{Z_2(s)E(s)}{Z_1(s) + Z_2(s)} = \frac{Z_2(s) - Z_1(s)}{Z_1(s) + Z_2(s)} E(s)$$

Thus

$$H(s) = \frac{V(s)}{I(s)} = \left\{\frac{Z_2(s) - Z_1(s)}{Z_1(s) + Z_2(s)} E(s)\right\} \Big/ \left\{\frac{2E(s)}{Z_1(s) + Z_2(s)}\right\} = \frac{1}{2}\{Z_2(s) - Z_1(s)\}$$

**7.40** In the lattice network of Problem 7.39 (Fig. 7-69), $Z_1(s)$ and $Z_2(s)$ are constructed as shown in Fig. 7-71. Find the poles and zeros of $H(s) = V(s)/I(s)$, and then evaluate $H(j\omega)$ graphically.

With reference to Fig. 7-71,

$$Z_1(s) = \frac{1}{s/4 + 1/2 + 1/2s} = \frac{4s}{s^2 + 2s + 2} \quad \text{and} \quad Z_2(s) = 1$$

Thus, using the solution of Problem 7.39,

$$H(s) = \frac{V(s)}{I(s)} = \frac{1}{2}\{Z_2(s) - Z_1(s)\} = \frac{1}{2}\left\{1 - \frac{4s}{s^2 + 2s + 2}\right\}$$

$$= \frac{1}{2}\left\{\frac{s^2 - 2s + 2}{s^2 + 2s + 2}\right\} = \frac{1}{2}\left\{\frac{(s - 1 - j)(s - 1 + j)}{(s + 1 - j)(s + 1 + j)}\right\}$$

Hence $H(s)$ has two poles at $s = -1 \pm j1$ and two zeros at $s = 1 \pm j1$.

Referring now to Fig. 7-72, where the poles and zeros have been plotted,

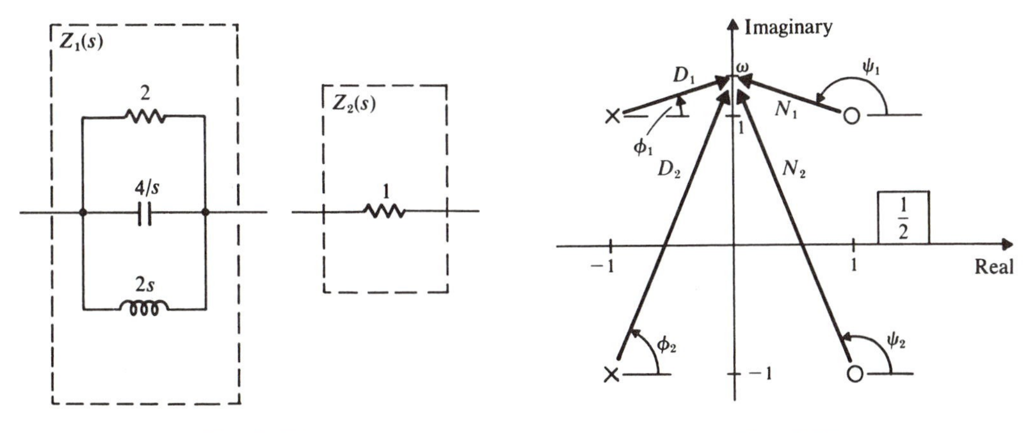


Fig. 7-71

Fig. 7-72

$$H(j\omega) = \frac{0.5(N_1 e^{j\psi_1})\,(N_2 e^{j\psi_2})}{(D_1 e^{j\phi_1})\,(D_2 e^{j\phi_2})}$$

But $N_1 = D_1$ and $N_2 = D_2$. Therefore $|H(j\omega)| = 0.5$ for all $\omega$. This is known as an all-pass network and it has very useful phase properties.

# PROBLEMS

**7.41** In the circuit of Fig. 7-73, find $i_{2,\text{sss}}(t)$ when $i_1(t) = 10 \sin (2t + \phi)$.

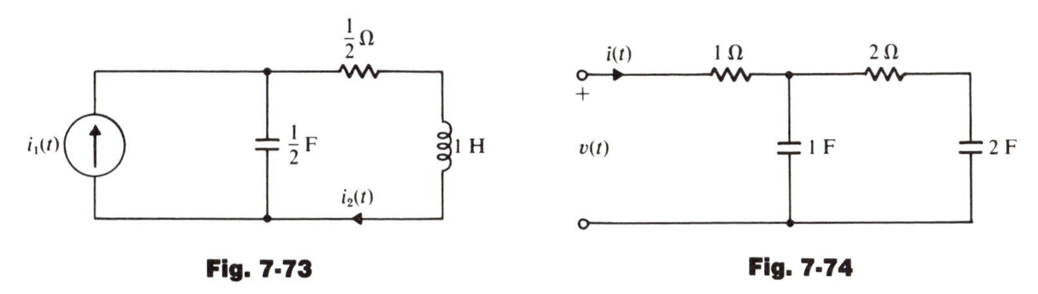


Fig. 7-73

Fig. 7-74

**7.42** Find the reactive part of the driving point impedance of the circuit in Fig. 7-74 if $i(t) = 17 \sin (t + 76°)$.

**7.43** Verify the following phasor matrix equation for the circuit shown in Fig. 7-75, where $v(t) = \frac{1}{2} \cos (5t + 76°)$:

$$\begin{bmatrix} 5 + j5 & -1 \\ -1 & 10 + j10 \end{bmatrix} \begin{bmatrix} \mathbf{V}_a \\ \mathbf{V}_b \end{bmatrix} = \begin{bmatrix} 2e^{j(5t+76°)} \\ 0 \end{bmatrix}$$

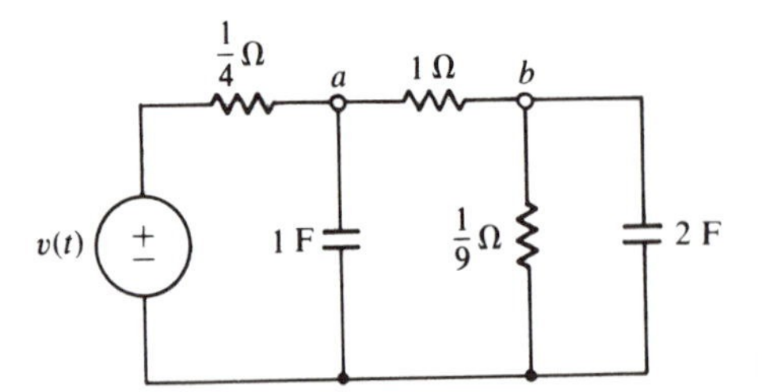


Fig. 7-75

**7.44** Find $v_{a,\,\text{sss}}(t)$ for the circuit of Problem 7.43.

**7.45** Show that for the circuit of Fig. 7-76, $Y(j1) = \mathbf{I}/\mathbf{V} = (\sqrt{40}/8)\, e^{-j18.4^\circ}$ and that $i_{\text{sss}}(t) = 5 \sin t$ when $v(t) = \sqrt{40} \sin (t + 18.4^\circ)$.

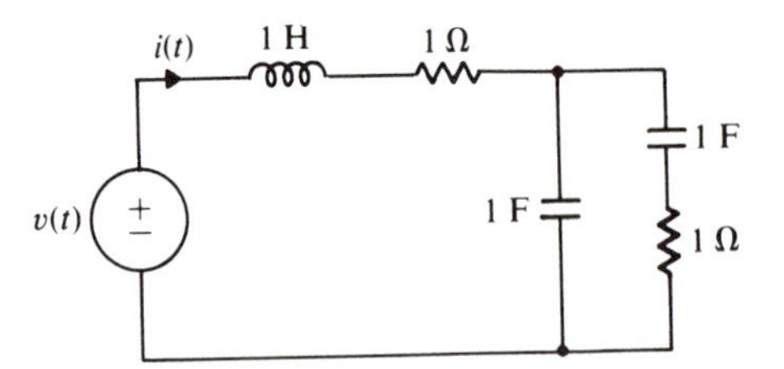


Fig. 7-76

**7.46** Confirm the following phasor matrix equation for the network shown in Fig. 7-77, where $i(t) = \frac{1}{2} \cos (5t + 67^\circ)$ and $v(t) = 4 \cos (5t - 17^\circ)$.

$$\begin{bmatrix} 5 + j5 & -1 \\ -1 & 10 + j10 \end{bmatrix} \begin{bmatrix} \mathbf{I}_1 \\ \mathbf{I}_2 \end{bmatrix} = \begin{bmatrix} 2e^{j(5t+67^\circ)} \\ 4e^{j(5t+163^\circ)} \end{bmatrix}$$

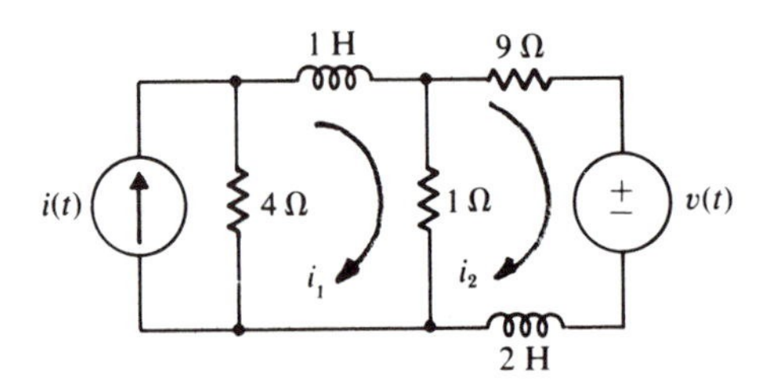


Fig. 7-77

**7.47** From oscillographic measurements on the circuit of Fig. 7-78 it is known that at some frequency $\omega$,

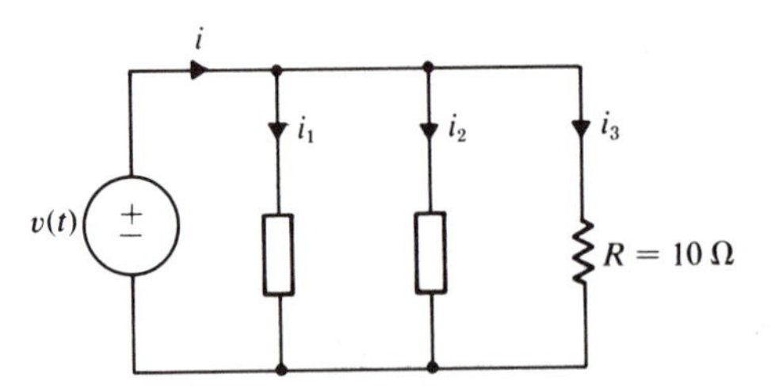


Fig. 7-78

$$i_1(t) = 2\cos(\omega t - 60°), \qquad i_2(t) = 5\cos(\omega t + 90°), \qquad i_3(t) = \cos\omega t$$

Calculate the driving point impedance $Z(j\omega) = \mathbf{V}/\mathbf{I}$.

**7.48** At what frequency does the reactive part of the driving point impedance become zero in the circuit of Fig. 7-79?

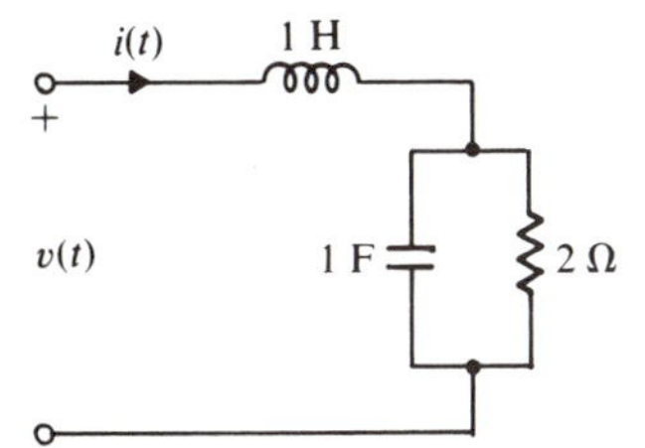


**Fig. 7-79**

**7.49** Use the graphical method to find $H(j1) = \mathbf{V}/\mathbf{E}$ for the circuit of Fig. 7-80.

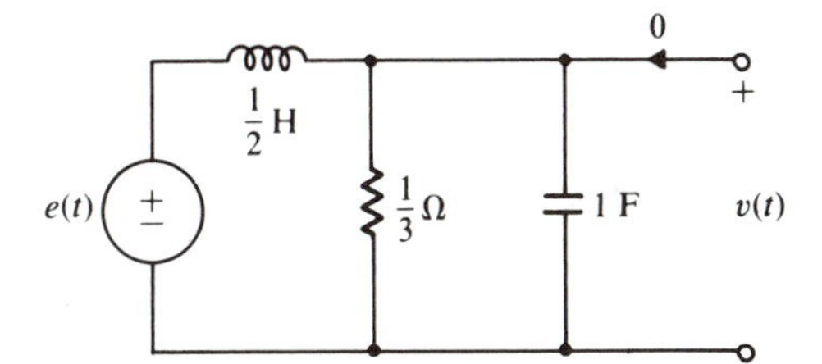


**Fig. 7-80**

**7.50** For $v(t) = A\cos t$ in Fig. 7-81, show that $i_{sss}(t) = \frac{1}{2}A\cos t$.

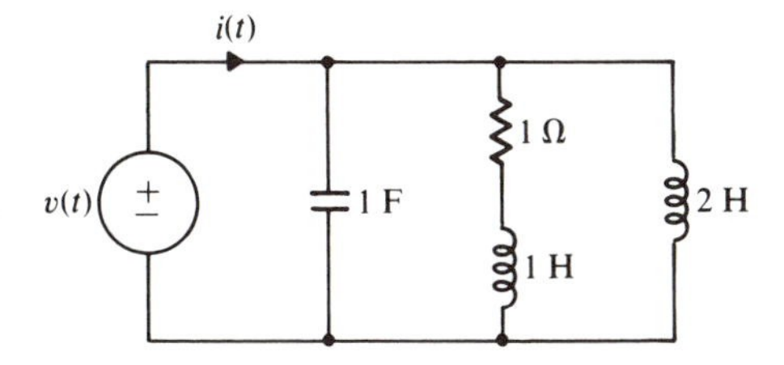


**Fig. 7-81**

**7.51** The following differential equation describes the input-output voltage relationship of the network in Fig. 7-82,

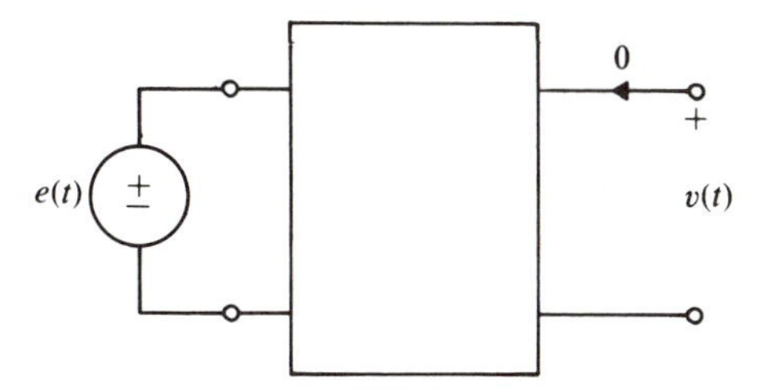


**Fig. 7-82**

$$\frac{d^3v}{dt^3} + 10\frac{d^2v}{dt^2} + 8\frac{dv}{dt} + 5v = 13\frac{de}{dt} + 7e$$

For what input $e(t)$ will the sinusoidal steady-state output be $v_{sss}(t) = 2 \sin t$?

**7.52** From oscillographic measurements in the sinusoidal steady-state it has been determined that in the circuit of Fig. 7-83, $v_{2,\,sss}(t) = 2 \sin 2t$. Show that $v_{1,\,sss}(t) = \sqrt{2} \sin (2t - 45°)$.

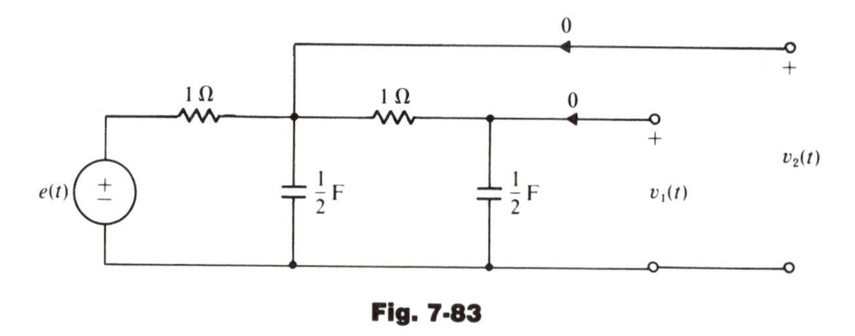


Fig. 7-83

**7.53** Find the sinusoidal output impedance at $\omega = 3$ rad/s for the network in Fig. 7-84.

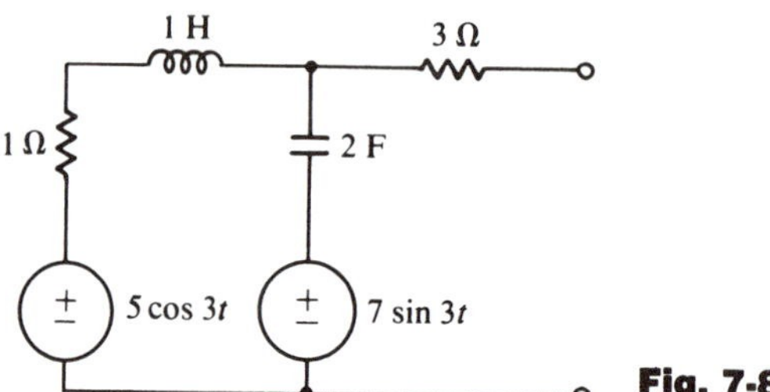


Fig. 7-84

**7.54** An electric circuit is described by the differential equation $\frac{1}{2}(di/dt) + \frac{3}{2}i = e(t)$. Find $i_{sss}(t)$ if $e(t) = 6\sqrt{2} \cos (3t + 60°)$.

**7.55** In the circuit of Fig. 7-85,

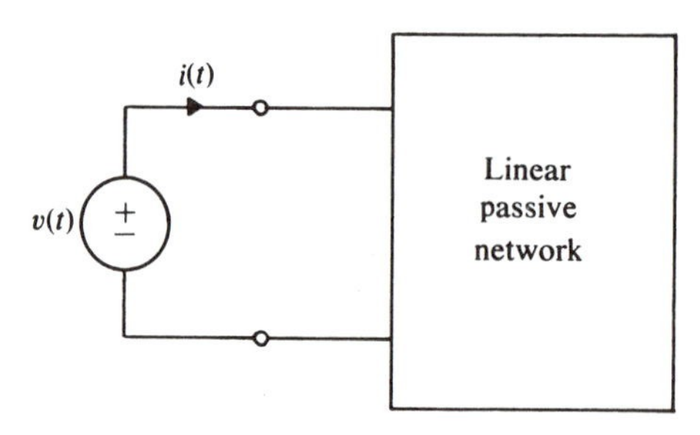


Fig. 7-85

$$\frac{d^4 i}{dt^4} + 20\frac{d^3 i}{dt^3} + 100\frac{d^2 i}{dt^2} + 1200\frac{di}{dt} + 12{,}000i = 100\frac{dv}{dt} + 900v$$

Find a simple two-element series network that would represent this circuit under sinusoidal steady-state conditions with $\omega = 10$ rad/s.

**7.56** For the circuit in Fig. 7-86, find $v_{sss}(t)$ when $e(t) = 100\cos(3t + 13°)$ and $i(t) = 0$.

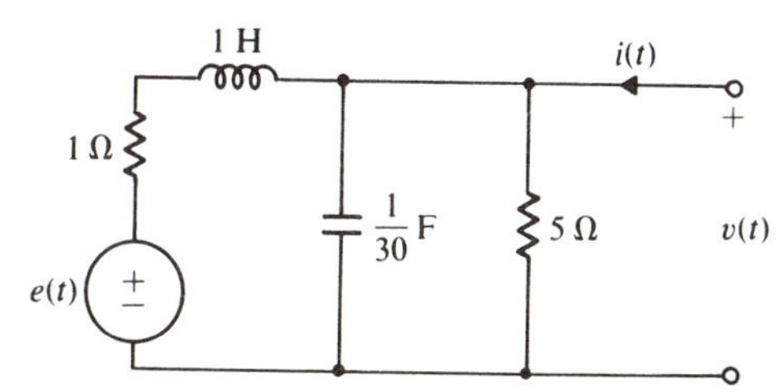


**Fig. 7-86**

**7.57** Find the output impedance $Z_0(j3) = \mathbf{V}/\mathbf{I}$ of the circuit in Fig. 7-86.

**7.58** It is known that for the circuit of Fig. 7-87, $I_{R,\text{eff}} = 4$ A and $I_{L,\text{eff}} = 5$ A. Use phasor diagrams to find the phase of $v_L(t)$ with repsect to $v_R(t)$.

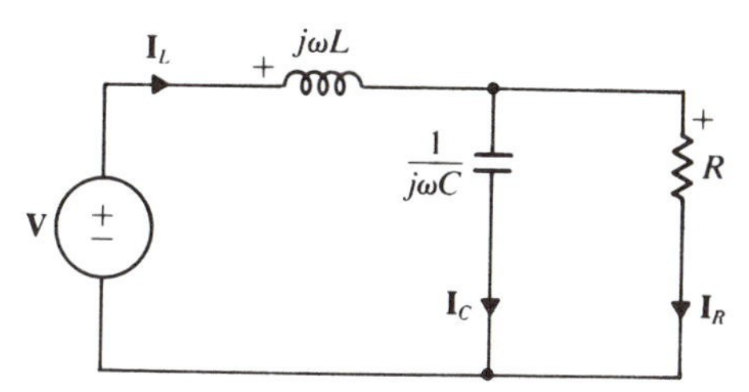


**Fig. 7-87**

**7.59** Find the Thevenin equivalent of the circuit of Fig. 7-88 as seen by the resistor R.

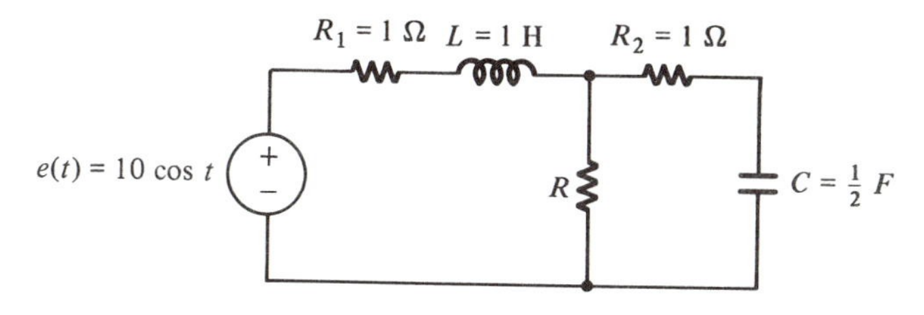


**Fig. 7-88**

Chapter

# 8

# FREQUENCY RESPONSE AND FILTERS

The calculation of a circuit's sinusoidal steady-state response to a sinusoidal input was developed in Chapter 7. The present chapter discusses the *application* of these techniques to the evaluation of a circuit's ac response as a function of frequency. That is, we will determine the circuit's *frequency response*.

Initially, we will be concerned with the various possibilities for *plotting* frequency response data in a meaningful and useful way. We will treat this in relation to network functions, but the same techniques may be applied to a network's *stationary* response phasor, which can be obtained experimentally as well as analytically.

Then we will go on to examine the properties of circuits which are designed to have specified frequency response characteristics. Such circuits are known as *filters*.

## THE RECTANGULAR PLOT

The most obvious frequency response plot is of $|H(j\omega)|$ and $\underline{/H(j\omega)}$ vs. $\omega$, as in Fig. 8-1. Suppose $H(j\omega)$ relates an amplifier's input and output, $H(j\omega) = \mathbf{V}_{out}/\mathbf{V}_{in}$. Then from a plot such as that in Fig. 8-1 we may read the *mid-frequency voltage gain* $|H(j\omega)|_{max}$ and the *bandwidth* $B = \omega_2 - \omega_1$, which is usually defined in terms of the *half-power frequencies* as in Fig. 8-1. (Since the power in a resistor, for example, is proportional to the voltage *squared*, the *power* will be halved when the voltage drops to $1/\sqrt{2} = 0.707$ of its maximum value.) In brief, the plot facilitates the examination of a circuit's characteristics as a parameter such as frequency is varied.

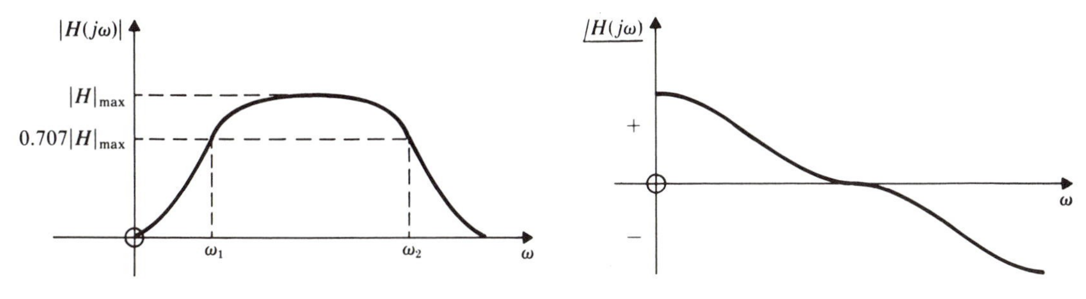


**Fig. 8-1**

## POLAR PLOT OR LOCUS DIAGRAM

If we plot $H(j\omega) = |H(j\omega)|e^{j\underline{/H(j\omega)}}$ as a frequency-dependent complex number, a *polar plot* or *locus diagram* results. As is usual when plotting complex numbers, we have a choice of plotting $|H(j\omega)|$ vs. $\underline{/H(j\omega)}$ on polar graph paper (see Fig. 8-2*a*) or Im $[H(j\omega)]$ vs. Re $[H(j\omega)]$ on rectangular graph paper (see Fig. 8-2*b*). Both procedures lead, of course, to the same graph or locus, as can be seen by comparing Figs. 8-2*a* and *b*.

It should be evident that the polar plot or locus diagram contains the same information as the *two* curves of the rectangular plot, and that we can extract the same information (e.g. bandwidth) *provided that* frequency values $\omega_a$, $\omega_b$, $\omega_c$, . . . have been written in beside each plotted point on the locus.

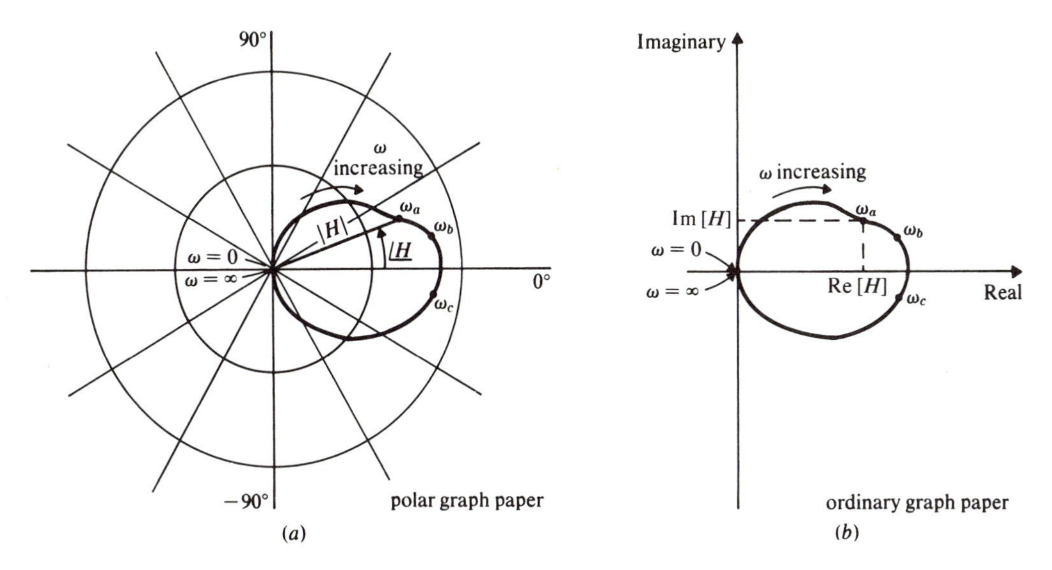


**Fig. 8-2**

## THE NYQUIST (POLAR) DIAGRAM

If the polar diagram for $H(j\omega)$ is plotted *for all* $\omega$, $-\infty \leq \omega \leq \infty$, the resulting curve is called the *Nyquist diagram*. (This assumes that the numerator polynomial of $H(s)$ is of equal or lower order than the denominator, and that there are no imaginary poles or zeros. Otherwise the Nyquist diagram must be defined more carefully.)

The Nyquist diagram plays an important role in stability theory, offering an alternative to the Routh procedure.

## LOGARITHMIC PLOTS AND THE BODE DIAGRAM

In practice the rectangular plots of Fig. 8-1 are not particularly useful, since both $|H(j\omega)|$ and $\omega$ are likely to vary through several orders of magnitude. The obvious move is to use log-log graph paper for the $|H(j\omega)|$ vs. $\omega$ curve and semilog paper for the $\underline{/H(j\omega)}$ vs. $\omega$ curve. This is illustrated in Fig. 8-3.

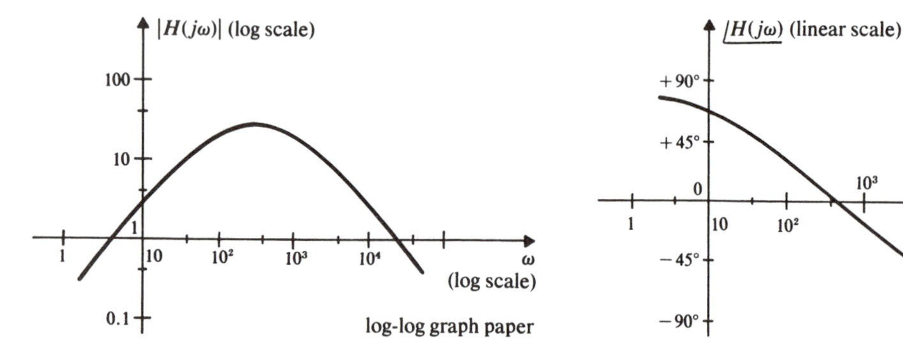


**Fig. 8-3**

There are a number of equivalent possibilities, the most common being to plot $20 \log |H(j\omega)|$ vs. $\omega$ and $\underline{/H(j\omega)}$ vs. $\omega$ on semilog graph paper, with $\omega$ along the log scale (see Fig. 8-4). The quantity $20 \log |H(j\omega)|$ is called the *decibel* or dB gain of the network.

These logarithmic plots are often referred to as Bode diagrams.

The logarithmic or Bode diagram of Fig. 8-3 or 8-4 contains the same information as the earlier graphs, so that characteristics such as bandwidth may still be read—if we observe that $0.707 \equiv 20 \log 0.707 \doteq -3$ dB in Fig. 8-4.

Like its near relative, the Nyquist diagram, the Bode diagram has applications in stability theory.

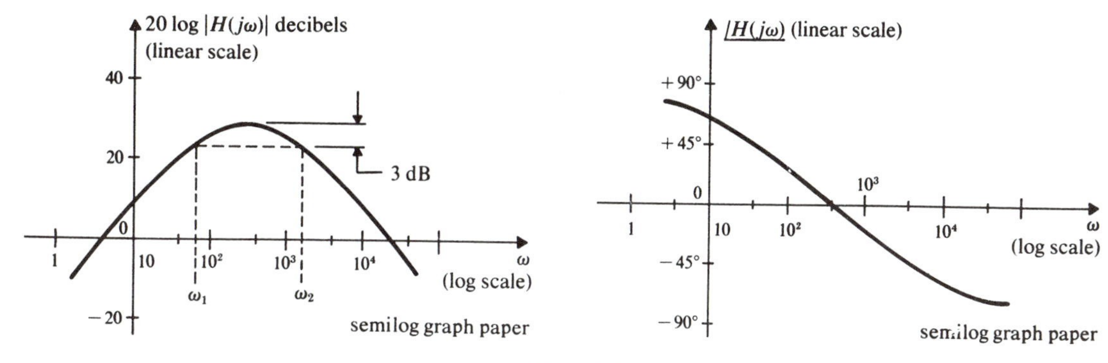


**Fig. 8-4**

## FREQUENCY RESPONSE CALCULATIONS

Given a network function $H(s)$, it is a relatively simple computer or calculator exercise to evaluate $|H(j\omega)|$ and $\underline{/H(j\omega)}$ for any specified value of $\omega$. (See, for example, Problems 7.2, 7.3, and 7.4.) And it is a simple matter to loop repeatedly through such a calculation to accumulate and plot frequency response data—that is, the value of $|H(j\omega)|$ and $\underline{/H(j\omega)}$ at each of several-to-many values of $\omega$. In fact, your computer center almost certainly has such a program available. Programs are also available for desktop computers and pocket calculators.

## ASYMPTOTIC FREQUENCY RESPONSE PLOTS

Here our objective is to develop a shortcut, graphical method for obtaining a network function's frequency response curves. It may reasonably be argued that this topic is an outdated artifact from the precomputer era. However, it is still a part of some feedback control courses.

We start with the assumption that the numerator and denominator of $H(s)$ come to us in factored form—which is often the case in control system analysis.

First, let us suppose that $H(s)$ is a network function whose poles and zeros are real,

$$H(s) = \frac{K(s - z_1)(s - z_2)\cdots}{(s - p_1)(s - p_2)\cdots} = \frac{K(s + 1/\tau_1)(s + 1/\tau_2)\cdots}{(s + 1/\tau_a)(s + 1/\tau_b)\cdots}$$

Then $H(j\omega) = H(s)\big|_{s=j\omega}$ can be written

$$H(j\omega) = \frac{K'(1 + j\omega\tau_1)(1 + j\omega\tau_2)\cdots}{(1 + j\omega\tau_a)(1 + j\omega\tau_b)\cdots} \tag{8.1}$$

which we will regard as the standard frequency response form, where $K' = K\tau_a\tau_b \ldots / \tau_1\tau_2 \ldots$.

It follows that we can calculate $H(j\omega)$ at any $\omega$ from

$$|H(j\omega)| = \frac{|K'|\sqrt{1 + \omega^2\tau_1^2}\,\sqrt{1 + \omega^2\tau_2^2}\cdots}{\sqrt{1 + \omega^2\tau_a^2}\,\sqrt{1 + \omega^2\tau_b^2}\quad\cdots} \tag{8.2}$$

and

$$\underline{/H(j\omega)} = \left.\begin{matrix}0^\circ\\ \pm 180^\circ\end{matrix}\right\} + \tan^{-1}\omega\tau_1 + \tan^{-1}\omega\tau_2 + \cdots$$
$$- (\tan^{-1}\omega\tau_a + \tan^{-1}\omega\tau_b + \cdots) \tag{8.3}$$

where the first term is $0^\circ$ if $K' > 0$ and $\pm 180^\circ$ if $K' < 0$.

We will now demonstrate that equations (*8.1*)–(*8.3*) lead to the desired graphical procedure. To this end we will consider a typical term in equation (*8.1*), namely $1 + j\omega\tau$.

Now $|1 + j\omega\tau| = \sqrt{1 + \omega^2\tau^2}$, which is graphed as the dashed curve in Fig. 8-5. Further,

$$|1 + j\omega\tau| = \sqrt{1 + \omega^2\tau^2} \quad \begin{cases} \to 1 & \text{for } \omega \ll 1/\tau \\ \to \omega\tau & \text{for } \omega \gg 1/\tau \end{cases} \tag{8.4}$$

and we see that the *asymptotic* or *straight-line approximation* plotted as the solid line in Fig. 8-5 is not too seriously in error, at least for values of $\omega$ remote from $\omega = 1/\tau$.

It should be clear that once the *break frequency* $\omega_b = 1/\tau$ has been located, *no further calculations are necessary*. The two straight-line segments can immediately be

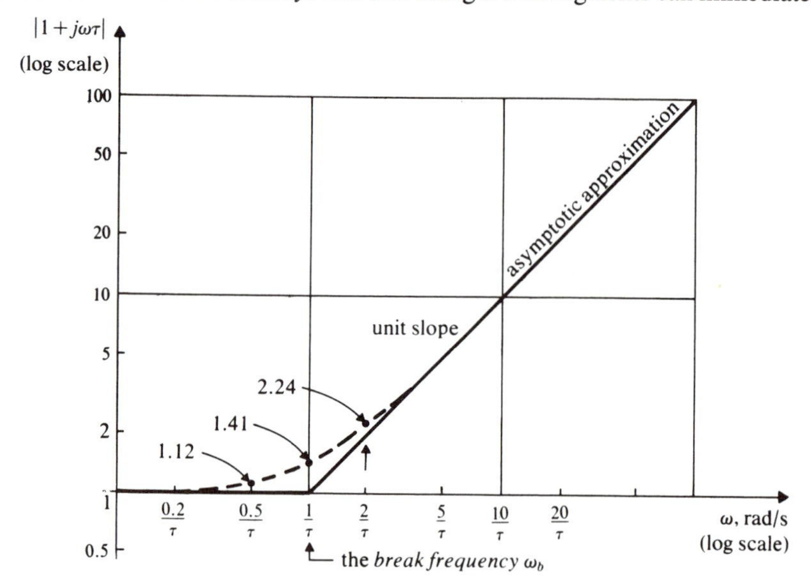


**Fig. 8-5**

drawn: one with zero slope at $|1 + j\omega\tau| = 1$, and one with unit slope meeting the first at the break frequency. If greater accuracy is warranted, we can read from Fig. 8-5 the correction factors of 1.12, 1.41, and $2.24/2 = 1.12$ at $\omega = 0.5/\tau$, $1/\tau$, and $2/\tau$; plot the corrected points; and then sketch the true curve.

There is one special case: the term $|j\omega|$ graphs as a straight line of unit slope through the point (1, 1).

And if any term is repeated $n$ times, all the corresponding slopes and magnitudes must be multiplied by $n$. Setting $n = -1$ we see that $|1/(1 + j\omega\tau)|$ breaks *downward* with an asymptotic slope of $-1$.

Now, since we work on logarithmic graph paper, the individual terms in equation (8.2) may be plotted separately and then *added* graphically to yield $|H(j\omega)|$. (In the logarithmic domain, multiplication becomes addition.)

## Example 8.1

We will obtain the frequency response curve of $|H(j\omega)|$ vs. $\omega$ for the network function

$$H(s) = \frac{4(s + 0.1)}{s(s + 2)}$$

The first step is to put $H(j\omega)$ into standard frequency response form:

$$H(j\omega) = \frac{4(0.1 + j\omega)}{j\omega(2 + j\omega)} = \frac{0.2(1 + j10\omega)}{j\omega(1 + j0.5\omega)}$$

Now we can plot 0.2, $1/|j\omega|$ and the asymptotic approximations for $|1 + j10\omega|$ and $1/|1 + j0.5\omega|$ as in Fig. 8-6. (The break frequencies for $|1 + j10\omega|$ and $1/|1 + j0.5\omega|$ are 0.1 and 2 rad/s respectively.)

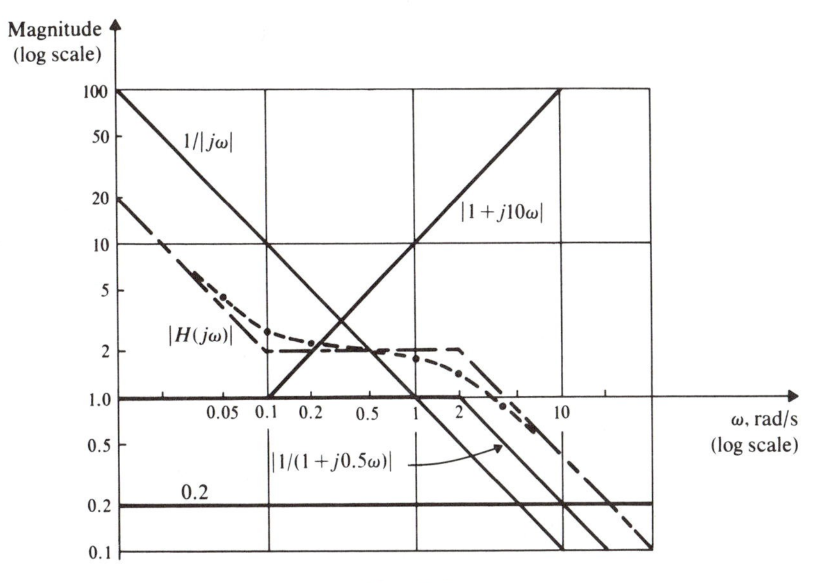


**Fig. 8-6**

These four graphs have been added in Fig. 8-6 to yield $|H(j\omega)|$ vs. $\omega$. Observe carefully that at each value of $\omega$ we add the vertical *distances* above (plus) and below (minus) the base line of unity magnitude. We have *not* added the *numbers* appearing along the magnitude axis. The chain line in Fig. 8-6 is the asymptotic approximation to $|H(j\omega)|$; the dashed line is the true curve which we obtain by multiplying the values on the asymptotic curve by 1.12, 1.41, and $2.24/2 = 1.12$ at $\omega = 0.05$, 0.1, and 0.2 rad/s and by $1/1.12 = 0.89$, $1/1.41 = 0.71$, and $2/2.24 = 0.89$ at $\omega = 1$, 2, and 4 rad/s. ■

The final asymptotic approximation to $|H(j\omega)|$ is even easier to plot once we realize that its *slope* at any $\omega$ is equal to the *sum* of the *slopes* of its components. It can only have slopes of 0, $\pm 1$, $\pm 2$, . . . and can "break" only at the break frequencies of its components. These characteristics can be observed in Fig. 8-6.

As an alternative we can plot the *decibel gain* vs. $\omega$ on *semilog* graph paper. There is no substantial difference; for example, the $|1 + j\omega\tau|_{\text{dB}}$ vs. $\omega$ curve of Fig. 8-7 is precisely the same as the $|1 + j\omega\tau|$ vs. $\omega$ curve of Fig. 8-5 except for the different vertical scale. (Slopes are often expressed in dB/octave. One *octave* represents a *doubling* of frequency.)

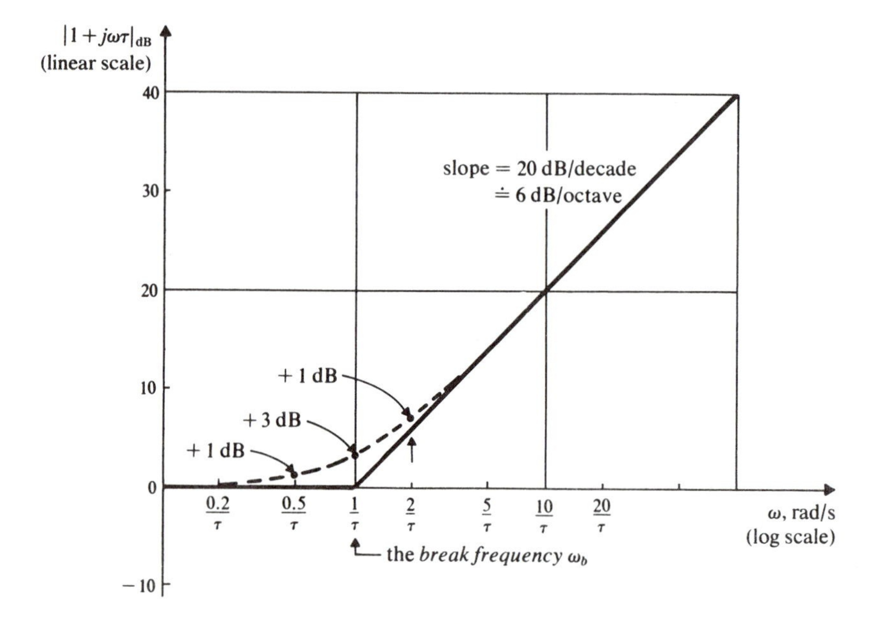


**Fig. 8-7**

## Example 8.2

Let us plot the curve of $|H(j\omega)|_{\text{dB}}$ vs. $\omega$ for the network function,

$$H(s) = \frac{4(s + 0.1)}{s(s + 2)} \quad \text{or} \quad H(j\omega) = \frac{0.2(1 + j10\omega)}{j\omega(1 + j0.5\omega)}$$

We must calculate $|0.2|_{\text{dB}} = 20 \log 0.2 = -14$ dB. Then we can plot $|0.2|_{\text{dB}}$, $|1/j\omega|_{\text{dB}}$, and

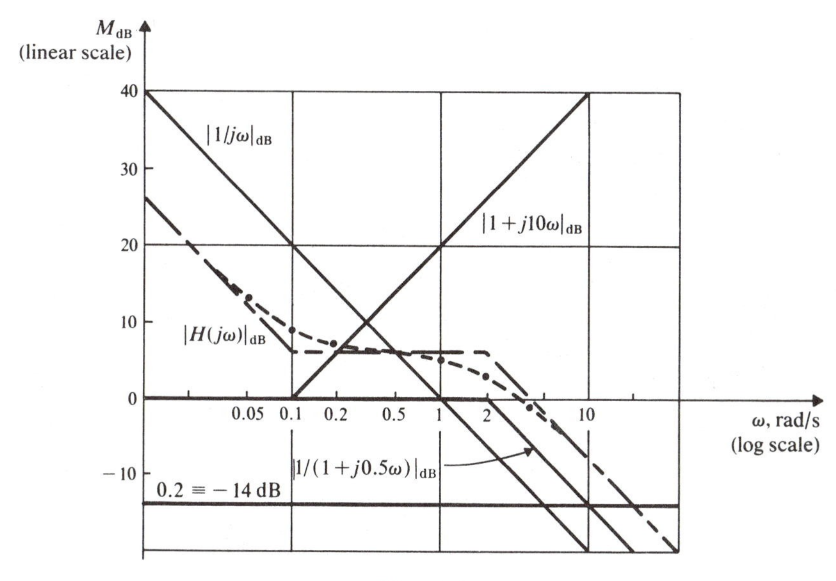


**Fig. 8-8**

the asymptotic approximations for $|1 + j10\omega|_{dB}$ and $|1/(1 + j0.5\omega)|_{dB}$ as in Fig. 8-8, where the four graphs have been added to yield $|H(j\omega)|_{dB}$ vs. $\omega$. Now, when working in decibels, we may add *either* vertical *distances* above (plus) and below (minus) the base line of 0 dB, *or* the corresponding dB values read from the magnitude scale. And the corrections needed to obtain the true curve are *added*; 1, 3, and 1 dB at $\omega = 0.05$, 0.1, and 0.2 rad/s and −1, −3, and −1 dB at $\omega = 1$, 2, and 4 rad/s. ■

Whether we choose to work with dB or with the actual magnitudes depends primarily on personal preference. The procedure is virtually the same.

Returning now to the *phase* of the typical term in equation (*8.1*), we have

$$\underline{/1 + j\omega\tau} = \tan^{-1} \omega\tau \tag{8.5}$$

which is plotted as the dashed line on the semilog paper of Fig. 8-9. This, too, can

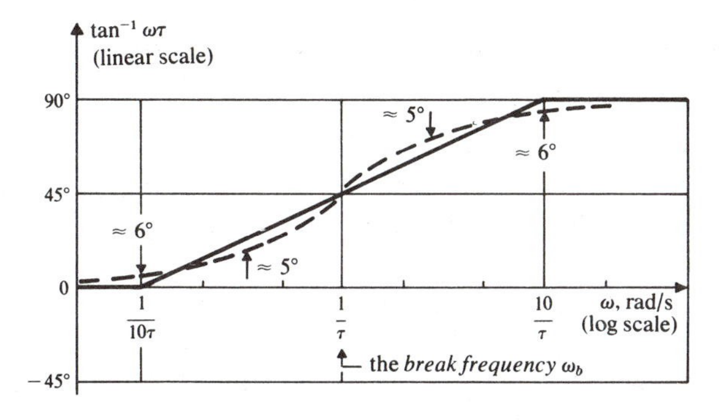


**Fig. 8-9**

reasonably be approximated by straight-line segments, as shown by the solid curve in Fig. 8-9.

Note that the center segment passes through the point $(\omega_b, 45°)$ with a slope of 45°/decade. This meets the low frequency asymptote at the point $(0.1\omega_b, 0)$ and the high frequency asymptote at $(10\omega_b, 90°)$. Both asymptotes have zero slope.

As a special case, $\angle j\omega$ is 90° for all $\omega$.

And if any term is repeated $n$ times, the corresponding phase at each value of $\omega$ must be multiplied by $n$.

Finally, equation (8.3) shows that if we plot the phase angle of the component terms on a linear scale, then $\angle H(j\omega)$ may be obtained by graphical addition.

## Example 8.3

Now we can obtain the frequency response curve $\angle H(j\omega)$ vs. $\omega$ for the sinusoidal network function of Examples 8.1 and 8.2, namely

$$H(j\omega) = \frac{0.2(1 + j10\omega)}{j\omega(1 + j0.5\omega)}$$

The straight-line approximations for $\angle 1 + j10\omega$, $\angle 1/(1 + j0.5\omega)$, and the constant $-90°$ phase of $1/j\omega$ have been graphed in Fig. 8-10, as has their sum. If we wished, we could sketch the true curve for $\angle H(j\omega)$.

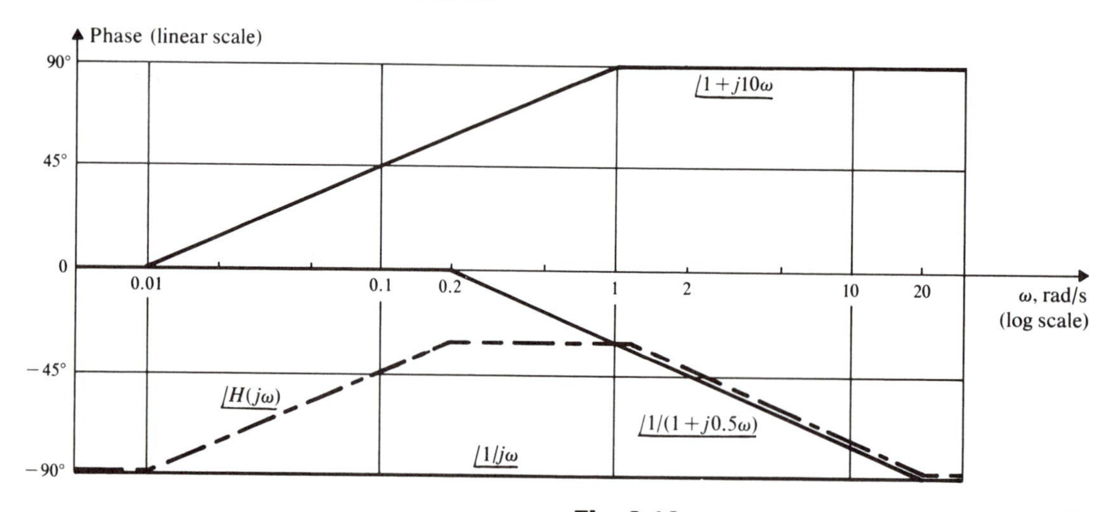


**Fig. 8-10** ■

If $H(s)$ is a network function having one or more pairs of complex conjugate poles or zeros, then we are faced with quadratic factors of the form

$$\frac{1}{\omega_n^2}(s^2 + 2\zeta\omega_n s + \omega_n^2)\big|_{s=j\omega} = \left(1 + j2\zeta\frac{\omega}{\omega_n} - \frac{\omega^2}{\omega_n^2}\right)$$

where we have divided through by $\omega_n^2$ to obtain the standard frequency response form.

The corresponding magnitude and phase are

$$M = \sqrt{\left(1 - \frac{\omega^2}{\omega_n^2}\right)^2 + \left(2\zeta\frac{\omega}{\omega_n}\right)^2} \quad \text{and} \quad \theta = \tan^{-1}\left\{\frac{2\zeta\omega}{\omega_n} \Big/ \left(1 - \frac{\omega^2}{\omega_n^2}\right)\right\} \tag{8.6}$$

which can be plotted against $\omega$ as in Fig. 8-11. As can be seen in that figure, the asymptotic magnitude curve can be seriously in error for $\omega/\omega_n \simeq 1$. In fact, straightforward calculations show that for $\zeta < \sqrt{2}/2$, $M$ has a minimum value at $\omega/\omega_n = \sqrt{1 - 2\zeta^2}$ and that value is $M = 2\zeta\sqrt{1 - \zeta^2}$.

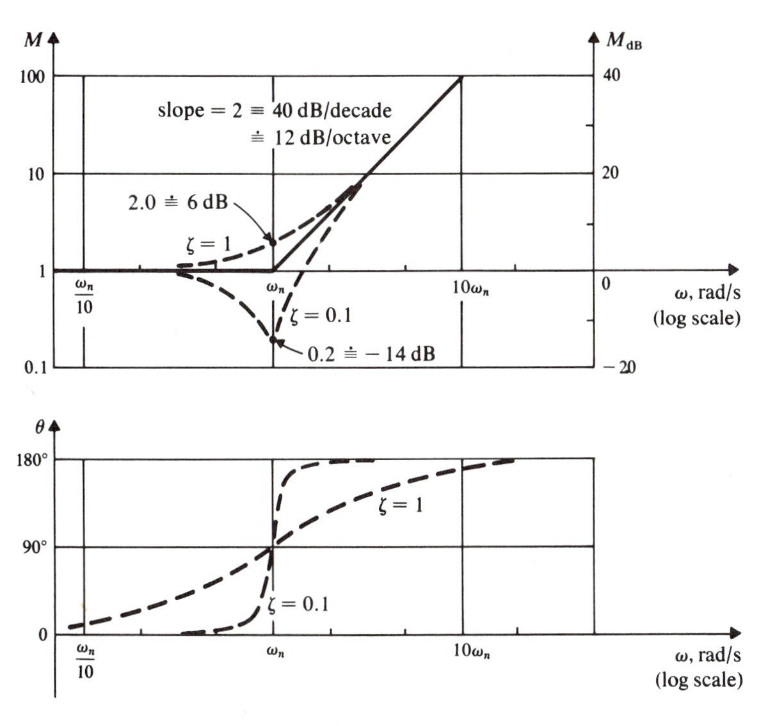


**Fig. 8-11**

Straight-line phase curves are sometimes used, but the slope of the center section must clearly be a function of $\zeta$.

## FILTERS

In virtually every branch of engineering we find requirements for circuits or devices which have gains—or attenuations—that depend, in an important way, on frequency. The bass and treble controls in a hi-fi, or a mechanical vibration damper, come to mind. These *filters,* so called, can conveniently be classified into the following categories.

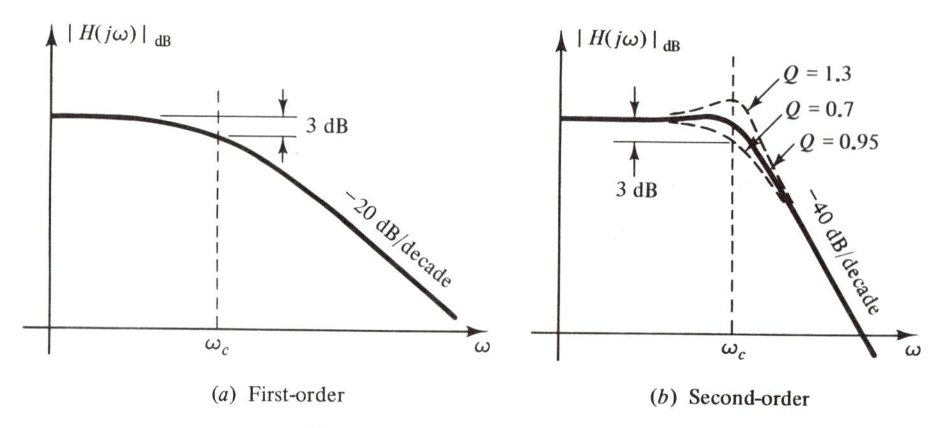


(a) First-order (b) Second-order

**Fig. 8-12 Low-pass filters**

*Low-pass filters* allow low-frequency signals to pass, while attenuating high frequencies—see Fig. 8-12. The simplest (first- and second-order) filters are described by

$$H(s) = \frac{K}{s + \omega_c} \quad \text{and} \quad H(s) = \frac{K}{s^2 + Bs + \omega_c^2} \tag{8.7}$$

where $\omega_c$ is called the *cut-off frequency*. $\omega_c$ and the parameter $B$ in the second-order transfer function are commonly combined to define another parameter, the circuit's $Q(=\omega_c/B)$, which controls the shape of the frequency response curve near cut off—see Fig. 8-12 again.

*High-pass filters* pass high frequency signals while attenuating low frequencies—see Fig. 8-13. The corresponding first- and second-order transfer functions are

$$H(s) = \frac{Ks}{s + \omega_c} \quad \text{and} \quad H(s) = \frac{Ks^2}{s^2 + Bs + \omega_c^2} \tag{8.8}$$

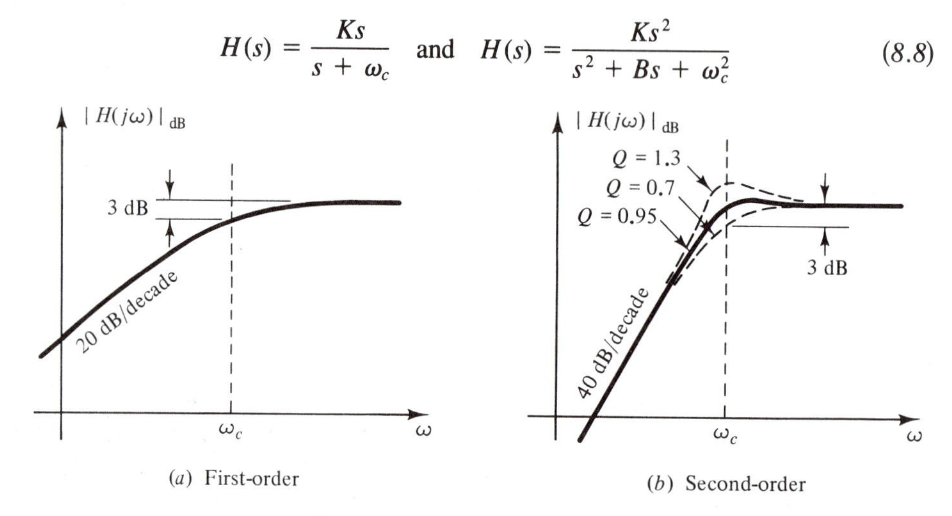


(a) First-order (b) Second-order

**Fig. 8-13 High-pass filters**

*Band-pass filters* pass a narrow band of frequencies, while attenuating those outside the band—see Fig. 8-14. The simplest such filter is second-order:

$$H(s) = \frac{Ks}{s^2 + Bs + \omega_o^2} \tag{8.9}$$

where $\omega_o$ is called the *resonant frequency* and $B$ is called the 3-dB *bandwidth*—shown on the $Q = 1$ curve of Fig. 8-14. The circuit's $Q(=\omega_o/B)$ determines the height and sharpness of the resonant peak. This filter is often called a resonant circuit.

*Band-reject* or *notch filters* attenuate a signal frequencies within the notch—see Fig. 8-15. The simplest notch filter is second-order:

$$H(s) = \frac{K(s^2 + \omega_o^2)}{s^2 + Bs + \omega_o^2} \tag{8.10}$$

This filter has zero gain ($-\infty$ dB) at the anti-resonant or notch frequency $\omega_o$. The 3-dB bandwidth is approximately equal to $B$ for $Q \gg 1$.

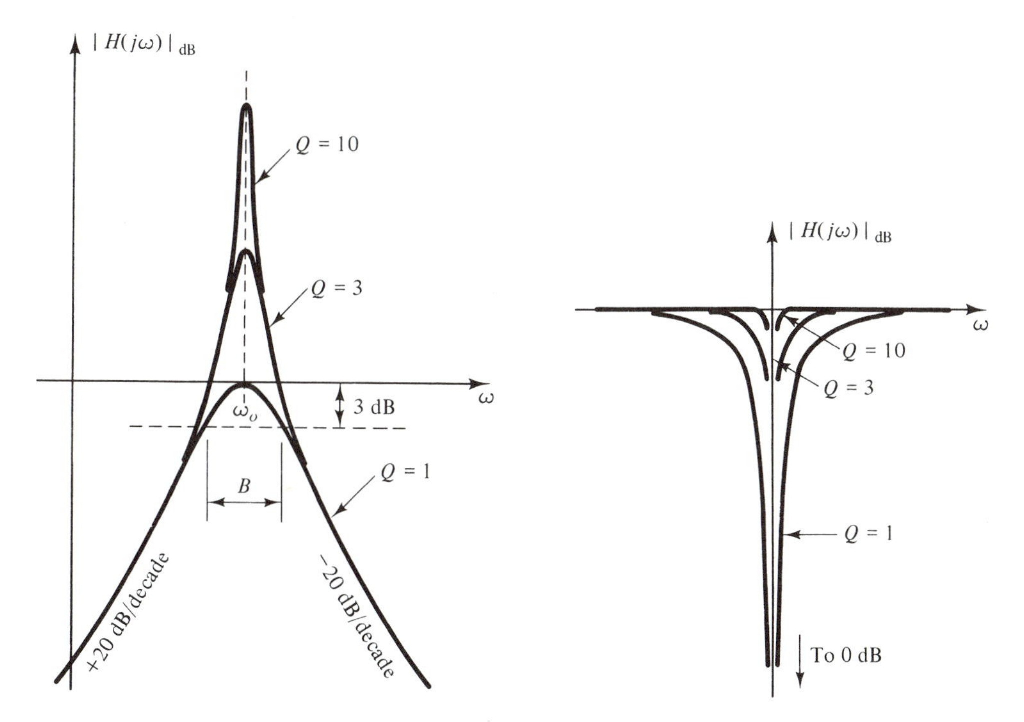


**Fig. 8-14 Band-pass filters**

**Fig. 8-15 Band-reject filters**

In practice, higher or more specialized performance than can be met by one of the above prototypical filters is often demanded. Filters of greater complexity are therefore commonplace.

## ACTIVE FILTERS

When we build a filter, our task is often simplified by designing a circuit around one or more op amps. Such a filter is said to be *active*. Op amps make it easy to avoid the unwanted effects of a load and the need for inductors (which are heavy, expensive, and usually far from ideal).

### Example 8.4

We will show that the active circuit of Fig. 8-16 is a second-order, low-pass filter. (Given: $R_1 = R_2 = R$.)

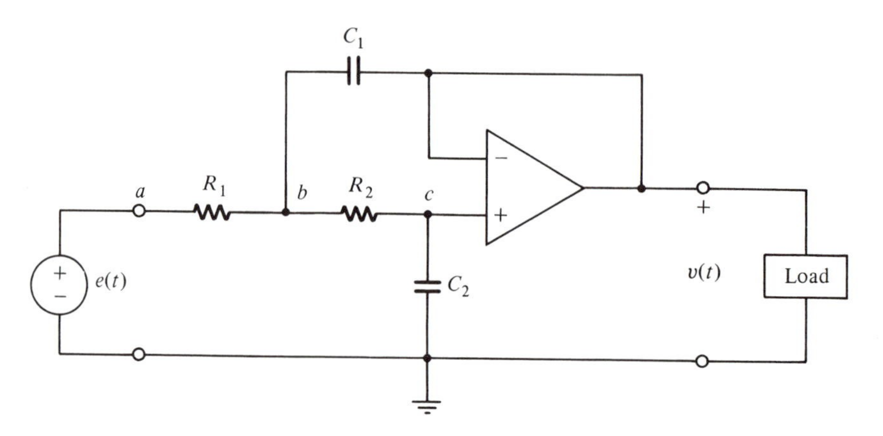


**Fig. 8-16**

Visualizing the circuit in the $s$-domain (with zero initial conditions), and making the usual op amp approximations, KCL at nodes $b$ and $c$ yields

$$\begin{bmatrix} \frac{2}{R} + C_1 s & -\frac{1}{R} - C_1 s \\ -\frac{1}{R} & \frac{1}{R} + C_2 s \end{bmatrix} \begin{bmatrix} V_b(s) \\ V(s) \end{bmatrix} = \begin{bmatrix} \frac{1}{R} E(s) \\ 0 \end{bmatrix}$$

Then, from Cramer's rule,

$$H(s) = \frac{V(s)}{E(s)} = \frac{1/R^2 C_1 C_2}{s^2 + (2/RC_1)s + 1/R^2 C_1 C_2}$$

which is of the same form as equation (8.7) for the second-order low-pass filter described above. Here $\omega_c = \frac{1}{R}\sqrt{1/C_1C_2}$, $B = 2/RC_1$, $Q = \omega_c/B = \frac{1}{2}\sqrt{C_1/C_2}$, and $K = 1/R^2C_1C_2$. The low frequency gain is $K/\omega_c^2 = 1$. ■

# ILLUSTRATIVE PROBLEMS

### *Frequency Response Plots*

**8.1** For the $RC$ network in Fig. 8-17, find $H(j\omega) = \mathbf{V}/\mathbf{I}$ and complete the table below.

| $\omega$ | $\lvert H(j\omega)\rvert$ | $\underline{/H(j\omega)}$ | $\text{Im}[H(j\omega)]$ | $\text{Re}[H(j\omega)]$ |
|---|---|---|---|---|
| 0<br>$1/2RC$<br>$1/RC$<br>$2/RC$<br>$\infty$ | | | | |

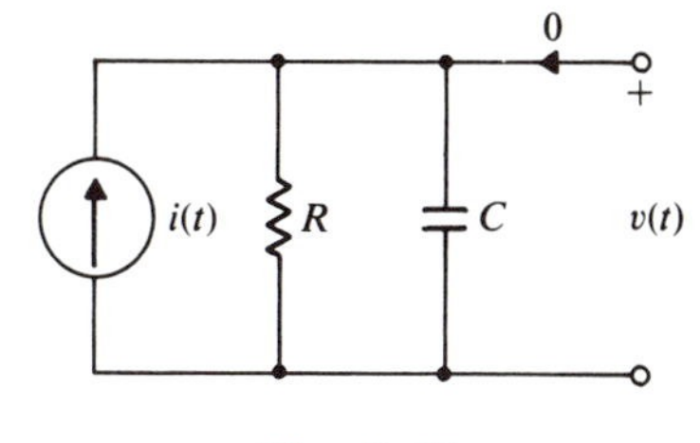


**Fig. 8-17**

Applying KCL, $(1/R)\mathbf{V} + j\omega C\mathbf{V} = \mathbf{I}$ and so

$$H(j\omega) = \frac{\mathbf{V}}{\mathbf{I}} = \frac{1}{1/R + j\omega C} = \frac{R}{1 + j\omega RC}$$

which is the standard form for $j\omega$-domain calculations. Here

$$|H(j\omega)| = \frac{R}{\sqrt{1 + (\omega RC)^2}} \quad \text{and} \quad \underline{/H(j\omega)} = -\tan^{-1} \omega RC$$

If we rationalize the expression for $H(j\omega)$,

$$H(j\omega) = \frac{R}{1 + (\omega RC)^2} - j\frac{\omega R^2 C}{1 + (\omega RC)^2}$$

or

$$\text{Re}[H(j\omega)] = \frac{R}{1 + (\omega RC)^2} \quad \text{and} \quad \text{Im}[H(j\omega)] = \frac{-\omega R^2 C}{1 + (\omega RC)^2}$$

The table can now be completed as follows:

| $\omega$ | $\lvert H(j\omega)\rvert$ | $\underline{/H(j\omega)}$ | $\text{Im}[H(j\omega)]$ | $\text{Re}[H(j\omega)]$ |
|---|---|---|---|---|
| 0 | $R$ | 0 | 0 | $R$ |
| $1/2RC$ | $0.894R$ | $-26.6°$ | $-0.4R$ | $0.8R$ |
| $1/RC$ | $0.707R$ | $-45°$ | $-0.5R$ | $0.5R$ |
| $2/RC$ | $0.448R$ | $-63.4$ | $-0.4R$ | $0.2R$ |
| $\infty$ | 0 | $-90°$ | 0 | 0 |

**8.2** From the data of Problem 8.1, plot the following curves: $|H(j\omega)|$ vs. $\omega$ and $\angle H(j\omega)$ vs. $\omega$; polar plot of $|H(j\omega)|$ vs. $\angle H(j\omega)$; Im $[H(j\omega)]$ vs. Re $[H(j\omega)]$.

The data from the completed table of Problem 8.1 are graphed in Fig. 8-18.

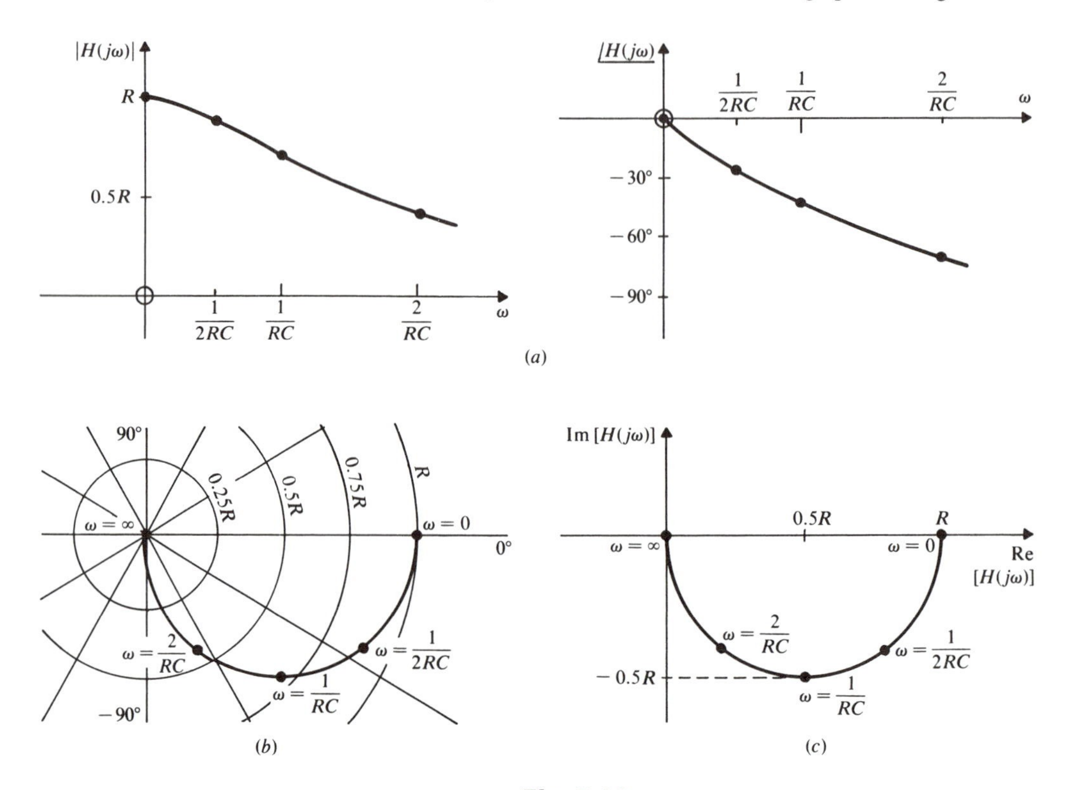


**Fig. 8-18**

**8.3** The network of Fig. 8-19 is sometimes called a *lag network* by control system engineers. Find $H(j\omega) = \mathbf{V}/\mathbf{E}$, complete the table below, sketch the polar plot of $H(j\omega)$, and find the *maximum phase shift* produced by this network.

| $\omega$ | $\lvert H(j\omega)\rvert$ | $\angle H(j\omega)$ |
|---|---|---|
| 0 | | |
| 5 | | |
| 7 | | |
| 10 | | |
| $\infty$ | | |

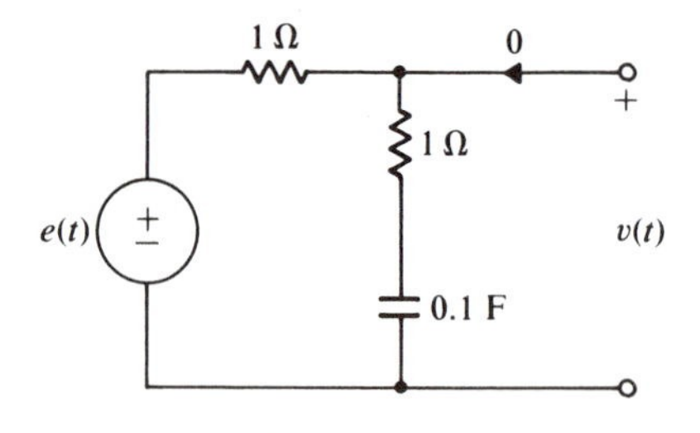


**Fig. 8-19**

Applying KCL,

$$\frac{\mathbf{V}}{1 + 10/j\omega} + \frac{\mathbf{V} - \mathbf{E}}{1} = 0$$

and, in standard form,

$$H(j\omega) = \frac{\mathbf{V}}{\mathbf{E}} = \frac{1 + j0.1\omega}{1 + j0.2\omega}$$

Therefore

$$|H(j\omega)| = \frac{\sqrt{1 + (0.1\omega)^2}}{\sqrt{1 + (0.2\omega)^2}} \quad \text{and} \quad \underline{/H(j\omega)} = \tan^{-1} 0.1\omega - \tan^{-1} 0.2\omega$$

from which the table can be completed:

| $\omega$ | $\lvert H(j\omega)\rvert$ | $\underline{/H(j\omega)}$ |
|---|---|---|
| 0 | 1 | 0 |
| 5 | 0.791 | −18.4° |
| 7 | 0.708 | −19° |
| 10 | 0.634 | −18.4° |
| ∞ | 0.500 | 0 |

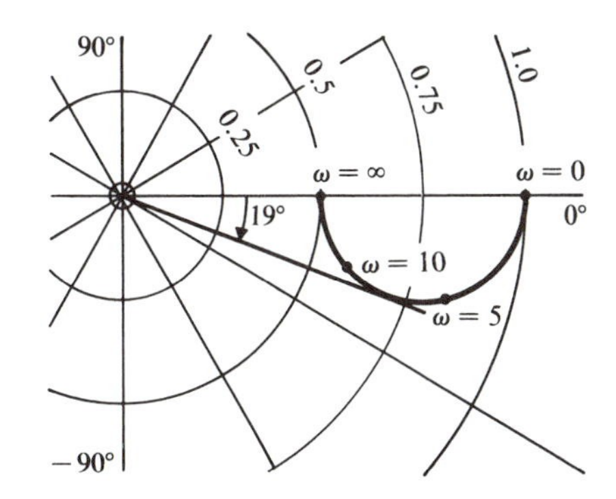


**Fig. 8-20**

These data are plotted in polar coordinates in Fig. 8-20. Again the graph is circular, and by drawing the tangent to the curve from the origin we can measure $\underline{/H(j\omega)}_{\text{min}} = -19°$. That is, **V** lags **E** by a maximum of 19°.

Alternatively we may proceed analytically. To find an extremum of $\underline{/H(j\omega)} = \tan^{-1} 0.1\omega - \tan^{-1} 0.2\omega$ with respect to $\omega$, we set the derivative equal to zero.

$$\frac{d\underline{/H(j\omega)}}{d\omega} = \frac{0.1}{1 + \omega^2/100} - \frac{0.2}{1 + \omega^2/25} = \frac{10}{100 + \omega^2} - \frac{5}{25 + \omega^2} = 0$$

If we solve for $\omega$, we find

$$\omega = \sqrt{50} = 7.07 \text{ rad/s}$$

From the graph we know this corresponds to the desired minimum of $\underline{/H(j\omega)}$, which is given by

$$\underline{/H(j\sqrt{50})} = \tan^{-1} 0.1\sqrt{50} - \tan^{-1} 0.2\sqrt{50} = -19.5°$$

**8.4** Find the poles and zeros of the transfer function $H(s) = V(s)/E(s)$ for the circuit of Fig. 8-19. Then evaluate $H(j5)$ graphically, in the pole-zero diagram.

Using the voltage divider concept,

$$H(s) = \frac{1 + 10/s}{1 + 1 + 10/s} = \frac{s + 10}{2s + 10} = \frac{0.5(s + 10)}{s + 5}$$

Thus $H(s)$ has a zero at $s = -10$ and a pole at $s = -5$. From Fig. 8-21,

$$H(j5) = \frac{0.5(N_1 e^{j\psi_1})}{D_1 e^{j\phi_1}} = \frac{0.5(\sqrt{125}e^{j26.6°})}{\sqrt{50}e^{j45°}} = 0.790e^{-j18.4°}$$

which agrees with the second entry in the table of Problem 8.3.

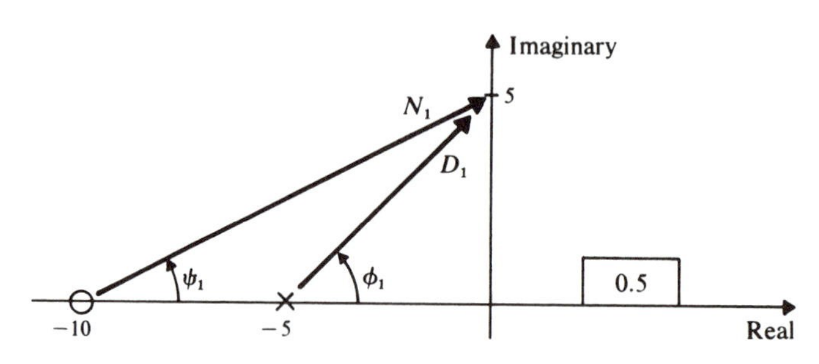


**Fig. 8-21**

**8.5** A polar plot of $Z(j\omega)$ for some (unspecified) circuit is shown in Fig. 8-22. Using these data, construct a polar plot of $Y(j\omega)$ for the same network.

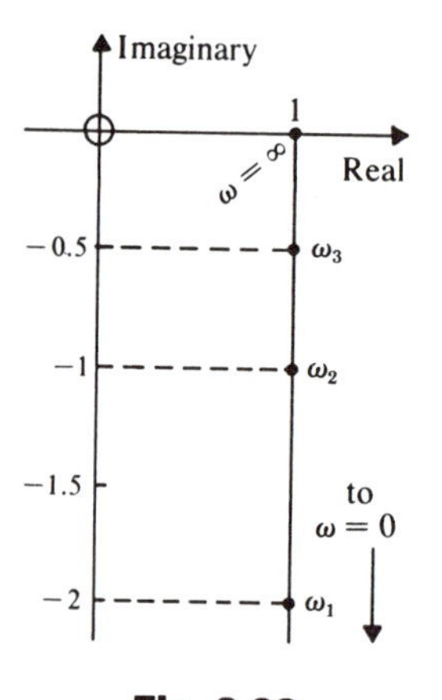


**Fig. 8-22**

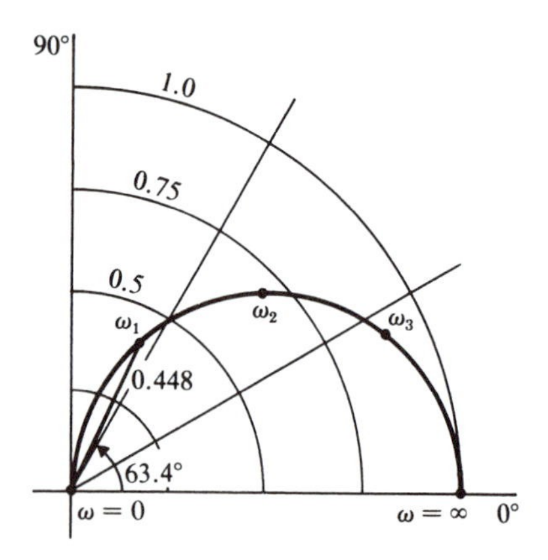


**Fig. 8-23**

Given $Z(j\omega) = |Z(j\omega)|e^{j\angle Z(j\omega)}$, it follows that

$$Y(j\omega) = \frac{1}{Z(j\omega)} = \frac{1}{|Z(j\omega)|}e^{-j\angle Z(j\omega)}$$

We can read values of $|Z(j\omega)|$ and $\angle Z(j\omega)$ from Fig. 8-22, from which $|Y(j\omega)|$ and $\angle Y(j\omega)$ can be computed. Thus:

| $\omega$ | $\lvert Z(j\omega)\rvert$ | $\underline{/Z(j\omega)}$ | $\lvert Y\rvert = 1/\lvert Z\rvert$ | $\underline{/Y} = -\underline{/Z}$ |
|---|---|---|---|---|
| 0 | $\infty$ | $-90°$ | 0 | $90°$ |
| $\omega_1$ | 2.24 | $-63.4°$ | 0.448 | $63.4°$ |
| $\omega_2$ | 1.41 | $-45°$ | 0.707 | $45°$ |
| $\omega_3$ | 1.12 | $-26.6°$ | 0.895 | $26.6°$ |
| $\infty$ | 1.00 | 0 | 1.000 | 0 |

The data from the last two columns are plotted in polar coordinates in Fig. 8-23.

*Comment*: If $Z(j\omega) = R(\omega) + jX(\omega)$, it is, of course, *incorrect* to write $Y(j\omega) = 1/R(\omega) \pm j\{1/X(\omega)\}$. To work in rectangular coordinates, we must rationalize $Y(j\omega) = 1/\{R(\omega) + jX(\omega)\}$, which is much more effort than that needed to find the reciprocal in polar coordinates, as above.

**8.6** It is often helpful to establish a circuit's ac characteristics at very low and at very high frequencies. For example:

(*a*) What is the frequency characteristic of $Z_C(j\omega) = 1/j\omega C$ and $Z_L(j\omega) = j\omega L$ as $\omega \to 0$?

(*b*) What is the frequency characteristic of these impedances as $\omega \to \infty$?

(*c*) For the lag network of Fig. 8-19, ascertain the low and high frequency approximations of the transfer function $H(j\omega) = \mathbf{V}/\mathbf{E}$.

(*d*) Repeat part (*c*) for the output impedance $Z_o(j\omega)$.

(*a*) $Z_C(j0) = \dfrac{1}{j0}$ or $\infty e^{-j90°}$ as $\omega \to 0$ (an open circuit)

$Z_L(j0) = j0$ as $\omega \to 0$ (a short circuit)

(*b*) $Z_C(j\omega) \to -j0$ as $\omega \to \infty$ (a short circuit)

$Z_L(j\omega) \to j\infty$ as $\omega \to \infty$ (an open circuit)

(*c*) At low frequency the capacitor in the circuit of Fig. 8-19 becomes an open circuit (infinite impedance). Thus the voltage divider ratio will be unity, i.e. $\mathbf{V}/\mathbf{E} \to 1$ as $\omega \to 0$. Similarly the ratio at high frequency will be 0.5, i.e. $\mathbf{V}/\mathbf{E} \to 0.5$ as $\omega \to \infty$.

Alternatively, from Problem 8.3,

$$H(j\omega) = \frac{\mathbf{V}}{\mathbf{E}} = \frac{1 + j0.1\omega}{1 + j0.2\omega} \begin{cases} \to 1 & \text{as } \omega \to 0 \\ \to 0.5 & \text{as } \omega \to \infty \end{cases}$$

(*d*) To find the output impedance, $\mathbf{E}$ must be set to zero, a short circuit. Then by inspection of the circuit in Fig. 8-19,

$$Z_o(j\omega) \to 1 \text{ as } \omega \to 0 \quad \text{and} \quad Z_o(j\omega) \to 0.5 \text{ as } \omega \to \infty$$

*Comment*: These asymptotic characteristics are so easily computed (often mentally, as in the above examples) that it is well worthwhile developing the habit of checking the "ends" of every frequency response plot in this way.

**8.7** The voltage gain of a ladder network is

$$\frac{V_o(s)}{V_i(s)} = H(s) = \frac{1}{(s + 1)(s + 10)}$$

Sketch the asymptotic (straight-line) Bode diagram for this network. That is, graph

$$|H(j\omega)| \text{ vs. } \omega \quad \text{and} \quad \underline{/H(j\omega)} \text{ vs. } \omega$$

The transfer function must be manipulated into the standard form for $j\omega$-domain calculations,

$$H(j\omega) = \frac{1}{(j\omega + 1)(j\omega + 10)} = \frac{0.1}{(1 + j\omega)(1 + j0.1\omega)}$$

The asymptotic plots are obtained by treating each term in $H(j\omega)$ separately. We note first that:

0.1 is a constant (independent of frequency),

$\dfrac{1}{1 + j\omega}$ has a break frequency of $1/\tau_a = 1/1 = 1$ rad/s.

$\dfrac{1}{1 + j0.1\omega}$ has a break frequency of $1/\tau_b = 1/0.1 = 10$ rad/s.

The magnitude of the first term is easily plotted—see $M_1$ in Fig. 8-24. The phase of this term is zero, since the constant is positive.

To plot the second term we merely "copy" the standard curves of Figs. 8-5 and 8-9. These must be properly located with respect to their break frequency of $\omega_b = 1$ rad/s, and they must be plotted *negatively*, since this is a denominator factor. See $M_2$

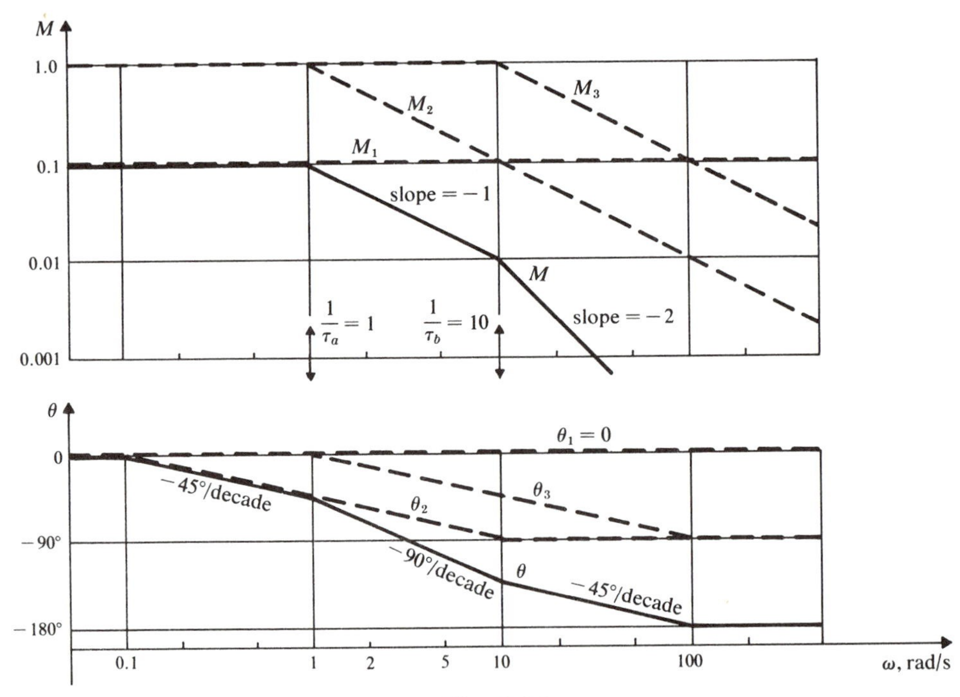


**Fig. 8-24**

and $\theta_2$ in Fig. 8-24. Observe that the magnitude curve is on log-log graph paper, while the phase curve is on semilog graph paper.

The third and final term is treated similarly, with its break frequency at 10 rad/s. See $M_3$ and $\theta_3$ in Fig. 8-24.

Now, summing the *distances* of the $M_1$, $M_2$, and $M_3$ curves from the unity reference base line, we obtain the overall $M$ curve shown in Fig. 8-24. Here, where all the distances are negative (i.e. below the base line), their algebraic sum is even further below unity magnitude.

Similarly the sum of the $\theta_1$, $\theta_2$, and $\theta_3$ curves yields the overall $\theta$ curve shown in Fig. 8-24.

### *Alternative Solution*

If we prefer to work in decibels on semilog paper, the magnitude plot is as shown in Fig. 8-25. The technique is unchanged, except that we "copy" the standard curve of Fig. 8-7 instead of that in Fig. 8-5.

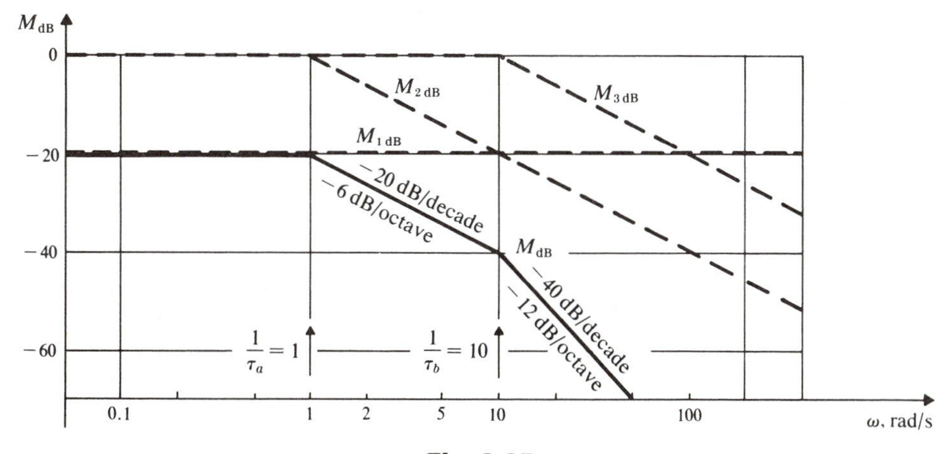


**Fig. 8-25**

**8.8** Asymptotic curves are approximate and need correcting when we want more accurate data. Use information from Fig. 8-5 and Fig. 8-9 to correct the gain and phase curves of Fig. 8-24.

It is almost always sufficiently accurate to estimate the corrections from graphs such as those of Figs. 8-5 and 8-9. But we must be careful since these curves were plotted for a typical *numerator* term, $1 + j\omega\tau$. Thus the magnitude corrections in this problem are the *reciprocal* of the corrections shown in Fig. 8-5. That is, we must correct by $1/1.12 = 0.89$ at $\omega = 0.5/\tau$, by $1/1.41 = 0.71$ at $\omega = 1/\tau$, and by $2/2.24 = 0.89$ at $\omega = 2/\tau$.

For the same reason, the phase corrections in this problem are the *negative* of those shown in Fig. 8-9.

In practice we can apply the corrections directly to the overall $M$ and $\theta$ curves. Thus

$M$ must be multiplied by 0.89 at $\omega = 0.5$ rad/s,

$M$ must be multiplied by 0.71 at $\omega = 1.0$ rad/s,

$M$ must be multiplied by 0.89 at $\omega = 2.0$ rad/s,
$\theta$ must be decreased by 6° at $\omega = 0.1$ rad/s,
$\theta$ must be increased by 5° at $\omega = 0.35$ rad/s,

and so on. The corrected points are plotted in Fig. 8-26.

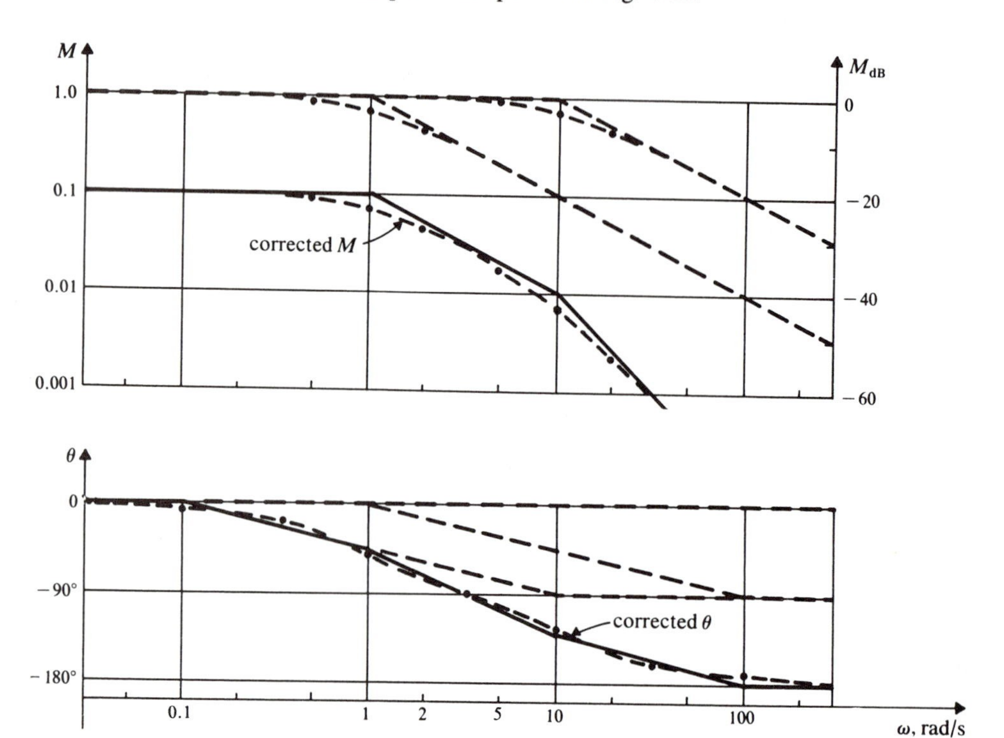


**Fig. 8-26**

*Alternative Solution*

If we work with decibels, the magnitude corrections in this problem are the *negative* of those shown in Fig. 8-7. That is, we must *subtract* 1 dB at $\omega = 0.5$ rad/s, 3 dB at $\omega = 1$ rad/s, etc. This can be seen in relation to the dB scale on the right of Fig. 8-26.

**8.9** In Problems 8.7 and 8.8 we generated the Bode diagram of Fig. 8-26 for the network function

$$H(j\omega) = \frac{0.1}{(1 + j\omega)(1 + j0.1\omega)}$$

Use data from the Bode diagram to plot the corresponding polar diagram.

From Fig. 8-26 we can read the following data:

| $\omega$ | $M$ | $\theta°$ |
|---|---|---|
| 0 | 0.1 | 0 |
| 0.1 | 0.1 | −6 |
| 0.5 | 0.09 | −30 |
| 1.0 | 0.071 | −50 |
| 2.0 | 0.045 | −72 |
| 5.0 | 0.018 | −106 |
| 10.0 | 0.007 | −130 |
| $\infty$ | 0 | −180 |

These can now be replotted in the polar coordinates of Fig. 8-27.

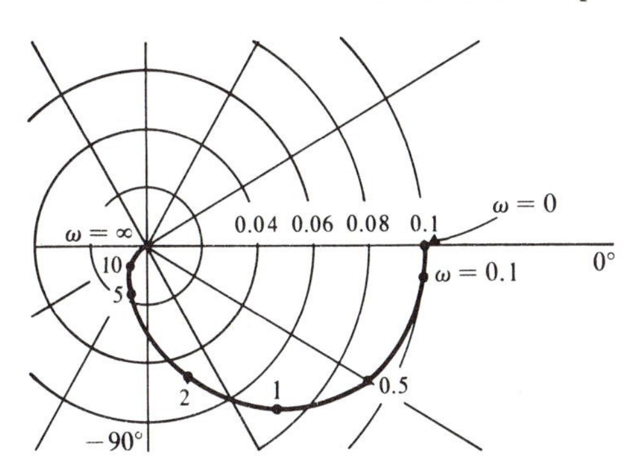


**Fig. 8-27**

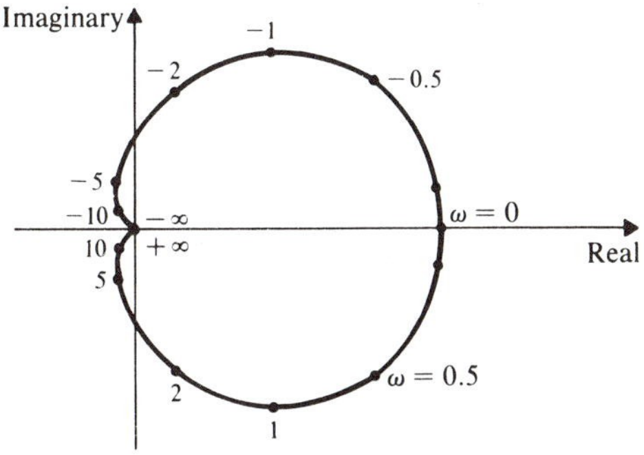

**Fig. 8-28**

**8.10** Given the polar diagram of Fig. 8-27 for the network function $H(j\omega) = 0.1/[(1 + j\omega)(1 + j0.1\omega)]$, draw the Nyquist diagram.

The Nyquist diagram is simply a polar diagram for *all* $\omega$, $-\infty \le \omega \le \infty$. We must therefore add to Fig. 8-27 the curve for $-\infty \le \omega \le 0$. This negative $\omega$ branch of the Nyquist diagram is found by setting $\omega = -\omega$ in $H(j\omega)$. The two branches may therefore be compared as follows:

$$H(j\omega) = H(s)\big|_{s=j\omega} = \alpha(\omega) + j\beta(\omega) \qquad H(-j\omega) = H(s)\big|_{s=j(-\omega)} = H(s)\big|_{s=(-j)\omega} = \alpha(\omega) - j\beta(\omega)$$

That is, for each point on the positive $\omega$ branch there will be a complex conjugate point on the negative $\omega$ branch. The latter is therefore the mirror image in the real axis of the ordinary polar diagram, as shown in Fig. 8-28.

*Comment:* The Nyquist diagram is always a closed curve.

**8.11** An active network is described by the transfer function $H(s) = 100(s + 1)/s^2(s + 10)$. Plot the asymptotic (straight-line) Bode diagram for $H(j\omega)$.

In standard form, $H(j\omega) = 10(1 + j\omega)/(j\omega)^2(1 + j0.1\omega)$.

The procedure follows that of Problem 8.7. Note that the magnitude curve for $1/(j\omega)^2$ passes through the point $(1, 1)$ with a slope of $-2$ and that the corresponding phase is a constant $-180°$. Otherwise the construction of Fig. 8-29 below is self-explanatory.

*Comment:* A decibel scale is provided on the right of Fig. 8-29 for those who prefer to work in this way.

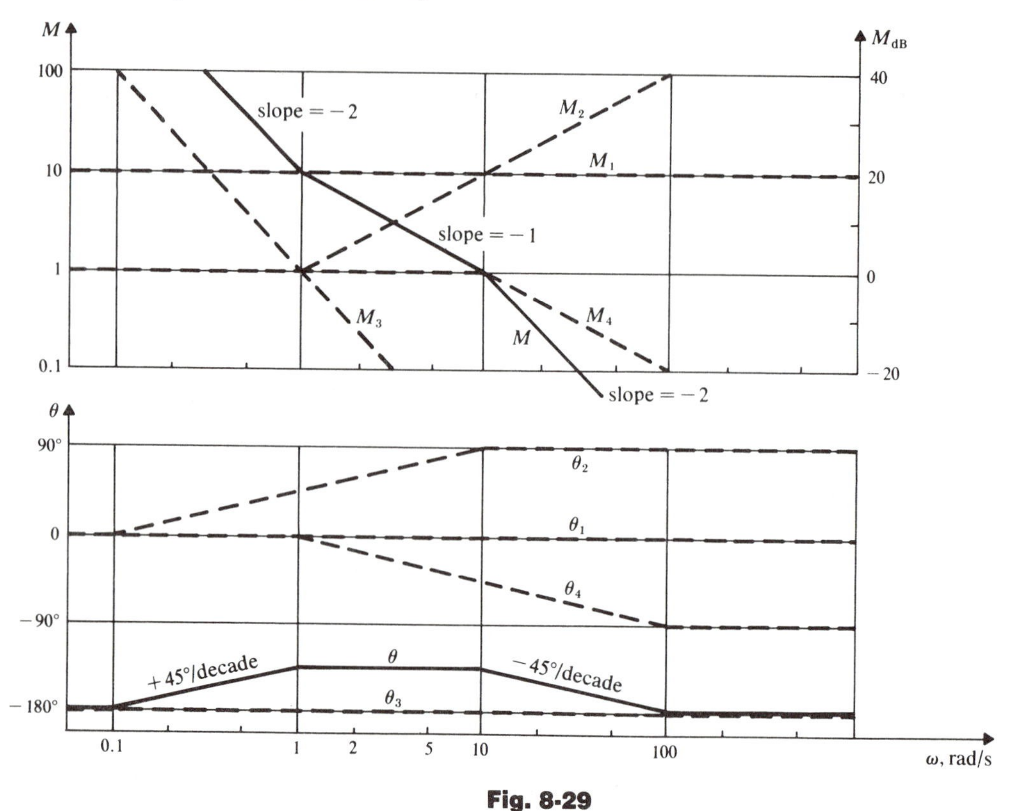


**Fig. 8-29**

**8.12** Sketch the straight-line Bode diagrams for $H_1(s) = V_L(s)/E(s)$ and $H_2(s) = V_C(s)/E(s)$ for the circuit in Fig. 8-30 below.

Using the voltage divider rationale,

$$H_1(s) = \frac{V_L(s)}{E(s)} = \frac{0.2s}{0.2s + 4.2 + 4/s} = \frac{s^2}{(s + 1)(s + 20)}$$

0.2 H, 4.2 Ω, $v_L$, $v_R$, $i(t)$, 0.25 F, $v_C$, $e(t)$

**Fig. 8-30**

Then in the standard $j\omega$-domain form,

$$H_1(j\omega) = \frac{0.05(j\omega)^2}{(1 + j\omega)(1 + j0.05\omega)}$$

The corresponding asymptotic Bode diagram is plotted in Fig. 8-31.

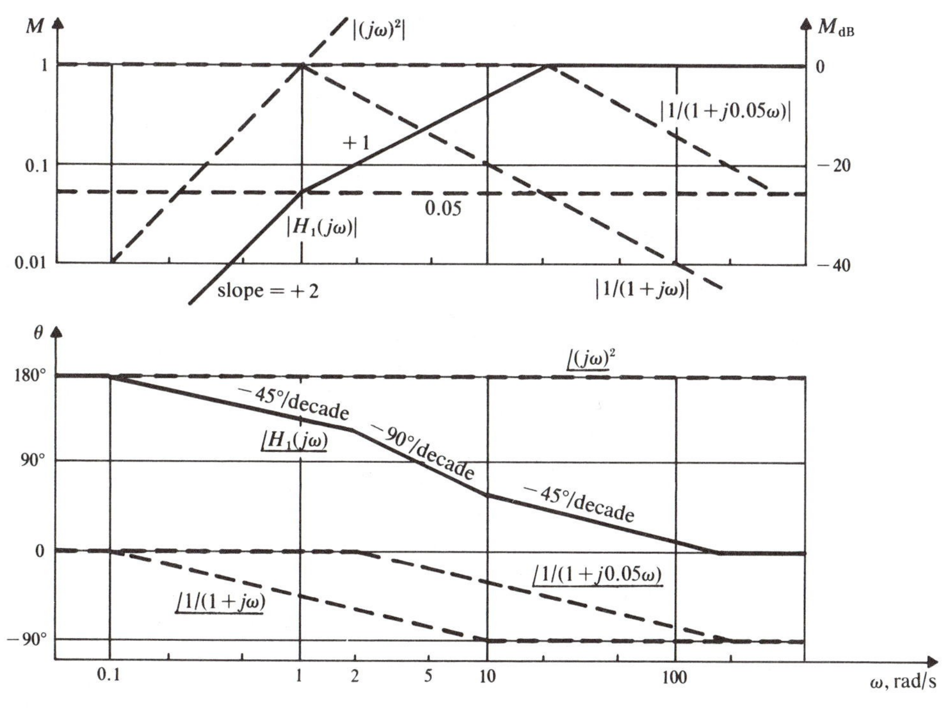


**Fig. 8-31**

Similarly,

$$H_2(s) = \frac{V_C(s)}{E(s)} = \frac{20}{(s + 1)(s + 20)}$$

and

$$H_2(j\omega) = \frac{1}{(1 + j\omega)(1 + j0.05\omega)}$$

The Bode diagram for $H_2(j\omega)$ is plotted in Fig. 8-32.

**8.13** Draw the Bode diagram corresponding to the transfer function

$$H(s) = \frac{s + 1}{s(s^2 + 2s + 100)}$$

Include all corrections to the straight-line approximations.

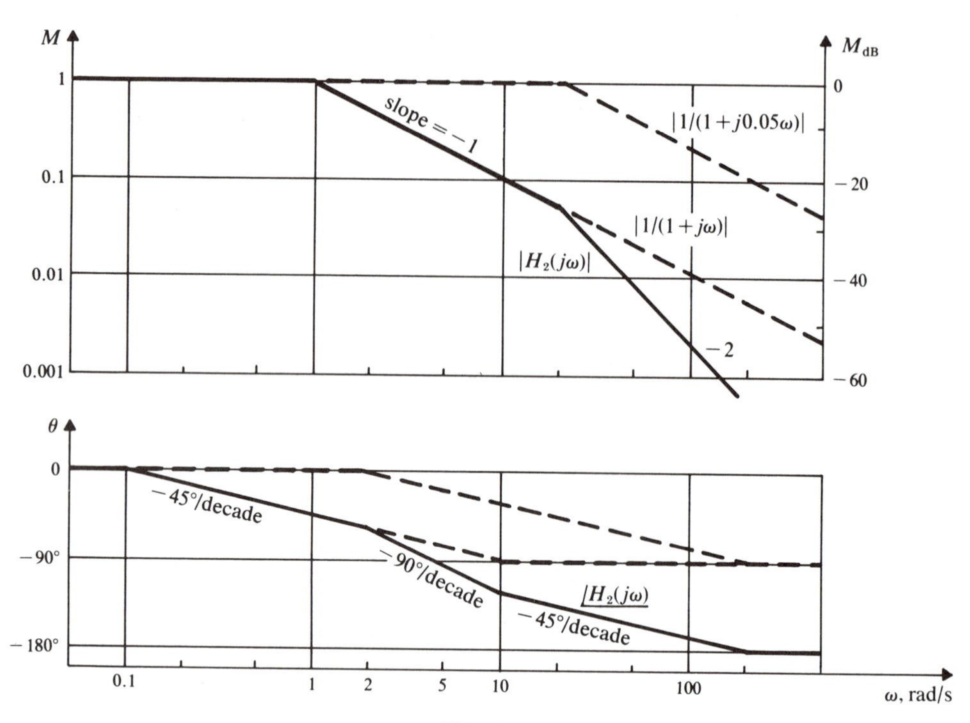


**Fig. 8-32**

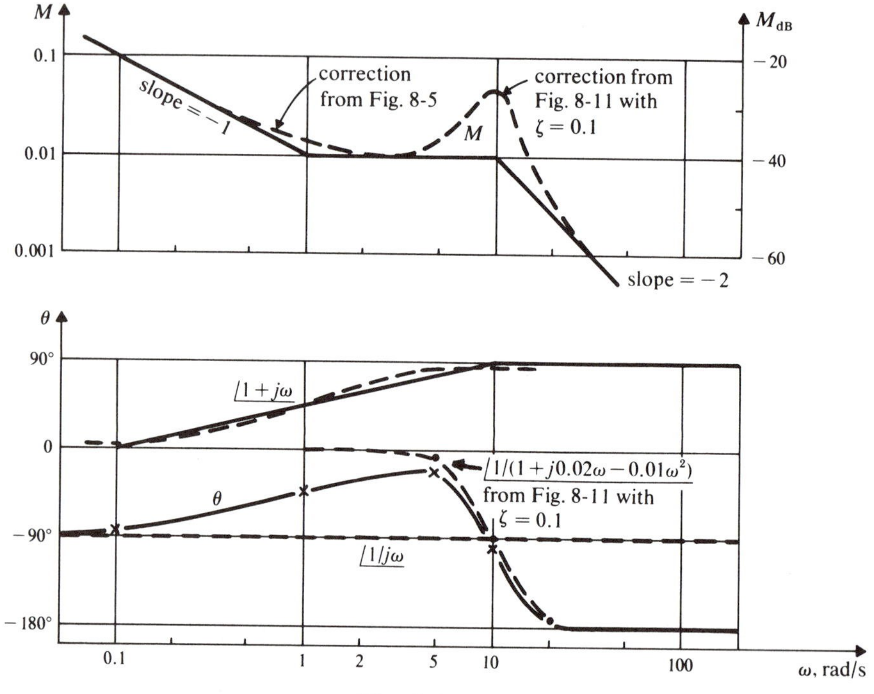


**Fig. 8-33**

In standard frequency-response form,

$$H(j\omega) = \frac{0.01(j\omega + 1)}{j\omega(1 + j0.02\omega - 0.01\omega^2)}$$

and we treat the quadratic term with the help of Fig. 8-11, or by calculator from equation (*8.6*), where in this instance

$$\omega_n = \sqrt{100} = 10 \text{ rad/s} \quad \text{and} \quad \zeta = 0.1$$

The rest of the procedure, shown in Fig. 8-33, follows the usual pattern.

**8.14** By definition, $N_{dB} = 20 \log N$. Graph this relation, i.e. $N_{dB}$ vs. $N$, on semilog graph paper, with $N_{dB}$ on the linear scale.

The above equation is of the form $y = mx + b$, with $x = \log N$ and $b = 0$. Therefore $N_{dB}$ plotted against $\log N$ on *ordinary* graph paper would be a straight line of slope 20 dB per unit of $\log N$ (i.e. per *decade* of $N$), passing through the point (0, 0).

If instead we plot $N_{dB}$ vs. $N$ with $N$ on a logarithmic scale, we get the same straight line, of slope 20 dB/decade of $N$ through the point (1, 0). This is illustrated in Fig. 8-34.

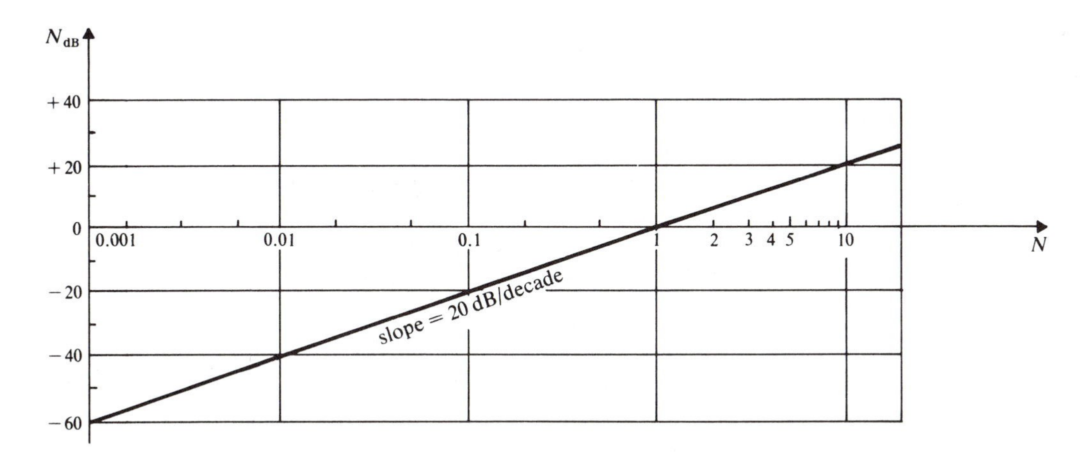


**Fig. 8-34**

This graph can be redrawn as in Fig. 8-35 to make the dB scale more readable. Although ordinary gains can easily be converted into dB gains with the help of a calculator, you should spend a little time learning to think in terms of decibels.

### *Filters*

**8.15** Show that the low frequency gain (i.e. the gain when $\omega \ll \omega_c$) is $K/\omega_c$ and $K/\omega_c^2$ for first- and second-order low-pass filters, respectively.

Setting $s = j\omega$ in equation (*8.7*),

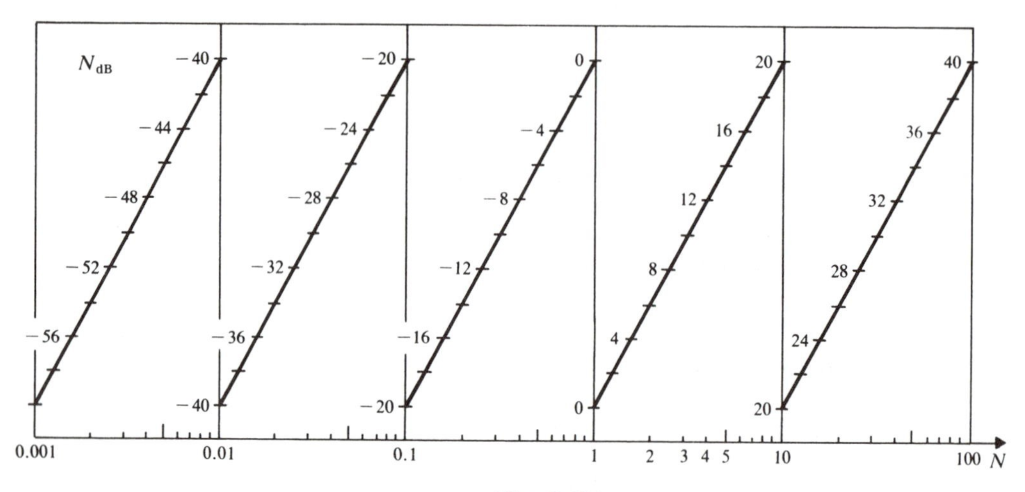


**Fig. 8-35**

$$H(j\omega) = \frac{K/\omega_c}{1 + j\omega/\omega_c} \quad \text{and} \quad H(j\omega) = \frac{K/\omega_c^2}{1 - \omega^2/\omega_c^2 + jB\omega/\omega_c}$$

Then for $\omega \ll \omega_c$, $H(j\omega) \to K/\omega_c$ and $K/\omega_c^2$, respectively.

**8.16** Show for the second-order band-pass filter that

(a) $|H(j\omega)|_{\max} = |H(j\omega_0)| = K/B$,

(b) $\omega_2 - \omega_1 = B$, where $\omega_1$ and $\omega_2$ are the frequencies where $|H(j\omega)| = (1/\sqrt{2})\,|H(j\omega)|_{\max}$, and

(c) the low- and high-frequency slopes of the $|H(j\omega)|$ vs. $\omega$ curve (on log-log graph paper) are $+1$ and $-1$, respectively.

(a) Setting $s = j\omega$ in equation (8.9),

$$H(j\omega) = \frac{K}{B + j\omega(1 - \omega_0^2/\omega^2)}$$

and

$$|H(j\omega)| = \frac{K}{\sqrt{B^2 + \omega^2(1 - \omega_0^2/\omega^2)^2}} \qquad (1)$$

Since $\omega$ occurs only in the denominator, it can be seen that $|H(j\omega)|$ will be a maximum when $\omega$ is chosen to make $\omega^2(1 - \omega_0^2/\omega^2)^2$ zero, that is, when $\omega = \omega_0$. The corresponding value of $|H(j\omega)|$ is $|H(j\omega)|_{\max} = |H(j\omega_0)| = K/\sqrt{B^2 + 0} = K/B$.

(b) We have just shown that $|H(j\omega)|_{\max} = K/\sqrt{B^2}$. So, to make $|H(j\omega)| = (1/\sqrt{2})\,|H(j\omega)|_{\max}$, we need

$$|H(j\omega)| = K/\sqrt{2B^2}$$

Thus by reference to equation (1), above, we require

$$\omega^2(1 - \omega_0^2/\omega^2)^2 = B^2$$

The rest is algebra. We want

$$\omega(1 - \omega_0^2/\omega^2) = \pm B$$

or

$$\omega^2 \mp B\omega - \omega_0^2 = 0$$

or

$$\omega_{1,2} = \frac{1}{2}\{\pm B \pm \sqrt{B^2 + 4\omega_0^2}\}$$

Only positive values of $\omega_{1,2}$ are legitimate, and so

$$\omega_{1,2} = \frac{1}{2}\{\pm B + \sqrt{B^2 + 4\omega_0^2}\}$$

Thus, $\omega_2 - \omega_1 = B$.
(*c*) When $\omega \ll \omega_0$, equation (*8.9*) shows that

$$H(j\omega) \rightarrow jK\omega/\omega_0^2$$

or

$$|H(j\omega)| \rightarrow K\omega/\omega_0^2$$

Thus $\log |H(j\omega)| \rightarrow \log K + \log \omega - \log \omega_0^2$ and

$$d \log |H(j\omega)|/d \log \omega = 1$$

which is the +1 slope we expected.
Similarly, when $\omega \gg \omega_0$, $H(j\omega) \rightarrow jK/(-\omega)$, which leads to a slope of −1.

**8.17** Use your calculator to verify the results of parts (*a*) and (*b*) of the previous problem for the filter described by

$$H(s) = 0.5s/(s^2 + 0.5s + 1)$$

That is, complete the following table, and use the tabulated data to determine $|H(j\omega)|_{max}$ and $\omega_2 - \omega_1$.

| $\omega$ | $\lvert H(j\omega)\rvert$ |
|---|---|
| 0.6 | |
| 0.7 | |
| 0.781 | |
| 0.99 | |
| 1.0 | |
| 1.01 | |
| 1.2 | |
| 1.281 | |

Here $K = 0.5$, $\omega_0 = 1$, and $B = 0.5$, so that from equation (*1*) in the previous problem,

$$|H(j\omega)| = 0.5/\sqrt{0.5^2 + \omega^2(1 - 1/\omega^2)^2}$$

from which the following data can be calculated.

| $\omega$ | $\lvert H(j\omega)\rvert$ | |
|---|---|---|
| 0.6 | 0.424 | |
| 0.7 | 0.566 | |
| 0.781 | 0.707 | $(1/\sqrt{2} = 0.707)$ |
| 0.99 | 0.999 | |
| 1.0 | 1.000 | |
| 1.01 | 0.999 | |
| 1.2 | 0.806 | |
| 1.281 | 0.707 | $(1/\sqrt{2} = 0.707)$ |

From this data, $|H(j\omega)|_{\max} = 1.00$, which is equal to $K/B$ as expected. The half-power frequencies correspond to $|H(j\omega)| = (1/\sqrt{2})|H(j\omega)|_{\max}$, whence by inspection of the table, $\omega_1 = 0.781$ rad/s and $\omega_2 = 1.281$ rad/s. It follows that $B = \omega_2 - \omega_1 = 0.500$ rad/s, as expected.

**8.18** Investigate the properties of $H(s) = V(s)/E(s)$ for the circuit of Fig. 8-36.

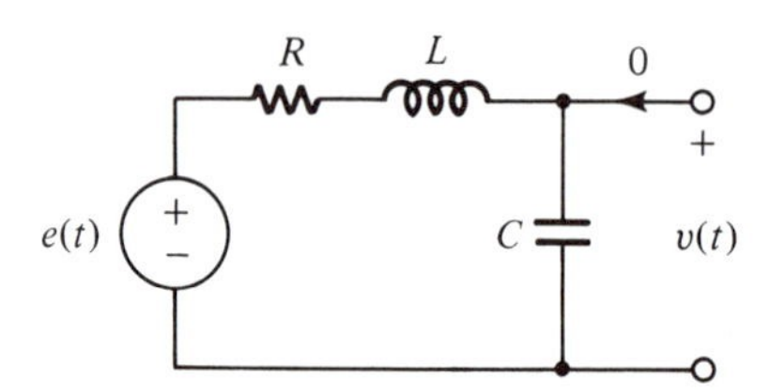


**Fig. 8-36**

From the voltage divider relation in the $s$-domain (where we visualize the circuit with zero initial conditions),

$$H(s) = \frac{1/Cs}{Ls + R + 1/Cs} = \frac{1/LC}{s^2 + (R/L)s + 1/LC}$$

which is of the form of equation (*8.7*) for a second-order low-pass filter with

$$K = 1/LC, \quad \omega_c = \sqrt{1/LC}, \quad B = R/L, \quad \text{and} \quad Q = \omega_c/B = \omega_c L/R.$$

This is a practical circuit if the use of an inductor is acceptable, but an output buffer such as an op amp voltage follower (see Problem 1.18) would be needed if any significant current were to be drawn from the circuit's terminals.

**8.19** Investigate the properties of $H(s) = V(s)/E(s)$ for the circuit of Fig. 8-37.

Here, by $s$-domain analysis (with zero initial conditions),

$$H(s) = \frac{s^2 + 1/LC}{s^2 + (R/L)s + 1/LC}$$

which is of the form of equation (*8.10*) for a second-order notch filter, with

$$K = 1, \quad \omega_0 = \sqrt{1/LC}, \quad B = R/L, \quad \text{and} \quad Q = \omega_0 L/R.$$

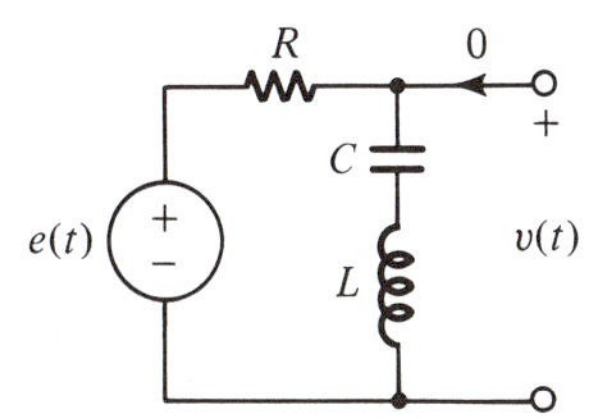


**Fig. 8-37**

In practice the gain of this circuit would not go to zero for $\omega = \omega_0$, since the inductor's inevitable resistance has been neglected in the model of Fig. 8-37.

**8.20** The $Q$ of a circuit or other dynamic system is sometimes defined under sinusoidal steady-state conditions as

$$Q = \frac{2\pi(\text{maximum stored energy})}{\text{energy dissipated per cycle}}$$

Show that this is equivalent to the definition $Q = \omega_0/B = \omega_0 L/R$ for the particular case of a series $L$-$R$-$C$ circuit (in the circuit of Fig. 8-36 or 8-37, for example).

The maximum stored energy in the inductor or in the capacitor could be chosen, since these are equal. Taking the former, the maximum stored energy during a cycle is

$$W_{\max} = \frac{1}{2}L\{i^2(t)\}_{\max} = \frac{1}{2}LI^2$$

where $I$ is the zero-to-peak amplitude of the sinusoid $i(t)$ *at resonance*. The energy dissipated per cycle is

$$W_d = P_{\text{av}}T = (RI_{\text{eff}}^2)T = \frac{1}{2}RI^2\left(\frac{2\pi}{\omega_0}\right)$$

Therefore, substituting into the above definition of $Q$,

$$Q = \frac{2\pi W_{\max}}{W_d} = \frac{2\pi(\frac{1}{2}LI^2)}{\pi RI^2/\omega_0} = \frac{\omega_0 L}{R} = Q$$

**8.21** Show that the circuit of Fig. 8-38 is a band-pass filter and find $\omega_0$, $B$, and $|H(j\omega)|_{\max}$ given that $C_1 = C_2 = 1\ F$, $R_1 = (1/2\alpha)\ \Omega$ and $R_2 = 2\alpha\ \Omega$.

Visualizing the circuit in the $s$-domain with zero initial conditions, and making the usual op amp assumptions, KCL at nodes $b$ and $c$ yields

$$\begin{bmatrix} \dfrac{1}{R_1} + (C_1 + C_2)s & -C_2 s \\ C_1 s & \dfrac{1}{R_2} \end{bmatrix} \begin{bmatrix} V_b(s) \\ V(s) \end{bmatrix} = \begin{bmatrix} \dfrac{1}{R_1}E(s) \\ 0 \end{bmatrix}$$

whence, from Cramer's rule,

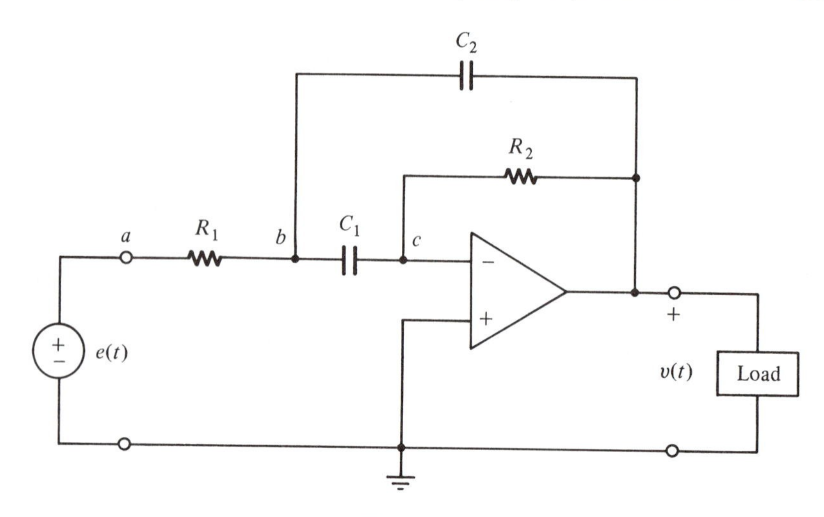


**Fig. 8-38**

$$\frac{V(s)}{E(s)} = \frac{-\dfrac{1}{R_1C_2}s}{s^2 + \dfrac{C_1 + C_2}{R_2C_1C_2}s + \dfrac{1}{R_1R_2C_1C_2}}$$

which is of the form of equation (8.9)—a band-pass filter. Now, substituting the data,

$$\frac{V(s)}{E(s)} = \frac{-2\alpha s}{s^2 + (1/\alpha)s + 1}$$

whence $\omega_0 = 1$ rad/s, $B = 1/\alpha$, and $|H(j\omega)|_{max} = |K/B| = 2\alpha^2$.

This circuit does not require any inductors, and the op amp buffers the output—the circuit can deliver a useful current to the load with little error. The minus sign in the transfer function does not, of course, affect $|H(j\omega)|$. It can be regarded as an additional phase shift of 180°, which—if objectionable—can be eliminated by inserting an op amp sign changer (see Problem 2.67) either ahead of or following the filter circuit itself.

**8.22** Show that the transfer function $H(s) = V(s)/E(s)$ for the op amp circuit of Fig. 8-39 corresponds to a notch filter.

*Hint:* You should recognize (see Problems 2.42, 2.43, and 2.67) in the four separate op amp circuits a summing integrator, a sign changer, an integrator, and a summer. Then you can relate the output of each of these circuits to its input(s). You should assume zero initial capacitor voltages. The rest involves somewhat messy $s$-domain algebra.

The equations describing the four op amp circuits are:

$$w(t) = -\int_0^t \{Bw(\tau) + \omega_0^2 y(\tau) + Ke(\tau)\}\, d\tau$$

$$x(t) = -w(t)$$

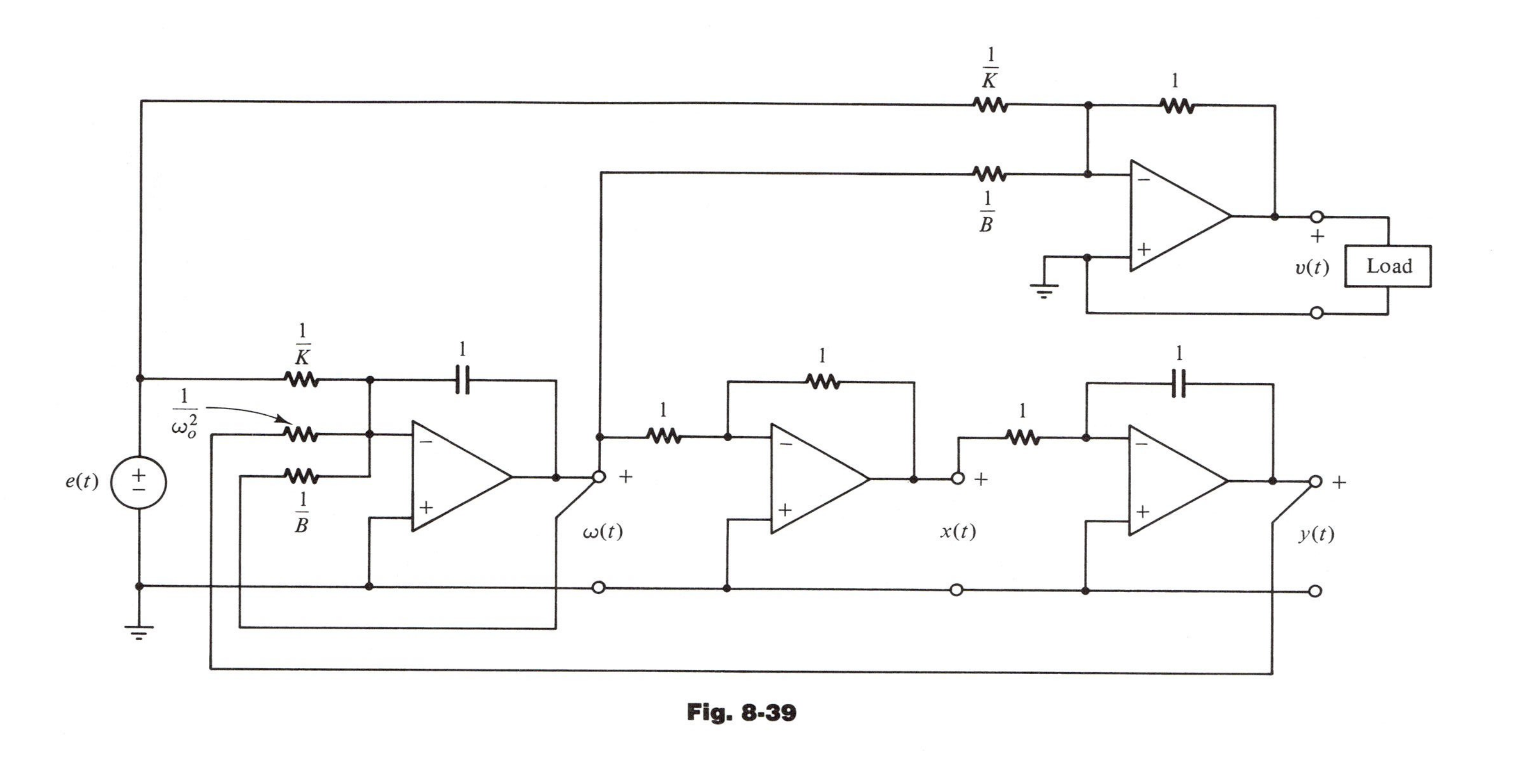


Fig. 8-39

$$y(t) = -\int_0^t x(\tau)\, d\tau$$

$$v(t) = -\{Bw(t) + Ke(t)\}$$

(Don't forget the sign change in each such operation!) Laplace transformation of these equations yields

$$W(s) = -\frac{1}{s}\{BW(s) + \omega_0^2 Y(s) + KE(s)\}$$

$$X(s) = -W(s)$$

$$Y(s) = -\frac{1}{s}X(s)$$

$$V(s) = -\{BW(s) + KE(s)\}$$

After some algebra to eliminate $W(s)$, $X(s)$, and $Y(s)$,

$$\frac{V(s)}{E(s)} = \frac{-K(s^2 + \omega_0^2)}{s^2 + Bs + \omega_0^2}$$

This is the transfer function of a notch filter. The minus sign has no effect on $|V/E|$. It can easily be eliminated by an additional sign changer if desired.

### *Reactance Plots and Locus Diagrams*

**8.23** With reference to the circuit of Fig. 8-40 and the definitions $Z(j\omega) = R(\omega) + jX(\omega)$, $Y(j\omega) = G(\omega) + jB(\omega)$, $Y_1(j\omega) = j\omega C_1 + 1/j\omega L_1$ and $Z_2(j\omega) = R + 1/j\omega C_2$, sketch the curves of $B_1(\omega)$, $X_2(\omega)$, $X_1(\omega)$, and $X(\omega)$ vs. $\omega$.

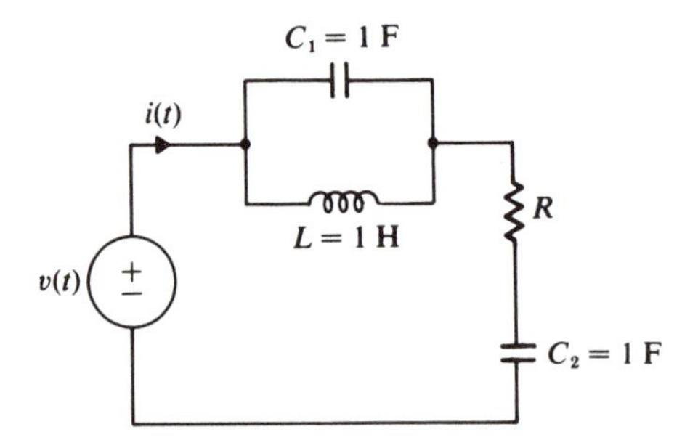


**Fig. 8-40**

$B_1(\omega) = \omega C_1 - 1/\omega L_1 = \omega - 1/\omega$, which is sketched in Fig. 8-41*a*. Next, $X_2(\omega) = -1/\omega C_2 = -1/\omega$ as in Fig. 8-41*b*.

Since $Z_1(j\omega)$ is purely imaginary, $X_1(\omega) = -1/B_1(\omega)$ and we may plot $X_1(\omega)$ by point-by-point inversion of Figs. 8-41*a*, with sign change. This is shown in Fig. 8-41*c*.

$X(\omega) = X_1(\omega) + X_2(\omega)$, and so Fig. 8-41*d* follows as the summation of the curves in Figs. 8-41*b* and *c*.

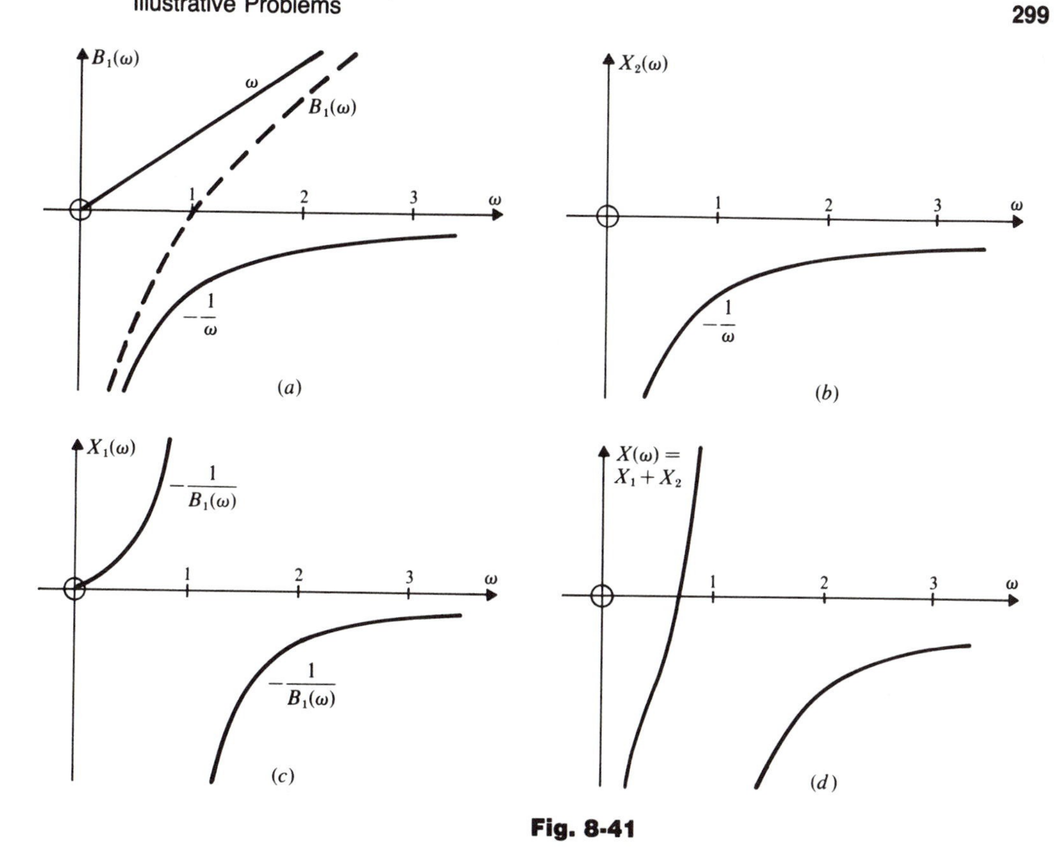


**Fig. 8-41**

**8.24** Sketch the locus of the tip of the frozen or stationary phasor **I** as $C$ is varied through the range $0 \leq C \leq \infty$ in the circuit of Fig. 8-42. The source voltage is $v(t) = 10 \cos 2t$.

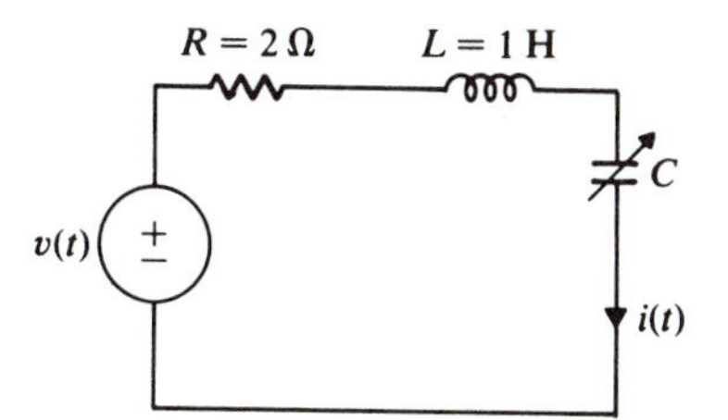


**Fig. 8-42**

Here, if we choose **V** as the reference and work with *stationary* phasors,

$$\mathbf{I} = \frac{\mathbf{V}}{R + j(\omega L - 1/\omega C)} = \frac{10}{2 + j(2 - 1/2C)}$$

Hence we can evaluate **I** for a number of values of $C$ as shown in the table on page 300. These data are plotted in Fig. 8-43.

| $C$ | $\lvert \mathbf{I} \rvert$ | $\angle \mathbf{I}$ |
|---|---|---|
| 0 | 0 | +90° |
| $\frac{1}{8}$ F | 3.54 | +45° |
| $\frac{1}{6}$ F | 4.46 | +26.6° |
| $\frac{1}{4}$ F | 5.0 | 0 |
| $\frac{1}{2}$ F | 4.46 | −26.6° |
| ∞ | 3.54 | −45° |

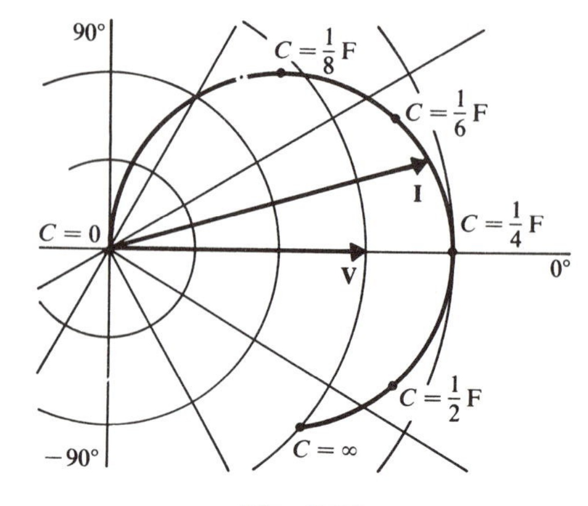


**Fig. 8-43**

# PROBLEMS

**8.25** For the circuit of Fig. 8-44, (*a*) find the transfer function $H(s) = I_L(s)/I(s)$, (*b*) sketch the polar plot of $H(j\omega)$, and (*c*) find the frequency which makes $\text{Re}[H(j\omega)] = 0$.

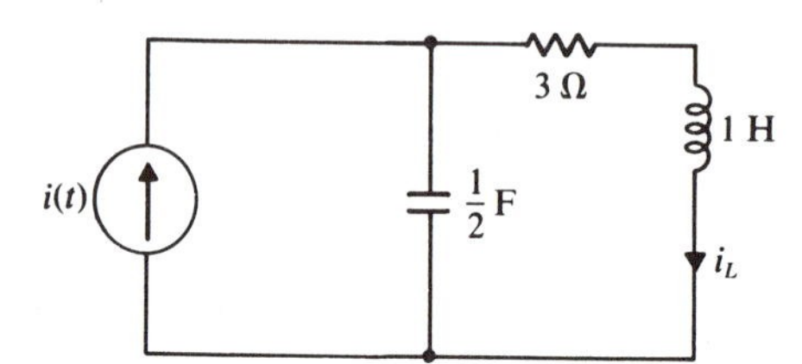


**Fig. 8-44**

**8.26** A frequency response test has been performed on the circuit in Fig. 8-45*a*. The data can be reasonably represented by the (asymptotic) straight lines of Fig. 8-45*b*. Find $H(s)$.

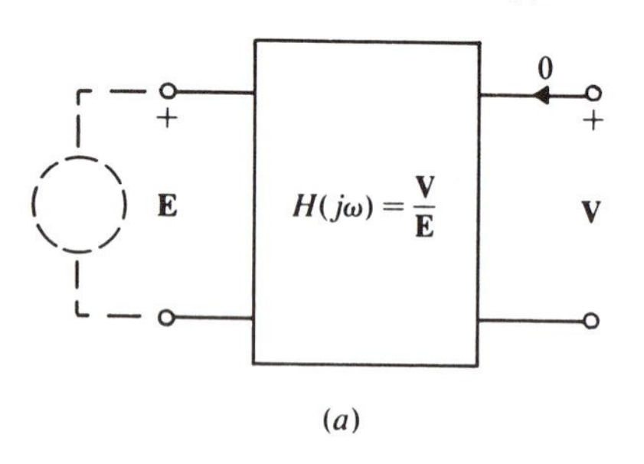


(*a*)

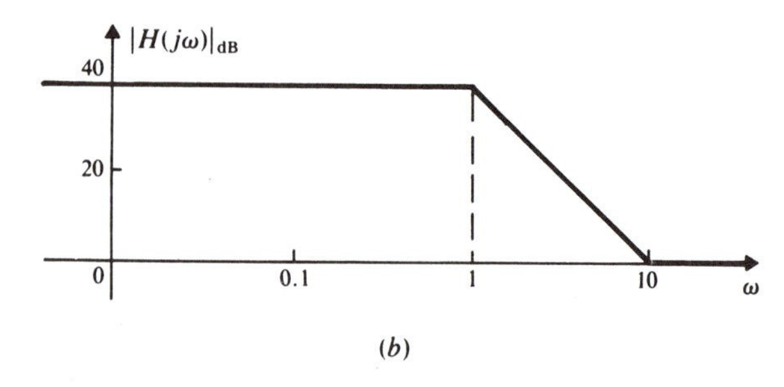


(*b*)

**Fig. 8-45**

**8.27** Sketch the asymptotic (straight-line) Bode plots (magnitude and phase) for the network function $H(s) = 4(s + 0.5)/(s + 20)$. From the graphs find (*a*) the "gain" at $\omega = 0.1$ rad/s, (*b*) the slope of the magnitude plot at $\omega = 2$ rad/s, (*c*) the slope of the magnitude plot at $\omega = 100$ rad/s, (*d*) the "gain" at 100 rad/s, (*e*) the phase at $\omega = 3$ rad/s, (*f*) the phase at $\omega = 220$ rad/s.

**8.28** Show that for a second-order band-pass filter,

$$\omega_1 \omega_2 = \omega_0^2$$

where $\omega_1$ and $\omega_2$ are the half-power frequencies (i.e. the frequencies where the gain is down to $1/\sqrt{2}$ of its maximum value), and $\omega_0$ is the resonant frequency.

**8.29** Show that $|H(j\omega_0)| = 0$ for the ideal second-order band-reject filter.

**8.30** Show that

$$H(s) = \frac{V(s)}{E(s)} = \frac{1}{2s^3 + 4s^2 + 4s + 2}$$

and that $|H(j\omega)| = 0.5/\sqrt{1 + \omega^6}$ for the circuit of Fig. 8-46. Then use your calculator to locate $\omega_c$ for this third-order low-pass filter.

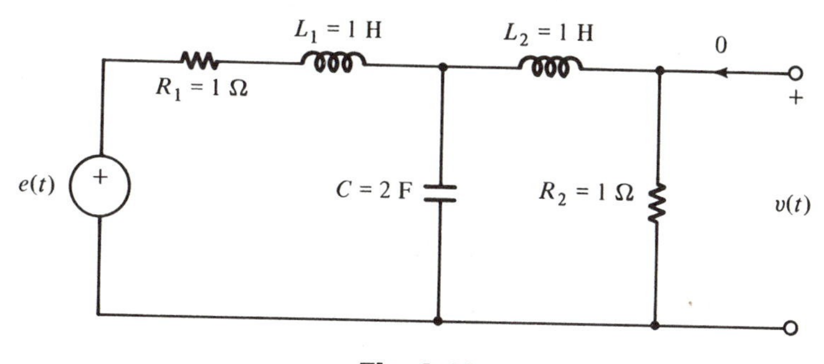


**Fig. 8-46**

**8.31** Show analytically that the high-frequency slope of the $|H(j\omega)|$ vs. $\omega$ graph (on log-log graph paper) is $-3$ for the filter of the previous problem. Use your calculator to confirm the result.

**8.32** Show that the transfer function $X(s)/E(s)$ for the op amp circuit of Fig. 8-39 corresponds to that of a second-order band-pass filter.

**8.33** Similarly, show that $Y(s)/E(s)$ for the circuit of Fig. 8-39 is the transfer function of a second-order low-pass filter. (This is a very versatile filter circuit. The availability of multiple op amps on a single integrated-circuit chip make it convenient and economical, too.)

**8.34** Use reactance plots (as in Problem 8.23) to show that the driving-point impedance of the circuit in Fig. 8-47 is zero at two positive values of $\omega$.

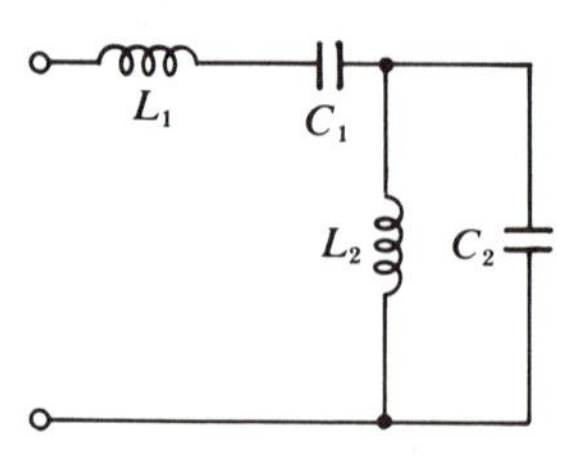


**Fig. 8-47**

**8.35** For how many positive finite frequencies does the driving-point impedance become infinite for the circuit of Fig. 8-47?

**8.36** Select the statement in the following list which best describes the plot in Fig. 8-48*a* for the circuit of Fig. 8-48*b*.

(*a*) $Y_{ab}(j\omega)$ for $0 \le C \le \infty$.
(*b*) $\text{Im}\,[Y_{ab}(j\omega)]$ for $0 \le \omega \le \infty$.
(*c*) $Z_{ab}(j\omega)$ for $0 \le \omega \le \infty$.
(*d*) $Z_{ab}(j\omega)$ for $0 \le R \le \infty$.

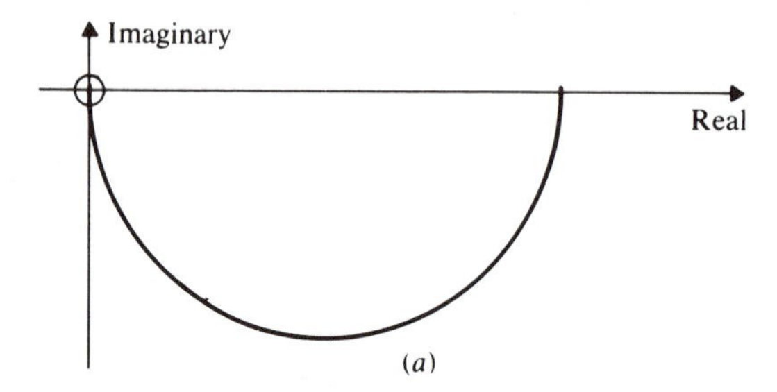


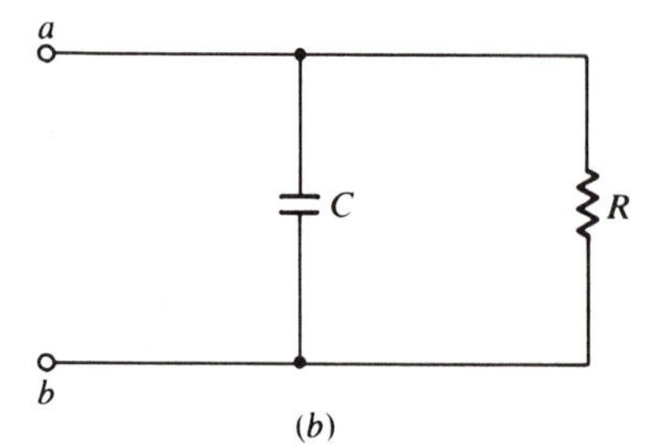


**Fig. 8-48**

**8.37** Sketch the polar plot of $Y_{ab}(j\omega)$ for the circuit of Fig. 8-48*b* as $R$ is varied in the range $0 \le R \le \infty$.

Chapter

9

# AC POWER, TRANSFORMERS, AND POLYPHASE CIRCUITS

The three topics discussed in this chapter are most commonly associated with the generation, transmission, and consumption of electrical power, although all three—and especially transformers—find their way into the design of electronic circuits.

## POWER IN THE SINUSOIDAL STEADY-STATE

Power and energy were defined in Chapter 1, as was the concept of effective values, and phasors were defined in Chapter 7. When working with power, we will usually assume that **V** and **I** are *stationary* phasors whose magnitudes are specified by their *effective* values.

If we assume sinusoidal steady-state conditions, the power delivered *to* the one port of Fig. 9-1 is

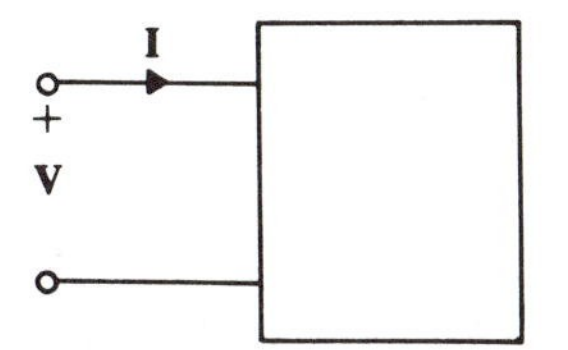


Fig. 9-1

$$
\begin{aligned}
p(t) &= V_m I_m \cos \omega t \cos(\omega t + \phi) \\
&= \tfrac{1}{2} V_m I_m \cos \phi + \tfrac{1}{2} V_m I_m \cos(2\omega t + \phi)
\end{aligned}
$$

so that the average power delivered is

$$P = \tfrac{1}{2} V_m I_m \cos \phi = V_{\text{eff}} I_{\text{eff}} \cos \phi \qquad \text{watts (W)} \tag{9.1}$$

where $P$ (strictly $P_{\text{av}}$) is called the *average active power* and where $\phi$ is *defined* to be the angle of $v(t)$ with respect to $i(t)$. The quantity $\cos \phi$ is called the *power factor*. $P$ is the *true*, *real*, or *useful* power delivered to the one port, and is the quantity measured by a wattmeter.

The product of RMS voltmeter and ammeter readings is called the *average apparent power* delivered to the one-port,

$$S = V_{\text{eff}} I_{\text{eff}} \qquad \text{volt amperes (VA)} \tag{9.2}$$

The *average reactive power*, which is just the reverse of "useful," is the product of $I_{\text{eff}}$ with the *out-of-phase* or *quadrature* component of $\mathbf{V}$, $V_{\text{eff}} \sin \phi$. That is,

$$Q = V_{\text{eff}} I_{\text{eff}} \sin \phi \qquad \text{volt amperes, reactive (VAR)} \tag{9.3}$$

The three quantities $P$, $S$, and $Q$ can be brought together in terms of the *complex power*,

$$
\begin{aligned}
\mathbf{S} = P + jQ &= V_{\text{eff}} I_{\text{eff}} (\cos \phi + j \sin \phi) \\
&= V_{\text{eff}} I_{\text{eff}} e^{j\phi} = S e^{j\phi} = \mathbf{V}\mathbf{I}^*
\end{aligned} \tag{9.4}
$$

where $S = |\mathbf{S}|$ and $\mathbf{I}^*$ is the complex conjugate of $\mathbf{I}$. This somewhat artificial relationship is expressed graphically in the *power triangle* of Fig. 9-2.

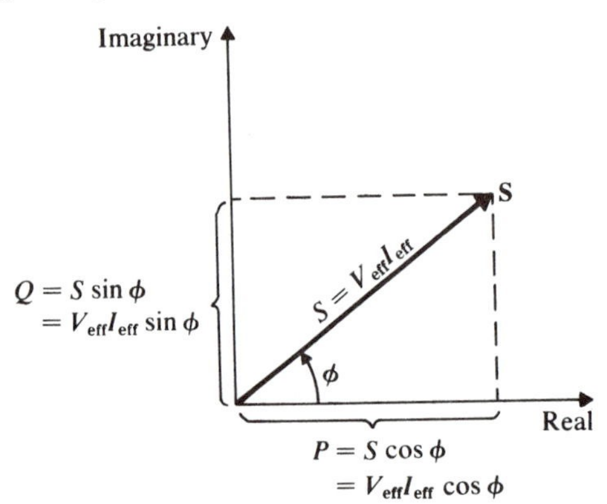


**Fig. 9-2**

If the one-port is *passive*, then it may be represented by an impedance $Z(j\omega) = R(\omega) + jX(\omega)$. Therefore

$$\phi = \underline{/\mathbf{V}} - \underline{/\mathbf{I}} = \underline{/Z(j\omega)}$$

The power factor $\cos\phi$ is said to be *inductive* if **V** leads **I** (positive $\phi$) and *capacitive* if **V** lags **I** ($\phi$ negative).

Also, for a passive one-port,

$$Z(j\omega) = \frac{\mathbf{V}}{\mathbf{I}} = \frac{V_{\text{eff}}}{I_{\text{eff}}} e^{j\phi}$$

so that equation (*9.4*) may be rewritten in the form

$$\mathbf{S} = V_{\text{eff}} I_{\text{eff}} e^{j\phi} = I_{\text{eff}}^2 Z(j\omega) \tag{9.5}$$

or

$$P = I_{\text{eff}}^2 R(\omega) \quad \text{and} \quad Q = I_{\text{eff}}^2 X(\omega) \tag{9.6}$$

Finally,

$$S = I_{\text{eff}}^2 |Z(j\omega)| = I_{\text{eff}}^2 \sqrt{R^2 + X^2} = \sqrt{P^2 + Q^2} \tag{9.7}$$

## Example 9.1

We will consider the three loads of Fig. 9-3 connected across a 2300-V power line. For the capacitive load of Fig. 9-3*a*,

$$\mathbf{I}_c = \frac{2300}{-j10} = 230e^{j90^\circ}$$

and therefore from equation (*9.4*),

$$\mathbf{S} = \mathbf{VI}^* = 2300(230e^{-j90^\circ}) = -j529 \times 10^3$$

or

$$P = 0 \text{ W}, \quad Q = -529 \text{ kVAR} \quad \text{and} \quad \cos\phi = 0$$

Thus the capacitor draws no average active power from the line, a result we anticipated in Chapter 1. That is, the capacitor draws energy as it is charged, but returns the same energy during the discharge part of each cycle. A wattmeter would read zero.

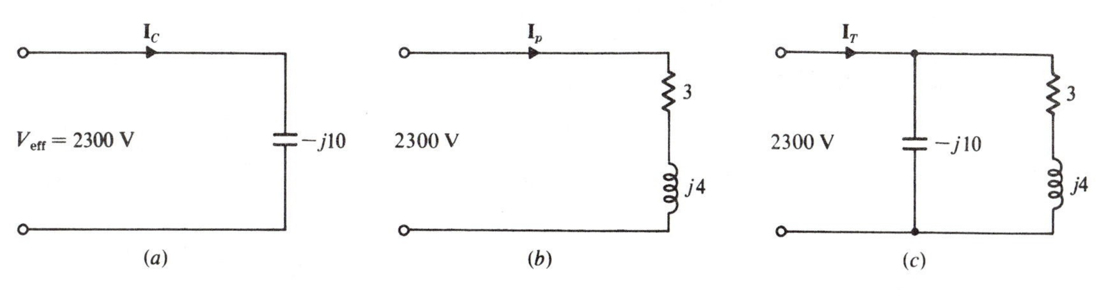


**Fig. 9-3**

This can be costly to a power company since the supply system must be sized to carry the current without unreasonable voltage drop or power loss. For this reason power factor meters are installed to monitor large industrial consumers.

In the second case (Fig. 9-3*b*), $I_{p,\text{eff}} = 2300/\sqrt{3^2 + 4^2} = 460$ A, and from equation (9.6),

$$P = 3(460)^2 = 635 \text{ kW}, \qquad Q = 4(460)^2 = 846 \text{ kVAR}$$

and

$$\cos \phi = \cos \underline{/Z(j\omega)} = \cos (\tan^{-1} 4/3) = 0.6$$

If this load is an industrial plant, $P$ corresponds to the "useful" power purchased from the power company to light and heat the plant and to run power tools, etc. The reactive power $Q$, which is usually due to the inductive nature of electric motors, has only nuisance value.

Finally, by combining the two loads as in Fig. 9-3*c* we can partially correct for the plant's poor power factor. Thus since the current through each of the elements is unchanged,

$$P = 0 + 635 = 635 \text{ kW}, \qquad Q = -529 + 846 = 317 \text{ kVAR}$$

and

$$\cos \phi = \cos (\tan^{-1} 317/635) = 0.89$$

The addition of the capacitor, which is common practice in large power systems, reduces the reactive power seen by the incoming line, and has brought the power factor much closer to unity.

## SOME PROPERTIES OF TRANSFORMERS

The transformer is a very common circuit element—from the large iron-cored power transformers used by the electrical power companies, to the miniature units used in electronic circuits for "coupling" or "tuning" purposes. To the extent that the inevitable nonlinearities and winding capacitances can be neglected, we will be concerned with the network of Fig. 9-4.

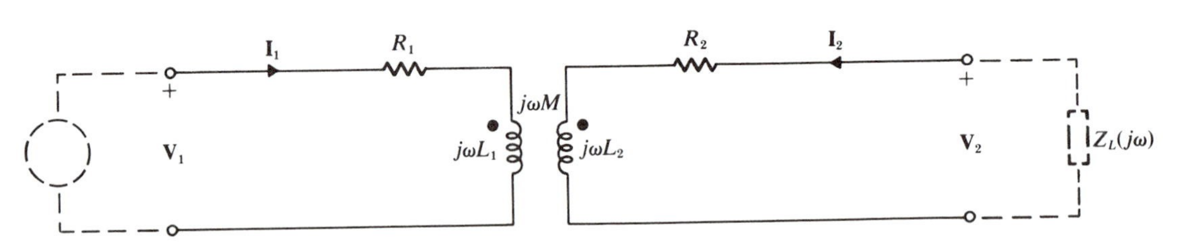


**Fig. 9-4**

It will be convenient to make further approximations, although we will find that we can later reintroduce the neglected factors. First, if we assume that $R_1$ and $R_2$ are negligible, then as shown in Problem 9.9,

$$\frac{\mathbf{V}_2}{\mathbf{V}_1} = \frac{j\omega M Z_L(j\omega)}{-\omega^2(L_1L_2 - M^2) + j\omega L_1 Z_L(j\omega)} \tag{9.8}$$

$$\frac{\mathbf{I}_2}{\mathbf{I}_1} = \frac{-j\omega M}{Z_L(j\omega) + j\omega L_2} \tag{9.9}$$

and

$$Z_{\text{in}}(j\omega) = \frac{-\omega^2(L_1L_2 - M^2) + j\omega L_1 Z_L(j\omega)}{Z_L(j\omega) + j\omega L_2} \tag{9.10}$$

Now the self and mutual inductances may be related through the *coefficient of coupling k*,

$$M = k\sqrt{L_1L_2} \tag{9.11}$$

where it can be shown that $k$ cannot exceed unity. However, for *close-coupled, iron-cored coils* it is reasonable to assume that $k = 1$. Upon making this second approximation the so-called *perfect transformer* results. Equation (*9.8*) now becomes

$$\frac{\mathbf{V}_2}{\mathbf{V}_1} = \sqrt{\frac{L_2}{L_1}} = \frac{1}{n} \tag{9.12}$$

where $n$ is called the *voltage transformation ratio* $|\mathbf{V}_1|/|\mathbf{V}_2|$.

Finally, if it is assumed that $\omega L_I \gg n^2|Z_L(j\omega)|$, corresponding to a transformer under load (i.e. delivering a "reasonably" large output current), then from equations (*9.9*), (*9.10*), and (*9.12*),

$$\frac{\mathbf{V}_2}{\mathbf{V}_1} = \frac{1}{n}, \qquad \frac{\mathbf{I}_2}{\mathbf{I}_1} = -n \tag{9.13}$$

and

$$Z_{\text{in}}(j\omega) = \frac{\mathbf{V}_1}{\mathbf{I}_1} = n^2 Z_L(j\omega) \tag{9.14}$$

Equations (*9.13*) and (*9.14*) describe the *ideal transformer*, which is useful as a conceptual building-block.

If the above approximations are *not* realistic (in a lightly loaded, air-cored, coupling transformer, for example), it can be shown (see Problems 9.10 and 9.11) that circuit models may still be based on the ideal transformer. This is illustrated in Fig. 9-5.

As a matter of terminology, the input side of a transformer is called the *primary*, and the output side is called the *secondary*.

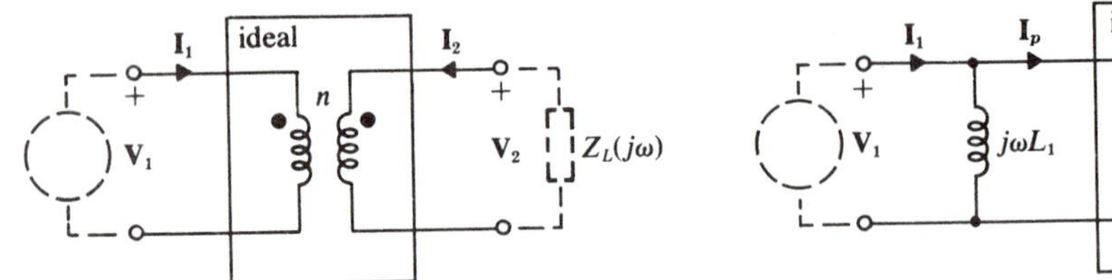


(*a*) *Ideal* transformer with $R_1$ and $R_2$ negligible; $k \doteq 1$; $\omega L_1 \gg n^2|Z_L(j\omega)|$

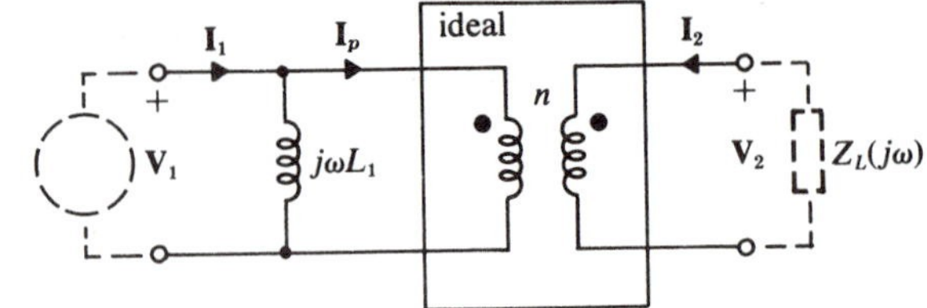


(*b*) *Perfect* transformer with $R_1$ and $R_2$ negligible; $k \doteq 1$

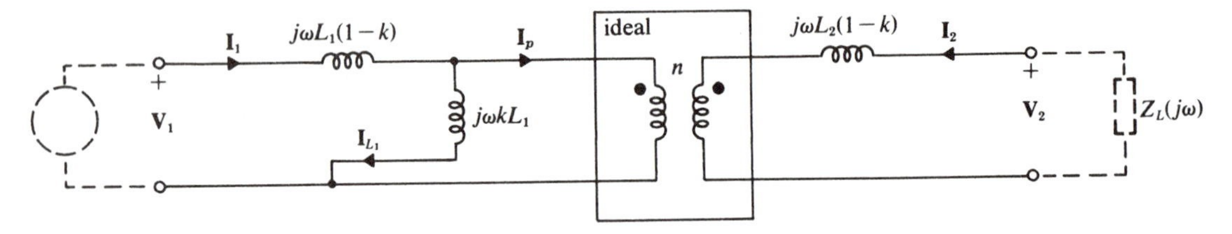


(*c*) Transformer with $R_1$ and $R_2$ negligible

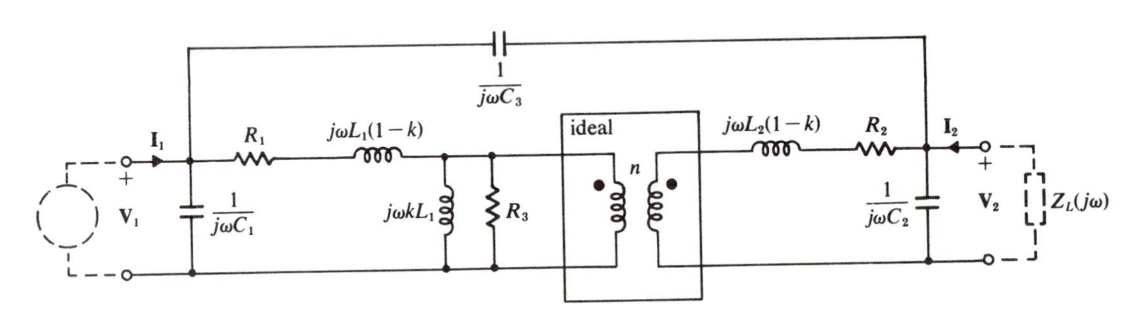


(*d*) Transformer with winding and core losses

**Fig. 9-5**

## Example 9.2

An electronic amplifier with an output impedance of $R_a$ is driving a loudspeaker whose impedance is $R_L$ through an ideal transformer (see Fig. 9-6). We shall find the transformation ratio $n$ which maximizes the power into the loudspeaker, given that $V$, $R_a$, and $R_L$ are fixed.

Equation (*9.14*) asserts that the amplifier will see an impedance of $n^2R_L$, as shown in Fig. 9-7. Then

$$\mathbf{I}_1 = \frac{\mathbf{V}}{R_a + n^2R_L}$$

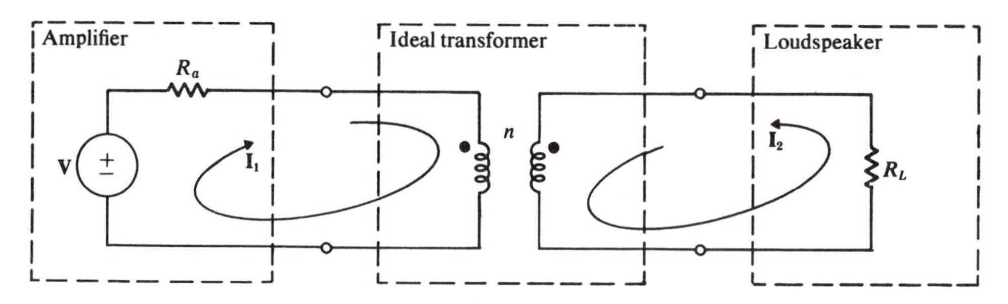


Fig. 9-6

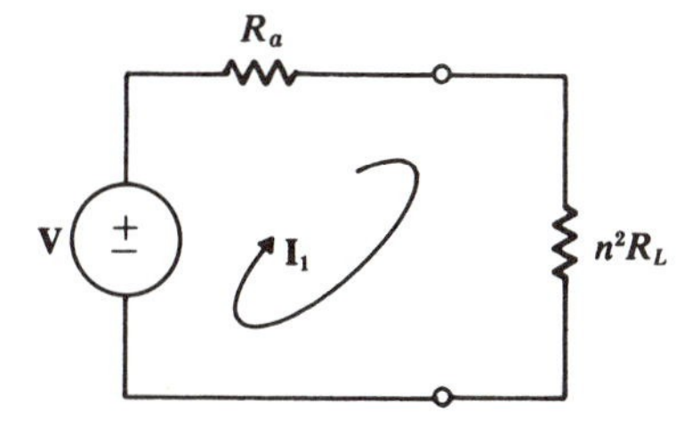


Fig. 9-7

But from equation (9.13), $\mathbf{I}_2 = -n\mathbf{I}_1$, so

$$P_L = R_L I_{2,\text{eff}}^2 = n^2 R_L I_{1,\text{eff}}^2 = \frac{n^2 R_L V_{\text{eff}}^2}{(R_a + n^2 R_L)^2}$$

To maximize $P_L$ with respect to $n$ we set $dP_L/dn$ equal to zero,

$$\frac{2nR_L V_{\text{eff}}^2 (R_a + n^2 R_L)^2 - n^2 R_L V_{\text{eff}}^2 \{4nR_L(R_a + n^2 R_L)\}}{(R_a + n^2 R_L)^4} = 0$$

from which, after a little algebraic simplification,

$$n^2 = \frac{R_a}{R_L}$$

Thus if we know the amplifier and loudspeaker impedances, we can choose a transformer with the proper transformation ratio, $n$.

## POLYPHASE CIRCUITS

There are economic and technical advantages to the generation, transmission, and consumption of *polyphase power*. Instead of a single rotating coil in an ac generator or *alternator*, two or more coils are set at definite angles to each other, resulting in a polyphase source. This is illustrated in Fig. 9-8.

Because of the fixed angle betweeen the nominally identical alternator coils, there will be a definite relationship between the source voltages of a polyphase source.

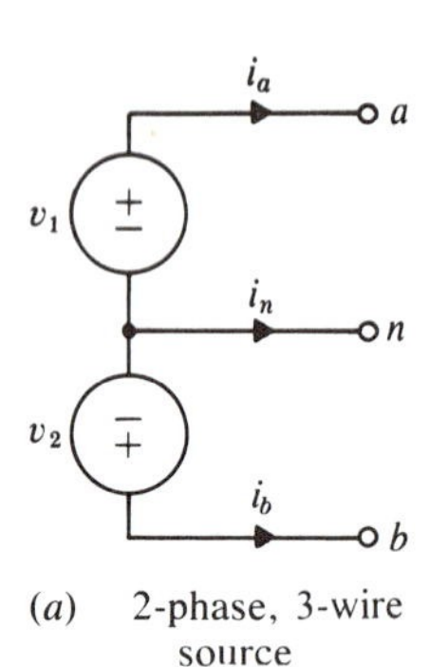


(*a*) 2-phase, 3-wire source

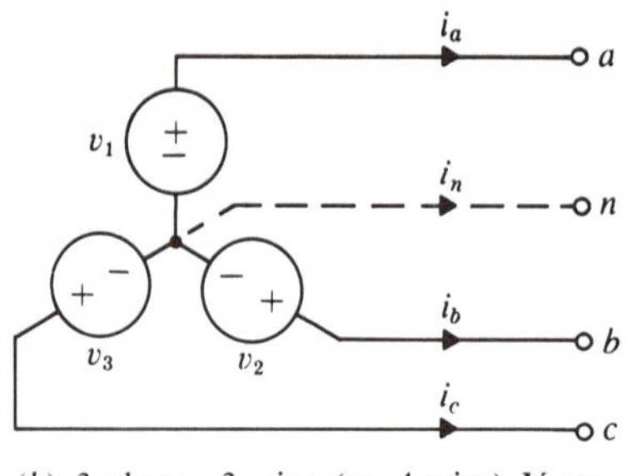


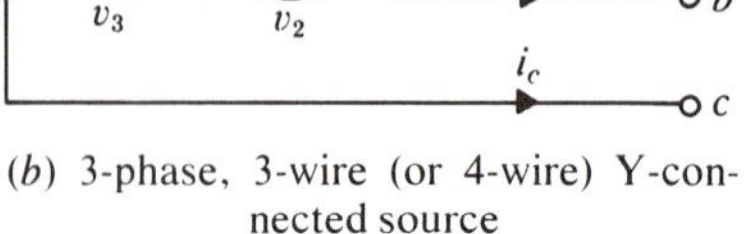

(*b*) 3-phase, 3-wire (or 4-wire) Y-connected source

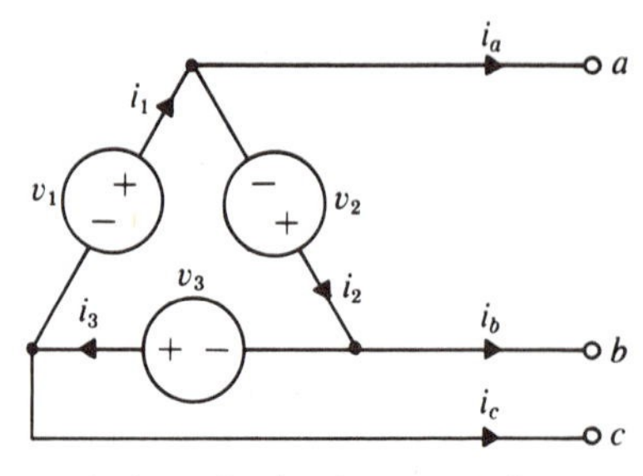


(*c*) 3-phase, 3-wire, Δ-connected source

**Fig. 9-8**

In the 2-phase system of Fig. 9-8*a*,

$$v_1(t) = A \cos \omega t$$
$$v_2(t) = A \cos (\omega t \pm 90^\circ) \tag{9.15}$$

The phase of $v_2$, $+90^\circ$, or $-90^\circ$ relative to $v_1$ will depend on the relative polarity of the source connections.

It is easily seen in the phasor diagrams of Fig. 9-9 that for the 2-phase, 3-wire system of Fig. 9-8*a*, we have the following:

1. The *line-to-line voltage* $v_{ab} = v_a - v_b = v_1 - v_2$ has a magnitude $\sqrt{2}$ times that of the *line-to-neutral voltage*.
2. If the system is *balanced* (i.e. has equal loads from each line to neutral), then the *line currents* will be

$$i_a(t) = B \cos (\omega t + \phi)$$
$$i_b(t) = B \cos (\omega t \pm 90^\circ + \phi)$$

and the *neutral current* $i_n = -i_a - i_b$ will have a magnitude $\sqrt{2}$ times that of either line current.

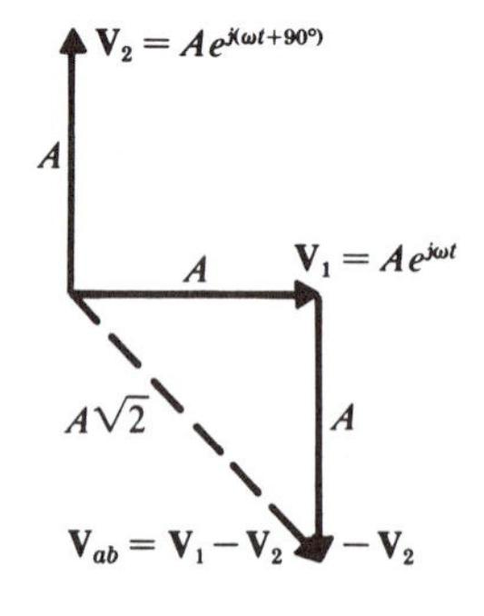


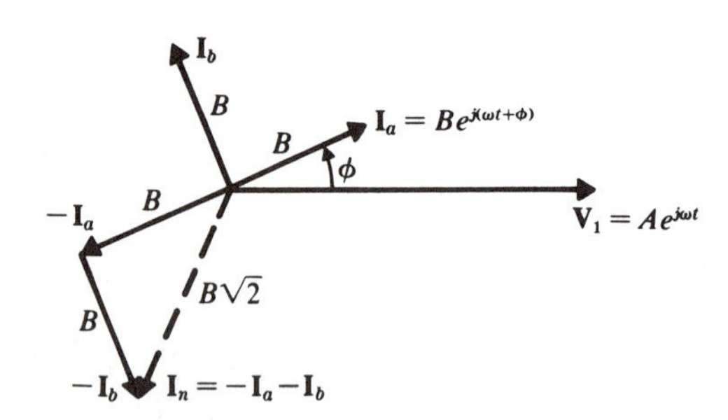


**Fig. 9-9**

In the case of the 3-phase system,

$$v_1(t) = A \cos \omega t$$
$$v_2(t) = A \cos (\omega t \pm 120°) \qquad (9.16)$$
$$v_3(t) = A \cos (\omega t \pm 240°)$$

Taking the plus signs, $v_3$ leads, followed by $v_2$ and then $v_1$. We say that the source has the *phase sequence* 3-2-1 or 1-3-2. If the minus signs are taken, the phase sequence would be 1-2-3. This concept is illustrated in Fig. 9-10.

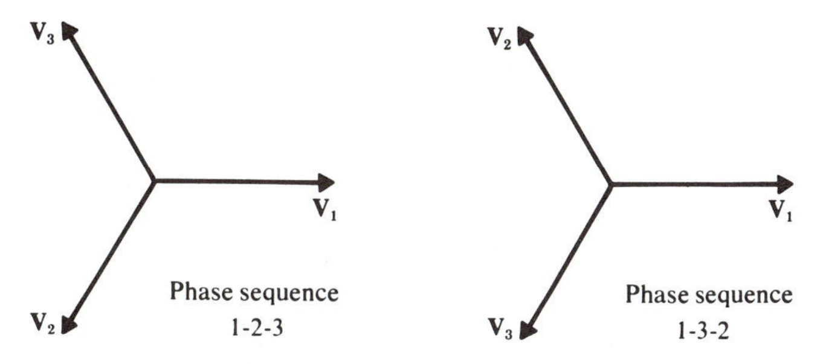


**Fig. 9-10**

Looking now at the phasor diagrams of Fig. 9-11 for the 3-phase, 4-wire system of Fig. 9-8*b*, we see the following:

1. The line-to-line voltage (for example, $v_{ab} = v_1 - v_2$) has a magnitude $\sqrt{3}$ times that of the line-to-neutral voltage.
2. If the system is balanced, so that

$$i_a = B \cos (\omega t + \phi) \qquad i_b = B \cos (\omega t \pm 120° + \phi)$$
$$i_c = B \cos (\omega t \pm 240° + \phi)$$

then the neutral current $i_n = -i_a - i_b - i_c$ will be identically zero. For this reason the neutral wire is often omitted.

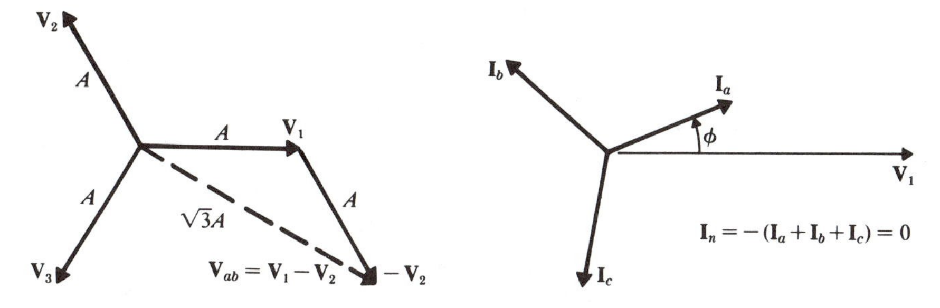


**Fig. 9-11**

Similar examination of the 3-phase, Δ-connected system (Fig. 9-8*c*) shows that under balanced conditions the magnitude of the line current is $\sqrt{3}$ times that of the *phase current* (for example, $i_1$).

Turning now to the 3-phase load—motors, heaters, lighting, etc.—it is clear that this, too, may be either Y- or Δ-connected as shown in Fig. 9-12.

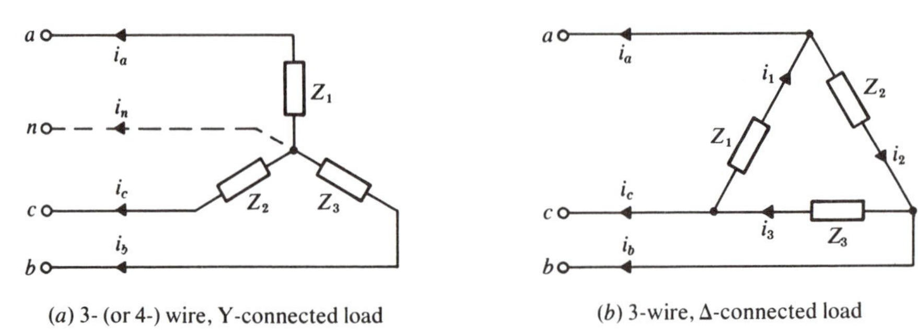


(*a*) 3- (or 4-) wire, Y-connected load (*b*) 3-wire, Δ-connected load

**Fig. 9-12**

The total active power delivered to a *balanced* 3-phase load, either Y- or Δ-connected, can be shown to be

$$P_T = \sqrt{3} V_{L,\text{eff}} I_{L,\text{eff}} \cos \phi \qquad (9.17)$$

where $V_{L,\text{eff}}$ is the line-to-line voltage and $I_{L,\text{eff}}$ is the line current. As in the case of single-phase power, $\cos \phi$ is the power factor, where $\phi$ is the angle of the impedance in one phase of the load.

Similarly, the total apparent power and the total reactive power are given by

$$S_T = \sqrt{3} V_{L,\text{eff}} I_{L,\text{eff}} \qquad \text{and} \qquad Q_T = \sqrt{3} V_{L,\text{eff}} I_{L,\text{eff}} \sin \phi \qquad (9.18)$$

Wattmeters may be used to measure the total power to a 3-phase load regardless of whether the system is balanced. Thus in Fig. 9-13*a* the sum of the three readings can be shown to equal $P_T$, while in Fig. 9-13*b* the sum of the *two* readings equals $P_T$.

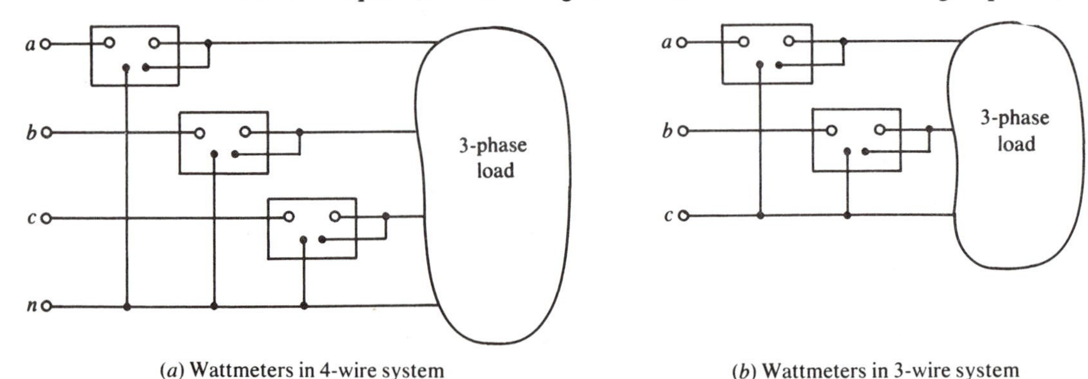


(*a*) Wattmeters in 4-wire system (*b*) Wattmeters in 3-wire system

**Fig. 9-13**

(A wattmeter has two sets of terminals: one heavily constructed pair through which the line current passes, and another more lightly constructed pair which senses voltage.)

# ILLUSTRATIVE PROBLEMS

## *AC Power*

**9.1** In the one-port network of Fig. 9-14,

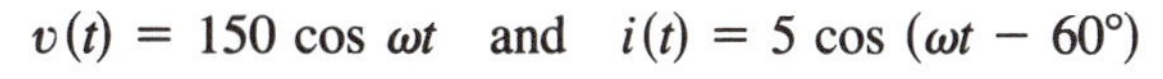

$$v(t) = 150 \cos \omega t \quad \text{and} \quad i(t) = 5 \cos (\omega t - 60^\circ)$$

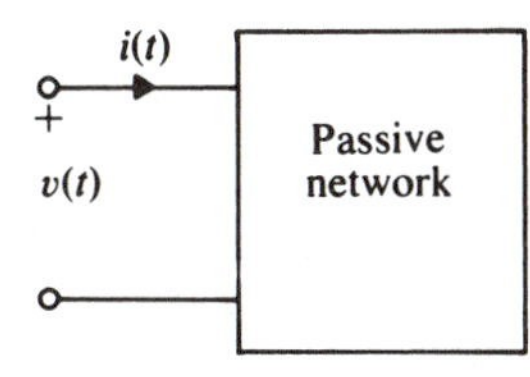


**Fig. 9-14**

Determine

(*a*) the average active (real) power $P$,
(*b*) the average reactive (quadrature) power $Q$,
(*c*) the average apparent power $S$, and
(*d*) the power factor of the network.

(*a*) With $V_{\text{eff}} = 150/\sqrt{2} = 106$ V, $I_{\text{eff}} = 5/\sqrt{2} = 3.53$ A and $\phi = +60^\circ$,

$$P = V_{\text{eff}} I_{\text{eff}} \cos \phi = (106)(3.53)(\cos 60^\circ) = 187 \text{ W}$$

(*b*)
$$Q = V_{\text{eff}} I_{\text{eff}} \sin \phi = (106)(3.53)(\sin 60^\circ) = 325 \text{ VAR}$$

(*c*)
$$S = V_{\text{eff}} I_{\text{eff}} = (106)(3.53) = 375 \text{ VA}$$

Or alternatively,

$$S = \sqrt{P^2 + Q^2} = \sqrt{187^2 + 325^2} = 375 \text{ VA}$$

(*d*) Power factor $= \cos \phi = \cos 60^\circ = 0.50$ inductive.

*Note:* The power factor $\cos \phi$ does not change sign as $\phi$ changes sign. The sense of the power factor should therefore always be stated—as *capacitive or inductive*.

### *Alternative Solution*

If we treat **V** and **I** as rotating phasors,

$$Z(j\omega) = R(\omega) + jX(\omega) = \frac{\mathbf{V}}{\mathbf{I}} = \frac{150e^{j\omega t}}{5e^{j(\omega t - 60^\circ)}} = 30e^{j60^\circ}$$

$$= 30(\cos 60^\circ + j\sin 60^\circ) = 15 + j26$$

Therefore $R(\omega) = 15\ \Omega$ and $X(\omega) = 26\ \Omega$. And as before, $I_{eff} = 3.53$ A. Then

(a) $P = I_{eff}^2 R(\omega) = (3.53)^2\,(15) = 187$ W,

(b) $Q = I_{eff}^2 X(\omega) = (3.53)^2\,(26) = 325$ VAR,

(c) $S = I_{eff}^2 |Z(j\omega)| = (3.53)^2\,(30) = 375$ VA,

(d) Power factor $= \cos\phi = \cos \angle Z(j\omega) = \cos 60° = 0.50$ inductive.

*Comment:* It is sometimes useful to represent the complex power $S = P + jQ$ graphically, as in the *power triangle* of Fig. 9-15.

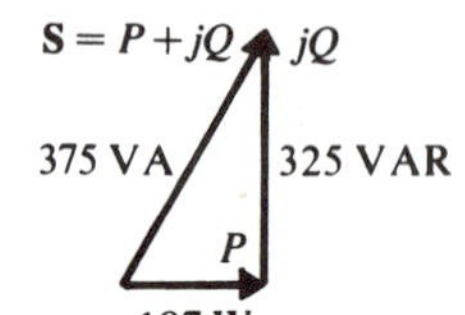


**Fig. 9-15**

**9.2** Assuming sinusoidal steady-state conditions and given $v(t) = 100\sqrt{2}\cos 377t$, $S = 500$ VA and power factor $= 0.866$ capacitive, find $i(t)$ in the circuit of Fig. 9-16.

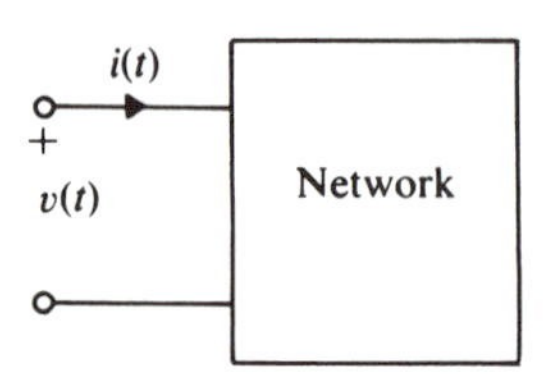


**Fig. 9-16**

Here $S = V_{eff} I_{eff}$ or $I_{eff} = S/V_{eff} = 500/100 = 5$ A. But the power factor is 0.866 capacitive. That is, the current leads the voltage by 30°. So $i(t) = 5\sqrt{2}\cos(377t + 30°)$.

**9.3** The system shown in Fig. 9-17 has the power demands tabulated:

| Load | $P$, kW | $Q$, kVAR | $S$, kVA |
|---|---|---|---|
| 1 | 5 | 5 | |
| 2 | 4 | | 5 |
| 3 | | −5 | 5 |
| Total | | +3 | |

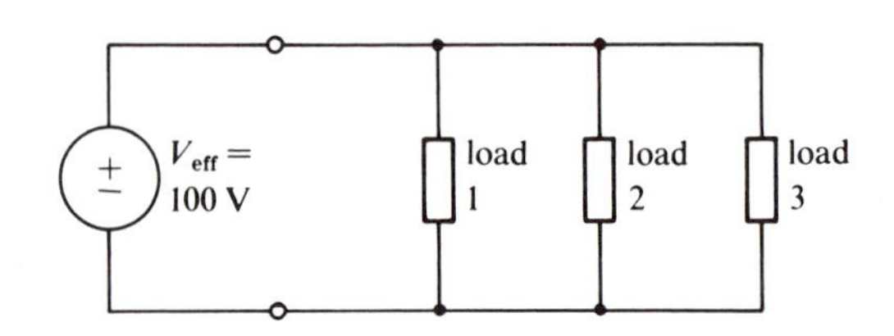


**Fig. 9-17**

Fill in the blanks in this table and find the power factor for the overall system.

By addition of "vectors,"

$$\mathbf{S}_T = P_T + jQ_T = (P_1 + P_2 + P_3) + j(Q_1 + Q_2 + Q_3)$$

so we must calculate $S_T$ from $S_T = \sqrt{P_T^2 + Q_T^2}$. Note that $S_T \neq S_1 + S_2 + S_3$.

To complete the table we may proceed as follows:

1. Since $S_3 = 5 = \sqrt{P_3^2 + Q_3^2} = \sqrt{P_3^2 + 5^2}$, then $P_3 = 0$ kW and $P_T = 5 + 4 + 0 = 9$ kW.
2. $Q_T = 3 = 5 + Q_2 + (-5)$, so $Q_2 = +3$ kVAR.
3. $S_1 = \sqrt{P_1^2 + Q_1^2} = \sqrt{5^2 + 5^2} = 7.07$ kVA.
4. $S_T = \sqrt{P_T^2 + Q_T^2} = \sqrt{9^2 + 3^2} = 9.5$ *kVA*.

Thus:

| Load | $P$, kW | $Q$, kVAR | $S$, kVA |
|---|---|---|---|
| 1 | 5 | 5 | 7.07 |
| 2 | 4 | 3 | 5 |
| 3 | 0 | −5 | 5 |
| Total | 3 | 3 | 9.5 |

Since $Q_T = I_{\text{eff}}^2 X_T(\omega) = +3$ kVAR, we know that the total load must be *inductive* and the system's overall power factor is therefore

$$\cos \phi_T = \frac{P_T}{S_T} = \frac{9}{9.5} = 0.946 \text{ inductive}$$

**9.4** In the circuit of Fig. 9-18 it is known that the industrial plant draws 250 kW with an inductive power factor of 0.707. A capacitor is to be added in parallel with the plant to improve the overall power factor to 0.866 inductive. Find the value of $C$ required to effect this change.

For the plant, in units of kW, kVAR, and kVA,

$$\mathbf{S}_p = 250 + j250$$

since the power factor is 0.707 and $\cos^{-1}(0.707) = 45°$. For the plant-capacitor combination, whose power factor is 0.866,

$$\mathbf{S}_T = 250 + j144$$

Both of these situations are shown in Fig. 9-19 and we see that the capacitor must draw

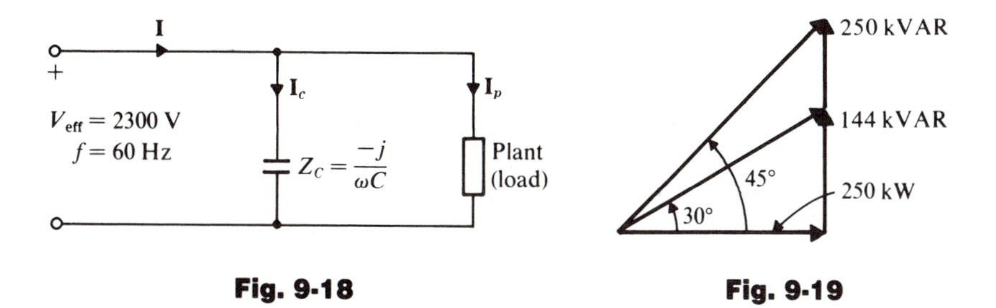


**Fig. 9-18** **Fig. 9-19**

$$Q_C = 144 - 250 = -106 \text{ kVAR}$$

of reactive power.

But for a capacitor, $\phi = \underline{/Z_C(j\omega)} = -90°$. Therefore

$$Q_C = V_{\text{eff}} I_{C,\text{eff}} \sin(-90°) = -\frac{V_{\text{eff}}^2}{1/\omega C} = -106 \times 10^3 \text{ VAR}$$

Thus

$$C = \frac{106 \times 10^3}{(2300)^2(2\pi \times 60)} = 53.1\ \mu\text{F}$$

### *Alternative Solution*

We may use a phasor diagram to solve this problem. For the plant,

$$P_p = 250 \times 10^3 = V_{\text{eff}} I_{p,\text{eff}} \cos\phi = (2300)(I_{p,\text{eff}})(0.707)$$

and thus $I_{p,\text{eff}} = 154$ A. If we choose $\mathbf{V} = 2300e^{j\omega t}$ as the reference, then $\mathbf{I}_p = 154e^{j(\omega t - 45°)}$.

Now the capacitor current $\mathbf{I}_C$ must lead $\mathbf{V}$ by 90° and must be of sufficient amplitude to reduce the angle between the *total* current $\mathbf{I}$ and $\mathbf{V}$ to 30°. This is shown in Fig. 9-20. From the geometry,

$$\mathbf{I}_{C,\text{eff}} = 109 - 109 \tan 30° = 46 \text{ A}$$

or $\mathbf{I}_C = 46e^{j(\omega t + 90°)}$. But we also know that

$$\mathbf{I}_C = j\omega C\mathbf{V} = j\omega C(2300e^{j\omega t}) = 2300\omega Ce^{j(\omega t + 90°)}$$

Equating these two expressions for $\mathbf{I}_C$ yields

$$C = \frac{46}{(2300)(2\pi \times 60)} = 53\ \mu\text{F}$$

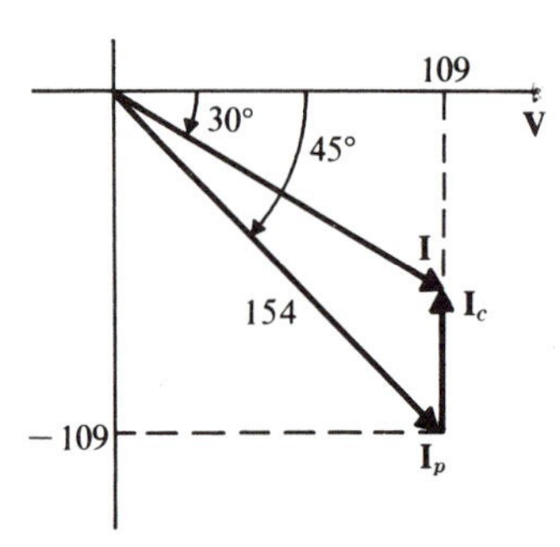


**Fig. 9-20**

**9.5** An ammeter, a voltmeter, and a wattmeter were connected as shown in Fig. 9-21 in an effort to determine $Z(j\omega)$. The readings were 5 A, RMS, 65 V, RMS, and 125 W. Determine the possible values of $Z(j\omega)$.

From the problem data,

$$|Z(j\omega)| = \frac{V_{\text{eff}}}{I_{\text{eff}}} = \frac{65}{5} = 13$$

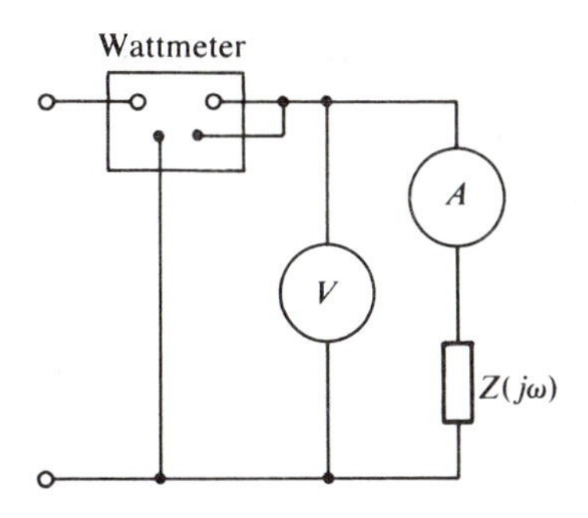


Fig. 9-21

And since $P = V_{eff} I_{eff} \cos \phi$,

$$\cos \phi = \frac{125}{(65)(5)} = 0.387$$

or

$$\phi = \pm 67.3°$$

Therefore since $\phi = \underline{/Z(j\omega)}$,

$$Z(j\omega) = 13e^{\pm j67.3°} = 5 \pm j12$$

Without additional data there is no way of establishing which sign applies.

**9.6** Using an RMS reading voltmeter, it has been found that the amplitudes of the three sinusoidal voltages $e(t)$, $v_L(t)$, and $v_R(t)$ in Fig. 9-22 are the same, namely 200 V. It is also known that the reactance of the inductor is $\omega L = 20\ \Omega$. Find the complex power **S** supplied by the source.

Choosing $\mathbf{V}_R$ as reference, the phasor diagram of Fig. 9-23 may be constructed. The triangle must be completed *above* $\mathbf{V}_R$ since, if it were not, $\mathbf{I}_L$ would *lag* $\mathbf{V}_R$ by 30°. This is not possible, since $\mathbf{I}_L$ is the current through the *R-C* combination, which must *lead* the voltage $\mathbf{V}_R$ across it.

It follows from the phasor diagram that

$$\mathbf{V}_L = 200e^{j(\omega t + 120°)} \quad \text{and} \quad \mathbf{E} = 200e^{j(\omega t + 60°)}$$

Thus

$$\mathbf{I}_L = \frac{\mathbf{V}_L}{j\omega L} = \frac{200e^{j(\omega t + 120°)}}{20e^{j90°}} = 10e^{j(\omega t + 30°)}$$

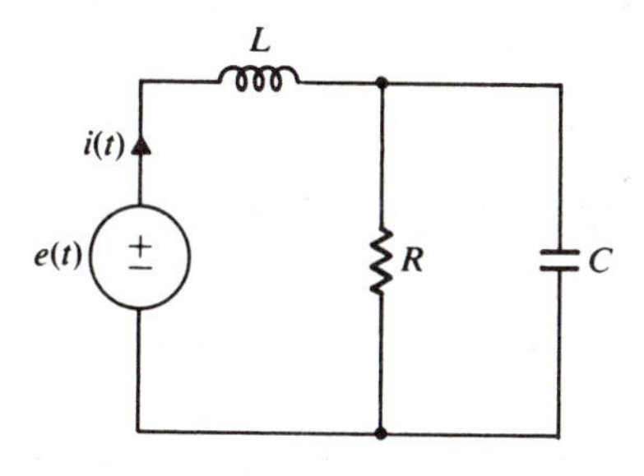


Fig. 9-22

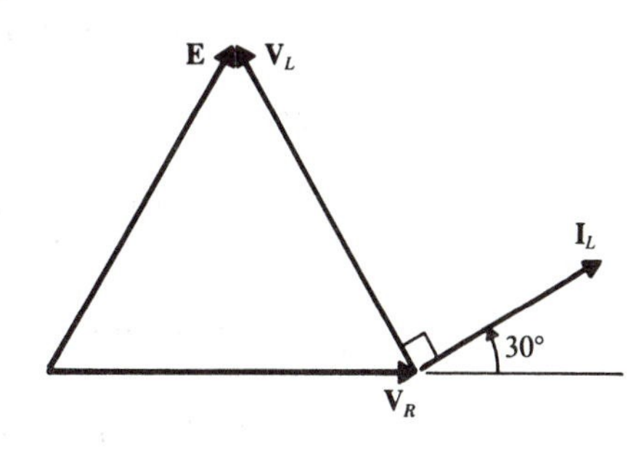


Fig. 9-23

But $\mathbf{I}_L = \mathbf{I}$, the phasor source current, and $\phi = \underline{/\mathbf{E}} - \underline{/\mathbf{I}} = 30°$. Hence

$$P = E_{\text{eff}} I_{\text{eff}} \cos \phi = (200)(10) \cos 30° = 1732 \ W$$

$$Q = E_{\text{eff}} I_{\text{eff}} \sin \phi = (200)(10) \sin 30° = 1000 \text{ VAR}$$

Finally,

$$\mathbf{S} = P + jQ = 1732 + j1000$$

### *Transformers*

**9.7** Find $v_{\text{sss}}(t)$ in the coupled circuit of Fig. 9-24, given $e(t) = A \cos \omega t$.

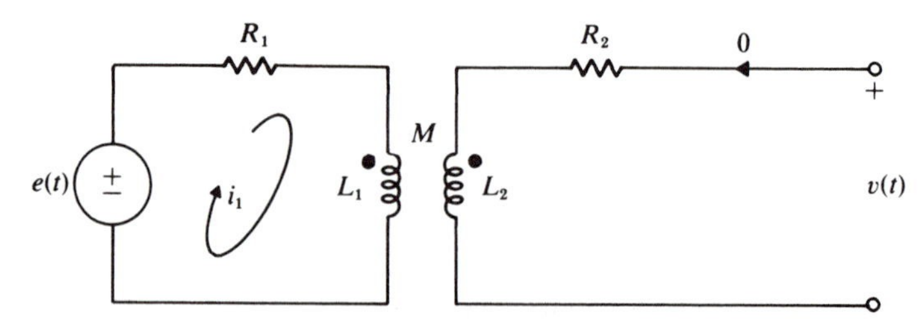


**Fig. 9-24**

The appropriate equations are

$$L_1 \frac{di_1}{dt} + R_1 i_1 = A \cos \omega t \quad \text{and} \quad M \frac{di_1}{dt} = v(t)$$

In phasor notation these become

$$j\omega L_1 \mathbf{I}_1 + R_1 \mathbf{I}_1 = Ae^{j\omega t} \quad \text{and} \quad j\omega M \mathbf{I}_1 = \mathbf{V}$$

from which we find

$$\mathbf{I}_1 = \frac{Ae^{j\omega t}}{R_1 + j\omega L_1} = \frac{A}{\sqrt{R_1^2 + (\omega L_1)^2}} e^{j\{\omega t - \tan^{-1}(\omega L_1/R_1)\}}$$

and

$$\mathbf{V} = j\omega M \mathbf{I}_1 = \omega M e^{j90°} \mathbf{I}_1 = \frac{A\omega M}{\sqrt{R_1^2 + (\omega L_1)^2}} e^{j\{\omega t + 90° - \tan^{-1}(\omega L_1/R_1)\}}$$

Finally,

$$v(t) = \frac{A\omega M}{\sqrt{R_1^2 + (\omega L_1)^2}} \cos \{\omega t + 90° - \tan^{-1} (\omega L_1/R_1)\}$$

**9.8** Find the reactive power supplied by the source in the circuit of Problem 9.7 (Fig. 9-24).

From Problem 9.7 the phase angle of $\mathbf{E}$ relative to $\mathbf{I}_1$ is $\phi = +\tan^{-1}(\omega L_1/R_1)$.

So

$$Q = E_{\text{eff}} I_{1,\text{eff}} \sin \phi = \frac{A}{\sqrt{2}} \frac{A/\sqrt{2}}{\sqrt{R_1^2 + (\omega L_1)^2}} \frac{\omega L_1}{\sqrt{R_1^2 + (\omega L_1)^2}} = \frac{A^2 \omega L_1}{2(R_1^2 + \omega^2 L_1^2)} \text{ VAR}$$

**9.9** Derive equations (*9.8*)–(*9.10*) for the transformer circuit of Fig. 9-4, page 306. Neglect the winding resistances $R_1$ and $R_2$.

The phasor loop equations are

$$j\omega L_1 \mathbf{I}_1 + j\omega M \mathbf{I}_2 = \mathbf{V}_1$$

$$j\omega M \mathbf{I}_1 + j\omega L_2 \mathbf{I}_2 = \mathbf{V}_2 = -Z_L(j\omega)\mathbf{I}_2$$

From the second equation,

$$\mathbf{I}_2 = \frac{-j\omega M \mathbf{I}_1}{Z_L(j\omega) + j\omega L_2} \quad \text{or} \quad \frac{\mathbf{I}_2}{\mathbf{I}_1} = \frac{-j\omega M}{Z_L(j\omega) + j\omega L_2} \tag{9.9}$$

Substituting this expression for $\mathbf{I}_2$ into both loop equations and dividing the second by the first yields

$$\frac{\mathbf{V}_2}{\mathbf{V}_1} = \frac{j\omega M Z_L(j\omega)}{-\omega^2(L_1 L_2 - M_2) + j\omega L_1 Z_L(j\omega)} \tag{9.8}$$

Finally, substituting for $\mathbf{I}_2$ in the first loop equation and rearranging,

$$\frac{\mathbf{V}_1}{\mathbf{I}_1} = Z_{\text{in}}(j\omega) = \frac{-\omega^2(L_1 L_2 - M^2) + j\omega L_1 Z_L(j\omega)}{Z_L(j\omega) + j\omega L_2} \tag{9.10}$$

**9.10** Obtain the ratios $\mathbf{V}_2/\mathbf{V}_1$, $\mathbf{I}_2/\mathbf{I}_1$, and $\mathbf{V}_1/\mathbf{I}_1 = Z_{\text{in}}(j\omega)$ for the *perfect transformer*. Then demonstrate that the circuit of Fig. 9-5*b*, page 308, is a valid representation.

To obtain the perfect transformer, we set the coefficient of coupling $k$ equal to unity. Then from equation (*9.11*), page 307,

$$M = \sqrt{L_1 L_2} \quad \text{or} \quad M^2 = L_1 L_2$$

If, further, we define the voltage ratio $\mathbf{V}_1/\mathbf{V}_2 = n$, equation (*9.8*) becomes

$$\frac{\mathbf{V}_2}{\mathbf{V}_1} = \frac{j\omega M Z_L(j\omega)}{j\omega L_1 Z_L(j\omega)} = \frac{M}{L_1} = \sqrt{\frac{L_2}{L_1}} = \frac{1}{n} \tag{9.12}$$

equation (*9.9*) is unchanged,

$$\frac{\mathbf{I}_2}{\mathbf{I}_1} = \frac{-j\omega M}{Z_L(j\omega) + j\omega L_2} \tag{9.9}$$

and equation (*9.10*) becomes

$$\frac{\mathbf{V}_1}{\mathbf{I}_1} = Z_{\text{in}}(j\omega) = \frac{j\omega L_1 Z_L(j\omega)}{Z_L(j\omega) + j\omega L_2}$$

It remains to be shown that the circuit of Fig. 9-5*b* correctly represents the perfect transformer; that is, we must show that the above equations properly describe the circuit.

With reference to Fig. 9-5*b* and to the characteristics of the *ideal* transformer, it follows from equation (*9.13*), page 307, that

$$\frac{\mathbf{V}_2}{\mathbf{V}_1} = \frac{1}{n} \quad \text{and} \quad \mathbf{I}_p = -\frac{\mathbf{I}_2}{n}$$

Now $\mathbf{I}_1 = \mathbf{I}_{L1} + \mathbf{I}_p$, so that

$$\mathbf{I}_1 = \frac{\mathbf{V}_1}{j\omega L_1} - \frac{\mathbf{I}_2}{n} = \frac{\mathbf{V}_1}{j\omega L_1} + \frac{\mathbf{V}_2}{nZ_L(j\omega)} = \frac{\mathbf{V}_1}{j\omega L_1} + \frac{\mathbf{V}_1}{n^2 Z_L(j\omega)}$$

and since $\sqrt{L_1/L_2} = n$,

$$\frac{\mathbf{V}_1}{\mathbf{I}_1} = Z_{\text{in}}(j\omega) = \frac{n^2 j\omega L_1 Z_L(j\omega)}{n^2 Z_L(j\omega) + j\omega L_1} = \frac{j\omega L_1 Z_L(j\omega)}{Z_L(j\omega) + j\omega L_2}$$

Finally,

$$\mathbf{I}_2 = -\frac{\mathbf{V}_2}{Z_L(j\omega)} = -\frac{\mathbf{V}_1}{nZ_L(j\omega)}$$

or, dividing through by $\mathbf{I}_1$,

$$\frac{\mathbf{I}_2}{\mathbf{I}_1} = -\frac{\mathbf{V}_1}{\mathbf{I}_1}\left(\frac{1}{nZ_L(j\omega)}\right) = -\frac{Z_{in}(j\omega)}{nZ_L(j\omega)} = -\frac{j\omega L_1/n}{Z_L(j\omega) + j\omega L_2} = \frac{-j\omega M}{Z_L(j\omega) + j\omega L_2}$$

Thus the circuit of Fig. 9-5*b* is described by the three equations that define the perfect transformer.

*Comment:* In any real transformer a primary current $\mathbf{I}_1$ will flow even if the output is open circuit ($\mathbf{I}_2 = 0$). This is called the magnetizing current and it is accounted for in the model by the inductance $L_1$ (see Fig. 9-5*b*).

**9.11** Show that the circuit of Fig. 9-5*c* is equivalent to Fig. 9-4 (with $R_1$ and $R_2$ neglected) as far as the terminal conditions are concerned.

Choosing the loop currents as shown in Fig. 9-25, the KVL equation around the left-hand mesh is

$$\{j\omega L_1(1 - k) + j\omega k L_1\}\mathbf{I}_1 - j\omega k L_1 \mathbf{I}_1' = 0$$

But $\mathbf{I}_1' = -\mathbf{I}_2/n$ and $kL_1/n = M$ (since $M = k\sqrt{L_1 L_2}$ and $\sqrt{L_2/L_1} = 1/n$). Therefore

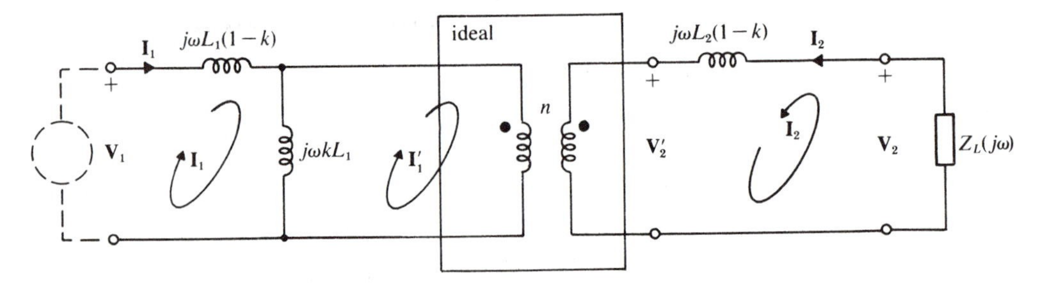


**Fig. 9-25**

$$j\omega L_1\mathbf{I}_1 + j\omega M\mathbf{I}_2 = 0$$

This is also the loop equation for the left-hand loop of the circuit of Fig. 9-4 if $R_1$ is neglected.

Next, the loop equation for the right-hand mesh of Fig. 9-25 is

$$\mathbf{V}_2' + j\omega L_2(1 - k)\mathbf{I}_2 = -Z_L(j\omega)\mathbf{I}_2 = \mathbf{V}_2$$

But $\mathbf{V}_2' = \mathbf{V}_1'/n$ and $\mathbf{V}_1' = j\omega kL_1(\mathbf{I}_1 - \mathbf{I}_1') = j\omega kL_1(\mathbf{I}_1 + \mathbf{I}_2/n)$. Therefore

$$\left(\frac{j\omega kL_1}{n}\right)\left(\mathbf{I}_1 + \frac{\mathbf{I}_2}{n}\right) + j\omega L_2(1 - k)\mathbf{I}_2 = -Z_L(j\omega)\mathbf{I}_2 = \mathbf{V}_2$$

or

$$j\omega M\mathbf{I}_1 + j\omega kL_2\mathbf{I}_2 + j\omega L_2(1 - k)\mathbf{I}_2 = -Z_L(j\omega)\mathbf{I}_2 = \mathbf{V}_2$$

or

$$j\omega M\mathbf{I}_1 + j\omega L_2\mathbf{I}_2 = -Z_L(j\omega)\mathbf{I}_2 = \mathbf{V}_2$$

which is also the loop equation for the right-hand mesh of Fig. 9-4 if $R_2$ is neglected.

### *Comments*

1. It follows that the circuit of Fig. 9-5*c* is a valid model of an ordinary transformer, i.e. neither perfect nor ideal, if the winding resistances can be neglected.
2. The inductance $kL_1$ in the model accounts for the *magnetizing flux*, while the reactances $j\omega L_1(1 - k)$ and $j\omega L_2(1 - k)$ take into account the fact that not all of the flux generated by coil 1 passes through coil 2, and vice versa. These are therefore called the *leakage reactances*.
3. The effects so far neglected can be incorporated as indicated in Fig. 9-5*d*. Although this model has been built up around the conceptual *ideal transformer*, it represents most real-world transformers with reasonable fidelity.

**9.12** Calculate the RMS amplitudes of the primary and secondary currents in the transformer circuit of Fig. 9-26, where $\mathbf{V}_2 = 4\mathbf{V}_1$ and $e(t) = 120\sqrt{2}\cos\omega t$.

The appropriate *stationary* phasor equations are

$$5\mathbf{I}_1 + \mathbf{V}_1 = 120 \quad \text{and} \quad \mathbf{V}_2 = 240(-\mathbf{I}_2)$$

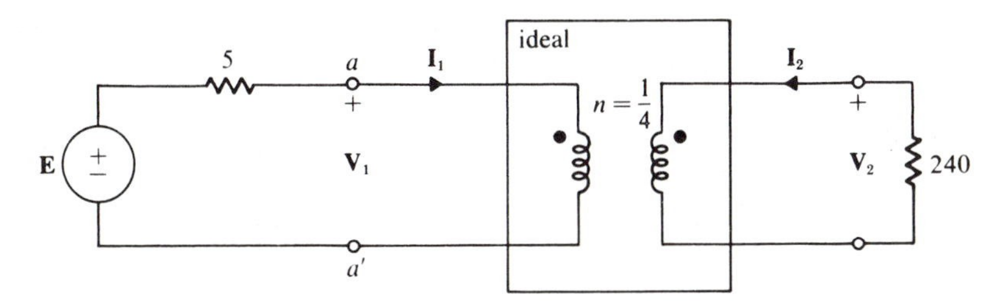


**Fig. 9-26**

But $n = \mathbf{V}_1/\mathbf{V}_2 = 1/4$. Hence from equation (*9.13*), $\mathbf{I}_2 = -\mathbf{I}_1/4$ and

$$5\mathbf{I}_1 + \mathbf{V}_2/4 = 120 \quad \text{or} \quad 5\mathbf{I}_1 - 60\mathbf{I}_2 = 120 \quad \text{or} \quad 5\mathbf{I}_1 + 15\mathbf{I}_1 = 120$$

So $\mathbf{I}_1 = 6$ and $\mathbf{I}_2 = -\mathbf{I}_1/4 = -1.5$. Thus the amplitude of the primary current is 6 A, RMS and that of the secondary is 1.5 A, RMS.

**9.13** The primary current $\mathbf{I}_1$ in the circuit of Fig. 9-27 is to be the same as that in Fig. 9-26. Find the value of $R$ required to make this possible. The source voltage $e(t)$ is unchanged.

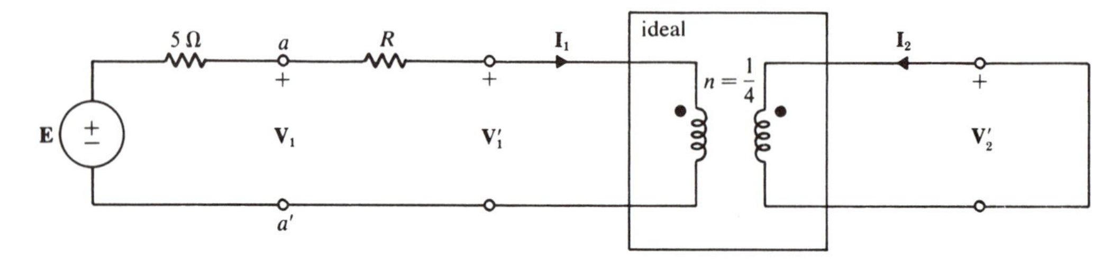


**Fig. 9-27**

If $\mathbf{I}_1$ is to be the same in both cases, the impedance $Z_{\text{in}}(j\omega)$ seen by the source must be the same. Then with the help of equation (*9.14*),

$$Z_{\text{in}}(j\omega) = \frac{\mathbf{E}}{\mathbf{I}_1} = 5 + n^2(240) \quad \text{(for Fig. 9-26)}$$

$$= 5 + R \qquad\qquad \text{(for Fig. 9-27)}$$

But $n = 1/4$, so that if $Z_{\text{in}}(j\omega)$ is to be the same in the two cases, $R = 240/16 = 15\ \Omega$.

### *Comments*

1. The secondary impedance $Z_L(j\omega)$ can be *reflected* into the primary circuit of an *ideal* transformer. Its value, when placed in the primary, must be $n^2 Z_L(j\omega)$.
2. In this example $E_{\text{eff}} = 120$ V and $I_{1,\text{eff}} = 6$ A. The voltage drop across the total of 20 Ω in the primary of Fig. 9-27 is also 120 V. And $\mathbf{E}$ and $\mathbf{I}_1$ are in phase. Therefore $\mathbf{V}_1' = 0$ and $\mathbf{V}_2' = (1/n)\mathbf{V}_1' = 0$. This makes the finite current $\mathbf{I}_2$ through the zero secondary impedance more palatable!
3. The transformer equivalents of Figs. 9-26 and 9-27 are equivalent only as seen through the terminals *a-a'*.

**9.14** Find $i_1(t)$ in the circuit of Fig. 9-28 if $v(t) = 2\cos 2t$ and the circuit is in the sinusoidal steady-state.

*Reflecting* the $\frac{1}{8}$-Ω resistor into the primary results in the circuit of Fig. 9-29. Then

$$\left(\frac{1}{2} + \frac{1}{j2}\right)\mathbf{I}_1 = \mathbf{V} = 2e^{j\omega t}$$

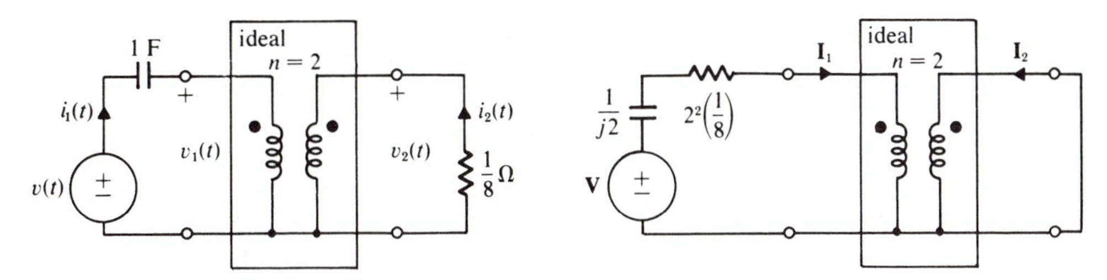


**Fig. 9-28** **Fig. 9-29**

$$\mathbf{I}_1 = \frac{2e^{j\omega t}}{\frac{1}{2} - \frac{1}{2}j} = \frac{2e^{j\omega t}}{\frac{1}{2}\sqrt{2}e^{-j45^\circ}} = 2.82e^{j(\omega t + 45^\circ)}$$

Therefore $i_1(t) = 2.82 \cos (2t + 45^\circ)$.

**9.15** Determine the real power supplied by the source $v_2(t)$ in the circuit of Fig. 9-30 given $v_1(t) = 10\sqrt{2} \sin t$ and $v_2(t) = 10\sqrt{2} \sin (t + 60^\circ)$.

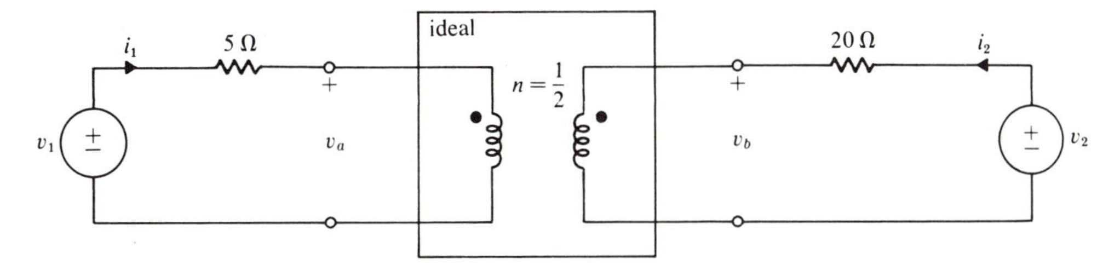


**Fig. 9-30**

The circuit equations in the $j\omega$-domain are

$$5\mathbf{I}_1 + \mathbf{V}_a = \mathbf{V}_1 = 10e^{j\omega t}$$

$$20\mathbf{I}_2 + \mathbf{V}_b = \mathbf{V}_2 = 10e^{j(\omega t + 60^\circ)}$$

$$\mathbf{V}_b = 2\mathbf{V}_a \quad \text{and} \quad \mathbf{I}_2 = -\tfrac{1}{2}\mathbf{I}_1$$

Thus in terms of $\mathbf{I}_2$ and $\mathbf{V}_b$,

$$5(-2\mathbf{I}_2) + \tfrac{1}{2}\mathbf{V}_b = 10e^{j\omega t}$$

$$20\mathbf{I}_2 + \mathbf{V}_b = 10e^{j(\omega t + 60^\circ)}$$

Solving for $\mathbf{I}_2$ by Cramer's rule,

$$\mathbf{I}_2 = \frac{\begin{vmatrix} 10e^{j\omega t} & 1/2 \\ 10e^{j(\omega t + 60^\circ)} & 1 \end{vmatrix}}{\begin{vmatrix} -10 & 1/2 \\ 20 & 1 \end{vmatrix}} = \frac{10e^{j\omega t} - 5e^{j(\omega t + 60^\circ)}}{-20}$$

$$= \frac{(10 - 2.5 - j4.33)e^{j\omega t}}{-20} = \frac{(7.5 - j4.33)e^{j\omega t}}{20e^{j180^\circ}}$$

$$= \frac{(8.66e^{-j30^\circ})e^{j\omega t}}{20e^{j180^\circ}} = 0.433e^{j(\omega t - 210^\circ)}$$

Now the power supplied by the source $v_2(t)$ will be

$$P = V_{2,\text{eff}} I_{2,\text{eff}} \cos\phi$$

where $\phi = \angle\mathbf{V}_2 - \angle\mathbf{I}_2 = 60^\circ - (-120^\circ) = 270^\circ$. Hence

$$P = (10)(0.433)(\cos 270^\circ) = 0 \text{ W}$$

## *Polyphase Circuits*

**9.16** The balanced set of 3-phase line-to-neutral voltages

$$v_{an} = V_m \cos \omega t$$

$$v_{bn} = V_m \cos(\omega t - 120^\circ)$$

$$v_{cn} = V_m \cos(\omega t - 240^\circ)$$

is applied to the $Y$-connected load of Fig. 9-31. Show that the instantaneous power delivered to the load is given by

$$p(t) = \tfrac{3}{2} V_m I_m \cos\phi$$

where $I_m = V_m/|Z(j\omega)|$ and $\phi = \angle Z(j\omega)$.

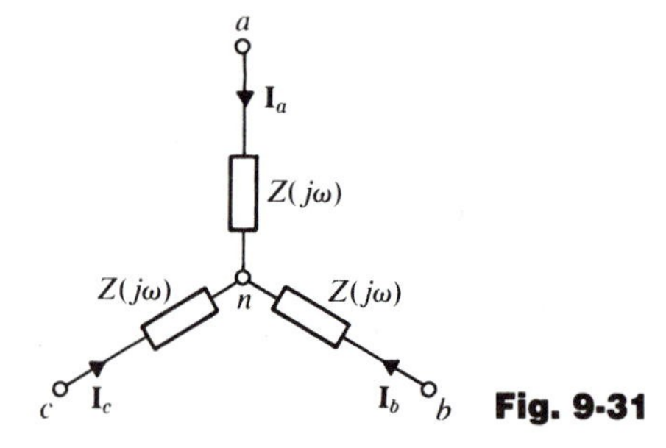


**Fig. 9-31**

For phase $a$,

$$p_a(t) = v_{an} i_a = \{V_m \cos \omega t\}\{I_m \cos(\omega t - \phi)\} = \frac{V_m I_m}{2}\{\cos\phi + \cos(2\omega t - \phi)\}$$

For phase $b$,

$$p_b(t) = v_{bn} i_b = \{V_m \cos(\omega t - 120^\circ)\}\{I_m \cos(\omega t - 120^\circ - \phi)\}$$

$$= \frac{V_m I_m}{2}\{\cos\phi + \cos(2\omega t - 240^\circ - \phi)\}$$

Similarly for phase $c$,

$$p_c(t) = \frac{V_m I_m}{2}\{\cos\phi + \cos(2\omega t - 120^\circ - \phi)\}$$

Adding the power in the three phases,

$$p(t) = p_a + p_b + p_c = \tfrac{3}{2}\, V_m I_m \cos \phi$$

since by phasor addition,

$$\cos (2\omega t - \phi) + \cos (2\omega t - \phi - 120°) + \cos (2\omega t - \phi - 240°) = 0$$

This demonstrates that the *instantaneous* power in a balanced, 3-phase system is *constant*—it does not vary with time.

**9.17** In the three-phase system of Fig. 9-32 the effective value of the balanced line-to-line voltage is 208 V and in the balanced Y-connected load $Z(j\omega) = 12e^{j30°}$.

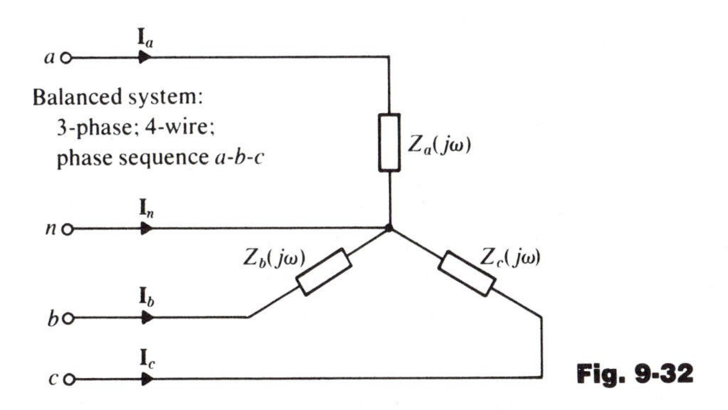


**Fig. 9-32**

(*a*) Draw a phasor diagram showing the line-to-neutral voltage phasors $\mathbf{V}_{an}$, $\mathbf{V}_{bn}$, and $\mathbf{V}_{cn}$.

(*b*) Find the line currents $\mathbf{I}_a$, $\mathbf{I}_b$, $\mathbf{I}_c$, and $\mathbf{I}_n$ and show these currents in a current phasor diagram.

(*c*) Calculate the total real power dissipated in this load.

(*a*) If the line-to-line voltage is 208 V, the line-to-neutral voltage will be $208/\sqrt{3} = 120$ V. Thus for the phase sequence *a-b-c*, the line-to-neutral voltages are as shown in Fig. 9-33*a*.

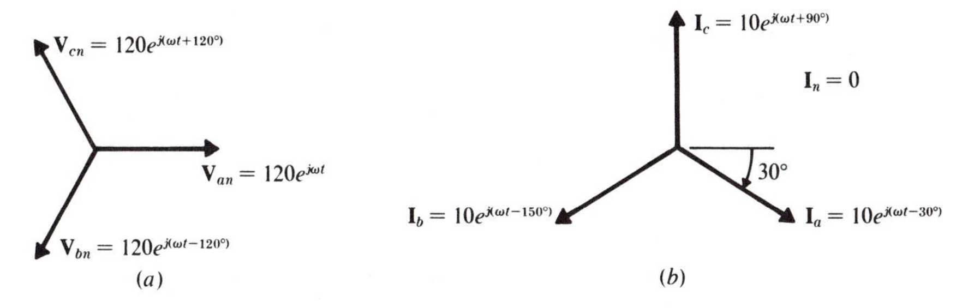


**Fig. 9-33**

(*b*) In terms of effective amplitudes,

$$\mathbf{I}_a = \frac{\mathbf{V}_{an}}{Z(j\omega)} = \frac{120e^{j\omega t}}{12e^{j30^\circ}} = 10e^{j(\omega t-30^\circ)}$$

$$\mathbf{I}_b = \frac{\mathbf{V}_{bn}}{Z(j\omega)} = \frac{120e^{j(\omega t-120^\circ)}}{12e^{j30^\circ}} = 10e^{j(\omega t-150^\circ)}$$

$$\mathbf{I}_c = \frac{\mathbf{V}_{cn}}{Z(j\omega)} = \frac{120e^{j(\omega t+120^\circ)}}{12e^{j30^\circ}} = 10e^{j(\omega t+90^\circ)}$$

These phasors are shown in Fig. 9-33*b*, where it can be seen that $\mathbf{I}_n = -(\mathbf{I}_a + \mathbf{I}_b + \mathbf{I}_c) = 0$.

*Comment:* When the load is balanced, the neutral wire carries no current. Therefore the same currents would have resulted if the supply were 3-wire.

(*c*) Using equation (*9.17*),

$$P_T = \sqrt{3}V_{L,\text{eff}}I_{L,\text{eff}} \cos \phi = \sqrt{3}(208)\ (10)\ (\cos 30^\circ) = 3120 \text{ W}$$

Alternatively, the power per phase is

$$P_\phi = \text{I}^2_{\text{eff}}\text{Re}[Z(j\omega)] = 10^2(10.4) = 1040 \text{ W} \quad \text{and} \quad P_T = 3P_\phi = 3120 \text{ W}$$

**9.18** In the balanced, $\Delta$-connected load of Fig. 9-34, $Z(j\omega) = 12e^{j30^\circ}$ and the line-to-line voltage is 208 V, RMS. Find the phasor line currents $\mathbf{I}_a$, $\mathbf{I}_b$, and $\mathbf{I}_c$ and calculate the total real power dissipated in the load.

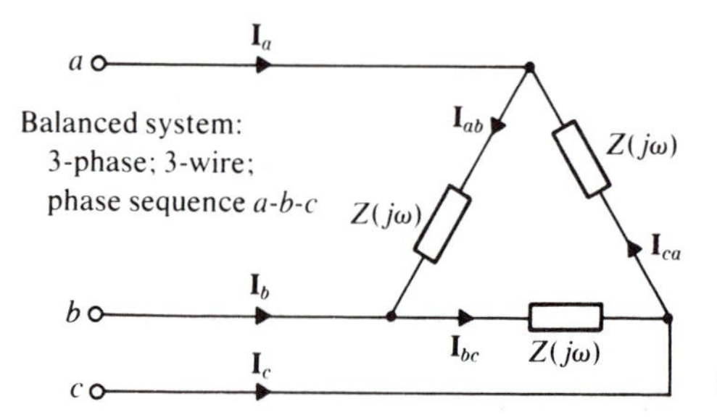


**Fig. 9-34**

One set of voltage phasors having a phase sequence *a-b-c* is shown in Fig. 9-35*a*. The set in Fig. 9-35*b* is equally correct, but not as symmetrical!

Now from KCL,

$$\mathbf{I}_a = \mathbf{I}_{ab} - \mathbf{I}_{ca}$$

where

$$\mathbf{I}_{ab} = \frac{\mathbf{V}_{ab}}{Z(j\omega)} = \frac{208e^{j\omega t}}{12e^{j30^\circ}} = 17.32e^{j(\omega t-30^\circ)}$$

and

$$\mathbf{I}_{ca} = 17.32e^{j(\omega t+90^\circ)}$$

Thus

$$\begin{aligned}\mathbf{I}_a &= 17.32e^{j(\omega t-30^\circ)} - 17.32e^{j(\omega t+90^\circ)} \\ &= (15.0 - j8.66 - j17.32)e^{j\omega t} = 30e^{j(\omega t-60^\circ)}\end{aligned}$$

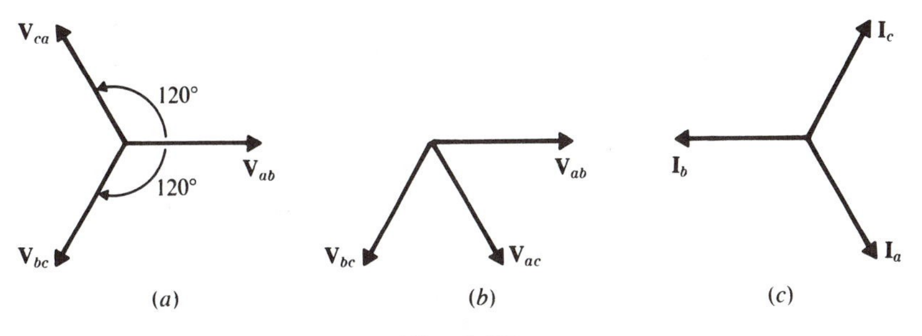


**Fig. 9-35**

As expected, the magnitude of the line current (30 A, RMS) is $\sqrt{3}$ times the magnitude of the phase current (17.32 A, RMS). Similarly,

$$\mathbf{I}_b = -\mathbf{I}_{ab} + \mathbf{I}_{bc} = 30e^{j(\omega t + 180°)}$$

and

$$\mathbf{I}_c = \mathbf{I}_{ca} - \mathbf{I}_{bc} = 30e^{j(\omega t + 60°)}$$

These three line current phasors are shown in Fig. 9-35*c*.

We can compute the total power from equation (*9.17*),

$$P_T = \sqrt{3}V_{L,\text{eff}}I_{L,\text{eff}} \cos\phi = \sqrt{3}(208)\,(30)\,(\cos 30°) = 9350 \text{ W}$$

Alternatively,

$$P_T = 3P_\phi = 3I^2_{\phi,\text{eff}}\text{Re}[Z(j\omega)] = 3(17.32)^2(10.4) = 9360 \text{ W}$$

*Comment:* By comparing the answers to Problems 9.17 and 9.18 we can see that if a load is switched from Y- to Δ-connection, then

1. the voltage across each $Z(j\omega)$ increases $\sqrt{3}$ times,
2. the current through each $Z(j\omega)$ increases $\sqrt{3}$ times,
3. the total power absorbed increases 3 times.

**9.19** Find the three line currents feeding the unbalanced Δ-connected load of Fig. 9-36, where $Z_1 = 24e^{j0°}$, $Z_2 = 24e^{j90°}$, and $Z_3 = 24e^{-j90°}$.

*Comment*: Here the current reference directions have been chosen as leaving the load. The choice is arbitrary and the reference directions need not even be the same for each of the three lines.

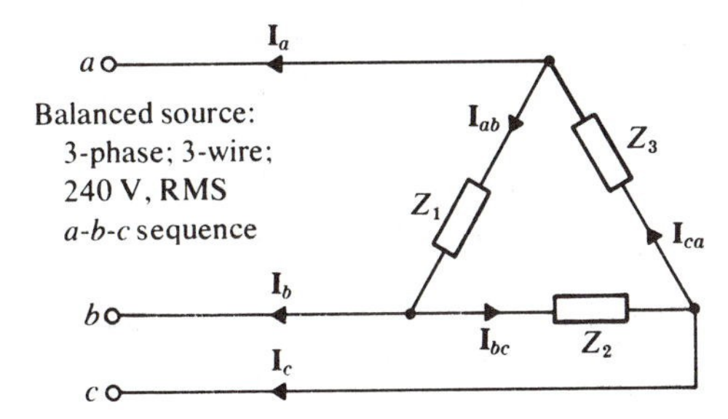


**Fig. 9-36**

A set of line voltage phasors with phase sequence *a-b-c* is shown in Fig. 9-37*a*. And

$$\mathbf{I}_a = \mathbf{I}_{ca} - \mathbf{I}_{ab} = \frac{\mathbf{V}_{ca}}{Z_3(j\omega)} - \frac{\mathbf{V}_{ab}}{Z_1(j\omega)} = \frac{240e^{j(\omega t+120°)}}{24e^{-j90°}} - \frac{240e^{j\omega t}}{24e^{j0°}}$$

$$= 10e^{j(\omega t+210°)} - 10e^{j\omega t} = (-8.66 - j5 - 10)e^{j\omega t} = 19.3e^{j(wt + 195°)}$$

Similarly,

$$\mathbf{I}_b = 19.3e^{j(\omega t-15°)} \quad \text{and} \quad \mathbf{I}_c = 10e^{j(\omega t+90°)}$$

These three line current phasors are shown in Fig. 9-37*b*, where it can be seen that $\mathbf{I}_a + \mathbf{I}_b + \mathbf{I}_c = 0$ as required by KCL.

The effective values of the line currents are $I_{a,\text{eff}} = 19.3$ A, $I_{b,\text{eff}} = 19.3$ A and $I_{c,\text{eff}} = 10$ A.

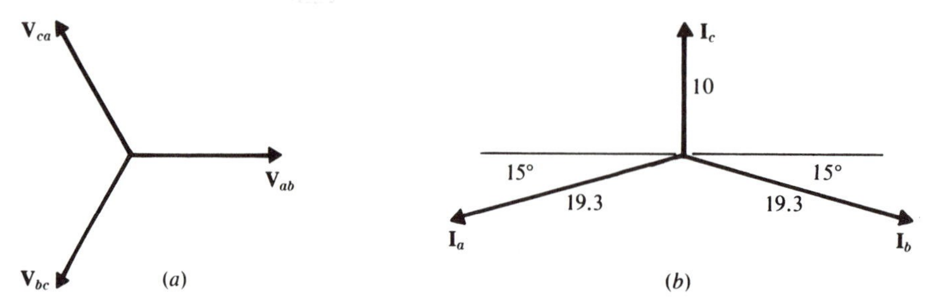


**Fig. 9-37**

**9.20** The two-wattmeter method, illustrated in Fig. 9-38, is to be used to determine the power drawn by a 3-phase, 440-V, 60-Hz induction motor, which may be regarded as a balanced load. The motor is developing 20 horsepower (1 Hp ≡ 746 W) at an efficiency of 74.6%. If the line current is 52.5 A, RMS, find the reading of each wattmeter.

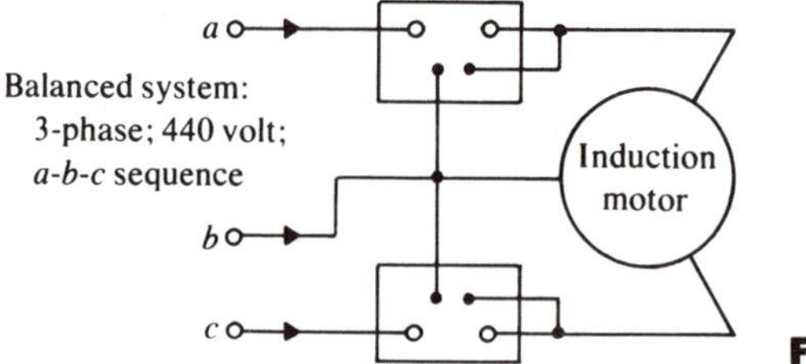


**Fig. 9-38**

From equation (*9.1*), the two wattmeters will read

$$W_a = V_{ab,\text{eff}} I_{a,\text{eff}} \cos\alpha \quad \text{and} \quad W_c = V_{cb,\text{eff}} I_{c,\text{eff}} \cos\beta$$

where $\alpha$ is the angle of $\mathbf{V}_{ab}$ with respect to $\mathbf{I}_a$ and $\beta$ is the angle of $\mathbf{V}_{cb}$ with respect to $\mathbf{I}_c$.

Now to ascertain the values of the angles $\alpha$ and $\beta$, we must determine the power factor of the motor load. The electrical power drawn by the motor is given by

$$P_{in} = \frac{P_{out}}{\text{efficiency}} = \frac{(20\text{ Hp})(746\text{ W/Hp})}{0.746} = 20\text{ kW}$$

Using equation (*9.17*), namely $P_T = \sqrt{3}\,V_{L,\text{eff}} I_{L,\text{eff}} \cos\phi$, it follows that

$$\cos\phi = \frac{20 \times 10^3}{\sqrt{3}(440)(52.5)} = 0.50 \quad \text{or} \quad \phi = +60°$$

($\phi$ will be positive since an induction motor is an inductive load.)

Now assuming the motor to be Y-connected (we could equally well assume Δ-connection), we can draw the phasor diagram of Fig. 9-39*a* where $\mathbf{V}_a$, $\mathbf{V}_b$, and $\mathbf{V}_c$ are the line-to-neutral voltages. From Fig. 9-39*b* the line-to-line voltage $\mathbf{V}_{ab}$ is

$$\mathbf{V}_{ab} = \mathbf{V}_a - \mathbf{V}_b = 440e^{j(\omega t + 30°)}$$

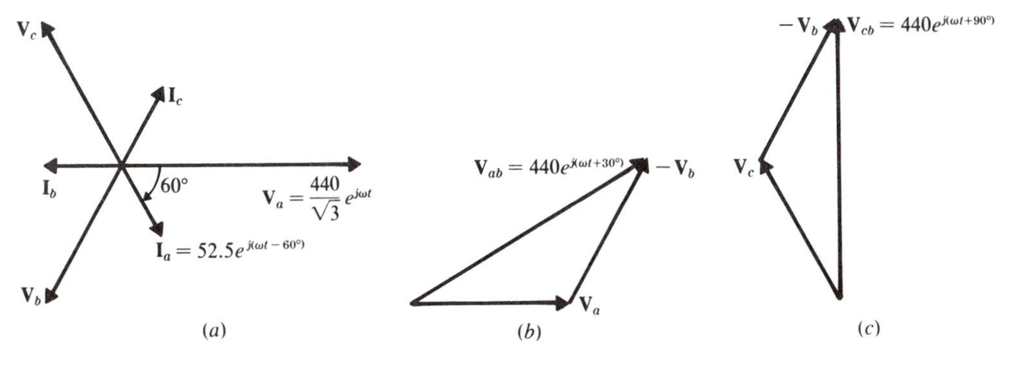


**Fig. 9-39**

Therefore $\alpha = \angle\mathbf{V}_{ab} - \angle\mathbf{I}_a = 30° - (-60°) = 90°$ and

$$W_a = (440)(52.5)(\cos 90°) = 0\text{ W}$$

From Fig. 9-39*c*,

$$\mathbf{V}_{cb} = \mathbf{V}_c - \mathbf{V}_b = 440e^{j(\omega t + 90°)}$$

making $\beta = \angle\mathbf{V}_{cb} - \angle\mathbf{I}_c = 90° - 60° = 30°$ and

$$W_c = (440)(52.5)(\cos 30°) = 20\text{ kW}$$

Note that the *sum* of the *two* wattmeter readings does equal the total 3-phase power of 20 kW.

**9.21** An industrial load draws 1000 kW with a power factor of 0.707 inductive from a balanced 3-phase, 4-kV source as shown in Fig. 9-40*a*. A capacitor bank is to be connected in parallel with the original load (see Fig. 9-40*b*) for the purpose of reducing the current drawn from the source to 160 A. How much reactive power must the capacitor bank draw? Compute the value of the capacitors that would be needed if they were connected in delta. What would be their value if they were connected in wye?

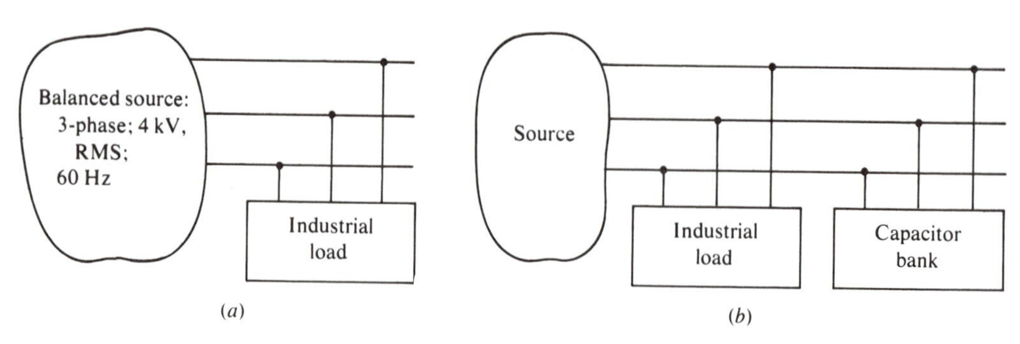


**Fig. 9-40**

From equation (*9.17*),

$$10^6 = \sqrt{3}(4000)(0.707)I_L \quad \text{or} \quad I_L = 204 \text{ A, RMS}$$

Applying the same equation to the load-capacitor combination, for which the real power is still to be 1000 kW at a new line current of 160 A, RMS,

$$10^6 = \sqrt{3}(4000)(160) \cos \phi \quad \text{or} \quad \cos \phi = 0.902$$

which will be the power factor of the combination.

Assuming that the power factor is not to be *over*corrected, we may draw the power triangles of Fig. 9-41, where the two angles are given by $\cos^{-1}(0.707) = 45°$ and $\cos^{-1}(0.902) = 25.5°$. After correction, the reactive power is

$$Q_T = P_T \tan \phi_T = 1000 \tan 25.5° = 476 \text{ kVAR}$$

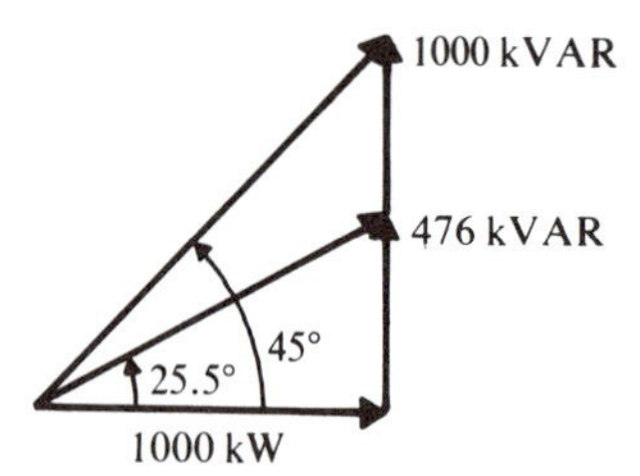


**Fig. 9-41**

The capacitor bank will have to draw 476 − 1000 = −524 kVAR of reactive power.

If we consider one phase of the system, *each* capacitor must draw −524/3 = −175 kVAR. So, considering both Y- and Δ-connected capacitors in relation to equation (*9.3*),

*Δ-Connection*

$$I_C = \frac{Q_C}{V_C \sin(-90°)} = \frac{-175 \times 10^3}{(4 \times 10^3)(-1)} = 43.8 \text{ A, RMS}$$

*Y-Connection*

$$I_C = \frac{Q_C}{V_C \sin(-90°)} = \frac{-175 \times 10^3}{(4 \times 10^3/\sqrt{3})(-1)} = 76 \text{ A, RMS}$$

Hence from $(1/j\omega C)\mathbf{I}_C = \mathbf{V}_C$,

$$\left(\frac{1}{377C}\right)43.8 = 4 \times 10^3 \qquad \left(\frac{1}{377C}\right)76 = \frac{4}{\sqrt{3}} \times 10^3$$

$$C = 29\,\mu\text{F} \qquad C = 87.2\,\mu\text{F}$$

**9.22** An *unbalanced*, Y-connected load is connected to a balanced, 3-phase, 240-V source as shown in Fig. 9-42. Find the magnitudes of the three line currents. *Hint:* Use the method of loop analysis.

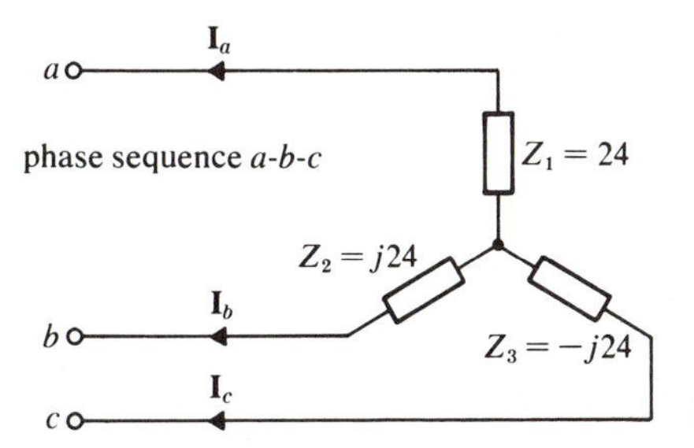


**Fig. 9-42**

It probably helps to visualize the loops if a 3-phase source is shown as in Fig. 9-43$a$. Then following the procedure developed in Chapter 7, we can write the loop equations by inspection.

$$\begin{bmatrix} Z_1 + Z_2 & -Z_2 \\ -Z_2 & Z_2 + Z_3 \end{bmatrix} \begin{bmatrix} \mathbf{I}_x \\ \mathbf{I}_y \end{bmatrix} = \begin{bmatrix} \mathbf{V}_b - \mathbf{V}_a \\ \mathbf{V}_c - \mathbf{V}_b \end{bmatrix} = \begin{bmatrix} \mathbf{V}_{ba} \\ \mathbf{V}_{cb} \end{bmatrix}$$

Using the voltages shown in the phasor diagram of Fig. 9-43$b$, and substituting the values of the load impedances,

$$\begin{bmatrix} 24 + j24 & -j24 \\ -j24 & 0 \end{bmatrix} \begin{bmatrix} \mathbf{I}_x \\ \mathbf{I}_y \end{bmatrix} = \begin{bmatrix} 240e^{j\omega t} \\ 240e^{j(\omega t - 120°)} \end{bmatrix}$$

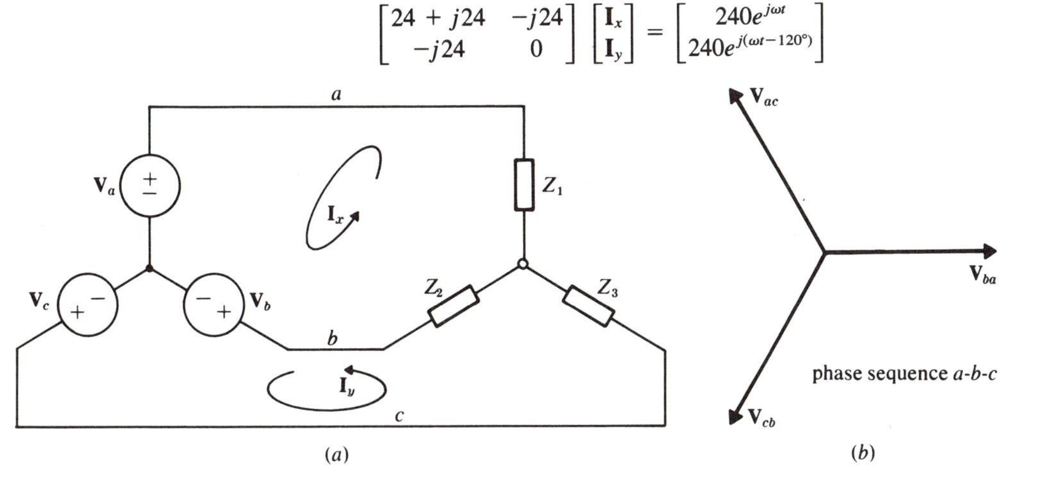


**Fig. 9-43**

This is easily solved for $\mathbf{I}_x$,

$$\mathbf{I}_x = \frac{240e^{j(\omega t - 120°)}}{-j24} = 10e^{j(\omega t - 30°)} = (8.66 - j5)e^{j\omega t}$$

Then substituting into the first equation,

$$\{24 + j24\}\{10e^{j(\omega t - 30°)}\} - j24\mathbf{I}_y = 240e^{j\omega t}$$

from which

$$\mathbf{I}_y = 5.2e^{j(\omega t - 45°)} = (3.66 - j3.66)e^{j\omega t}$$

Finally, from Figs. 9-42 and 9-43*a*,

$$\mathbf{I}_a = \mathbf{I}_x = 10e^{j(\omega t - 30°)}$$

$$\mathbf{I}_b = \mathbf{I}_y - \mathbf{I}_x = (3.66 - j3.66 - 8.66 + j5)e^{j\omega t} = (-5 + j1.34)e^{j\omega t}$$

$$= 5.2e^{j(\omega t + 165°)}$$

$$\mathbf{I}_c = -\mathbf{I}_y = 5.2e^{j(\omega t + 135°)}$$

Thus the three line currents have effective values of 10 A, 5.2 A, and 5.2 A.

*Comment:* The analysis of this unbalanced, Y-connected circuit is rather messier than that for an unbalanced delta. In Chapter 10 we will solve this problem by converting the Y-connected load into an equivalent delta configuration.

**9.23** Three ideal transformers are connected in the wye-delta arrangement of Fig. 9-44 and are supplying a balanced, 3-phase load of 100 kVA. If the line-to-line primary voltage (lines *a*, *b*, *c*) is 4150 V, RMS and the transformation ratio *n* is 10, find the effective values of the primary and the secondary voltages and currents for each transformer.

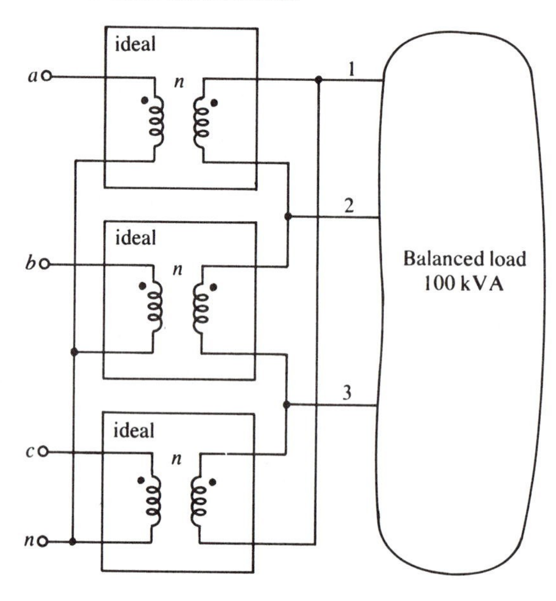


**Fig. 9-44**

For the wye-connected primary,

$$V_\phi = \frac{V_L}{\sqrt{3}} = \frac{4150}{\sqrt{3}} = 2400 \text{ V}$$

which is the primary voltage of each transformer. With a transformation ratio of $n = 10$, the secondary voltage of each transformer is

$$V_{\text{sec}} = \frac{V_{\text{pri}}}{10} = 240 \text{ V}$$

Now the line current of the secondary may be obtained from equation (*9.18*),

$$S_T = 3V_L I_L$$

or

$$I_L = \frac{100 \times 10^3}{(\sqrt{3})(240)} = 241 \text{ A}$$

But the transformer secondary is $\Delta$-connected, so that

$$I_{\text{sec}} = \frac{241}{\sqrt{3}} = 139 \text{ A}$$

But $n = 10$ and therefore

$$I_{\text{pri}} = \frac{I_{\text{sec}}}{10} = 13.9 \text{ A}$$

The apparent power supplied by each transformer may be calculated from

$$V_{\text{pri}} I_{\text{pri}} = (2400)(13.9) = 33.3 \text{ kVA}$$

or alternatively from

$$V_{\text{sec}} I_{\text{sec}} = (240)(139) = 33.3 \text{ kVA}$$

Thus each transformer supplies one-third of the apparent power.

# PROBLEMS

**9.24** For the linear network of Fig. 9-45, in the sinusoidal steady-state,

$$i(t) = 20\sqrt{2} \cos (100t + 30°)$$
$$P = 2400 \text{ W}$$
$$Q = 3200 \text{ VAR}$$

Find the effective value of the source voltage $v(t)$.

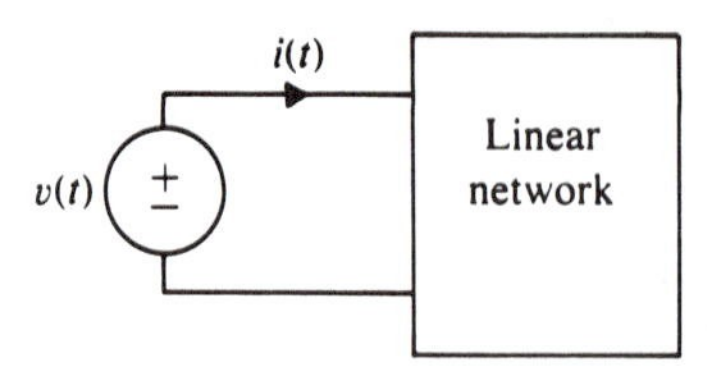


Fig. 9-45

**9.25** For the circuit of Problem 9.24, find the value of the capacitance or inductance (as the case may be) which must be placed in *series* with the network so that the reactive power will become zero.

**9.26** Assume that the series element found in Problem 9.25 has been added to the circuit of Fig. 9-45. Calculate the new value of $P$.

**9.27** The plant in the system of Fig. 9-46 draws 250 kW at an inductive power factor of 0.707 from the single-phase, 2300-V, 60-Hz supply. A capacitor is to be placed in parallel with the plant to change the overall power factor to 0.866 capacitive. How much reactive power must this capacitor provide?

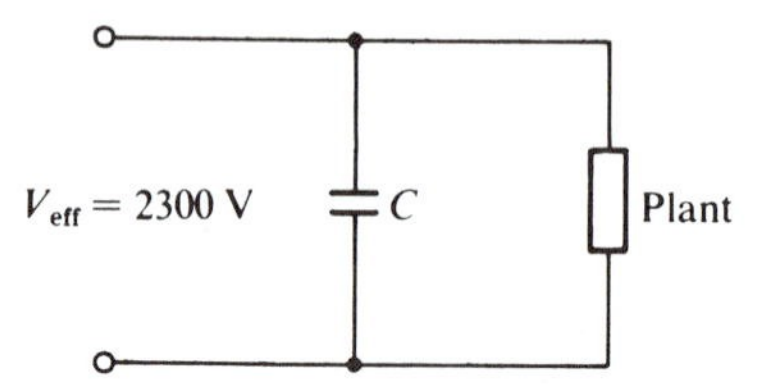


Fig. 9-46

**9.28** In Fig. 9-47,

$$v(t) = 200 \cos (866t - 10°)$$

$$i(t) = 2.0 \sin (866t + 20°)$$

(*a*) Replace the network by *two* passive elements *in series*.
(*b*) Find the active and reactive power supplied to the network.

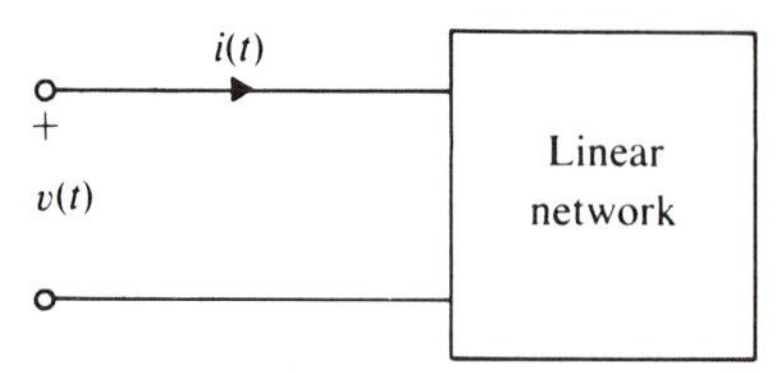


Fig. 9-47

**9.29** Determine the element which when placed across the terminals of the network of Problem 9.28 will make the power factor of the combination equal to unity.

**9.30** The circuit of Fig. 9-48 contains an ideal transformer for which $\mathbf{V}_1/\mathbf{V}_2 = \frac{1}{2}$. Find $v_{sss}(t)$ if $i_1(t) = 10 \sin t$.

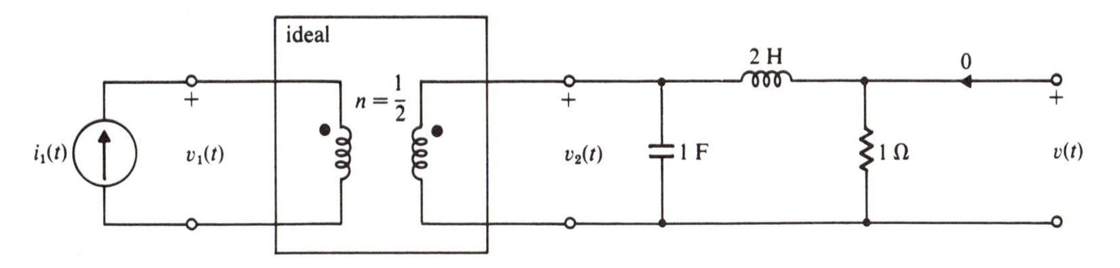


Fig. 9-48

**9.31** Show that the sinusoidal input admittance of the circuit in Fig. 9-49 is

$$\frac{\mathbf{I}}{\mathbf{V}} = Y_{\text{in}}(j\omega) = \frac{j\omega}{-\omega^2 + n^2/2 + 2j\omega}$$

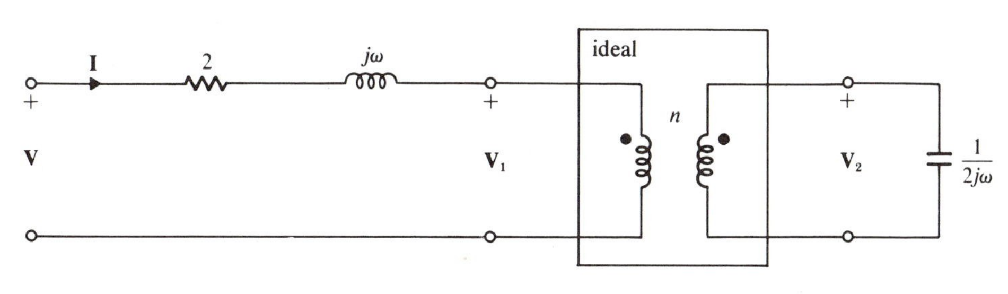


Fig. 9-49

**9.32** Show that the transformerless circuit of Fig. 9-50 is equivalent to the transformer circuit of Fig. 9-4.

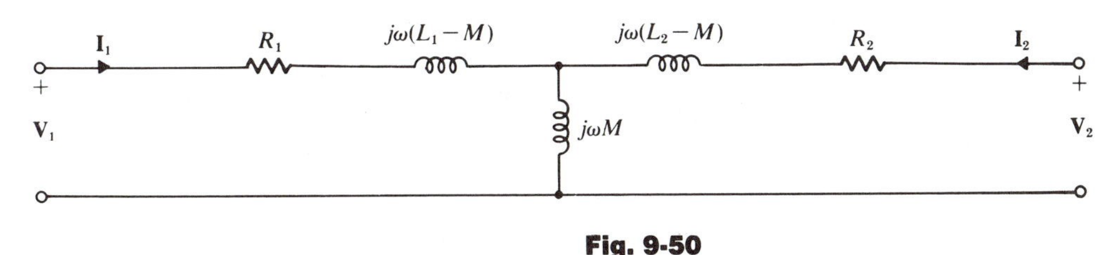


Fig. 9-50

**9.33** Find the total real power and the total reactive power delivered to the balanced, wye-connected load of Fig. 9-51.

**9.34** The same impedances ($Z(j\omega) = 10e^{-j53.2^\circ}$ in Fig. 9-51) are reconnected in delta to the same 208-V, 3-phase source. Find the line current and the total real power.

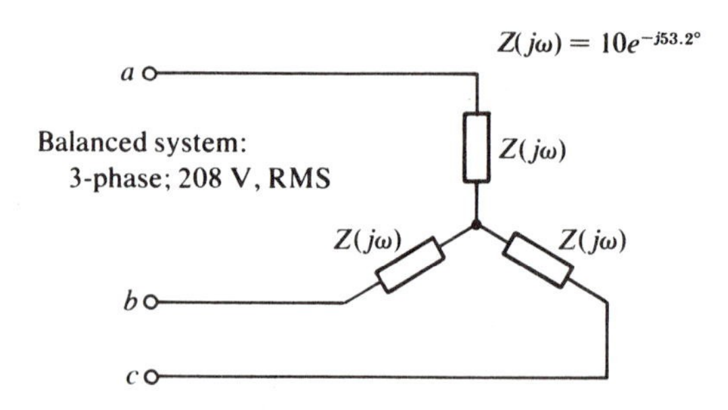


**Fig. 9-51**

**9.35** Three capacitors, connected as in Fig. 9.52*a*, are used to make the system power factor equal to unity. If the connection of Fig. 9-52*b* is to effect the same change in the overall power factor, how must $C$ and $C'$ be related?

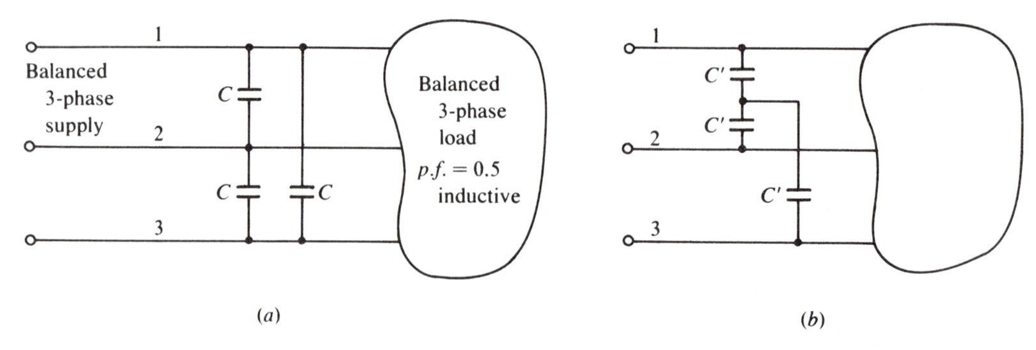


**Fig. 9-52**

**9.36** A resistance heating oven is to be designed to operate from a balanced, three-phase, 240-V line. It is to draw 10 kW when delta-connected. Find the value of the resistor required in each leg of the delta.

**9.37** In Fig. 9-53 the transformation ratio is 10 and the secondary winding is center-tapped. Find the magnitude of the current in the neutral line.

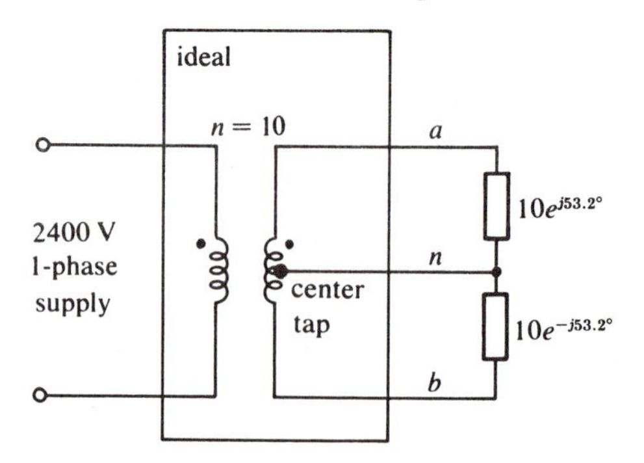


**Fig. 9-53**

**9.38** Find the effective value of the neutral current in the system of Fig. 9-54 if the phase sequence is *a-b-c*.

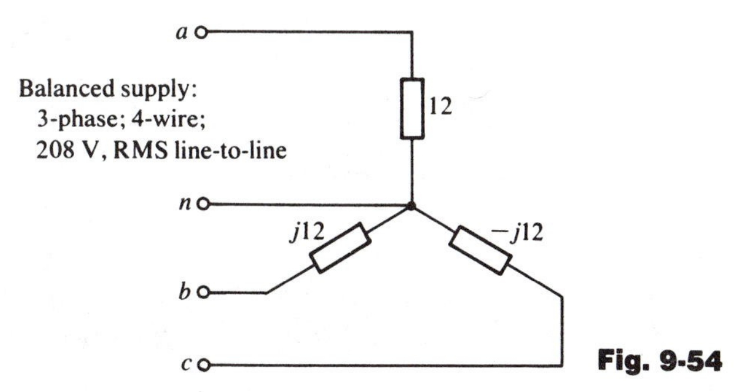


**Fig. 9-54**

**9.39** Repeat Problem 9.38 for the phase sequence *a-c-b*.

**9.40** Calculate the power read by each of the two wattmeters in Fig. 9-55 and show that this is equal to the total real power dissipated in the load.

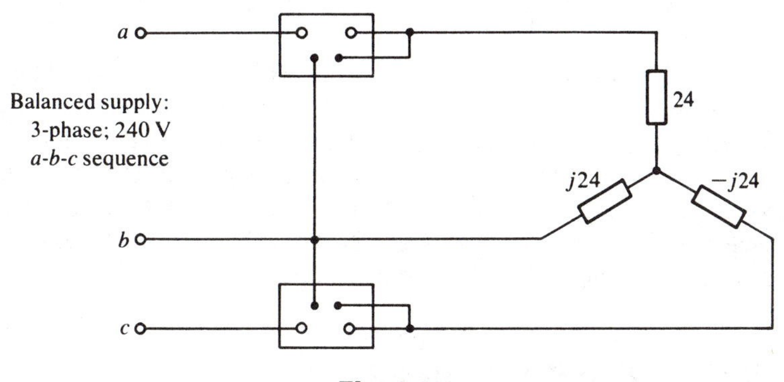


**Fig. 9-55**

## Chapter 10

# NETWORK THEOREMS AND NETWORK GRAPHS

Here some of the more theoretical aspects of circuit analysis will be introduced. First, the network theorems: these can often simplify the solution of a circuit problem by reducing a circuit to a simpler equivalent. They are also the basis for the development of some of the more advanced and/or more theoretical features of circuit analysis and synthesis. The latter is also true of the abstract topological approach to network analysis which is based on graph theory and which is the basis of modern digital computer methods of analysis and synthesis.

### THE NETWORK THEOREMS

**Definition 10.1:** **A Linear Element** An element (resistor, inductor, capacitor, transformer, or dependent source) is linear if its parameter ($R$, $L$, $C$, etc.) is either a constant or a function of time.

**Definition 10.2:** **A Linear, Time-Invariant Element** An element whose parameter is a constant is linear and time-invariant.

**Definition 10.3:** **A Linear Circuit** A circuit is linear if every element is either linear or an independent source.

**Definition 10.4:** **A Linear, Time-Invariant Circuit** A circuit in which every element is either linear and time-invariant or an independent source is linear and time-invariant.

**Definition 10.5:** **A Passive Circuit** A circuit containing neither dependent nor independent sources. Zero initial conditions are generally implied, also.

**Theorem 10.1: Homogeneity** Suppose an input $u(t)$ to a linear circuit causes the output $y(t)$. If the input is changed to $ku(t)$, then the output will become $ky(t)$.

**Theorem 10.2: Additivity** If an input $u_1(t)$ to a linear circuit causes an output $y_1(t)$, and an input $u_2(t)$ causes an output $y_2(t)$, then if $u_1(t)$ and $u_2(t)$ are applied simultaneously, the resulting output will be $y_1(t) + y_2(t)$.

The additivity or *superposition* theorem permits us to find the response due to each of several inputs and/or initial conditions *separately,* and to then add or *superimpose* the several responses to find the total response. (A circuit's nonzero initial conditions may be replaced by equivalent sources, as in Chapter 5, to facilitate the application of this theorem.)

## Example 10.1

In the circuit of Fig. 10-1, $f(t) = A\cos(2t + \alpha)$ and $e(t) = B\cos(4t + \beta)$. Find $i_{sss}(t)$.

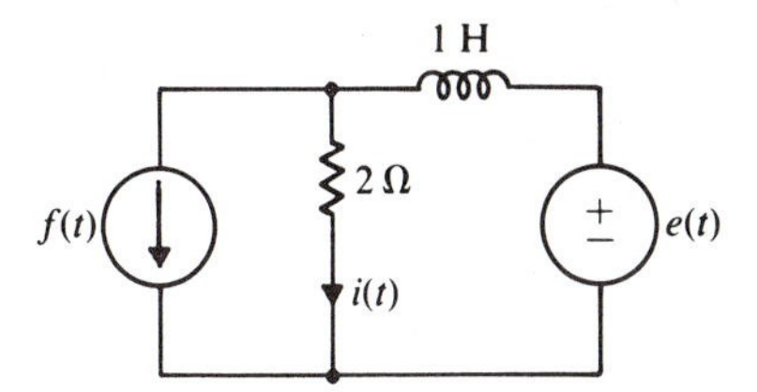


**Fig. 10-1**

We cannot use the methods of Chapter 7 *directly* since the inputs are of two different frequencies. However, we *can* use these methods to find the ac current through the resistor due to each input *separately*. Then we can apply Theorem 10.2.

From the *s*-domain diagram of Fig. 10-2*a*,

$$H_1(s) = \frac{I_1(s)}{F(s)} = \frac{-1/2}{1/2 + 1/s} = \frac{-s}{s + 2}$$

so for $\omega = 2$ rad/s,

$$H_1(j2) = \frac{-j2}{j2 + 2} = \frac{1}{\sqrt{2}}e^{-j135°}$$

Therefore from equation (*7.2*), page 216,

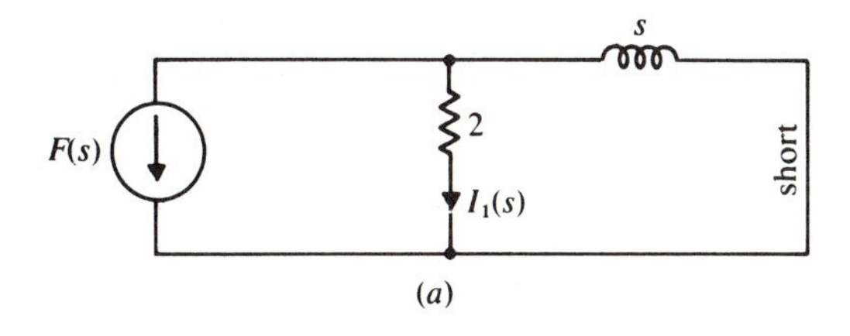


(*a*)

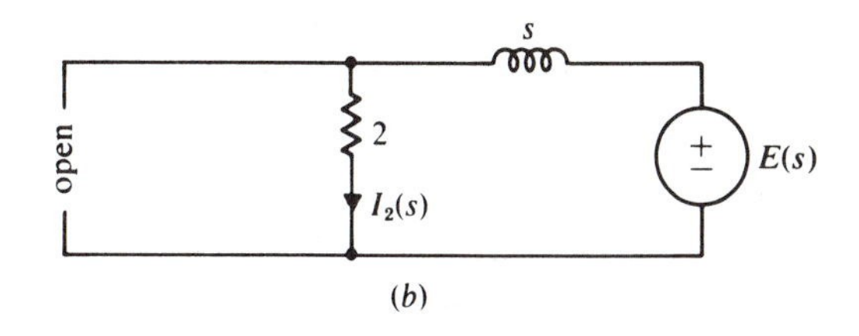


(*b*)

**Fig. 10-2**

$$i_{1,\text{sss}}(t) = \frac{A}{\sqrt{2}} \cos (2t + \alpha - 135°)$$

Similarly, from Fig. 10-2*b*,

$$H_2(s) = \frac{I_2(s)}{E(s)} = \frac{1}{s + 2} \quad \text{and} \quad H_2(j4) = \frac{1}{\sqrt{20}} e^{-j63.4°}$$

Thus from equation (7.2),

$$i_{2,\text{sss}}(t) = \frac{B}{\sqrt{20}} \cos (4t + \beta - 63.4°)$$

Finally, applying the superposition or additivity theorem,

$$i_{\text{sss}}(t) = \frac{A}{\sqrt{2}} \cos (2t + \alpha - 135°) + \frac{B}{\sqrt{20}} \cos (4t + \beta - 63.4°)$$ ■

**Theorem 10.3: Reciprocity** Any passive, linear, two-port network with zero initial conditions is said to be *reciprocal,* and the following results apply:

1. $v_2 = v_1'$ for all $t$, in the circuit of Fig. 10-3*a*,

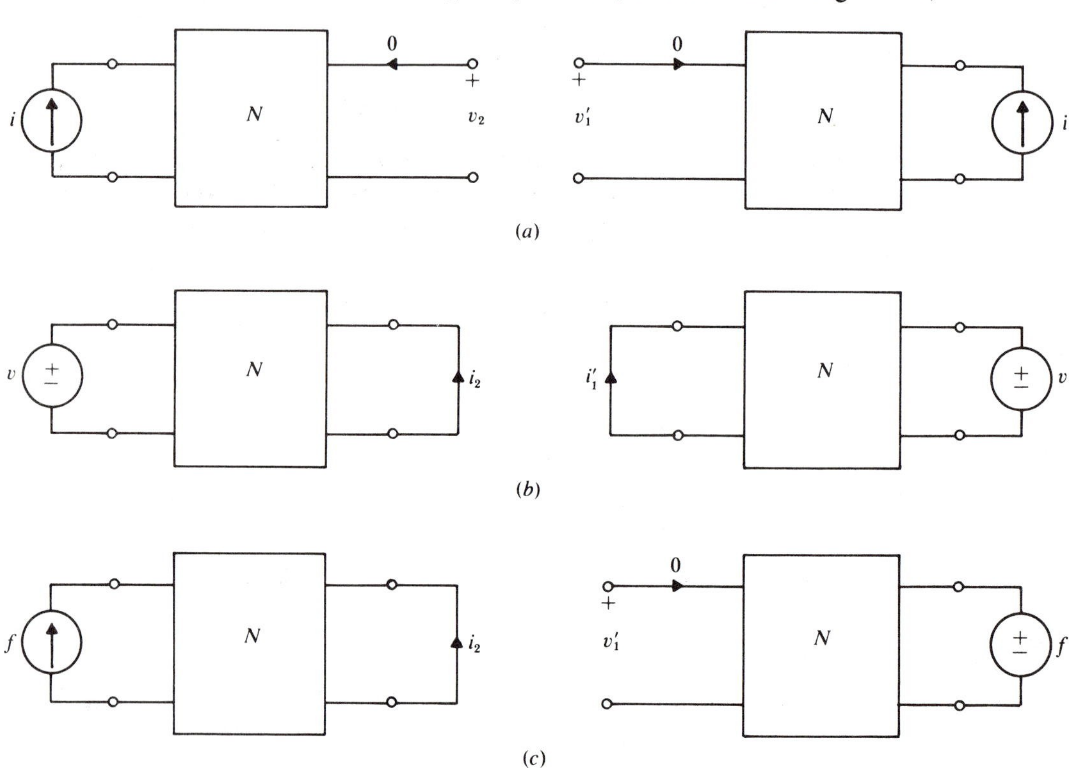


**Fig. 10-3**

2. $i_2 = i_1'$ for all $t$, in the circuit of Fig. 10-3*b*, and
3. $i_2 = -v_1'$ for all $t$, in the circuit of Fig. 10-3*c*.

An alternative statement of the reciprocity theorem will be derived in Problem 10.5, and a more general statement will be deduced as a consequence of Tellegen's theorem in Problem 10.22.

### *Preliminary to Thévenin's and Norton's Theorems*

You are already familiar with the Thévenin equivalent one port (see Chapters 2, 6, and 7), and you have briefly met the Norton equivalent (see Problem 6.28). Here we will be slightly more formal, restating the concept—the ultimate reduction of a one port—in the form of two theorems.

The *open-circuit voltage* $v_{oc}$ and the *short-circuit current* $i_{sc}$ due to the sources and initial conditions in $N$ are defined in Figs. 10-4*b* and *c*. And the *Thévenin network* $N_{Th}$ is defined to be the network $N$ with all *independent* sources and initial conditions set to zero (see Fig. 10-4*d*).

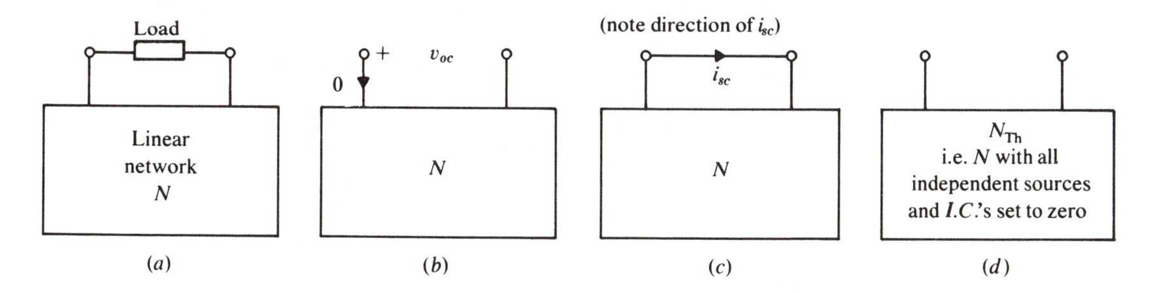


**Fig. 10-4**

**Theorem 10.4: Thévenin's Theorem** The theorem states that, *as far as the load is concerned, N* may be replaced by $N_{Th}$ and $v_{oc}$ in series, as shown in Fig. 10-5.

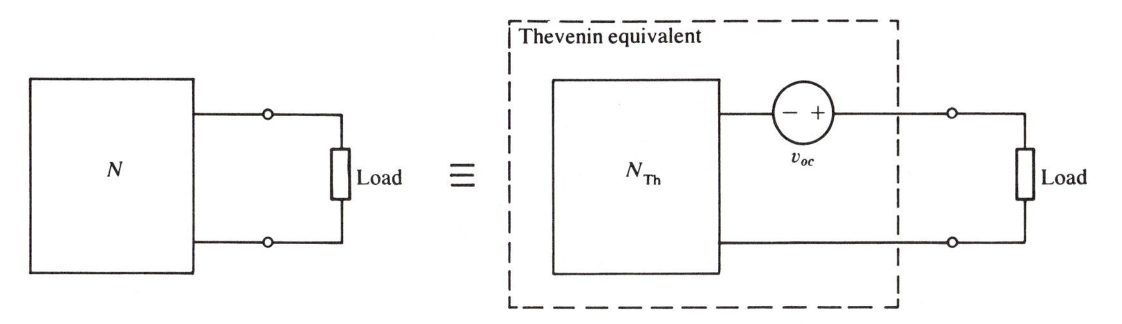


**Fig. 10-5**

**Theorem 10.5: Norton's Theorem** This theorem states that, *as far as the load is concerned,* $N$ may be replaced by $N_{\text{Th}}$ and $i_{sc}$ in parallel, as shown in Fig. 10-6.

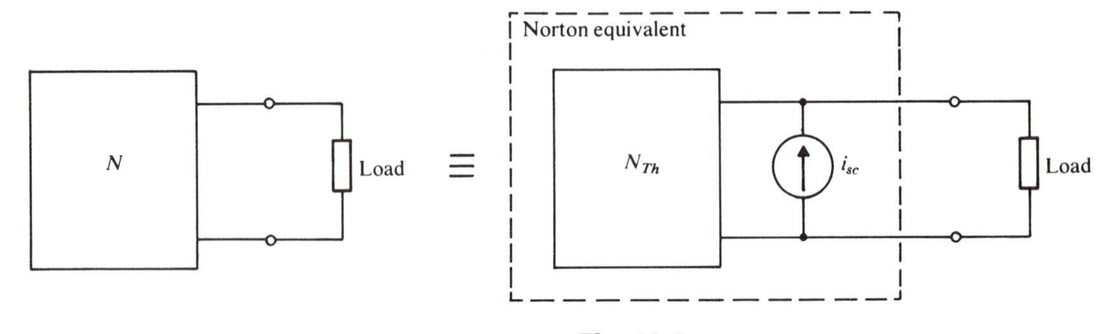


**Fig. 10-6**

*Postscript to Thévenin and Norton*

$v_{oc}$ (also written $v_{\text{Th}}$), $i_{sc}$ (also written $i_N$), and the resistance of $N_{\text{Th}}$ (written $R_{\text{Th}}$) are found by any of the methods with which you are already familiar.

If the network $N$ is not resistive, as has so far been implied, then the theorems must be restated in $s$- or $j\omega$-domain terminology, whichever may be appropriate, and the calculations performed in that domain.

The theorems of Thévenin and Norton establish a state of mind as much as or more than a method of calculation. Thus any linear circuit, *no matter how complicated,* will always be seen *by the load* as a single impedance in series with a single voltage source *or* as a single impedance in parallel with a current source.

**Theorem 10.6: Maximum Power** If a given phasor-domain source $\mathbf{V}_s$ of impedance $Z_s(j\omega)$ is to deliver the maximum possible average active power into a freely adjustable load $Z_L(j\omega)$, then we must set $Z_L(j\omega) = Z_s^*(j\omega)$, where $Z_s^*(j\omega)$ is the complex conjugate of $Z_s(j\omega)$.

This is the most common of several maximum power theorems. The others deal with situations where parameters other than $Z_L(j\omega)$ are adjustable, or where there are constraints on the adjustment of $Z_L(j\omega)$.

**Theorem 10.7: Tellegen's Theorem** If a set of voltages $v_1', v_2', \ldots$ satisfy KVL around all loops in a network, and if a set of currents $i_1'', i_2'', \ldots$ satisfy KCL at all nodes, then

$$\sum_{\substack{\text{all} \\ \text{branches}}} v_k' i_k'' = 0$$

where $v'_k$ and $i''_k$ refer to the voltage across and current through the $k$th branch. However, the $v'_k$ and $i''_k$ need not refer to the values at the same instant of time, nor need the branches be the same at the times that the $v'_k$ and $i''_k$ are "measured." The *only* requirements are that the $v'_k$ satisfy KVL, that the $i''_k$ satisfy KCL, and that the associated sign convention be adopted.*

## Example 10.2

In the circuit of Fig. 10-7$a$, at some time $t'$, $v'_1 = 2$ V, $v'_2 = 3$ V, $v'_3 = 1$ V, and $v'_4 = -1$ V. And in the circuit of Fig. 10-7$b$, at some time $t''$, $i''_1 = -2$ A, $i''_2 = 2$ A, $i''_3 = 5$ A, and $i''_4 = 7$ A. Show that the $v'_k$ satisfy KVL, the $i''_k$ satisfy KCL, and $\Sigma\, v'_k i''_k = 0$.

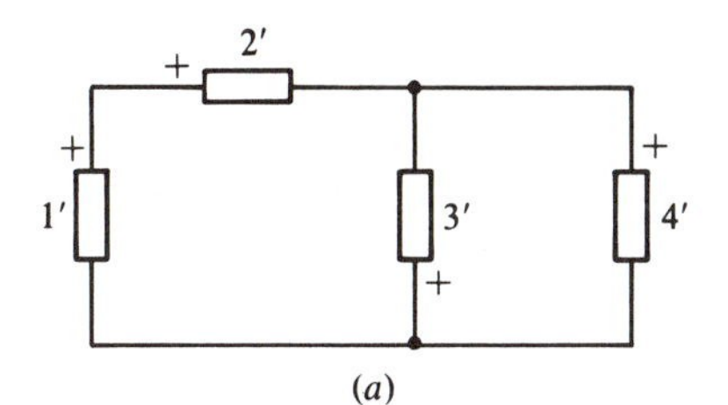

(a)

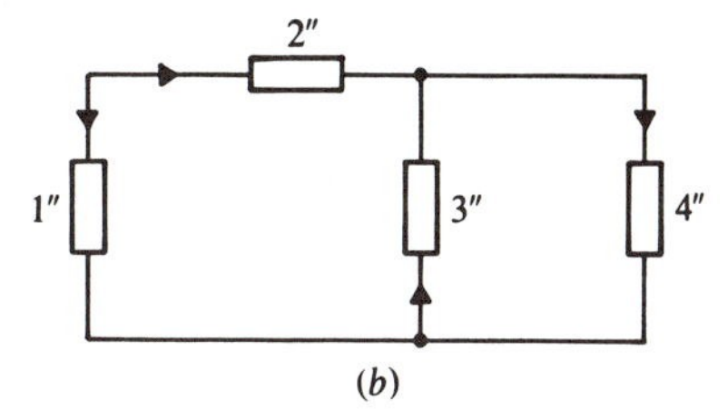

(b)

**Fig. 10-7**

From Fig. 10-7$a$ and KVL,

$$-v'_1 + v'_2 - v'_3 = -2 + 3 - 1 = 0$$

$$v'_3 + v'_4 = 1 - 1 = 0$$

which demonstrates that the $v'_k$ satisfy KVL.

Next, from Fig. 10-7$b$ and KCL,

$$i''_1 + i''_2 = -2 + 2 = 0$$

$$i''_2 + i''_3 - i''_4 = 2 + 5 - 7 = 0$$

which shows that the $i''_k$ satisfy KCL.

Now we can test Tellegen's theorem. Thus

$$\sum v'_k i''_k = (2)(-2) + (3)(2) + (1)(5) + (-1)(7) = 0$$ ■

Note that we have required only that the associated sign convention be adopted (compare Figs. 10-7$a$ and $b$) and that KVL and KCL be satisfied. *Nothing* has been said about the nature of the circuit branches. There is no reason, for example, why the branch 1′ should not be a voltage source in series with a resistor, while 1″ is a diode. This is a *very* general theorem, and it applies in the $s$- and $j\omega$-domains as well.

* This general and powerful theorem is stated somewhat more generally, is proved, and is widely applied in *Tellegen's Theorem and Electrical Networks,* Penfield, Spence and Duinker, Research Monograph No. 58 (Cambridge, Mass.: M.I.T. Press, 1970).

**Theorem 10.8: Substitution** Any branch of a network with a branch voltage $v_k$ and branch current $i_k$ may be replaced by a substitute branch provided it also has a voltage $v_k$ when carrying a current $i_k$.

**Theorem 10.9: Compensation** If the impedance of a network branch carrying a current **I** is changed from $Z(j\omega)$ to $Z + \Delta Z$, then the corresponding change in any current or voltage in the (linear) network will be equal to that due to a *compensation* voltage source $\Delta Z\mathbf{I}$ placed in series with $Z + \Delta Z$ and *opposing* the current **I**. The change $\Delta Z$ need not be small.

This theorem can be restated for application in the $s$-domain.

**Theorem 10.10: The Y-Δ Transformation** The Δ and the Y configurations of Fig. 10-8*a* (or the tee-pi configurations of Fig. 10-8*b*) are equivalent, as far as the circuit outside their terminals is concerned, provided

$$Z_a = \frac{Z_{ab}Z_{ca}}{Z_{ab} + Z_{bc} + Z_{ca}}$$

or

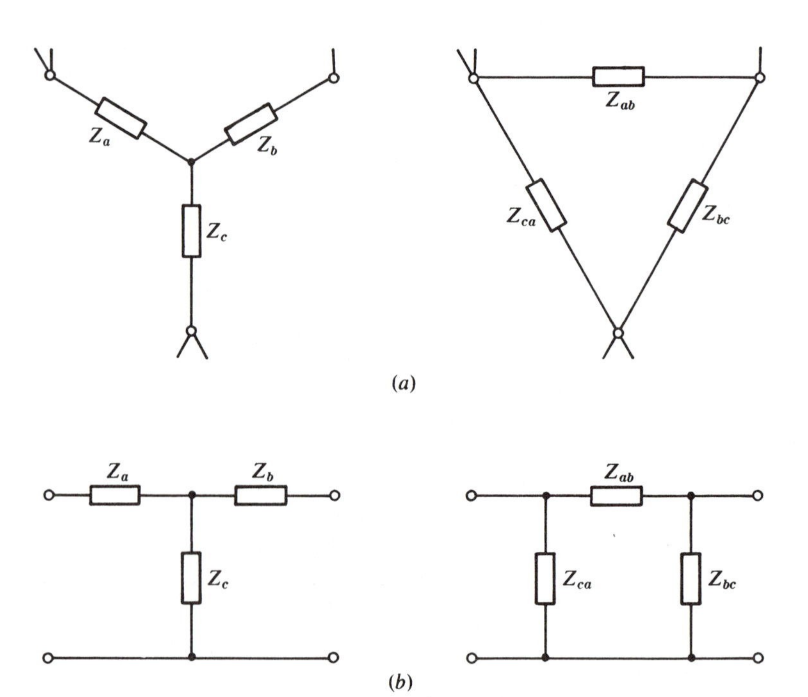


**Fig. 10-8**

$$Z_{ab} = \frac{Z_a Z_b + Z_b Z_c + Z_c Z_a}{Z_c}$$

and similarly for $Z_b$, $Z_c$, $Z_{bc}$, and $Z_{ca}$.

These transformations are useful when we wish to eliminate a node (Y-Δ or T-Π) or a mesh (Δ-Y or Π-T). The Y-Δ transformation is also directly useful in the analysis of many problems in 3-phase power systems.

**Theorem 10.11: Scaling or Normalization** To avoid unwieldy powers of 10 it is often convenient to *scale* or *normalize* our calculations. Scale factors $k$ and $K$ are defined by

$$Z(j\omega) = kZ'(j\omega)$$

$$\omega = K\omega'$$

where $Z'(j\omega)$ and $\omega'$ are the scaled equivalents of $Z(j\omega)$ and $\omega$. Every $L$, $R$, and $C$ in a circuit must be scaled,

$$R = kR', \quad L = \frac{k}{K}L', \quad \text{and} \quad C = \frac{1}{kK}C'$$

This procedure applies only to linear circuits. Most of the problems in this book have, in fact, been scaled to simplify the calculations.

## NETWORK GRAPHS AND TOPOLOGY

The topological arrangement of a network's elements and the validity of KVL, KCL, and Tellegen's theorem are independent of the nature of the elements. Here, some of the properties of a network's *oriented graph* will be introduced as a prerequisite to the formal and systematic methods of network analysis to be developed in Chapter 14.

The oriented graph of Fig. 10-9*b* consists of two *separate parts;* that is, it is *not*

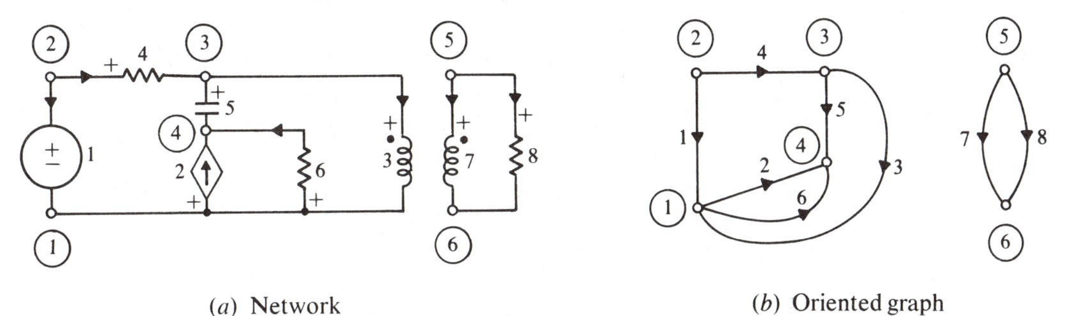

(*a*) Network (*b*) Oriented graph

**Fig. 10-9**

*connected*. If there were a branch joining nodes ① and ⑥ and/or ③ and ⑥, say, then the graph would be connected, but *hinged*. A graph that can be drawn with no intersections except at nodes is said to be *planar*.

Any *loop* of a planar graph with no interior branches is called a *mesh* and the loop of a planar, connected, unhinged graph with no exterior branches is called the *outer mesh* (see Fig. 10-10*a*). The number of meshes in a connected, planar graph with $b$ branches and $n_t$ nodes is $l = b - n_t + 1$ (not counting the outer mesh).

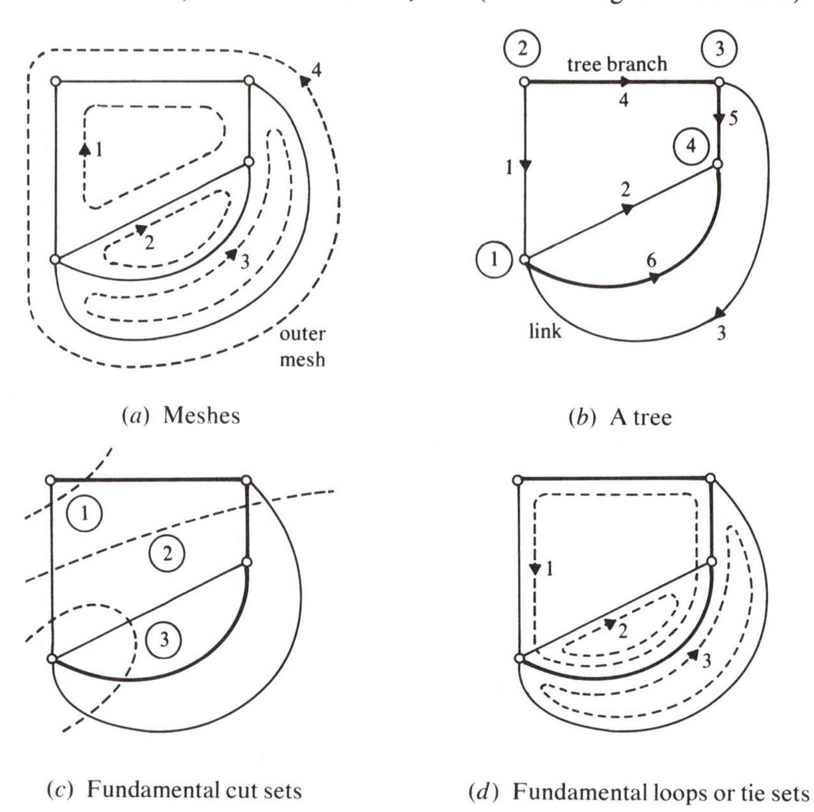


(*a*) Meshes

(*b*) A tree

(*c*) Fundamental cut sets

(*d*) Fundamental loops or tie sets

**Fig. 10-10**

Any set of branches in a connected graph which includes all nodes and forms no closed paths is called a *tree* (see Fig. 10-12*b*). The number of *tree branches* in a connecterd graph is $n = n_t - 1$. Any branch not in the tree is a *link*.

If a connected graph is divided into two separate parts by the removal of a set of branches, and if the removal of all but any one of the set leaves the graph connected, then this set of branches is called a *cut set*. If, further, a tree has been specified for a connected graph, then a cut set containing exactly one tree branch is called a *fundamental cut set* (for example, the set of branches cut by any one of the dashed lines in Fig. 10-10*c*). If the total number of nodes in a graph is $n_t = n + 1$, then the number of fundamental cut sets will be $n$; one for each tree branch.

A loop formed by exactly one link and one or more tree branches in a connected graph is called a *fundamental loop*, and the set of branches in the fundamental loop is called a *tie set* (see Fig. 10-10*d*). The number of fundamental loops is $l = b - n_t + 1$.

The connections between the nodes of a graph completely define the graph, and these connections can be conveniently specified by an $n_t \times b$ *augmented incidence matrix* $\mathbf{A}_a$ for which

$$a_{ij} = \begin{cases} 1 & \text{if branch } j \text{ } \textit{leaves} \text{ node } i \\ -1 & \text{if branch } j \text{ } \textit{enters} \text{ node } i \\ 0 & \text{if branch } j \text{ is not incident upon } i \end{cases}$$

For the graph of Fig. 10-10*b*,

$$\mathbf{A}_a = \begin{bmatrix} -1 & 1 & -1 & 0 & 0 & 1 \\ 1 & 0 & 0 & 1 & 0 & 0 \\ 0 & 0 & 1 & -1 & 1 & 0 \\ 0 & -1 & 0 & 0 & -1 & -1 \end{bmatrix}$$

## DUALITY

A *dual graph* $\mathcal{G}'$ of a given connected, unhinged, planar graph $\mathcal{G}$ is defined by the following algorithm:

1. Number the nodes and meshes of $\mathcal{G}$ in such a way that the reference node and the outer mesh have the same number (see Fig. 10-11*a*).
2. Associate a *node* of $\mathcal{G}'$ with every *mesh* of $\mathcal{G}$ as shown in Fig. 10-11*b*. The numbering should correspond.
3. Wherever $\mathcal{G}$ has a branch common to two *meshes*, draw a branch of $\mathcal{G}'$ between the corresponding *nodes*. This is illustrated in Fig. 10-11*b*.

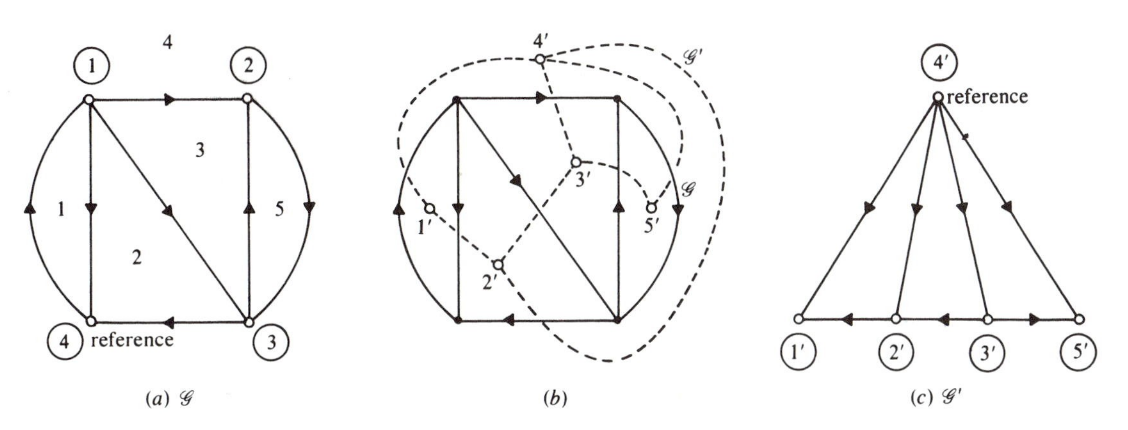


(*a*) $\mathcal{G}$ (*b*) (*c*) $\mathcal{G}'$

**Fig. 10-11**

4. Imagine a directional vector in each branch of $\mathcal{G}$ turned clockwise (by convention) into alignment with the intersecting branch of $\mathcal{G}'$. This will establish the reference direction in the latter (see Fig. 10-11*c*).

A network $N'$ is the *dual* of $N$ if the oriented graphs are duals, and if the corresponding branch equations are related by the following substitutions:

$$v \to i'$$

$$i \to v'$$

$$q \to \lambda' \text{ ($q$ is charge, $\lambda$ is flux linkage)}$$

$$\lambda \to q'$$

Consider a resistive branch $k$ for which $v_k = Ki_k$. The dual relationship for the "intersecting" branch $k'$ must be $i'_k = Kv'_k$ or $v'_k = (1/K)i'_k$. That is, a resistor $R_k$ in $N$ corresponds to a resistor $R'_k$ of value $1/R_k$ in $N'$. Similarly, an inductor $L_k$ corresponds to a capacitor $C'_k$ of value $L_k$, and a capacitor $C_k$ corresponds to an inductor $L'_k$ of value $C_k$.

Dual networks have the same equations and the same solutions—except that the "names" of the variables will differ, according to the above schedule of substitutions.

More generally, if $S$ is a true statement about $N$, and if $S'$ is the dual statement formed by substituting the dual of every *graph-theoretic* or "electrical" word in $S$ (see Table 10-1), then $S'$ is a true statement about $N'$.

**Table 10-1 Dual Concepts.**

| | | | |
|---|---|---|---|
| Voltage | Current | Short circuit | Open circuit |
| Charge | Flux linkage | Series | Parallel |
| Resistor | Resistor | Node | Mesh |
| Resistance | Conductance | Reference node | Outer mesh |
| Inductor | Capacitor | Cut set | Loop (tie set) |
| Inductance | Capacitance | Fundamental cut set | Fundamental loop |
| Impedance | Admittance | Tree branch | Link |

# ILLUSTRATIVE PROBLEMS

### *Superposition and Reciprocity*

**10.1** Find the steady-state current $i(t)$ in the circuit of Fig. 10-12, where $f(t) = 10 \cos t$ and $v(t) = 10$.

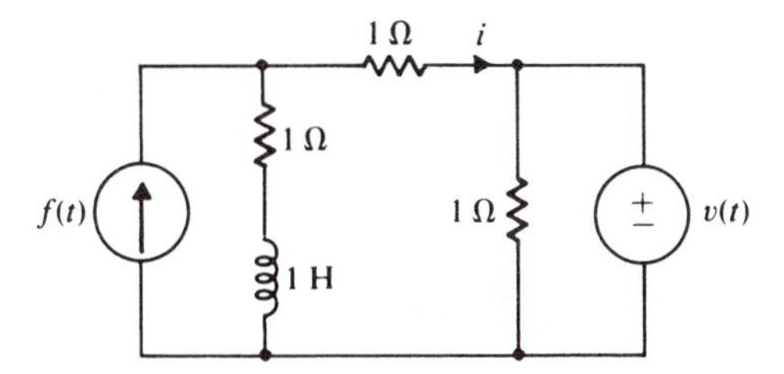


Fig. 10-12

First we determine the steady-state current due to each of the sources separately. Thus from Fig. 10-13$a$ and the current divider concept,

$$\frac{\mathbf{I}_1}{\mathbf{F}} = \frac{1}{1 + 1/(1 + j)} = \frac{1 + j}{2 + j} = \frac{\sqrt{2}e^{j45°}}{\sqrt{5}e^{j26.6°}}$$
$$= 0.634e^{j18.4°}$$

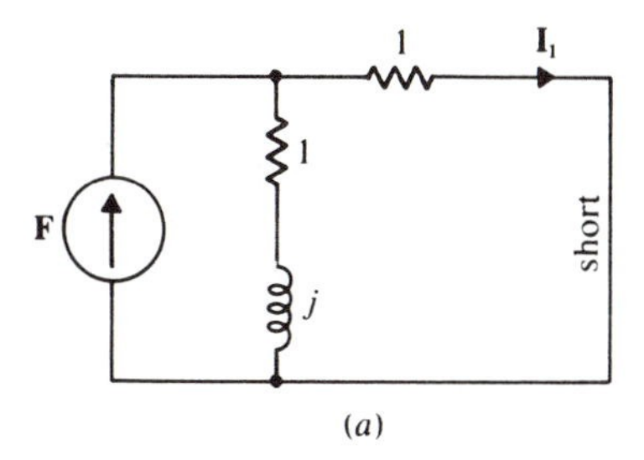


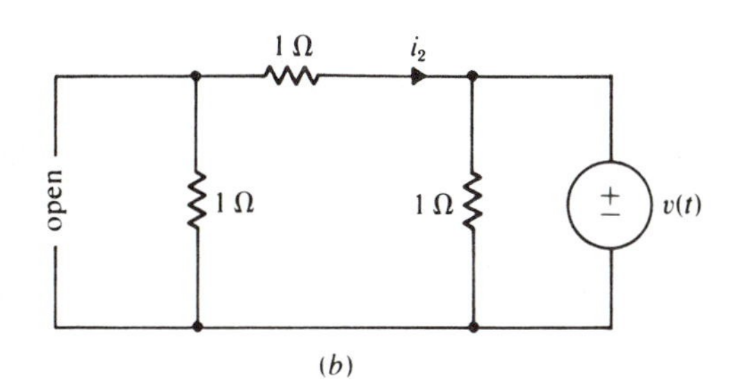


Fig. 10-13

Therefore the phasor current due to the source phasor **F** is

$$\mathbf{I}_1 = (0.634e^{j18.4°})(10e^{jt}) = 6.34e^{j(t+18.4°)}$$

and the ac current is

$$i_1(t) = 6.34 \cos (t + 18.4°)$$

Next we turn to Fig. 10-13$b$ to find the dc steady-state current due to $v(t)$. Combining the two left-most 1-Ω resistors in series and applying Ohm's law,

$$i_2(t) = -\tfrac{1}{2}v(t) = -5$$

Finally, from Theorem 10.2,

$$i(t) = i_1(t) + i_2(t) = 6.34\ cos\ (t + 18.4°) - 5$$

in the steady-state.

**10.2** Figure 10-14$a$ is the circuit diagram of a two-port network that has three series inputs of *different* frequency. It is assumed that Fig. 10-14$b$ adequately represents $H(j\omega) = \mathbf{E}/\mathbf{V}$ over the frequency range of interest. Find $e_{sss}(t)$ when $v_1 = 10 \cos 100t$, $v_2 = 20 \cos (200t + 10°)$, and $v_3 = 30 \cos (300t + 20°)$.

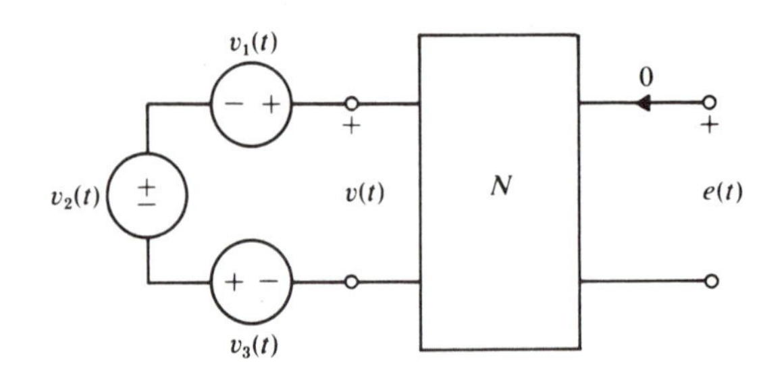


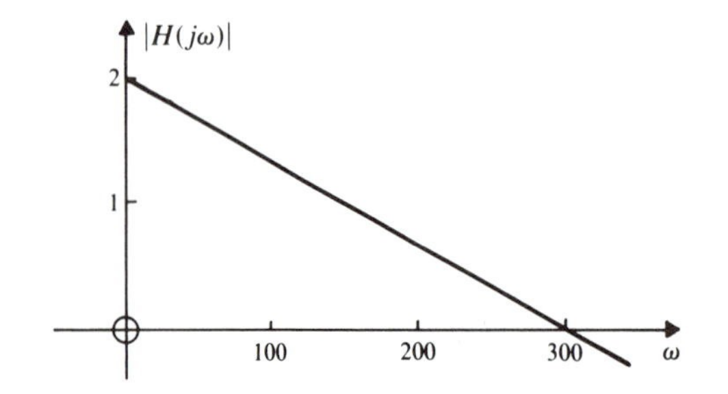


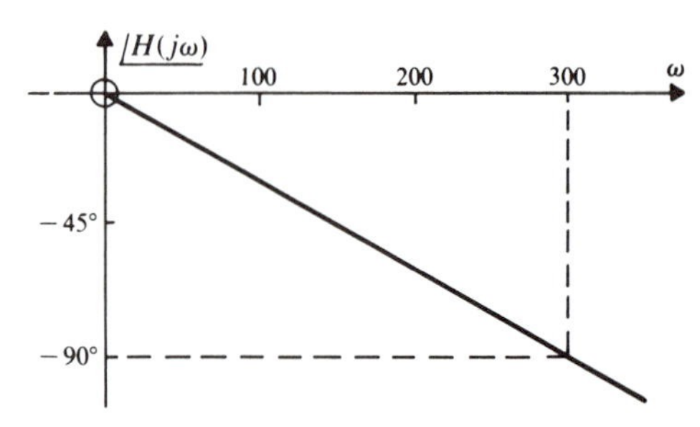


**Fig. 10-14**

The aim must be to find $e_1(t)$, $e_2(t)$, and $e_3(t)$, the ac voltages due to $v_1$, $v_2$, and $v_3$, separately. Then by the principle of superposition we can determine the total sinusoidal steady-state response.

For $\omega = 100$ rad/s, $H(j100) = \frac{4}{3}e^{-j30°}$.
For $\omega = 200$ rad/s, $H(j200) = \frac{2}{3}e^{-j60°}$.
For $\omega = 300$ rad/s, $H(j300) = 0e^{-j90°}$.
But $\mathbf{V}_1 = 10e^{j100t}$, $\mathbf{V}_2 = 20e^{j(200t+10°)}$, and $\mathbf{V}_3 = 30e^{j(300t+20°)}$. Thus

$$\mathbf{E}_1 = H(j100)\mathbf{V}_1 = (\tfrac{4}{3}e^{-j30°})(10e^{j100t}) = \tfrac{40}{3}e^{j(100t-30°)}$$

$$\mathbf{E}_2 = H(j200)\mathbf{V}_2 = (\tfrac{2}{3}e^{-j60°})(20e^{j(200t+10°)}) = \tfrac{40}{3}e^{j(200t-50°)}$$

$$\mathbf{E}_3 = H(j300)\mathbf{V}_3 = 0$$

Using the principle of superposition, that is, Theorem 10.2,

$$\begin{aligned} e_{\text{sss}}(t) &= \text{Re}\,[\mathbf{E}_1] + \text{Re}\,[\mathbf{E}_2] + \text{Re}\,[\mathbf{E}_3] \\ &= e_{1,\text{sss}}(t) + e_{2,\text{sss}}(t) + e_{3,\text{sss}}(t) \\ &= \tfrac{40}{3}\cos(100t - 30°) + \tfrac{40}{3}\cos(200t - 50°) \end{aligned}$$

**10.3** Draw an $s$-domain diagram for the network of Fig. 10-15 showing the initial conditions $i(0)$ and $v(0)$ as equivalent sources. Then find $V_{L_1}(s)$, the component of $V_L(s)$ which is due to the initial capacitor voltage $v(0)$.

Following the method of Chapter 5, the initial conditions may be replaced by erquivalent sources as in Fig. 10-16$a$. Then setting $e(t)$ and $i(0)$ to zero we obtain the $s$-domain circuit of Fig. 10-16$b$, and from KCL at node $a$,

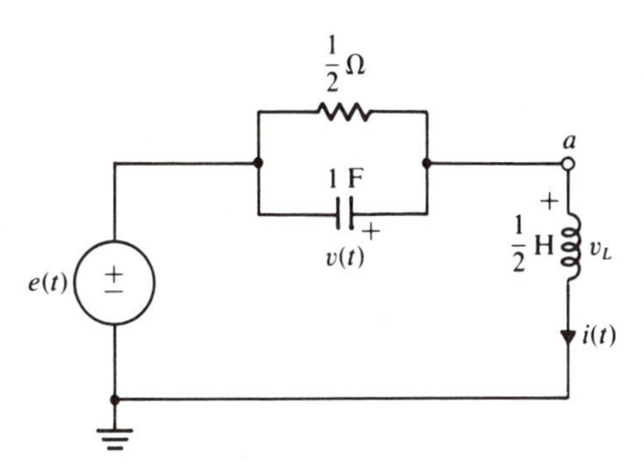


Fig. 10-15

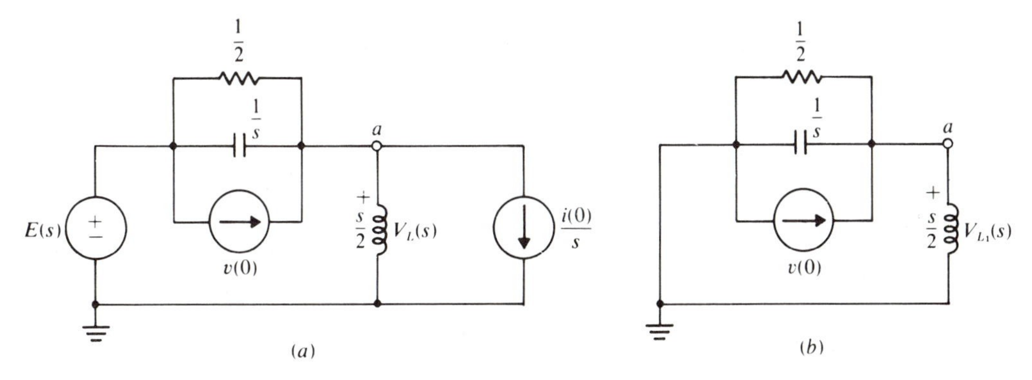


Fig. 10-16

$$\frac{V_{L_1}(s)}{1/2} + \frac{V_{L_1}(s)}{1/s} + \frac{V_{L_1}(s)}{s/2} = v(0)$$

from which

$$V_{L_1}(s) = \frac{v(0)}{2 + s + 2/s} = \frac{sv(0)}{s^2 + 2s + 2}$$

*Comment:* Similarly, we could find $V_{L_2}(s)$ and $V_{L_3}(s)$, due to $i(0)$ and $e(t)$:

$$V_{L_2}(s) = \frac{i(0)}{s^2 + 2s + 2} \quad \text{and} \quad V_{L_3}(s) = \frac{s(s + 2)E(s)}{s^2 + 2s + 2}$$

Finally, by superposition,

$$V_L(s) = V_{L_1}(s) + V_{L_2}(s) + V_{L_3}(s)$$

**10.4** The reciprocal network of Fig. 10-17 is in the sinusoidal steady-state. From laboratory tests it is known that

$$i_1(t) = 1.0 \cos(\omega t + 18.4°) \quad \text{and} \quad i_2(t) = -\frac{1}{\sqrt{2}} \cos(\omega t - 26.6°)$$

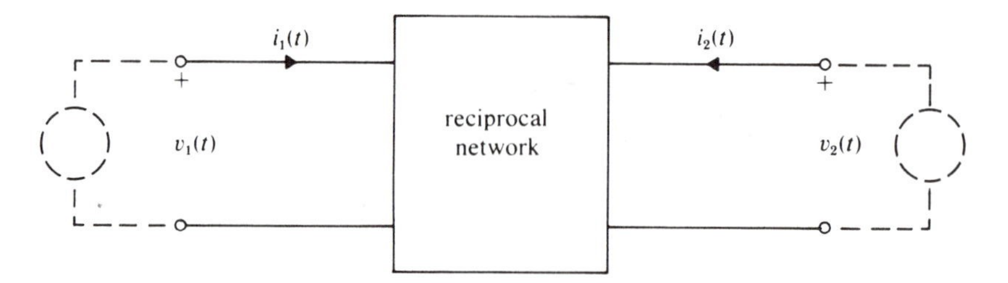


**Fig. 10-17**

when $v_1(t) = (5/\sqrt{2}) \cos \omega t$ and $v_2(t) = 0$. Find $i_1(t)$ when $v_1(t) = 5 \cos (\omega t - 90°)$ and $v_2(t) = 5 \cos \omega t$.

Here we will utilize homogeneity (Theorem 10.1), additivity (Theorem 10.2), and reciprocity (Theorem 10.3) to analyze the situation. It is easier to keep track of the strategy if we work in a table, and the notation can be abbreviated by electing to use stationary phasors. Thus

| Step | $\mathbf{V}_1$ | $\mathbf{V}_2$ | $\mathbf{I}_1$ | $\mathbf{I}_2$ | Remarks |
|---|---|---|---|---|---|
| 1 | $\frac{5}{\sqrt{2}} e^{j0°}$ | 0 | $1e^{j18.4°}$ | $-\frac{1}{\sqrt{2}} e^{-j26.6°}$ | Given data |
| 2 | $5e^{-j90°}$ | 0 | $\sqrt{2}e^{-j71.6°}$ | $-1e^{-j116.6°}$ | Homogeneity applied to step 1 data |
| 3 | 0 | $\frac{5}{\sqrt{2}} e^{j0°}$ | $-\frac{1}{\sqrt{2}} e^{-j26.6°}$ | — | Reciprocity (see Fig. 10-3*b*) applied to step 1 data. |
| 4 | 0 | $5e^{j0°}$ | $-1e^{-j26.6°}$ | — | Homogeneity applied to step 3 data |
| 5 | $5e^{-j90°} + 0$ | $0 + 5e^{j0°}$ | $\sqrt{2}e^{-j71.6°} - 1e^{-j26.6°}$ | — | Superposition (i.e. additivity) applied to the data of steps 2 and 4 |

Finally,

$$i_1(t) = \sqrt{2} \cos (\omega t - 71.6°) - \cos (\omega t - 26.6°)$$

when $v_1(t) = 5 \cos (\omega t - 90°)$ and $v_2(t) = 5 \cos \omega t$.

*Comments:* In step 2 we applied the homogeneity theorem in its complex form. That is, we multiplied by the complex constant $\sqrt{2}e^{-j90°}$. Notice, too, that reciprocity as applied in step 3 yields no information about $\mathbf{I}_2$.

**10.5** Show that in the circuit of Fig. 10-18*a*, the positions of the excitation $E(s)$ and

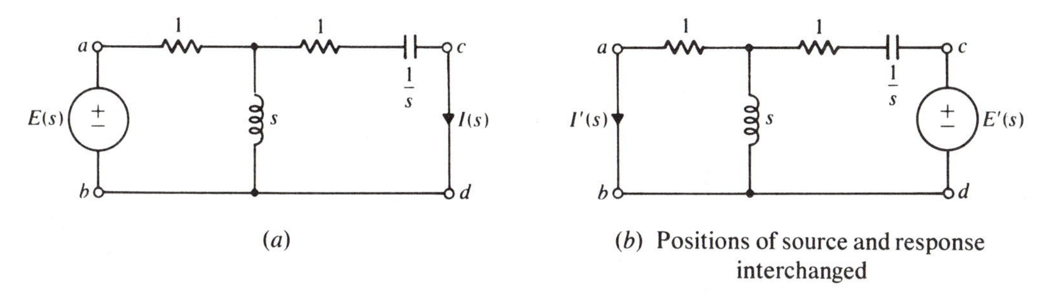


(a) (b) Positions of source and response interchanged

**Fig. 10-18**

response $I(s)$ may be interchanged without affecting the ratio $I(s)/E(s)$. That is, show that $I(s)/E(s)$ in Fig. 10-18a equals $I'(s)/E'(s)$ in Fig. 10-18b.

We want the responses due to $E(s)$ and $E'(s)$, so we set all the initial conditions equal to zero. Then the circuit "inside" the terminals *a-b* and *c-d* is a passive, linear, time-invariant two port for which reciprocity is guaranteed.

Therefore if $E'(s)$ in Fig. 10-18b equals $E(s)$ in Fig. 10-18a, reciprocity (see Fig. 10-3b) yields

$$I(s) = I'(s) \quad \text{and thus} \quad \frac{I(s)}{E(s)} = \frac{I'(s)}{E(s)} = \frac{I'(s)}{E'(s)}$$

*Comments:* The circuit of Fig. 10-18 need not be regarded as a two port with an input source $E(s)$ and an output short circuit. Thinking of it as a *single source* circuit, we can say that it is reciprocal if the positions of the excitation and response may be interchanged without affecting the ratio of the response to the excitation. This is an alternative statement of the reciprocity theorem, and all three interchanges implied by Fig. 10-3 are admissible.

When interchanging source and response the associated sign convention of Chapter 1 must be followed. For example, when a current response takes the place of a voltage source as in Fig. 10-18, its reference direction is that of the current associated with the original source.

**10.6** The network of Fig. 10-19 is a low-frequency circuit model of a common-emitter transistor amplifier. Show that it does *not* satisfy the reciprocity relations.

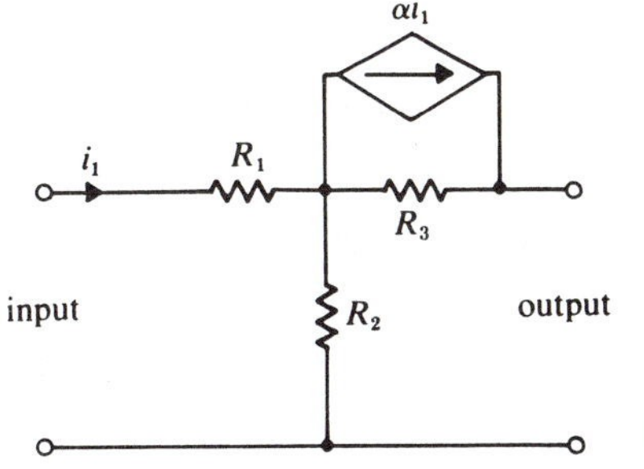


**Fig. 10-19**

If we refer to the reciprocity relationships of Fig. 10-3, then we can, for example, test the present circuit by comparing $v_2$ in Fig. 10-20*a* with $v_1'$ in Fig. 10-20*b*.

In the circuit of Fig. 10-20*a*, $i_1 = i = i_{R_2}$ and $v_2 = v_{R_2} + v_{R_3}$. Thus

$$v_2 = R_2 i + R_3 \alpha i = (R_2 + \alpha R_3) i$$

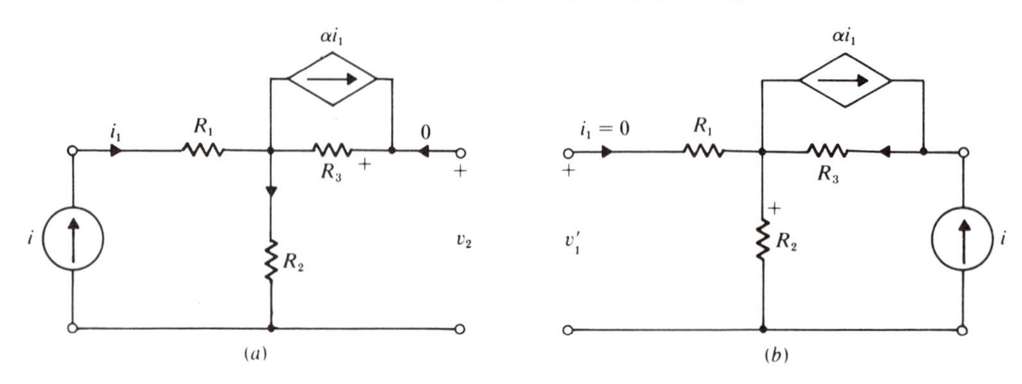


**Fig. 10-20**

Now, in the circuit of Fig. 10-20*b*, $i_1 = 0$, $i_{R_2} = i_{R_3} = i$, and $v_1' = v_{R_2}$, or

$$v_1' = R_2 i$$

This is patently *not* equal to $v_2$.

We could, of course, have tested the network according to the relationship of Fig. 10-3*b* or *c* rather than that of Fig. 10-3*a*, as we have done here.

### *Thévenin and Norton*

**10.7** The network of Fig. 10-21 is to be analyzed under sinusoidal steady-state conditions for a variety of load networks. We are given $i(t) = 2 \cos 4t$ and $e(t) = 2 \cos 4t$. As a first step, find $\mathbf{V}_{oc}$ and draw the $j\omega$-domain Thévenin network $N_{\text{Th}}$.

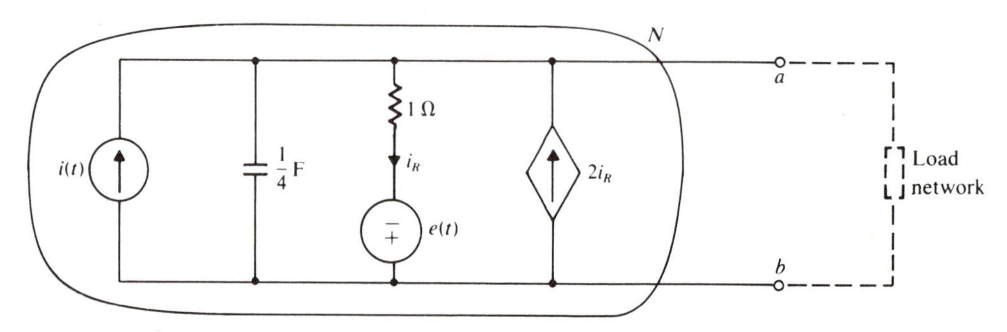


**Fig. 10-21**

The $j\omega$-domain network is shown with its output open-circuited in Fig. 10-22*a*. From KCL,

$$-2\mathbf{I}_R + \mathbf{I}_R + \mathbf{I}_C - 2e^{j4t} = 0 \quad \text{or} \quad -\mathbf{I}_R + \mathbf{I}_C = 2e^{j4t}$$

But from KVL, $\mathbf{V}_{oc} = \mathbf{I}_R - 2e^{j4t}$ and $\mathbf{I}_C = j\mathbf{V}_{oc}$. Substituting into the KCL equation to eliminate $\mathbf{I}_R$ and $\mathbf{I}_C$,

$$-\mathbf{V}_{oc} - 2e^{j4t} + j\mathbf{V}_{oc} = 2e^{j4t}$$

or

$$\mathbf{V}_{oc} = \frac{4e^{j4t}}{-1+j} = \frac{4e^{j4t}}{\sqrt{2}e^{j135°}} = 2\sqrt{2}e^{j(4t-135°)}$$

To obtain $N_{\text{Th}}$ we simply "kill" all the *independent* sources in $N$ as shown in Fig. 10-22*b*.

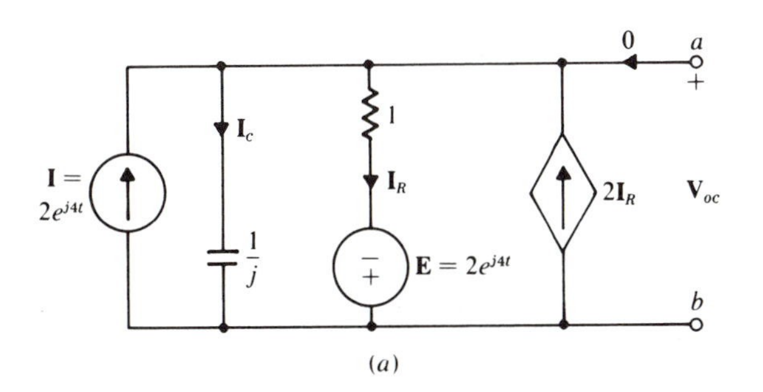


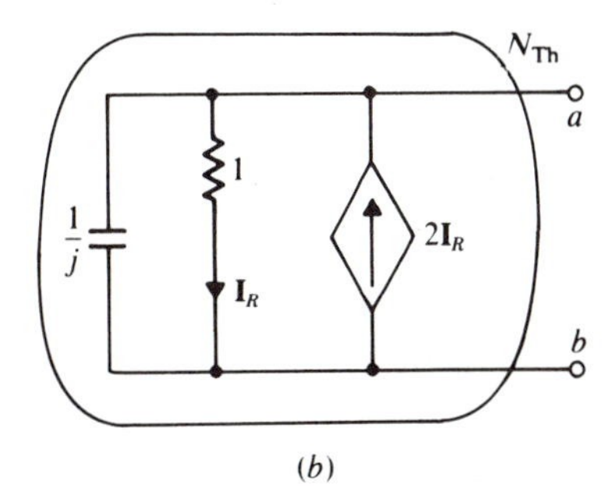


**Fig. 10-22**

**10.8** The network $N_{\text{Th}}$ of Fig. 10-22*b* is to be represented by an impedance $Z_{\text{Th}}(j\omega)$. Use two different methods to find this impedance.

*Method 1*

To find $Z_{\text{Th}}(j\omega)$ we can "connect" a phasor source $\mathbf{V}$ to $N_{\text{Th}}$ and calculate the resulting $\mathbf{I}$, or vice versa, as suggested by Fig. 10-23*a*. Then $Z_{\text{Th}}(j\omega) = \mathbf{V}/\mathbf{I}$.

Writing a nodal equation for the circuit of Fig. 10-23*a*,

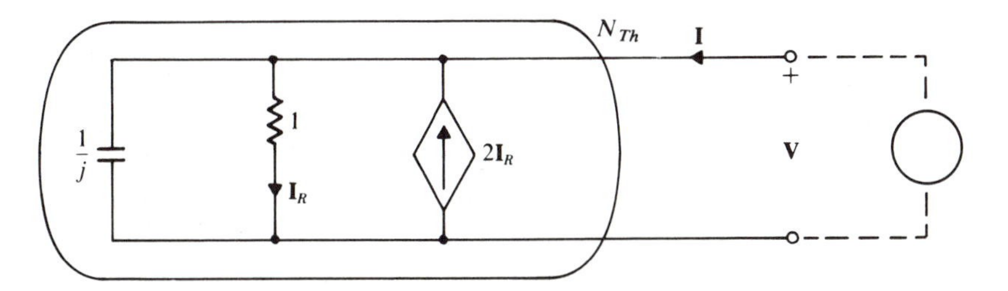


**Fig. 10-23a**

$$-2\mathbf{I}_R + \mathbf{I}_R + j\mathbf{V} - \mathbf{I} = 0$$

where $\mathbf{I}_R = 1\mathbf{V} = \mathbf{V}$. Therefore $-\mathbf{V} + j\mathbf{V} - \mathbf{I} = 0$ and

$$Z_{\text{Th}}(j\omega) = \frac{\mathbf{V}}{\mathbf{I}} = \frac{1}{-1 + j} = \frac{1}{\sqrt{2}} e^{-j135°}$$

*Method 2*

$Z_{\text{Th}}(j\omega) = \mathbf{V}_{oc}/\mathbf{I}_{sc}$ and we have already found $\mathbf{V}_{oc}$ in Problem 10.7. Now we must find $\mathbf{I}_{sc}$ at the shorted terminals of $N$, as indicated in Fig. 10-23*b*.

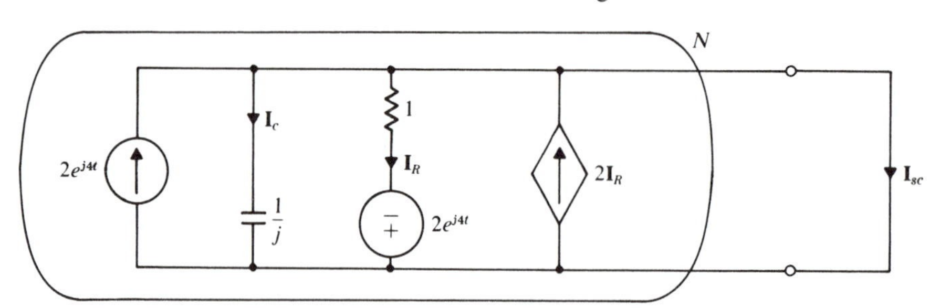


**Fig. 10-23*b***

Here $\mathbf{I}_C = 0$ since $\mathbf{V}_C = 0$; and thus from KCL,

$$\mathbf{I}_{sc} - 2\mathbf{I}_R + \mathbf{I}_R = 2e^{j4t}$$

But from KVL, $\mathbf{I}_R - 2e^{j4t} = 0$. Therefore $\mathbf{I}_{sc} = 4e^{j4t}$ and

$$Z_{\text{Th}}(j\omega) = \frac{\mathbf{V}_{oc}}{\mathbf{I}_{sc}} = \frac{2\sqrt{2}e^{j(4t-135°)}}{4e^{j4t}} = \frac{1}{\sqrt{2}} e^{-j135°}$$

where the expression for $\mathbf{V}_{oc}$ was obtained from Problem 10.7.

**10.9** To summarize the results of Problems 10.7 and 10.8, Thévenin's theorem states that, *as far as the load is concerned,* the circuit of Fig. 10-24*a* is equivalent to the original circuit of Fig. 10-21. Solve for the ac voltage $v_{ab}(t)$ for each of the loads in Fig. 10-24*b*.

For the first load, we must analyze the circuit of Fig. 10-25*a*. From KVL,

$$(-\tfrac{1}{2} - j\tfrac{1}{2} + \tfrac{1}{2} - j\tfrac{1}{2})\mathbf{I} - 2\sqrt{2}e^{j(4t-135°)} = 0$$

or

$$\mathbf{I} = \frac{2\sqrt{2}e^{j(4t-135°)}}{-j} = 2\sqrt{2}e^{j(4t-45°)}$$

Thus

$$\mathbf{V}_{ab} = (\tfrac{1}{2} - j\tfrac{1}{2})\mathbf{I} = \left(\frac{1}{\sqrt{2}} e^{-j45°}\right)(2\sqrt{2}e^{j(4t-45°)}) = 2e^{j(4t-90°)}$$

and $v_{ab}(t) = 2 \cos (4t - 90°)$.

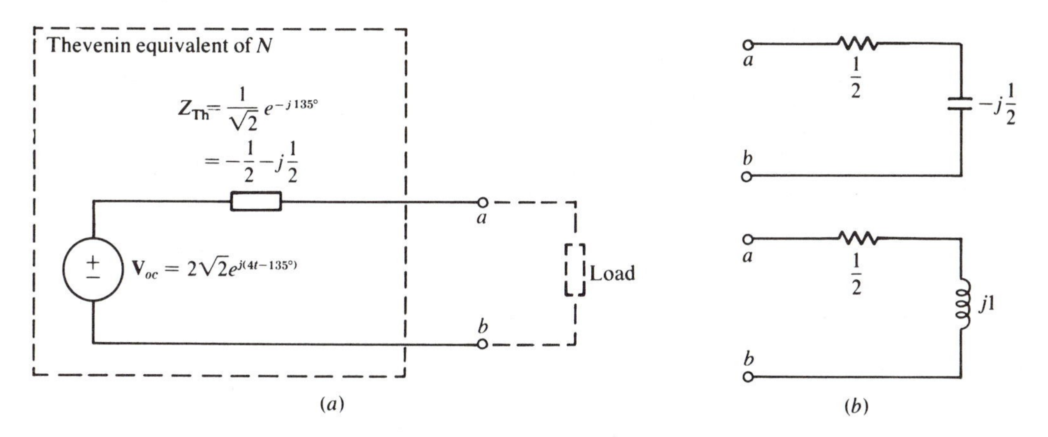


**Fig. 10-24**

For the second load network we go to Fig. 10-25*b*,

$$(-\tfrac{1}{2} - j\tfrac{1}{2} + \tfrac{1}{2} + j1)\mathbf{I} - 2\sqrt{2}e^{j(4t-135°)} = 0$$

If we proceed as in the first half of this problem, we find after the same two steps,

$$v_{ab}(t) = 6.34 \cos (4t - 161.6°)$$

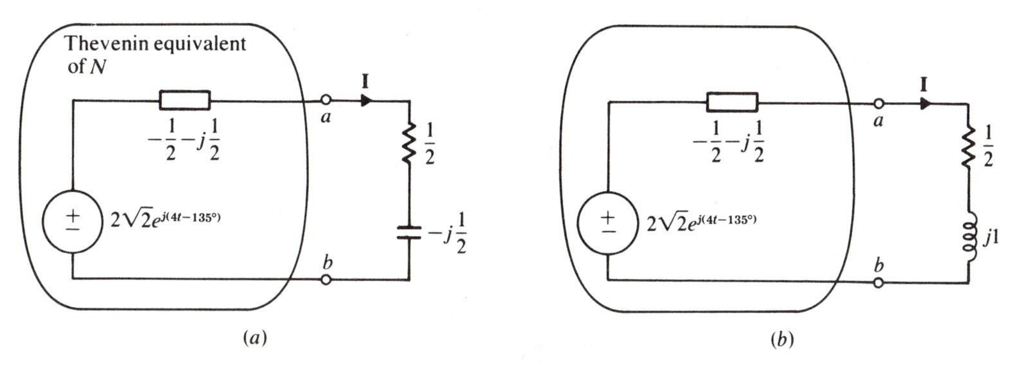


**Fig. 10-25**

**10.10** A circuit model of a transistor driven by a current source $i(t)$ is shown in Fig. 10-26 where $R_s$ is the source resistance and $h_i$, $h_r$, $h_f$, and $1/h_o$ are transistor parameters. Find the Thévenin and Norton equivalents of this circuit with respect to the terminals *a*-*g*.

First, under open-circuit conditions (see Fig. 10-27*a*),

$$h_i i_1 + h_r v_{oc} + R_s\{i_1 - i(t)\} = 0 \quad \text{and} \quad \frac{v_{oc}}{1/h_o} + h_f i_1 = 0$$

Rearranging,

$$(h_i + R_s)i_1 + h_r v_{oc} = R_s i(t)$$

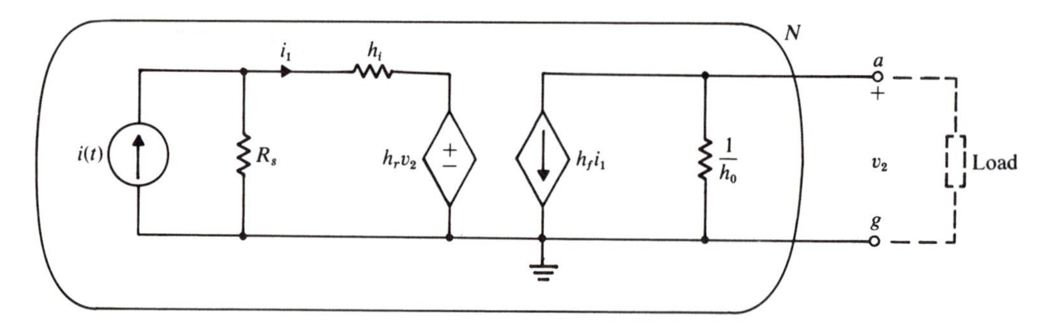


**Fig. 10-26**

$$h_f i_1 + h_O v_{oc} = 0$$

and using Cramer's rule,

$$v_{oc} = \frac{\begin{vmatrix} h_i + R_s & R_s i(t) \\ h_f & 0 \end{vmatrix}}{\begin{vmatrix} h_i + R_s & h_r \\ h_f & h_O \end{vmatrix}} = \frac{-h_f R_s i(t)}{h_i h_O + R_s h_O - h_r h_f}$$

Next, under short-circuit conditions (see Fig. 10-27*b*),

$$h_i i_1 + \cancel{h_r v_2} + R_s\{i_1 - i(t)\} = 0 \quad \text{and} \quad i_{sc} + h_f i_1 = 0$$

from which $i_1 = R_s i(t)/(h_i + R_s)$ since $v_2 = 0$, and

$$i_{sc} = -h_f i_1 = -\frac{h_f R_s i(t)}{h_i + R_s}$$

Finally,

$$R_{\text{Th}} = \frac{v_{oc}}{i_{sc}} = \left\{\frac{-h_f R_s i(t)}{h_i h_O + R_s h_O - h_r h_f}\right\}\left\{-\frac{h_i + R_s}{h_f R_s i(t)}\right\} = \frac{h_i + R_s}{h_i h_O + R_s h_O - h_r h_f}$$

Now, with $R_{\text{Th}}$, $v_{oc}$, and $i_{sc}$ determined, the Thévenin and Norton equivalents of Fig. 10-27*c* follow at once from Theorems 10.4 and 10.5.

### *Alternative Evaluation of $R_{\text{Th}}$*

To find $R_{\text{Th}}$ we can assume a voltage at the terminals of $N_{\text{Th}}$ and calculate the resulting current (see Fig. 10-27*d*). Then $R_{\text{Th}} = v_2/i_2$. Here

$$\frac{v_2}{1/h_O} + h_f i_1 = i_2 \quad \text{and} \quad (h_i + R_s)i_1 + h_r v_2 = 0$$

Solving the second equation for $i_1$ and substituting this into the first,

$$\left(h_O - \frac{h_r h_f}{h_i + R_s}\right)v_2 = i_2 \quad \text{and} \quad R_{\text{Th}} = \frac{v_2}{i_2} = \frac{h_i + R_s}{h_i h_O + R_s h_O - h_r h_f}$$

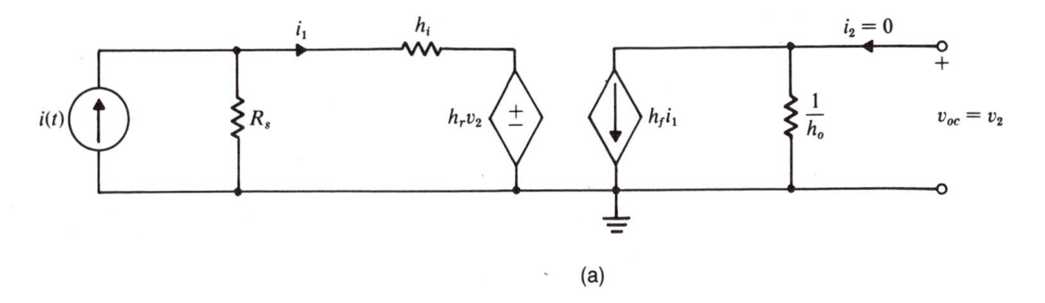
$i_1$
$h_i$
$i_2 = 0$
$i(t)$
$R_s$
$h_r v_2$
$h_f i_1$
$\frac{1}{h_o}$
$v_{oc} = v_2$

(a)

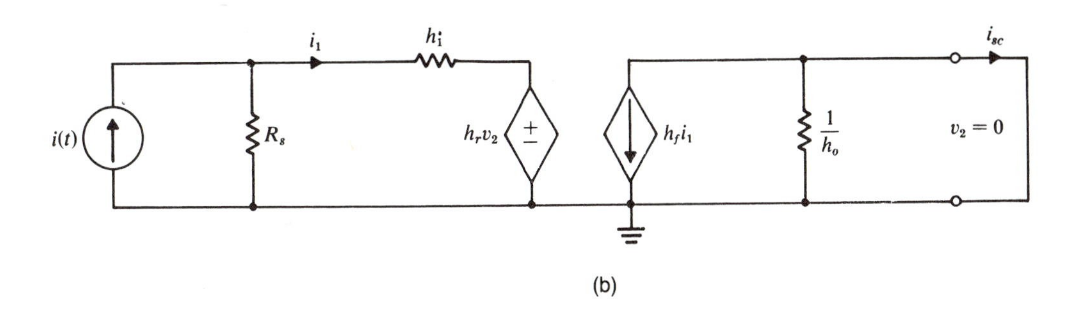
$i_1$
$h_i$
$i_{sc}$
$i(t)$
$R_s$
$h_r v_2$
$h_f i_1$
$\frac{1}{h_o}$
$v_2 = 0$

(b)

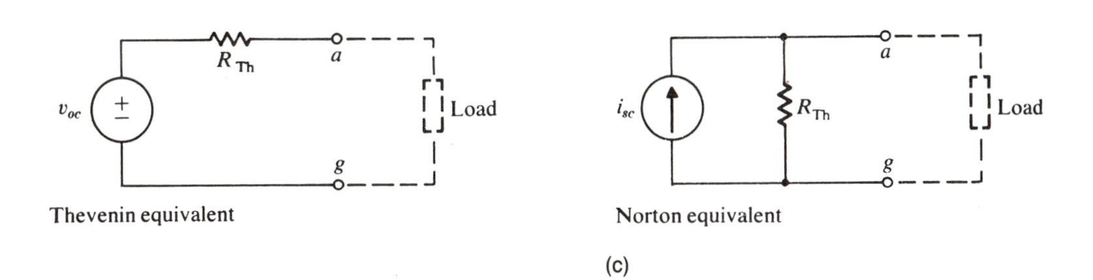
$R_{Th}$
a
$v_{oc}$
Load
g
Thevenin equivalent
a
$i_{sc}$
$R_{Th}$
Load
g
Norton equivalent

(c)

**Fig. 10-27**

### *Yet Another Method*

The most direct method, and possibly the simplest one, is to find the original circuit's *v-i* or *i-v* terminal relation by loop or nodal analysis. Then $v_{\text{Th}}$ (i.e. $v_{oc}$) and $R_{\text{Th}}$ follow from equation (2.8) by inspection.

The circuit has been redrawn for loop analysis in Fig. 10-28. (Nodal analysis would be equally appropriate.) Note that to avoid confusion with the terminal voltage and current $v$ and $i$, the source current has been renamed $i_s(t)$ and the output voltage $v_2$ has been labeled $v$.

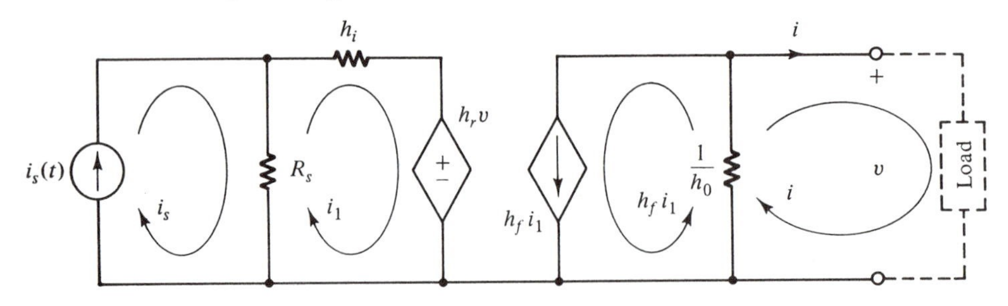


**Fig. 10-28**

Here in the usual way we obtain the following loop equations, where $v$ has been treated as a known.

$$\begin{bmatrix} R_s + h_i & 0 \\ h_f/h_o & 1/h_o \end{bmatrix} \begin{bmatrix} i_1 \\ i \end{bmatrix} = \begin{bmatrix} R_s i_s - h_r v \\ -v \end{bmatrix}$$

Cramer's rule now yields

$$i = \frac{\begin{vmatrix} R_s + h_i & R_s i_s - h_r v \\ h_f/h_o & -v \end{vmatrix}}{\begin{vmatrix} R_s + h_i & 0 \\ h_f/h_o & 1/h_o \end{vmatrix}}$$

$$= -\frac{h_i h_o + R_s h_o - h_r h_f}{h_i + R_s} v - \frac{h_f R_s i_s(t)}{h_i + R_s}$$

Comparison with equation (2.8) leads to the same values of $R_{\text{Th}}$ and $v_{\text{Th}}$ (i.e. $v_{oc}$) as were obtained above by the other two methods.

**10.11** Find the Norton equivalent of the circuit of Fig. 10-29 with respect to the terminals *x-y*. Use this equivalent to find $H(s) = I_{xy}(s)/I(s)$ when $Z(s) = 1/s$. Then if $i(t) = 10 \cos t$, find the steady-state value of $i_{xy}(t)$.

The network $N_{\text{Th}}$, which is shown in Fig. 10-30*a*, is passive and we may therefore find $Z_{\text{Th}}$ by impedance reduction techniques. Thus

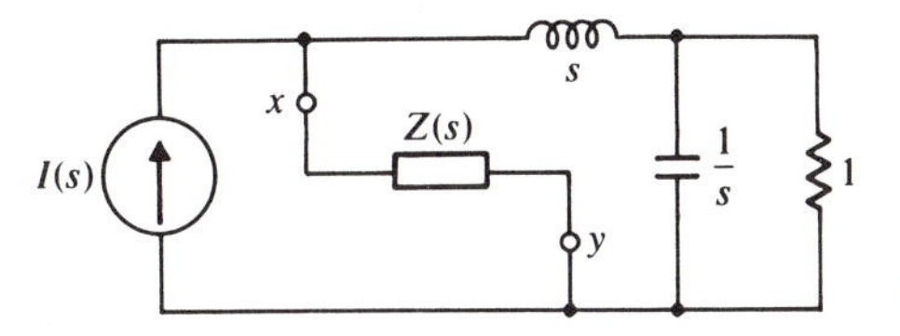


Fig. 10-29

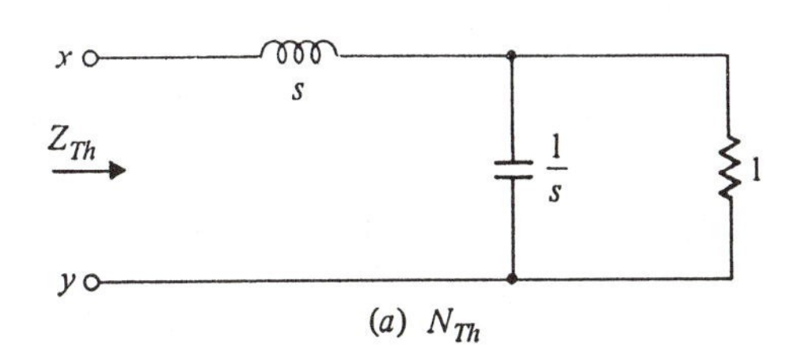


(a) $N_{Th}$

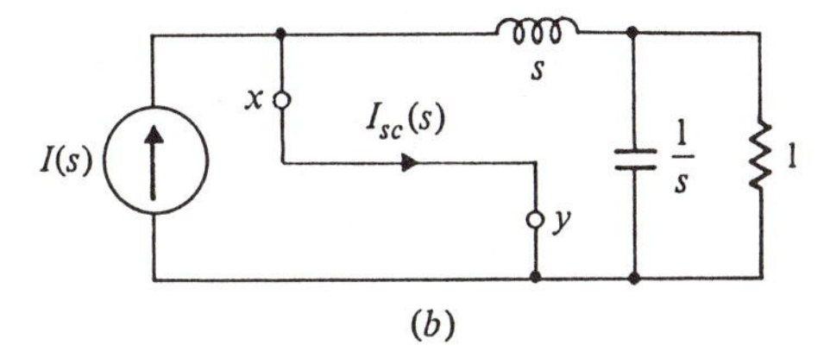


(b)

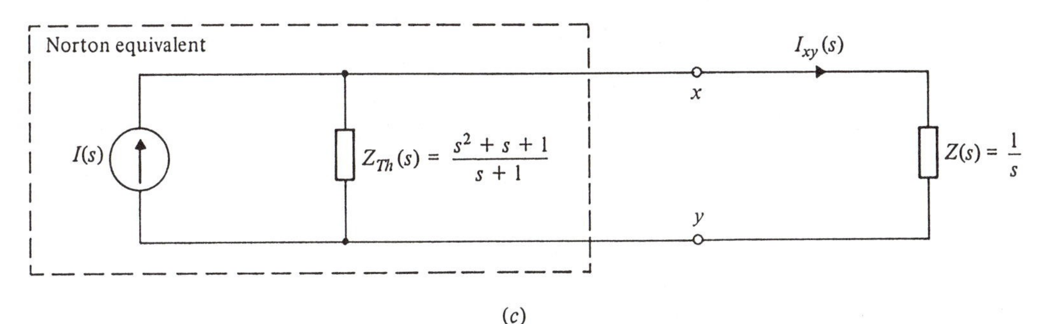


(c)

Fig. 10-30

$$Z_{\text{Th}}(s) = s + \frac{1}{s+1} = \frac{s^2 + s + 1}{s + 1}$$

To find $I_{sc}(s)$ we consider Fig. 10-30*b*, where it is clear that $I_{sc}(s) = I(s)$. The Norton equivalent of Fig. 10-30*c* follows from Theorem 10.5. The rest is easy! From the current divider rule,

$$I_{xy}(s) = \frac{sI(s)}{s + (s+1)/(s^2 + s + 1)} \quad \text{or} \quad H(s) = \frac{s(s^2 + s + 1)}{s^3 + s^2 + 2s + 1}$$

Moving into the $j\omega$-domain, with $\omega = 1$ rad/s,

$$\mathbf{I}_{xy} = H(j1)\mathbf{I} = \left(\frac{-j - 1 + j}{-j - 1 + j2 + 1}\right)(10e^{jt}) = \frac{1e^{j180^\circ}}{1e^{j90^\circ}} 10e^{jt} = 10e^{j(t+90^\circ)}$$

Thus in the sinusoidal steady-state,

$$i_{xy}(t) = 10 \cos (t + 90^\circ)$$

**10.12** The open-circuit voltage of the network in Fig. 10-31 is 12 V. When a resistor

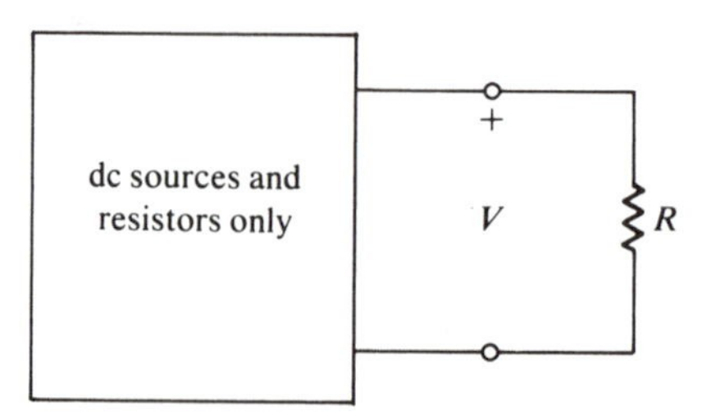


**Fig. 10-31**

$R = 4\,\Omega$ is connected, the terminal voltage drops to 8 V. Find the Thévenin and Norton equivalents of this network.

Since we know that $V_{oc} = 12$ V, we can draw the Thévenin equivalent of Fig. 10-32*a*, where $V = 8$ V when $R = 4\,\Omega$. To find $R_{\mathrm{Th}}$ we write the equations

$$(R_{\mathrm{Th}} + 4)I - 12 = 0 \quad \text{and} \quad V = 8 = 4I$$

which give $I = 2$ A and $R_{\mathrm{Th}} = 2\,\Omega$. The Thévenin and Norton equivalents of Fig. 10-32*b* then follow from Theorems 10.4 and 10.5.

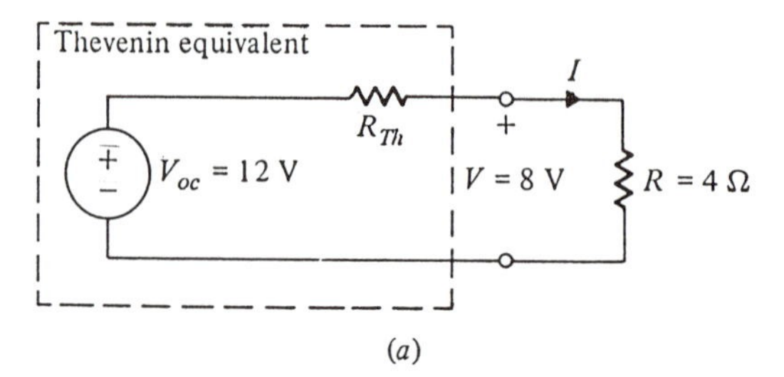


(*a*)

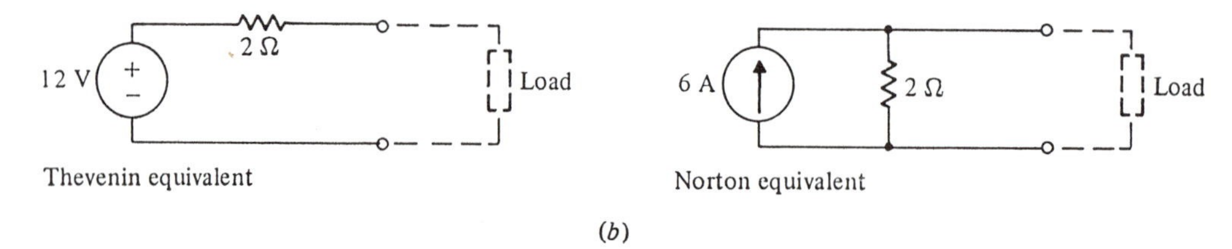


(*b*)

**Fig. 10-32**

*Comment*: $V_{oc}$ and/or $I_{sc}$ can often be obtained by direct measurement. Sometimes, however, the attempt may lead to the overloading or destruction of the circuit. Under these circumstances the technique suggested by this problem would be appropriate.

**10.13** The active network in Fig. 10-33 contains sinusoidal independent sources which are all of the same frequency, in addition to resistors, capacitors, and inductors. A test was performed to determine the Thévenin equivalent network seen by the load, and the following data were obtained:

| $Z_L(j\omega)$ | $\infty$ | $-j8$ | $-j4$ |
|---|---|---|---|
| $V_{\text{eff}}$, V | 100 | 160 | 133.3 |

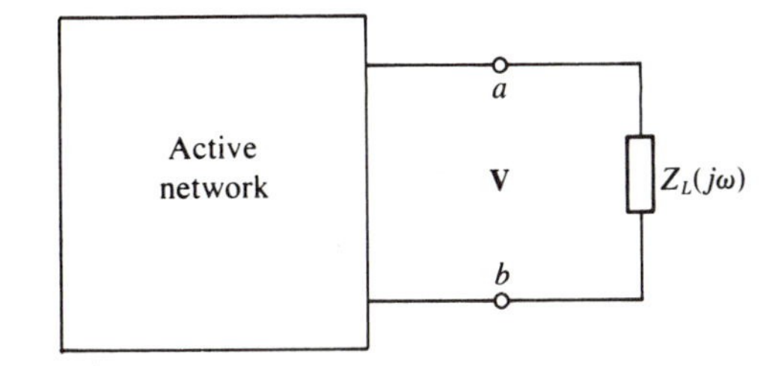


**Fig. 10-33**

Find the Thévenin equivalent with respect to the terminals *a*-*b*.

First we postulate a Thévenin network as in Fig. 10-34*a*. Then since $Z_L = \infty$ corresponds to an open circuit, we know that $V_{oc,\text{eff}} = 100$ V. And we will choose $\mathbf{V}_{oc}$ to be the stationary RMS reference phasor $\mathbf{V}_{oc} = 100$, as indicated in Fig. 10-34*b*.

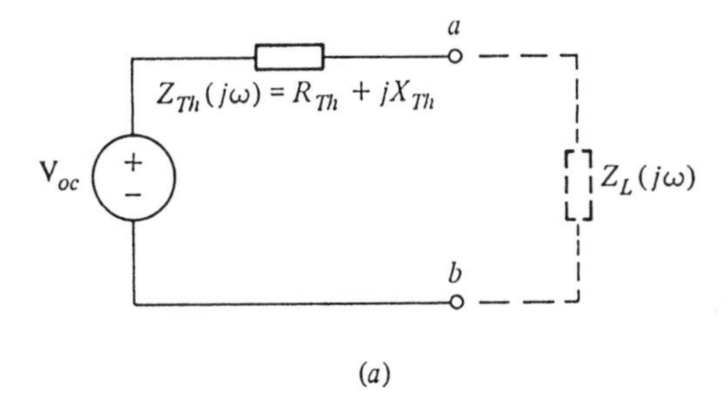


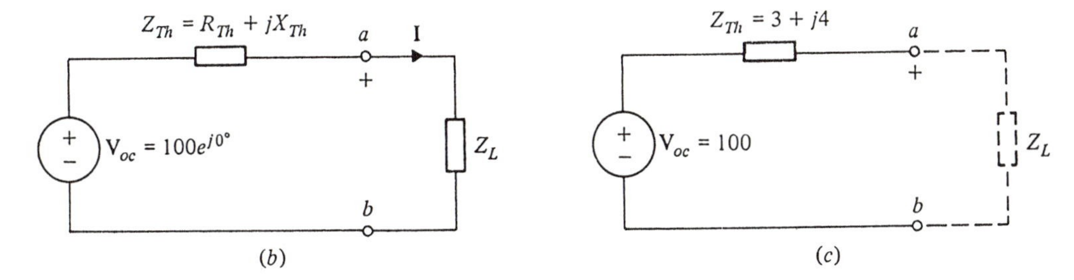


**Fig. 10-34**

When $Z_L(j\omega) = -j8$,

$$|\mathbf{I}| = \frac{V_{\text{eff}}}{|Z_L(j\omega)|} = \frac{160}{8} = 20 \text{ A, RMS}$$

and the total circuit impedance is

$$|Z(j\omega)| = \frac{|\mathbf{V}_{oc}|}{|\mathbf{I}|} = \frac{100}{33.3} = 3\,\Omega$$

But from Fig. 10-34*b*, with $Z_L(j\omega) = -j8$,

$$|Z(j\omega)| = 5 = \sqrt{R_{\mathrm{Th}}^2 + (X_{\mathrm{Th}} - 8)^2}$$

or

$$25 = R_{\mathrm{Th}}^2 + (X_{\mathrm{Th}} - 8)^2 \tag{1}$$

Similarly when $Z_L(j\omega) = -j4$,

$$|\mathbf{I}| = \frac{V_{\mathrm{eff}}}{|Z_L(j\omega)|} = \frac{133.3}{4} = 33.3 \text{ A, RMS}$$

and the total circuit impedance is

$$|Z(j\omega)| = \frac{|\mathbf{V}_{oc}|}{|\mathbf{I}|} = \frac{100}{33.3} = 3\,\Omega$$

Thus from Fig. 10-34*b*, with $Z_L(j\omega) = -j4$,

$$|Z(j\omega)| = 3 = \sqrt{R_{\mathrm{Th}}^2 + (X_{\mathrm{Th}} - 4)^2}$$

or

$$9 = R_{\mathrm{Th}}^2 + (X_{\mathrm{Th}} - 4)^2 \tag{2}$$

Equations (*1*) and (*2*) now yield $R_{\mathrm{Th}} = 3\,\Omega$ and $X_{\mathrm{Th}} = 4\,\Omega$, and the Thévenin equivalent of Fig. 10-34*c* results.

**10.14** The switch in the circuit of Fig. 10-35 has been closed for a long time, and is then opened at $t = 0$. Find $v_b(t)$ for $t \geq 0$ with the help of Thévenin's (or Norton's) theorem.

First the initial conditions $v_a(0)$ and $v_b(0)$ must be determined. The dc steady-state circuit applicable at $t = 0^-$ is shown in Fig. 10-36. Using the current divider idea, $I_1 = 8$ A and $I_2 = 2$ A, making $v_a(0) = 0.4$ V and $v_b(0) = 0.2$ V.

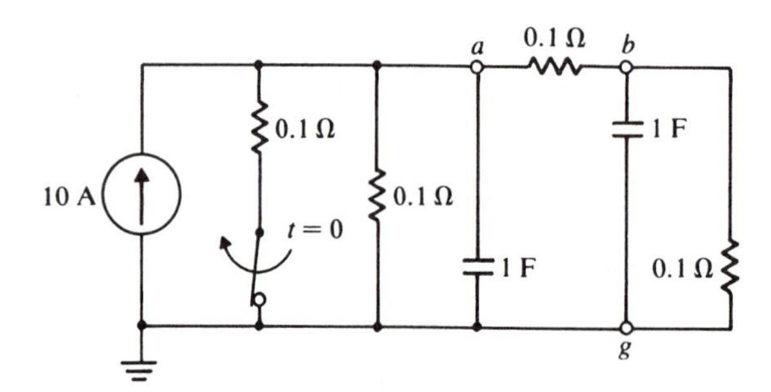


**Fig. 10-35**

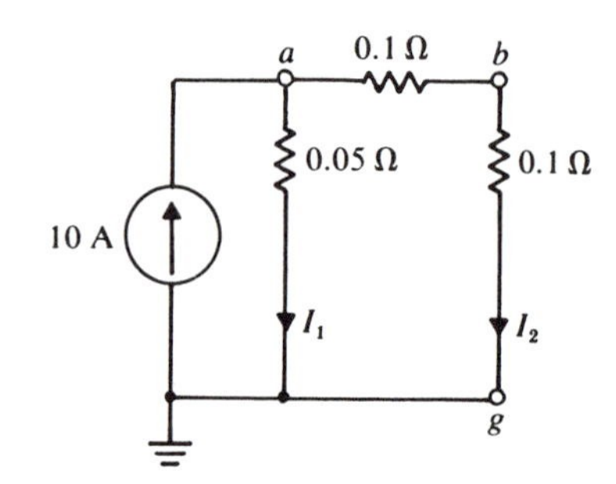


**Fig. 10-36**

An *s*-domain diagram for $t \geq 0$ is shown in Fig. 10-37*a*. To calcualte $V_{oc}(s)$ for the terminals *b*-*g*, we remove the load as in Fig. 10-37*b*. Then, applying KCL,

$$V_a(s)\{10 + s\} = \frac{10}{s} + 0.4 \quad \text{or} \quad V_a(s) = V_{oc}(s) = \frac{0.4(s + 25)}{s(s + 10)}$$

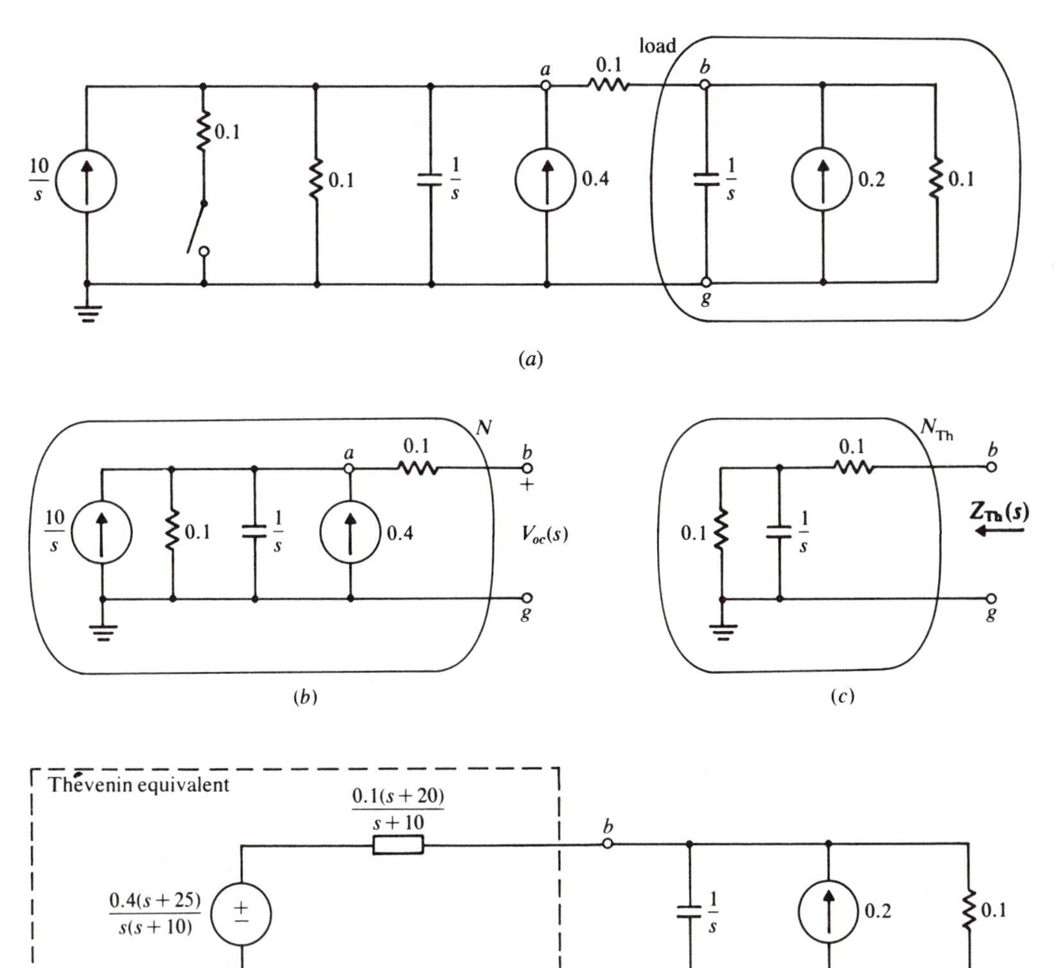


**Fig. 10-37**

Next, if we "kill" the sources in $N$ we obtain $N_{Th}$ as in Fig. 10-37$c$, and

$$Z_{Th}(s) = 0.1 + \frac{1}{s + 10} = \frac{0.1s + 1 + 1}{s + 10} = \frac{0.1(s + 20)}{s + 10}$$

The Thévenin equivalent is shown in Fig. 10-37$d$.

To find $V_b(s)$ we write a nodal equation at $b$,

$$\{s + 10\}V_b(s) - 0.2 + \frac{V_b(s) - 0.4(s + 25)/s(s + 10)}{0.1(s + 20)/(s + 10)} = 0$$

from which, after rearrangement,

$$V_b(s) = \frac{0.2(s^2 + 40s + 500)}{s(s^2 + 40s + 300)}$$

Expanding in partial fractions,

$$V_b(s) = \frac{0.2(s^2 + 40s + 500)}{s(s + 10)(s + 30)} \equiv \frac{C_1}{s} + \frac{C_2}{s + 10} + \frac{C_3}{s + 30}$$

where

$$C_1 = \left.\frac{0.2(s^2 + 40s + 500)}{(s + 10)(s + 30)}\right|_{s=0} = \frac{100}{300} = \frac{1}{3}$$

$$C_2 = \left.\frac{0.2(s^2 + 40s + 500)}{s(s + 30)}\right|_{s=-10} = \frac{20 - 80 + 100}{-10(20)} = -\frac{1}{5}$$

and

$$C_3 = \left.\frac{0.2(s^2 + 40s + 500)}{s(s + 10)}\right|_{s=-30} = \frac{180 - 240 + 100}{-30(-20)} = \frac{1}{15}$$

Therefore

$$v_b(t) = \tfrac{1}{3} - \tfrac{1}{5}e^{-10t} + \tfrac{1}{15}e^{-30t}, \qquad t \geq 0$$

### *Alternative Solution*

We may, of course, use Norton's theorem to find $V_b(s)$. From the above expressions for $Z_{Th}(s)$ and $V_{oc}(s)$,

$$I_{sc}(s) = \frac{V_{oc}(s)}{Z_{Th}(s)} = \left\{\frac{0.4(s + 25)}{s(s + 10)}\right\}\left\{\frac{10(s + 10)}{s + 20}\right\} = \frac{4(s + 25)}{s(s + 20)}$$

The corresponding Norton equivalent circuit is shown in Fig. 10-38.

At node $b$,

$$\left\{\frac{10(s + 10)}{s + 20} + s + 10\right\}V_b(s) = \frac{4(s + 25)}{s(s + 20)} + 0.2$$

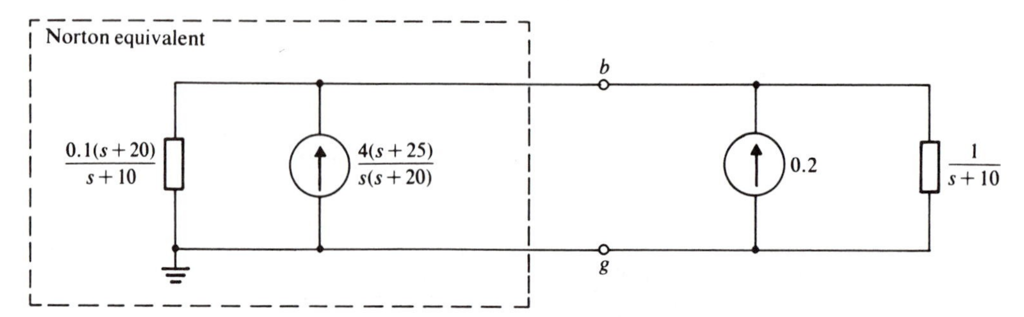


**Fig. 10-38**

or

$$V_b(s) = \frac{0.2(s^2 + 40s + 500)}{s(s + 20)} \frac{s + 20}{s^2 + 40s + 300} = \frac{0.2(s^2 + 40s + 500)}{s(s + 10)(s + 30)}$$

### *Maximum Power*

**10.15** Figure 10-39*b* is the Thévenin equivalent of the circuit of Fig. 10-39*a*. Determine the value of $Z_L(j\omega)$ which will result in the maximum active power being delivered by the network to the load.

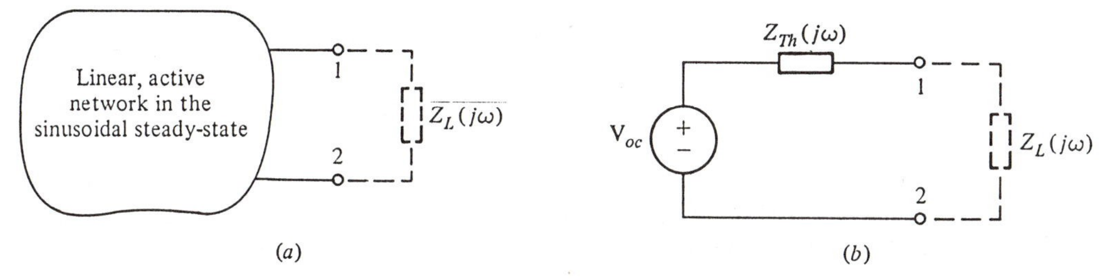


**Fig. 10-39**

Consider the circuit in Fig. 10-40, where the Thévenin phasor voltage $\mathbf{V}_{oc}$ is stationary and of magnitude $V_{\text{eff}}$. Then the loop current is

$$I_{\text{eff}} = \frac{V_{\text{eff}}}{\sqrt{(R_{\text{Th}} + R_L)^2 + (X_{\text{Th}} + X_L)^2}}$$

and the real power dissipated in the load becomes

$$P = I_{\text{eff}}^2 \text{Re}[Z_L(j\omega)] = I_{\text{eff}}^2 R_L$$

$$= \frac{V_{\text{eff}}^2 R_L}{(R_{\text{Th}} + R_L)^2 + (X_{\text{Th}} + X_L)^2}$$

As the first step in maximizing the above expression for load power, we can set $X_L = -X_{\text{Th}}$. This is both logical and legal since no real power can be dissipated in a reactance.

At this point, $P = V_{\text{eff}}^2 R_L/(R_{\text{Th}} + R_L)^2$; and to establish the conditions for an extremum as $R_L$ varies, we equate the derivative $dP/dR_L$ to zero. That is,

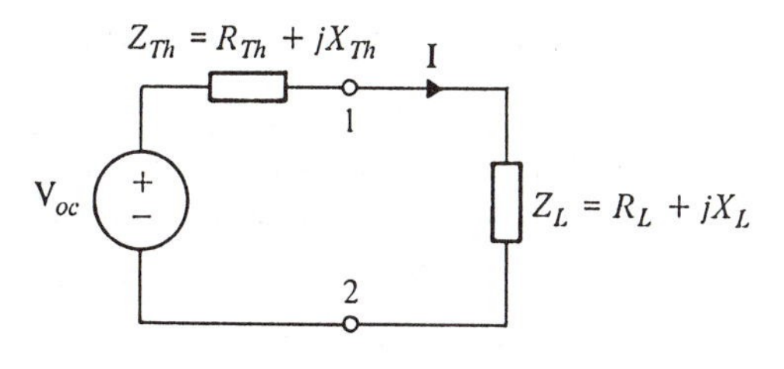


**Fig. 10-40**

$$\frac{dP}{dR_L} = \frac{(R_{Th} + R_L)^2 V_{eff}^2 - 2V_{eff}^2 R_L (R_{Th} + R_L)}{(R_{Th} + R_L)^4} = 0$$

or

$$(R_{Th} + R_L)^2 V_{eff}^2 - 2V_{eff}^2 R_L (R_{Th} + R_L) = 0$$

from which

$$R_{Th} + R_L - 2R_L = 0 \quad \text{or} \quad R_L = R_{Th}$$

which we can show, by further differentiation, to be the condition for a *maximum* of $P$.

Summarizing, the maximum possible power will be transferred to $Z_L(j\omega)$ from the original network when

$$Z_L(j\omega) = R_{Th} - jX_{Th} = Z_{Th}^*(j\omega)$$

where $Z_{Th}^*(j\omega)$ is the complex conjugate of $Z_{Th}(j\omega)$. Thus Theorem 10.6 is proved. Notice that this assumes $Z_L(j\omega)$ and only $Z_L(j\omega)$ is variable.

**10.16** Find the value of $R_L$ such that maximum possible power will be extracted from the network $A$ in Fig. 10-41. Find also the value of $P_{max}$ and the amount of power supplied by the source under these conditions.

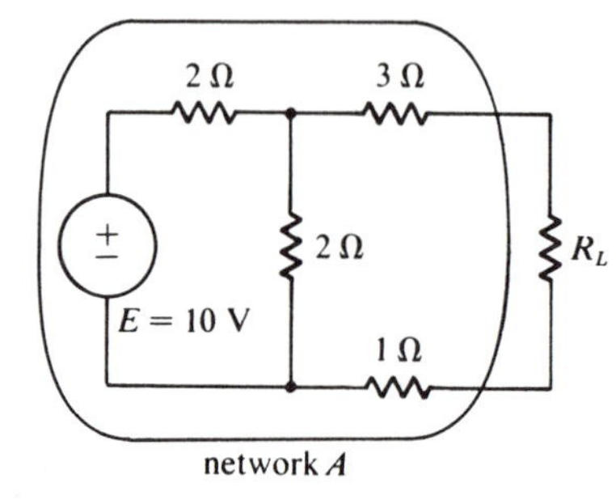


**Fig. 10-41**

The first move is to find the Thévenin equivalent insofar as the resistor $R_L$ is concerned. Using Figs. 10-42*a* and *b*,

$$V_{oc} = \frac{2}{2+2} E = 5 \text{ V}$$

and

$$R_{Th} = 3 + 1 + \frac{1}{\frac{1}{2} + \frac{1}{2}} = 5\,\Omega$$

Thus we arrive at the Thévenin equivalent of Fig. 10-42*c*, and for maximum power transfer, $R_L$ should equal 5 Ω (see Theorem 10.6).

From Fig. 10-42*c*, $I = 5/10 = 0.5$ A and the power in $R_L$ is

$$P_{max} = I^2 R_L = (0.5)^2 5 = 1.25 \text{ W}$$

To find the power delivered by the *source*, we *must* work with the *original* network since the Thévenin equivalent of Fig. 10-42*c* is valid *only* in relation to the

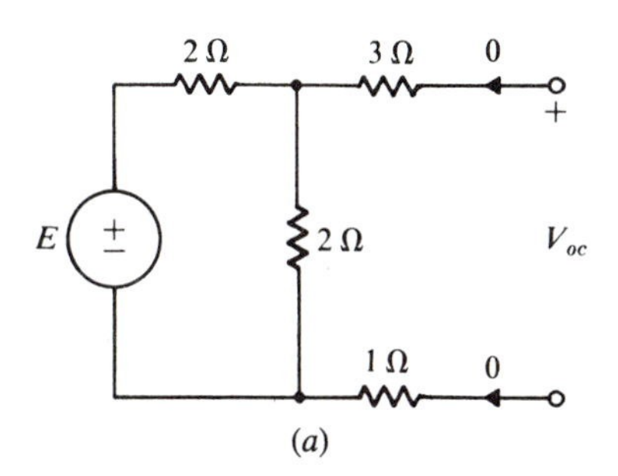


(a)

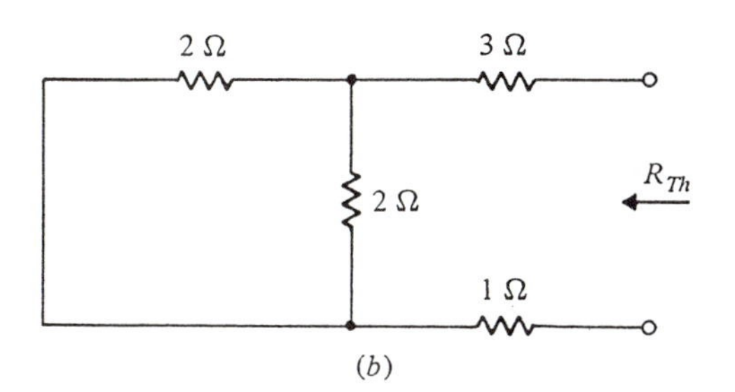


(b)

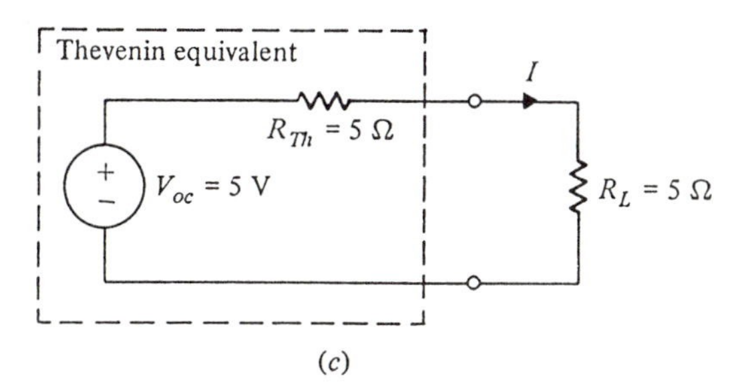


(c)

**Fig. 10-42**

*load*. Referring to Fig. 10-41, with $R_L = 5\,\Omega$, the impedance seen by the source is

$$R_{\text{in}} = 2 + \frac{1}{\frac{1}{2} + \frac{1}{9}} = 2 + 1.64 = 3.64\ \Omega$$

and

$$P_s = \frac{V_s^2}{R_{\text{in}}} = \frac{10^2}{3.64} = 27.5\ W$$

*Comment:* The transmission efficiency $P_{\max}/P_s$ is usually low in a maximum-power situation.

**10.17** The Norton equivalent of a linear active circuit is shown in Fig. 10-43. $\mathbf{I}_{sc}$ is

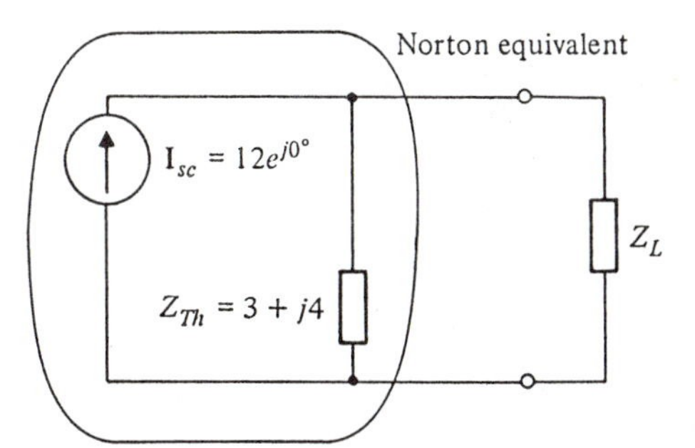


**Fig. 10-43**

a stationary phasor, and we are given $I_{sc,\text{eff}} = 12$ A. Find $Z_L(j\omega)$ for maximum real power to the load and the corresponding value of $P_{\max}$.

In order to apply the maximum power theorem we must convert Fig. 10-43 to the Thévenin equivalent of Fig. 10-44. This is no more than a source transformation.

Thus for maximum power, $Z_L(j\omega) = Z^*_{\text{Th}}(j\omega) = 3 - j4$. Under these conditions,

$$\mathbf{I} = \frac{60e^{j53.2^\circ}}{3 + j4 + 3 - j4} = 10e^{j53.2^\circ} \quad \text{or} \quad I_{\text{eff}} = 10 \text{ A}$$

Therefore the maximum real power to the load is $P_{\max} = I^2_{\text{eff}} R_L = (10^2)3 = 300$ W.

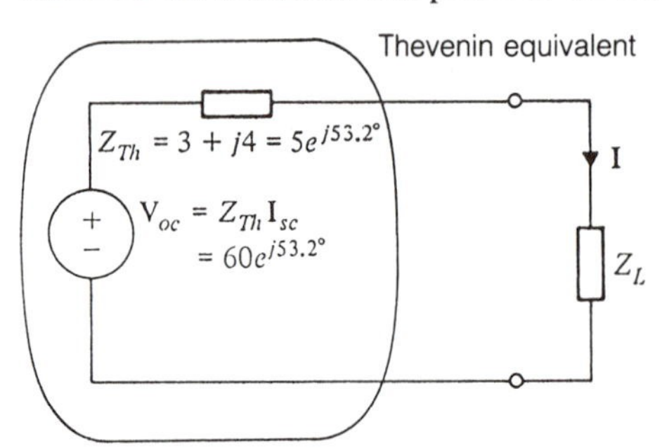


**Fig. 10-44**

**10.18** Suppose $Z_L(j\omega)$ in Problem 10.17 (see Fig. 10-43) is restricted to *resistive* values. What is then the condition for maximum real power into the load?

The appropriate Thévenin equivalent is shown in Fig. 10-45. We *cannot* apply the maximum power theorem since $Z_L(j\omega)$ is not freely adjustable. It is necessary to start afresh. Thus the real power in the load is

$$P = I^2_{\text{eff}} R_L = \left\{\frac{|\mathbf{V}_{oc}|}{|Z_{\text{Th}}(j\omega) + R_L|}\right\}^2 R_L = \frac{(60)^2 R_L}{(3 + R_L)^2 + 16}$$

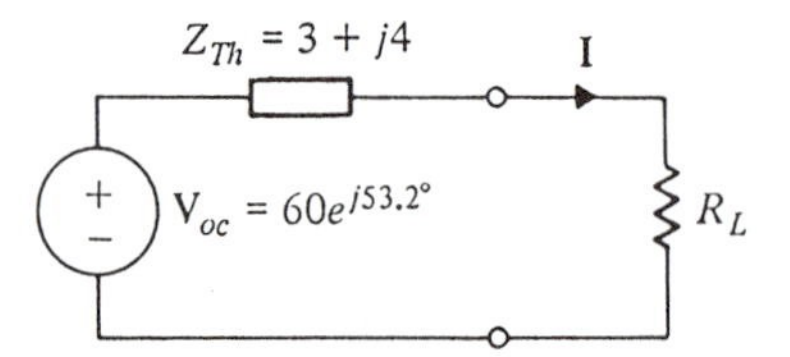


**Fig. 10-45**

And, proceeding as in Problem 10.15,

$$\frac{dP}{dR_L} = \frac{\{(3 + R_L)^2 + 16\}(60)^2 - (60)^2 R_L\{2(3 + R_L)\}}{\{(3 + R_L)^2 + 16\}^2} = 0$$

or

$$(3 + R_L)^2 + 16 - R_L(6 + 2R_L) = 0$$

from which for $P_{\max}$, $R_L = 5\ \Omega$.

*Comment:* In general, if the variation of $Z_L(j\omega)$ is restricted such that $\underline{/Z_L(j\omega)}$ is constant, then we must make $|Z_L(j\omega)| = |Z_{Th}(j\omega)|$ in order to achieve maximum active power transfer.

**10.19** Figure 10-46 shows the Thévenin equivalent of an amplifier driving a loudspeaker through an ideal transformer. Find the value of the transformation ratio $n$ which will result in maximum power being transferred to the loudspeaker.

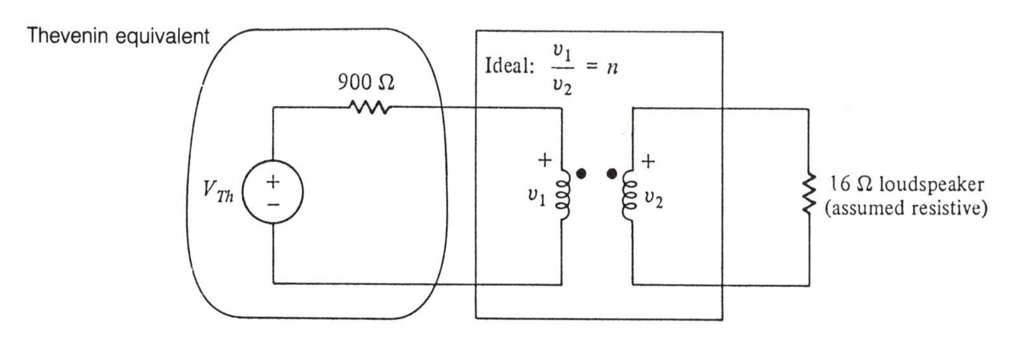


**Fig. 10-46**

Using the results of the previous problems, the idea is to *match* the 900-Ω Thévenin impedance. Now from equation (*9.14*), page 307, we know that the impedance seen looking into the primary of the ideal transformer is $16n^2$. So for a maximum power match we require

$$16n^2 = 900 \quad \text{or} \quad n = \sqrt{900/16} = 7.5$$

*Comment:* The concept of matching source and load for maximum power transfer finds its applications almost entirely in communications circuits. The low efficiency (high losses in the source network) is usually not critical in communications, but would be completely unacceptable in the electric power industry.

### *Tellegen's Theorem*

**10.20** In the circuit of Fig. 10-47 *associated* reference polarities and directions have been assigned. Choose a set of branch voltages $v_k'$ and currents $i_k''$ such that

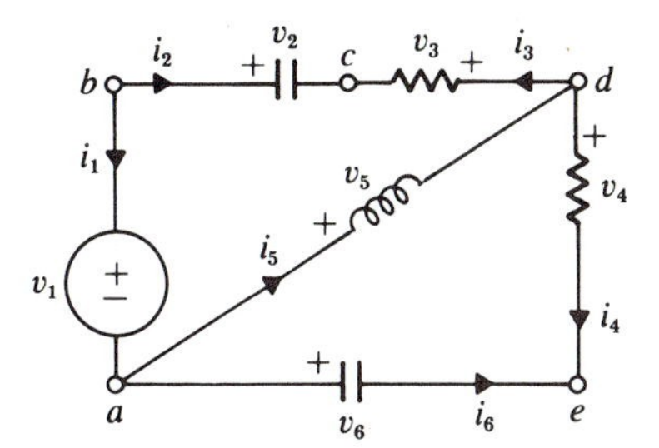


**Fig. 10-47**

Kirchhoff's laws are satisfied. Then show that these voltages and currents satisfy Tellegen's theorem.

We could, for example, choose $v_1' = 1$ V, $v_2' = 2$ V, and $v_3' = 3$ V. Then KVL requires $v_5' = -v_1' + v_2' - v_3' = -2$ V.

Next, if we choose $v_4' = 4$ V, then from KVL, $v_6' = v_5' + v_4' = 2$ V.

Turning now to the currents, if we choose to let $i_1'' = 1$ A, then by KCL, $i_2'' = -1$ A and $i_3'' = 1$ A. Next, choosing $i_6'' = 6$ A, KCL requires $i_4'' = -6$ A and $i_5'' = i_3'' + i_4'' = -5$ A.

Summarizing, the following voltages and currents satisfy the constraints of KVL and KCL:

$$v_1' = 1\text{ V},\quad v_2' = 2\text{ V},\quad v_3' = 3\text{ V},\quad v_4' = 4\text{ V},\quad v_5' = -2\text{ V},\quad v_6' = 2\text{ V}$$

$$i_1'' = 1\text{ A},\quad i_2'' = -1\text{ A},\quad i_3'' = 1\text{ A},\quad i_4'' = -6\text{ A},\quad i_5'' = -5\text{ A},\quad i_6'' = 6\text{ A}$$

If we apply Tellegen's theorem to this set,

$$\sum_{k=1}^{k=6} v_k' i_k'' = (1)(1) + (2)(-1) + (3)(1) + (4)(-6) + (-2)(-5) + (2)(6) = 0$$

*Comment:* This is a graphic illustration of the fact that Tellegen's theorem depends only on the Kirchhoff laws and *not* in any way on the nature of the circuit elements themselves. Note, too, that there was no special relationship between the set of $v_k'$ and the set of $i_k''$. They were chosen independently.

**10.21** The network $N$ in Fig. 10-48 has several ports where other networks $N_1$, $N_2$, . . . are connected, Show that

$$\sum_{\substack{\text{all}\\ \text{ports}}} v_p' i_p'' = \sum_{\substack{\text{network}\\ N}} v_n' i_n''$$

Tellegen's theorem states that

$$\sum_{\substack{\text{all}\\ \text{branches}}} v_k' i_k'' = 0$$

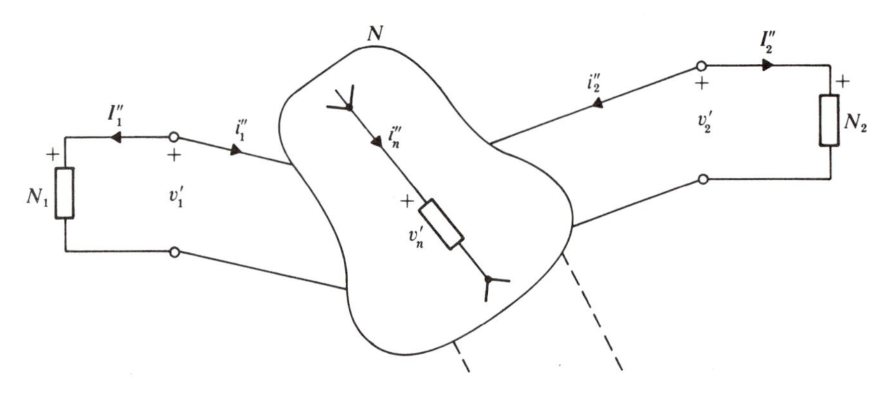


**Fig. 10-48**

It is convenient to consider the branches of Fig. 10-48 in two groups: those in $N$, and the branches $N_1, N_2, \ldots$ . Then expanding the summation of Tellegen's theorem,

$$v_1'I_1'' + v_2'I_2'' + \cdots + \sum_{\substack{\text{network}\\ N}} v_n'i_n'' = 0$$

But by the associated sign convention, $I_1'' = -i_1''$, $I_2'' = -i_2''$, . . . and so

$$-v_1'i_1'' - v_2'i_2'' - \cdots + \sum_{\substack{\text{network}\\ N}} v_n'i_n'' = 0$$

or

$$\sum_{\substack{\text{all}\\ \text{ports}}} v_p'i_p'' = \sum_{\substack{\text{network}\\ N}} v_n'i_n''$$

The primes and double primes simply indicate as usual that the set of voltages and the set of currents need not be related.

**10.22** In Fig. 10-49 a linear, time-invariant, passive, two-port network $N$ is shown with two different external connections and two corresponding sets of port currents and voltages. Use Tellegen's theorem to show that

$$\mathbf{V}_1'\mathbf{I}_1'' + \mathbf{V}_2'\mathbf{I}_2'' = \mathbf{V}_1''\mathbf{I}_1' + \mathbf{V}_2''\mathbf{I}_2'$$

From Problem 10.21,

$$\mathbf{V}_1'\mathbf{I}_1'' + \mathbf{V}_2'\mathbf{I}_2'' = \sum_N \mathbf{V}_n'\mathbf{I}_n'' \tag{1}$$

$$\mathbf{V}_1''\mathbf{I}_1' + \mathbf{V}_2''\mathbf{I}_2' = \sum_N \mathbf{V}_n''\mathbf{I}_n'$$

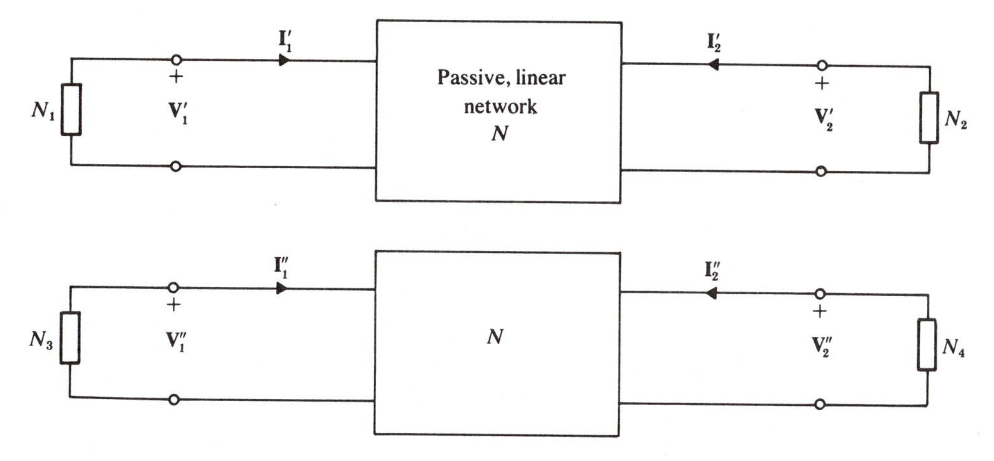


**Fig. 10-49**

Now $N$ has been defined so that each branch in $N$ has a fixed impedance $Z_n(j\omega)$,

$$\mathbf{V}'_n = Z_n(j\omega)\mathbf{I}'_n \quad \text{or} \quad \mathbf{V}''_n = Z_n(j\omega)\mathbf{I}''_n$$

and therefore

$$\mathbf{V}'_n\mathbf{I}''_n = \mathbf{V}''_n\mathbf{I}'_n \quad \text{or} \quad \sum_N \mathbf{V}'_n\mathbf{I}''_n = \sum_N \mathbf{V}''_n\mathbf{I}'_n$$

Given this equality, we may conclude from equation (*1*) that

$$\mathbf{V}'_1\mathbf{I}''_1 + \mathbf{V}'_2\mathbf{I}''_2 = \mathbf{V}''_1\mathbf{I}'_1 + \mathbf{V}''_2\mathbf{I}'_2$$

*Comment:* This is a more general statement of reciprocity than those considered earlier. In fact, the earlier reciprocity relationships can easily be deduced as special cases of the above equation.

**10.23** The following measurements have been made on the resistive network of Fig. 10-50.

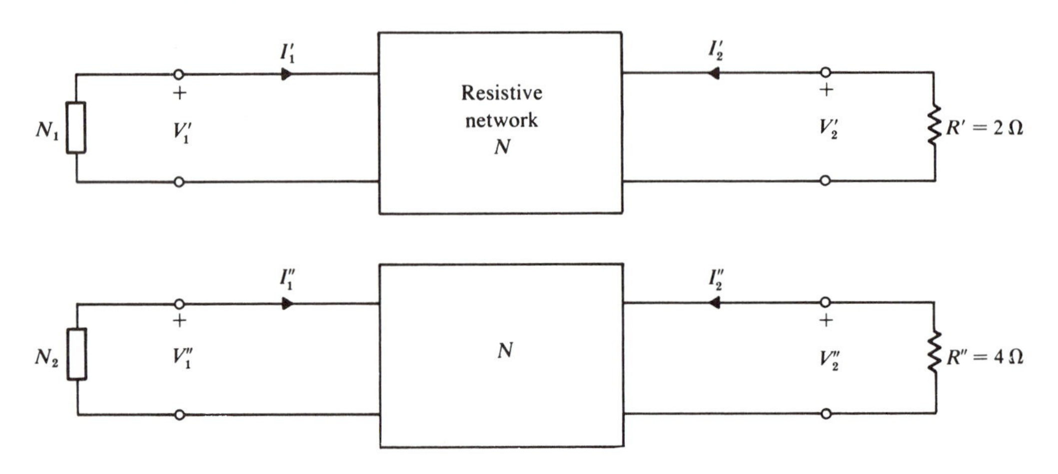


**Fig. 10-50**

$$V'_1 = 12\text{ V} \qquad I'_1 = 3\text{ A} \qquad R' = 2\ \Omega$$

$$V'_2 = 3\text{ V} \qquad I'_2 = -1.5\text{ A}$$

Use the reciprocity relation of Problem 10.22 to find the power dissipated in $R''$ when the conditions are changed to $V''_1 = 22$ V, $I''_1 = 5$ A, and $R'' = 4\ \Omega$.

From problem 10.22,

$$V'_1I''_1 + V'_2I''_2 = V''_1I'_1 + V''_2I'_2$$

from which

$$(12)(5) + (3)I''_2 = (22)(3) + V''_2(-1.5)$$

But we also know that $V''_2 = -R''I''_2 = -4I''_2$. Thus

$$60 + 3I''_2 = 66 + 6I''_2 \quad \text{or} \quad I''_2 = -2\text{ A}$$

and consequently $V_2'' = -4I_2'' = 8$ V. Therefore

$$P'' = V_2''(-I_2'') = 16 \text{ W}$$

### *Substitution and Compensation*

**10.24** Use the substitution theorem (Theorem 10.8, page 344) to replace the effects of mutual inductance in the circuit of Fig. 10-51 by equivalent dependent sources.

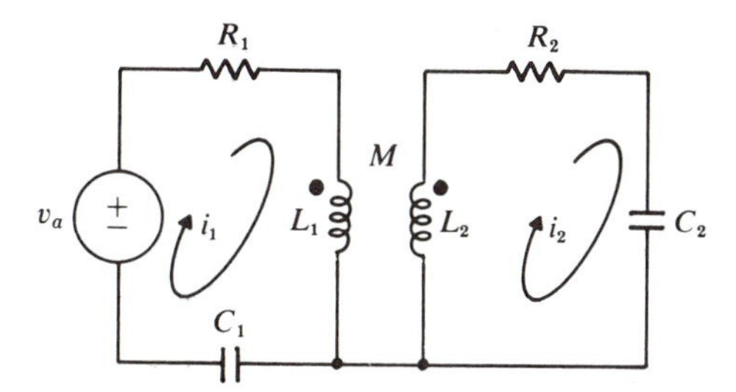


**Fig. 10-51**

The phasor loop equations for the circuit in Fig. 10-51 are

$$\left(R_1 + j\omega L_1 + \frac{1}{j\omega C_1}\right)\mathbf{I}_1 - j\omega M\mathbf{I}_2 = \mathbf{V}_a$$

$$-j\omega M\mathbf{I}_1 + \left(R_2 + j\omega L_2 + \frac{1}{j\omega C_2}\right)\mathbf{I}_2 = 0$$

Theorem 10.8 states that we may replace the "branches" whose voltages are $-j\omega M\mathbf{I}_2$ and $-j\omega M\mathbf{I}_1$ by voltage sources of the same value. We have done this in Fig. 10-52.

It is easy to see that this substitution has in no way altered the circuit's phasor loop equations.

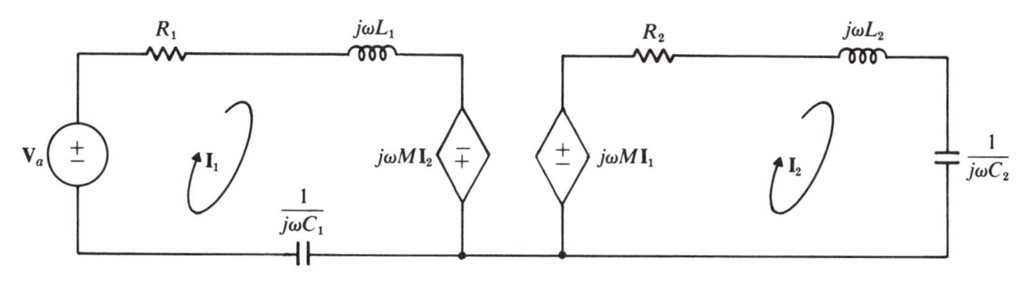


**Fig. 10-52**

**10.25** The circuit of Fig. 10-53$a$ is a linear active network in the sinusoidal steady-state. Its Thévenin equivalent with respect to terminals $a$-$b$ is shown in Fig. 10-53$b$. Determine the change $\Delta\mathbf{I}$ in the current $\mathbf{I}$ which would be caused by a small change in $Z$ (see Fig. 10-53$c$).

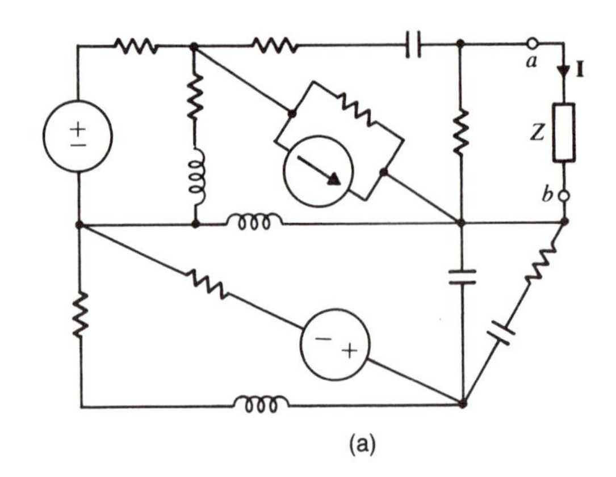


(a)

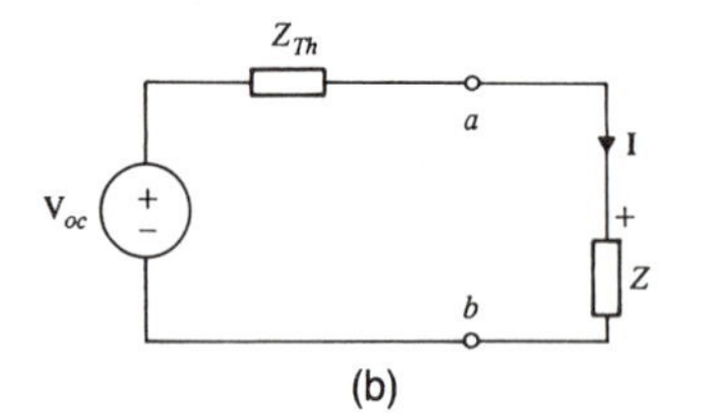


(b)

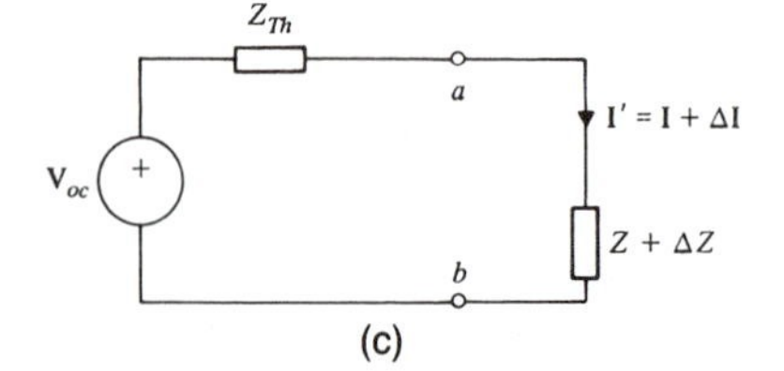


(c)

**Fig. 10-53**

From Figs. 10-53*b* and *c*,

$$(Z_{\text{Th}} + Z)\mathbf{I} = \mathbf{V}_{oc}$$

and

$$(Z_{\text{Th}} + Z + \Delta Z)\ (\mathbf{I} + \Delta\mathbf{I}) = \mathbf{V}_{oc}$$

Eliminating $\mathbf{V}_{oc}$,

$$(Z_{\text{Th}} + Z + \Delta Z)\ (\mathbf{I} + \Delta\mathbf{I}) = (Z_{\text{Th}} + Z)\mathbf{I}$$

or

$$\Delta Z\mathbf{I} + (Z_{\text{Th}} + Z + \Delta Z)\Delta\mathbf{I} = 0$$

and

$$\Delta\mathbf{I} = \frac{-\Delta Z\mathbf{I}}{Z_{\text{Th}} + Z + \Delta Z} \doteq \frac{-\Delta Z\mathbf{I}}{Z_{\text{Th}} + Z} \quad \text{if} \quad |\Delta Z| \ll |Z + Z_{\text{Th}}|$$

*Alternative Solution*

Theorem 10.9 (the compensation theorem) states that the effect of changing $Z$ to $Z + \Delta Z$ can be found by calculating the effect of the voltage source $\Delta Z\mathbf{I}$ as shown in Fig. 10-54. Therefore

$$\Delta\mathbf{I} = \frac{-\Delta Z\mathbf{I}}{Z_{\text{Th}} + Z + \Delta Z} \doteq \frac{-\Delta Z\mathbf{I}}{Z_{\text{Th}} + Z} \qquad \text{if } |\Delta Z| \ll |Z + Z_{\text{Th}}|$$

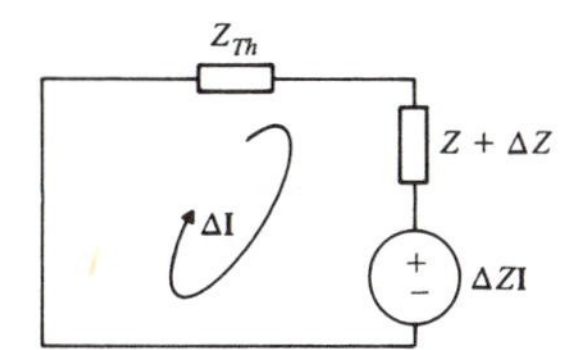


**Fig. 10-54**

**10.26** The voltage $V_{ab}$ across a resistor $R$ in a linear, active, resistive circuit is to be read by a voltmeter of resistance $r$. The Thévenin equivalent of the remainder of the circuit is assumed known (see Fig. 10-55), and it may be assumed that $r \gg R$. Find the fractional error $\Delta V_{ab}/V_{ab}$ caused by the *loading* effect of the voltmeter on the circuit.

Before the voltmeter is connected, $I = V_{oc}/(R_{\text{Th}} + R)$ and $V_{ab} = RV_{oc}/(R_{\text{Th}} + R)$. When the voltmeter is connected, the resistance between the terminals $a$-$b$ will change by

$$\Delta Z = \Delta R = \frac{1}{1/R + 1/r} - R \doteq -\frac{R^2}{r} \qquad \text{for } r \gg R$$

We can now use Theorem 10.9 to find $\Delta I$. Thus from Fig. 10-56,

$$\Delta I = \left\{\frac{R^2 V_{oc}}{r(R_{\text{Th}} + R)}\right\} \Big/ (R_{\text{Th}} + R) = \frac{R^2 V_{oc}}{r(R_{\text{Th}} + R)^2}$$

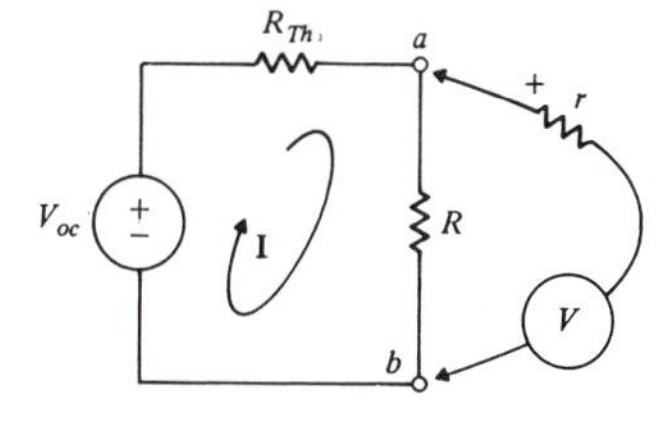


**Fig. 10-55**

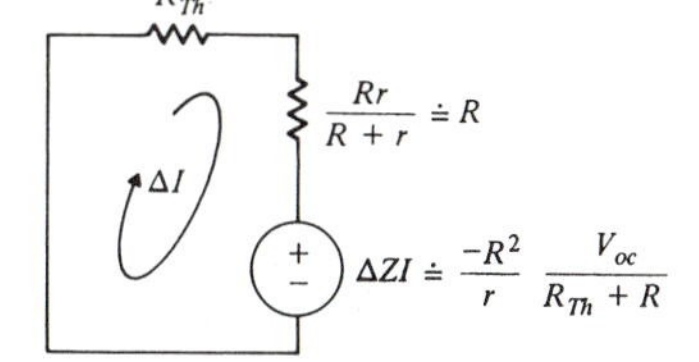


**Fig. 10-56**

Now from KVL,

$$\Delta V_{ab} = -\Delta V_{R_{\text{Th}}} = -R_{\text{Th}}\Delta I = \frac{-R_{\text{Th}} R^2 V_{oc}}{r(R_{\text{Th}} + R)^2}$$

and so the fractional error is

$$\frac{\Delta V_{ab}}{V_{ab}} = \frac{-R_{\text{Th}} R}{r(R_{\text{Th}} + R)}$$

### *Y-Δ Transformations*

**10.27** Find $Z_a(s)$, $Z_b(s)$, and $Z_c(s)$ in the tee-equivalent of the pi-network of Fig. 10-57$a$. Hence find the tee-equivalent in the $j\omega$-domain when $\omega = 2$ rad/s and when $\omega = 1$ rad/s.

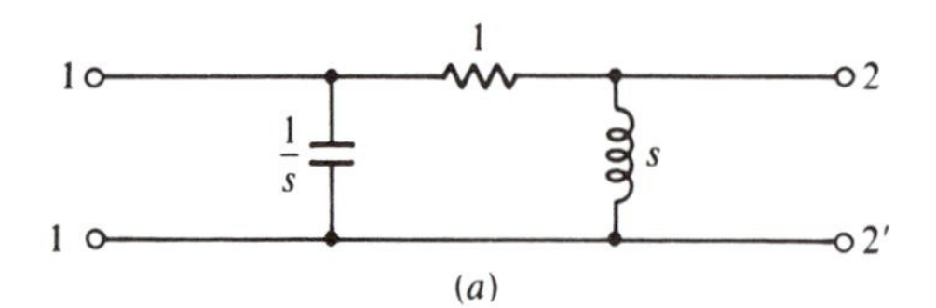


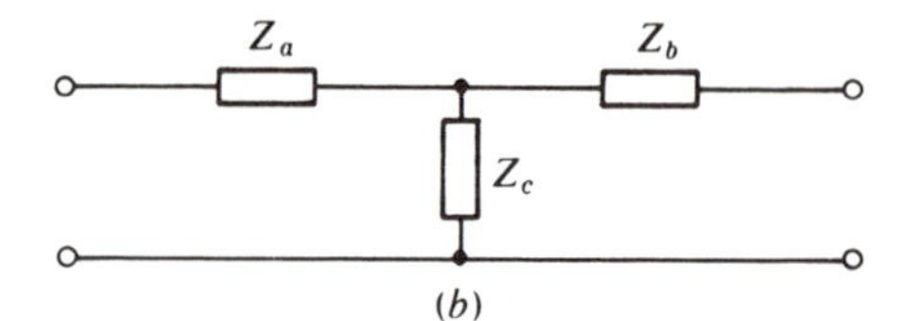


**Fig. 10-57**

From Theorem 10.10, page 344 (and Fig. 10-8*b*),

$$Z_a(s) = \frac{(1)\,(1/s)}{s + 1 + 1/s} = \frac{1}{s^2 + s + 1}$$

$$Z_b(s) = \frac{(1)\,(s)}{s + 1 + 1/s} = \frac{s^2}{s^2 + s + 1}$$

$$Z_c(s) = \frac{(s)\,(1/s)}{s + 1 + 1/s} = \frac{s}{s^2 + s + 1}$$

If we now set $s = j2$,

$$Z_a(j2) = \frac{1}{-4 + j2 + 1} = \frac{1}{-3 + j2} = \frac{-3 - j2}{13}$$

$$Z_b(j2) = \frac{-4}{-3 + j2} = \frac{12 + j8}{13}$$

$$Z_c(j2) = \frac{j2}{-3 + j2} = \frac{4 - j6}{13}$$

*Note:* It is not necessarily possible to build the equivalent circuit from passive elements. Here, for example, $Z_a(j2)$ has a negative real part—a negative resistance!

Finally, setting $s = j1$,

$$Z_a(j1) = \frac{1}{j}, \quad Z_b(j1) = j, \quad \text{and} \quad Z_c(j1) = 1$$

In this instance the tee-equivalent is easily built (see Fig. 10-58). But the equivalency is valid only at the one frequency $\omega = 1$ rad/s.

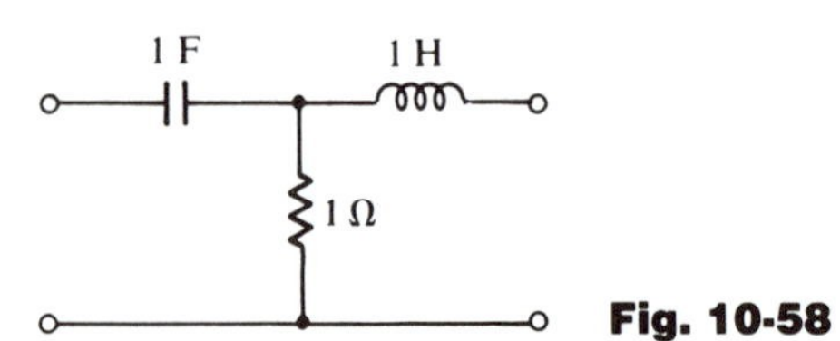


**Fig. 10-58**

**10.28** Use the wye-delta transformation to convert the unbalanced wye load of Fig. 10-59 to a delta load, and then solve for the line currents.

The equivalent Δ-network is shown in Fig. 10-60*a*, where from Theorem 10.10,

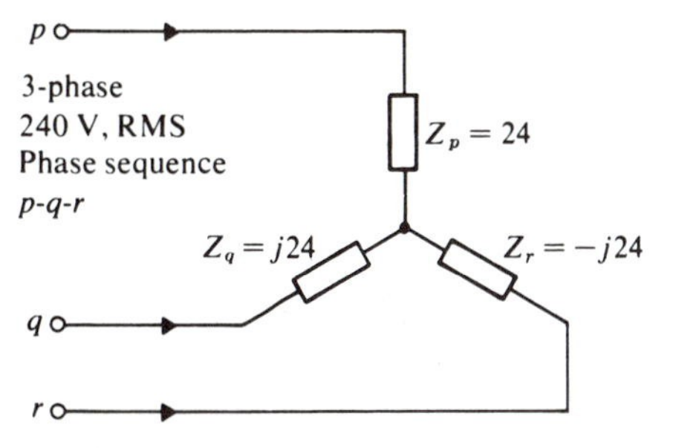


Fig. 10-59

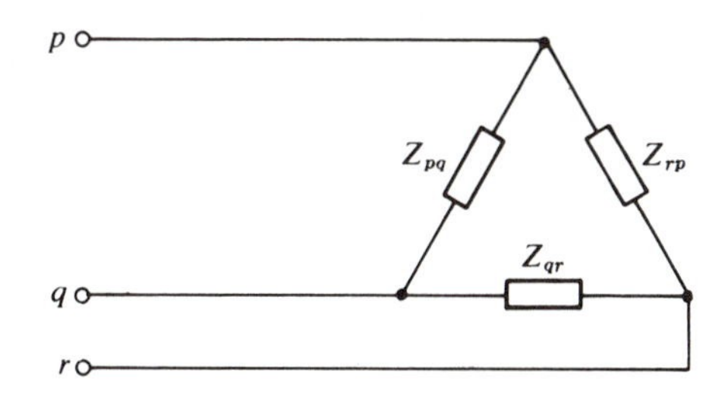


(a)

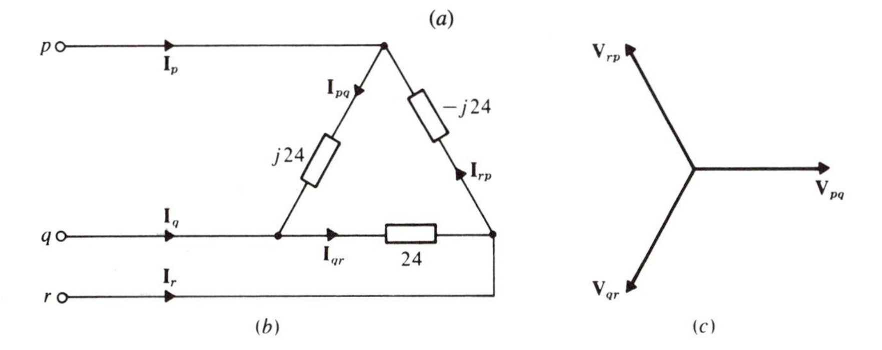


(b) (c)

Fig. 10-60

$$Z_{pq} = \frac{Z_pZ_q + Z_qZ_r + Z_rZ_p}{Z_r} = \frac{(24)\,(j24) + (j24)(-j24) + (-j24)\,(24)}{-j24} = j24$$

$$Z_{qr} = \frac{24^2}{Z_p} = \frac{24^2}{24} = 24 \quad \text{and} \quad Z_{rp} = \frac{24^2}{Z_q} = \frac{24^2}{j24} = -j24$$

as shown in Fig. 10-60*b*.

Choosing $\mathbf{V}_{pq}$ as the stationary reference phasor, as in Fig. 10-60*c*,

$$\mathbf{I}_p = \mathbf{I}_{pq} - \mathbf{I}_{rp} = \frac{240e^{j0°}}{j24} - \frac{240e^{j120°}}{-j24} = -j10 - (-8.66 - j5) = 10e^{-j30°}$$

$$\mathbf{I}_q = \mathbf{I}_{qr} - \mathbf{I}_{pq} = \frac{240e^{-j120°}}{24} - \frac{240e^{j0°}}{j24} = 5.17e^{j165°}$$

$$\mathbf{I}_r = \mathbf{I}_{rp} - \mathbf{I}_{qr} = \frac{240e^{j120°}}{-j24} - \frac{240e^{-j120°}}{24} = 5.17e^{j135°}$$

*Comment:* This is a simpler way to treat a 3-phase unbalanced-Y than was the direct approach of Problem 9.22.

### *Scaling and Normalization*

**10.29** The driving-point impedance of the network in Fig. 10-61*a* is $Z(j\omega) = R + j\omega L + 1/j\omega C$. Suppose we want a network with the same general characteristics (resonant frequency, bandwidth, etc.), but at an impedance level $\alpha$ times the original. Find the appropriate values of $R'$, $L'$, and $C'$ (see Fig. 10-61*b*).

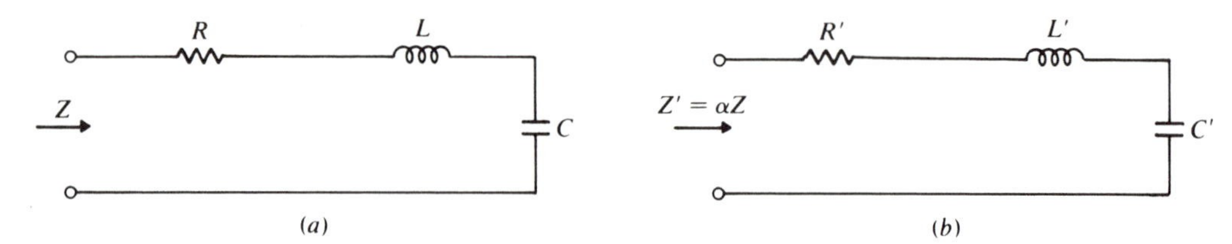


**Fig. 10-61**

The elements must be *magnitude scaled* according to the rules stated in Theorem 10.11, page 345, where $k$ is now $1/\alpha$. Thus

$$R' = \frac{1}{k}R = \alpha R, \qquad L' = \frac{1}{k}L = \alpha L, \qquad C' = kC = \frac{1}{\alpha}C$$

This is also evident from the fact that each term in $Z(j\omega)$ is to be scaled by a factor $\alpha$.

If we were to plot $|Z(j\omega)|$ vs. $\omega$, and $|Z'(j\omega)|$ vs. $\omega$, the curves would be related by a vertical scale factor of $\alpha$ as shown in Fig. 10-62, where it has been assumed that $\alpha > 1$.

Magnitude scaling has no effect on phase. That is, $\angle Z(j\omega) = \angle Z'(j\omega)$.

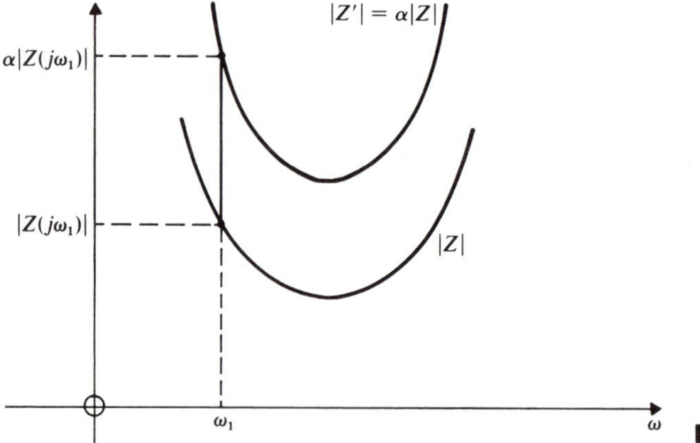


**Fig. 10-62**

**10.30** Suppose the network of Fig. 10-61$a$ is to be *frequency scaled*. That is, we want an impedance $Z'$ which, at a frequency $\omega' = \beta\omega$, will equal the original impedance $Z$ at a frequency $\omega$. In other words, find the values of $R'$, $L'$, and $C'$ which will make $Z'(j\beta\omega) = Z(j\omega)$.

Referring to Theorem 10.11, $\beta = 1/K$ and

$$R' = R, \qquad L' = KL = \frac{1}{\beta}L, \qquad C' = KC = \frac{1}{\beta}C$$

This is also evident from the fact that each term in $Z$ is to take on the same value at $\beta$ times the frequency.

As can be seen in Fig. 10-63, the effect of frequency scaling is to move the impedance curve horizontally. The impedance level at each new frequency is unchanged, and the same is true of the phase.

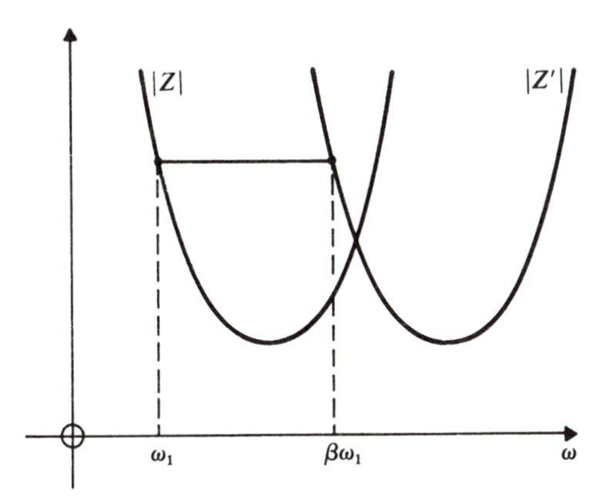


**Fig. 10-63**

**10.31** Figure 10-64$a$ shows a low-pass filter having a transfer function

$$H(s) = \frac{V_2(s)}{V_1(s)} = \frac{1}{s^3 + 2s^2 + 2s + 1}$$

$|H(j\omega)|$ is plotted vs. $\omega$ in Fig. 10-64$b$. It is desired to change this circuit so

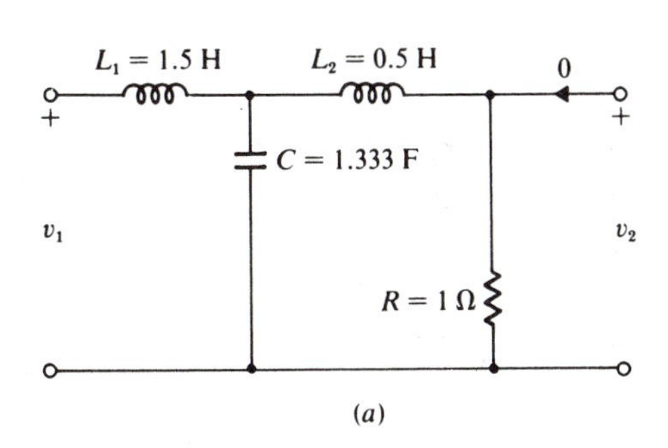


(a)

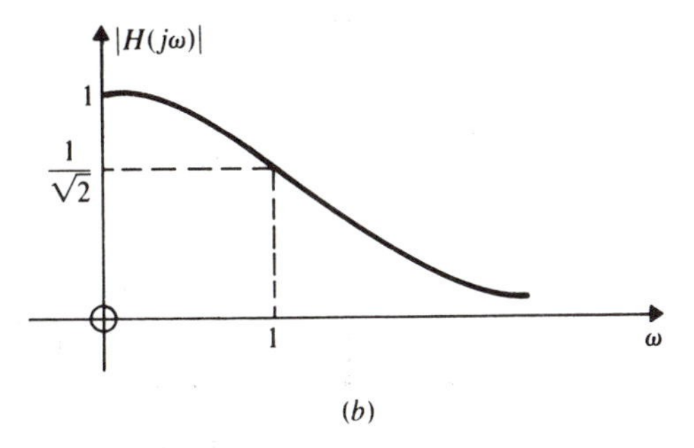


(b)

**Fig. 10-64**

that the impedance level will be 1000 Ω at $\omega = 0$ and so that the gain will be down by 3 dB at $10/2\pi$ kHz. Determine the new element values.

Since the low frequency impedance of the circuit in Fig. 10-64 is 1 Ω, and the gain is down by 3 dB at $\omega = 1$ rad/s or $f = 1/2\pi$ Hz, the required scale factors are

$$\alpha = \frac{1}{k} = 1000 \qquad \text{and} \qquad \beta = \frac{1}{K} = 10{,}000$$

for magnitude and frequency, respectively. It then follows from Theorem 10.11 that

$$R' = \frac{1}{K}R = 1000R = 1000\ \Omega \qquad\qquad L_2' = \frac{K}{k}L_2 = \frac{1000}{10{,}000}L_2 = 0.05\ \text{H}$$

$$L_1' = \frac{K}{k}L_1 = \frac{1000}{10{,}000}L_1 = 0.1\ L_1 = 0.15\ \text{H}$$

$$C' = KkC = (10^{-4})\ (10^{-3})C = 0.1333\mu\text{F}$$

The new circuit is shown in Fig. 10-65$a$ and its response is sketched in Fig. 10-65$b$.

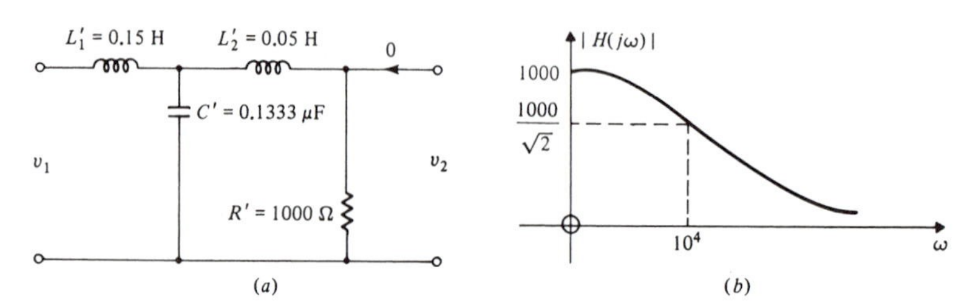


**Fig. 10-65**

## *Network Graphs and Duality*

**10.32** (*a*) Draw the oriented graph for the network of Fig. 10-66.
(*b*) Write the augmented incidence matrix $\mathbf{A}_a$.
(*c*) Draw at least three trees for this network, and on each graph show how the fundamental cut sets would be determined.
(*d*) Find the dual graph.

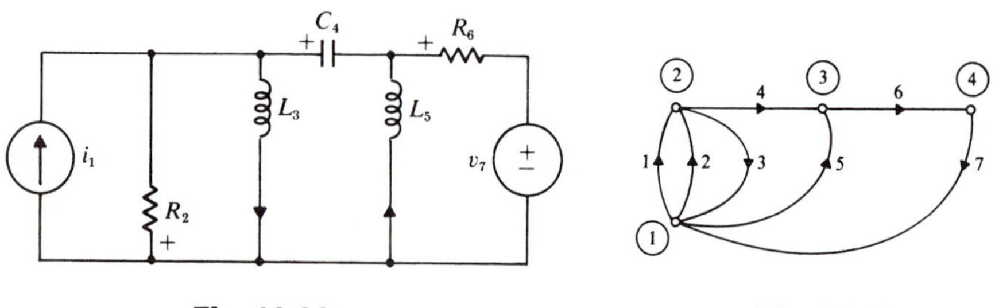


**Fig. 10-66** **Fig. 10-67**

(*a*) See Fig. 10-67. The orientation of the branches has been derived from the given reference voltage polarities and/or reference current directions.

(*b*) From the definition of the augmented incidence matrix (see page 347),

$$\mathbf{A}_a = \begin{bmatrix} 1 & 1 & -1 & 0 & 1 & 0 & -1 \\ -1 & -1 & 1 & 1 & 0 & 0 & 0 \\ 0 & 0 & 0 & -1 & -1 & 1 & 0 \\ 0 & 0 & 0 & 0 & 0 & -1 & 1 \end{bmatrix}$$

(*c*) Three possible trees—there are many others—are shown in Fig. 10-68. The dashed lines define (i.e. cut) the branches in each fundamental cut set.

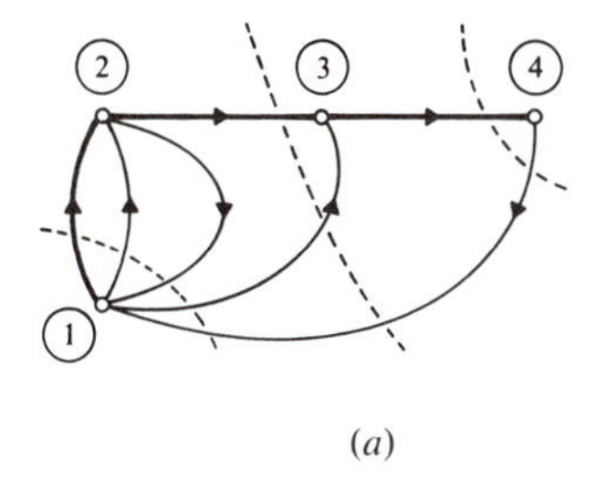


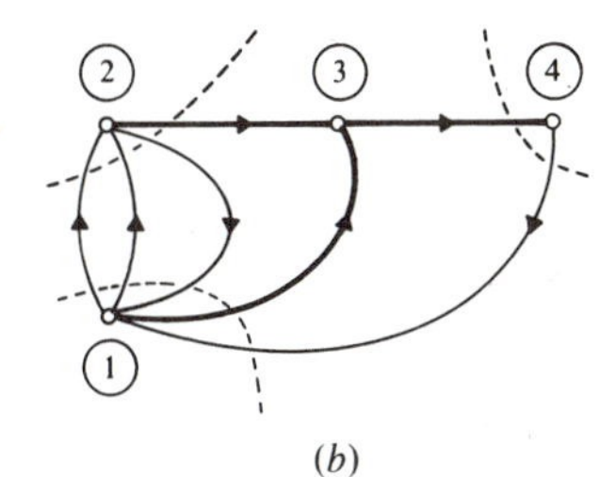


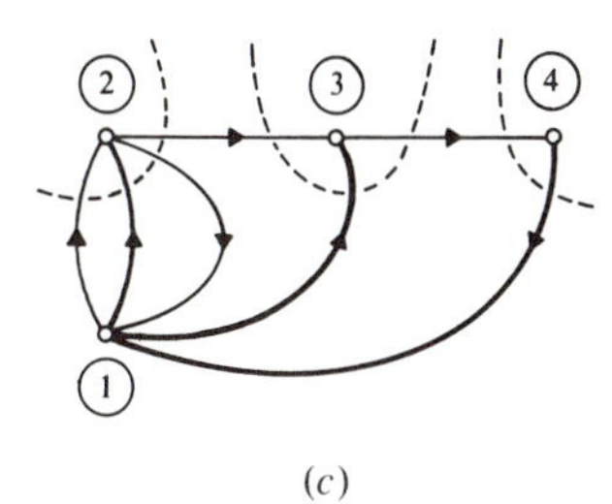


**Fig. 10-68**

(*d*) Applying the algorithm for obtaining the dual graph, we first number the meshes of $\mathscr{G}$ as shown in Fig. 10-69*a*. Note that the reference node and outer mesh have been given the same number.

Next we establish the *nodes* of $\mathscr{G}'$, one for each *mesh* of $\mathscr{G}$, as in Fig. 10-69*b*. Then, wherever $\mathscr{G}$ has a branch common to two *meshes*, we draw a branch between the two correspoding *nodes* of $\mathscr{G}'$ as shown by the dashed lines in Fig. 10-69*b*.

Finally, by visualizing a directional vector in each branch of $\mathscr{G}$, and by turning each of these *clockwise* into alignment with the intersecting branch of $\mathscr{G}'$, we obtain the branch orientations of $\mathscr{G}'$ as in Fig. 10-69*c*.

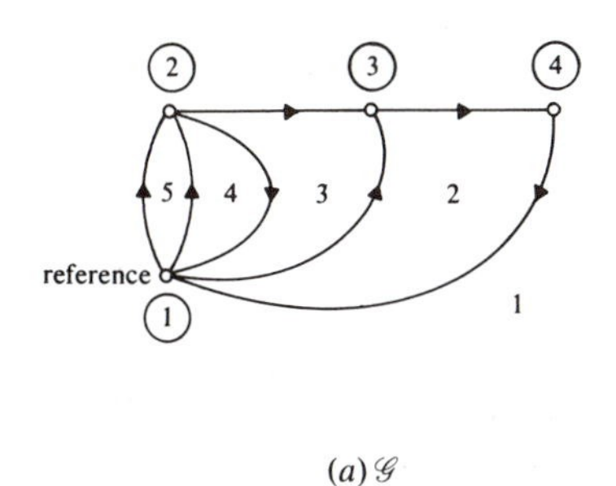


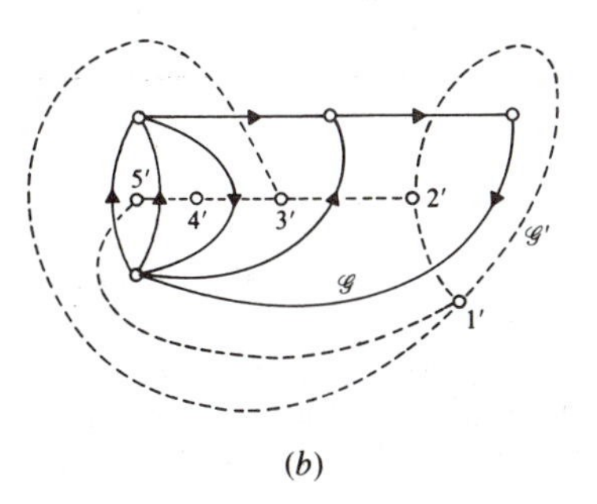


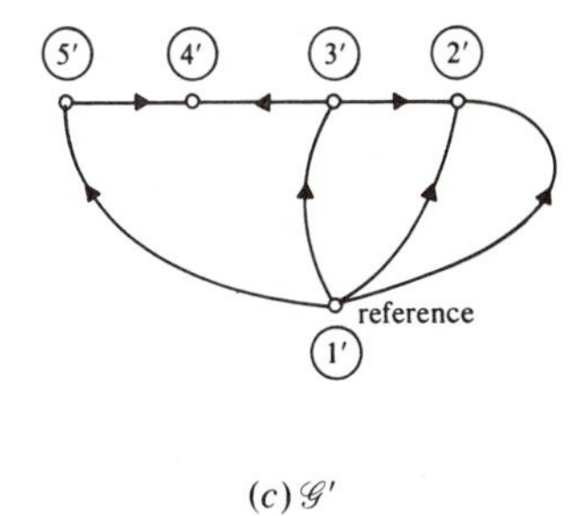


**Fig. 10-69**

**10.33** Construct the dual of the network in Fig. 10-70.

The preliminary steps in the construction of the dual *graph* are shown in Fig. 10-71*a*. Then, identifying the dual elements and their values from Table 10-1,

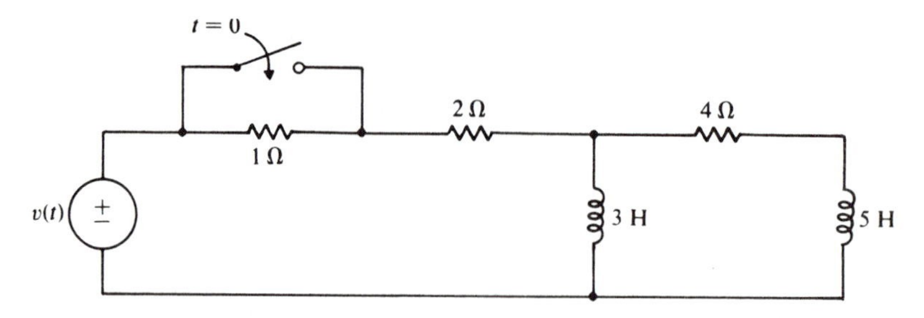


**Fig. 10-70**

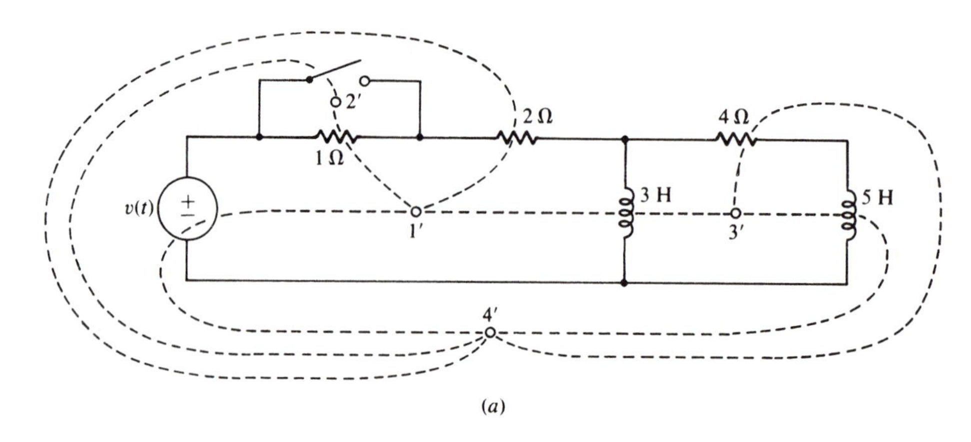


(a)

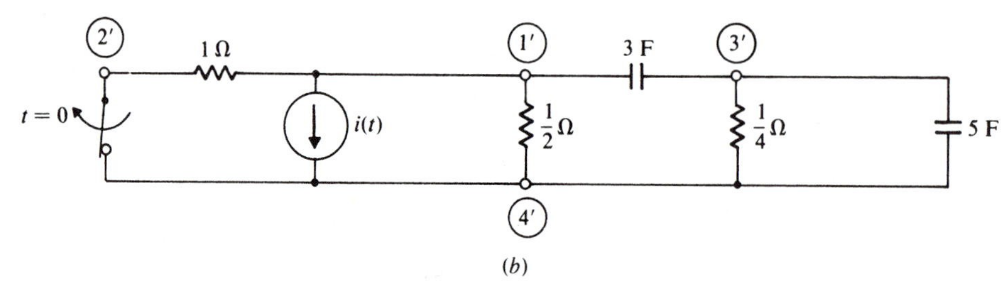


(b)

**Fig. 10-71**

page 348, we can complete the dual as in Fig. 10-71*b*.

Note the duality between resistance and conductance, inductance and capacitance, current source and voltage source, and "open" and "closed."

**10.34** The driving-point impedance of the network in Fig. 10-72 is

$$Z(s) = \frac{6s^2 + 2s + 1}{6s^2 + 1}$$

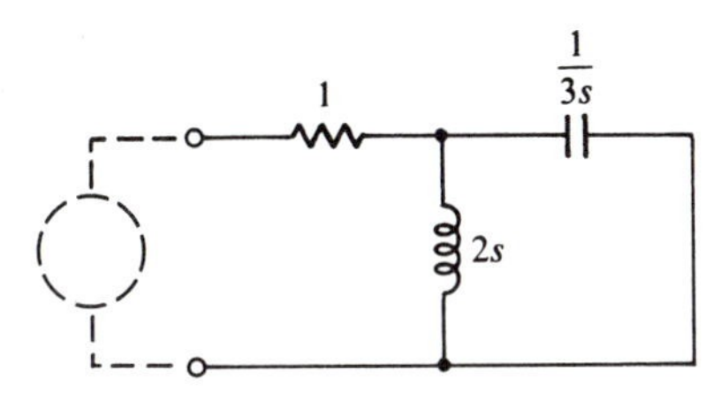

Fig. 10-72

Construct a network having a driving-point impedance of

$$Z(s) = \frac{6s^2 + 1}{6s^2 + 2s + 1}$$

Admittance and impedance are duals. Therefore the *dual* of the circuit in Fig. 10-72 will have the impedance desired.

If we proceed as in Problem 10.33, the dual network in Fig. 10-73*b* results.

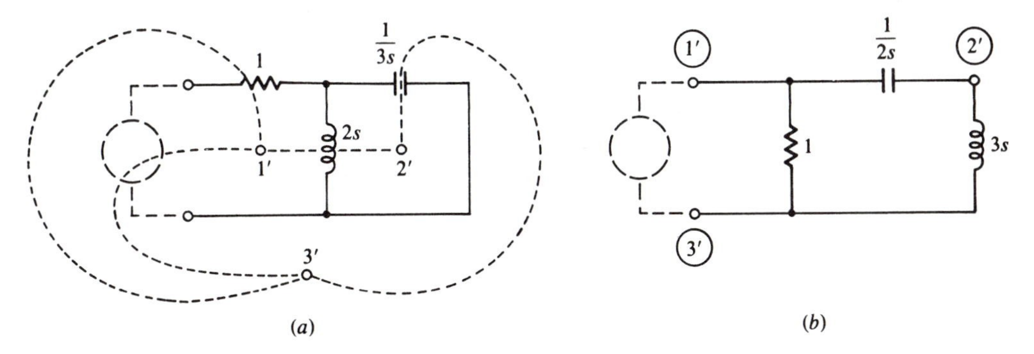

Fig. 10-73

# PROBLEMS

**10.35** In the resistive network of Fig. 10-74, $i(t) = 2$ A when $v(t) = 20$ V. Find $i(t)$ if $v(t) = 40 + 30e^{-t} - 20t$.

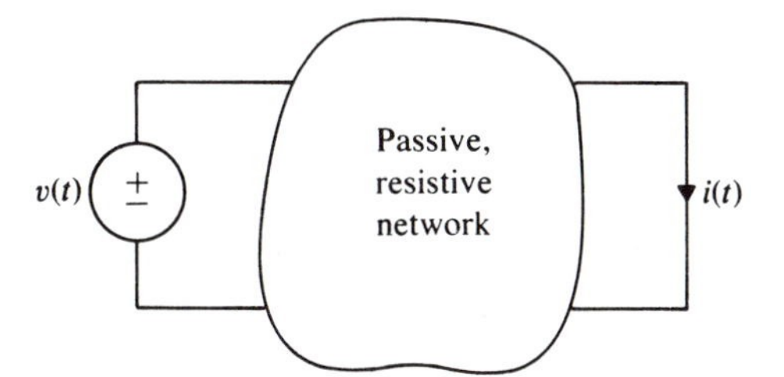

Fig. 10-74

**10.36** Find the contribution of $i(t)$ to the voltage $v(t)$, for $t \geq 0$, in the network of Fig. 10-75.

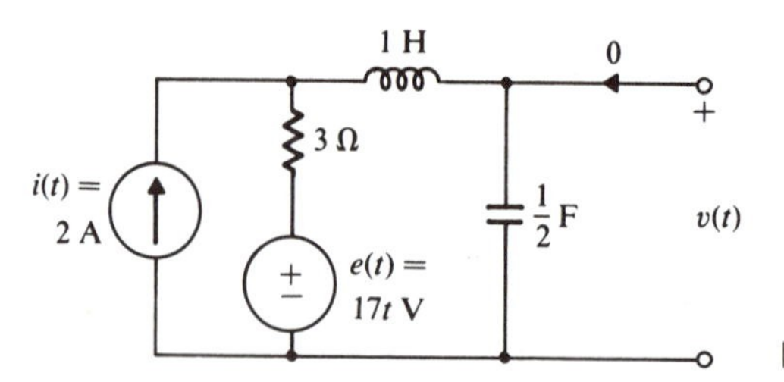


**Fig. 10-75**

**10.37** The active network of Fig. 10-76 contains sinusoidal sources of the same frequency, in addition to resistors, capacitors, and inductors. The following test data are available:

| $R$ | $\infty$ | 11 | 4 |
|---|---|---|---|
| $V_{\text{eff}}$ | 100 | 55 | 26.7 |

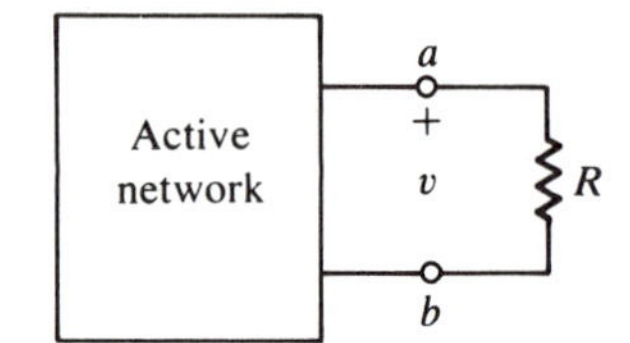


**Fig. 10-76**

Find the Thévenin and Norton equivalents for the terminals $a$-$b$.

**10.38** Find the value of the resistor $R$ in the circuit of Fig. 10-77 for maximum real power to be dissipated in that resistor. Calculate the value of this power.

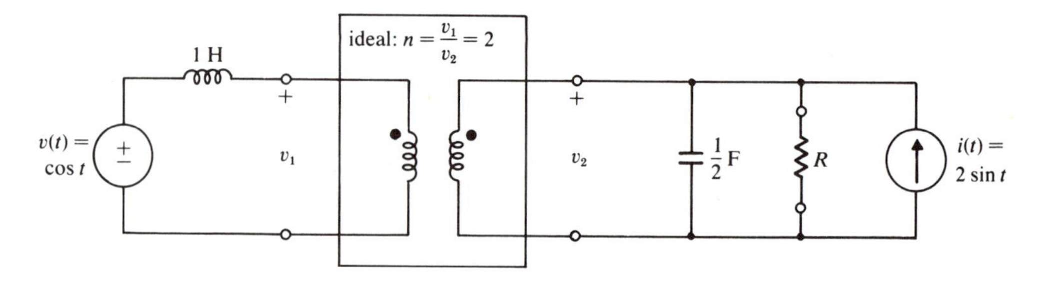


**Fig. 10-77**

**10.39** In the circuit of Fig. 10-78, the load impedance $Z_L$ may take any value. What value of $Z_L$ will extract maximum real power from the network $N$? How much power will this be?

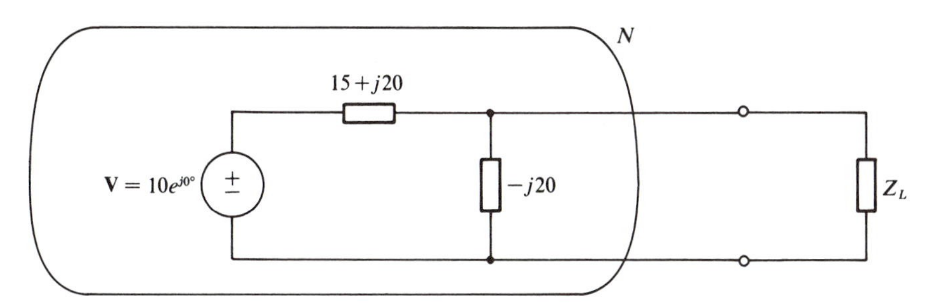


**Fig. 10-78**

**10.40** A series *L-R-C* circuit has characteristic roots $s_{1,2} = -0.5 \pm j0.866$ when $R = 1\ \Omega$. Calculate the resonant frequency. The circuit elements are to be changed to make the impedance 600 Ω at a resonant frequency of $200/\pi$ kHz. Find the new element values.

**10.41** In the network of Fig. 10-79*a* below, a unit step of voltage $v(t) = \mathbb{1}(t)$ causes a current $i(t)$ to flow. If, in Fig. 10-79*b*, the source voltage is a unit impulse $v'(t) = \delta(t)$, find $i'(t)$.

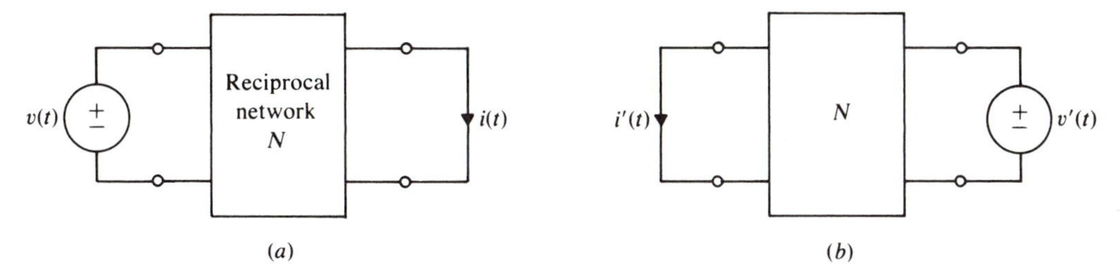


**Fig. 10-79**

**10.42** Show that an ideal transformer is a reciprocal two port.

**10.43** Show that the two networks of Fig. 10-80 have equivalent terminal conditions when $\omega = 1/\sqrt{LC}$.

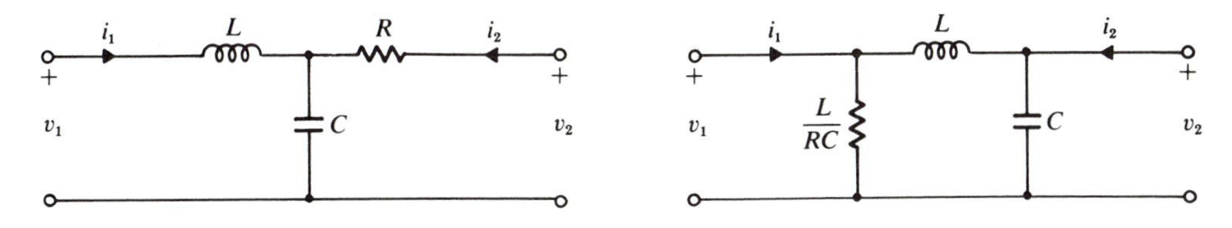


**Fig. 10-80**

**10.44** Show that the two $j\omega$-domain circuits of Fig. 10-81 are equivalent.

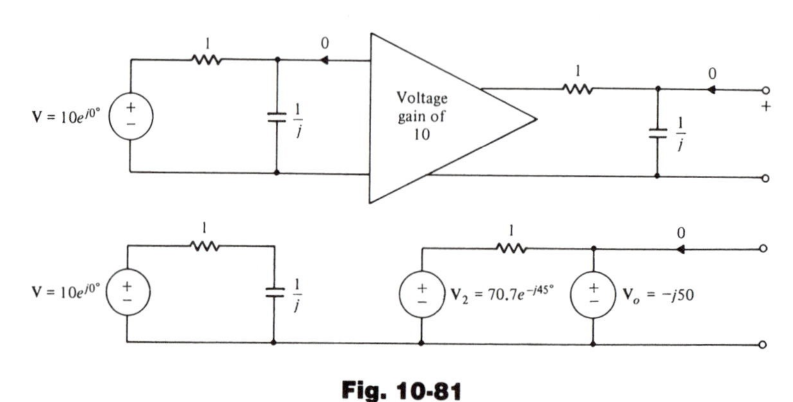


**Fig. 10-81**

**10.45** Prove that the two $j\omega$-domain networks of Fig. 10-82 are equivalent.

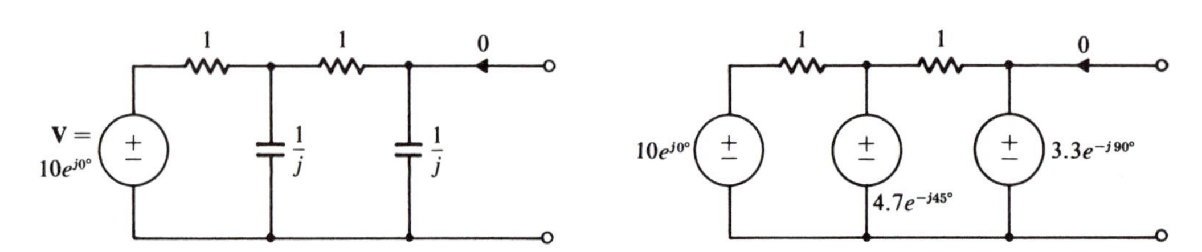


**Fig. 10-82**

**10.46** For the circuit of Fig. 10-83, construct the corresponding oriented graph, the dual graph, and the dual network.

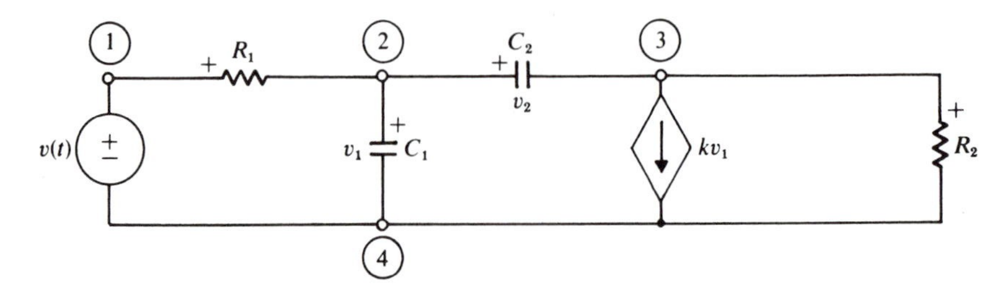


**Fig. 10-83**

**10.47** Two tests have been run on a reciprocal network as indicated in Fig. 10-84. Find $Z(j\omega)$ at the frequency of the tests.

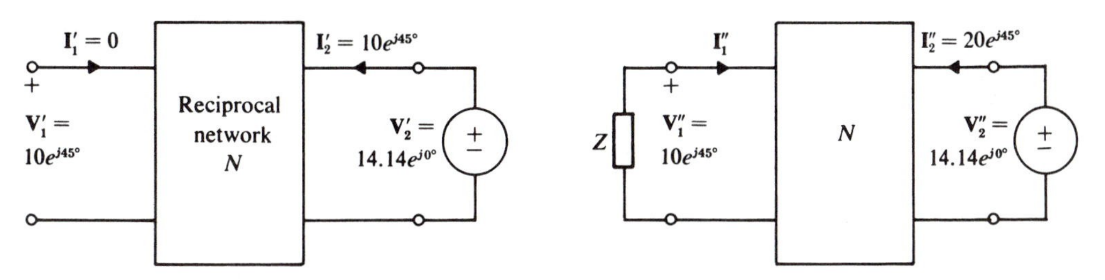


**Fig. 10-84**

Chapter

# 11

# TWO-PORT NETWORKS

Many practical circuits have just two *ports* of access: two places where signals may be input or output. For example, a coaxial cable between New York and San Francisco has two ports, one at each of those cities. The object here is to analyze such networks in terms of their terminal characteristics without particular regard to the internal composition of the network. That is, the network will be described by relationships between the port voltages and currents.

It is convenient and conventional to represent two-port networks as in Fig. 11-1. Such a four-terminal network is said to be a *two port* if the *same* current enters and leaves each *terminal-pair,* as shown. The terms *two port* and *two-terminal-pair* are synonymous.

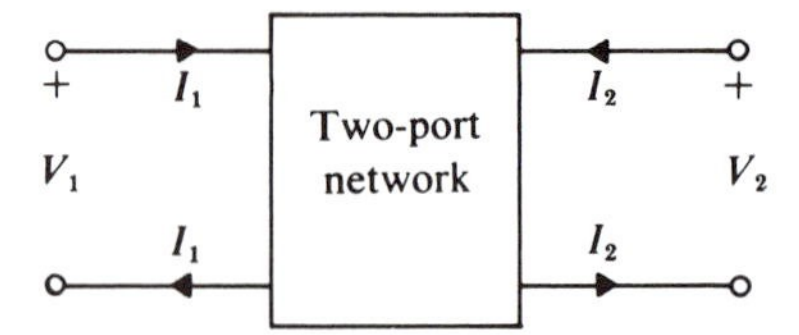


**Fig. 11-1**

It will be assumed throughout this chapter that there are no *independent* sources or nonzero initial conditions within the linear two port. And the normal associated sign convention will apply. Further, the two port equations will be written in a general form, since they are applicable in both the $s$- and $j\omega$-domains, as well as in the time-domain if the two port is resistive.

## THE z-PARAMETERS OF THE TWO PORT

There are several ways in which the variables $V_1$, $V_2$, $I_1$, and $I_2$ (see Fig. 11-1) may be grouped. One arrangement, for example, is

$$V_1 = z_{11}I_1 + z_{12}I_2$$
$$V_2 = z_{21}I_1 + z_{22}I_2 \qquad (11.1)$$

Here the $z$'s are parameters which characterize the two port, and two of the variables will be dependent upon the other two independent (given or known) variables.

The $z$'s, which have units of ohms, are called the *z-parameters*. This representation of the two port is particularly convenient when $I_1$ and $I_2$ are known and $V_1$ and $V_2$ are to be found.

### Example 11.1

Let us find the $z$-parameters of the network of Fig. 11-2.

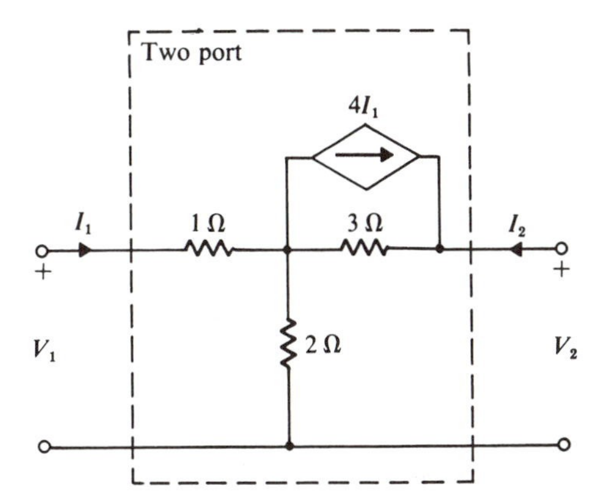


**Fig. 11-2**

The format of equation (*11.1*) should remind us of loop equations. Consequently we start out by defining the loop currents $I_1$ and $I_2$ as in Fig. 11-3. These are equal to the port currents. Then by KVL,

$$V_1 = 3I_1 + 2I_2$$
$$V_2 = 2I_1 + 5I_2 + 3(4I_1) = 14I_1 + 5I_2$$

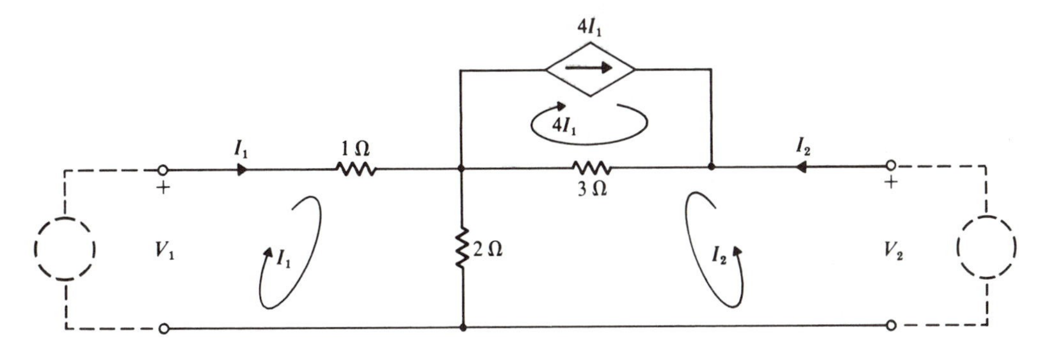


**Fig. 11-3**

And so by comparison with equation (*11.1*) we see that the $z$-parameters are

$$\mathbf{z} = \begin{bmatrix} 3 & 2 \\ 14 & 5 \end{bmatrix}$$

■

## THE y-PARAMETERS

If $I_1$ and $I_2$ are the unknowns, another grouping of the variables is more convenient:

$$\begin{aligned} I_1 &= y_{11}V_1 + y_{12}V_2 \\ I_2 &= y_{21}V_1 + y_{22}V_2 \end{aligned} \tag{11.2}$$

The $y$'s, which also characterize the two port, have units of siemens (S) and are called the *y-parameters*. They are related to the $z$-parameters and may be calculated from them.

## THE HYBRID PARAMETERS

The *hybrid* or *h-parameters* are particularly useful when $V_1$ and $I_2$ are the unknowns, since

$$\begin{aligned} V_1 &= h_{11}I_1 + h_{12}V_2 \\ I_2 &= h_{21}I_1 + h_{22}V_2 \end{aligned} \tag{11.3}$$

These equations are very frequently and conveniently used to represent a transistor. It is clear that $h_{11}$ has units of ohms, $h_{22}$ has units of siemens, and that $h_{12}$ and $h_{21}$ are dimensionless. Hence the term "hybrid."

### Example 11.2

A two-port "black box" is completely defined by any one of its parameter sets. To illustrate, we will solve for the output voltage $V_2$ in the circuit of Fig. 11-4.

From KVL and equation (*11.3*),

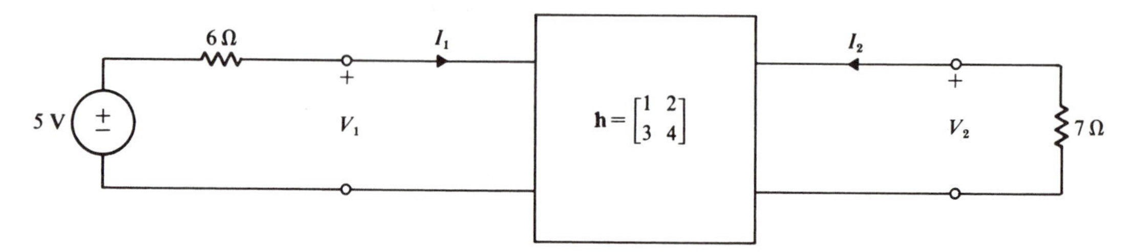


**Fig. 11-4**

$$V_1 = 5 - 6I_1 = 1I_1 + 2V_2 = I_1 + 2(-7I_2)$$

$$I_2 = 3I_1 + 4V_2 = 3I_1 + 4(-7I_2)$$

or, after rearrangement,

$$7I_1 - 14I_2 = 5$$

$$3I_1 - 29I_2 = 0$$

Then from Cramer's rule we find $I_2 = 0.093$ A, and so

$$V_2 = -7I_2 = -7(0.093) = -0.65 \text{ V}$$

■

## MEASUREMENT OR CALCULATION OF PARAMETERS

We have seen one method for calculating a two port's $z$-parameters in Example 11.1. Duality suggests, correctly, that we should be able to obtain the $y$-parameters by setting up the nodal equations for the two port.

There is another, quite different, approach which is especially useful for the *experimental* determination of one or more parameters. Suppose, for example, that we wish to measure or calcualte $h_{11}$. If we set $V_2 = 0$ in the first relation of equation (*11.3*), which corresponds to shorting the output terminals, then

$$h_{11} = \left.\frac{V_1}{I_1}\right|_{V_2=0}$$

Experimentally we would apply a source to the input terminals, with the output shorted. Then the ratio $V_1/I_1$, the input impedance with the output shorted, would yield $h_{11}$. But note that the ratio $V_1/I_1$, like any other network function, is meaningless unless the circuit is resistive or unless we work in the $s$- or $j\omega$-domain.

Similarly,

$$z_{12} = \left.\frac{V_1}{I_2}\right|_{I_1=0}$$

$$y_{22} = \left.\frac{I_2}{V_2}\right|_{V_1=0}$$

and so on.

## THE SIX FORMS OF A TWO PORT'S EQUATIONS

We have seen three of the six ways in which the four two-port variables may be grouped. The six possibilities are summarized in Table 11-1.

**Table 11-1 Two-Port Parameters.**

| Parameters | Equations | Parameters | Equations |
|---|---|---|---|
| $z$ (impedance) | $V_1 = z_{11}I_1 + z_{12}I_2$<br>$V_2 = z_{21}I_1 + z_{22}I_2$ | $g$ (inverse hybrid) | $I_1 = g_{11}V_1 + g_{12}I_2$<br>$V_2 = g_{21}V_1 + g_{22}I_2$ |
| $y$ (admittance) | $I_1 = y_{11}V_1 + y_{12}V_2$<br>$I_2 = y_{21}V_1 + y_{22}V_2$ | $T$ (transmission) | $V_1 = AV_2 - BI_2$<br>$I_1 = CV_2 - DI_2$ |
| $h$ (hybrid) | $V_1 = h_{11}I_1 + h_{12}V_2$<br>$I_2 = h_{21}I_1 + h_{12}V_2$ | $T'$ (inverse transmission) | $V_2 = A'V_1 - B'I_1$<br>$I_2 = C'V_1 - D'I_1$ |

## RELATIONS BETWEEN PARAMETER SETS

There are definite relationships between each of the parameter sets. These are summarized in Table 11-2.

**Table 11-2 Parameter Conversions.**

| | **z** | **y** | **h** | **g** | **T** | **T′** |
|---|---|---|---|---|---|---|
| **z** | $\begin{matrix} z_{11} & z_{12} \\ z_{21} & z_{22} \end{matrix}$ | $\begin{matrix} \frac{y_{22}}{\Delta_y} & -\frac{y_{12}}{\Delta_y} \\ -\frac{y_{21}}{\Delta_y} & \frac{y_{11}}{\Delta_y} \end{matrix}$ | $\begin{matrix} \frac{\Delta_h}{h_{22}} & \frac{h_{12}}{h_{22}} \\ -\frac{h_{21}}{h_{22}} & \frac{1}{h_{22}} \end{matrix}$ | $\begin{matrix} \frac{1}{g_{11}} & -\frac{g_{12}}{g_{11}} \\ \frac{g_{21}}{g_{11}} & \frac{\Delta_g}{g_{11}} \end{matrix}$ | $\begin{matrix} \frac{A}{C} & \frac{\Delta_T}{C} \\ \frac{1}{C} & \frac{D}{C} \end{matrix}$ | $\begin{matrix} \frac{D'}{C'} & \frac{1}{C'} \\ \frac{\Delta_{T'}}{C'} & \frac{A'}{C'} \end{matrix}$ |
| **y** | $\begin{matrix} \frac{z_{22}}{\Delta_z} & -\frac{z_{12}}{\Delta_z} \\ -\frac{z_{21}}{\Delta_z} & \frac{z_{11}}{\Delta_z} \end{matrix}$ | $\begin{matrix} y_{11} & y_{12} \\ y_{21} & y_{22} \end{matrix}$ | $\begin{matrix} \frac{1}{h_{11}} & -\frac{h_{12}}{h_{11}} \\ \frac{h_{21}}{h_{11}} & \frac{\Delta_h}{h_{11}} \end{matrix}$ | $\begin{matrix} \frac{\Delta_g}{g_{22}} & \frac{g_{12}}{g_{22}} \\ -\frac{g_{21}}{g_{22}} & \frac{1}{g_{22}} \end{matrix}$ | $\begin{matrix} \frac{D}{B} & -\frac{\Delta_T}{B} \\ -\frac{1}{B} & \frac{A}{B} \end{matrix}$ | $\begin{matrix} \frac{A'}{B'} & -\frac{1}{B'} \\ -\frac{\Delta_{T'}}{B'} & \frac{D'}{B'} \end{matrix}$ |
| **h** | $\begin{matrix} \frac{\Delta_z}{z_{22}} & \frac{z_{12}}{z_{22}} \\ -\frac{z_{21}}{z_{22}} & \frac{1}{z_{22}} \end{matrix}$ | $\begin{matrix} \frac{1}{y_{11}} & -\frac{y_{12}}{y_{11}} \\ \frac{y_{21}}{y_{11}} & \frac{\Delta_y}{y_{11}} \end{matrix}$ | $\begin{matrix} h_{11} & h_{12} \\ h_{21} & h_{22} \end{matrix}$ | $\begin{matrix} \frac{g_{22}}{\Delta_g} & -\frac{g_{12}}{\Delta_g} \\ -\frac{g_{21}}{\Delta_g} & \frac{g_{11}}{\Delta_g} \end{matrix}$ | $\begin{matrix} \frac{B}{D} & \frac{\Delta_T}{D} \\ -\frac{1}{D} & \frac{C}{D} \end{matrix}$ | $\begin{matrix} \frac{B'}{A'} & \frac{1}{A'} \\ -\frac{\Delta_{T'}}{A'} & \frac{C'}{A'} \end{matrix}$ |
| **g** | $\begin{matrix} \frac{1}{z_{11}} & -\frac{z_{12}}{z_{11}} \\ \frac{z_{21}}{z_{11}} & \frac{\Delta_z}{z_{11}} \end{matrix}$ | $\begin{matrix} \frac{\Delta_y}{y_{22}} & \frac{y_{12}}{y_{22}} \\ -\frac{y_{21}}{y_{22}} & \frac{1}{y_{22}} \end{matrix}$ | $\begin{matrix} \frac{h_{22}}{\Delta_h} & -\frac{h_{12}}{\Delta_h} \\ -\frac{h_{21}}{\Delta_h} & \frac{h_{11}}{\Delta_h} \end{matrix}$ | $\begin{matrix} g_{11} & g_{12} \\ g_{21} & g_{22} \end{matrix}$ | $\begin{matrix} \frac{C}{A} & -\frac{\Delta_T}{A} \\ \frac{1}{A} & \frac{B}{A} \end{matrix}$ | $\begin{matrix} \frac{C'}{D'} & -\frac{1}{D'} \\ \frac{\Delta_{T'}}{D'} & \frac{B'}{D'} \end{matrix}$ |
| **T** | $\begin{matrix} \frac{z_{11}}{z_{21}} & \frac{\Delta_z}{z_{21}} \\ \frac{1}{z_{21}} & \frac{z_{22}}{z_{21}} \end{matrix}$ | $\begin{matrix} -\frac{y_{22}}{y_{21}} & -\frac{1}{y_{21}} \\ -\frac{\Delta_y}{y_{21}} & -\frac{y_{11}}{y_{21}} \end{matrix}$ | $\begin{matrix} -\frac{\Delta_h}{h_{21}} & -\frac{h_{11}}{h_{21}} \\ -\frac{h_{22}}{h_{21}} & -\frac{1}{h_{21}} \end{matrix}$ | $\begin{matrix} \frac{1}{g_{21}} & \frac{g_{22}}{g_{21}} \\ \frac{g_{11}}{g_{21}} & \frac{\Delta_g}{g_{21}} \end{matrix}$ | $\begin{matrix} A & B \\ C & D \end{matrix}$ | $\begin{matrix} \frac{D'}{\Delta_{T'}} & \frac{B'}{\Delta_{T'}} \\ \frac{C'}{\Delta_{T'}} & \frac{A'}{\Delta_{T'}} \end{matrix}$ |
| **T′** | $\begin{matrix} \frac{z_{22}}{z_{12}} & \frac{\Delta_z}{z_{12}} \\ \frac{1}{z_{12}} & \frac{z_{11}}{z_{12}} \end{matrix}$ | $\begin{matrix} -\frac{y_{11}}{y_{12}} & -\frac{1}{y_{12}} \\ -\frac{\Delta_y}{y_{12}} & -\frac{y_{22}}{y_{12}} \end{matrix}$ | $\begin{matrix} \frac{1}{h_{12}} & \frac{h_{11}}{h_{12}} \\ \frac{h_{22}}{h_{12}} & \frac{\Delta_h}{h_{12}} \end{matrix}$ | $\begin{matrix} -\frac{\Delta_g}{g_{12}} & -\frac{g_{22}}{g_{12}} \\ -\frac{g_{11}}{g_{12}} & -\frac{1}{g_{12}} \end{matrix}$ | $\begin{matrix} \frac{D}{\Delta_T} & \frac{B}{\Delta_T} \\ \frac{C}{\Delta_T} & \frac{A}{\Delta_T} \end{matrix}$ | $\begin{matrix} A' & B' \\ C' & D' \end{matrix}$ |

This table asserts, for example, that

$$\mathbf{z} = \begin{bmatrix} z_{11} & z_{12} \\ z_{21} & z_{22} \end{bmatrix} = \frac{1}{\Delta_y}\begin{bmatrix} y_{22} & -y_{12} \\ -y_{21} & y_{11} \end{bmatrix} = \frac{1}{h_{22}}\begin{bmatrix} \Delta_h & h_{12} \\ -h_{21} & 1 \end{bmatrix} = \cdots$$

where $\Delta_y = |\mathbf{y}| = y_{11}y_{22} - y_{12}y_{21}$, $\Delta_h = |\mathbf{h}| = h_{11}h_{22} - h_{12}h_{21}$, . . . . Thus given any parameter set for a two port, we may easily compute any other set for the same network. *However,* for some networks one or more parameter sets *may not exist.*

### RECIPROCAL TWO-PORT NETWORKS

The reciprocity relations for a reciprocal two port have been stated in Chapter 10.

## ILLUSTRATIVE PROBLEMS

**11.1** Identify the two-port networks in Fig. 11-5.

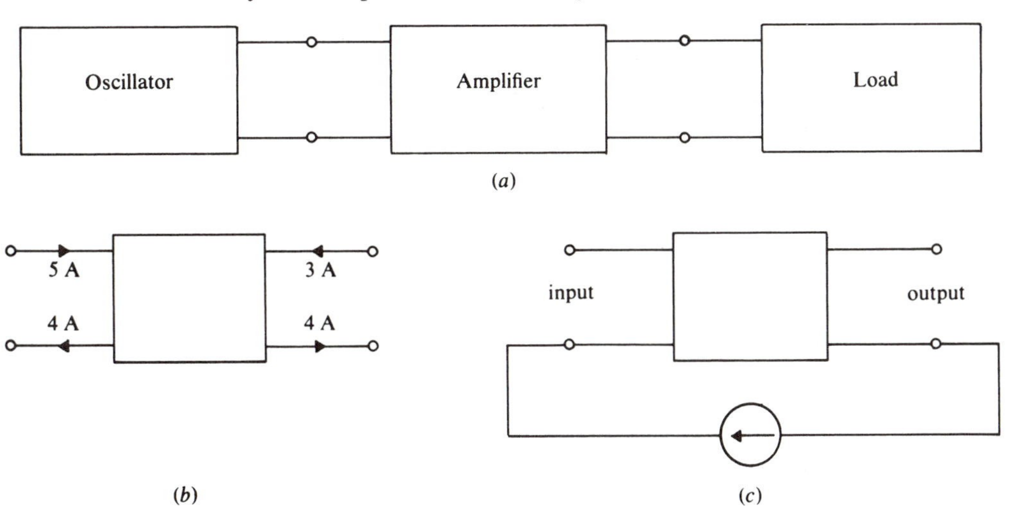


**Fig. 11-5**

The amplifier in Fig. 11-5*a* is the only two port. Kirchhoff's current law requires that the same current enters and leaves the oscillator. Therefore the port current condition is satisfied at the amplifier input. Similarly, the current condition is met at the amplifier output.

The current condition is *not* met in Fig. 11-5*b*. This situation is always a possibility if the two port is imbedded in a circuit which provides one or more paths between the input and output ports. This is illustrated in the simple circuit of Fig. 11-5*c*. For this reason, connections (or measurements) between input and output ports are usually "forbidden."

### *z*- and *y*-Parameters

**11.2** Given: the parameters $z_{11} = 20\ \Omega$, $z_{12} = z_{21} = 15\ \Omega$, $z_{22} = 25\ \Omega$, and the independent variables $I_1 = 2$ A and $I_2 = -1$ A as in Fig. 11-6. Solve for the dependent variables $V_1$ and $V_2$.

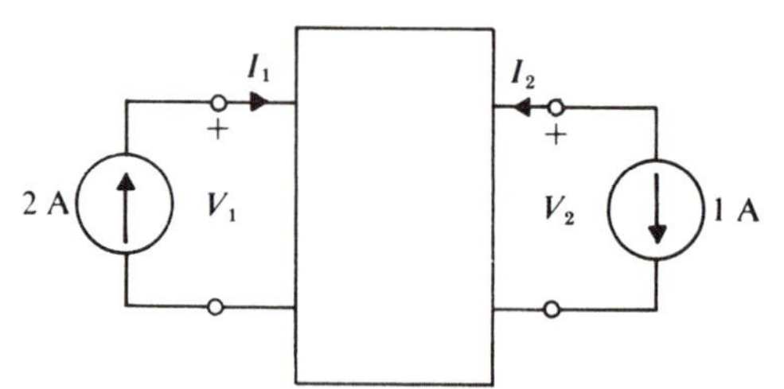


**Fig. 11-6**

Substituting known values into equation (*11.1*),

$$V_1 = 20(2) + 15(-1) = 25\ \text{V}$$

$$V_2 = 15(2) + 25(-1) = 5\ \text{V}$$

**11.3** The same two port as in Problem 11.2 is connected as shown in Fig. 11-7. The measured values of $V_1$ and $I_1$ are 25 V and 2 A. Find $V_2$ and $I_2$, and hence the value of $R$.

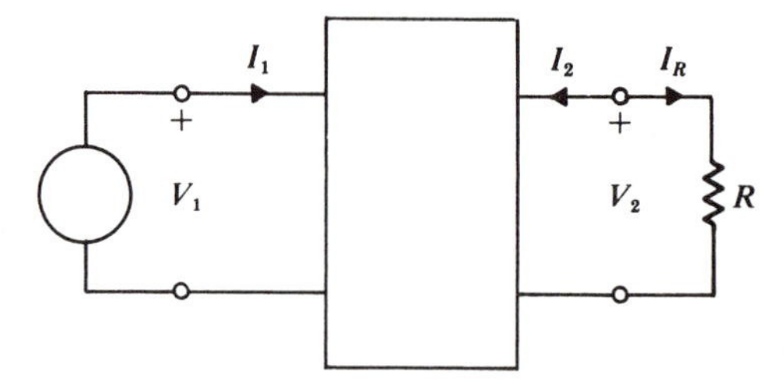


**Fig. 11-7**

In this case $V_2$ and $I_2$ are the dependent (unknown) variables. Substituting the data into equation (*11.1*),

$$25 = 20(2) + 15I_2$$

$$V_2 = 15(2) + 25I_2$$

from which $I_2 = -1$ A and $V_2 = 5$ V. Therefore

$$R = \frac{V_2}{I_R} = \frac{V_2}{-I_2} = 5\ \Omega$$

**11.4** Find $I_1$ and $I_2$ given $V_1 = 10$ V and $V_2 = -5$ V, the $z$'s remaining as in Problem 11.2.

Here $I_1$ and $I_2$ are the dependent variables, and from equation (*11.1*),

$$10 = 20I_1 + 15I_2$$
$$-5 = 15I_1 + 25I_2$$

yielding $I_1 = 1.18$ A, $I_2 = -0.91$ A.

**11.5** Write the loop equations for the tee network of Fig. 11-8 and hence find the $z$-parameters. Then determine the values of $R_1$, $R_2$, and $R_3$ needed to obtain $z_{11} = 20\ \Omega$, $z_{12} = z_{21} = 15\ \Omega$, and $z_{22} = 25\ \Omega$.

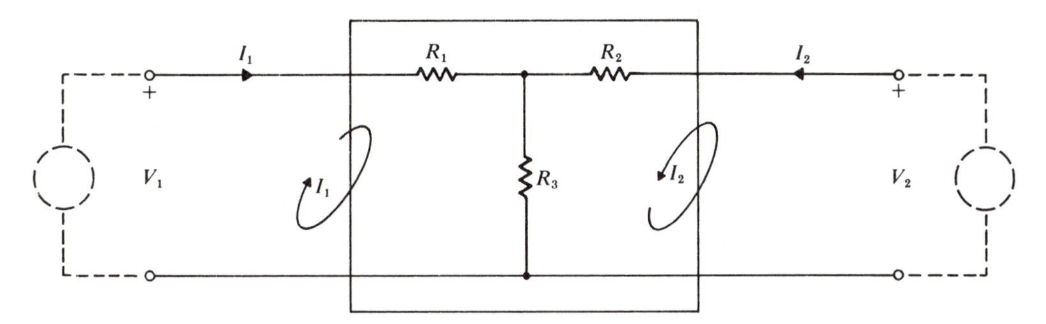


**Fig. 11-8**

The loop equations are

$$V_1 = (R_1 + R_3)I_1 + R_3 I_2$$
$$V_2 = R_3 I_1 + (R_2 + R_3)I_2$$

Now the loop currents are also the port currents (see Fig. 11-8), and so by comparison with equation (*11.1*),

$$\mathbf{z} = \begin{bmatrix} R_1 + R_3 & R_3 \\ R_3 & R_2 + R_3 \end{bmatrix}$$

In order to obtain the desired values for the parameters, we must have

$$R_3 = z_{12} = z_{21} = 15\ \Omega$$
$$R_1 + R_3 = 20\ \Omega \quad \text{or} \quad R_1 = 5\ \Omega$$
$$R_2 + R_3 = 25\ \Omega \quad \text{or} \quad R_2 = 10\ \Omega$$

**11.6** Given $z_{11} = 20\ \Omega$, $z_{12} = z_{21} = 15\ \Omega$, and $z_{22} = 25\ \Omega$, use Table 11-2 to find the $y$-parameters. Then if $V_1 = 10$ V and $V_2 = -5$ V, find $I_1$ and $I_2$.

From Table 11-2,

$$\mathbf{y} = \frac{1}{\Delta_z}\begin{bmatrix} z_{22} & -z_{12} \\ -z_{21} & z_{11} \end{bmatrix}$$

where $\Delta_z = z_{11}z_{22} - z_{12}z_{21} = (20)(25) - (15)(15) = 275$. Therefore

$$\mathbf{y} = \frac{1}{275}\begin{bmatrix} 25 & -15 \\ -15 & 20 \end{bmatrix} = \begin{bmatrix} 0.091 & -0.0545 \\ -0.0545 & 0.0728 \end{bmatrix}$$

Then from equation (*11.2*),

$$I_1 = (0.091)(10) + (-0.0545)(-5) = 1.183 \text{ A}$$

$$I_2 = (-0.0545)(10) + (0.0728)(-5) = -0.909 \text{ A}$$

which we obtained in Problem 11.4 from the $z$-parameter equations.

**11.7** Write the nodal equations for the pi network of Fig. 11-9 and hence find the $y$-parameters. Then determine the values of $R_a$, $R_b$, and $R_c$ needed to obtain $y_{11} = 0.091$ S, $y_{12} = y_{21} = -0.0545$ S, and $y_{22} = 0.0728$ S.

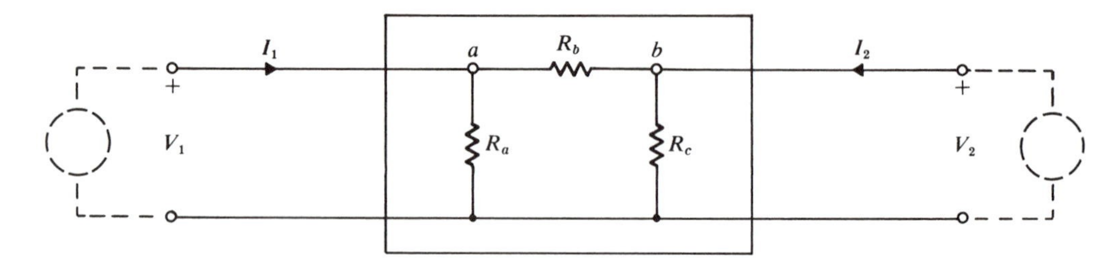


**Fig. 11-9**

The nodal equations are

$$I_1 = \left(\frac{1}{R_a} + \frac{1}{R_b}\right)V_1 - \frac{1}{R_b}V_2$$

$$I_2 = -\frac{1}{R_b}V_1 + \left(\frac{1}{R_b} + \frac{1}{R_c}\right)V_2$$

Therefore by comparison with equation (*11.2*),

$$\mathbf{y} = \begin{bmatrix} \dfrac{1}{R_a} + \dfrac{1}{R_b} & -\dfrac{1}{R_b} \\ -\dfrac{1}{R_b} & \dfrac{1}{R_b} + \dfrac{1}{R_c} \end{bmatrix}$$

In order to obtain the desired values for the parameters, we require that

$$-\frac{1}{R_b} = -0.0545 \quad \text{or} \quad R_b = 18.4\ \Omega$$

$$\frac{1}{R_a} + \frac{1}{R_b} = 0.091 \quad \text{or} \quad R_a = 27.4\ \Omega$$

$$\frac{1}{R_b} + \frac{1}{R_c} = 0.0728 \quad \text{or} \quad R_c = 55\ \Omega$$

*Note:* Since we obtained the $y$-parameters of this problem from the $z$-parameters of Problems 11.4–11.6, the tee network of Problem 11.5 has the same port characteristics as those of the above pi network (see Fig. 11-10). These two "black boxes" are identical as far as any port measurements are concerned.

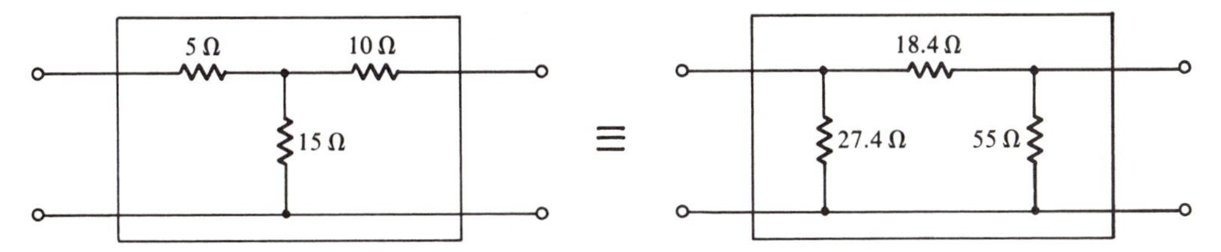


**Fig. 11-10**

### *Hybrid or h-Parameters*

**11.8** Write a loop and a nodal equation for the transistor model in Fig. 11-11. Then by comparison with equation (*11.3*), find **h**.

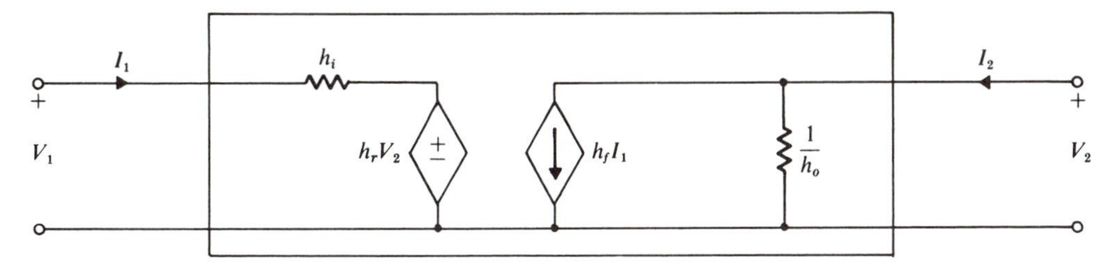


**Fig. 11-11**

The input loop equation and the output nodal equation are

$$V_1 = h_i I_1 + h_r V_2$$

$$I_2 = h_f I_1 + h_o V_2$$

Thus by comparison with equation (*11.3*),

$$\mathbf{h} = \begin{bmatrix} h_i & h_r \\ h_f & h_o \end{bmatrix}$$

That is, the symbols $h_i$, $h_r$, $h_f$, and $h_o$, which are commonly used in the analysis of transistor circuits, are equivalent to $h_{11}$, $h_{12}$, $h_{21}$, and $h_{22}$, respectively.

**11.9** Given the two-port circuit of Fig. 11-12, find the current gain $A_I = -I_2/I_1$.

We can at once write

$$I_2 = h_{21} I_1 + h_{22} V_2 \quad \text{and} \quad V_2 = R_L(-I_2)$$

Therefore $I_2 = h_{21} I_1 + h_{22}(-R_L I_2)$ or

$$A_I = -\frac{I_2}{I_1} = -\frac{h_{21}}{1 + h_{22} R_L}$$

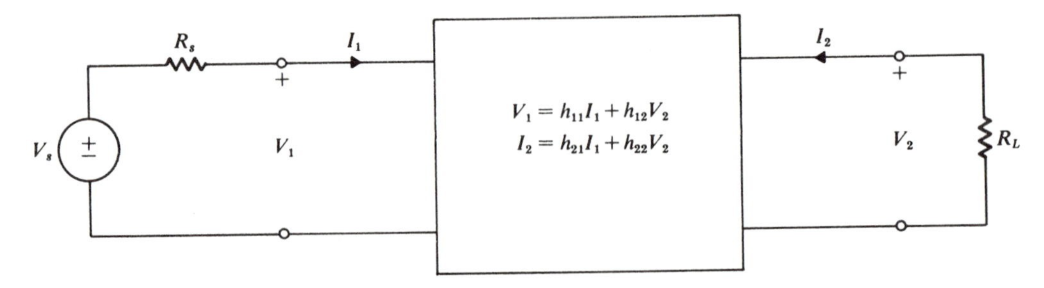


**Fig. 11-12**

*Note:* Recall that the $h$-parameters are usually written $h_i$, $h_r$, $h_f$, and $h_o$ when we are describing a transistor. For a typical transistor in the common-emitter configuration, $h_{11} = 1000\ \Omega$, $h_{12} = 2.5 \times 10^{-4}$, $h_{21} = 50$, and $h_{22} = 25 \times 10^{-6}$ S. If $R_L = 100\ \text{k}\Omega$, then by substitution, $A_I = -14.3$.

**11.10** Show that for the two-port network of Fig. 11-12,

$$Z_i = \frac{V_1}{I_1} = h_{11} - \frac{h_{12}h_{21}R_L}{1 + h_{22}R_L}$$

and evaluate this expression for the numerical data given in Problem 11.9. Similarly, calculate the numerical value of the voltage gain $A_V = V_2/V_1$ and the output admittance $Y_o = I_2/V_2$, given $R_s = 500\ \Omega$.

We know that

$$V_1 = h_{11}I_1 + h_{12}V_2 \quad \text{and} \quad V_2 = R_L(-I_2)$$

Hence $V_1 = h_{11}I_1 - h_{12}R_LI_2$ or

$$Z_i = \frac{V_1}{I_1} = h_{11} - h_{12}R_L\frac{I_2}{I_1} = h_{11} + h_{12}R_LA_I$$

Then substituting the expression for $A_I$ in Problem 11.9,

$$Z_i = h_{11} - \frac{h_{12}h_{21}R_L}{1 + h_{22}R_L}$$

and if we substitute the numerical data, $Z_i = 643\ \Omega$.

To find $A_V$ we write

$$V_1 = h_{11}I_1 + h_{12}V_2 = h_{11}\frac{V_1}{Z_i} + h_{12}V_2$$

Therefore

$$A_V = \frac{V_2}{V_1} = \frac{1 - h_{11}/Z_i}{h_{12}}$$

or, substituting the numerical data, $A_V = -2200$.

In order to find the output admittance $Y_o$ we must calculate $I_2$ due to $V_2$, or vice versa, *with the input source $V_s$ set to zero as in Fig. 11-13*. The circuit has now been

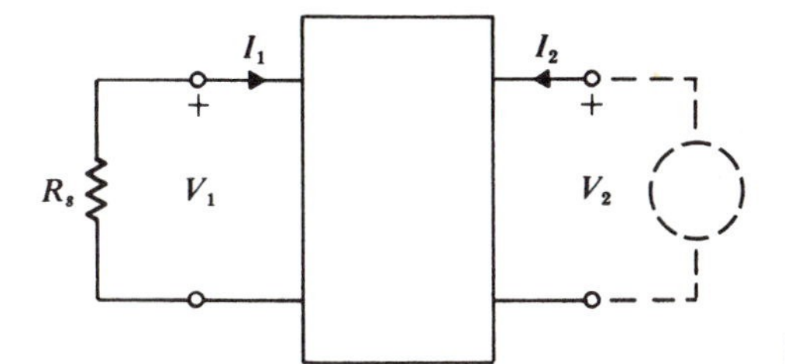


**Fig. 11-13**

changed, and so the previously calculated results ($A_I$, $A_V$, and $Z_i$) *are not relevant* to this calculation.

From equation *(11.3)*,

$$I_2 = h_{21} I_1 + h_{22} V_2$$

or

$$Y_o = \frac{I_2}{V_2} = h_{21} \frac{I_1}{V_2} + h_{22}$$

Also,

$$V_1 = R_s(-I_1) = h_{11} I_1 + h_{12} V_2 \quad \text{or} \quad \frac{I_1}{V_2} = \frac{-h_{12}}{h_{11} + R_s}$$

which when put into the expression for $Y_o$ yields

$$Y_o = -\frac{h_{12} h_{21}}{h_{11} + R_s} + h_{22}$$

When we substitute the numerical data, $Y_o = 16.7\ \mu S$.

If we wish to compute the output admittance *including the effect of* $R_L$, we have only to add the parallel admittances $Y_o$ and $1/R_L = 10\ \mu S$. This yields $Y_o' = 16.7 + 10 = 26.7\ \mu S$.

**11.11** Show, *without* using Table 11-2, that if we are given the $h$-parameters of a two port, then the $y$-parameters may be calculated from

$$\mathbf{y} = \frac{1}{h_{11}} \begin{bmatrix} 1 & -h_{12} \\ h_{21} & \Delta_h \end{bmatrix}$$

where

$$\Delta_h = \begin{vmatrix} h_{11} & h_{12} \\ h_{21} & h_{22} \end{vmatrix} = h_{11} h_{22} - h_{12} h_{21}$$

The two sets of two-port equations are

$$V_1 = h_{11} I_1 + h_{12} V_2 \qquad I_1 = y_{11} V_1 + y_{12} V_2$$

$$I_2 = h_{21} I_1 + h_{22} V_2 \qquad I_2 = y_{21} V_1 + y_{22} V_2$$

The idea is to put the $h$-equations into the same form as the $y$-equations, and then to equate coefficients. Rearranging the first $h$-equation yields

$$I_1 = \frac{1}{h_{11}} V_1 - \frac{h_{12}}{h_{11}} V_2$$

and comparing with the first $y$-equation we see that

$$y_{11} = \frac{1}{h_{11}} \quad \text{and} \quad y_{12} = -\frac{h_{12}}{h_{11}}$$

It is necessary to eliminate $I_1$ from the second $h$-equation before it can be put into the same form as the second $y$-equation. Thus

$$h_{21}\left\{\frac{1}{h_{11}} V_1 - \frac{h_{12}}{h_{11}} V_2\right\} + h_{22}V_2 = I_2 \quad \text{or} \quad \frac{h_{21}}{h_{11}} V_1 + \frac{h_{11}h_{22} - h_{21}h_{12}}{h_{11}} V_2 = I_2$$

But $h_{11}h_{22} - h_{21}h_{12} = \Delta_h$, so

$$y_{21} = h_{21}/h_{11} \quad \text{and} \quad y_{22} = \Delta_h/h_{11}$$

thus completing the derivation. Clearly the $y$-parameters will not exist if $h_{11} = 0$.

Similarly we can derive the other entries in the conversion table (see Table 11-2).

### *Measurement of Parameters*

**11.12** Show how we might go about measuring $h_{22}$, $y_{11}$, $y_{12}$, and $z_{21}$ for some given resistive two port.

*To find $h_{22}$:*

The second relation of equation (*11.3*) is $I_2 = h_{21}I_1 + h_{22}V_2$. Thus by setting $I_1 = 0$, that is, by open-circuiting the input of the two port,

$$h_{22} = \left.\frac{I_2}{V_2}\right|_{I_1=0}$$

$h_{22}$ may therefore be measured as suggested by Fig. 11-14*a*. In words, this parameter is the output admittance of the two port *when the input is open*.

*To find $y_{11}$:*

If we set $V_2 = 0$ in the first relation of equation (*11.2*),

$$y_{11} = \left.\frac{I_1}{V_1}\right|_{V_2=0}$$

Thus $y_{11}$ is the input admittance *with the output shorted*. (See Fig. 11-14*b*.)

*To find $y_{12}$:*

Setting $V_1 = 0$ in equation (*11.2*),

$$y_{12} = \left.\frac{I_1}{V_2}\right|_{V_1=0}$$

This is illustrated in Fig. 11-14*c*.

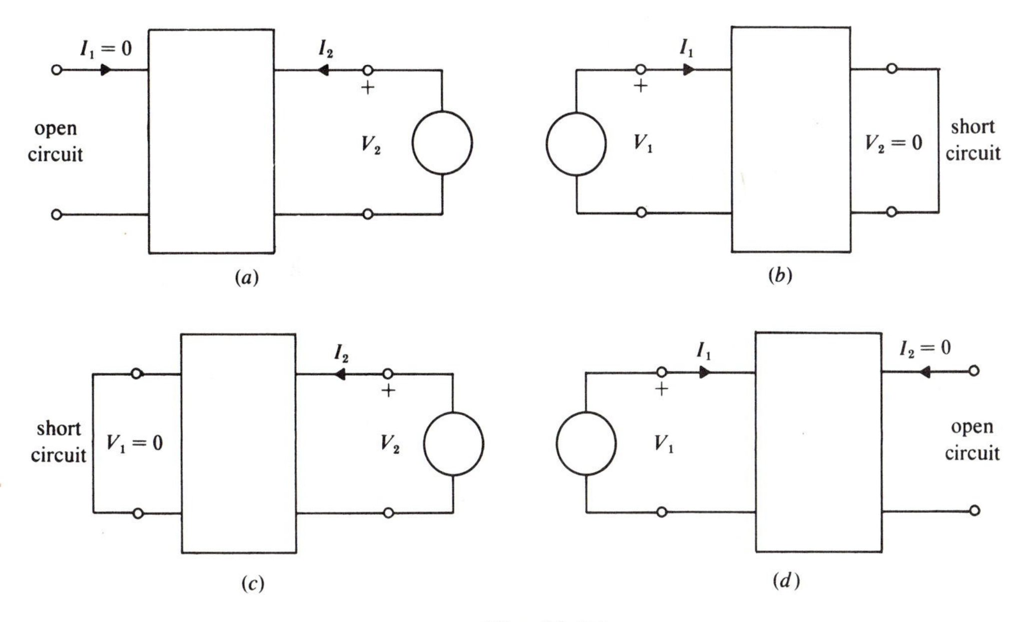


**Fig. 11-14**

*To find $z_{21}$:*

Setting $I_2 = 0$ in equation (*11.1*),

$$z_{21} = \left.\frac{V_2}{I_1}\right|_{I_2=0}$$

This is illustrated in Fig. 11-14*d*.

### Comments

1. The $y$-parameters are often called the *short-circuit admittance parameters*.
2. The $z$-parameters are often called the *open-circuit impedance parameters*.

**11.13** Over a "middle" range of frequencies a transistor may be represented as shown in Fig. 11-15. Find the $h$-parameters in terms of $\alpha$, $R_1$, $R_2$, $C$, and $\omega$, assuming sinusoidal steady-state conditions.

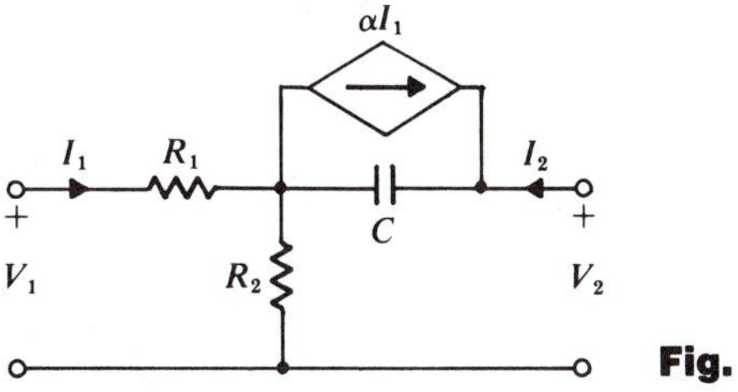


**Fig. 11-15**

Here in terms of the appropriate phasors,

$$h_{11} = \left.\frac{\mathbf{V}_1}{\mathbf{I}_1}\right|_{\mathbf{V}_2=0}$$

and we must consider the corresponding circuit of Fig. 11-16*a*.

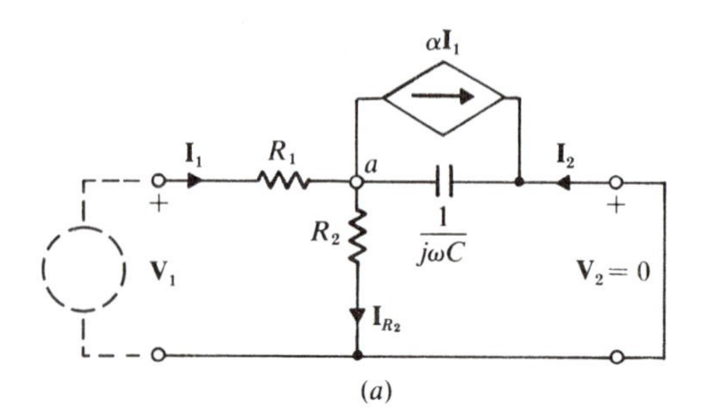

(*a*)

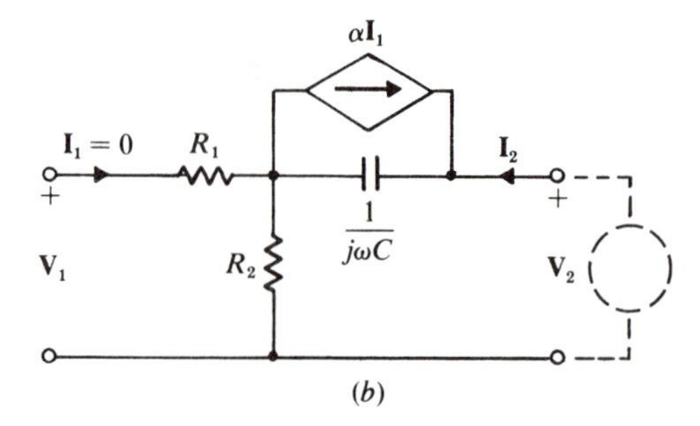

(*b*)

**Fig. 11-16**

Writing a nodal equation at *a*.

$$\mathbf{I}_1 = \alpha\mathbf{I}_1 + j\omega C\mathbf{V}_a + \frac{1}{R_2}\mathbf{V}_a$$

But $\mathbf{V}_a = \mathbf{V}_1 - R_1\mathbf{I}_1$ and therefore

$$\mathbf{I}_1 = \alpha\mathbf{I}_1 + j\omega C(V_1 - R_1\mathbf{I}_1) + \frac{1}{R_2}(\mathbf{V}_1 - R_1\mathbf{I}_1)$$

from which

$$h_{11} = \frac{\mathbf{V}_1}{\mathbf{I}_1} = \frac{1 - \alpha + R_1/R_2 + j\omega R_1 C}{1/R_2 + j\omega C} = \frac{R_1 + (1 - \alpha)R_2 + j\omega R_1 R_2 C}{1 + j\omega R_2 C}$$

Similarly,

$$h_{12} = \left.\frac{\mathbf{V}_1}{\mathbf{V}_2}\right|_{\mathbf{I}_1=0}$$

which we can evaluate from Fig. 11-16*b* where the input is open. Note that the dependent source current is zero. Thus

$$\mathbf{I}_2 = \frac{\mathbf{V}_2}{R_2 + 1/j\omega C} \quad \text{and} \quad \mathbf{V}_1 = R_2\mathbf{I}_2 = \frac{R_2\mathbf{V}_2}{R_2 + 1/j\omega C}$$

Hence

$$h_{12} = \frac{\mathbf{V}_1}{\mathbf{V}_2} = \frac{R_2}{R_2 + 1/j\omega C} = \frac{j\omega R_2 C}{1 + j\omega R_2 C}$$

Next we can return to Fig. 11-16*a* to evaluate

$$h_{21} = \left.\frac{\mathbf{I}_2}{\mathbf{I}_1}\right|_{\mathbf{V}_2=0}$$

At node $a$,

$$\mathbf{I}_1 = \alpha\mathbf{I}_1 + j\omega C\mathbf{V}_a + \mathbf{I}_{R2}$$

But $\mathbf{V}_a = R_2\mathbf{I}_{R2}$ and $\mathbf{I}_{R2} = \mathbf{I}_1 + \mathbf{I}_2$. Thus

$$\mathbf{I}_1 = \alpha\mathbf{I}_1 + j\omega CR_2(\mathbf{I}_1 + \mathbf{I}_2) + (\mathbf{I}_1 + \mathbf{I}_2)$$

and

$$h_{21} = \frac{\mathbf{I}_2}{\mathbf{I}_1} = -\frac{\alpha + j\omega R_2C}{1 + j\omega R_2C}$$

Finally,

$$h_{22} = \left.\frac{\mathbf{I}_2}{\mathbf{V}_2}\right|_{\mathbf{I}_1=0}$$

and from Fig. 11-16$b$, where $\mathbf{I}_1$ and the dependent source current are zero,

$$\mathbf{I}_2 = \frac{\mathbf{V}_2}{R_2 + 1/j\omega C} \quad \text{and} \quad h_{22} = \frac{j\omega C}{1 + j\omega R_2C}$$

### *Alternative Solution*

For many networks one particular set of parameters will be easier to evaluate than the others. For example, any tee network lends itself to loop analysis, from which the $z$-parameters are easily found. Then we can use the conversion table to determine the desired parameter set.

Here, referring to Fig. 11-15, we can write the loop equations

$$(R_1 + R_2)\mathbf{I}_1 + R_2\mathbf{I}_2 = \mathbf{V}_1$$

$$\left(R_2 + \frac{\alpha}{j\omega C}\right)\mathbf{I}_1 + \left(R_2 + \frac{1}{j\omega C}\right)\mathbf{I}_2 = \mathbf{V}_2$$

By comparison with equation (11.1) we find

$$\mathbf{z} = \begin{bmatrix} R_1 + R_2 & R_2 \\ R_2 + \dfrac{\alpha}{j\omega C} & R_2 + \dfrac{1}{j\omega C} \end{bmatrix}$$

We can see from Table 11-2 that we will need to calculate $\Delta_z$. Thus

$$\Delta_z = |\mathbf{z}| = R_1R_2 + (R_1 + R_2 - \alpha R_2)/j\omega C$$

Now from Table 11-2,

$$h_{11} = \frac{\Delta_z}{z_{22}} = \frac{R_1R_2 + (R_1 + R_2 - \alpha R_2)/j\omega C}{R_2 + 1/j\omega C} = \frac{R_1 + (1 - \alpha)R_2 + j\omega R_1R_2C}{1 + j\omega R_2C}$$

And, using the same table,

$$h_{12} = \frac{z_{12}}{z_{22}} = \frac{R_2}{R_2 + 1/j\omega C} = \frac{j\omega R_2 C}{1 + j\omega R_2 C}$$

$$h_{21} = -\frac{z_{21}}{z_{22}} = -\frac{R_2 + \alpha/j\omega C}{R_2 + 1/j\omega C} = -\frac{\alpha + j\omega R_2 C}{1 + j\omega R_2 C}$$

$$h_{22} = \frac{1}{z_{22}} = \frac{j\omega C}{1 + j\omega R_2 C}$$

### *Transmission Parameters*

**11.14** Given the transmission parameters $A = 2$, $B = 20\ \Omega$, $C = 1/12$ S, and $D = 4/3$, find the two port's input impedance $Z_i = V_1/I_1$ when

(*a*) the output terminals are open circuited,

(*b*) the output terminals are short circuited,

(*c*) $R_L = 8\ \Omega$ is connected across the output terminals.

(*a*) When $I_2 = 0$ the transmission equations are

$$V_1 = 2V_2 - \cancel{20I_2} \quad \text{and} \quad I_1 = \tfrac{1}{12}V_2 - \cancel{\tfrac{4}{3}I_2}$$

Then dividing one equation by the other,

$$Z_i = \frac{V_1}{I_1} = \frac{2V_2}{\frac{1}{12}V_2} = 24\ \Omega$$

(*b*) When $V_2 = 0$,

$$V_1 = \cancel{2V_2} - 20I_2 \quad \text{and} \quad I_1 = \cancel{\tfrac{1}{12}V_2} - \tfrac{4}{3}I_2$$

and so

$$Z_i = \frac{V_1}{I_1} = \frac{-20I_2}{-\frac{4}{3}I_2} = 15\ \Omega$$

(*c*) When $R_L = 8\ \Omega$, as in Fig. 11-17 below,

$$V_1 = 2V_2 - 20I_2, \quad I_1 = \tfrac{1}{12}V_2 - \tfrac{4}{3}I_2, \quad \text{and} \quad V_2 = -8I_2$$

**Fig. 11-17**

Thus

$$V_1 = -16I_2 - 20I_2 = -36I_2, \qquad I_1 = -\tfrac{8}{12}I_2 - \tfrac{4}{3}I_2 = -2I_2,$$

and

$$Z_i = \frac{V_1}{I_1} = \frac{-36I_2}{-2I_2} = 18\ \Omega$$

**11.15** Given the two port shown in Fig. 11-18, find the transmission parameters $A$, $B$, $C$, and $D$ in terms of $L$, $R$, $C$, and $\omega$, assuming sinusoidal steady-state conditions. (Note: $C$ is used both for a parameter and a capacitance.)

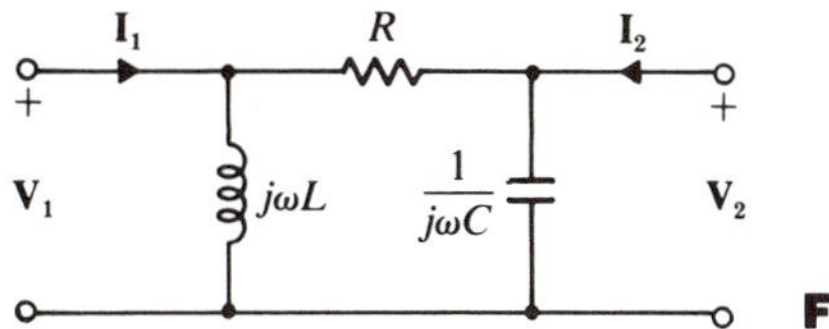


Fig. 11-18

The transmission equations are

$$\mathbf{V}_1 = A\mathbf{V}_2 - B\mathbf{I}_2 \quad \text{and} \quad \mathbf{I}_1 = C\mathbf{V}_2 - D\mathbf{I}_2$$

To solve for $A$ and $C$ we set $\mathbf{I}_2 = 0$ as in Figure 11-19a. Then the two nodal equations are

$$\left(\frac{1}{R} + \frac{1}{j\omega L}\right)\mathbf{V}_1 + \left(-\frac{1}{R}\right)\mathbf{V}_2 = \mathbf{I}_1$$

$$\left(-\frac{1}{R}\right)\mathbf{V}_1 + \left(\frac{1}{R} + j\omega C\right)\mathbf{V}_2 = 0$$

Solving for $\mathbf{V}_1$ and $\mathbf{V}_2$ using Cramer's rule,

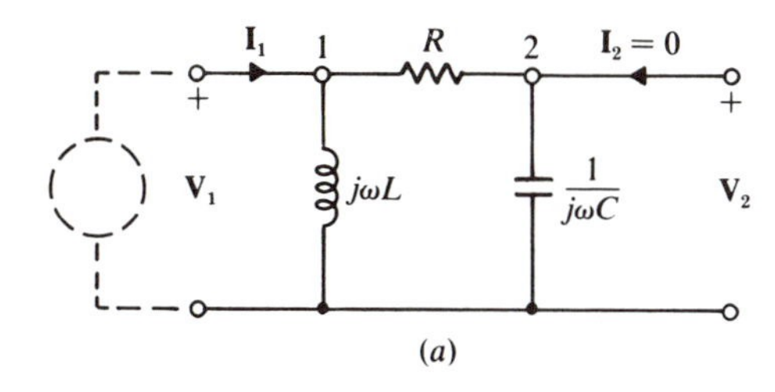


(a)

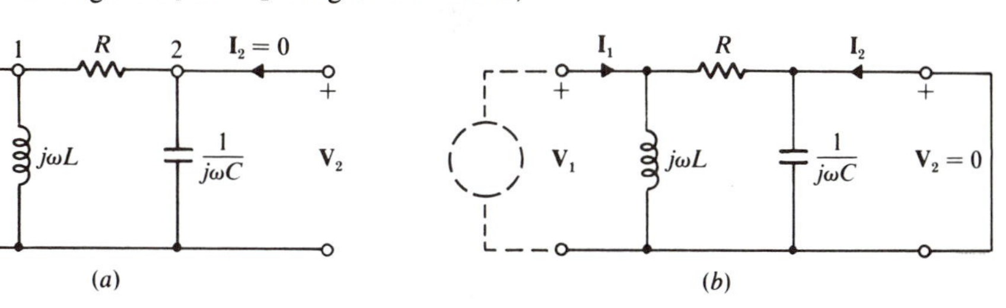


(b)

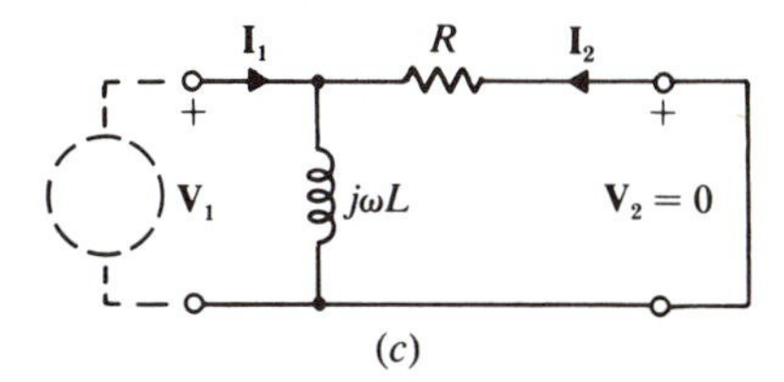


(c)

Fig. 11-19

$$\mathbf{V}_1 = \frac{(1/R + j\omega C)\mathbf{I}_1}{(1/R + 1/j\omega L)(1/R + j\omega C) - 1/R^2} \quad \text{and}$$

$$\mathbf{V}_2 = \frac{(1/R)\mathbf{I}_1}{(1/R + 1/j\omega L)(1/R + j\omega C) - 1/R^2}$$

Thus

$$A = \left.\frac{\mathbf{V}_1}{\mathbf{V}_2}\right|_{\mathbf{I}_2=0} = \frac{1/R + j\omega C}{1/R} = 1 + j\omega RC$$

and

$$C = \left.\frac{\mathbf{I}_1}{\mathbf{V}_2}\right|_{\mathbf{I}_2=0} = R\left(\frac{1}{R} + \frac{1}{j\omega L}\right)\left(\frac{1}{R} + j\omega C\right) - \frac{1}{R}$$

To solve for $B$ and $D$ we set $\mathbf{V}_2 = 0$ as in Fig. 11-19*b*. Note that since the capacitor has been short circuited we can redraw this circuit as in Fig. 11-19*c*. Then

$$\mathbf{I}_1 = \left(\frac{1}{R} + \frac{1}{j\omega L}\right)\mathbf{V}_1 \quad \text{and} \quad \mathbf{V}_1 = (-R)\mathbf{I}_2$$

Thus

$$B = \left.\frac{\mathbf{V}_1}{-\mathbf{I}_2}\right|_{\mathbf{V}_2=0} = R$$

$$D = \left.\frac{\mathbf{I}_1}{-\mathbf{I}_2}\right|_{\mathbf{V}_2=0} = \frac{(1/R + 1/j\omega L)\mathbf{V}_1}{(1/R)\mathbf{V}_1} = 1 + \frac{R}{j\omega L}$$

**11.16** The transmission matrix $\mathbf{T}$ is defined by

$$\begin{bmatrix} V_1 \\ I_1 \end{bmatrix} = \begin{bmatrix} A & B \\ C & D \end{bmatrix}\begin{bmatrix} V_2 \\ -I_2 \end{bmatrix} = \mathbf{T}\begin{bmatrix} V_2 \\ -I_2 \end{bmatrix}$$

Find $\mathbf{T}$ for the network shown in Fig. 11-20.

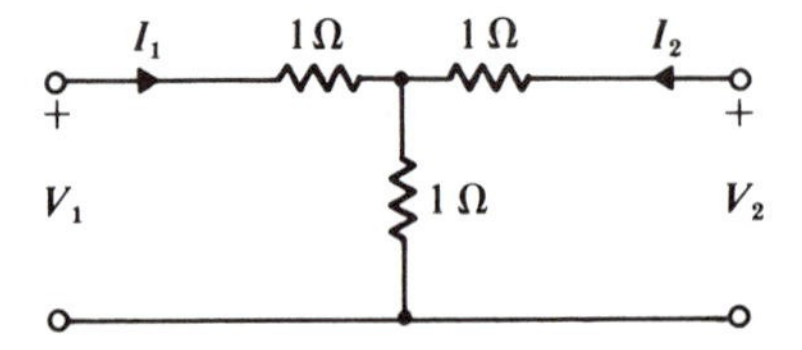


**Fig. 11-20**

Since this is a simple tee network we would be well advised to calculate the $z$-parameters, and then to convert them to the desired $T$-parameters. The loop equations are

$$\begin{aligned} 2I_1 + I_2 &= V_1 \\ I_1 + 2I_2 &= V_2 \end{aligned} \quad \text{or} \quad \mathbf{z} = \begin{bmatrix} 2 & 1 \\ 1 & 2 \end{bmatrix}$$

Then from Table 11-2,

$$\mathbf{T} = \begin{bmatrix} z_{11}/z_{21} & \Delta_z/z_{21} \\ 1/z_{21} & z_{22}/z_{21} \end{bmatrix} = \begin{bmatrix} 2 & 3 \\ 1 & 2 \end{bmatrix}$$

### *Alternative Solution*

To find $A$ and $C$ we set $I_2 = 0$ as in Fig. 11-21*a*. Then

$$V_2 = 1I_1 \quad \text{and} \quad I_1 = V_1/2$$

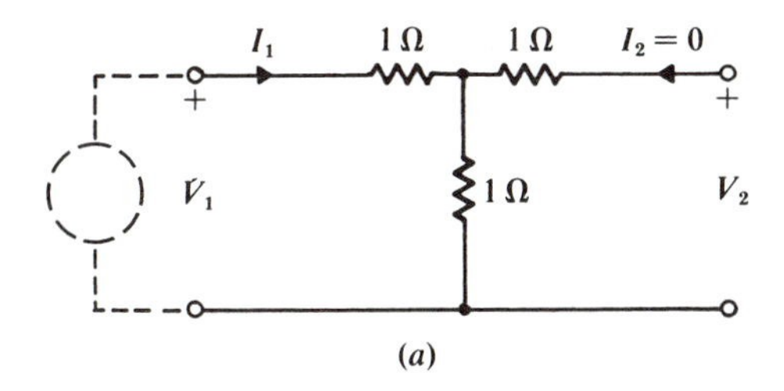

(*a*)

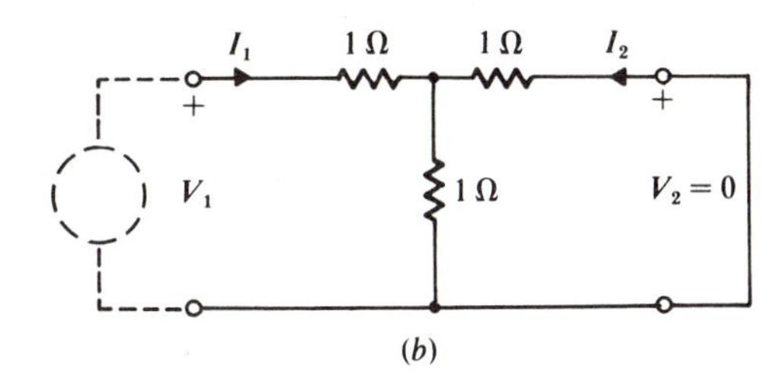

(*b*)

**Fig. 11-21**

Thus

$$A = \left.\frac{V_1}{V_2}\right|_{I_2=0} = 2 \quad \text{and} \quad C = \left.\frac{I_1}{V_2}\right|_{I_2=0} = 1$$

Then setting $V_2 = 0$ as in Fig. 11-21*b*, $-I_2 = I_1/2$ and

$$D = \left.\frac{I_1}{-I_2}\right|_{V_2=0} = 2$$

Finally, by loop analysis, $V_1 = \frac{3}{2}I_1$ and

$$B = \left.\frac{V_1}{-I_2}\right|_{V_2=0} = \frac{\frac{3}{2}I_1}{\frac{1}{2}I_1} = 3$$

**11.17** The transmission equations are particularly useful when two or more two ports are connected in *cascade* (see Fig. 11-22). This is a very common configuration, typified by a multistage amplifier or a ladder network. Prove that the two networks represented by $\mathbf{T}_a$ and $\mathbf{T}_b$ in Fig. 11-22 may be treated as a single two port whose transmission matrix is $\mathbf{T} = \mathbf{T}_a\mathbf{T}_b$.

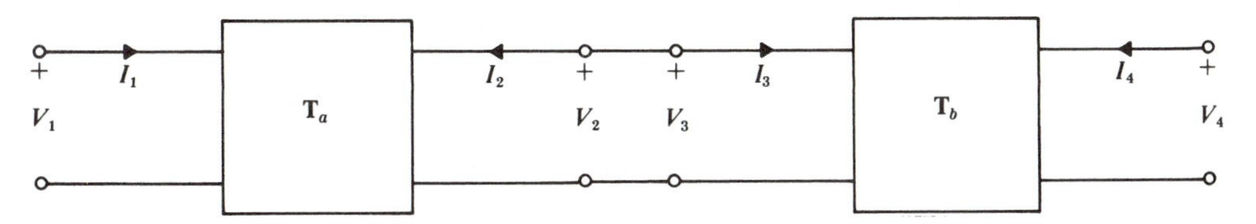


**Fig. 11-22**

By definition,

$$\begin{bmatrix} V_1 \\ I_1 \end{bmatrix} = \begin{bmatrix} A_a & B_a \\ C_a & D_a \end{bmatrix} \begin{bmatrix} V_2 \\ -I_2 \end{bmatrix} \quad \text{and} \quad \begin{bmatrix} V_3 \\ I_3 \end{bmatrix} = \begin{bmatrix} A_b & B_b \\ C_b & D_b \end{bmatrix} \begin{bmatrix} V_4 \\ -I_4 \end{bmatrix}$$

But from Fig. 11-22, $V_2 = V_3$ and $-I_2 = I_3$. That is,

$$\begin{bmatrix} V_2 \\ -I_2 \end{bmatrix} = \begin{bmatrix} V_3 \\ I_3 \end{bmatrix}$$

By substitution,

$$\begin{bmatrix} V_1 \\ I_1 \end{bmatrix} = \begin{bmatrix} A_a & B_a \\ C_a & D_a \end{bmatrix} \begin{bmatrix} A_b & B_b \\ C_b & D_b \end{bmatrix} \begin{bmatrix} V_4 \\ -I_4 \end{bmatrix} = \mathbf{T}_a \mathbf{T}_b \begin{bmatrix} V_4 \\ -I_4 \end{bmatrix} = \mathbf{T} \begin{bmatrix} V_4 \\ -I_4 \end{bmatrix}$$

**11.18** Using the results of Problems 11.16 and 11.17, show that for the network of Fig. 11-23,

$$\begin{bmatrix} V_1 \\ I_1 \end{bmatrix} = \begin{bmatrix} 7 & 12 \\ 4 & 7 \end{bmatrix} \begin{bmatrix} V_4 \\ -I_4 \end{bmatrix}$$

Then find the input impedance $Z_i$ when the output terminals are short circuited.

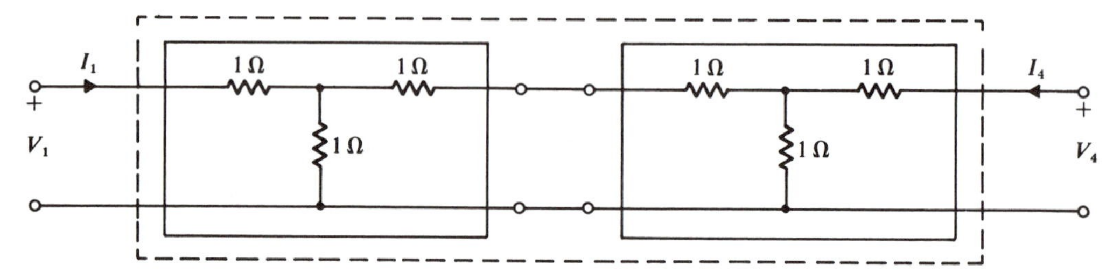


**Fig. 11-23**

From Problem 11.16 the transmission matrix for *each* tee network is

$$\mathbf{T} = \begin{bmatrix} 2 & 3 \\ 1 & 2 \end{bmatrix}$$

Therefore from Problem 11.17,

$$\begin{bmatrix} V_1 \\ I_1 \end{bmatrix} = \begin{bmatrix} 2 & 3 \\ 1 & 2 \end{bmatrix} \begin{bmatrix} 2 & 3 \\ 1 & 2 \end{bmatrix} \begin{bmatrix} V_4 \\ -I_4 \end{bmatrix} = \begin{bmatrix} 7 & 12 \\ 4 & 7 \end{bmatrix} \begin{bmatrix} V_4 \\ -I_4 \end{bmatrix}$$

With the output short circuited, $V_4 = 0$. Thus

$$\begin{bmatrix} V_1 \\ I_1 \end{bmatrix} = \begin{bmatrix} 7 & 12 \\ 4 & 7 \end{bmatrix} \begin{bmatrix} 0 \\ -I_4 \end{bmatrix} = \begin{bmatrix} -12I_4 \\ -7I_4 \end{bmatrix}$$

and

$$Z_i = \frac{V_1}{I_1} = \frac{-12I_4}{-7I_4} = \frac{12}{7}\ \Omega$$

**11.19** Show that

$$\mathbf{T} = \begin{bmatrix} 1 & Z(j\omega) \\ 0 & 1 \end{bmatrix}$$

for the network of Fig. 11-24. Then use Table 11-2 to find the $y$-parameters, and show that the $z$-parameters do not exist.

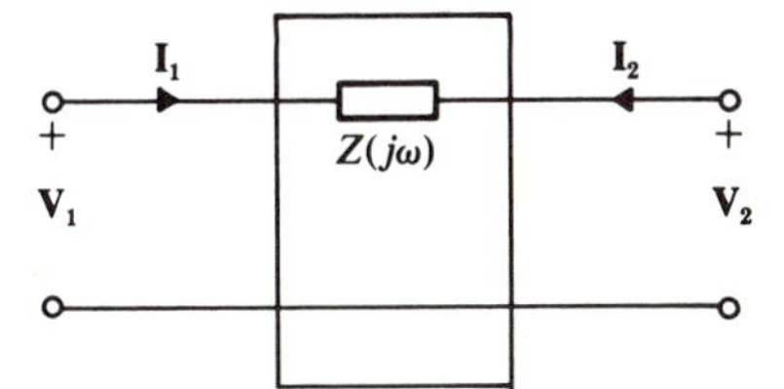


**Fig. 11-24**

In the usual way, referring to Fig. 11-25$a$,

$$A = \left.\frac{\mathbf{V}_1}{\mathbf{V}_2}\right|_{\mathbf{I}_2=0} = 1 \quad \text{and} \quad C = \left.\frac{\mathbf{I}_1}{\mathbf{V}_2}\right|_{\mathbf{I}_2=0} = 0$$

Similarly, referring to Fig. 11-25$b$,

$$B = \left.\frac{\mathbf{V}_1}{-\mathbf{I}_2}\right|_{\mathbf{V}_2=0} = Z(j\omega) \quad \text{and} \quad D = \left.\frac{\mathbf{I}_1}{-\mathbf{I}_2}\right|_{\mathbf{V}_2=0} = 1$$

That is,

$$\mathbf{T} = \begin{bmatrix} 1 & Z(j\omega) \\ 0 & 1 \end{bmatrix} \quad \text{and} \quad \Delta_\mathbf{T} = \begin{vmatrix} 1 & Z(j\omega) \\ 0 & 1 \end{vmatrix} = 1$$

Then from the conversion table,

$$\mathbf{y} = \begin{bmatrix} D/B & -\Delta_T/B \\ -1/B & A/B \end{bmatrix} = \frac{1}{Z(j\omega)} \begin{bmatrix} 1 & -1 \\ -1 & 1 \end{bmatrix}$$

*Check:* We can determine the $y$-parameters directly by applying KCL at the input and output. Thus from Fig. 11-24,

$$\mathbf{I}_1 = \frac{\mathbf{V}_1 - \mathbf{V}_2}{Z(j\omega)} \quad \text{and} \quad \mathbf{I}_2 = \frac{\mathbf{V}_2 - \mathbf{V}_1}{Z(j\omega)}$$

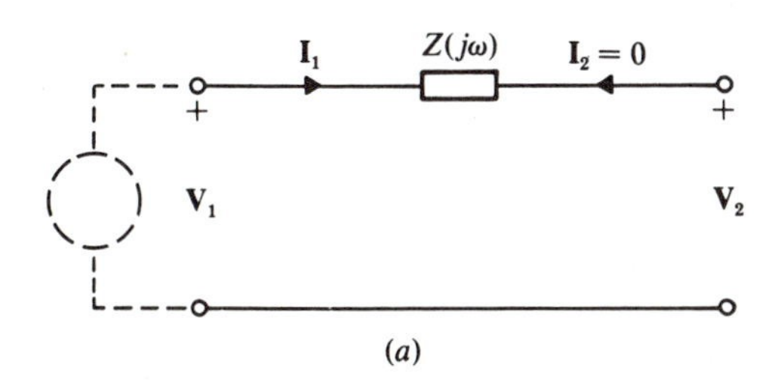


($a$)

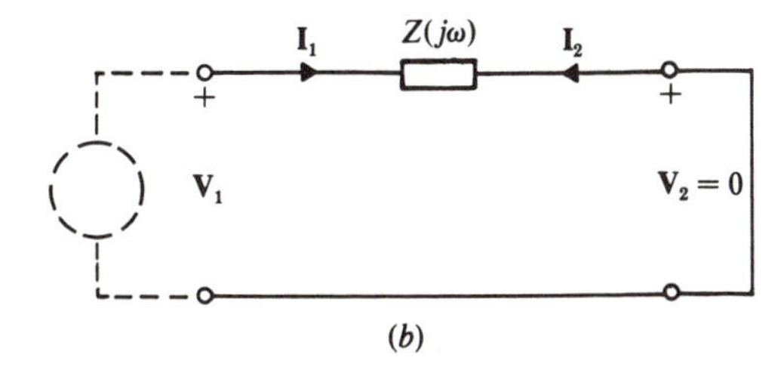


($b$)

**Fig. 11-25**

And by comparison with equation (11.2),

$$\mathbf{y} = \frac{1}{Z(j\omega)}\begin{bmatrix} 1 & -1 \\ -1 & 1 \end{bmatrix}$$

From Table 11-2 we see that we must divide by $C$ to obtain the $z$-parameters. Since $C = 0$, the $z$-parameters do not exist.

**11.20** Show that

$$\mathbf{T} = \begin{bmatrix} 1 & 0 \\ 1/Z(j\omega) & 1 \end{bmatrix}$$

for the two port of Fig. 11-26.

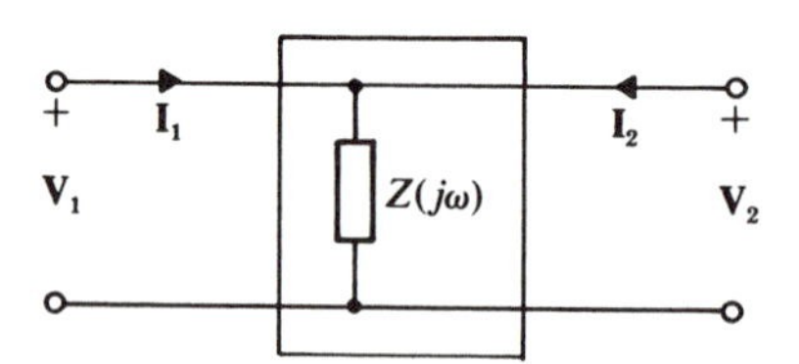


**Fig. 11-26**

Writing the loop equations for the circuit of Fig. 11-26,

$$Z(j\omega)\mathbf{I}_1 + Z(j\omega)\mathbf{I}_2 = \mathbf{V}_1$$

$$Z(j\omega)\mathbf{I}_1 + Z(j\omega)\mathbf{I}_2 = \mathbf{V}_2$$

and so by comparison with equation (11.1),

$$\mathbf{z} = Z(j\omega)\begin{bmatrix} 1 & 1 \\ 1 & 1 \end{bmatrix} \quad \text{and} \quad \Delta_z = |\mathbf{z}| = 0$$

Thus with the help of Table 11-2,

$$\mathbf{T} = \frac{1}{z_{21}}\begin{bmatrix} z_{11} & \Delta_z \\ 1 & z_{22} \end{bmatrix} = \begin{bmatrix} 1 & 0 \\ 1/Z(j\omega) & 1 \end{bmatrix}$$

**11.21** Use the results of Problems 11.17, 11.19, and 11.20 to find the transmission parameters for the ladder network in Fig. 11-27 under sinusoidal steady-state conditions. *Hint:* partition the ladder as shown.

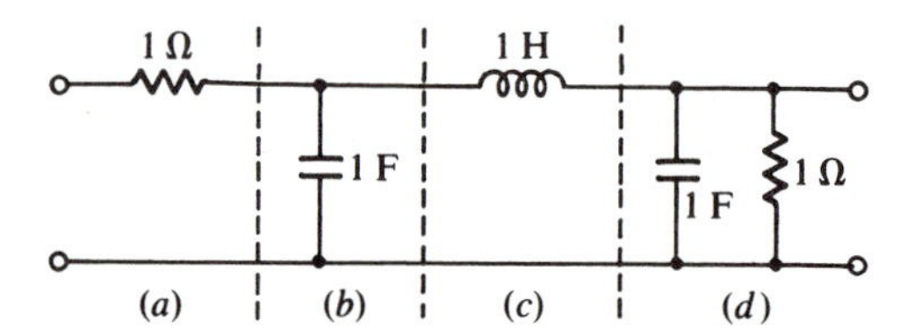


**Fig. 11-27**

Applying the results of Problems 11.19 and 11.20 to the circuit of Fig. 11-27,

$$\mathbf{T}_a = \begin{bmatrix} 1 & 1 \\ 0 & 1 \end{bmatrix} \qquad \mathbf{T}_b = \begin{bmatrix} 1 & 0 \\ j\omega & 1 \end{bmatrix}$$

$$\mathbf{T}_c = \begin{bmatrix} 1 & j\omega \\ 0 & 1 \end{bmatrix} \qquad \mathbf{T}_d = \begin{bmatrix} 1 & 0 \\ 1 + j\omega & 1 \end{bmatrix}$$

Then extending the result of Problem 11.17 to four cascaded two ports,

$$\mathbf{T} = \mathbf{T}_a\mathbf{T}_b\mathbf{T}_c\mathbf{T}_d$$

After a little hard labor, multiplying out these four matrices,

$$\mathbf{T} = \begin{bmatrix} 2(1 - \omega^2) + j(3\omega - \omega^3) & (1 - \omega^2) + j\omega \\ (1 - \omega^2) + j(2\omega - \omega^3) & (1 - \omega^2) \end{bmatrix}$$

### *Reciprocity Revisited*

**11.22** Find the relationship among the $z$-parameters which must be satisfied if a two port is to be reciprocal.

We will consider the statement of reciprocity in Theorem 10.3 and Fig. 10-3*b*, page 340.

First, let us assume an input source $\mathbf{V}_1 = \mathbf{V}$ with the output shorted. Then from equation (11.1), page 391, and Cramer's rule,

$$\mathbf{I}_2 = \frac{1}{\Delta_z}\begin{bmatrix} z_{11} & \mathbf{V} \\ z_{21} & 0 \end{bmatrix} = -\frac{z_{21}\mathbf{V}}{\Delta_z}$$

Next we will assume an output source $\mathbf{V}_2 = \mathbf{V}$ with the input shorted. Then

$$\mathbf{I}_1' = \frac{1}{\Delta_z}\begin{bmatrix} 0 & z_{12} \\ \mathbf{V} & z_{22} \end{bmatrix} = -\frac{z_{12}\mathbf{V}}{\Delta_z}$$

These two currents will be equal and the network reciprocal if $z_{12} = z_{21}$.

#### *Alternative Solution*

Referring instead to Fig. 10-3*a*, we will assume an input current source $\mathbf{I}_1 = \mathbf{I}$ with the output open. Equation (11.1) now yields

$$\mathbf{V}_2 = z_{21}\mathbf{I}_1$$

Then with the input open and $\mathbf{I}_2 = \mathbf{I}$,

$$\mathbf{V}_1' = z_{12}\mathbf{I}$$

and these two voltages will be equal and the network reciprocal if $z_{12} = z_{21}$.

**11.23** Show that $y_{12} = y_{21}$ for a reciprocal two port, and find the reciprocity condition in terms of the $h$- and $T$-parameters.

From Table 11-2,

$$\mathbf{z} = \frac{1}{\Delta_y}\begin{bmatrix} y_{22} & -y_{12} \\ -y_{21} & y_{11} \end{bmatrix} = \frac{1}{h_{22}}\begin{bmatrix} \Delta_h & h_{12} \\ -h_{21} & 1 \end{bmatrix} = \frac{1}{C}\begin{bmatrix} A & \Delta_T \\ 1 & D \end{bmatrix}$$

Therefore for $z_{12} = z_{21}$,

$$y_{12} = y_{21}, \quad h_{12} = -h_{21}, \quad \text{and} \quad \Delta_T = AD - BC = 1$$

These are the three (equivalent) reciprocity conditions required.

### *Some Two-Port Applications*

**11.24** The two port of Fig. 11-28 is resistive. Find the Thévenin equivalent with respect to the output terminals, and hence find $I_2$.

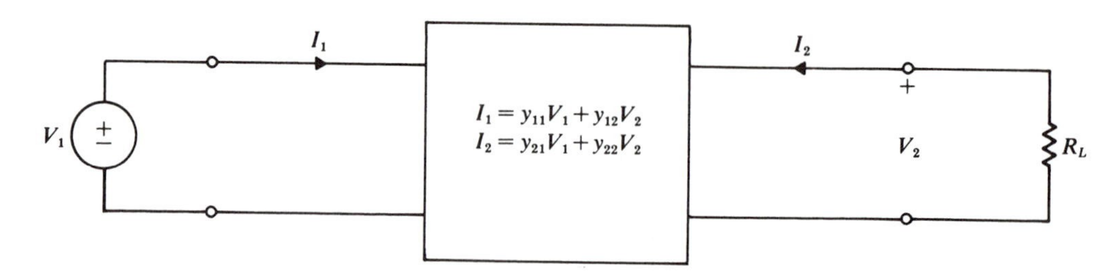


**Fig. 11-28**

From the second relation of equation *(11.2)*, with $I_2 = 0$ and hence $V_2 = V_{oc}$,

$$0 = y_{21}V_1 + y_{22}V_{oc} \quad \text{or} \quad V_{oc} = -\frac{y_{21}}{y_{22}}V_1$$

Similarly, with $V_2 = 0$ and hence $I_2 = -I_{sc}$,

$$-I_{sc} = y_{21}V_1 + y_{22}(0) \quad \text{or} \quad I_{sc} = -y_{21}V_1$$

Thus the Thévenin impedance is

$$Z_{Th} = R_{Th} = \frac{V_{oc}}{I_{sc}} = \frac{1}{y_{22}}$$

The Thévenin equivalent of Fig. 11-29 then follows at once from Thévenin's theorem. And from Fig. 11-29,

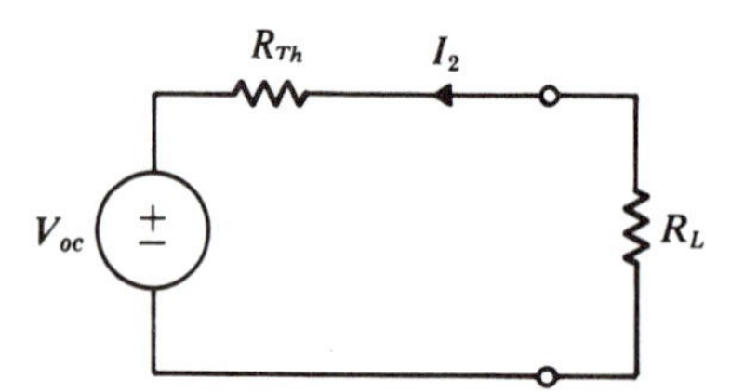


**Fig. 11-29**

$$I_2 = -\frac{V_{oc}}{R_{Th} + R_L} = -\frac{-y_{21}/y_{22}}{1/y_{22} + R_L}V_1 = \frac{y_{21}V_1}{1 + y_{22}R_L}$$

**11.25** The ideal transformer in Fig. 11-30 is described by the inverse transmission parameters

$$\mathbf{T}' = \begin{bmatrix} 0.5 & 0 \\ 0 & 2 \end{bmatrix}$$

Also, $L = 1/8$ H, $C = 1/8$ F, and $R = 1/12$ Ω. Find the value of $\omega$ which will cause the maximum average real power to be dissipated in $R$ under sinusoidal steady-state conditions. Calculate this power given $v_s(t) = \sqrt{2}\cos \omega t$.

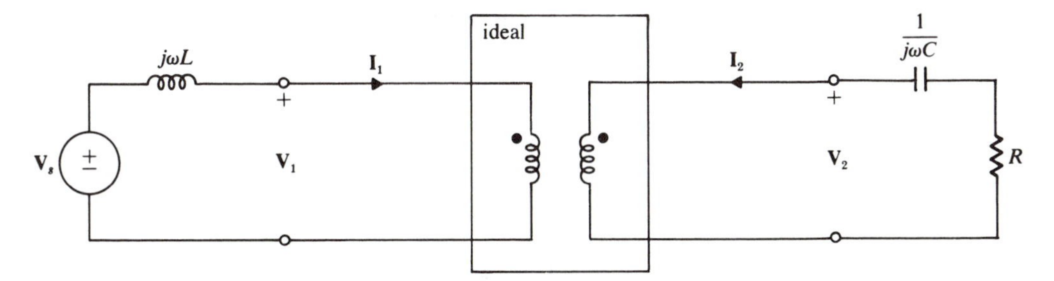


**Fig. 11-30**

We must first solve for $\mathbf{I}_2$. Then $P = RI^2_{2,\text{eff}}$ can be maximized with respect to $\omega$. The governing phasor loop equations are

$$-\mathbf{V}_s + \frac{j\omega}{8}\mathbf{I}_1 + \mathbf{V}_1 = 0$$

$$\frac{1}{12}\mathbf{I}_2 + \frac{8}{j\omega}\mathbf{I}_2 + \mathbf{V}_2 = 0$$

and from Table 11-1, page 394, and the given data,

$$\mathbf{V}_2 = 0.5\mathbf{V}_1 \quad \text{and} \quad \mathbf{I}_2 = -2\mathbf{I}_1$$

After substitution, the loop equations become

$$-\frac{j\omega}{16}\mathbf{I}_2 + 2\mathbf{V}_2 = \mathbf{V}_s$$

$$\left(\frac{1}{12} + \frac{8}{j\omega}\right)\mathbf{I}_2 + \mathbf{V}_2 = 0$$

Using Cramer's rule, $\mathbf{I}_2 = \mathbf{V}_s/(-j\omega/16 - 2/12 - 16/j\omega)$ and so $i_2(t)$ will have maximum amplitude when $-j\omega/16 - 16/j\omega = 0$ or $\omega = 16$ rad/s. Then $\mathbf{I}_2 = 6\mathbf{V}_s$ and

$$P_{\text{max}} = (6V_{s,\text{eff}})^2\frac{1}{12} = \frac{36}{12} = 3 \text{ W}$$

# PROBLEMS

**11.26** For the network of Fig. 11-31, show that

$$z_{11}(j\omega) = \frac{2j\omega - \omega^2 + 1}{j\omega} \quad \text{and} \quad y_{11}(j\omega) = \frac{3j\omega}{5j\omega - 3\omega^2 + 3}$$

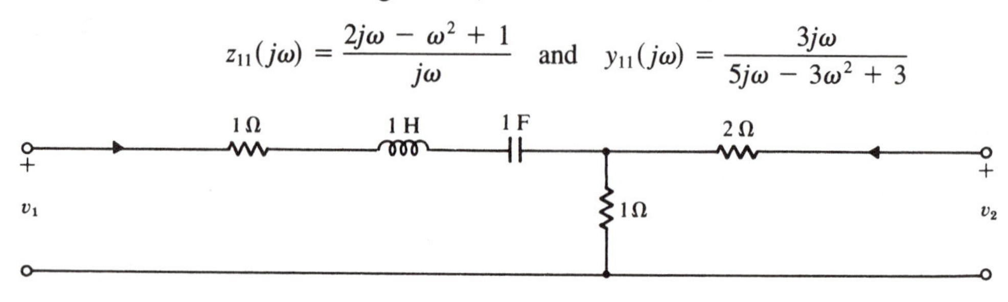


**Fig. 11-31**

**11.27** In Fig. 11-32,

$$\mathbf{z}_x = \begin{bmatrix} 1 & 2 \\ 2 & 3 \end{bmatrix} \quad \text{and} \quad \mathbf{z}_y = \begin{bmatrix} 5 & 6 \\ 6 & 7 \end{bmatrix}$$

Show that

$$\begin{bmatrix} v_1 \\ v_2 \end{bmatrix} = \begin{bmatrix} 6 & 8 \\ 8 & 10 \end{bmatrix} \begin{bmatrix} i_1 \\ i_2 \end{bmatrix}$$

In other words, two two ports in *series* act as a single two port with $\mathbf{z}_{eq} = \mathbf{z}_x + \mathbf{z}_y$, provided that the port conditions are not violated.

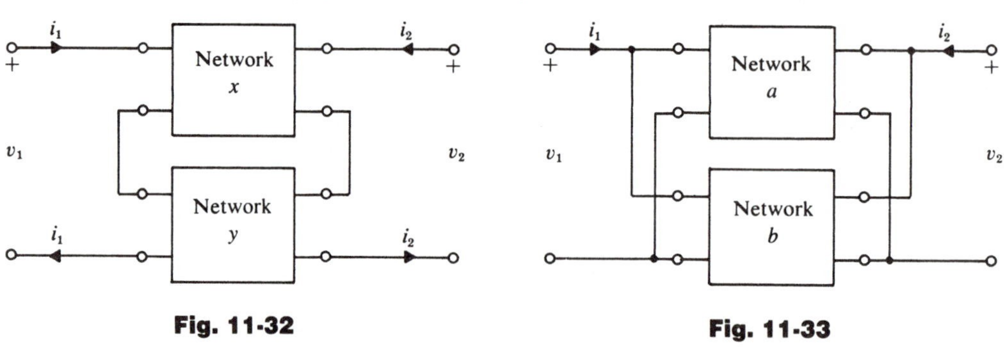


**Fig. 11-32** **Fig. 11-33**

**11.28** Show that for two two ports in *parallel* (see Fig. 11-33 above), $\mathbf{y}_{eq} = \mathbf{y}_a + \mathbf{y}_b$ provided that the port conditions are not violated.

**11.29** Show that the *h*-parameters will not exist for any network having $z_{22} = 0$.

**11.30** Find the *h*-parameters for the pi network of Fig. 11-34. *Hint:* It is very easy to find the *y*-parameters for this network.

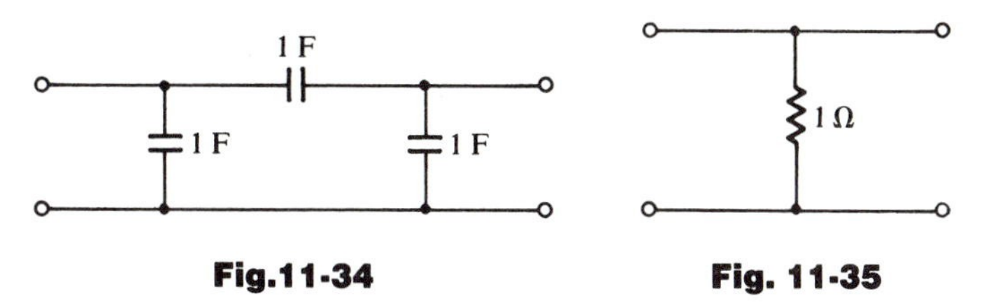


**Fig.11-34** **Fig. 11-35**

**11.31** Find the $z$-, $y$-, and $h$-parameters for the network of Fig. 11-35.

**11.32** Find $h_{21}$ for the circuit of Fig. 11-36.

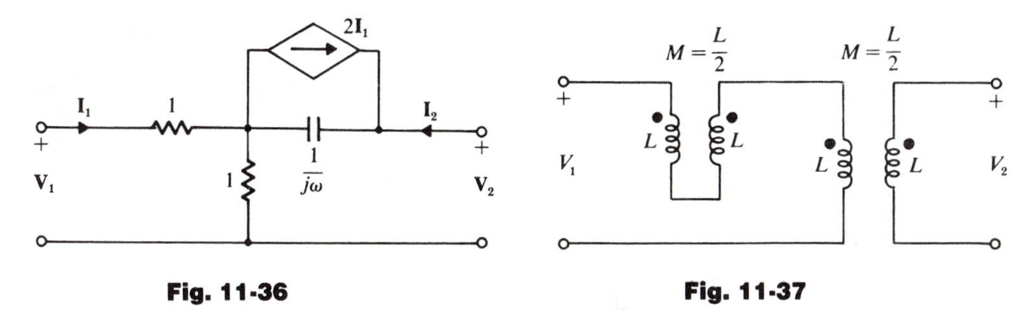


**Fig. 11-36** **Fig. 11-37**

**11.33** Two identical transformers are connected as shown in Fig. 11-37. There is no coupling from one transformer to the other. Find **z** for the entire network, working in the $j\omega$-domain.

**11.34** The $z$-parameters of the network $N$ in Fig. 11-38 are known. Find $\mathbf{V}_2/\mathbf{I}$ in terms of $R_1$, $R_2$, and the $z$-parameters.

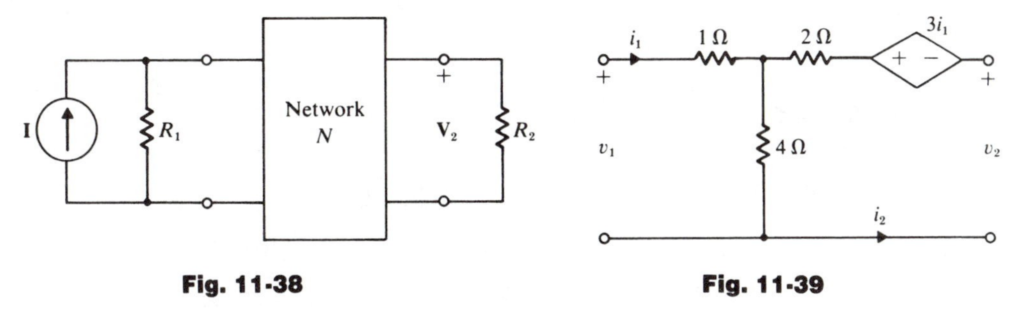


**Fig. 11-38** **Fig. 11-39**

**11.35** Show that the following equations correctly represent the network of Fig. 11-39:

$$v_2 = 1.5v_1 - 6.5i_1 \quad \text{and} \quad i_2 = 0.25v_1 - 1.25i_1$$

Chapter

# 12

# FOURIER METHODS OF ANALYSIS

The calculation of the steady-state response of a linear network to a sinusoidal input was developed in Chapter 7. Now almost all practical *periodic* waveforms may be expressed as the sum of a series of sine and cosine terms—the so-called *Fourier series*. And since the response to each sinusoidal term is easily computed, it is equally easy to find the *total* response with the help of superposition.

For most waveforms, periodic or nonperiodic, we will find that a limiting process leads to the *Fourier transform,* which is heavily used in communication theory. It has properties analogous to the Laplace transform, to which it is closely related.

## THE SINE–COSINE SERIES

A periodic function $g(t)$ can be written as the sum of an infinite series of sine and cosine terms, as follows:

$$g(t) = C_0 + \sum_{k=1}^{\infty} A_k \cos k\omega_0 t + \sum_{k=1}^{\infty} B_k \sin k\omega_0 t \qquad (12.1)$$

where

$$C_0 = \frac{1}{T}\int_{t_0}^{t_0+T} g(t)\, dt$$

$$A_k = \frac{2}{T}\int_{t_0}^{t_0+T} g(t) \cos k\omega_0 t\, dt \qquad (k = 1, 2, \ldots)$$

$$B_k = \frac{2}{T}\int_{t_0}^{t_0+T} g(t)\sin k\omega_0 t\,dt \qquad (k = 1, 2, \ldots)$$

This is the most common form of the Fourier series.

If, alternatively, we write the constant term as $A_0/2$, then $A_0 = A_k|_{k=0}$. Note too that the constant term $C_0$ or $A_0/2$ is the *average value* of $g(t)$.

As a matter of nomenclature, $\omega_0$ is called the *fundamental frequency* of the periodic waveform $g(t)$, while $k\omega_0$ is called the *kth harmonic*. The *fundamental period* is $T = 2\pi/\omega_0$.

## Example 12.1

Let us calculate the coefficients of the sine-cosine series for the periodic pulse-train shown in Fig. 12-1, given $\alpha = 10$ and $\omega_0 = 2\pi/T = 2$ rad/s.

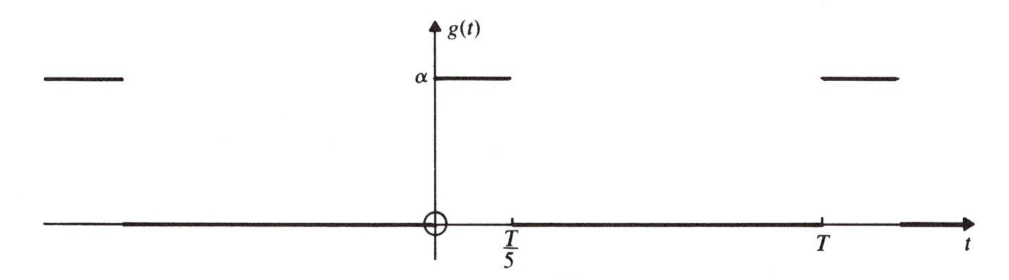


**Fig. 12-1**

From equation (*12.1*),

$$C_0 = \frac{1}{T}\int_{t_0}^{t_0+T} g(t)\,dt = \frac{1}{T}\int_0^{T/5}\alpha\,dt + \frac{1}{T}\int_{T/5}^{T} 0\,dt = \frac{\alpha}{5}$$

Similarly

$$A_k = \frac{2}{T}\int_0^{T/5}\alpha\cos k\omega_0 t\,dt = \frac{2\alpha}{Tk\omega_0}\sin\frac{k\omega_0 T}{5}$$

But $\omega_0 = 2\pi/T$, and so

$$A_k = \frac{\alpha}{\pi k}\sin\frac{2\pi k}{5}$$

After a similar calculation,

$$B_k = \frac{\alpha}{\pi k}\left\{1 - \cos\frac{2\pi k}{5}\right\}$$

Now we can substitute the given data and evaluate the first few terms in the series, as follows:

$$g(t) = C_0 + \sum_{k=1}^{\infty} A_k\cos k\omega_0 t + \sum_{k=1}^{\infty} B_k\sin k\omega_0 t$$

$$= 2 + 3.02 \cos 2t + 0.94 \cos 4t - 0.62 \cos 6t + \cdots$$
$$+ 2.20 \sin 2t + 2.88 \sin 4t + 1.92 \sin 6t + \cdots$$ ■

Needless to say, the integrations for $A_k$ and $B_k$ can become rather unpleasant for all but a few simple periodic functions. However, the integrations can easily be performed numerically on a digital computer. The computer could also be used to sum the first few terms of the series to see how well a finite number of terms approximates the function.

## THE SYMMETRY RULES

A function $g(t)$ is said to be *even* if

$$g(t) = g(-t)$$

and *odd* if

$$g(t) = -g(-t)$$

This concept is illustrated in Fig. 12-2. Of course many functions are *neither* even *nor* odd.

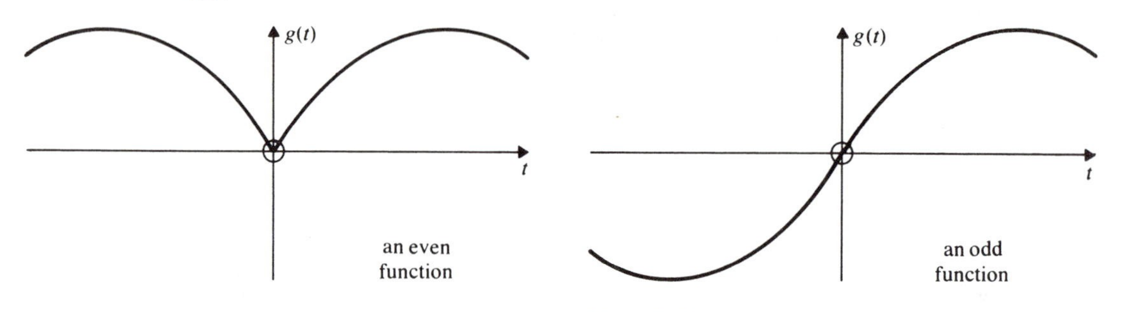


**Fig. 12-2**

If a function is *even,* there will be no *sine* terms in the sine-cosine series. That is,

$$B_k = 0 \qquad (k = 1, 2, \ldots)$$

And for an *odd* function there will be no *constant* and no *cosine* terms. That is,

$$C_0 = 0 \quad \text{and} \quad A_k = 0 \qquad (k = 1, 2, \ldots)$$

## THE EXPONENTIAL SERIES

The Fourier series can also be written as the sum of an infinite number of exponential terms, namely

$$g(t) = \sum_{k=-\infty}^{\infty} \mathbf{C}_k e^{jk\omega_0 t} \tag{12.2}$$

where

$$\mathbf{C}_k = \frac{1}{T}\int_{t_0}^{t_0+T} g(t)e^{-jk\omega_0 t}\,dt \qquad (k = \ldots, -2, -1, 0, 1, 2, \ldots)$$

Observe that once a $\mathbf{C}_k$ is evaluated, then the coefficient $\mathbf{C}_{-k}$ follows at once as the complex conjugate of $\mathbf{C}_k$,

$$\mathbf{C}_{-k} = \mathbf{C}_k^*$$

Further, the coefficient $\mathbf{C}_0$ is identical to the constant term $C_0$ in the sine-cosine series. It is still the average value of $g(t)$,

$$\mathbf{C}_0 = C_0 = \frac{1}{T}\int_{t_0}^{t_0+T} g(t)\,dt$$

### Example 12.2

Here we will compute the coefficients of the exponential series for the function considered in Example 12.1 (see Fig. 12-1).

From equation (*12.2*),

$$\mathbf{C}_k = \frac{1}{T}\int_0^{T/5} \alpha e^{-jk\omega_0 t}\,dt + \frac{1}{T}\int_{T/5}^{T} 0e^{-jk\omega_0 t}\,dt = \frac{1}{T}\left[\frac{\alpha e^{-jk\omega_0 t}}{-jk\omega_0}\right]_0^{T/5} = \frac{\alpha}{Tk\omega_0}\left\{\frac{e^{-jk\omega_0 T/5} - 1}{-j}\right\}$$

$$= \frac{\alpha}{2\pi k}\left\{\frac{1 - e^{-j2\pi k/5}}{j}\right\} = \frac{\alpha}{2\pi k}\left\{\frac{1 - \cos(2\pi k/5) + j\sin(2\pi k/5)}{j}\right\}$$

$$= \frac{\alpha}{2\pi k}\left\{\sin\frac{2\pi k}{5} + j\left(\cos\frac{2\pi k}{5} - 1\right)\right\}$$

and, as in Example 12.1, $\mathbf{C}_0 = C_0 =$ average value of $g(t) = \alpha/5$.

Given $\alpha = 10$ and $\omega_0 = 2$ rad/s, we can compute a few terms in the series,

$$g(t) = \sum_{-\infty}^{\infty} \mathbf{C}_k e^{jk\omega_0 t} = \cdots + (-0.312 + j0.960)e^{-j6t} + (0.468 + j1.44)e^{-j4t}$$
$$+ (1.51 + j1.10)e^{-j2t} + 2 + (1.51 - j1.10)e^{j2t}$$
$$+ (0.468 - j1.44)e^{j4t} + (-0.312 - j0.960)e^{j6t} + \cdots$$

■

Once again, the aid of a digital computer is recommended for the evaluation of the integral.

## SOME RELATIONSHIPS

As can be seen in Example 12.2, the $\mathbf{C}_k$ are, in general, complex numbers. We will now show that the sine-cosine and the exponential forms of the Fourier series are closely related. Thus from equation (*12.2*) and the Euler relation,

$$\mathbf{C}_k = \frac{1}{T}\int_{t_0}^{t_0+T} g(t)e^{-jk\omega_0 t}\,dt = \frac{1}{T}\int_{t_0}^{t_0+T} g(t)\{\cos k\omega_0 t - j\sin k\omega_0 t\}\,dt$$

which, from equation (*12.1*), can be written

$$\mathbf{C}_k = \tfrac{1}{2}(A_k - jB_k) \qquad (12.3)$$

or conversely,

$$A_k = 2\,\mathrm{Re}[\mathbf{C}_k] \quad \text{and} \quad B_k = -2\,\mathrm{Im}[\mathbf{C}_k]$$

That is, given the $\mathbf{C}_k$, we can find the $A_k$ and $B_k$, or vice versa. Comparing the results of Examples 12.1 and 12.2 for $k = 2$, say,

$$\mathbf{C}_2 = \tfrac{1}{2}(A_2 - jB_2) = \tfrac{1}{2}(0.94 - j2.88) = 0.47 - j1.44$$

which checks.

From a slightly different point of view we can write

$$\begin{aligned} g(t) &= \sum_{-\infty}^{\infty} |\mathbf{C}_k|\, e^{j\angle\mathbf{C}_k}\, e^{jk\omega_0 t} \\ &= \sum_{-\infty}^{\infty} \{|\mathbf{C}_k| \cos(k\omega_0 t + \angle\mathbf{C}_k) + j|\mathbf{C}_k|\sin(k\omega_0 t + \angle\mathbf{C}_k)\} \\ &= \sum_{-\infty}^{-1} |\mathbf{C}_k|\cos(k\omega_0 t + \angle\mathbf{C}_k) + |\mathbf{C}_0|\cos\angle\mathbf{C}_0 + \sum_{1}^{\infty}|\mathbf{C}_k|\cos(k\omega_0 t + \angle\mathbf{C}_k) \\ &\quad + j\sum_{-\infty}^{-1}|\mathbf{C}_k|\sin(k\omega_0 t + \angle\mathbf{C}_k) + j|\mathbf{C}_0|\sin\angle\mathbf{C}_0 + j\sum_{1}^{\infty}|\mathbf{C}_k|\sin(k\omega_0 t + \angle\mathbf{C}_k) \end{aligned}$$

Now we have seen that $\mathbf{C}_{-k} = \mathbf{C}_k^*$. Therefore $|\mathbf{C}_{-k}| = |\mathbf{C}_k|$ and $\angle\mathbf{C}_{-k} = -\angle\mathbf{C}_k$. Thus

$$\begin{aligned} \sum_{-\infty}^{-1}|\mathbf{C}_k|\cos(k\omega_0 t + \angle\mathbf{C}_k) &= \sum_{1}^{\infty}|\mathbf{C}_{-k}|\cos(-k\omega_0 t + \angle\mathbf{C}_{-k}) \\ &= \sum_{1}^{\infty}|\mathbf{C}_k|\cos(k\omega_0 t + \angle\mathbf{C}_k) \end{aligned}$$

and similarly,

$$\sum_{-\infty}^{-1}|\mathbf{C}_k|\sin(k\omega_0 t + \angle\mathbf{C}_k) = -\sum_{1}^{\infty}|\mathbf{C}_k|\sin(k\omega_0 t + \angle\mathbf{C}_k)$$

Finally, substituting these results and the facts that $\cos\angle\mathbf{C}_0 = 1$ and $\sin\angle\mathbf{C}_0 = 0$,

$$g(t) = C_0 + \sum_{k=1}^{\infty} M_k\cos(k\omega_0 t + \theta_k) \qquad (12.4)$$

where

$$M_k = 2|\mathbf{C}_k| \quad \text{and} \quad \theta_k = \angle\mathbf{C}_k$$

This is a third form of the Fourier series. Similarly we could write a sine series.

The exponential series is probably the most useful, theoretically, but it leaves us with a series whose coefficients are complex. What we have just shown is that by judiciously combining terms we can arrive at a real cosine series.

Sometimes we plot $|\mathbf{C}_k|$ and $\underline{/\mathbf{C}_k}$ or $M_k$ and $\theta_k$ vs. $k$. These are called the line or discrete *amplitude and phase spectra*.

### Example 12.3

Now we can obtain the cosine series corresponding to the pulse-train of Fig. 12-1.

In Example 12.2 we found $C_0$ and $\mathbf{C}_k$. Making use of these results,

$$C_0 = \frac{\alpha}{5}, \qquad M_k = 2|\mathbf{C}_k| = \frac{\alpha}{\pi k}\sqrt{\sin^2\frac{2\pi k}{5} + \left(\cos\frac{2\pi k}{5} - 1\right)^2} = \frac{\alpha}{\pi k}\sqrt{2 - 2\cos\frac{2\pi k}{5}}$$

and

$$\theta_k = \underline{/\mathbf{C}_k} = \tan^{-1}\left\{\left(\cos\frac{2\pi k}{5} - 1\right)\Big/\sin\frac{2\pi k}{5}\right\}$$

If, once again, we are given $\alpha = 10$ and $\omega_0 = 2$ rad/s, we can evaluate a few terms in the series. This yields

$$\begin{aligned} g(t) &= C_0 + \sum_{1}^{\infty} M_k \cos(k\omega_0 t + \theta_k) \\ &= 2 + 3.74\cos(2t - 36°) + 3.03\cos(4t - 72°) + 2.02\cos(6t - 108°) + \cdots \end{aligned}$$

The amplitude and phase spectra for $g(t)$ are graphed in Fig. 12-3.

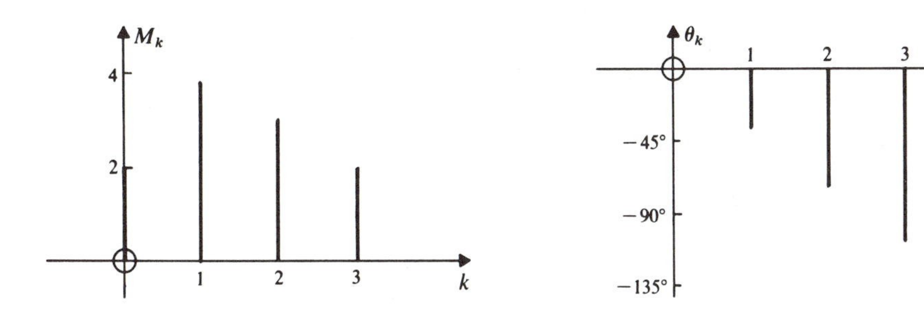


**Fig. 12-3** ■

As a check on our work, and to consolidate our understanding of the relationships between the three forms of the Fourier series, it is worthwhile comparing the answers to Examples 12.1, 12.2, and 12.3. For example, $A_2 = 0.94$ and $B_2 = 2.88$ in Example 12.1 checks with $\mathbf{C}_2 = \frac{1}{2}(A_2 - jB_2) = 0.47 - j1.44$ in Example 12.2. Also, $M_2 = 2|\mathbf{C}_2| = 2\sqrt{0.47^2 + 1.44^2} = 3.03$ and $\theta_2 = \underline{/\mathbf{C}_2} = \tan^{-1}(-1.44/0.47) = -72°$ in Example 12.3.

### RESPONSE OF LINEAR NETWORKS

Given the sinusoidal network function $H(j\omega) = \mathbf{Y}/\mathbf{U}$ and the periodic input function

$$u(t) = \sum_{k=-\infty}^{\infty} \mathbf{U}_k e^{jk\omega_0 t}$$

the steady-state response will be

$$y_{ss}(t) = \sum_{k=-\infty}^{\infty} \mathbf{Y}_k e^{jk\omega_0 t} = Y_0 + \sum_{k=1}^{\infty} 2|\mathbf{Y}_k| \cos(k\omega_0 t + \angle\mathbf{Y}_k) \qquad (12.5)$$

where $\mathbf{Y}_k = \mathbf{U}_k H(jk\omega_0)$. That is, given the Fourier coefficients $\mathbf{U}_k$ for the input, we can easily compute the corresponding coefficients $\mathbf{Y}_k$ for the output.

#### Example 12.4

We will suppose that the pulse-train of Fig. 12-1 is applied as an input to the circuit of Fig. 12-4. The objective is to find a Fourier series representation of the output.

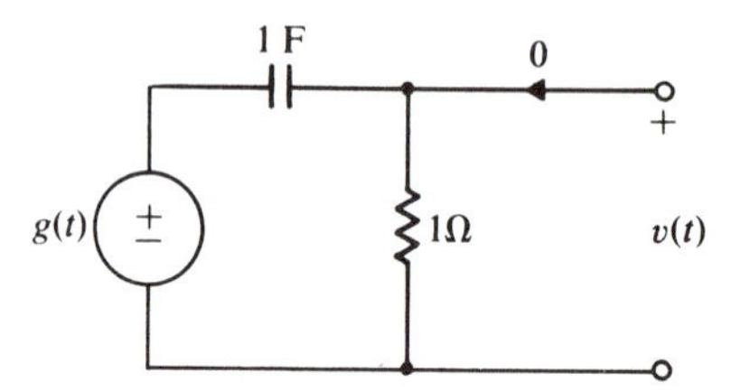


**Fig. 12-4**

By inspection of the circuit,

$$H(j\omega) = \frac{\mathbf{V}}{\mathbf{G}} = \frac{1}{1 + 1/j\omega} = \frac{j\omega}{1 + j\omega}$$

Now from equation (12.5),

$$\mathbf{V}_k = \mathbf{G}_k H(jk\omega_0) = \mathbf{G}_k \frac{jk\omega_0}{1 + jk\omega_0}$$

and

$$v_{ss}(t) = V_0 + \sum_{k=1}^{\infty} 2|\mathbf{V}_k| \cos(k\omega_0 t + \angle\mathbf{V}_k)$$

But $\mathbf{G}_k$ is simply another symbol for the $\mathbf{C}_k$ found in Example 12.2. Thus

$$V_0 = G_0 H(j0) = \frac{\alpha}{5} 0 = 0$$

$$2|\mathbf{V}_k| = 2|\mathbf{G}_k|\,|jk\omega_0|/|1 + jk\omega_0| = \frac{\alpha k\omega_0}{\pi k\sqrt{1 + k^2\omega_0^2}} \sqrt{2 - 2\cos\frac{2\pi k}{5}}$$

and

$$\underline{/\mathbf{V}_k} = \underline{/\mathbf{G}_k} + 90° - \tan^{-1} k\omega_0 = \tan^{-1}\left\{\left(\cos\frac{2\pi k}{5} - 1\right)\Big/\sin\frac{2\pi k}{5}\right\} + 90° - \tan^{-1} k\omega_0$$

If we wished, we could substitute $\alpha = 10$ and $\omega_0 = 2$ rad/s as before, and then evaluate the first few terms of the cosine series. ■

## EFFECTIVE VALUE AND POWER

Given the periodic waveform

$$g(t) = \sum_{k=-\infty}^{\infty} \mathbf{G}_k e^{jk\omega_0 t} = G_0 + \sum_{k=1}^{\infty} 2|\mathbf{G}_k| \cos(k\omega_0 t + \underline{/\mathbf{G}_k})$$

the harmonics will have RMS amplitudes of $G_{1,\text{eff}} = 2|\mathbf{G}_1|/\sqrt{2}$, $G_{2,\text{eff}} = 2|\mathbf{G}_2|/\sqrt{2}$, . . . and the RMS value of $g(t)$ will be given by

$$G_{\text{eff}}^2 = G_0^2 + G_{1,\text{eff}}^2 + G_{2,\text{eff}}^2 + \cdots \tag{12.6}$$

And if in some network branch $v(t)$ and $i(t)$ are periodic, with the same period, then the average active power associated with the $k$th harmonic is

$$P_k = V_{k,\text{eff}} I_{k,\text{eff}} \cos \phi_k$$

where $\phi_k$ is the angle of the harmonic sinusoid $v_k(t)$ relative to $i_k(t)$. The *total* average power is

$$P = P_0 + P_1 + P_2 + \cdots$$

The plot of the terms in this series against $k$ is called the line or discrete *power spectrum*.

## THE FOURIER TRANSFORM

It is possible to go through a limiting process of the Fourier series to arrive at an integral representation of the function $g(t)$, namely,

$$g(t) = \frac{1}{2\pi}\int_{-\infty}^{\infty} \mathbf{G}(\omega)e^{j\omega t}\, d\omega \tag{12.7}$$

where

$$\mathbf{G}(\omega) = \int_{-\infty}^{\infty} g(t)e^{-j\omega t}\, dt \tag{12.8}$$

The latter integral is called the *Fourier integral* or the *Fourier transform* of $g(t)$, and is closely related to the Laplace transform of $g(t)$. The similar notation

$$\mathcal{F}[g(t)] = \mathbf{G}(\omega) \quad \text{and} \quad \mathcal{F}^{-1}[\mathbf{G}(\omega)] = g(t)$$

is often used, and for many functions the substitution of $j\omega$ for $s$ relates the two transforms. As in the case of the Laplace transform, tables of transform pairs and properties may be set up as in Tables 12-1 and 12-2.

Note carefully, however, that the factor $1/2\pi$ is sometimes assigned to equation (*12.8*) rather than (*12.7*), and that sometimes the factor $1/\sqrt{2\pi}$ is assigned to *both* equations. Here we will consistently use equations (*12.7*) and (*12.8*) as stated above.

The transform pair of equations (*12.7*) and (*12.8*) is often written in terms of the true frequency $f$ as

$$g(t) = \int_{-\infty}^{\infty} \mathbf{G}(f)e^{j2\pi ft}\, df \tag{12.9}$$

where

$$\mathbf{G}(f) = \int_{-\infty}^{\infty} g(t)e^{-j2\pi ft}\, dt \tag{12.10}$$

Some relationships between $f$- and $\omega$-transforms are explored in Problem 12.27.

**Table 12-1 Fourier Transform Properties.**

| Property | $g(t)$ | $\mathbf{G}(\omega)$ |
|---|---|---|
| 1. Definition | $g(t)$ | $\int_{-\infty}^{\infty} g(t)e^{-j\omega t}\, dt$ |
| 2. Inversion | $\frac{1}{2\pi}\int_{-\infty}^{\infty} \mathbf{G}(\omega)e^{j\omega t}\, d\omega$ | $\mathbf{G}(\omega)$ |
| 3. Linearity | $ag_1(t) + bg_2(t)$ | $a\mathbf{G}_1(\omega) + b\mathbf{G}_2(\omega)$ |
| 4. Symmetry | $g(t)$<br>$\mathbf{G}(t)$ | $\mathbf{G}(\omega)$<br>$2\pi g(-\omega)$ |
| 5. Time shift | $g(t-t_0)$ | $e^{-j\omega t_0}\mathbf{G}(\omega)$ |
| 6. Frequency shift (modulation) | $e^{j\omega_0 t}g(t)$ | $\mathbf{G}(\omega-\omega_0)$ |
| 7. Scaling | $g(at)$ | $\frac{1}{\lvert a\rvert}\mathbf{G}\left(\frac{\omega}{a}\right)$ |
| 8. Differentiation | $\frac{d^n g(t)}{dt^n}$ | $(j\omega)^n\, \mathbf{G}(\omega)$ |
| 9. Time convolution† | $g(t) = g_1(t) * g_2(t)$<br>$= \int_{-\infty}^{\infty} g_1(\tau)g_2(t-\tau)\, d\tau$ | $\mathbf{G}(\omega) = \mathbf{G}_1(\omega)\mathbf{G}_2(\omega)$ |
| 10. Frequency convolution† | $g(t) = g_1(t)g_2(t)$ | $\mathbf{G}(\omega) = \frac{1}{2\pi}[\mathbf{G}_1(\omega) * \mathbf{G}_2(\omega)]$<br>$= \frac{1}{2\pi}\int_{-\infty}^{\infty} \mathbf{G}_1(y)\mathbf{G}_2(\omega-y)\, dy$ |

†The concept of convolution is the subject of the next chapter.

**Table 12-2 Some Fourier Transform Pairs.**

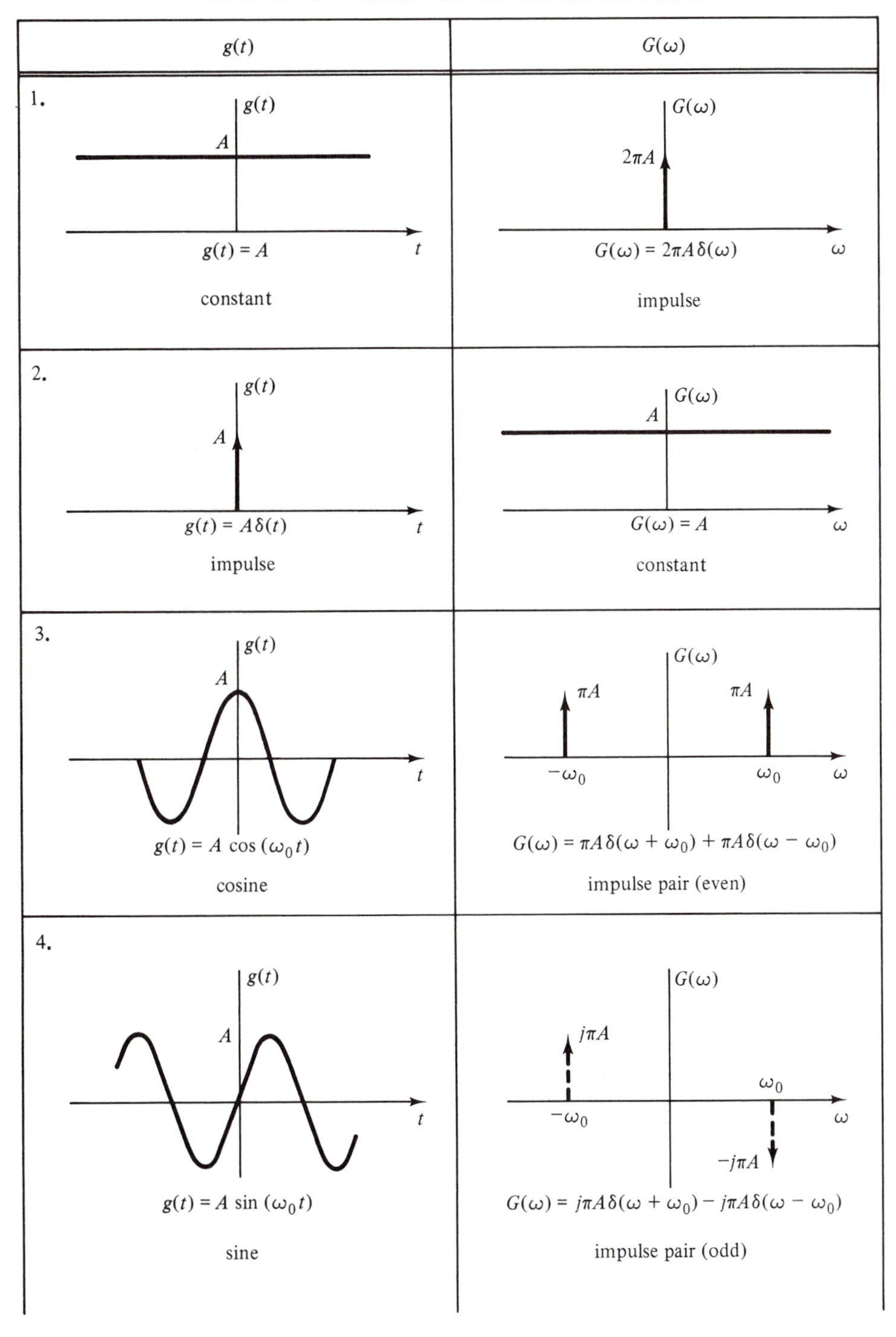

**Table 12-2 (*cont.*)**

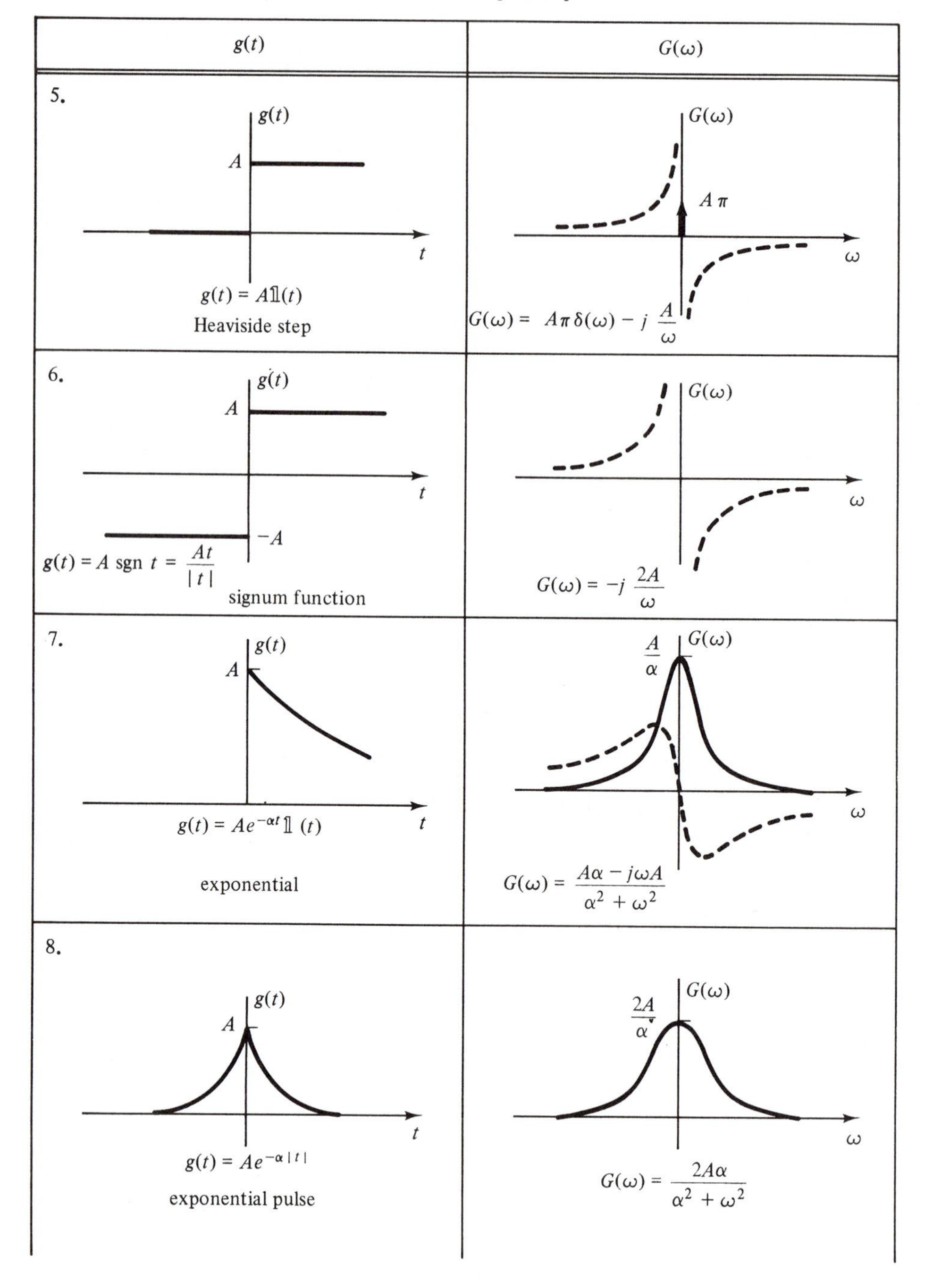

**Table 12-2 (*cont.*)**

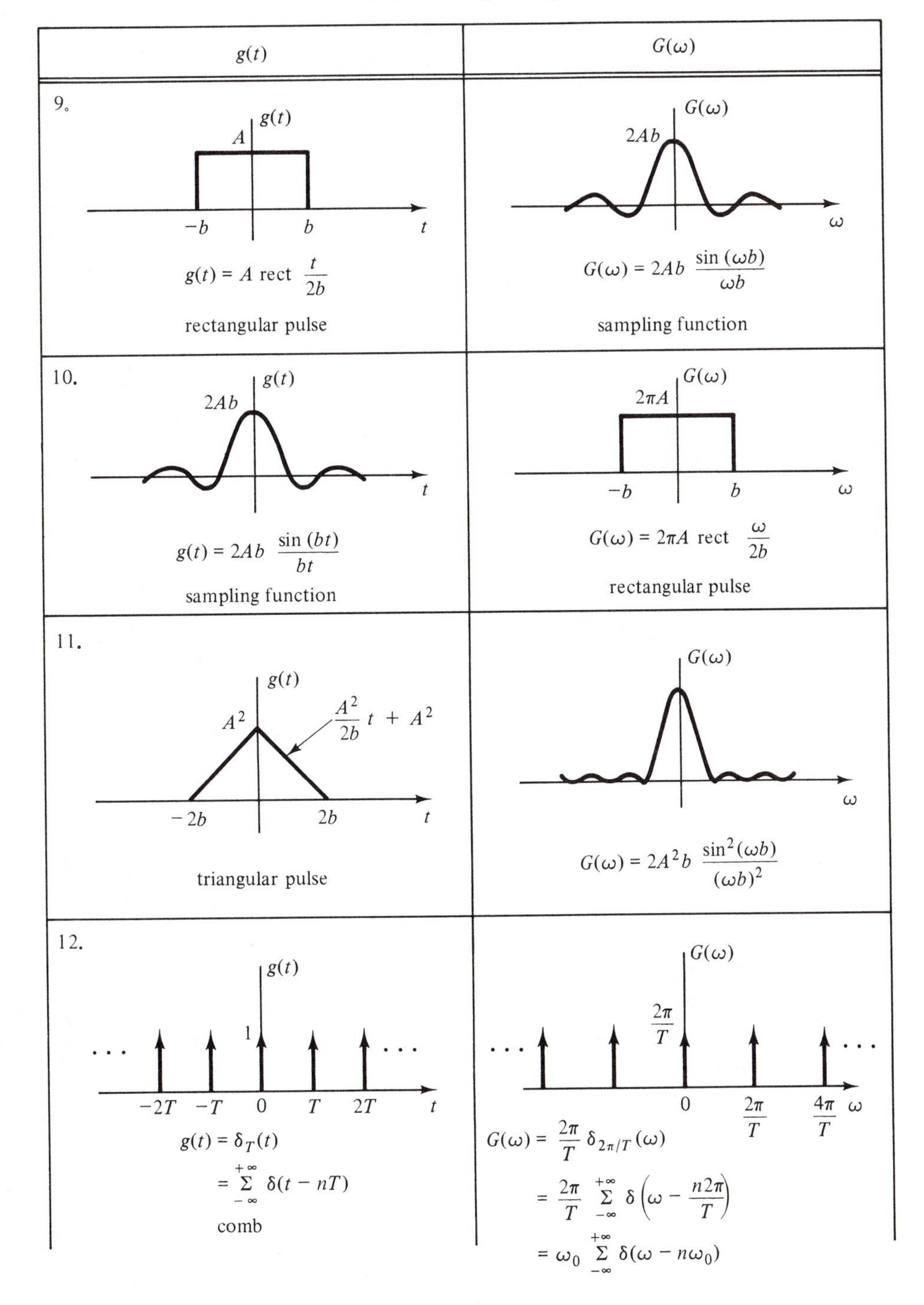

The Fourier transform approach is likely to be valuable whenever a signal's *spectrum* and/or a circuit's *frequency response* are important. Under these conditions the Fourier transform can take on meaning in its own right, in contrast to the Laplace transform, which is usually just a mathematical step on the way to a time-domain solution.

$\mathbf{G}(\omega)$, the Fourier transform of $g(t)$, may be written

$$\mathbf{G}(\omega) = |\mathbf{G}(\omega)| e^{j\underline{/\mathbf{G}(\omega)}}$$

and to help us visualize the properties of the transform we often plot the continuous *amplitude and phase spectra,* $|\mathbf{G}(\omega)|$ vs. $\omega$ and $\underline{/\mathbf{G}(\omega)}$ vs. $\omega$ for $-\infty \leq \omega \leq \infty$.

## RESPONSE OF LINEAR NETWORKS

The sinusoidal network function $H(j\omega)$ may now be defined as

$$H(j\omega) = \frac{\text{Fourier transform of response}}{\text{Fourier transform of input}} = \frac{\mathbf{Y}(\omega)}{\mathbf{U}(\omega)}$$

and the transform of the output must be

$$\mathbf{Y}(\omega) = H(j\omega)\mathbf{U}(\omega) \tag{12.11}$$

The output spectra follow at once from

$$|\mathbf{Y}(\omega)| = |H(j\omega)|\,|\mathbf{U}(\omega)| \quad \text{and} \quad \underline{/\mathbf{Y}(\omega)} = \underline{/H(j\omega)} + \underline{/\mathbf{U}(\omega)}$$

### Example 12.5

The single exponential pulse $e(t) = 10e^{-t}\mathbb{1}(t)$ of Fig. 12-5*b* is the input to the circuit of Fig. 12-5*a*. Find the Fourier transform of the output $v(t)$ and sketch the corresponding amplitude and phase spectra.

From equation (*12.11*),

$$\mathbf{V}(\omega) = H(j\omega)\mathbf{E}(\omega)$$

where

$$H(j\omega) = \frac{1/j\omega}{j\omega + 1 + 1/j\omega} = \frac{1}{1 - \omega^2 - j\omega}$$

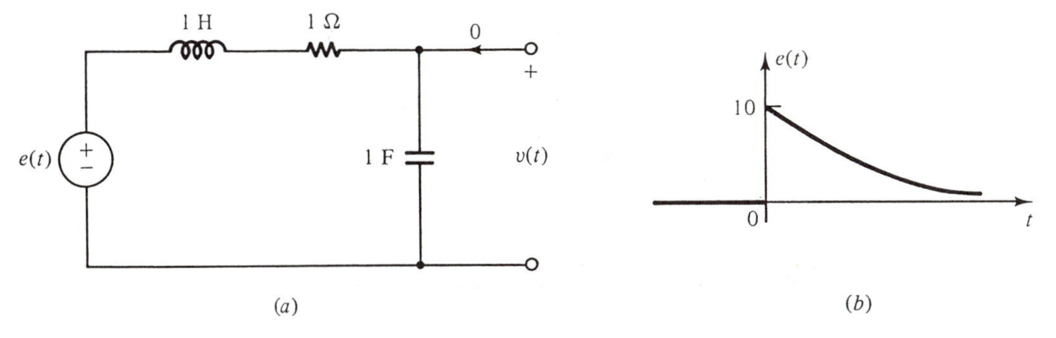


**Fig. 12-5**

and

$$\mathbf{E}(\omega) = \mathscr{F}[10e^{-t}1(t)]$$

$$= \int_{-\infty}^{0} 0e^{-j\omega t}\,dt + \int_{0}^{\infty} 10e^{-t}e^{-j\omega t}\,dt = \left[\frac{-10e^{-(1+j\omega)t}}{1+j\omega}\right]_0^\infty = \frac{10}{1+j\omega}$$

Therefore the transform of the output is

$$\mathbf{V}(\omega) = \frac{1}{1-\omega^2+j\omega}\,\frac{10}{1+j\omega} = \frac{10}{1-2\omega^2+j(2\omega-\omega^3)}$$

We can now evaluate the magnitude and phase of $\mathbf{V}(\omega)$ for a few values of $\omega$ (see table below) and hence plot the spectra, as shown in Fig. 12-6. Note that $|\mathbf{V}(-\omega)| = |\mathbf{V}(\omega)|$ and $\underline{/\mathbf{V}(-\omega)} = -\underline{/\mathbf{V}(\omega)}$.

| $\omega$ | $1-2\omega^2$ | $2\omega-\omega^3$ | $\lvert\mathbf{V}(\omega)\rvert$ | $\underline{/\mathbf{V}(\omega)}$ |
|---|---|---|---|---|
| 0 | 1 | 0 | 10 | 0 |
| 0.5 | 0.5 | 0.875 | 9.92 | −60° |
| 1 | −1 | 1 | 7.07 | −135° |
| 1.41 | −3 | 0 | 3.33 | −180° |
| 2 | −7 | −4 | 1.24 | −210° |
| 5 | −49 | −105 | 0.09 | −245° |
| ∞ | −∞ | −∞ | 0 | −270° |

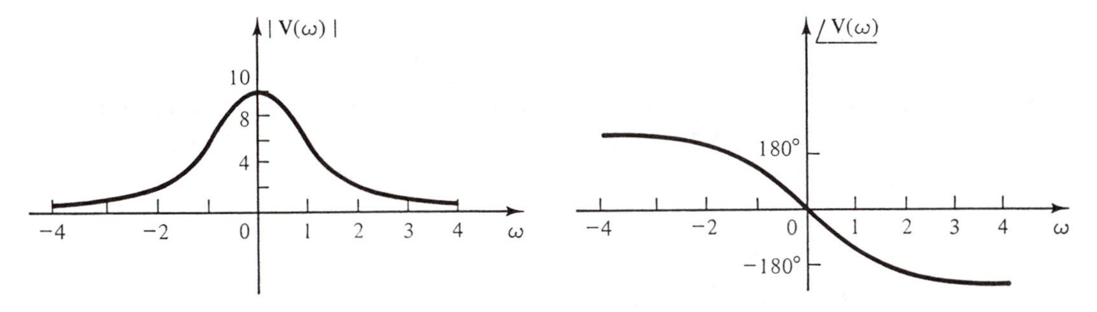


**Fig. 12-6**

*Comment:* The evaluation of the definite integral is worthy of a second glance. We assumed

$$[e^{-(1+j\omega)t}]_0^\infty = 0 - 1$$

The lower limit presents no difficulty, but we should justify the value at the upper limit of infinity. Thus

$$\lim_{t\to\infty} e^{-(1+j\omega)t} = \lim_{t\to\infty} e^{-t}e^{-j\omega t} = 0$$

since $|e^{-j\omega t}| = 1$ for *any* positive or negative value of $\omega t$.

### POWER (OR ENERGY) SPECTRA

If we take $\{g(t)\}^2$ as representative of the instantaneous power in a signal $g(t)$, it follows that the total signal energy can be represented by

$$E = \int_{-\infty}^{\infty} \{g(t)\}^2\, dt$$

Further, it can be shown that

$$\int_{-\infty}^{\infty} \{g(t)\}^2\, dt = \frac{1}{2\pi}\int_{-\infty}^{\infty} |\mathbf{G}(\omega)|^2\, d\omega \qquad (12.12)$$

which is known as Parseval's theorem. The graph of $(1/2\pi)|\mathbf{G}(\omega)|^2$ vs. $\omega$ is known as the *energy spectrum* or *energy density spectrum* of $g(t)$. That is, Parseval's theorem allows us to look at a signal's energy content in the transform domain.

## ILLUSTRATIVE PROBLEMS

### Fourier Series

**12.1** Indicate which of the following functions are periodic. For those that *are* periodic, determine the period and the fundamental radian frequency.

(*a*) $v(t) = 1\cos(628t + 17°) + 2\cos(1256t - 76°) + 3\cos(1884t - 55°)$,

(*b*) $i(t) = 0.5 \sin\sqrt{2}t + 0.25\sin\sqrt{4}t$,

(*c*) $p(t) = 1.23\cos(2\pi t + 45°) + 0.678\cos(2.2\pi t + 90°)$,

(*d*) $\phi(t) = (2\sin \pi t)^2$.

In Chapter 1 we defined a function $f(t)$ to be periodic if

$$f(t) = f(t - T) \qquad \text{for all } t$$

where the period is the smallest positive value of $T$ satisfying this relation.

(*a*) Here, since $1256/628 = 2$ and $1884/628 = 3$, $\omega_0 = 628$ rad/s is the fundamental frequency and the other sinusoids are the second and third harmonics. The fundamental period will be $T = 2\pi/\omega_0 = 0.01$ s.

(*b*) In this case the frequency ratio is $\sqrt{4}/\sqrt{2}$, an irrational number, so although each component of $i(t)$ is periodic, their sum is not.

(*c*) Here the frequency ratio is $2.2\pi/2\pi = 1.1$, making $p(t)$ periodic. For these frequencies the highest common factor is $0.2\pi$, which must be the fundamental frequency. Then $\omega = 2\pi$ is the 10th harmonic and $\omega = 2.2\pi$ is the 11th harmonic. That is,

$$\omega_0 = 0.2\pi \text{ rad/s} \quad \text{and} \quad T = 2\pi/\omega_0 = 10 \text{ s}$$

(*d*)
$$\phi(t) = (2 \sin \pi t)^2 = 4 \sin^2 \pi t = 4\left(\frac{1 - \cos 2\pi t}{2}\right)$$

But the sum of a constant and a periodic function is also periodic. Therefore $\phi(t)$ is periodic, with

$$\omega_0 = 2\pi \text{ rad/s} \quad \text{and} \quad T = 2\pi/\omega_0 = 1 \text{ s}$$

**12.2** A voltage $v(t) = A \cos(2\pi t + \alpha) + B \cos(4\pi t + \beta) + \Gamma \cos(6\pi t + \gamma)$ is applied to the $R$-$C$ circuit of Fig. 12-7. Use the principle of superposition to find the steady-state current $i_{ss}(t)$.

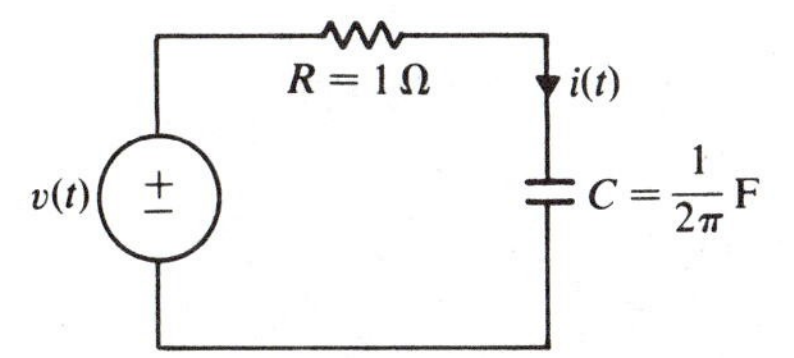


**Fig. 12-7**

If we let $v(t) = v_1(t) + v_2(t) + v_3(t)$, then the phasor currents due to $v_1(t)$, $v_2(t)$, and $v_3(t)$ separately will be

$$\mathbf{I}_1 = \frac{\mathbf{V}_1}{Z(j2\pi)} = \frac{Ae^{j(2\pi t + \alpha)}}{1 + 1/j} = \frac{A}{1.41} e^{j(2\pi t + \alpha + 45°)}$$

$$\mathbf{I}_2 = \frac{\mathbf{V}_2}{Z(j4\pi)} = \frac{Be^{j(4\pi t + \beta)}}{1 + 1/j2} = \frac{B}{1.12} e^{j(4\pi t + \beta + 26.6°)}$$

$$\mathbf{I}_3 = \frac{\mathbf{V}_3}{Z(j6\pi)} = \frac{\Gamma}{1.05} e^{j(6\pi t + \gamma\, +\, 18.4°)}$$

and by superposition (the additivity theorem),

$$i_{ss}(t) = i_{1,\text{sss}} + i_{2,\text{sss}} + i_{3,\text{sss}}$$

$$= \frac{A}{1.41}\cos(2\pi t + \alpha + 45°) + \frac{B}{1.12}\cos(4\pi t + \beta + 26.6°)$$

$$+ \frac{\Gamma}{1.05}\cos(6\pi t + \gamma + 18.4°)$$

**12.3** The periodic square wave of Fig. 12-8 is to be represented by a Fourier sine-cosine series. Calculate $C_0$, $A_k$, and $B_k$. Then write out a few terms in the series.

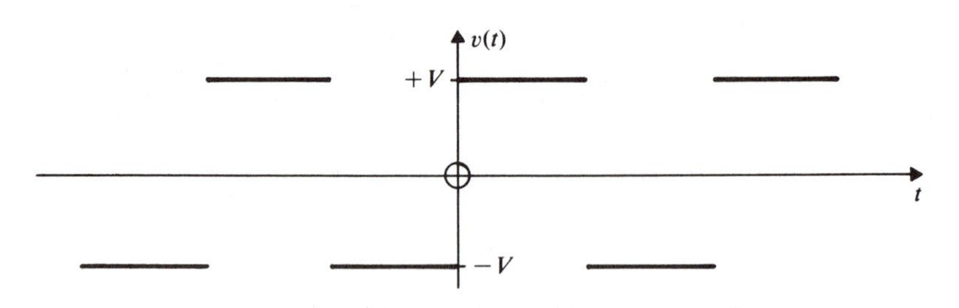


**Fig. 12-8**

If we choose to let $t_0 = 0$ in equation (*12.1*), then

$$C_0 = \frac{1}{T}\int_0^T v(t)\,dt = \frac{1}{T}\int_0^{T/2} V\,dt + \frac{1}{T}\int_{T/2}^{T} -V\,dt = 0$$

$$A_k = \frac{2}{T}\int_0^{T/2} V\cos k\omega_0 t\,dt + \frac{2}{T}\int_{T/2}^{T} -V\cos k\omega_0 t\,dt$$

$$= \frac{2V}{T}\left[\frac{\sin k\omega_0 t}{k\omega_0}\right]_0^{T/2} - \frac{2V}{T}\left[\frac{\sin k\omega_0 t}{k\omega_0}\right]_{T/2}^{T}$$

$$= \frac{2V}{k\omega_0 T}\left\{\sin k\omega_0\frac{T}{2} - \sin 0\right\} - \frac{2V}{k\omega_0 T}\left\{\sin k\omega_0 T - \sin k\omega_0\frac{T}{2}\right\}$$

$$= \frac{V}{k\pi}\{\sin k\pi - \sin 0 - \sin k2\pi + \sin k\pi\} = 0$$

$$B_k = \frac{2}{T}\int_0^{T/2} V\sin k\omega_0 t\,dt + \frac{2}{T}\int_{T/2}^{T} -V\sin k\omega_0 t\,dt$$

$$= \frac{V}{k\pi}\{-\cos k\pi + \cos 0 + \cos k2\pi - \cos k\pi\} = \frac{2V}{k\pi}(1 - \cos k\pi)$$

The first two results should not come as a surprise. $C_0$ is the *average* of $v(t)$, which by inspection of Fig. 12-8 is zero. Also, the function $v(t)$ is *odd*. Therefore by the symmetry rules, $C_0$ and $A_k$ are zero.

Substituting the first few values of $k$,

$$B_1 = \frac{2V}{\pi}\{1 - (-1)\} = \frac{4V}{\pi} \qquad B_2 = \frac{2V}{2\pi}\{1 - 1\} = 0$$

$$B_3 = \frac{2V}{3\pi}\{1 - (-1)\} = \frac{4V}{3\pi} \qquad B_4 = \frac{2V}{4\pi}\{1 - 1\} = 0$$

or, in general,

$$B_k = \begin{cases} 4V/k\pi & \text{for } k \text{ odd} \\ 0 & \text{for } k \text{ even} \end{cases}$$

Thus

$$v(t) = \frac{4V}{\pi}\sin\omega_0 t + \frac{4V}{3\pi}\sin 3\omega_0 t + \cdots$$

### *Alternative Solution*

The choice of $t_0$ is quite arbitrary. If we choose to make $t_0 = -T/2$, then

$$A_k = \frac{2}{T}\int_{-T/2}^{0} -V\cos k\omega_0 t\, dt + \frac{2}{T}\int_{0}^{T/2} V\cos k\omega_0 t\, dt$$

$$= \frac{2V}{k\omega_0 T}[-\sin k\omega_0 t]_{-T/2}^{0} + \frac{2V}{k\omega_0 T}[\sin k\omega_0 t]_{0}^{T/2} = 0$$

Similarly, the same values of $B_k$ follow from the appropriate integration between the limits $-T/2$ to 0 and 0 to $T/2$.

This demonstrates the fact that the choice of $t_0$ does *not* affect the value of the Fourier coefficients.

**12.4** Redo Problem 12.3 for the function $v(t)$ of Fig. 12-9. Note that the time origin has been changed from that in Fig. 12-8.

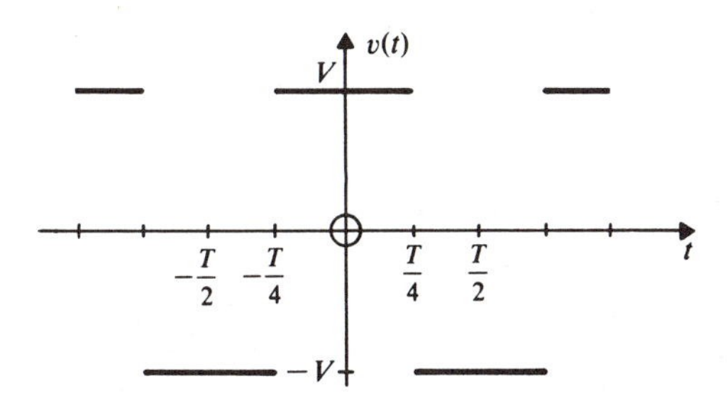


**Fig. 12-9**

$v(t)$ in Fig. 12-9 is an *even* function, and so from the symmetry rules, $B_k = 0$. Also, by inspection of the waveform the average is still zero, and thus $C_0 = 0$. Finally, choosing $t_0 = -T/2$,

$$A_k = \frac{2}{T}\left\{\int_{-T/2}^{-T/4} -V\cos k\omega_0 t\, dt + \int_{-T/4}^{T/4} V\cos k\omega_0 t\, dt + \int_{T/4}^{T/2} -V\cos k\omega_0 t\, dt\right\}$$

$$= \frac{V}{k\pi}\left\{2\sin\frac{k\pi}{2} - 2\sin\left(-\frac{k\pi}{2}\right)\right\} = \frac{4V}{k\pi}\sin\frac{k\pi}{2}$$

or

$$A_k = \begin{cases} 4V/k\pi & \text{for } k = 1, 5, 9, \ldots \\ 0 & \text{for } k \text{ even} \\ -4V/k\pi & \text{for } k = 3, 7, 11, \ldots \end{cases}$$

Hence

$$v(t) = \frac{4V}{\pi}\cos\omega_0 t - \frac{4V}{3\pi}\cos 3\omega_0 t + \frac{4V}{5\pi}\cos 5\omega_0 t - \cdots$$

which can be written alternatively as

$$v(t) = \frac{4V}{\pi}\cos\omega_0 t + \frac{4V}{3\pi}\cos(3\omega_0 t + 180°) + \frac{4V}{5\pi}\cos 5\omega_0 t + \cdots$$

The corresponding line spectra are plotted in Fig. 12-10.

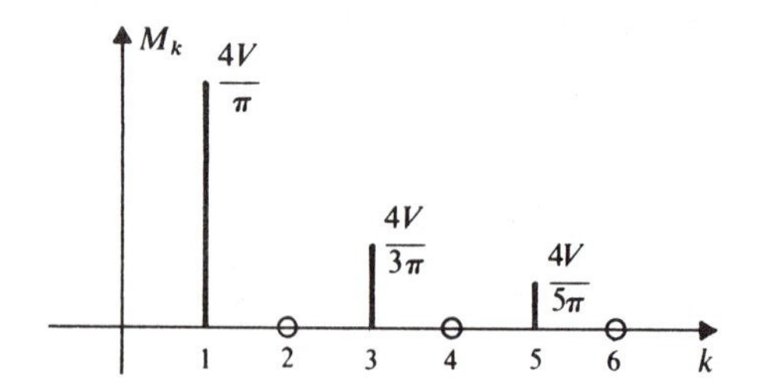


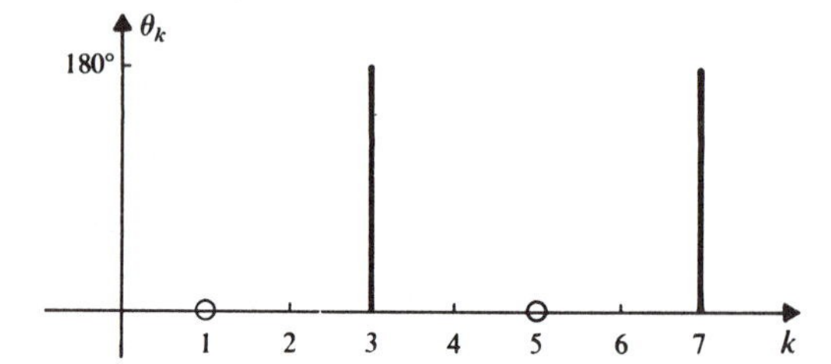


**Fig. 12-10**

**12.5** Derive the exponential Fourier series

$$f(t) = \sum_{k=-\infty}^{\infty} \mathbf{C}_k e^{jk\omega_0 t}$$

from the sine-cosine series of equation (*12.1*), page 418.

First, applying the Euler relations to equation (*12.1*),

$$f(t) = C_0 + \sum_{k=1}^{\infty}\left\{A_k\frac{e^{jk\omega_0 t} + e^{-jk\omega_0 t}}{2} + B_k\frac{e^{jk\omega_0 t} - e^{-jk\omega_0 t}}{2j}\right\}$$

$$= C_0 + \sum_{k=1}^{\infty}\{\tfrac{1}{2}(A_k - jB_k)e^{jk\omega_0 t} + \tfrac{1}{2}(A_k + jB_k)e^{-jk\omega_0 t}\}$$

Now let

$$\mathbf{C}_k = \tfrac{1}{2}(A_k - jB_k), \quad \mathbf{C}_{-k} = \tfrac{1}{2}(A_k + jB_k), \quad \text{and} \quad \mathbf{C}_0 = C_0$$

With these substitutions the trigonometric series becomes

$$f(t) = \mathbf{C}_0 + \sum_{k=1}^{\infty}\{\mathbf{C}_k e^{jk\omega_0 t} + \mathbf{C}_{-k}e^{-jk\omega_0 t}\}$$

and if we let $k$ range from $-\infty$ through zero to $+\infty$,

$$f(t) = \sum_{k=-\infty}^{\infty} \mathbf{C}_k e^{jk\omega_0 t}$$

Now we have only to derive an integral expression for $\mathbf{C}_k$, given $A_k$ and $B_k$ from equation (*12.1*). Thus

$$\begin{aligned}\mathbf{C}_k &= \tfrac{1}{2}(A_k - jB_k) \qquad (k = 1, 2, \ldots) \\ &= \frac{1}{T}\int_{t_0}^{t_0+T} f(t)\cos k\omega_0 t\,dt - \frac{j}{T}\int_{t_0}^{t_0+T} f(t)\sin k\omega_0 t\,dt \\ &= \frac{1}{T}\int_{t_0}^{t_0+T} f(t)\,\{\cos k\omega_0 t - j\sin k\omega_0 t\}\,dt \\ &= \frac{1}{T}\int_{t_0}^{t_0+T} f(t)\,e^{-jk\omega_0 t}\,dt \qquad (k = 1, 2, \ldots)\end{aligned}$$

Similarly,

$$\begin{aligned}\mathbf{C}_{-k} &= \tfrac{1}{2}(A_k + jB_k) \qquad (k = 1, 2, \ldots) \\ &= \frac{1}{T}\int_{t_0}^{t_0+T} f(t)\,\{\cos k\omega_0 t + j\sin k\omega_0 t\}\,dt = \frac{1}{T}\int_{t_0}^{t_0+T} f(t)\,e^{jk\omega_0 t}\,dt\end{aligned}$$

which may be written

$$\mathbf{C}_k = \frac{1}{T}\int_{t_0}^{t_0+T} f(t)\,e^{-jk\omega_0 t}\,dt \qquad (k = \ldots, -2, -1)$$

Both of these expressions for $\mathbf{C}_k$ can be combined into the one statement

$$\mathbf{C}_k = \frac{1}{T}\int_{t_0}^{t_0+T} f(t)\,e^{-jk\omega_0 t}\,dt \qquad (k = \ldots, -2, -1, 0, 1, 2, \ldots)$$

Note that the special case of $k = 0$ correctly yields

$$\mathbf{C}_0 = \frac{1}{T}\int_{t_0}^{t_0+T} f(t)\,dt = C_0$$

In summary, the exponential series of equation (*12.2*) is equivalent to and can be derived from the sine-cosine series of equation (*12.1*).

**12.6** Find $\mathbf{C}_k$ for the pulse-train of Fig. 12-11 and express $v(t)$ as an exponential series.

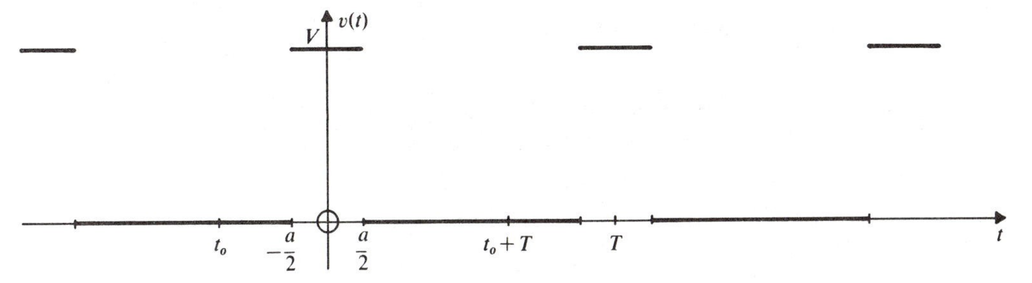


**Fig. 12-11**

As before, we may choose $t_0$ arbitrarily, as shown in Fig. 12-11. Then from equation (*12.2*), page 420,

$$\mathbf{C}_k = \frac{1}{T}\int_{-a/2}^{a/2} Ve^{-jk\omega_0 t}\,dt = \frac{V}{T}\left[\frac{e^{-jk\omega_0 t}}{-jk\omega_0}\right]_{-a/2}^{a/2}$$

$$= \frac{V}{T}\left\{\frac{e^{-jk\omega_0 a/2} - e^{jk\omega_0 a/2}}{-jk\omega_0}\right\} = \frac{V}{\pi k}\left\{\frac{e^{jk\omega_0 a/2} - e^{-jk\omega_0 a/2}}{2j}\right\}$$

or

$$\mathbf{C}_k = \frac{V}{\pi k}\sin\frac{k\omega_0 a}{2} \qquad (k = \ldots, -2, -1, 0, 1, 2, \ldots)$$

Thus the exponential series is

$$v(t) = \sum_{k=-\infty}^{\infty}\left\{\frac{V}{\pi k}\sin\frac{k\omega_0 a}{2}\right\}e^{jk\omega_0 t}$$

As a special case of $\mathbf{C}_k$, with $k = 0$, we obtain

$$\mathbf{C}_0 = \frac{V}{\pi 0}\sin 0$$

and so we must use l'Hopital's rule,

$$\mathbf{C}_0 = \frac{V}{\pi}\frac{\omega_0 a}{2}\cos\frac{k\omega_0 a}{2}\bigg|_{k=0} = \frac{V\omega_0 a}{2\pi} = \frac{Va}{T}$$

This is easily checked by calculating the average value of the waveform $v(t)$ in Fig. 12-11.

**12.7** Express $v(t)$ in Fig. 12-11 as a cosine series, using the results of Problem 12.6.

From Problem 12.6,

$$\mathbf{C}_k = \frac{V}{\pi k}\sin\frac{k\omega_0 a}{2}$$

which is real. Therefore, from equation (*12.4*), page 422,

$$M_k = 2|\mathbf{C}_k| \quad \text{and} \quad \theta_k = 0° \text{ or } 180°$$

Thus

$$v(t) = C_0 + \sum_{k=1}^{\infty} M_k\cos(k\omega_0 t + \theta_k) = \frac{Va}{T} + \sum_{k=1}^{\infty}\left|\frac{2V}{\pi k}\sin\frac{k\omega_0 a}{2}\right|\cos(k\omega_0 t + \theta_k)$$

**12.8** Plot the amplitude and phase spectra for the waveform of Fig. 12-11, using the results of Problems 12.6 and 12.7. Given: $a = T/4$.

From Problem 12.7, with $a = T/4$,

$$v(t) = \frac{V}{4} + \sum_{k=1}^{\infty}\left|\frac{2V}{k\pi}\sin\frac{k\pi}{4}\right|\cos(k\omega_0 t + \theta_k)$$

and we can tabulate the amplitude and phase of the harmonics as follows:

| Component of $v(t)$ | Amplitude, $M_k$ | Phase, $\theta_k$ |
|---|---|---|
| dc or average | $\dfrac{V}{4} = 0.25V$ | — |
| fundamental ($k = 1$) or 1st harmonic | $\dfrac{2V}{\pi} \sin \dfrac{\pi}{4} = 0.45V$ | 0° |
| 2nd harmonic | $\dfrac{2V}{2\pi} \sin \dfrac{2\pi}{4} = 0.318V$ | 0° |
| 3rd harmonic | $\dfrac{2V}{3\pi} \sin \dfrac{3\pi}{4} = 0.150V$ | 0° |
| 4th harmonic | $\dfrac{2V}{4\pi} \sin \dfrac{4\pi}{4} = 0$ | — |
| 5th harmonic | $\left\lvert \dfrac{2V}{5\pi} \sin \dfrac{5\pi}{4} \right\rvert = 0.090V$ | 180° |
| 6th harmonic | $\left\lvert \dfrac{2V}{6\pi} \sin \dfrac{6\pi}{4} \right\rvert = 0.106V$ | 180° |
| 7th harmonic | $\left\lvert \dfrac{2V}{7\pi} \sin \dfrac{7\pi}{4} \right\rvert = 0.064V$ | 180° |
| 8th harmonic | $\dfrac{2V}{8\pi} \sin \dfrac{8\pi}{4} = 0$ | — |

The corresponding spectra are plotted in Fig. 12-12*a*.

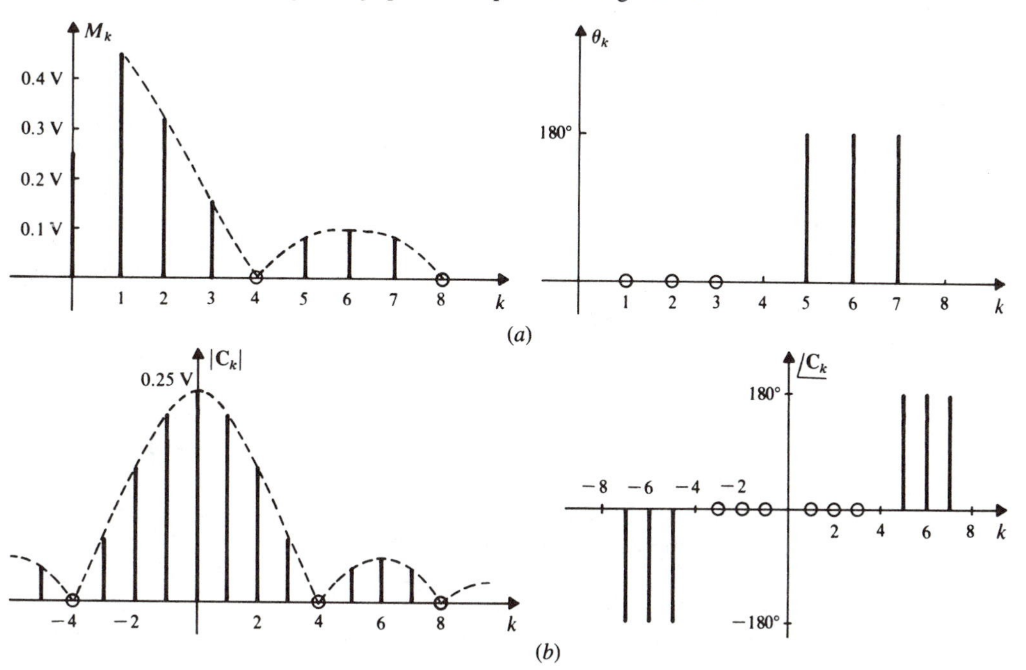


**Fig. 12-12**

Alternatively we can plot the spectra corresponding to $|\mathbf{C}_k|$ and $\underline{/\mathbf{C}_k}$. Now we know that

$$|\mathbf{C}_k| = |\mathbf{C}_{-k}| = \tfrac{1}{2}M_k \quad \text{and} \quad \underline{/\mathbf{C}_k} = -\underline{/\mathbf{C}_{-k}} = \theta_k$$

for $k = 1, 2, \ldots$. And of course $\mathbf{C}_0 = C_0$. Thus we can easily plot these spectra as shown in Fig. 12-12*b*.

The *envelope* of the $|\mathbf{C}_k|$ spectrum, shown dashed in Fig. 12-12*b*, takes the shape of $|\sin x/x|$, since

$$|\mathbf{C}_k| = \frac{V}{4}\left|\frac{\sin(k\pi/4)}{k\pi/4}\right| \qquad \text{for all integer } k$$

**12.9** The voltage waveform of Fig. 12-11 is impressed on a 1-Ω resistor. Calculate the total real power dissipated in the resistor, and the percentage of this power contained in the dc term and the first four harmonics. Given: $a = T/4$.

The average power is

$$P = \frac{1}{T}\int_{t_0}^{t_0+T} p(t)\,dt = \frac{1}{T}\int_{-a/2}^{a/2} \frac{V^2}{1}\,dt = \frac{V^2}{T}\int_{-T/8}^{T/8} dt = 0.25V^2$$

Now from Ohm's law, $i(t) = v(t)/1$, and, making use of the tabulation in Problem 12.8,

$$i(t) = \frac{v(t)}{1} \doteq V\{0.25 + 0.45\cos\omega_0 t + 0.318\cos 2\omega_0 t + 0.150\cos 3\omega_0 t + 0\}$$

Then from equation (*12.6*), page 425,

$$P = I_{\text{eff}}^2 R \doteq I_0^2 + I_{1,\text{eff}}^2 + I_{2,\text{eff}}^2 + I_{3,\text{eff}}^2 + I_{4,\text{eff}}^2$$

$$= V^2\left\{(0.25)^2 + \left(\frac{0.45}{\sqrt{2}}\right)^2 + \left(\frac{0.318}{\sqrt{2}}\right)^2 + \left(\frac{0.150}{\sqrt{2}}\right)^2 + 0\right\}$$

$$= V^2(0.0625 + 0.100 + 0.050 + 0.0112) = 0.224V^2$$

Thus the percentage of the real power contained in the dc term plus the first four harmonics is $(0.224V^2/0.25V^2)100 = 89.5$ percent.

**12.10** The saw-tooth voltage $v(t)$ of Fig. 12-13*a* is the input to the circuit of Fig. 12-13*b*, and it is known that

$$v(t) = \frac{V\pi}{2} + \sum_{k=1}^{\infty} \frac{V}{k}\cos(kt + 90^\circ)$$

The objective is to find the approximate value of $i_{ss}(t)$ using Fourier series and superposition. To this end,

(*a*) plot the amplitude and phase spectra for the input $v(t)$,
(*b*) plot the amplitude and phase characteristics of $Y(j\omega) = \mathbf{I}/\mathbf{V}$,
(*c*) from the plots of (*a*) and (*b*), construct the amplitude and phase spectra for the output $i_{ss}(t)$, and
(*d*) using data from (*c*), write down a few terms in the cosine series of the output signal $i_{ss}(t)$.

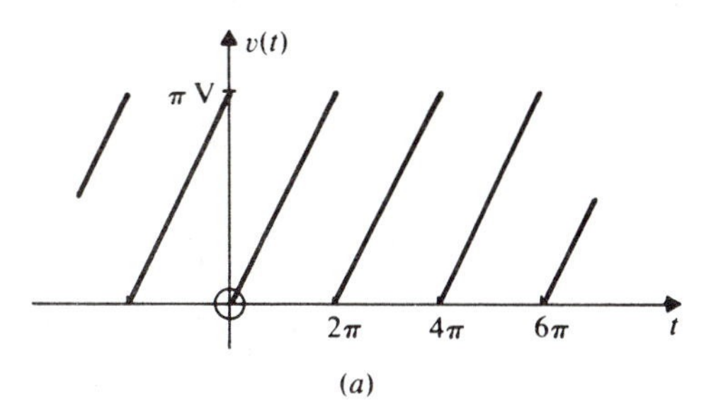

(a)

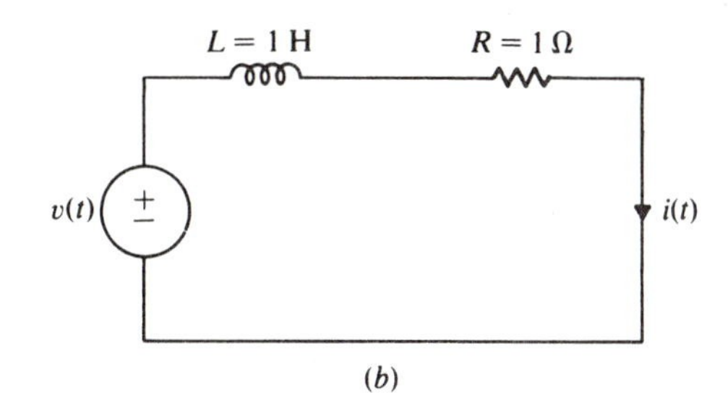

(b)

**Fig. 12-13**

(a) From Fig. 12-13a, $T = 2\pi$ s and so $\omega_0 = 1$ rad/s. And from the given cosine series we may at once plot the spectra of $v(t)$ as in Fig. 12-14a.

(b) By inspection of Fig. 12-13b,

$$Y(j\omega) = \frac{1}{R + j\omega L} = \frac{1}{1 + j\omega} = \frac{1}{\sqrt{1 + \omega^2}} e^{-j\tan^{-1}\omega}$$

from which follow the frequency response plots of Fig. 12-14b.

(c) From equation (*12.5*), page 424, $\mathbf{I}_k = Y(jk\omega_0)\mathbf{V}_k$. Or, when $\omega_0 = 1$,

$$|\mathbf{I}_k| = |Y(jk)|\,|\mathbf{V}_k| \quad \text{and} \quad \angle\mathbf{I}_k = \angle Y(jk) + \angle\mathbf{V}_k$$

Thus we have only to *multiply* the magnitude curves of Figs. 12-14a and b and to *add* the phase curves to obtain the output spectra of $N_k = 2|\mathbf{I}_k|$ and $\phi_k = \angle\mathbf{I}_k$ vs. $k$, as shown in Fig. 12-14c.

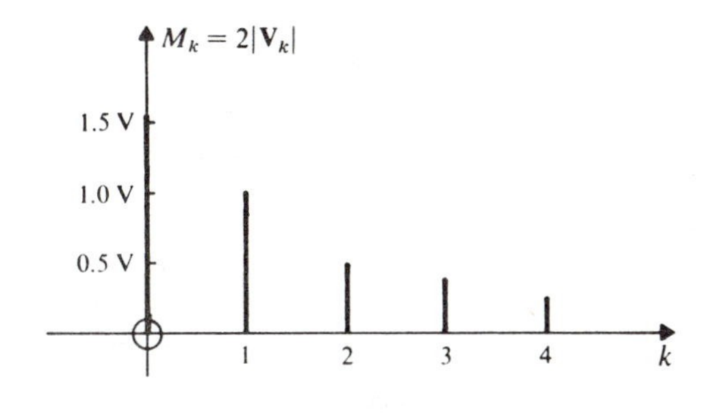


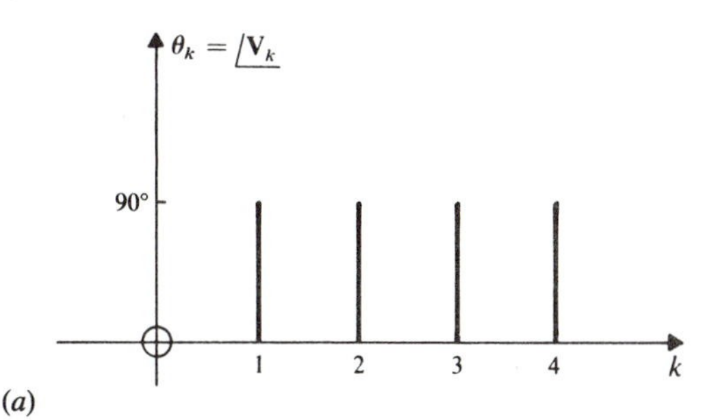

(a)

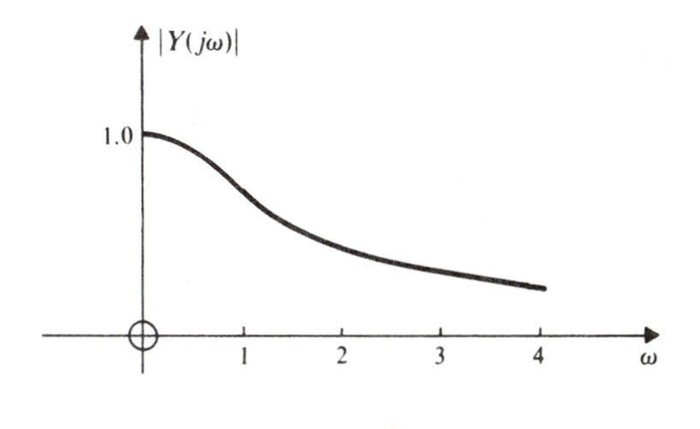


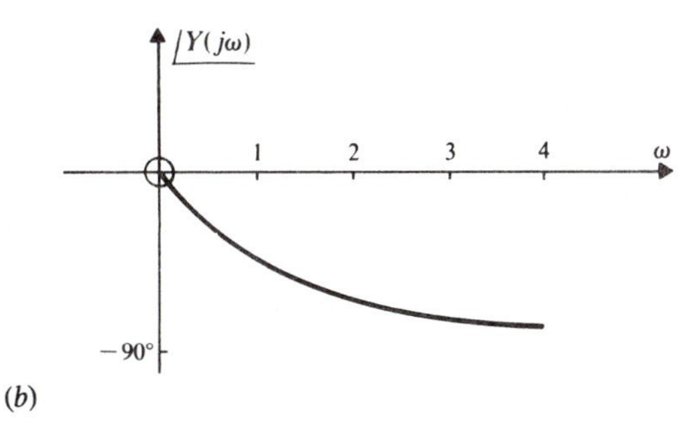

(b)

**Fig. 12-14**

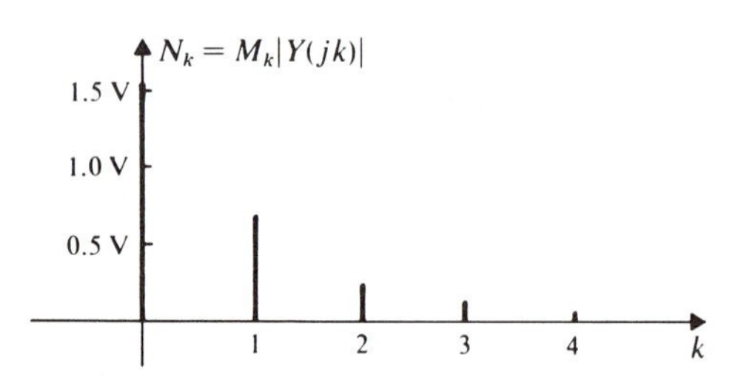


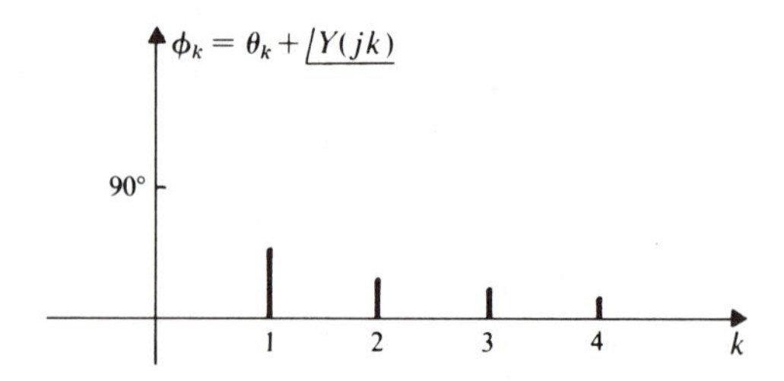


**Fig. 12-14c**

(*d*) Here, where $\omega_0 = 1$, equation (*12.4*) becomes

$$i_{ss}(t) = \mathbf{I}_0 + \sum_{k=1}^{\infty} N_k \cos(kt + \phi_k)$$

where $\mathbf{I}_0 = I_0 = N_0$ which can be read off with the other $N_k$ and the $\phi_k$ from Fig. 12-14*c*. This yields

$$i_{ss}(t) \doteq V\{1.57 + 0.707 \cos(t + 45°) + 0.223 \cos(2t + 26.6°) + 0.105 \cos(3t + 18.5°) + \cdots\}$$

**12.11** Find $\mathbf{C}_k$ and hence the exponential Fourier series for the *unit impulse train* of Fig. 12-15*a*. If this waveform, with $T = 2\pi$, is the input to the circuit of Fig. 12-15*b*, find and plot the amplitude spectrum of $i_{ss}(t)$. *Hint:* Use the result of Problem 12.6.

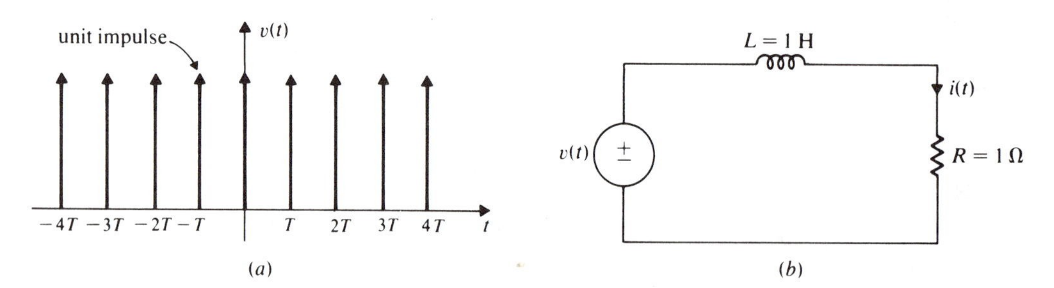


**Fig. 12-15**

For the pulse train of Fig. 12-11 (see Problem 12.6),

$$\mathbf{C}_k = \frac{V}{\pi k} \sin \frac{k\omega_0 a}{2}$$

If we make $V = 1/a$, each pulse will be of unit area, and

$$\mathbf{C}_k = \frac{1}{\pi k a} \sin \frac{k\omega_0 a}{2}$$

Finally, if $a \to 0$, the unit pulse train becomes a unit *impulse* train, and

$$\mathbf{C}_k = \lim_{a\to 0}\left\{\frac{1}{\pi k a}\sin\frac{k\omega_0 a}{2}\right\} = \lim_{a\to 0}\left\{\frac{\omega_0}{2\pi}\frac{\sin k\omega_0 a/2}{k\omega_0 a/2}\right\} = \frac{\omega_0}{2\pi} = \frac{1}{T}$$

since $\lim_{x\to 0}[(\sin x)/x] = 1$.

Thus each harmonic has the same amplitude, a rather remarkable result. The exponential series is

$$v(t) = \sum_{k=-\infty}^{\infty} \frac{1}{T} e^{jk\omega_0 t}$$

while the trigonometric series is

$$v(t) = \frac{1}{T} + \sum_{k=1}^{\infty} \frac{2}{T}\cos k\omega_0 t$$

We considered the circuit of Fig. 12-15*b* in Problem 12.10, where $Y(j\omega) = 1/(1 + j\omega)$. Referring to equation (*12.5*), page 424, with $\omega_0 = 1$,

$$|\mathbf{I}_k| = |Y(jk\omega_0)|\,|\mathbf{V}_k| = |Y(jk)|\,|\mathbf{V}_k|$$

The appropriate graphs are plotted in Fig. 12-16, where the final spectrum is obtained by multiplying together the graphs of $|\mathbf{V}_k|$ and $|Y(jk)|$.

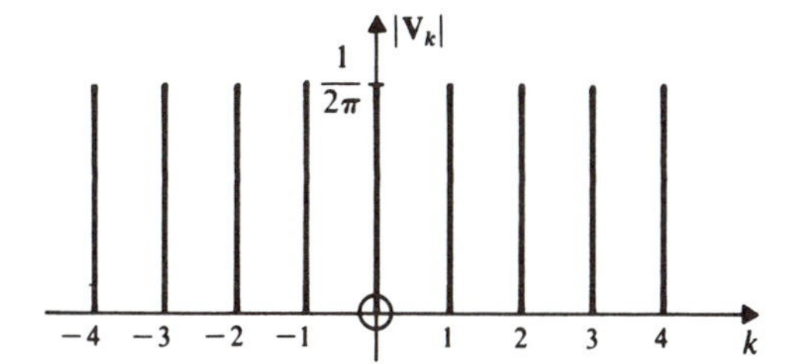


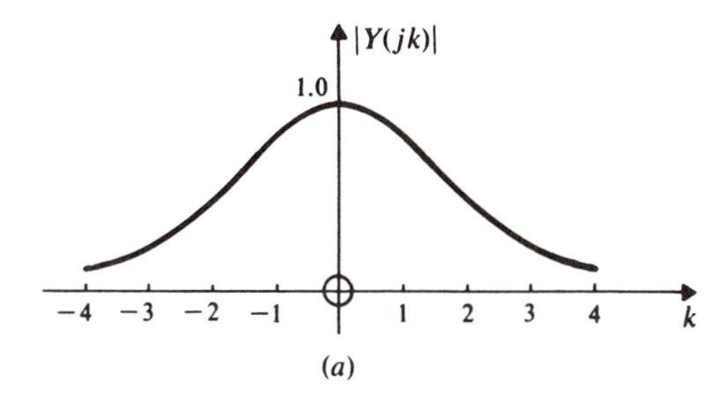


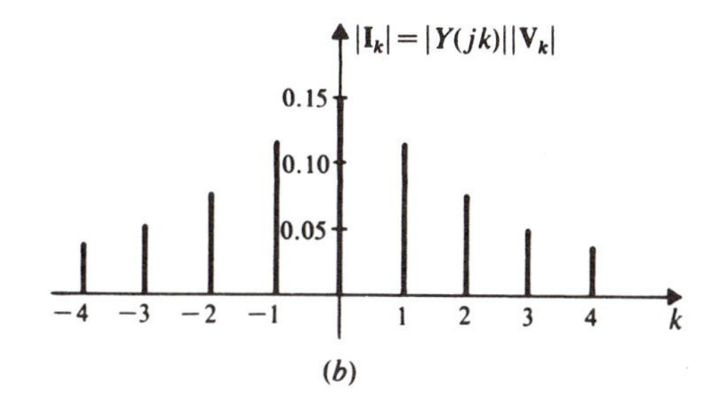


**Fig. 12-16**

If we also plotted the phase spectrum,

$$\begin{aligned} \underline{/\mathbf{I}_k} &= \underline{/Y(jk)} + \underline{/\mathbf{V}_k} \\ &= \underline{/Y(jk)} \end{aligned}$$

we could obtain the steady-state current from

$$i_{ss}(t) = \mathbf{I}_0 + \sum_{k=1}^{\infty} 2|\mathbf{I}_k|\cos(kt + \underline{/\mathbf{I}_k})$$

## Fourier Transforms

**12.12** $\mathbf{G}(\omega)$ is the Fourier transform of $g(t)$ as defined by equations (*12.7*) and (*12.8*), page 425. Place check marks in the appropriate boxes in the following table:

| If $g(t)$ is | Then $\mathbf{G}(\omega)$ will be in general | | | | |
|---|---|---|---|---|---|
| | Real | Imaginary | Complex | Even | Odd |
| Real | | | | | |
| Real and even | | | | | |
| Real and odd | | | | | |

First, for $g(t)$ real and odd *or* even,

$$\mathscr{F}[g(t)] = \mathbf{G}(\omega) = \int_{-\infty}^{\infty} g(t)e^{-j\omega t}\,dt = \int_{-\infty}^{\infty} g(t)\{\cos\omega t - j\sin\omega t\}\,dt$$

$$= \int_{-\infty}^{\infty} g(t)\cos\omega t\,dt - j\int_{-\infty}^{\infty} g(t)\sin\omega t\,dt$$

Thus, in general, $\mathbf{G}(\omega)$ is complex.

Let us now consider some consequences of evenness and oddness. If $e_1(t)$, $e_2(t)$, and $e_3(t)$ are even functions and $o_1(t)$ and $o_2(t)$ are odd,

$$e_1(t)e_2(t) = e_3(t), \qquad e_1(t)o_1(t) = o_2(t), \qquad o_1(t)o_2(t) = e_1(t)$$

$$\int_{-\infty}^{\infty} e_1(t)\,dt = 2\int_0^{\infty} e_1(t)\,dt, \qquad \int_{-\infty}^{\infty} o_1(t)\,dt = 0$$

If, now, $g(t)$ is real *and even,*

$$g(t)\cos\omega t \text{ is even} \quad \text{and} \quad \int_{-\infty}^{\infty} g(t)\cos\omega t\,dt = 2\int_0^{\infty} g(t)\cos\omega t\,dt$$

$$g(t)\sin\omega t \text{ is odd} \quad \text{and} \quad \int_{-\infty}^{\infty} g(t)\sin\omega t\,dt = 0$$

Therefore for $g(t)$ real and even,

$$\mathbf{G}(\omega) = 2\int_0^{\infty} g(t)\cos\omega t\,dt + 0$$

which is *real*. To test for evenness or oddness *with respect to* $\omega$, we must substitute $-\omega$ for $\omega$,

$$\mathbf{G}(-\omega) = 2\int_0^{\infty} g(t)\cos(-\omega t)\,dt = 2\int_0^{\infty} g(t)\cos\omega t\,dt = \mathbf{G}(\omega)$$

Therefore $\mathbf{G}(\omega)$ is also *even*.

Similarly, for $g(t)$ real *and odd,*

$$\mathbf{G}(\omega) = 0 - 2j\int_0^{\infty} g(t)\sin\omega t\,dt$$

and

$$\mathbf{G}(-\omega) = -2j \int_0^{\infty} g(t) \sin(-\omega t)\, dt = 2j \int_0^{\infty} g(t) \sin \omega t\, dt = -\mathbf{G}(\omega)$$

Thus $\mathbf{G}(\omega)$ is *imaginary* and *odd*.

*Summary:*

$g(t)$ real; $\mathbf{G}(\omega)$ complex

$g(t)$ real and even; $\mathbf{G}(\omega)$ real and even

$g(t)$ real and odd; $\mathbf{G}(\omega)$ imaginary and odd

**12.13** Find the Fourier transform of the *single* pulse shown in Fig. 12-17 and plot the amplitude spectrum.

From equation (*12.8*), page 445,

$$\mathbf{V}(\omega) = \mathcal{F}[v(t)] = \int_{-\infty}^{\infty} v(t)e^{-j\omega t}\, dt$$

$$= \int_{-a/2}^{a/2} Ve^{-j\omega t}\, dt = \left[\frac{Ve^{-j\omega t}}{-j\omega}\right]_{-a/2}^{a/2} = \frac{2V}{\omega}\left\{\frac{e^{j\omega a/2} - e^{-j\omega a/2}}{2j}\right\} = \frac{2V}{\omega} \sin \frac{\omega a}{2}$$

This is a sin $x/x$ type of function, since it can be rewritten as

$$\mathbf{V}(\omega) = aV\left\{\frac{\sin(\omega a/2)}{\omega a/2}\right\}$$

The corresponding amplitude spectrum is graphed in Fig. 12-18. Since the Fourier transform is defined *for all* $\omega$, the spectrum must be plotted for both positive and negative $\omega$.

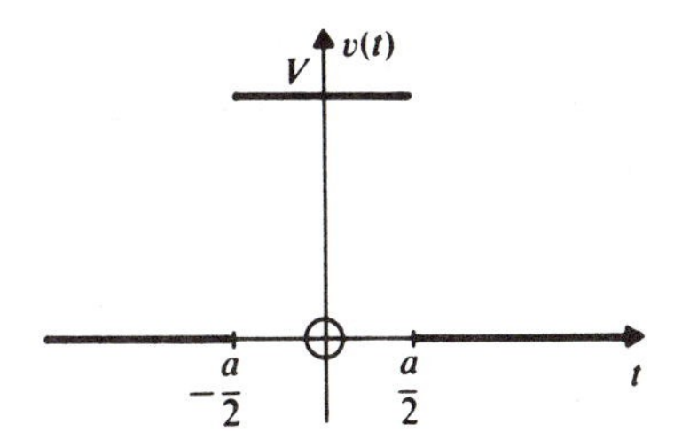


**Fig. 12-17**

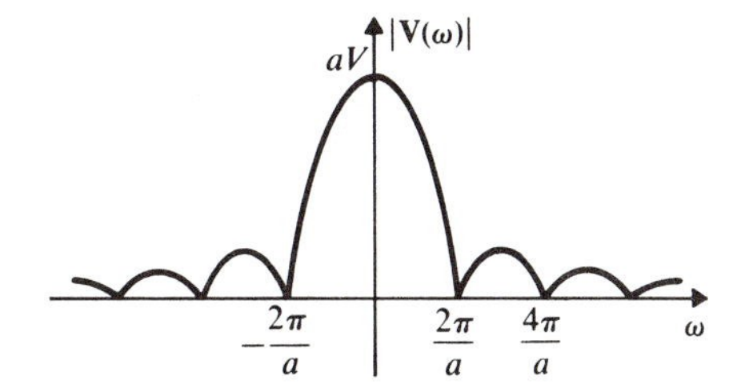


**Fig. 12-18**

Note that the spectrum is continuous rather than discrete, as was the case for periodic signals, and that with $v(t)$ real and even, $\mathbf{V}(\omega)$ is real and even, as predicted in Problem 12.12.

**12.14** Use the results of Problem 12.13 to find the Fourier transform of a *unit* impulse function.

First, the pulse of Fig. 12-17 can be made into a pulse $p(t)$ of unit area by making the amplitude $V = 1/a$. Then from Problem 12.13,

$$\mathscr{F}[p(t)] = a\left(\frac{1}{a}\right)\frac{\sin(\omega a/2)}{\omega a/2} = \frac{\sin(\omega a/2)}{\omega a/2}$$

Finally, to obtain the unit impulse $\delta(t)$ from $p(t)$, we let $a \to 0$,

$$\mathscr{F}[\delta(t)] = \lim_{a\to 0} \frac{\sin(\omega a/2)}{\omega a/2} = 1$$

This pair, multiplied through by $A$, is entry 2 in Table 12-2, page 427.

We have avoided some rather difficult mathematics. Strictly speaking, we should have taken the limit *before* computing the transform.

**12.15** Use the symmetry property (see entry 4 in Table 12-1, page 426) and the result of the previous problem to find $\mathscr{F}[1]$.

The symmetry property works this way: If

$$g(t) \Leftrightarrow \mathbf{G}(\omega)$$

then

$$\mathbf{G}(t) \Leftrightarrow 2\pi g(-\omega)$$

where $\mathbf{G}(t) = \mathbf{G}(\omega)|_{\omega=t}$ and $g(-\omega) = g(t)|_{t=-\omega}$.

In this case, from Problem 12.14,

$$g(t) = \delta(t) \quad \text{and} \quad \mathbf{G}(\omega) = 1$$

so

$$\mathbf{G}(t) = \mathbf{G}(\omega)|_{\omega=t} = 1|_{\omega=t} = 1$$

and

$$g(-\omega) = g(t)|_{t=-\omega} = \delta(-\omega) = \delta(\omega)$$

since the $\delta$-function is even. Therefore from the symmetry property,

$$\mathscr{F}[\mathbf{G}(t)] = \mathscr{F}[1] = 2\pi g(-\omega) = 2\pi\delta(\omega)$$

This pair, multiplied through by $A$, is the first entry in Table 12-2.
The concept of symmetry is illustrated in Fig. 12-19.

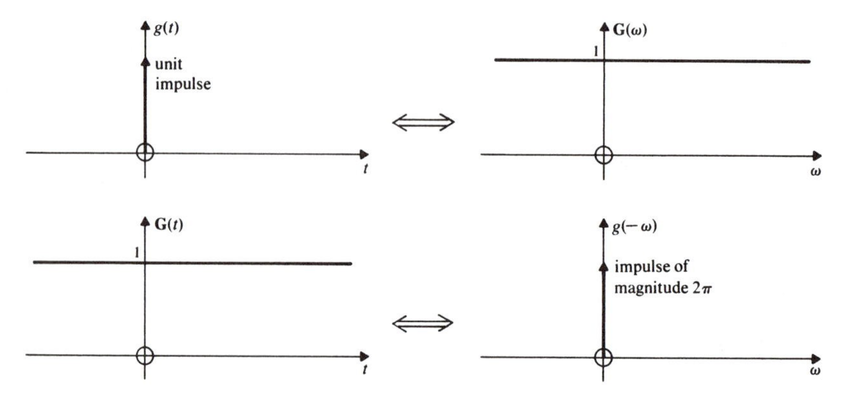


**Fig. 12-19**

**12.16** Find the Fourier transform of the voltage $v(t) = \mathbb{1}(t)Ae^{-\alpha t}$ shown in Fig. 12-20. Then sketch the amplitude and phase spectra for $v(t)$. Given: $\alpha > 0$.

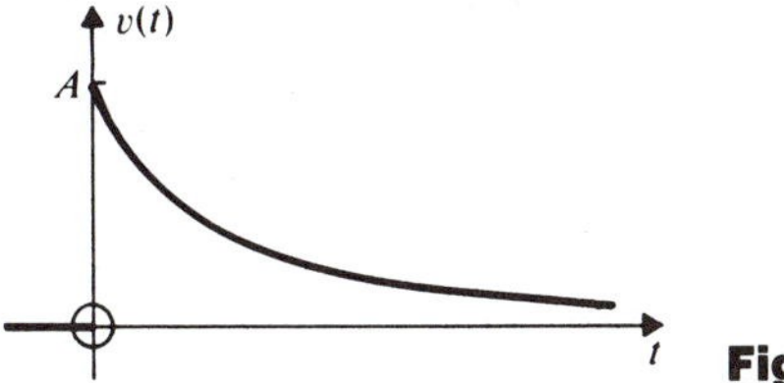


**Fig. 12-20**

From equation *(12.8)*,

$$\mathcal{F}[v(t)] = \int_{-\infty}^{0} 0e^{-j\omega t}\,dt + \int_{0}^{\infty} Ae^{-\alpha t}e^{-j\omega t}\,dt$$

$$= \int_{0}^{\infty} Ae^{-(\alpha+j\omega)t}\,dt = \left[\frac{Ae^{-\alpha t}e^{-j\omega t}}{-(\alpha + j\omega)}\right]_{0}^{\infty}$$

Now $e^{-j\omega t}$ *always* has unit magnitude, regardless of the value of $\omega t$. Therefore with $\alpha > 0$,

$$Ae^{-\alpha t}e^{-j\omega t} \to 0 \quad \text{as } t \to \infty$$

Thus

$$\mathcal{F}[v(t)] = \left[\frac{Ae^{-\alpha t}e^{-j\omega t}}{-(\alpha + j\omega)}\right]_{0}^{\infty} = \frac{A}{\alpha + j\omega} = \frac{A\alpha - Aj\omega}{\alpha^2 + \omega^2}$$

This pair is entry 7 in Table 12-2. In polar form,

$$\mathbf{V}(\omega) = \frac{A}{\sqrt{\alpha^2 + \omega^2}}\,e^{-j\tan^{-1}\omega/\alpha}$$

That is,

$$|\mathbf{V}(\omega)| = \frac{A}{\sqrt{\alpha^2 + \omega^2}} \quad \text{and} \quad \underline{/\mathbf{V}(\omega)} = -\tan^{-1}\omega/\alpha$$

The corresponding amplitude and phase spectra are plotted in Fig. 12-21. Again, since $\mathbf{V}(\omega)$ is defined for all $\omega$, the spectra must be plotted for both positive and negative $\omega$.

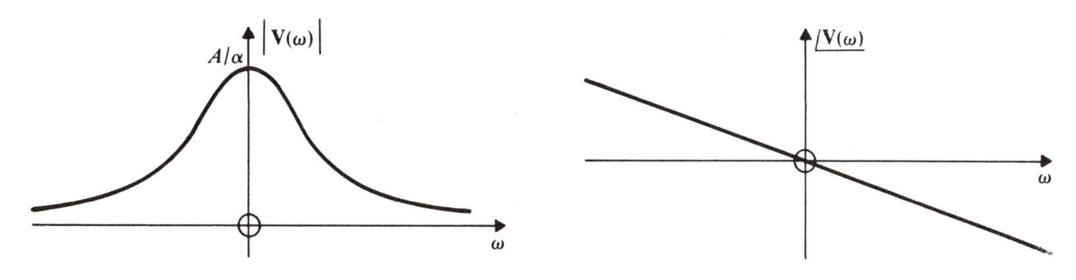


**Fig. 12-21**

**12.17** Find $\mathbf{V}(\omega)$ for the voltage waveform

$$v(t) = \begin{cases} Ae^{\alpha t} & \text{for } t < 0 \\ Ae^{-\alpha t} & \text{for } t \geq 0 \end{cases} \qquad \alpha > 0$$

which is sketched in Fig. 12-22.

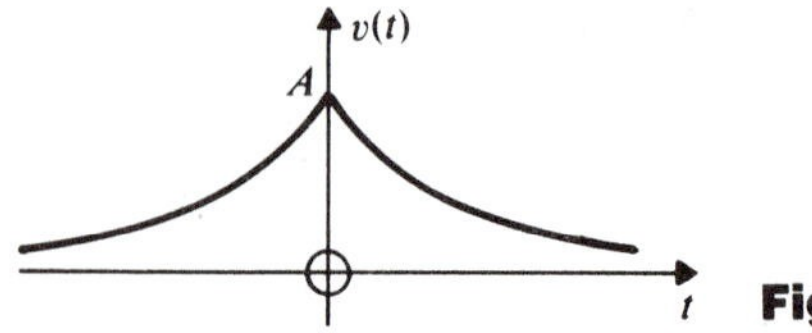


**Fig. 12-22**

From equation (*12.8*),

$$\mathcal{F}[v(t)] = \int_{-\infty}^{0} Ae^{\alpha t}e^{-j\omega t}\,dt + \int_{0}^{\infty} Ae^{-\alpha t}e^{-j\omega t}\,dt$$

$$= \left[\frac{Ae^{(\alpha - j\omega)t}}{\alpha - j\omega}\right]_{-\infty}^{0} + \left[\frac{Ae^{-(\alpha + j\omega)t}}{-(\alpha + j\omega)}\right]_{0}^{\infty}$$

$$= \frac{A}{\alpha - j\omega} + \frac{A}{\alpha + j\omega} = \frac{2\alpha A}{\alpha^2 + \omega^2}$$

As predicted in Problem 12.12, $\mathbf{V}(\omega)$ is real and even since $v(t)$ is real and even.

**12.18** The input to the circuit of Fig. 12-23 is

$$e(t) = 10e^{-2t}\mathbb{1}(t)$$

Find and sketch the amplitude and phase spectra for the output $v(t)$.

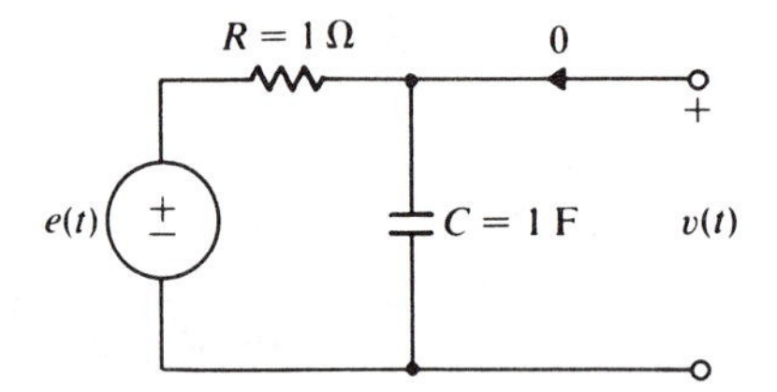


**Fig. 12-23**

From the result of Problem 12.16, with $A = 10$ and $\alpha = 2$,

$$\mathbf{E}(\omega) = \frac{10}{2 + j\omega} = \frac{10}{\sqrt{4 + \omega^2}}e^{-j\tan^{-1}\omega/2}$$

And by inspection of Fig. 12-23,

$$H(j\omega) = \frac{\mathbf{V}(\omega)}{\mathbf{E}(\omega)} = \frac{1/j\omega C}{R + 1/j\omega C} = \frac{1}{1 + j\omega} = \frac{1}{\sqrt{1 + \omega^2}}e^{-j\tan^{-1}\omega}$$

Finally, from equation (*12.11*), page 430,

$$|\mathbf{V}(\omega)| = \frac{10}{\sqrt{4+\omega^2}} \frac{1}{\sqrt{1+\omega^2}}$$

$$\underline{/\mathbf{V}(\omega)} = -\tan^{-1}\frac{\omega}{2} - \tan^{-1}\omega$$

These quantities can be evaluated for plotting as follows:

| $\omega$ | $\lvert\mathbf{V}(\omega)\rvert$ | $\underline{/\mathbf{V}(\omega)}$ |
|---|---|---|
| 0 | 5 | 0° |
| 1 | $\frac{10}{\sqrt{5}}\frac{1}{\sqrt{2}} = 3.16$ | $-45° - 26.6° = -71.6°$ |
| 2 | $\frac{10}{\sqrt{8}}\frac{1}{\sqrt{5}} = 1.58$ | $-63.4° - 45° = -108.4°$ |
| 3 | $\frac{10}{\sqrt{13}}\frac{1}{\sqrt{10}} = 0.88$ | $-56.3° - 71.5° = -127.8°$ |
| $\infty$ | 0 | $-180°$ |

For negative $\omega$, $|\mathbf{V}(\omega)|$ is unchanged and $\underline{/\mathbf{V}(\omega)}$ changes sign. Thus the spectra take the form shown in Fig. 12-24.

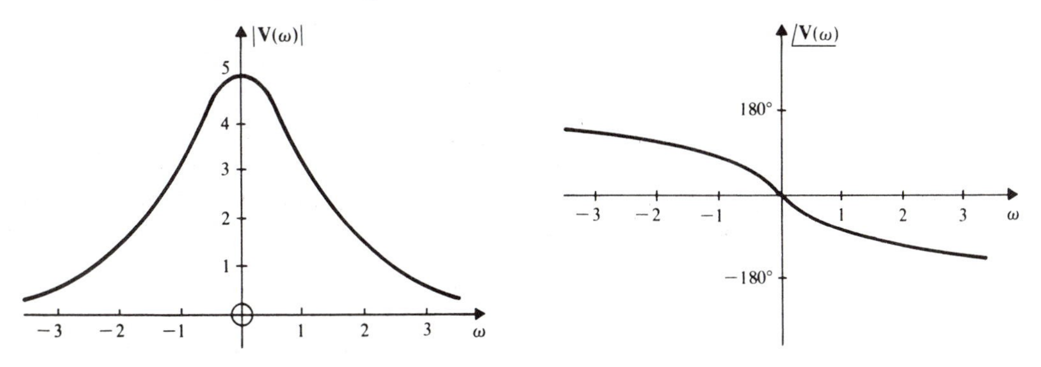


**Fig. 12-24**

**12.19** A linear circuit has the sinusoidal transfer function

$$H(j\omega) = Ke^{-j\omega t_0}$$

where $K$ and $t_0$ are positive real numbers. Determine the time response of this circuit to an arbitrary input $u(t)$.

If we call the output $y(t)$, then

$$\mathbf{Y}(\omega) = H(j\omega)\mathbf{U}(\omega) = Ke^{-j\omega t_0}\mathbf{U}(\omega)$$

Using the time shift property of Fourier transforms (entry 5 in Table 12-1, page 426), we obtain the result,

$$y(t) = Ku(t - t_0)$$

This circuit is a *time-delayer*. The output is the same *shape* as the input, but is *delayed* by $t_0$ seconds. The output is also amplified by a factor $K$.

**12.20** A plot of the (real) Fourier transform of a time signal $v(t)$ is shown in Fig. 12-25. Find an expression for $v(t)$.

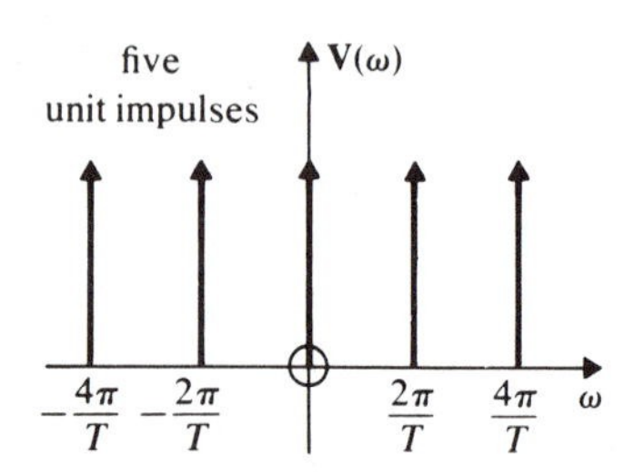


**Fig. 12-25**

From Fig. 12-25,

$$\mathbf{V}(\omega) = 1\delta(\omega) + \left\{1\delta\left(\omega - \frac{2\pi}{T}\right) + 1\delta\left(\omega + \frac{2\pi}{T}\right)\right\} + \left\{1\delta\left(\omega - \frac{4\pi}{T}\right) + 1\delta\left(\omega + \frac{4\pi}{T}\right)\right\}$$

Then, using entries 1 and 3 in Table 12-2, page 427,

$$v(t) = \frac{1}{2\pi} + \frac{1}{\pi}\cos\frac{2\pi}{T}t + \frac{1}{\pi}\cos\frac{4\pi}{T}t$$

Alternatively, we could use entry 6 in Table 12-1 to find the inverse of each of the "shifted" delta functions. Thus

$$\mathscr{F}^{-1}\left[\delta\left(\omega - \frac{2\pi}{T}\right)\right] = \frac{1}{2\pi}e^{j\omega_0 t} \quad \text{and} \quad \mathscr{F}^{-1}\left[\delta\left(\omega + \frac{2\pi}{T}\right)\right] = \frac{1}{2\pi}e^{-j\omega_0 t}$$

Combining these terms gives $(1/\pi)\cos(2\pi/T)t$ as before. A similar maneuver would lead to the term $(1/\pi)\cos(4\pi/T)t$.

**12.21** Sketch $\mathbf{P}(\omega)$ vs. $\omega$ given

$$p(t) = K\{1 + m(t)\}\cos\omega_c t$$

where

$$m(t) = \tfrac{1}{2}\cos\omega_m t, \qquad \omega_c > \omega_m$$

(*Comment: $m(t)$ is a sinusoidal modulation signal* impressed upon the *carrier signal $K\cos\omega_c t$*. This is *amplitude modulation*. $\omega_c$ is the carrier frequency and $\omega_m$ the modulation frequency.)

$$p(t) = K\{1 + \tfrac{1}{2}\cos\omega_m t\}\cos\omega_c t$$
$$= K\cos\omega_c t + \tfrac{1}{4}K\{\cos(\omega_c - \omega_m)t + \cos(\omega_c + \omega_m)t\}$$

Thus from entry 5 in Table 12-2,

$$\mathbf{P}(\omega) = K\pi\{\delta(\omega - \omega_c) + \delta(\omega + \omega_c)\}$$
$$+ \tfrac{1}{4}K\pi\{\delta(\omega - \omega_c + \omega_m) + \delta(\omega + \omega_c - \omega_m)\}$$
$$+ \tfrac{1}{4}K\pi\{\delta(\omega - \omega_c - \omega_m) + \delta(\omega + \omega_c + \omega_m)\}$$

$\mathbf{P}(\omega)$ is plotted vs. $\omega$ in Fig. 12-26. Since $p(t)$ is real and even, $\mathbf{P}(\omega)$ is also real and even.

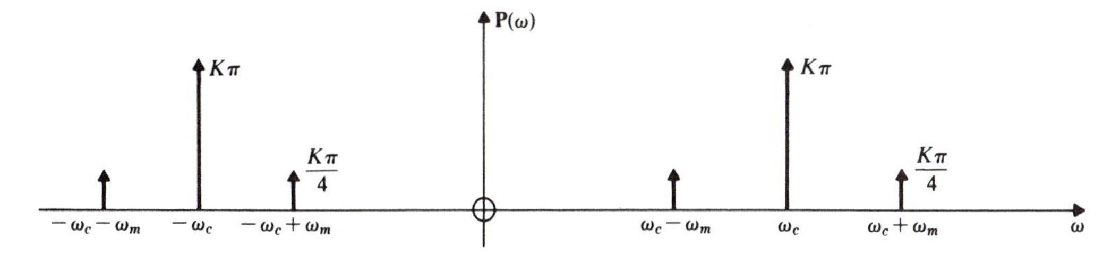


**Fig. 12-26**

**12.22** A *periodic* function can be represented by the Fourier *series*,

$$g(t) = \sum_{k=-\infty}^{\infty} \mathbf{C}_k e^{jk\omega_0 t}$$

Find $\mathbf{G}(\omega)$, the Fourier *transform* of $g(t)$.

$$\mathbf{G}(\omega) = \mathcal{F}\left[\sum_{k=-\infty}^{\infty} \mathbf{C}_k e^{jk\omega_0 t}\right]$$

But from entry 1 of Table 12-2 and entry 6 of Table 12-1, with $g(t) = 1$ and $\mathbf{C}_k$ a complex constant,

$$\mathcal{F}[1e^{jk\omega_0 t}] = 2\pi\delta(\omega - k\omega_0)$$

Thus

$$\mathbf{G}(\omega) = 2\pi \sum_{k=-\infty}^{\infty} \mathbf{C}_k \delta(\omega - k\omega_0)$$

So the amplitude spectrum of a periodic function consists of a sequence of impulse functions located at the harmonic frequencies of the function and of strength $2\pi|\mathbf{C}_k|$. The phase spectrum will consist of lines at the same harmonic frequencies with values of $\angle\mathbf{C}_k$.

**12.23** Use the results of Problems 12.11 and 12.22 to determine the Fourier transform of a unit impulse train,

$$\delta_T(t) = \sum_{k=-\infty}^{\infty} \delta(t - kT)$$

From Problem 12.11, the coefficients for the exponential Fourier series for a unit impulse train are given by

$$\mathbf{C}_k = \frac{1}{T}$$

Therefore from the result of Problem 12.22,

$$\mathscr{F}[\delta_T(t)] = \frac{2\pi}{T} \sum_{k=-\infty}^{\infty} \delta(\omega - k\omega_0) = \omega_0 \sum_{k=-\infty}^{\infty} \delta(\omega - k\omega_0)$$

Again, this is a remarkable result. The Fourier transform of an impulse train in the time-domain becomes an impulse train of $\omega_0$ times the amplitude in the frequency-domain.

This is entry 12 in Table 12-2.

**12.24** The input to the linear network of Fig. 12-27*a* is

$$u(t) = A\,\frac{\sin \pi t}{\pi t}$$

while the system transfer function is

$$H(j\omega) = \frac{\mathbf{Y}(\omega)}{\mathbf{U}(\omega)} = \begin{cases} 2 & \text{for } |\omega| \le 8\pi \\ 0 & \text{for } |\omega| > 8\pi \end{cases}$$

as shown in Fig. 12-27*b*. This would be described as an ideal *low-pass* characteristic. First, find $\mathbf{U}(\omega)$ and plot its spectrum. Then find $\mathbf{Y}(\omega)$ and $y(t)$ and sketch $|\mathbf{Y}(\omega)|$ vs. $\omega$.

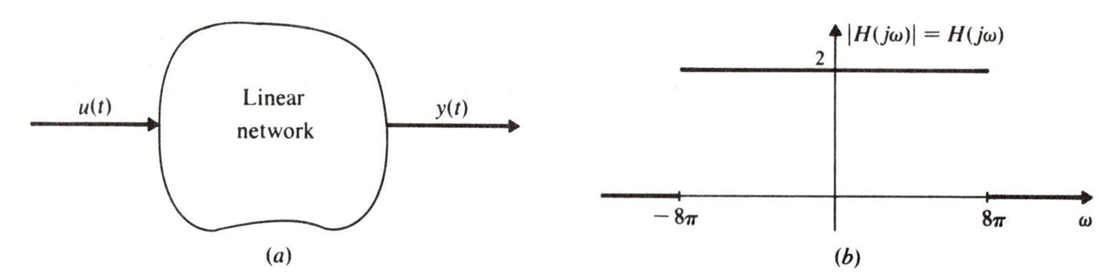


**Fig. 12-27**

Here it is helpful to make use of the symmetry property to find $\mathbf{U}(\omega)$. In Problem 12.13 the Fourier transform of a voltage pulse was found to be

$$\mathbf{V}(\omega) = aV\left\{\frac{\sin(\omega a/2)}{\omega a/2}\right\}$$

where

$$v(t) = V\{\mathbb{1}(t + a/2) - \mathbb{1}(t - a/2)\}$$

If we elect to set $a/2 = \pi$ and $V = A/2\pi$, then

$$\frac{A}{2\pi}\{\mathbb{1}(t+\pi) - \mathbb{1}(t-\pi)\} \Leftrightarrow A\frac{\sin \pi\omega}{\pi\omega}$$

and so from entry 4 of Table 12-1, page 426,

$$A\frac{\sin \pi t}{\pi t} \Leftrightarrow A\{\mathbb{1}(-\omega+\pi) - \mathbb{1}(-\omega-\pi)\}$$

The expression on the right is even, so that we may substitute $-\omega$ for $\omega$. Thus

$$\mathbf{U}(\omega) = \mathcal{F}\left[A\frac{\sin \pi t}{\pi t}\right] = A\{\mathbb{1}(\omega+\pi) - \mathbb{1}(\omega-\pi)\}$$

which is plotted in Fig. 12-28$a$.

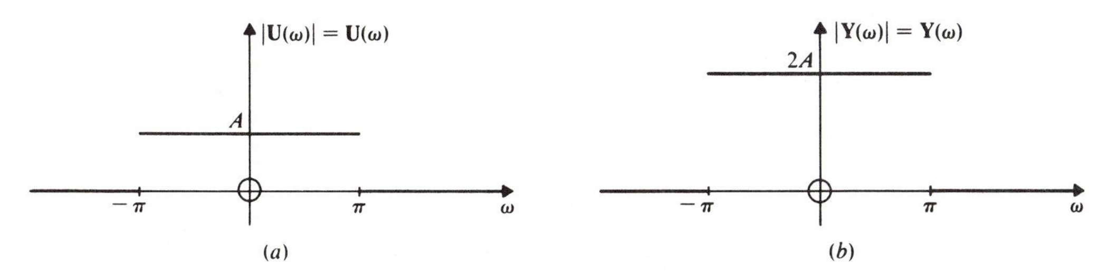


**Fig. 12-28**

Now from equation (*12.11*), page 430,

$$\mathbf{Y}(\omega) = H(j\omega)\mathbf{U}(\omega)$$

which is plotted in Fig. 12-28$b$.

From Fig. 12-28$b$, and knowing that $\underline{/H(j\omega)}$ and $\underline{/\mathbf{U}(\omega)}$ are both zero,

$$\mathbf{Y}(\omega) = 2A\{\mathbb{1}(\omega+\pi) - \mathbb{1}(\omega-\pi)\}$$

Finally, applying the symmetry property again,

$$y(t) = 2A\frac{\sin \pi t}{\pi t}$$

**12.25** Find the transform $\mathbf{V}(\omega)$ of the function $v(t)$ shown in Fig. 12-29.

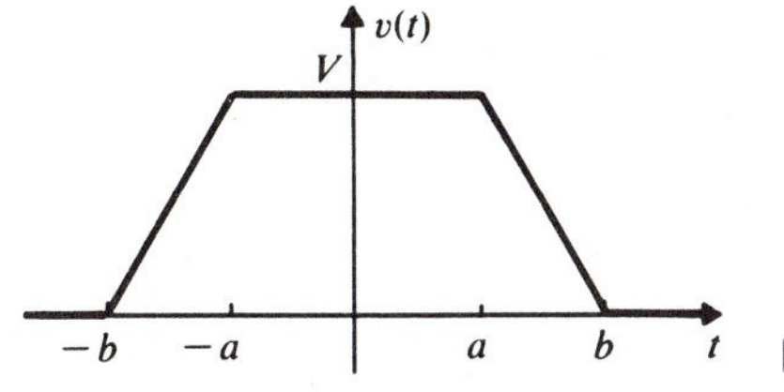


**Fig. 12-29**

When a function is composed of straight-line segments, we may differentiate twice to obtain a set of impulse functions. These impulses are easily transformed, and then from entry 8 of Table 12-1 the transform follows.

In this problem the first and second derivatives are graphed in Fig. 12-30. Then

$$\mathcal{F}[\ddot{v}(t)] = \frac{V}{b-a}[e^{j\omega b} - e^{j\omega a} - e^{-j\omega a} + e^{-j\omega b}] = \frac{2V}{b-a}\{\cos b\omega - \cos a\omega\}$$

Therefore from entry 8 in Table 12-1,

$$\mathcal{F}[v(t)] = \frac{1}{(j\omega)^2}\mathcal{F}[\ddot{v}(t)] = \frac{2V(\cos a\omega - \cos b\omega)}{\omega^2(b-a)}$$

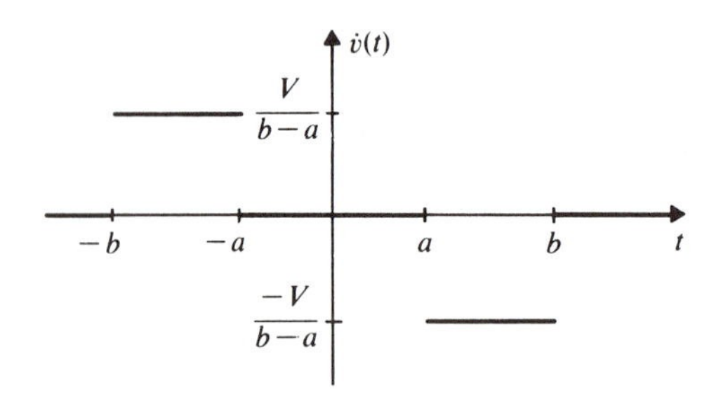


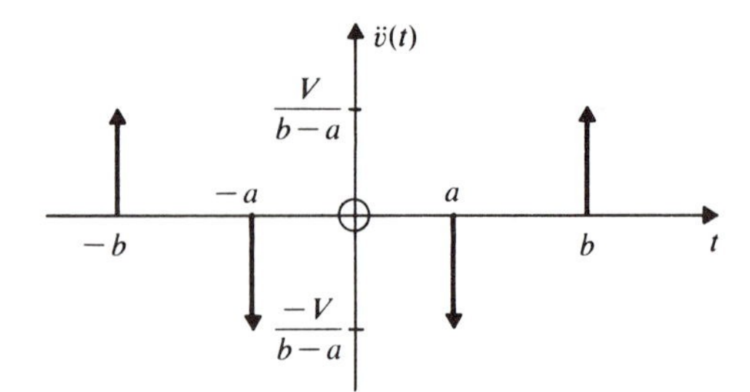


**Fig. 12-30**

**12.26** The signal multiplier of Fig. 12-31$a$ is defined by $c(t) = a(t)b(t)$. Given $\mathbf{A}(\omega)$ and $\mathbf{B}(\omega)$ as in Fig. 12-31$b$, find $c(t)$.

From Fig. 12-31$b$ we see that

$$\mathbf{A}(\omega) = 6\pi\{\delta(\omega - 1400\pi) + \delta(\omega + 1400\pi)\}$$

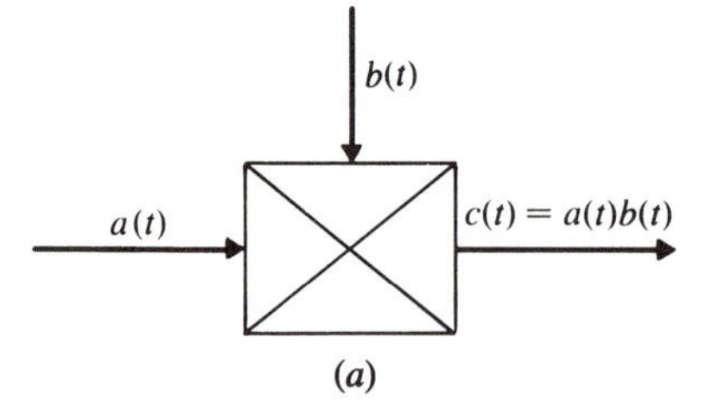


($a$)

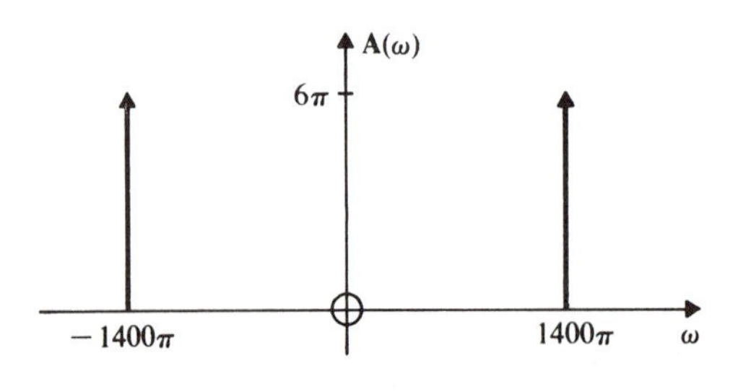


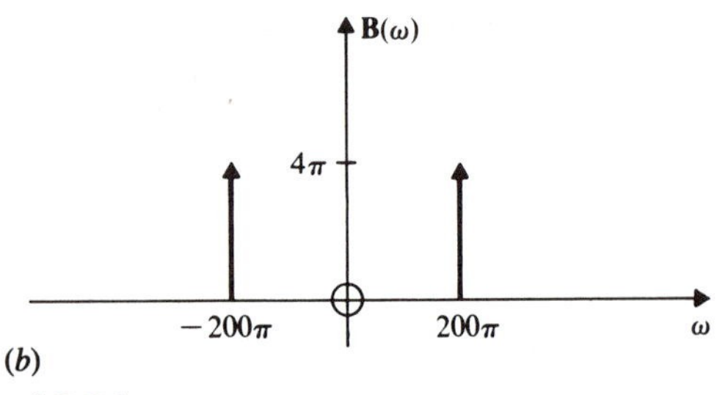


($b$)

**Fig. 12-31**

and so from entry 3 in Table 12-2, page 427,

$$a(t) = 6 \cos 1400\pi t$$

and similarly

$$b(t) = 4 \cos 200\pi t$$

Hence

$$c(t) = a(t)b(t) = 24\{\cos 1400\pi t\}\{\cos 200\pi t\} = 12\{\cos 1200\pi t + \cos 1600\pi t\}$$

**12.27** It is often helpful to work in terms of the frequency $f$ rather than the angular frequency $\omega$. To this end, write the following in terms of $f$:

(*a*) The definition of $\mathcal{F}[g(t)]$.
(*b*) The definition of the inverse transform.
(*c*) The symmetry property.
(*d*) $\mathcal{F}[\delta(t)]$.
(*e*) $\mathcal{F}[A]$.
(*f*) $\mathcal{F}[\cos 2\pi f_0 t]$.
(*g*) $H(j\omega) = \dfrac{1}{1 + j\omega}$.

(*a*) Substituting $2\pi f$ for $\omega$ in entry 1 of Table 12-1,

$$\mathcal{F}[g(t)] = \mathbf{G}(f) = \int_{-\infty}^{\infty} g(t)e^{-j2\pi ft}\,dt \tag{1}$$

Strictly speaking, we should write $\mathbf{G}(2\pi f)$, but just as we write $\mathbf{G}(\omega)$ instead of $\mathbf{G}(j\omega)$, it is common practice to drop the $2\pi$.

(*b*) Similarly, from the second entry in the table,

$$g(t) = \frac{1}{2\pi}\int_{-\infty}^{\infty} \mathbf{G}(f)e^{j2\pi ft}\,d(2\pi f) = \int_{-\infty}^{\infty} \mathbf{G}(f)e^{j2\pi ft}\,df \tag{2}$$

(*c*) Care must be taken with the abbreviated notation. Thus it is conventional to write

$$\mathbf{G}(f) = \mathbf{G}(\omega)\big|_{\omega=2\pi f}$$

But

$$\mathbf{G}(t) = \mathbf{G}(f)\big|_{f=t} \quad \text{and} \quad g(-f) = g(t)\big|_{t=-f}$$

Here, if we substitute $-f$ for $t$ and $-t$ for $f$ in equation (2),

$$g(-f) = \int_{-\infty}^{\infty} \mathbf{G}(-t)e^{j2\pi ft}\,d(-t) = \int_{-\infty}^{\infty} \mathbf{G}(t)e^{-j2\pi ft}\,dt = \mathcal{F}[\mathbf{G}(t)] \tag{3}$$

where we have replaced the "dummy" variable of integration $-t$ by $t$.

Thus from equations (2) and (3) the symmetry property may be expressed as

$$g(t) \Leftrightarrow \mathbf{G}(f)$$
$$\mathbf{G}(t) \Leftrightarrow g(-f)$$

(*d*) From Problem 12.14, the Fourier transform of a unit *pulse* is

$$\mathcal{F}[p(t)] = \frac{\sin(\omega a/2)}{\omega a/2} = \frac{\sin \pi fa}{\pi fa}$$

and so

$$\mathcal{F}[\delta(t)] = \lim_{a\to 0} \frac{\sin \pi fa}{\pi fa} = 1$$

(*e*) Using the symmetry property,

$$\delta(t) \Leftrightarrow 1$$
$$1 \Leftrightarrow \delta(-f) = \delta(f)$$

Therefore

$$\mathcal{F}[A] = A\delta(f)$$

(*f*) Rewriting the frequency shift property in terms of $f$,

$$e^{j2\pi f_0 t} g(t) \Leftrightarrow \mathbf{G}(f - f_0)$$

Thus when $g(t) = 1$,

$$\mathcal{F}[e^{j2\pi f_0 t}] = \delta(f - f_0) \quad \text{and} \quad \mathcal{F}[e^{-j2\pi f_0 t}] = \delta(f + f_0)$$

from which

$$\mathcal{F}[\cos 2\pi f_0 t] = \tfrac{1}{2}\{\delta(f - f_0) + \delta(f + f_0)\}$$

(*g*) Substituting $2\pi f$ for $\omega$,

$$H(f) = \frac{1}{1 + j2\pi f}$$

**12.28** The energy in a signal $g(t)$ is conventionally defined to be $E = \int_{-\infty}^{\infty} \{g(t)\}^2\, dt$. However, the integration can be quite messy and often it is easier to work in the transform-domain with the help of Parseval's theorem. Find the energy $E$ in the signal $g(t) = A(\sin \pi t)/\pi t$.

From equation (*12.12*), page 432,

$$\int_{-\infty}^{\infty} \{g(t)\}^2\, dt = \frac{1}{2\pi}\int_{-\infty}^{\infty} |\mathbf{G}(\omega)|^2\, d\omega$$

where from Problem 12.24 and Fig. 12-28*a*,

$$\mathbf{G}(\omega) = \begin{cases} 0 & \text{for } \omega < -\pi \\ A & \text{for } -\pi \le \omega \le \pi \\ 0 & \text{for } \omega > \pi \end{cases}$$

Thus

$$E = \frac{1}{2\pi}\int_{-\pi}^{\pi} A^2\, d\omega = \frac{1}{2\pi}[A^2\omega]_{-\pi}^{\pi} = A^2$$

*Check*: From a table of integrals, $\int_0^{\infty} [(\sin^2 ax)/x^2]\, dx = |a|\,\pi/2$. Hence since $(\sin^2 \pi t)/\pi^2 t^2$ is even,

$$E = \int_{-\infty}^{\infty} \frac{A^2}{\pi^2} \frac{\sin^2 \pi t}{t^2}\, dt = 2\int_{0}^{\infty} \frac{A^2}{\pi^2} \frac{\sin^2 \pi t}{t^2}\, dt = \frac{2A^2}{\pi^2} \,|\pi|\, \frac{\pi}{2} = A^2$$

*Comment*: The energy spectrum of $g(t)$, namely $(1/2\pi)\,|\mathbf{G}(\omega)|^2$ vs. $\omega$, is plotted in Fig. 12-32. The area under this graph is clearly $A^2$.

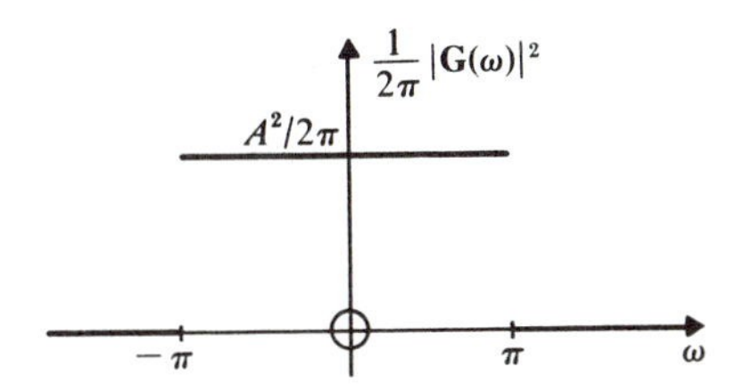


Fig. 12-32

# PROBLEMS

**12.29** Find the Fourier coefficient $\mathbf{C}_k$ for the periodic triangular current waveform shown in Fig. 12-33. Then write the *trigonometric* Fourier series.

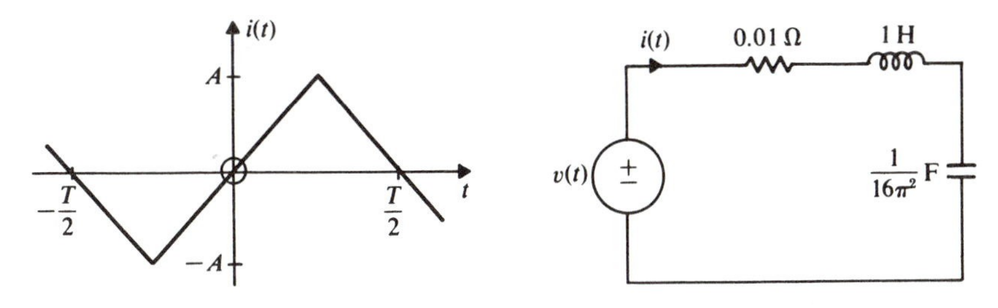


Fig. 12-33

Fig. 12-34

**12.30** The Fourier series for $v(t)$ in the circuit of Fig. 12-34 is

$$v(t) = \frac{V}{2} + \sum_{k=1}^{\infty} \frac{A}{\pi k} \cos\left(2\pi k t + \frac{\pi}{2}\right)$$

Find the first three terms in the cosine series for $i(t)$.

**12.31** In the circuit of Fig. 12-35,

$$v(t) = 10 \sin \omega_1 t + 5 \sin 3\omega_1 t + 2 \sin 5\omega_1 t$$

$$i(t) = 2 \sin\left(\omega_1 t + \frac{\pi}{4}\right) + 4 \sin\left(5\omega_1 t + \frac{\pi}{6}\right)$$

Calculate the average real power supplied to the circuit.

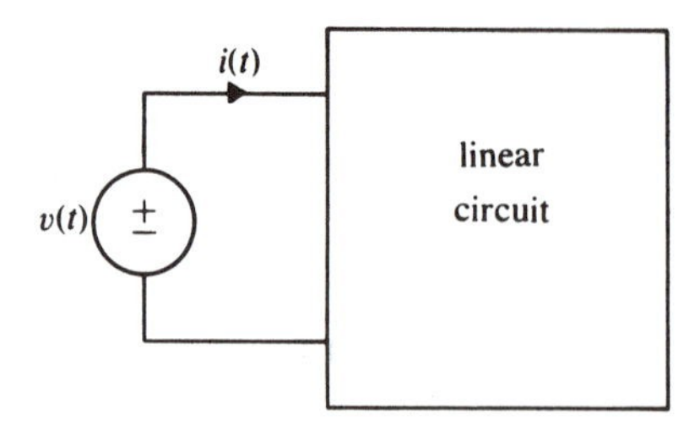


**Fig. 12-35**

**12.32** The input to a linear network is

$$u(t) = \sum_{k=-\infty}^{\infty} \frac{1}{k^2} e^{jkt}$$

and the appropriate network function is $H(s) = s/(s + 1)$. Find the steady-state output $y_{ss}(t)$.

**12.33** Find the Fourier transform of the semi-infinite sinusoidal function in Fig. 12-36.

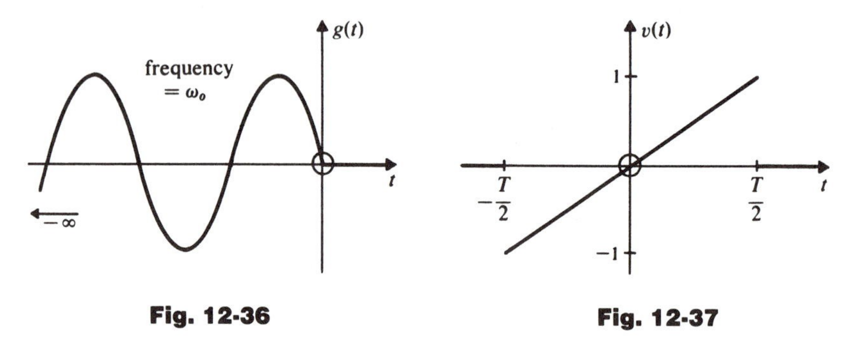


**Fig. 12-36** **Fig. 12-37**

**12.34** Find $\mathbf{V}(\omega)$ for the function $v(t)$ shown in Fig. 12-37.

**12.35** The *periodic* current of Fig. 12-38 flows through a resistor of $R$ ohms. Find the percentage of the total power which is contained in the dc and fundamental terms.

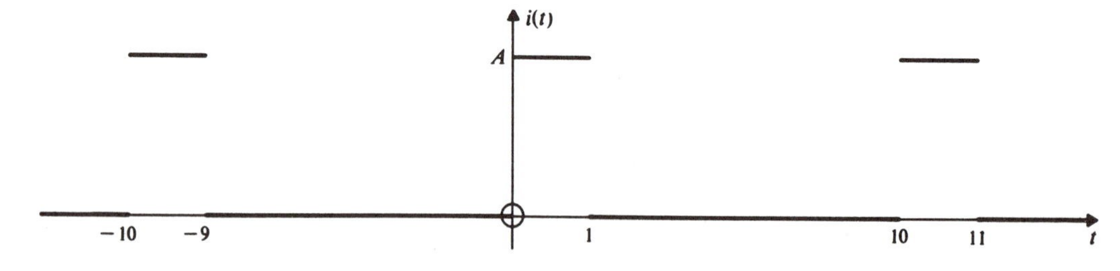


**Fig. 12-38**

**12.36** Derive the result given as entry 6 in Table 12-2, page 428. That is, show that

$$\mathcal{F}[A \operatorname{sgn} t] = \mathcal{F}\left[\frac{At}{|t|}\right] = \frac{2A}{j\omega}$$

*Hint*: First find the transform of the function

$$q(t) = \begin{cases} e^{-\alpha t} & \text{for } t \geq 0 \\ -e^{\alpha t} & \text{for } t < 0 \end{cases} \quad \alpha > 0$$

which is graphed in Fig. 12-39.

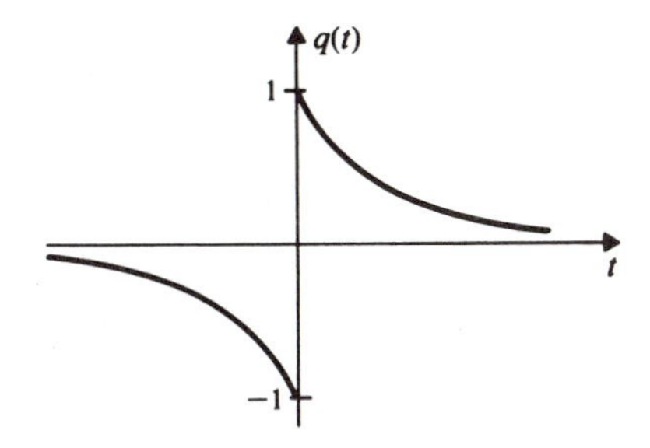


**Fig. 12-39**

**12.37** Starting with $\mathcal{F}[\operatorname{sgn} t] = 2/j\omega$, show that

$$\mathcal{F}[\mathbb{1}(t)] = \frac{1}{j\omega} + \pi\delta(\omega)$$

**12.38** Show that $\underline{/\mathbf{Q}(f)} = -\frac{1}{2}\pi \operatorname{sgn} f$ for the function $q(t)$ defined in Problem 12.36.

**12.39** Given $g(t) = 2e^{-3t}\mathbb{1}(t) + 5e^{-4|t|}$, find $\mathbf{G}(f)$.

**12.40** Find $\mathbf{S}(f)$ and $\mathbf{S}(\omega)$ for the signal of Fig. 12-40.

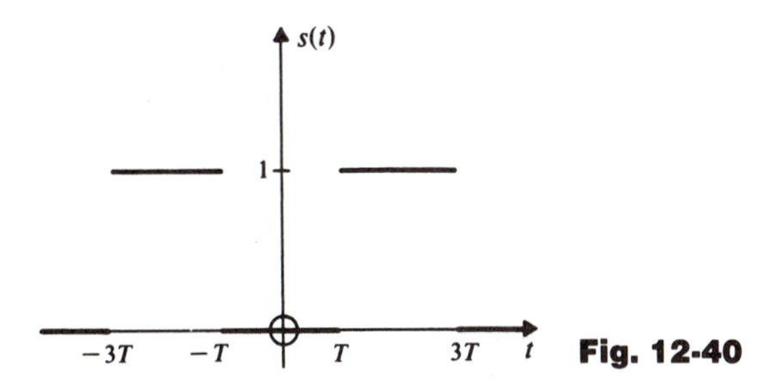


**Fig. 12-40**

**12.41** If $\mathbf{G}(\omega) = \dfrac{-\omega^2 + 1}{(j\omega + 1)(j\omega - 1)(j\omega - 3)}$, find $g(t)$.

Chapter

13

# CONVOLUTION

There are a number of ways of solving a linear network's differential equations—the classical method; transform methods; and superposition methods, such as the Fourier series approach of the previous chapter and the convolution technique of this chapter. Basically, the superposition methods depend upon decomposing an arbitrary signal into an infinite number or continuum of elementary functions, such as the sinusoid or the impulse. For example, we can show that a function $u(t)$ can be written as the integral of a continuum of impulse functions.

$$u(t) = \int_{-\infty}^{\infty} u(\tau)\delta(t - \tau)\, d\tau \qquad (13.1)$$

This equation is often referred to as the *sifting property* of the impulse function since it "sifts out" one particular value $u(\tau)|_{\tau=t}$ from the function $u(\tau)$.

If the network's response to a *single* unit impulse is known, then by superposition the response to the *total* signal should be calculable. It can be shown, in fact, that if $h(t)$ is the unit impulse response of a linear network, then an arbitrary input $u(t)$ will cause a response $y(t)$ given by the *convolution integral,*

$$y(t) = \int_{-\infty}^{\infty} h(\tau)u(t - \tau)\, d\tau = \int_{-\infty}^{\infty} u(\tau)h(t - \tau)\, d\tau \qquad (13.2)$$

Equation (*13.2*) is also called the *Faltung, Duhamel,* or *superposition integral.* It is often expressed operationally as

$$y(t) = h(t) * u(t) = u(t) * h(t) \qquad (13.3)$$

Convolution can be applied to any linear system, but we will restrict ourselves to linear, time-invariant circuits.

## CONVOLUTION AND THE TRANSFORM DOMAINS

First, it is well to recall that the Laplace and Fourier transforms of a unit impulse are both unity. Therefore if $H(s)$ is the network function relating an input $u(t)$ and an output $y(t)$, then the network's impulse response is given by

$$h(t) = \mathcal{L}^{-1}[H(s)] = \mathcal{F}^{-1}[H(\omega)] = \mathcal{F}^{-1}[H(f)] \tag{13.4}$$

Secondly, if $p$, $q$, and $r$ are three functions of time such that

$$p(t) = q(t) * r(t) \tag{13.5}$$

then it can be shown that

$$\begin{aligned} P(s) &= Q(s)R(s) \\ \mathbf{P}(\omega) &= \mathbf{Q}(\omega)\mathbf{R}(\omega) \\ \mathbf{P}(f) &= \mathbf{Q}(f)\mathbf{R}(f) \end{aligned} \tag{13.6}$$

That is, convolution in time corresponds to simple multiplication in any of the transform domains. Conversely, if

$$p(t) = q(t)r(t) \tag{13.7}$$

then

$$\begin{aligned} P(s) &= \frac{1}{2\pi j} Q(s) * R(s) \\ \mathbf{P}(\omega) &= \frac{1}{2\pi}\mathbf{Q}(\omega) * \mathbf{R}(\omega) = \frac{1}{2\pi}\int_{-\infty}^{\infty} \mathbf{Q}(y)\mathbf{R}(\omega - y)\, dy \\ \mathbf{P}(f) &= \mathbf{Q}(f) * \mathbf{R}(f) \end{aligned} \tag{13.8}$$

## EVALUATION OF THE CONVOLUTION INTEGRAL

The convolution integral can be evaluated in three distinct ways.

1. *Graphical convolution:* If $h(t)$ and $u(t)$ are available in graphical form, then by *folding, shifting*, multiplication, and graphical integration we may obtain the graph of $y(t) = h(t) * u(t)$.
2. *Numerical convolution:* If $h(t)$ and $u(t)$ are available as numerical sequences (most conveniently the values of $h$ and $u$ at equal intervals of time), then $y(t) = h(t) * u(t)$ may be found in the form of a third number sequence by purely numerical techniques. This is ideally suited to digital computer implementation.

3. *Analytical convolution:* If $h(t)$ and $u(t)$ can be expressed analytically, it is usually possible to integrate equation (*13.2*) formally. However, transform methods are almost always simpler, since convolution is transformed into multiplication.

### Example 13.1

The meaning of convolution is best illustrated by the process of graphical evaluation. Suppose that the impulse response of a linear network has been measured, with the result shown in Fig. 13-1. We will now use graphical convolution to find the output $y(t)$ at $t = 2$ and 4 s due to the input of Fig. 13-2.

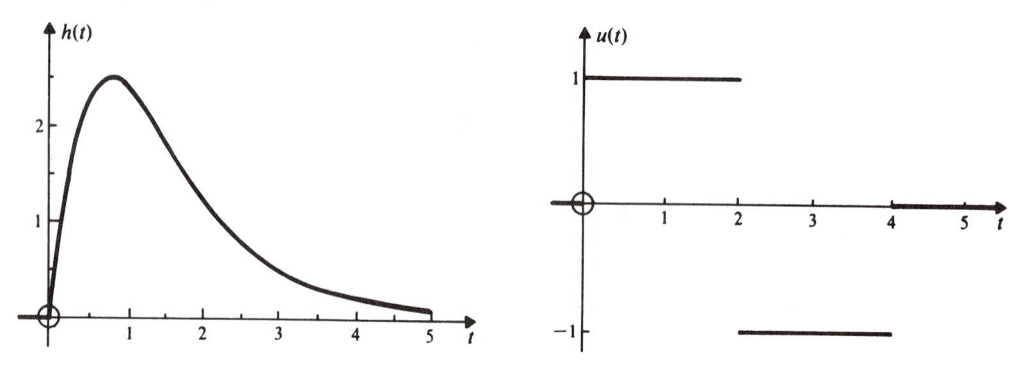


**Fig. 13-1** **Fig. 13-2**

Our first objective is to evaluate equation (*13.2*), namely

$$y(t) = h(t) * u(t) = \int_{-\infty}^{\infty} h(\tau)u(t - \tau)\, d\tau$$

for $t = 2$ s. We have been given a plot of $h(t)$ vs. $t$, and of course $h(\tau)$ vs. $\tau$ is identical, as in Fig. 13-3*a*. To obtain $u(-\tau)$ from the given graph of $u(t)$, we first imagine changing the variable from $t$ to $\tau$ as we have just done to obtain $h(\tau)$. Then $u(-\tau)$ is obtained by *folding* this curve about the $\tau = 0$ axis. This is illustrated in Fig. 13-3*b*. Finally, to find $u(2 - \tau)$ we *shift* the curve of $u(-\tau)$ two units to the *right*, as in Fig. 13-3*c*.

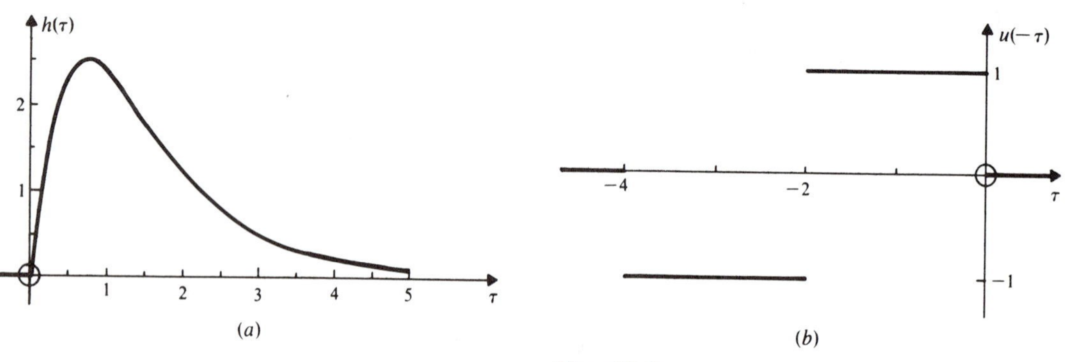


**Fig. 13-3**

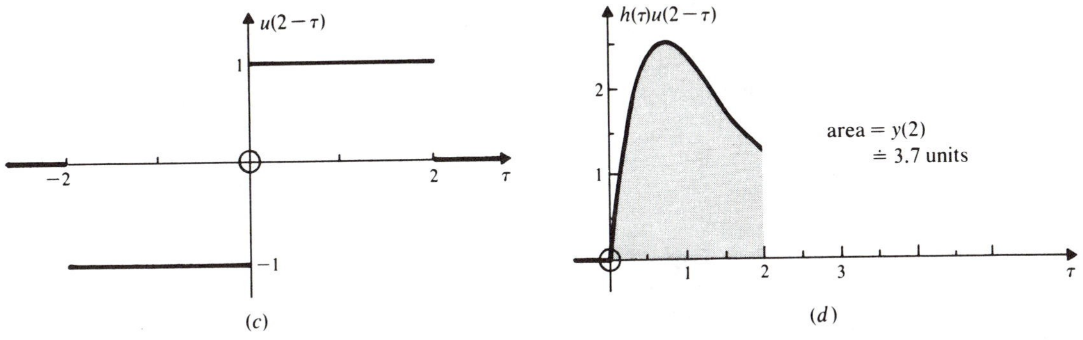


**Fig. 13-3 (*continued*)**

The *product* $h(\tau)u(2-\tau)$ is now easily deduced as in Fig. 13-3*d*, and the area can be found by counting squares, or by any convenient equivalent. Equation (*13.2*) indicates that this area is $y(2)$.

To evaluate $y(4)$ we follow the same procedure, except that we now shift the curve of $u(-\tau)$ *four* units to the right. This is shown in Fig. 13-4.

If we were to compute $y(t)$ for several more values of $t$ we would obtain the curve graphed in Fig. 13-5, where our two calculated points are ringed.

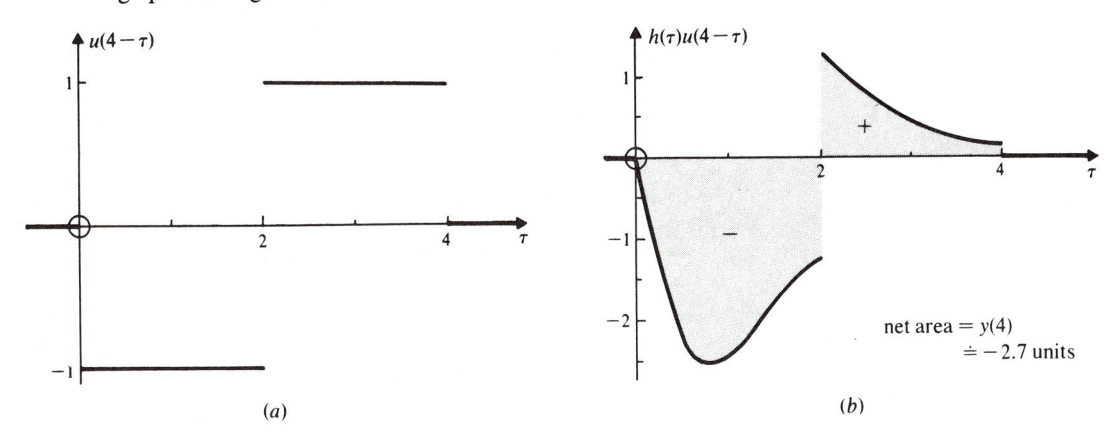


**Fig. 13-4**

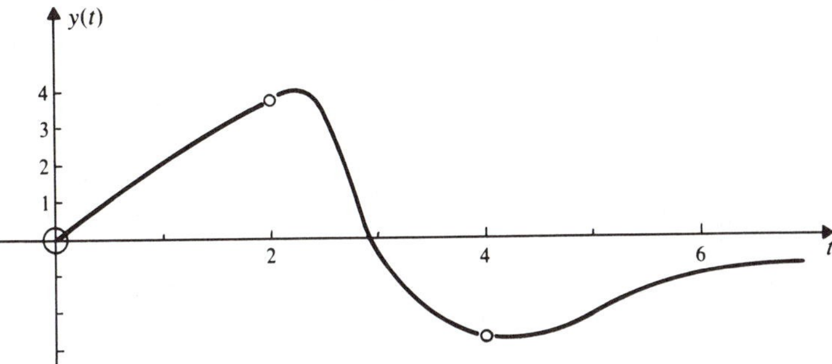


**Fig. 13-5**

# ILLUSTRATIVE PROBLEMS

## *Graphical Convolution*

**13.1** Figure 13-6 is a graph of $u(t)$ vs. $t$. Sketch the following:
(*a*) $u(\tau)$ vs. $\tau$
(*b*) $u(-\tau)$ vs. $\tau$
(*c*) $u(2-\tau)$ vs. $\tau$
(*d*) $u(-2-\tau)$ vs. $\tau$

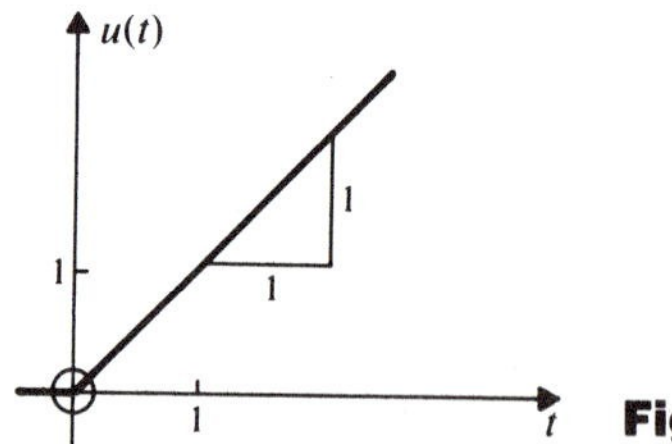


**Fig. 13-6**

(*a*) $u(\tau)$ vs. $\tau$ is the same as $u(t)$ vs. $t$ with $\tau$ substituted for $t$ (see Fig. 13-7*a*).
(*b*) $u(-\tau)$ is obtained form $u(\tau)$ by substituting $-\tau$ for $\tau$ as in Fig. 13-7*b*. This is often referred to as *folding* $u(\tau)$ about the $\tau = 0$ axis.
(*c*) $u(2-\tau)$ vs. $\tau$ is $u(-\tau)$ vs. $\tau$ *shifted* two units in the *positive* $\tau$ direction (see Fig. 13-7*c*). If this is not clear, try substituting $\tau = 1$, say, into $u(2-\tau)$. That is,

$$u(2-\tau)\big|_{\tau=1} = u(1)$$

which from Fig. 13-6 or 13-7*a* is equal to 1. This checks with the value read from Fig. 13-7*c* when $\tau = 1$.
(*d*) $u(-2-\tau)$ vs. $\tau$ is $u(-\tau)$ vs. $\tau$ shifted two units in the negative $\tau$ direction (see Fig. 13-7*d*).

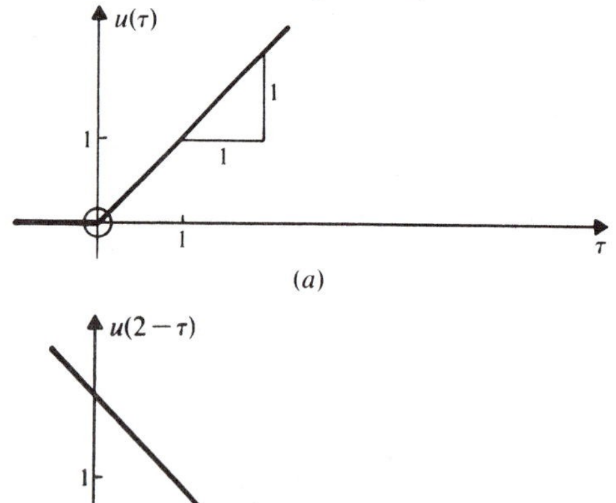


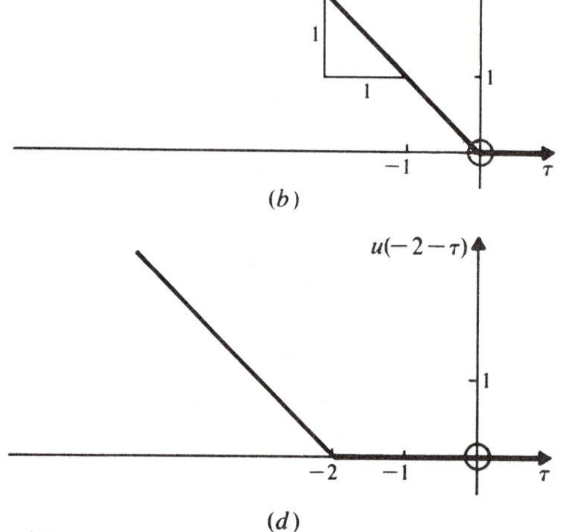


**Fig. 13-7**

**13.2** The linear network of Fig. 13-8$a$ has the impulse response $h(t)$ given in Fig. 13-8$b$. Use graphical convolution to determine the forced output $y(t)$ due to the input $u(t)$ of Fig. 13-8$c$.

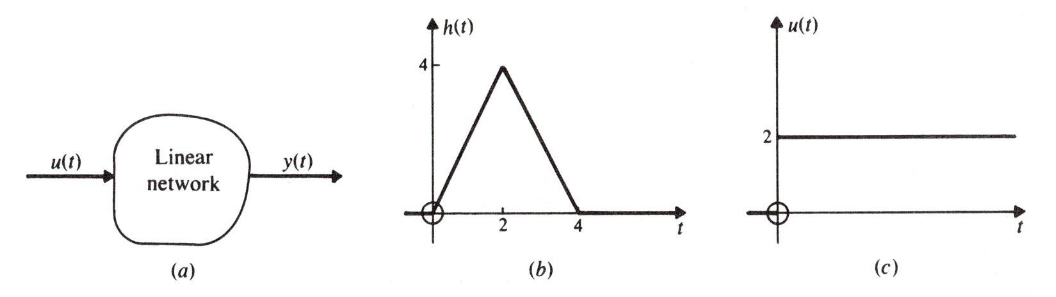


**Fig. 13-8**

Considering the convolution integral

$$y(t) = \int_{-\infty}^{\infty} h(\tau)u(t - \tau)\, d\tau$$

the first move is to plot $h(\tau)$ and $u(-\tau)$ as in Fig. 13-9.

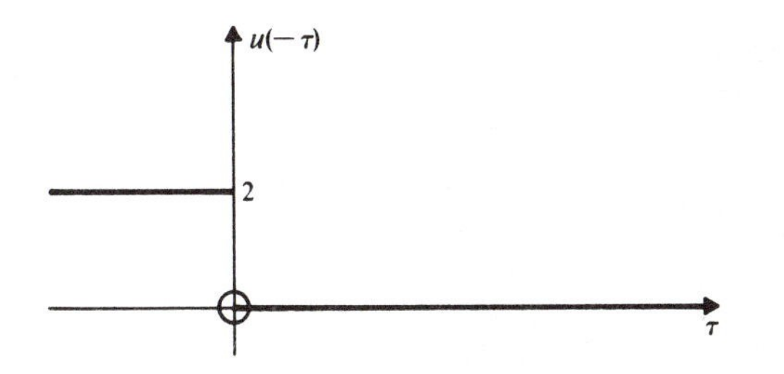


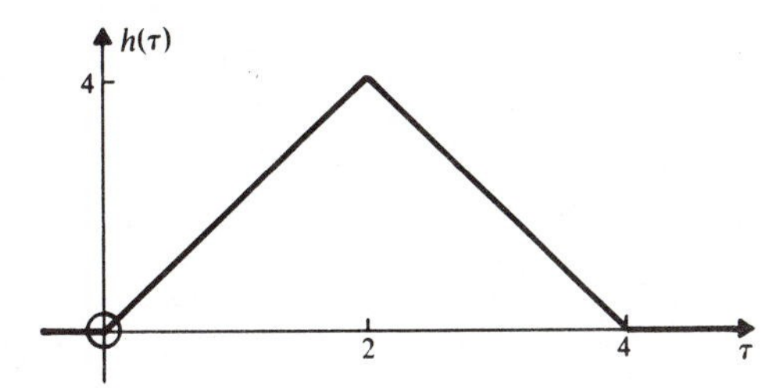


**Fig. 13-9**

Then if we elect to evaluate $y(t)$ at $t = 1$, say, the corresponding $u(1 - \tau)$ and $h(\tau)u(1 - \tau)$ may be plotted as in Fig. 13-10$a$.

Then by graphical integration,

$$y(1) = \int_{-\infty}^{\infty} h(\tau)u(1 - \tau)\, d\tau = 2$$

We must repeat the above process for other values of $t$ in order to define the output function $y(t)$. Thus with the aid of Figs. 13-10$b$-$e$ we find

$$y(2) = 8, \quad y(3) = 14, \quad y(t) = 16 \quad \text{for } t \geq 4, \quad \text{and} \quad y(t) = 0 \quad \text{for } t \leq 0$$

The function $y(t)$ defined by these data is graphed in Fig. 13-11.

## *Comments*

1. The above impulse response is not that of a real, physical network, but was chosen to simplify the computations.

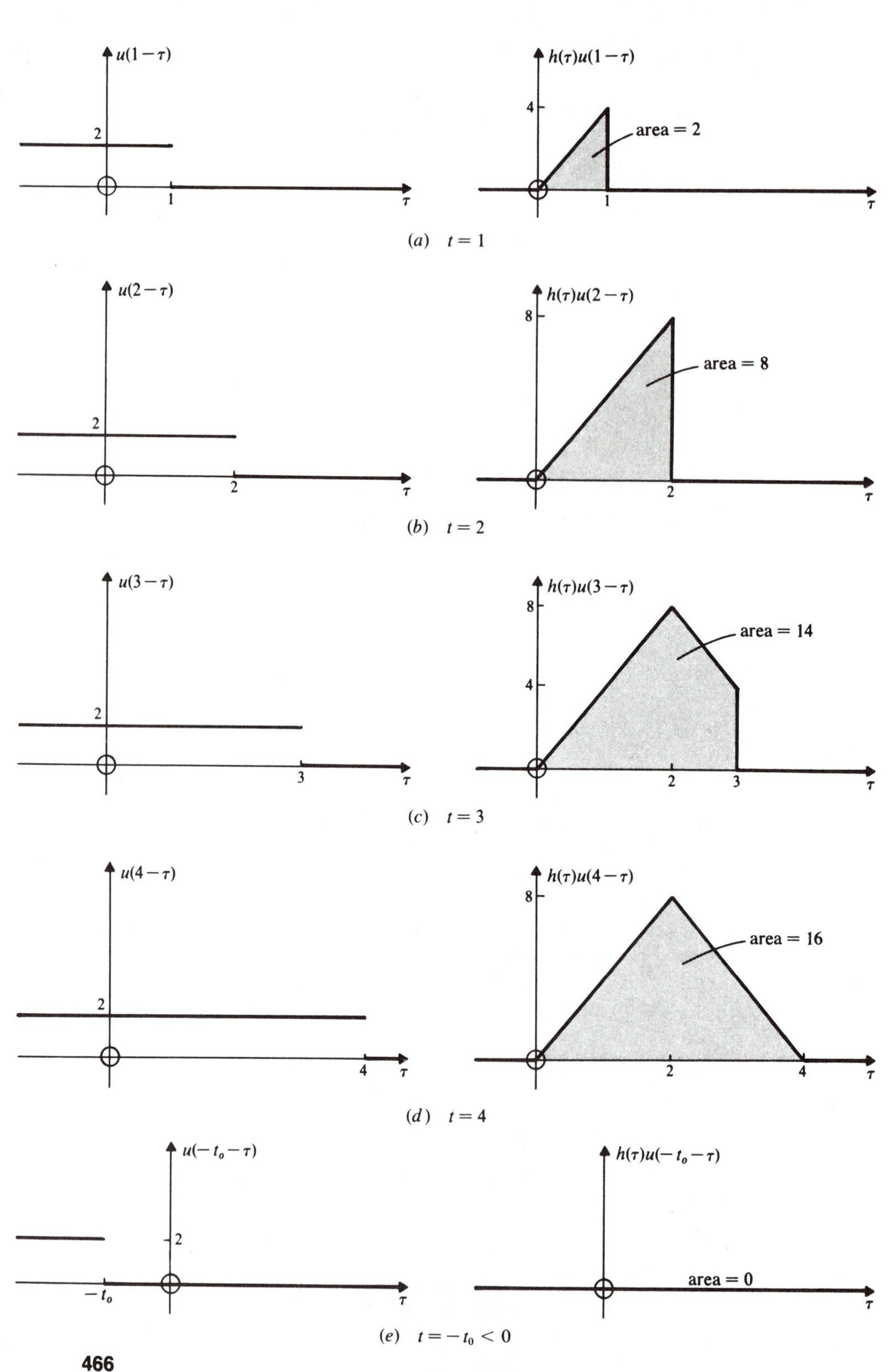

$u(1-\tau)$
2
1
$\tau$
$h(\tau)u(1-\tau)$
4
area = 2
1
$\tau$
(a) $t = 1$
$u(2-\tau)$
2
2
$\tau$
$h(\tau)u(2-\tau)$
8
area = 8
2
$\tau$
(b) $t = 2$
$u(3-\tau)$
2
3
$\tau$
$h(\tau)u(3-\tau)$
8
4
area = 14
2
3
$\tau$
(c) $t = 3$
$u(4-\tau)$
2
4
$\tau$
$h(\tau)u(4-\tau)$
8
area = 16
2
4
$\tau$
(d) $t = 4$
$u(-t_o-\tau)$
2
$-t_o$
$\tau$
$h(\tau)u(-t_o-\tau)$
area = 0
$\tau$
(e) $t = -t_0 < 0$




**Fig. 13-10**

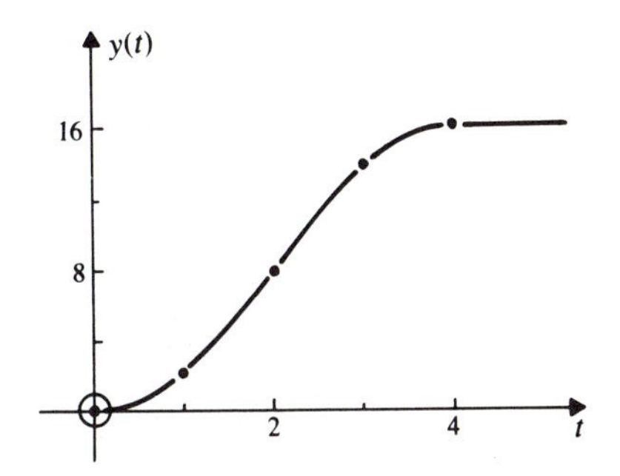


**Fig. 13-11**

2. A unit step function is the integral of a unit impulse function. Hence we can, in this instance, calculate $y(t)$ by integrating $h(t)$ and then multiplying by 2. The result checks the outcome of the convolution.

**13.3** A linear circuit whose impulse response is shown in Fig. 13-12*a* has an input $u(t)$ shown in Fig. 13-12*b*. Find the corresponding output $y(t)$.

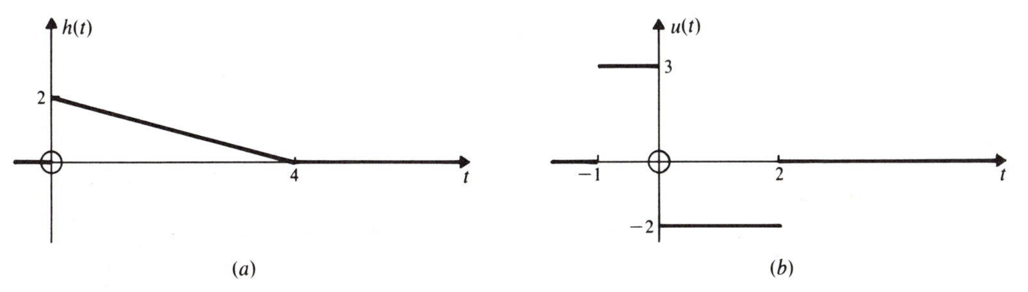


**Fig. 13-12**

Here we will find it simpler to fold and shift $h(t)$. Thus our objective will be the evaluation of

$$y(t) = \int_{-\infty}^{\infty} u(\tau)h(t - \tau)\, d\tau$$

where $u(\tau)$ and $h(-\tau)$ are as graphed in Fig. 13-13.

First, by mentally shifting $h(-\tau)$ one or more units to the left, we can see that the product $u(\tau)h(t - \tau)$ is zero for $t \leq -1$. That is,

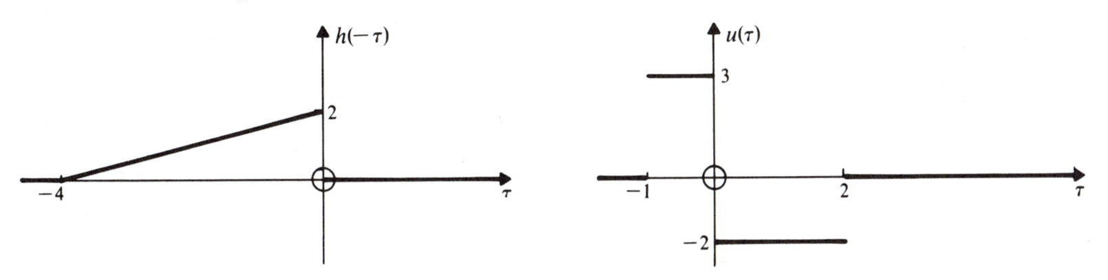


**Fig. 13-13**

$$y(t) = 0 \qquad \text{for } t \leq -1$$

Similarly, by mentally shifting $h(-\tau)$ to the right, we conclude that

$$y(t) = 0 \qquad \text{for } t \geq 6$$

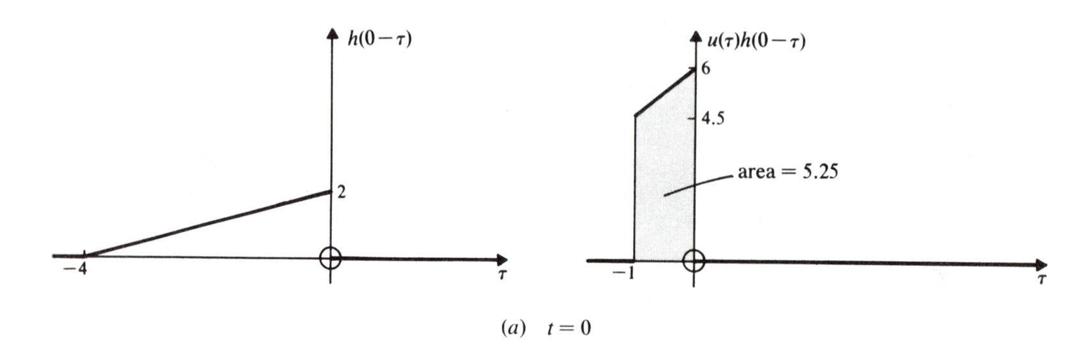


(a) t = 0

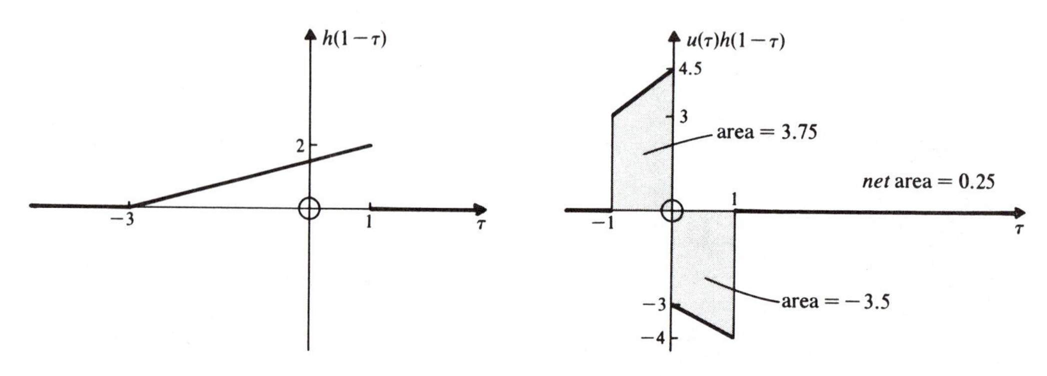


(b) t = 1

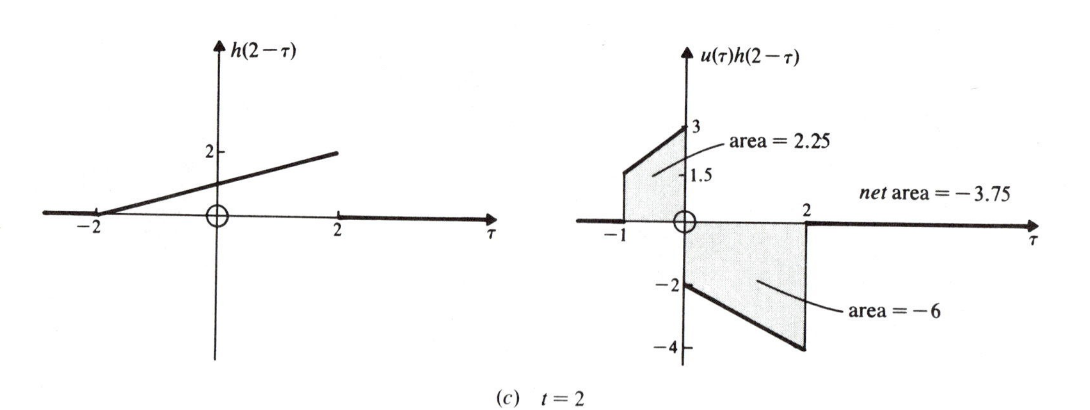


(c) t = 2

**Fig. 13-14**

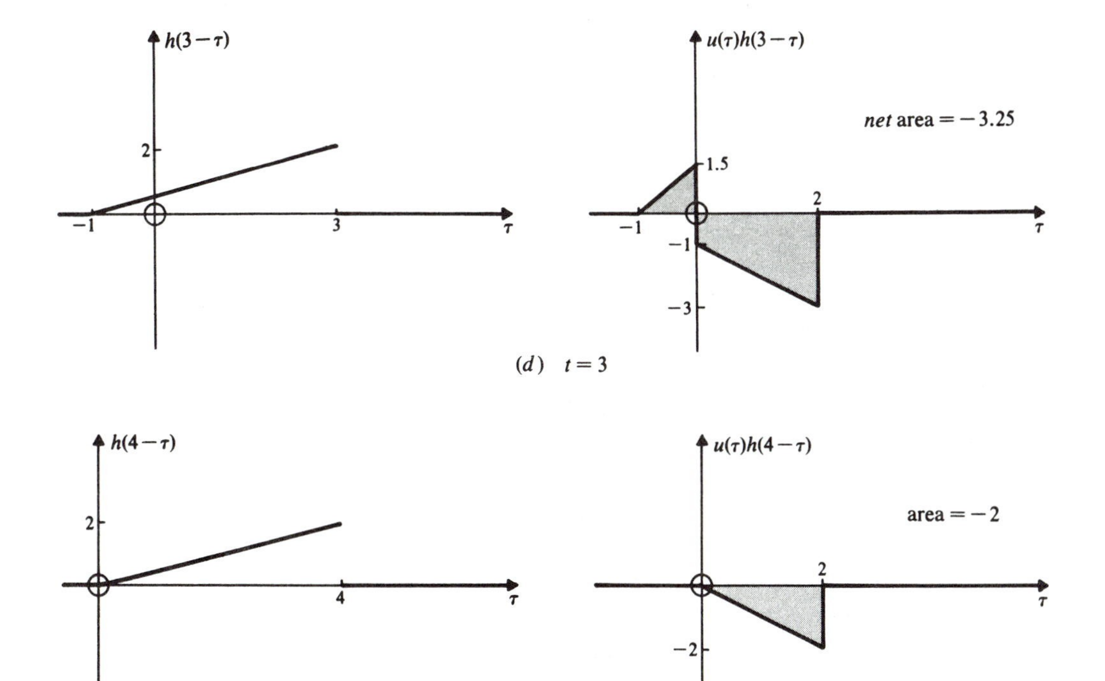


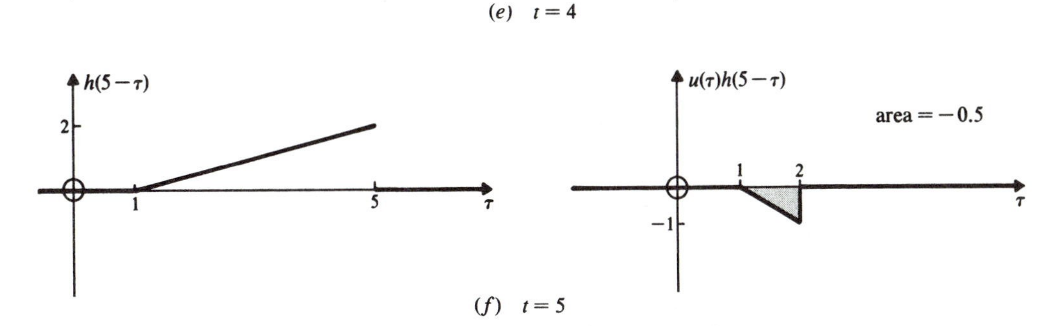


**Fig. 13-14 (*continued*)**

Finally, with the aid of Figs. 13-14*a-f* we can generate the following data:

| $t$ | 0 | 1 | 2 | 3 | 4 | 5 |
|---|---|---|---|---|---|---|
| $y(t)$ | 5.25 | 0.25 | −3.75 | −3.25 | −2 | −0.5 |

The data from the above table are plotted in Fig. 13-15. Calculations would have to be carried out for some intermediate values of $t$ if a reasonably accurate graph of $y(t)$ were wanted—particularly in the range $-1 < t < 2$.

**13.4** Use graphical convolution to obtain a *rough* sketch of the response $v(t)$ of the circuit in Fig. 13-16*a* to the voltage pulse $e(t)$ of Fig. 13-16*b*.

We can easily show (for example, from the voltage divider concept) that

$$H(s) = \frac{V(s)}{E(s)} = \frac{1}{1+s}$$

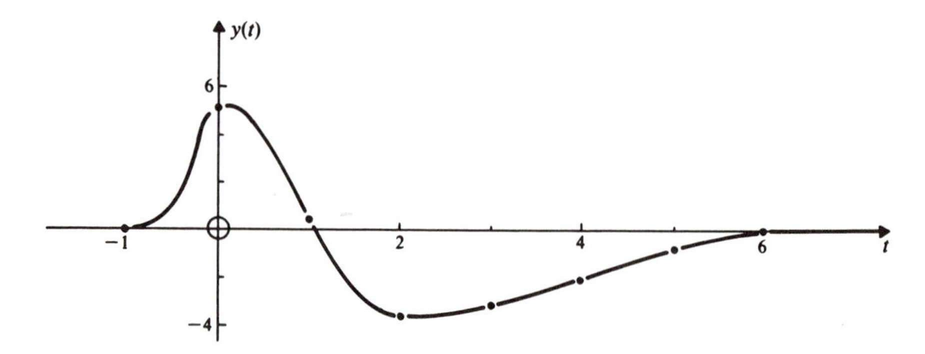


**Fig. 13-15**

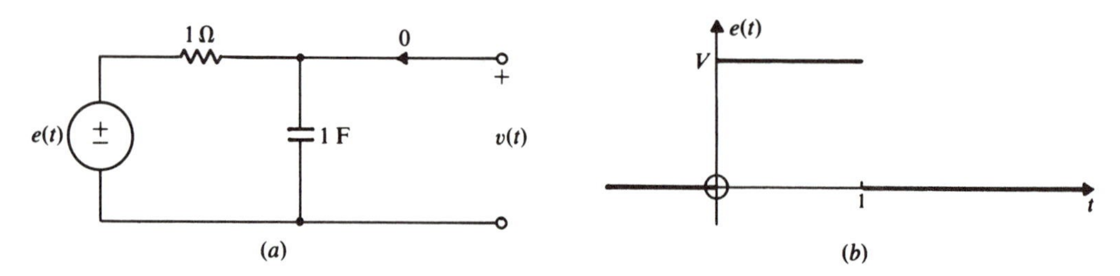


**Fig. 13-16**

Therefore, taking the inverse Laplace transform we find the unit impulse response to be

$$h(t) = \mathscr{L}^{-1}[H(s)] = e^{-t}, \qquad t \geq 0$$

Naturally the impulse response $h(t)$ is zero for $t < 0$, that is, before the impulse has been applied. The function $h(t)$ has been sketched in Fig. 13-17*a*.

Here we will fold and shift $e(\tau)$ and hence evaluate

$$v(t) = \int_{-\infty}^{\infty} h(\tau)e(t - \tau)\, d\tau$$

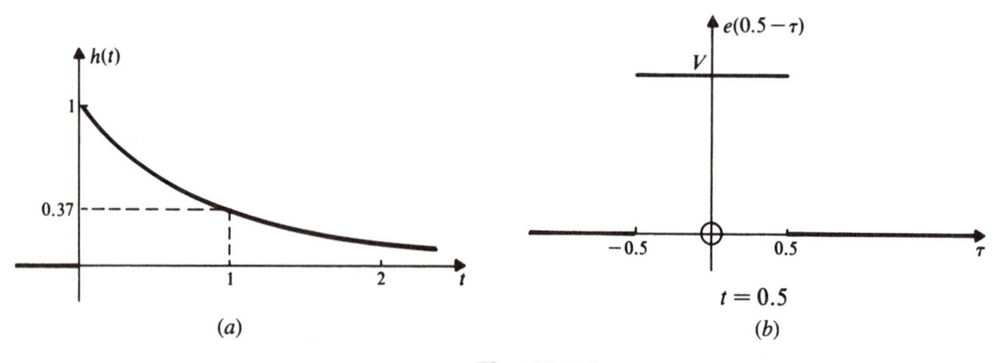


**Fig. 13-17**

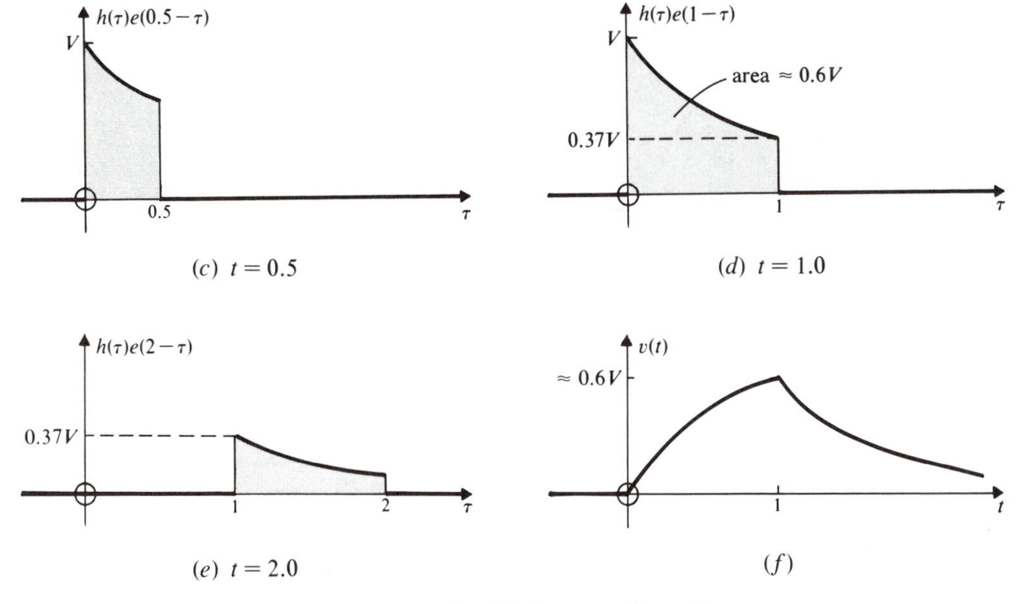


(c) $t = 0.5$

(d) $t = 1.0$

(e) $t = 2.0$

(f)

**Fig. 13-17 (*continued*)**

First, it is clear that

$$v(t) = 0 \quad \text{for} \quad t \leq 0 \qquad \text{and} \qquad v(t) \to 0 \quad \text{for } t \to \infty$$

Next, as the curve of $e(-\tau)$ is shifted to the right, the area under the graph of $h(\tau)e(t - \tau)$ will *increase* until $t = 1$ and will then *decrease* toward its asymptotic value of zero (see Figs. 13-17*b-e*). A little further thought will show that the curve is initially convex upward, and then, beyond $t = 1$, concave upward. It follows that $v(t)$ must be of the form indicated in Fig. 13-17*f*.

## *Numerical Convolution*

In the next five problems the concept of numerical convolution will be developed in a form appropriate for solution on a computer or calculator.

**13.5** We will use the circuit of Problem 13.2 to illustrate this numerical procedure. For a start, prepare a number sequence that will adequately represent the unit impulse response, $h(t)$, and the input, $u(t)$, of Figs. 13-8*b* and 13-8*c*.

The accuracy with which we wish to evaluate the convolution integral will determine how often we must *sample* the continuous functions. It is convenient to sample at equal intervals of time, $\Delta t$. Here, for a rather crude evaluation, we will choose $\Delta t = 0.5$. Were we to use a digital computer, we would sample much more finely, since the additional calculations would be performed painlessly.

We must also decide upon a convention at points of discontinuity. We will *define* the value of a function $p(t)$ at $t = t_0$ as

$$p(t_0) = \frac{p(t_0^-) + p(t_0^+)}{2}$$

With these preliminaries behind us we can write the sequences,

$$\begin{array}{l}
h(t) \sim [\cdots \;\; 0 \;\; 0 \;\; 0 \;\; 1 \;\; 2 \;\; 3 \;\; 4 \;\; 3 \;\; 2 \;\; 1 \;\; 0 \;\; 0 \;\; \cdots] \\
\qquad\qquad \updownarrow t=-1 \;\; \updownarrow t=0 \;\; \updownarrow t=1 \;\; \updownarrow t=2 \;\; \updownarrow t=3 \;\; \updownarrow t=4 \\
u(t) \sim [\cdots \;\; 0 \;\; 0 \;\; 1 \;\; 2 \;\; 2 \;\; 2 \;\; 2 \;\; 2 \;\; 2 \;\; 2 \;\; 2 \;\; 2 \;\; \cdots]
\end{array}$$

**13.6** Write the numerical sequence for $u(\tau)$ and *fold* it to obtain the sequence for $u(0 - \tau)$.

The sequence for $u(\tau)$ is the same as for $u(t)$, while $u(0 - \tau)$ is $u(\tau)$ folded about the point $\tau = 0$. Thus

$$\begin{array}{rl}
u(\tau) \sim & [\cdots \;\; 0 \;\; 0 \;\; 0 \;\; 1 \;\; 2 \;\; 2 \;\; 2 \;\; \cdots] \\
& \qquad\qquad\quad \updownarrow \tau=0 \\
u(0-\tau) \sim & [\cdots \;\; 2 \;\; 2 \;\; 2 \;\; 1 \;\; 0 \;\; 0 \;\; 0 \;\; \cdots]
\end{array}$$

**13.7** To find $y(t)$ for $t = 0$, namely $y(0) = \int_{-\infty}^{\infty} h(\tau)u(0 - \tau)\, d\tau$, we must obtain the sequence $h(\tau)u(0 - \tau)$ and then integrate numerically. From the data below it is clear that $h(\tau)u(0 - \tau) \equiv 0$, and so $y(0) = 0$.

$$\begin{array}{rl}
h(\tau) \sim & [\cdots \;\; 0 \;\; 0 \;\; 0 \;\; 1 \;\; 2 \;\; 3 \;\; 4 \;\; 3 \;\; 2 \;\; 1 \;\; 0 \;\; \cdots] \\
& \qquad\qquad \updownarrow \tau=0 \\
u(0-\tau) \sim & [\cdots \;\; 2 \;\; 2 \;\; 1 \;\; 0 \;\; 0 \;\; 0 \;\; 0 \;\; 0 \;\; 0 \;\; 0 \;\; 0 \;\; \cdots]
\end{array}$$

Now write the sequences for $h(\tau)$, $u(0.5 - \tau)$, and the product $h(\tau)u(0.5 - \tau)$. Then integrate numerically to find $y(0.5)$.

Since $\Delta t = 0.5$, the sequence for $u(0.5 - \tau)$ is obtained by *shifting* the $u(-\tau)$ sequence one column to the right. Thus

$$\begin{array}{rl}
h(\tau) \sim & [\cdots \;\; 0 \;\; 0 \;\; 0 \;\; 1 \;\; 2 \;\; 3 \;\; 4 \;\; 3 \;\; 2 \;\; \cdots] \\
& \qquad\qquad \updownarrow \tau=0 \\
u(0.5-\tau) \sim & [\cdots \;\; 2 \;\; 2 \;\; 2 \;\; 1 \;\; 0 \;\; 0 \;\; 0 \;\; 0 \;\; 0 \;\; \cdots] \\
& \qquad\qquad \updownarrow \tau=0 \\
h(\tau)u(0.5-\tau) \sim & [\cdots \;\; 0 \;\; 0 \;\; 0 \;\; 1 \;\; 0 \;\; 0 \;\; 0 \;\; 0 \;\; 0 \;\; \cdots]
\end{array}$$

and

The product sequence is plotted in Fig. 13-18*a*.

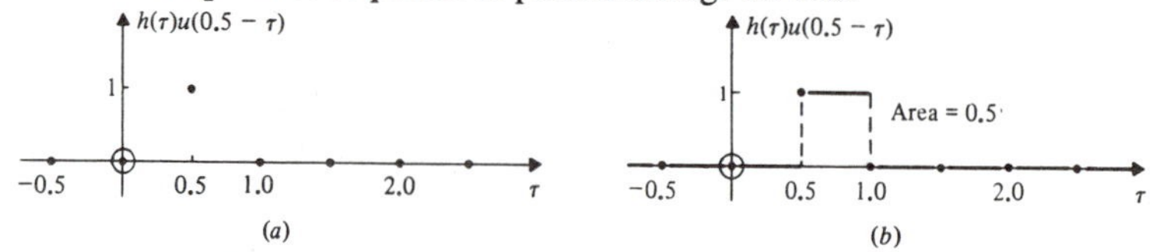


**Fig. 13-18**

The simplest method of numerical integration *assumes* that the function to be integrated is *piecewise constant* (see Fig. 13-18*b*). The integral is then easily evaluated! Here

$$y(0.5) = \int_{-\infty}^{\infty} h(\tau)u(0.5 - \tau)\, d\tau \doteq 1(0.5) = 0.5$$

**13.8** Find $y(1)$ and $y(1.5)$ using the method of Problem 13.7.

The sequence for $u(1-\tau)$ is obtained by shifting that of $u(0.5-\tau)$ one column to the right. Thus

$$h(\tau) \sim [\cdots \quad 0 \quad 0 \quad \underset{\updownarrow\,\tau=0}{0} \quad 1 \quad 2 \quad 3 \quad 4 \quad 3 \quad 2 \quad \cdots]$$

$$u(0.5-\tau) \sim [\cdots \quad 2 \quad 2 \quad \underset{\updownarrow\,\tau=0}{2} \quad 1 \quad 0 \quad 0 \quad 0 \quad 0 \quad 0 \quad \cdots]$$

and

$$h(\tau)u(0.5-\tau) \sim [\cdots \quad 0 \quad 0 \quad 0 \quad 1 \quad 0 \quad 0 \quad 0 \quad 0 \quad 0 \quad \cdots]$$

This product is plotted in Fig. 13-19*a*. Integrating numerically, with the piecewise-constant approximation, we find

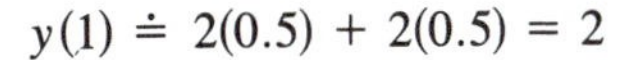

$$y(1) \doteq 2(0.5) + 2(0.5) = 2$$

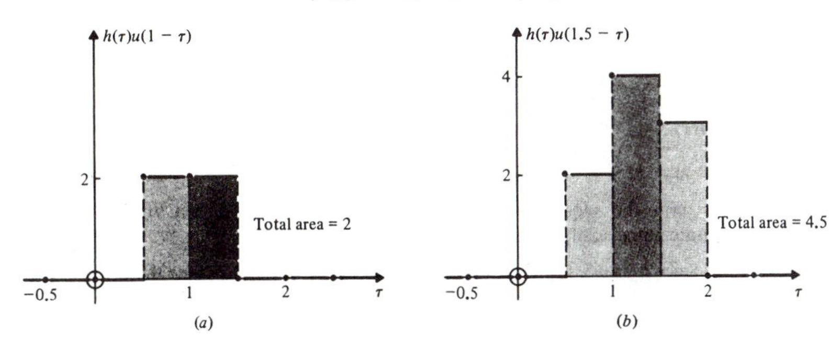


**Fig. 13-19**

Similarly, to find $y(1.5)$,

$$h(\tau) \sim [\cdots \quad 0 \quad 0 \quad \underset{\updownarrow\,\tau=0}{0} \quad 1 \quad 2 \quad 3 \quad 4 \quad 3 \quad 2 \quad \cdots]$$

$$u(1.5-\tau) \sim [\cdots \quad 2 \quad 2 \quad \underset{\updownarrow\,\tau=0}{2} \quad 2 \quad 2 \quad 1 \quad 0 \quad 0 \quad 0 \quad \cdots]$$

and

$$h(\tau)u(1.5-\tau) \sim [\cdots \quad 0 \quad 0 \quad 0 \quad 2 \quad 4 \quad 3 \quad 0 \quad 0 \quad 0 \quad \cdots]$$

The piecewise-constant approximation of $h(\tau)u(1.5-\tau)$ is sketched in Fig. 13-19*b*. Clearly,

$$y(1.5) \doteq 3(0.5) + 4(0.5) + 2(0.5) = 4.5$$

If we continued in the same way, we would find

$$y(2) = (4+6+4+2)(0.5) = 8$$

$$y(2.5) = (3+8+6+4+2)(0.5) = 11.5$$

$$y(3) = (2+6+8+6+4+2)(0.5) = 14$$

$$y(3.5) = (1+4+6+8+6+4+2)(0.5) = 15.5$$

$$y(4) = (2+4+6+8+6+4+2)(0.5) = 16$$

$$y(4.5) = (2+4+6+8+6+4+2)(0.5) = 16$$

and $y(t)$ does not change thereafter.

The function $y(t)$ as *evaluated* is plotted in Fig. 13-20. But note that we have made two approximations: in defining the value of a function at a discontinuity; and by assuming a piecewise-constant function for integration.

If we compare our results here with the solution of Problem 13-2, we find that the ringed points in Fig. 13-20 are *exact*. But this is "luck." The method is approximate.

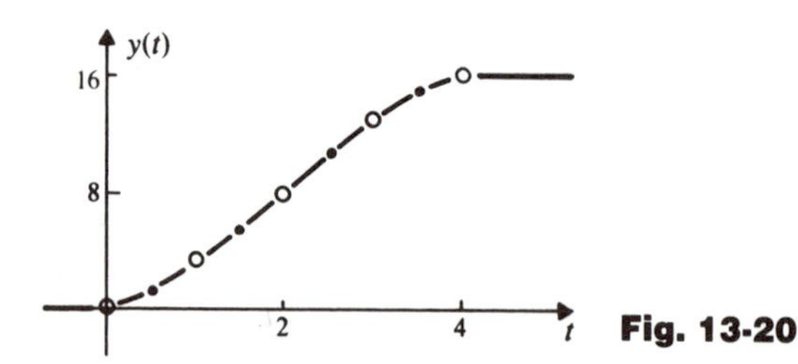

**Fig. 13-20**

**13.9** The numerical procedure of Problems 13.7 and 13.8 may be put into a general form suitable for digital computation. If the sequences that represent $h(t)$ and $u(t)$ at equal intervals of time $\Delta t$ are

$$h(t) \sim [h_{-\infty} \quad \cdots \quad h_{-2} \quad h_{-1} \quad \underset{\updownarrow t=0}{h_0} \quad h_1 \quad h_2 \quad \cdots \quad h_\infty]$$

$$u(t) \sim [u_{-\infty} \quad \cdots \quad u_{-2} \quad u_{-1} \quad u_0 \quad u_1 \quad u_2 \quad \cdots \quad u_\infty]$$

write the convolution of $h(t)$ and $u(t)$ at $t = 0$, $t = \Delta t$, $t = 2\Delta t$, and $t = -\Delta t$. Then write a general expression for $y_k = y(k\Delta t)$.

The $h(\tau)$ and the folded $u(\tau)$ sequences are

$$h(\tau) \sim [h_{-\infty} \quad \cdots \quad h_{-2} \quad h_{-1} \quad h_0 \quad h_1 \quad h_2 \quad \cdots \quad h_\infty]$$

$$u(-\tau) \sim [u_\infty \quad \cdots \quad u_2 \quad u_1 \quad u_0 \quad u_{-1} \quad u_{-2} \quad \cdots \quad u_{-\infty}]$$

Thus if we integrate the piecewise-constant function $h(\tau)u(-\tau)$, following the procedure of Problem 13.8,

$$y(0) = (h_{-\infty}u_\infty + \cdots + h_{-2}u_2 + h_{-1}u_1 + h_0u_0 + h_1u_{-1} + h_2u_{-2} + \cdots + h_\infty u_{-\infty})(\Delta t)$$

Similarly, after shifting the $u(-\tau)$ sequence one interval to the right,

$$y(\Delta t) = (h_{-\infty+1}u_\infty + \cdots + h_{-2}u_3 + h_{-1}u_2 + h_0u_1 + h_1u_0 + h_2u_{-1} + \cdots + h_\infty u_{-\infty+1})(\Delta t)$$

And

$$y(2\Delta t) = (h_{-\infty+2}u_\infty + \cdots + h_{-2}u_4 + h_{-1}u_3 + h_0u_2 + h_1u_1 + h_2u_0 + h_3u_{-1} + \cdots + h_\infty u_{-\infty+2})(\Delta t)$$

Finally, if we shift the $u(-\tau)$ sequence one interval to the *left*,

$$y(-\Delta t) = (h_{-\infty-1}u_\infty + \cdots + h_{-2}u_1 + h_{-1}u_0 + h_0u_{-1} + h_1u_{-2} + h_2u_{-3} + \cdots + h_\infty u_{-\infty-1})(\Delta t)$$

The foregoing results can be written as

$$y(-\Delta t) = y_{-1} = \Delta t \sum_{n=-\infty}^{\infty} h_n u_{-n-1}$$

$$y(0) = y_0 = \Delta t \sum_{n=-\infty}^{\infty} h_n u_{-n}$$

$$y(\Delta t) = y_1 = \Delta t \sum_{n=-\infty}^{\infty} h_n u_{-n+1}$$

$$y(2\Delta t) = y_2 = \Delta t \sum_{n=-\infty}^{\infty} h_n u_{-n+2}$$

or, in general,

$$y(k\Delta t) = y_k = \Delta t \sum_{n=-\infty}^{\infty} h_n u_{-n+k} \qquad \text{for all integer } k$$

If the sequences for $h(t)$ and $u(t)$ are entered into a computer as arrays, then repeated indexing (i.e. "shifting") of the subscripts, multiplication, and addition will lead successively to the elements in the sequence for $y(t)$.

**13.10** Given $h(t) = e^{-t}\mathbb{1}(t)$ for the circuit of Fig. 13-16*a* (see page 470), and the input $e(t) = t\mathbb{1}(t)$, use the result of Problem 13.9 to evaluate the output $v(t)$, when $t = 1$ s.

To obtain a reasonably accurate result, $\Delta t$ should be small compared with the circuit's "response time": of the order of 0.02 s, perhaps. On a computer this would present no difficulty (roughly 50 multiplications and additions), and here . . . ! At the cost of some inaccuracy, then, we will choose $t = 0.2$ s. Then from Problem 13.9,

$$v_k = \Delta t \sum_{n=-\infty}^{\infty} h_n e_{-n+k}$$

where $k = 5$, corresponding to $t = 1$ s, $\Delta t = 0.2$ s, and $t = k\Delta t$.

If we mentally graph $h(\tau)$ and $e(1 - \tau)$, it is clear that $h(\tau)e(1 - \tau) = 0$ for all $\tau$ except $0 < t \leq 1$. Thus we need only consider $h_n$ in the range $0 \leq n \leq 5$ and $e_{-n+5}$ in the range $0 \leq (-n + 5) \leq 5$. That is, $n$ need only take on the values 0 through 5. Proceeding as systematically as is possible without a computer, we generate the following tabulation:

| | $n = 0$ | $n = 1$ | $n = 2$ | $n = 3$ | $n = 4$ | $n = 5$ |
|---|---|---|---|---|---|---|
| 1. $h_n = e^{-0.2nt}$ | $\dfrac{0+1}{2} = 0.5$ | 0.819 | 0.670 | 0.549 | 0.449 | 0.368 |
| 2. $e_n = 0.2n$ | 0 | 0.2 | 0.4 | 0.6 | 0.8 | 1 |
| 3. $-n+5$ | 5 | 4 | 3 | 2 | 1 | 0 |
| 4. $e_{-n+5}$ | 1 | 0.8 | 0.6 | 0.4 | 0.2 | 0 |
| 5. $h_n e_{-n+5}$ | 0.5 | 0.655 | 0.402 | 0.220 | 0.090 | 0 |

$$\sum_{n=0}^{5} h_n e_{-n+5} = 1.867 \quad \text{and} \quad v_5 = \Delta t \sum_{n=0}^{5} h_n e_{-n+5} = (0.2)(1.867) = 0.373$$

The last figure, $v_5 = 0.373$, is the required approximate value of $v(1)$. The *exact* value, obtained analytically, is $v(1) = 0.368$.

If we were working with a computer, the first two lines in the table would correspond to data entry. The sequences for $h(t)$ and $e(t)$ could be quite arbitrary, based on laboratory measurements, perhaps, and need not be analytic as they were in this problem. The *average* value should be used at every discontinuity, as in the entry for $h_0$ in the above table.

Lines 3 and 4 correspond to an indexing operation, which is followed by the multiplications of line 5 and the running summation leading to the answer.

**13.11** Given a linear, time-invariant, causal circuit that is subjected to an input $u(t)$ which is zero for $t < 0$, show that the output is given by

$$y(t) = \int_0^t u(\tau)h(t - \tau)\, d\tau, \qquad t \geq 0$$

(*Note:* In a *causal* circuit, no response due to an input can occur before the input is applied. Causality is guaranteed for any real physical system.)

As shown in Fig. 13-21, $h(t) = 0$ for $t < 0$ if the circuit is causal. And we are given $u(t) = 0$ for $t < 0$. We can therefore deduce from Fig. 13-21 that $u(\tau)h(t - \tau) = 0$ everywhere except for $0 \leq \tau < t$. Therefore

$$y(t) = \int_{-\infty}^{\infty} u(\tau)h(t - \tau)\, d\tau = \int_0^t u(\tau)h(t - \tau)\, d\tau, \qquad t \geq 0$$

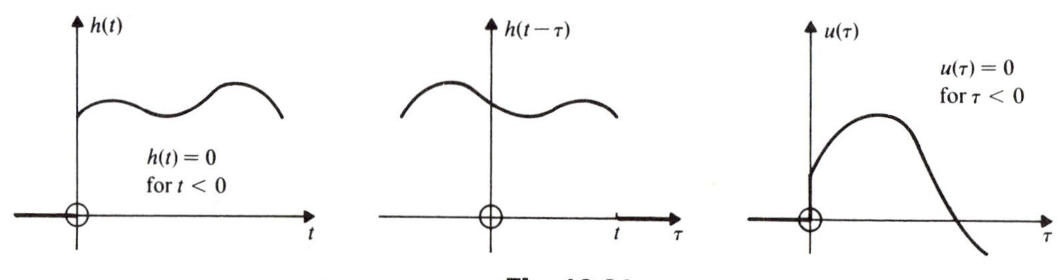


**Fig. 13-21**

A similar argument shows that the convolution operation is still commutative when expressed in this way. That is, it is also true that

$$y(t) = \int_0^t h(\tau)u(t - \tau)\, d\tau, \qquad t \geq 0$$

Note that under the stated conditions, $y(t) = 0$ for $t < 0$.

### *Analytical Convolution*

**13.12** The impulse response for a linear circuit is $h(t) = (e^{-t} - e^{-2t})\, \mathbb{1}(t)$. Use analytical convolution to find the output $y(t)$ due to the input $u(t) = 2e^{-3t}\, \mathbb{1}(t)$.

Using the result of Problem 13.11,

$$y(t) = \int_0^t u(\tau)h(t - \tau)\, d\tau, \qquad t \geq 0$$

$$= \int_0^t \{2e^{-3\tau}\}\{e^{-(t-\tau)} - e^{-2(t-\tau)}\}\, d\tau$$

$$= \int_0^t \{2e^{-t}e^{-2\tau} - 2e^{-2t}e^{-\tau}\}\, d\tau = 2e^{-t}\int_0^t e^{-2\tau}\, d\tau - 2e^{-2t}\int_0^t e^{-\tau}\, d\tau$$

$$= 2e^{-t}\left[\frac{-e^{-2\tau}}{2}\right]_0^t - 2e^{-2t}[-e^{-\tau}]_0^t = e^{-t} - 2e^{-2t} + e^{-3t}$$

And from Problem 13.11 we know that $y(t) = 0$ for $t < 0$. Hence the solution is

$$y(t) = (e^{-t} - 2e^{-2t} + e^{-3t})\mathbb{1}(t)$$

**13.13** Use the fact that convolution in the time-domain is equivalent to multiplication in the $s$-domain to solve Problem 13.12.

From equation (*13.5*) and (*13.6*), page 461, we can write

$$y(t) = h(t) * u(t) = \mathcal{L}^{-1}[H(s)U(s)], \qquad t \geq 0$$

where

$$H(s) = \mathcal{L}[h(t)] = \mathcal{L}[e^{-t} - e^{-2t}] = \frac{1}{(s+1)(s+2)}$$

and

$$U(s) = \mathcal{L}[u(t)] = \mathcal{L}[2e^{-3t}] = \frac{2}{s+3}$$

Thus

$$Y(s) = \frac{2}{(s+1)(s+2)(s+3)} = \frac{1}{s+1} - \frac{2}{s+2} + \frac{1}{s+3}$$

and

$$y(t) = \mathcal{L}^{-1}[Y(s)] = e^{-t} - 2e^{-2t} + e^{-3t}, \qquad t \geq 0$$

This is of course exactly what we have done in $s$-domain computations from Chapter 5 onwards. Convolution merely offers us an opportuinty to remain firmly in the time-domain.

**13.14** The unit impulse response of current in the circuit of Fig. 13-22 is $h(t) = \delta(t) - e^{-t}\mathbb{1}(t)$. Find $i(t)$ when

$$v(t) = (1 - e^{-2t})\mathbb{1}(t)$$

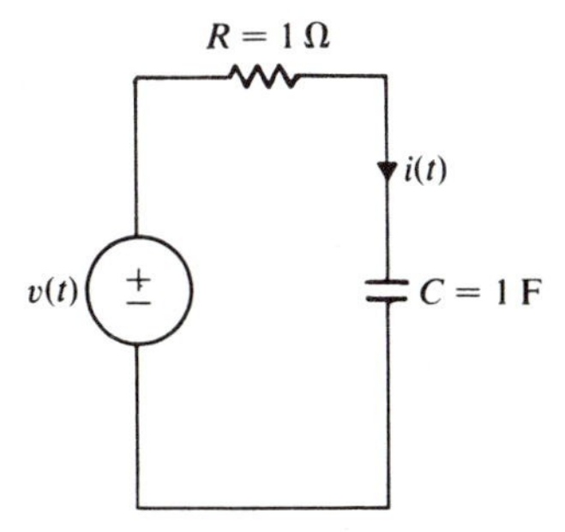


**Fig. 13-22**

*Hint:* Make use of equation (*13.1*), namely,

$$u(t) = \int_{-\infty}^{\infty} u(\tau)\delta(t-\tau)\,d\tau = u(t) * \delta(t)$$

which, if $u(t) = 0$ for $t < 0$, can be written

$$u(t) = \int_0^t u(\tau)\delta(t-\tau)\,d\tau, \qquad t \geq 0$$

Using the result of Problem 13.11,

$$i(t) = \int_0^t v(\tau)h(t-\tau)\,d\tau, \qquad t \geq 0$$

$$= \int_0^t \{1 - e^{-2\tau}\}\{\delta(t-\tau) - e^{-(t-\tau)}\}\,d\tau$$

$$= \int_0^t 1\delta(t-\tau)\,d\tau - \int_0^t e^{-(t-\tau)}\,d\tau$$

$$- \int_0^t e^{-2\tau}\delta(t-\tau)d\tau + \int_0^t e^{-2\tau}e^{-(t-\tau)}\,d\tau$$

from which, making use of the given mathematical identity,

$$i(t) = 1 - e^{-t}\Big[e^{\tau}\Big]_0^t - e^{-2t} + e^{-t}\Big[-e^{-\tau}\Big]_0^t = 2e^{-t} - 2e^{-2t}, \qquad t \geq 0$$

Further, we know from Problem 13.11 that $i(t) = 0$ for $t < 0$. Therefore

$$i(t) = 2(e^{-t} - e^{-2t})\mathbb{1}(t)$$

*Alternative Solution*

Since $i(t) = h(t) * v(t)$, it follows that

$$I(s) = H(s)V(s)$$

where

$$H(s) = \mathcal{L}[h(t)] = 1 - \frac{1}{s+1} = \frac{s}{s+1}$$

and

$$V(s) = \mathcal{L}[v(t)] = \frac{1}{s} - \frac{1}{s+2} = \frac{2}{s(s+2)}$$

Therefore

$$I(s) = H(s)V(s) = \frac{2}{(s+1)(s+2)} = \frac{2}{s+1} - \frac{2}{s+2}$$

and so

$$i(t) = 2e^{-t} - 2e^{-2t}, \qquad t \geq 0$$

**13.15** Rework Problem 13.4 by analytical convolution. That is, given that $h(t) = e^{-t}\mathbb{1}(t)$ and that the input $e(t)$ is the pulse sketched in Fig. 13-23, find the output $v(t) = h(t) * e(t)$.

First, $e(t)$ may be synthesized as the sum of $e_1(t)$ and $e_2(t)$ shown in Fig. 13-24. Then the solution $v(t)$ will be the sum of

$v_1(t)$, due to the step function at $t = 0$

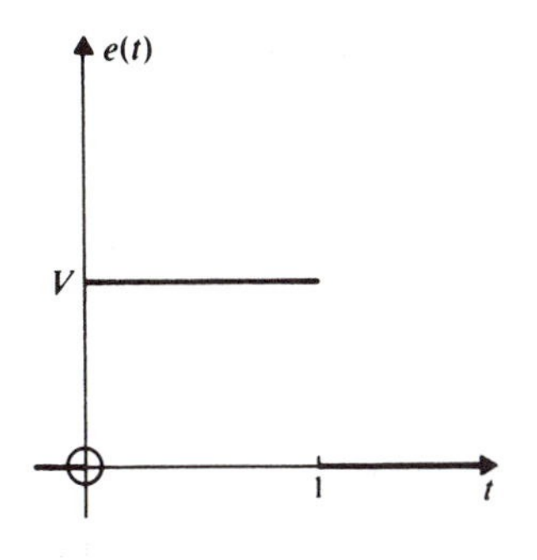


Fig. 13-23

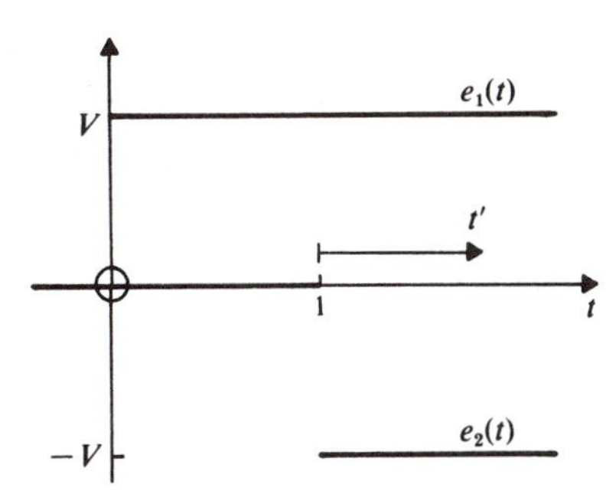


Fig. 13-24

and

$$v_2(t), \text{ due to the negative step function at } t = 1$$

In the usual way,

$$v_1(t) = \int_0^t h(\tau)e(t - \tau)\, d\tau = V\int_0^t e^{-\tau}1\, d\tau = -Ve^{-\tau}\Big|_0^t = -Ve^{-t} + V, \qquad t \geq 0$$

and

$$v_1(t) = 0 \qquad \text{for } t < 0$$

The negative step function and the corresponding response are best described relative to a new independent variable $t'$ defined in Fig. 13-24. Then $e_2(t') = -\mathbb{1}(t')$ and

$$v_2(t') = \int_0^{t'} h(\tau)e(t' - \tau)\, d\tau = V\int_0^{t'} e^{-\tau}(-1)\, d\tau = Ve^{-\tau}\Big|_0^{t'} = Ve^{-t'} - V, \qquad t' \geq 0$$

But $t' = t - 1$, and therefore $v_2(t')$ may be written

$$v_2(t) = \begin{cases} 0 & \text{for } t < 1 \\ Ve^{-(t-1)} - V & \text{for } t \geq 1 \end{cases}$$

Summarizing,

$$v(t) = v_1(t) + v_2(t) = \begin{cases} 0 & \text{for } t < 0 \\ V\{1 - e^{-t}\} & \text{for } 0 \leq t \leq 1 \\ V\{e^{-(t-1)} - e^{-t}\} & \text{for } t \geq 1 \end{cases}$$

This function is sketched (roughly) in Fig. 13-17*f*, page 471.

**13.16** If $f_1(t) = A\delta(t - a)$ and $f_2(t) = B\delta(t - b)$, show that

$$f_3(t) = f_1(t) * f_2(t) = AB\delta\{t - (a + b)\}$$

This is most easily proved by moving into a transform-domain, where convolution becomes multiplication. Thus taking the Fourier transform of $f_1(t)$ and $f_2(t)$ with the help of entry 5 in Table 12-1, page 426, and entry 2 in Table 12-2, page 427,

$$\mathbf{F}_1(\omega) = Ae^{-j\omega a} \quad \text{and} \quad \mathbf{F}_2(\omega) = Be^{-j\omega b}$$

Thus

$$\mathbf{F}_3(\omega) = \mathcal{F}[f_1(t) * f_2(t)] = \mathbf{F}_1(\omega)\mathbf{F}_2(\omega) = ABe^{-j\omega(a+b)}$$

Returning to the time-domain with the help of the same entries in the Fourier transform tables, we find that

$$f_3(t) = AB\delta\{t - (a + b)\}$$

In words, a $\delta$-function of amplitude $A$ at $t = a$, convolved with one of amplitude $B$ at $t = b$, results in a $\delta$-function of amplitude $AB$ at $t = a + b$.

### Comments

1. This problem can be approached by graphical convolution, but this involves the integration of a $\delta$-function whose *amplitude* is also a $\delta$-function. However, if we perform this integration, we arrive at the same result.
2. Similarly if $\mathbf{A}(\omega) = A\delta(\omega - a)$ and $\mathbf{B}(\omega) = B\delta(\omega - b)$, then

$$\mathbf{A}(\omega) * \mathbf{B}(\omega) = AB\delta\{\omega - (a + b)\}$$

In words, a $\delta$-function of amplitude $A$ at $\omega = a$, convolved with one of amplitude $B$ at $\omega = b$, results in a $\delta$-function of amplitude $AB$ at $\omega = a + b$.

**13.17** Given the electronic multiplier of Fig. 13-25$a$ and the frequency spectra $\mathbf{A}(\omega)$ vs. $\omega$ and $\mathbf{B}(\omega)$ vs. $\omega$ of Fig. 13-25$b$, find $\mathbf{C}(\omega)$, and from $\mathbf{C}(\omega)$, deduce $c(t)$. Use the result given in Comment 2 following Problem 13.16.

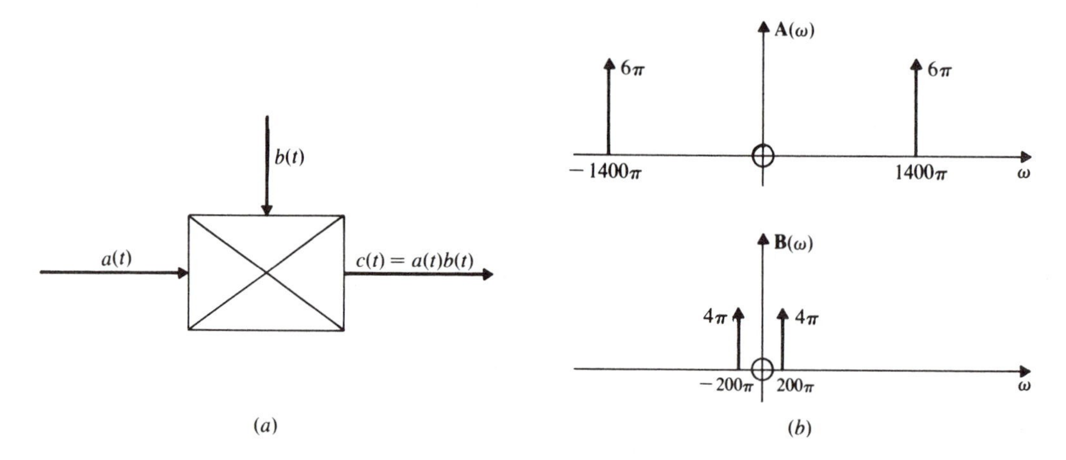


**Fig. 13-25**

Multiplication in time corresponds to convolution in frequency. That is, from equations (*13.7*) and (*13.8*), page 461,

$$\mathbf{C}(\omega) = \mathcal{F}[a(t)b(t)] = \frac{1}{2\pi}\mathbf{A}(\omega) * \mathbf{B}(\omega)$$

Now $\mathbf{A}(\omega)$ and $\mathbf{B}(\omega)$ consist of impulses, and in Problem 13.16 we saw how we convolve two impulses in the frequency-domain.

If therefore we convolve the impulses in Fig. 13-25$b$ one pair at a time, we see, for example, that the impulse at $-1400\pi$ convolved with the one at $-200\pi$, leads to an impulse of amplitude $24\pi^2$ at $-1600\pi$, as shown at the left of the graph in Fig. 13-26.

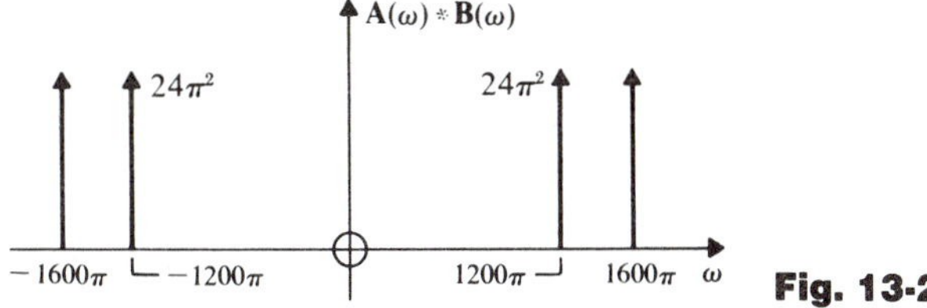


**Fig. 13-26**

Similarly, upon convolving the impulse at $-1400\pi$ with that at $200\pi$ we obtain an impulse of magnitude $24\pi^2$ at $-1200\pi$. Finally, the convolution can be completed by convolving the remaining impulse at $1400\pi$ with those at $\pm 200\pi$. This result is shown in Fig. 13-26.

Now we have

$$\mathbf{C}(\omega) = \mathcal{F}[a(t)b(t)] = \frac{1}{2\pi}\mathbf{A}(\omega) * \mathbf{B}(\omega)$$

$$= \frac{24\pi^2}{2\pi}\{\delta(\omega - 1600\pi) + \delta(\omega + 1600\pi) + \delta(\omega - 1200\pi) + \delta(\omega + 1200\pi)\}$$

and from entry 3 in Table 12-2, page 427,

$$c(t) = 12 \cos 1600\pi t + 12 \cos 1200\pi t$$

which agrees with the result of Problem 12.26.

**13.18** Given a signal

$$p(t) = K\{1 + m(t)\} \cos \omega_c t$$

where $m(t) = \frac{1}{2} \cos \omega_m t$, $\omega_c > \omega_m$, use convolution to obtain a plot of $\mathbf{P}(\omega)$ vs. $\omega$.

Combining the two equations,

$$p(t) = K \cos \omega_c t + \frac{K}{2} \cos \omega_m t \cos \omega_c t$$

$$= p_1(t) + p_2(t) = p_1(t) + p_a(t)p_b(t)$$

and hence

$$\mathbf{P}(\omega) = \mathbf{P}_1(\omega) + \mathbf{P}_2(\omega)$$

$$= \mathbf{P}_1(\omega) + \frac{1}{2\pi}\mathbf{P}_a(\omega) * \mathbf{P}_b(\omega)$$

where the spectra of $\mathbf{P}_1(\omega)$, $\mathbf{P}_a(\omega)$, and $\mathbf{P}_b(\omega)$ are as shown in Fig. 13-27.

We can now compute $\mathbf{P}_a(\omega) * \mathbf{P}_b(\omega)$ as in Problem 13.17 (see Fig. 13-28*a*). Then $\mathbf{P}_2(\omega)$ and finally $\mathbf{P}(\omega)$ follow as in Figs. 13-28*b* and *c*. The result agrees with that of Problem 12.21.

### *Comment*

When a *carrier* sinusoid with a frequency $\omega_c$ is *amplitude modulated* by a sinusoid of frequency $\omega_m$, we obtain a waveform which contains the original carrier frequency and

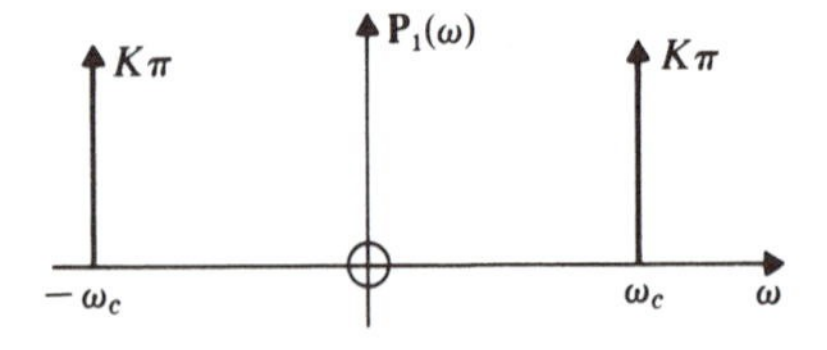


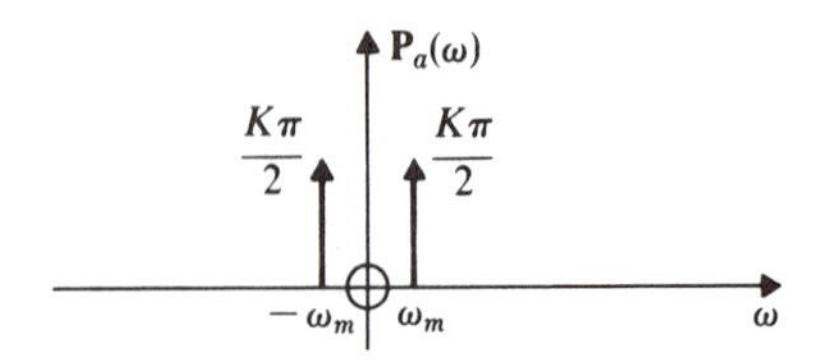


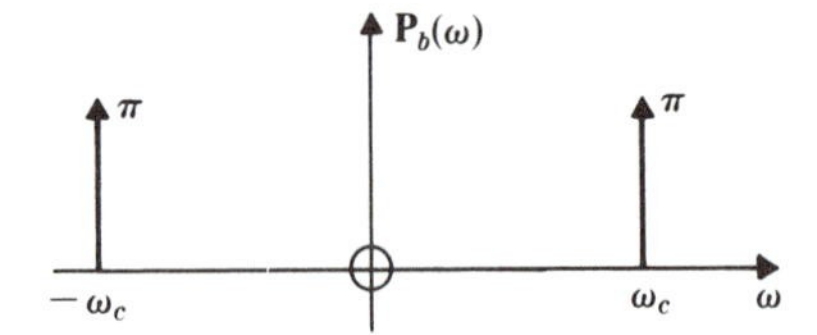


Fig. 13-27

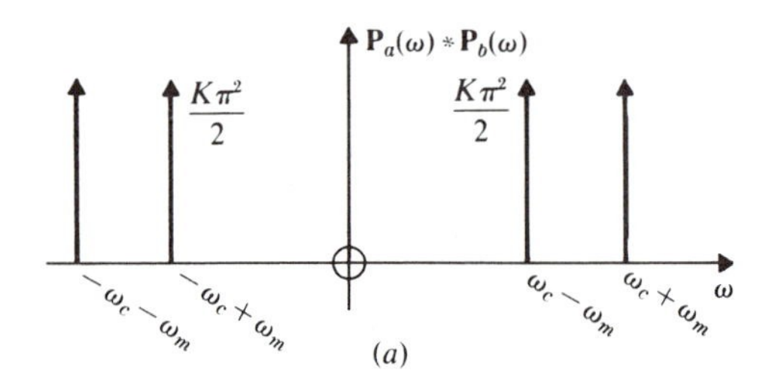


(*a*)

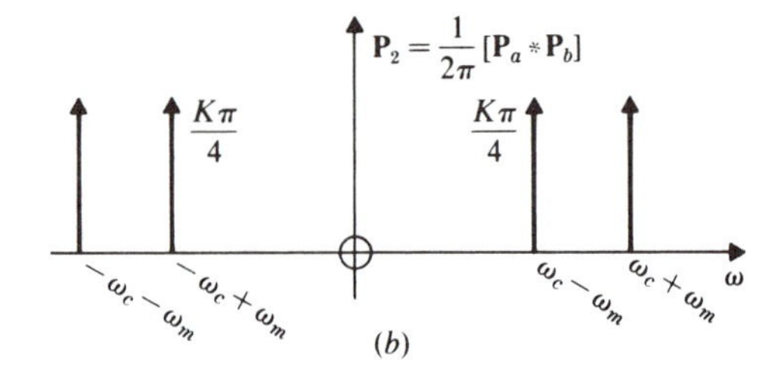


(*b*)

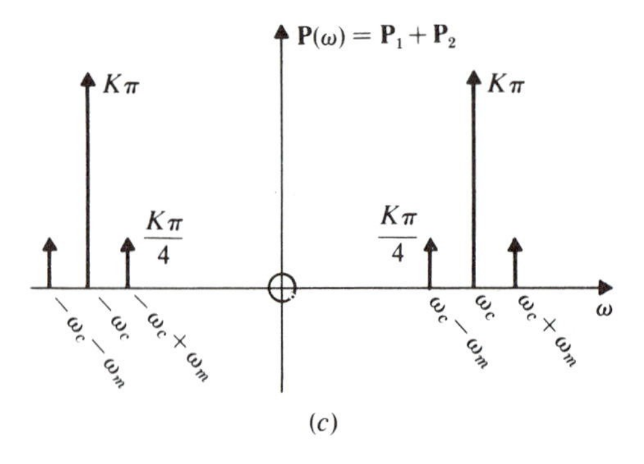


(*c*)

Fig. 13-28

the upper and lower *sideband* frequencies $\omega_c + \omega_m$ and $\omega_c - \omega_m$. A radio receiver tuned to $\omega_c$ must therefore have a sufficient bandwidth to receive the adjacent sidebands. Modern *single-sideband* (SSB) systems suppress one of the sidebands and all or part of the carrier.

# PROBLEMS

**13.19** Use graphical convolution to find $y(t)$ given the information in Fig. 13-29.

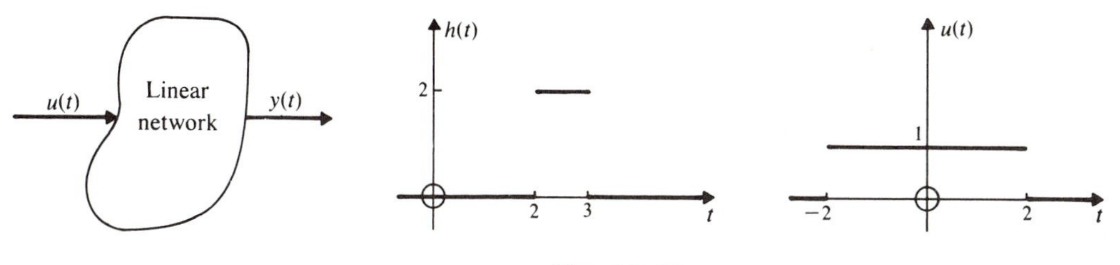


**Fig. 13-29**

**13.20** The impulse response of a linear system is

$$h(t) = \tfrac{3}{4}\{e^{-t} + e^{-3t}\}\mathbb{1}(t)$$

Determine the output $y(t)$ for the following inputs: (*a*) $u(t) = 2\delta(t - 5)$, (*b*) $u(t) = 4\mathbb{1}(t)$.

**13.21** Make use of frequency convolution to find the Fourier transform of the signal shown in Fig. 13-30. *Hint:* $g(t) = (-\sin \omega_0 t)\mathbb{1}(-t)$.

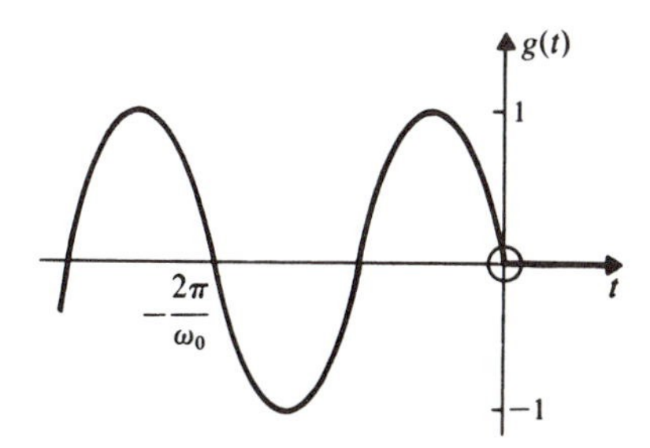


**Fig. 13-30**

**13.22** Find $f_1(t) * f_2(t)$ at $t = -3, -2, -1$, and 0 for $f_1(t)$ and $f_2(t)$ as given in Fig. 13-31.

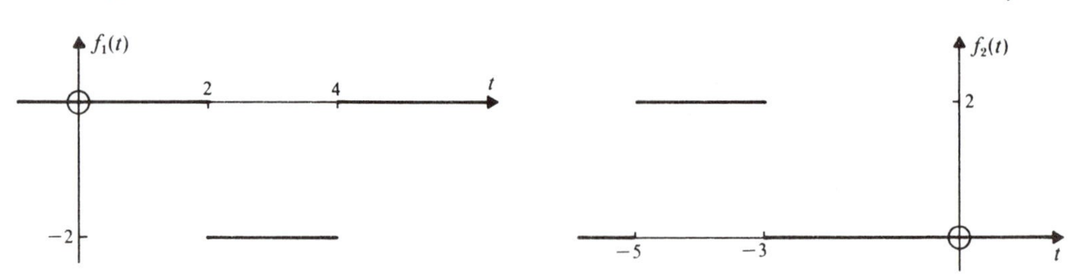


**Fig. 13-31**

**13.23** A linear system has the unit *step* response $w(t)$ pictured in Fig. 13-32. Write an expression for the unit *impulse* response $h(t)$.

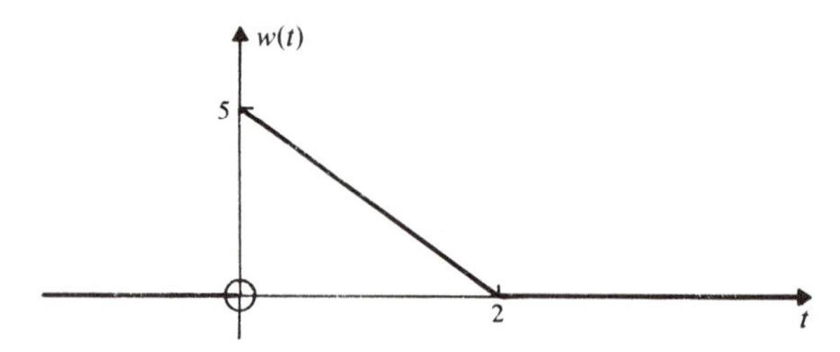


**Fig. 13-32**

**13.24** For a linear system,

$$\mathbf{X}(f) = \mathbf{Y}(f)\mathbf{Z}(f)$$

where $\mathbf{Y}(f)$ vs. $f$ and $z(t)$ vs. $t$ are given in Fig. 13-33. Sketch $x(t)$ vs. $t$.

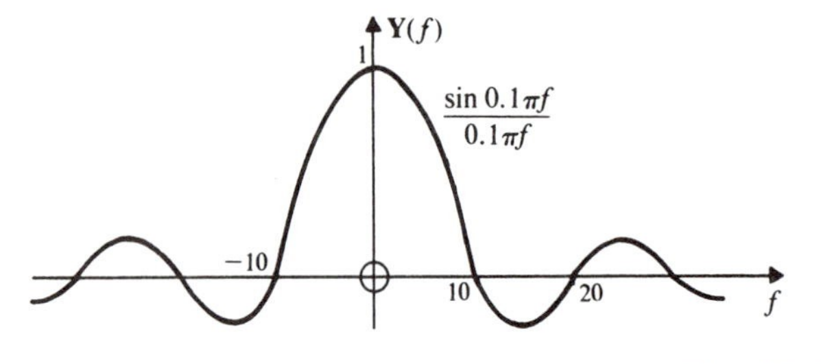


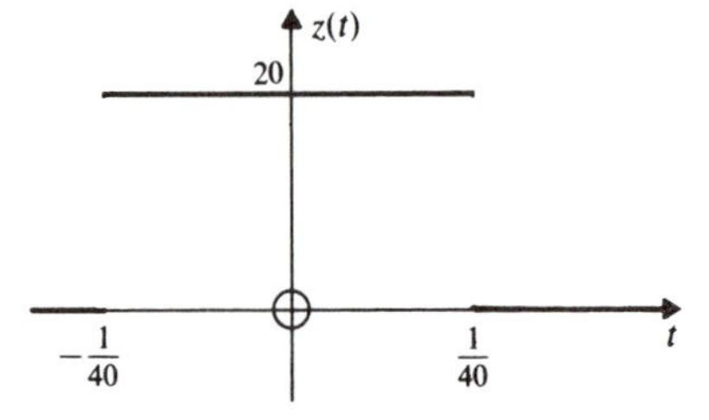


**Fig. 13-33**

**13.25** Two signals $u(t)$ and $h(t)$ may be represented by the sequences below, where the sampling interval is $\Delta t = 0.5$ s,

$$u(t) \sim [\cdots \; 0 \; 0 \; 0 \; 0 \; 0 \; 1 \; 2 \; 3 \; 4 \; 0 \; 0 \; \cdots]$$

$$\updownarrow \; t=0$$

$$h(t) \sim [\cdots \; 4 \; 3 \; 2 \; 1 \; 0 \; 0 \; 0 \; 0 \; 0 \; 0 \; 0 \; \cdots]$$

Find $u(t) * h(t)$ at $t = -1$ and $+1$.

**13.26** Use convolution to prove the trigonometric identity

$$\cos \omega_1 t \cos \omega_2 t = \tfrac{1}{2}\{\cos (\omega_2 + \omega_1)t + \cos (\omega_2 - \omega_1)t\}$$

# Chapter 14

# MATRIX EQUATION FORMULATION AND THE STATE EQUATION

Up to this point the emphasis has been on analytical methods of circuit analysis. But when we are faced with large, complicated and/or nonlinear circuits, the analytical approaches become unattractive or untenable, and we are likely to consider numerical methods, backed up by a digital computer. In fact, elaborate computer programs have been specially developed to meet this need.

The objective here is to obtain a circuit's equations in as systematic a form as possible, which is what the computer prefers!

## ORIENTED GRAPHS

The concept of an *oriented graph* (see Fig. 14-1) was introduced in Chapter 10. In review, Fig. 14-2 exemplifies *fundamental loops, fundamental cut sets,* and *meshes.*

It is usual to adopt a number of conventions. Thus after choosing a *tree,* as in Fig. 14-2*a*, we use the following procedure:

1. The *links* are numbered first, from 1 to $l$. Then the *tree branches* are numbered from $l + 1$ to $b$ (see Fig. 14-2*a*).
2. Each fundamental loop (see Fig. 14-2*b*) takes its direction and identification number from its defining link.

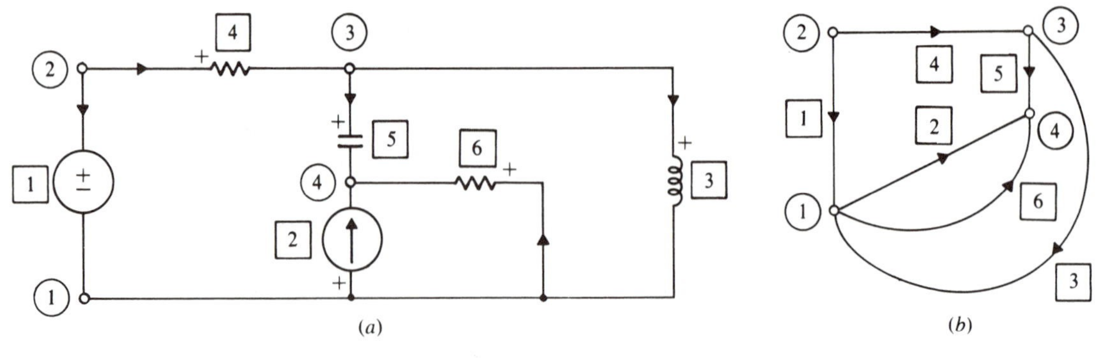


Fig. 14-1

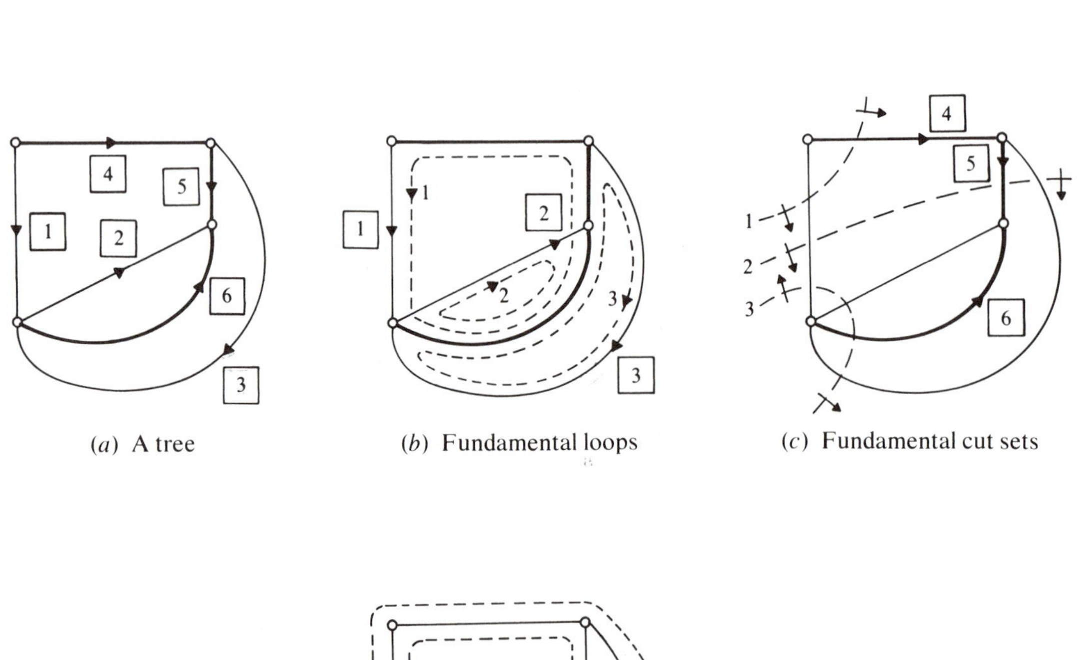


(a) A tree (b) Fundamental loops (c) Fundamental cut sets

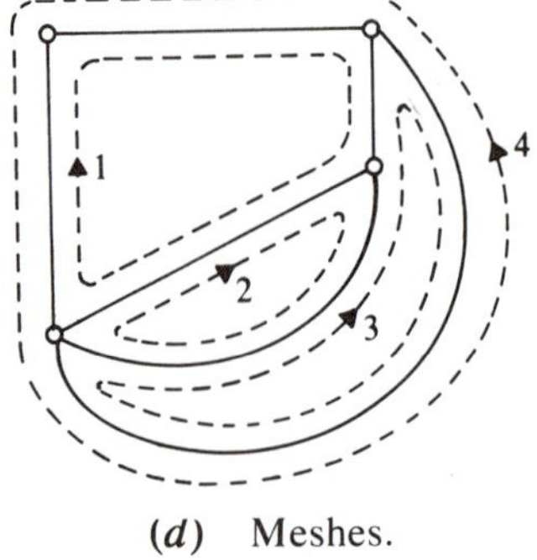


(d) Meshes. Fig. 14-2

3. The direction of a fundamental cut set is taken from its defining tree branch and the cut sets are numbered 1, 2, 3, . . . corresponding to the defining tree branches $l + 1$, $l + 2$, $l + 3$, . . . as shown in Fig. 14-2*c*.
4. Meshes are all clockwise except for the outer mesh which is counterclockwise (see Fig. 14-2*d*). Their numbering is arbitrary.

Further, we will assume throughout this chapter that the graph is connected, and, when meshes are involved, that it is also planar and unhinged. Then if

$n_t$ = number of nodes (including the reference node)
$n$ = number of nodes (*not* including the reference node)
$l$ = number of links
= number of meshes (*not* including the outer mesh)
$b$ = number of branches

it follows that

$$n_t = n + 1 \quad \text{and} \quad l = b - n_t + 1$$

## THE GRAPH'S MATRICES

The $n_t \times b$ augmented incidence matrix $\mathbf{A}_a$ is defined by

$$a_{ij} = \begin{cases} 1 & \text{if branch } j \text{ *leaves* node } i \\ -1 & \text{if branch } j \text{ *enters* node } i \\ 0 & \text{if branch } j \text{ is not incident upon } i \end{cases} \tag{14.1}$$

To obtain the $n \times b$ *reduced incidence matrix* $\mathbf{A}$ we simply delete the row in $\mathbf{A}_a$ corresponding to the reference node.

The $l \times b$ *fundamental loop matrix* $\mathbf{B}$ is defined by

$$b_{ij} = \begin{cases} 1 & \text{if branch } j \text{ is in loop } i \text{ and is in the same direction} \\ -1 & \text{if branch } j \text{ is in loop } i \text{ and is in the opposite direction} \\ 0 & \text{if branch } j \text{ is not in loop } i \end{cases} \tag{14.2}$$

The $n \times b$ *fundamental cut set matrix* $\mathbf{Q}$ is defined by

$$q_{ij} = \begin{cases} 1 & \text{if branch } j \text{ is in cut set } i \text{ and is in the same direction} \\ -1 & \text{if branch } j \text{ is in cut set } i \text{ and is in the opposite direction} \\ 0 & \text{if branch } j \text{ is not in cut set } i \end{cases} \tag{14.3}$$

Finally, the $(l + 1) \times b$ *augmented mesh matrix* $\mathbf{M}_a$ is defined by

$$m_{ij} = \begin{cases} 1 & \text{if branch } j \text{ is in mesh } i \text{ and is in the same direction} \\ -1 & \text{if branch } j \text{ is in mesh } i \text{ and is in the opposite direction} \\ 0 & \text{if branch } j \text{ is not in mesh } i \end{cases} \tag{14.4}$$

To obtain the *reduced mesh matrix* **M** we simply omit the row in $\mathbf{M}_a$ corresponding to the outer mesh.

### Example 14.1

Find **A**, **B**, **Q**, and **M** for the graph of Fig. 14-3.

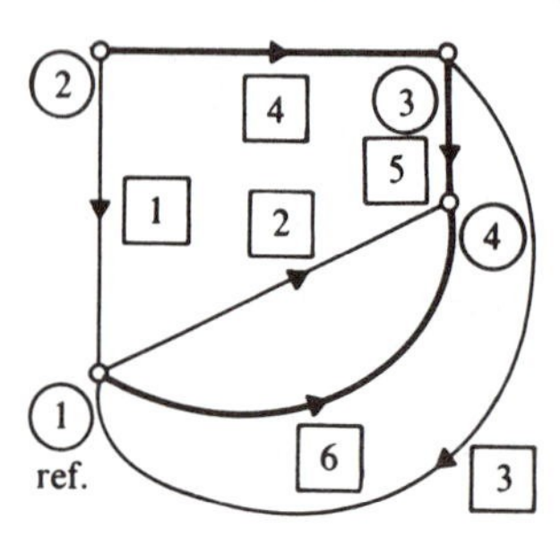


**Fig. 14-3**

Referring to the rule defining the elements $a_{ij}$, the augmented incidence matrix is

$$\mathbf{A}_a = \begin{bmatrix} -1 & 1 & -1 & 0 & 0 & 1 \\ 1 & 0 & 0 & 1 & 0 & 0 \\ 0 & 0 & 1 & -1 & 1 & 0 \\ 0 & -1 & 0 & 0 & -1 & -1 \end{bmatrix}$$

where, for example, $a_{24} = 1$ because branch $\boxed{4}$ *leaves* node ②.

To obtain the reduced incidence matrix we have only to strike out the row corresponding to the reference node, in this case the first, leaving

$$\mathbf{A} = \begin{bmatrix} 1 & 0 & 0 & 1 & 0 & 0 \\ 0 & 0 & 1 & -1 & 1 & 0 \\ 0 & -1 & 0 & 0 & -1 & -1 \end{bmatrix}$$

Recalling that the fundamental loops take their reference directions and identification numbers from their defining links, and referring to the definition of **B**, the fundamental loop matrix is

$$\mathbf{B} = \begin{bmatrix} 1 & 0 & 0 & -1 & -1 & 1 \\ 0 & 1 & 0 & 0 & 0 & -1 \\ 0 & 0 & 1 & 0 & -1 & 1 \end{bmatrix}$$

where $b_{35} = -1$ because branch $\boxed{5}$ is in loop ③, but is in the opposite direction.

Similarly, the fundamental cut set matrix is

$$\mathbf{Q} = \begin{bmatrix} 1 & 0 & 0 & 1 & 0 & 0 \\ 1 & 0 & 1 & 0 & 1 & 0 \\ -1 & 1 & -1 & 0 & 0 & 1 \end{bmatrix}$$

where the cut set numbering and directions were defined in Fig. 14-2*c*.

Finally, choosing to number the meshes as in Fig. 14-2*d*, and omitting the outer mesh, the reduced mesh matrix is

$$\mathbf{M} = \begin{bmatrix} -1 & -1 & 0 & 1 & 1 & 0 \\ 0 & 1 & 0 & 0 & 0 & -1 \\ 0 & 0 & 1 & 0 & -1 & 1 \end{bmatrix}$$

## NETWORK VOLTAGES AND CURRENTS

We now define the following column vectors:

$\mathbf{v}_b \equiv$ the $b$ branch potential differences,
$\mathbf{v}_n \equiv$ the $n$ node-to-reference voltages,
$\mathbf{v}_t \equiv$ the $n$ tree-branch potential differences,
$\mathbf{i}_b \equiv$ the $b$ branch currents,
$\mathbf{i}_m \equiv$ the $l$ mesh currents,
$\mathbf{i}_l \equiv$ the $l$ fundamental loop currents.

These vectors are related. We can show with the help of Kirchhoff's laws that

$$\mathbf{v}_b = \mathbf{A}'\mathbf{v}_n \qquad \mathbf{i}_b = \mathbf{M}'\mathbf{i}_m$$
$$\mathbf{v}_b = \mathbf{Q}'\mathbf{v}_t \qquad \mathbf{i}_b = \mathbf{B}'\mathbf{i}_l \tag{14.5}$$

where a prime indicates the *transpose* of a matrix.

## KIRCHHOFF'S LAWS

In matrix form Kirchhoff's laws can be written

$$\mathbf{A}\mathbf{i}_b = \mathbf{0} \qquad \mathbf{M}\mathbf{v}_b = \mathbf{0}$$
$$\mathbf{Q}\mathbf{i}_b = \mathbf{0} \qquad \mathbf{B}\mathbf{v}_b = \mathbf{0} \tag{14.6}$$

## THE BRANCH EQUATIONS

We will assume, initially, that every branch in a circuit is uncoupled, linear and time-invariant, and that each can be reduced to one of the equivalent standard forms of Fig. 14-4, with $Z_b(s)$ neither zero nor infinite.

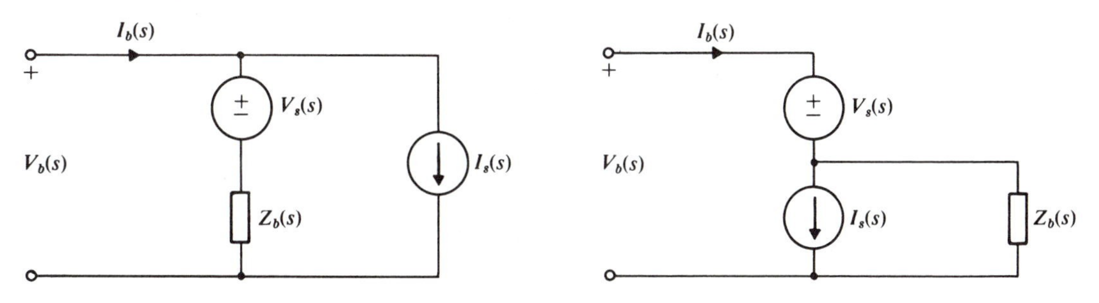


**Fig. 14-4**

Note carefully the sign convention relating the source reference directions to the reference directions for $V_b(s)$ and $I_b(s)$ in Fig. 14-4. Any initial conditions are included appropriately in $V_s(s)$ and $I_s(s)$.

Now we can write the matrix *v-i* and *i-v* relations for a circuit,

$$\mathbf{V}_b(s) = \mathbf{Z}_b(s)\mathbf{I}_b(s) + \mathbf{V}_s(s) - \mathbf{Z}_b(s)\mathbf{I}_s(s) \tag{14.7}$$

or

$$\mathbf{I}_b(s) = \mathbf{Y}_b(s)\mathbf{V}_b(s) + \mathbf{I}_s(s) - \mathbf{Y}_b(s)\mathbf{V}_s(s) \tag{14.8}$$

where $\mathbf{Z}_b(s)$ and $\mathbf{Y}_b(s)$ are the branch impedance and admittance matrices.

Of course these relations can also be written in $j\omega$-domain form.

## THE NETWORK EQUATIONS

Five distinct ways of writing circuit equations will be described here, namely,

1. Node equations
2. Mesh equations
3. Fundamental cut set equations
4. Fundamental loop equations
5. State equation

### I The Node Equations

From equations (*14.5*), (*14.6*), and (*14.8*),

$$\begin{aligned} \mathbf{A}\mathbf{I}_b(s) &= \mathbf{0} && \text{(KCL)} \\ \mathbf{V}_b(s) &= \mathbf{A}'\mathbf{V}_n(s) && \text{(KVL)} \\ \mathbf{I}_b(s) &= \mathbf{Y}_b(s)\mathbf{V}_b(s) + \mathbf{I}_s(s) - \mathbf{Y}_b(s)\mathbf{V}_s(s) && \text{($i$-$v$ relations)} \end{aligned}$$

Here $\mathbf{A}$, which defines the network graph's topology, and $\mathbf{Y}_b(s)$, $\mathbf{V}_s(s)$, and $\mathbf{I}_s(s)$, which define the branches, would be given. The aim is to solve for the nodal voltages $\mathbf{V}_n(s)$.

Premultiplying the third equation by $\mathbf{A}$,

$$\mathbf{A}\mathbf{I}_b(s) = \mathbf{A}\mathbf{Y}_b(s)\mathbf{V}_b(s) + \mathbf{A}\mathbf{I}_s(s) - \mathbf{A}\mathbf{Y}_b(s)\mathbf{V}_s(s)$$

Then substituting the first two equations and rearranging, we obtain the single matrix node equation,

$$\{\mathbf{A}\mathbf{Y}_b(s)\mathbf{A}'\}\mathbf{V}_n(s) = \mathbf{A}\mathbf{Y}_b(s)\mathbf{V}_s(s) - \mathbf{A}\mathbf{I}_s(s)$$

which is often written

$$\mathbf{Y}_n(s)\mathbf{V}_n(s) = \mathbf{I}_{sn}(s) \tag{14.9}$$

where the *node admittance matrix* and *node current source matrix* are

$$\mathbf{Y}_n(s) = \mathbf{A}\mathbf{Y}_b(s)\mathbf{A}' \quad \text{and} \quad \mathbf{I}_{sn}(s) = \mathbf{A}\mathbf{Y}_b(s)\mathbf{V}_s(s) - \mathbf{A}\mathbf{I}_s(s)$$

The matrix node equation (*14.9*) represents $n$ scalar equations which may be shown to be independent. They may therefore be solved for the $n$ node voltages,

$$\mathbf{V}_n(s) = \mathbf{Y}_n^{-1}(s)\mathbf{I}_{sn}(s)$$

Finally, $\mathbf{V}_b(s)$ and $\mathbf{I}_b(s)$ follow in turn from

$$\mathbf{V}_b(s) = \mathbf{A}'\mathbf{V}_n(s) \quad \text{and} \quad \mathbf{I}_b(s) = \mathbf{Y}_b(s)\mathbf{V}_b(s) + \mathbf{I}_s(s) - \mathbf{Y}_b(s)\mathbf{V}_s(s)$$

## II The Mesh Equations

From equations (*14.5*), (*14.6*), and (*14.7*),

$$\mathbf{M}\mathbf{V}_b(s) = \mathbf{0} \qquad \text{(KVL)}$$

$$\mathbf{I}_b(s) = \mathbf{M}'\mathbf{I}_m(s) \qquad \text{(KCL)}$$

$$\mathbf{V}_b(s) = \mathbf{Z}_b(s)\mathbf{I}_b(s) + \mathbf{V}_s(s) - \mathbf{Z}_b(s)\mathbf{I}_s(s) \qquad (v\text{-}i \text{ relations})$$

from which we can eliminate $\mathbf{I}_b(s)$ and $\mathbf{V}_b(s)$ to obtain the matrix mesh equation in the mesh currents $\mathbf{I}_m(s)$,

$$\{\mathbf{M}\mathbf{Z}_b(s)\mathbf{M}'\}\mathbf{I}_m(s) = \mathbf{M}\mathbf{Z}_b(s)\mathbf{I}_s(s) - \mathbf{M}\mathbf{V}_s(s)$$

or

$$\mathbf{Z}_m(s)\mathbf{I}_m(s) = \mathbf{V}_{sm}(s) \tag{14.10}$$

## III The Fundamental Cut Set Equations

In like manner

$$\mathbf{Q}\mathbf{I}_b(s) = \mathbf{0} \qquad \text{(KCL)}$$

$$\mathbf{V}_b(s) = \mathbf{Q}'\mathbf{V}_t(s) \qquad \text{(KVL)}$$

$$\mathbf{I}_b(s) = \mathbf{Y}_b(s)\mathbf{V}_b(s) + \mathbf{I}_s(s) - \mathbf{Y}_b(s)\mathbf{V}_s(s) \qquad (i\text{-}v \text{ relations})$$

Here $\mathbf{I}_b(s)$ and $\mathbf{V}_b(s)$ may be eliminated, yielding an equation in the tree-branch voltages $\mathbf{V}_t(s)$,

$$\{\mathbf{Q}\mathbf{Y}_b(s)\mathbf{Q}'\}\mathbf{V}_t(s) = \mathbf{Q}\mathbf{Y}_b(s)\mathbf{V}_s(s) - \mathbf{Q}\mathbf{I}_s(s)$$

or

$$\mathbf{Y}_t(s)\mathbf{V}_t(s) = \mathbf{I}_{st}(s) \tag{14.11}$$

### IV The Fundamental Loop Equations

$$\mathbf{B}\mathbf{V}_b(s) = \mathbf{0} \qquad \text{(KVL)}$$

$$\mathbf{I}_b(s) = \mathbf{B}'\mathbf{I}_l(s) \qquad \text{(KCL)}$$

$$\mathbf{V}_b(s) = \mathbf{Z}_b(s)\mathbf{I}_b(s) + \mathbf{V}_s(s) - \mathbf{Z}_b(s)\mathbf{I}_s(s) \qquad (v\text{-}i \text{ relations})$$

and the fundamental loop currents may be found from

$$\{\mathbf{B}\mathbf{Z}_b(s)\mathbf{B}'\}\mathbf{I}_l(s) = \mathbf{B}\mathbf{Z}_b(s)\mathbf{I}_s(s) - \mathbf{B}\mathbf{V}_s(s)$$

or

$$\mathbf{Z}_l(s)\mathbf{I}_l(s) = \mathbf{V}_{sl}(s) \qquad (14.12)$$

*Comments*

1. The node and mesh approaches are duals, as are the loop and cut set methods.
2. The nodal method is not necessarily a special case of the cut set method. Neither is the mesh method necessarily a special case of the loop method. It may not, for example, be possible to choose a tree such that the meshes are fundamental loops.

### V The State Equation

A set of *first-order* simultaneous equations of the form

$$\dot{\mathbf{x}} = \mathbf{f}(\mathbf{x}, \mathbf{u}, t) \qquad (14.13)$$

is said to be a *state equation* where $\mathbf{x}$ represents the system's *state variables* and $\mathbf{u}$ represents the inputs. $\mathbf{x}$ is said to be the *state* of the system.

This form of a network's equations is extremely valuable when we wish to employ numerical (i.e. computer) methods of analysis, especially since it is relatively easy to incorporate nonlinear and time-varying characteristics. With this intent in mind, it is usually advantageous to remain in the time-domain.

*The Linear, Time-Invariant Case*

The following procedure may be used to obtain a circuit's state equations:

1. Choose a tree containing *all* the capacitors and *none* of the inductors. This will always be possible as long as there is no loop of capacitors or cut set of inductors. A tree chosen in this way is known as a *proper tree*.
2. Choose the capacitor voltages and inductor currents as the state variables. (We may also choose charges and fluxes, or more strictly, flux linkages.)

3. Write the fundamental cut set equation for each capacitor, and the fundamental loop equation for each inductor in terms of the chosen state variables.
4. Eliminate all variables which are not state variables. This step can be involved. There *is* a systematic method, but it will not be included here.

The resulting equations can be put in the form

$$\dot{\mathbf{x}} = \mathbf{A}\mathbf{x} + \mathbf{B}\mathbf{u} \tag{14.14}$$

where $\mathbf{x}$ represents the states (i.e. capacitor voltages and inductor currents, or charges and fluxes), $\mathbf{A}$ and $\mathbf{B}$ are matrices of constants, and $\mathbf{u}$ is the vector of inputs.

Further, if $\mathbf{y}$ is a vector of *outputs*, such as branch currents and voltages, loop currents and/or nodal voltages, then we can write an *output equation,*

$$\mathbf{y} = \mathbf{C}\mathbf{x} + \mathbf{D}\mathbf{u} \tag{14.15}$$

where $\mathbf{C}$ and $\mathbf{D}$ are matrices of constants.

### *The Linear, Time-Varying Case*

It will usually be advantageous to choose capacitor charges and inductor fluxes as the state variables. Then the procedure outlined above leads to

$$\dot{\mathbf{x}} = \mathbf{A}(t)\mathbf{x} + \mathbf{B}(t)\mathbf{u} \tag{14.16}$$

where the matrices $\mathbf{A}(t)$ and $\mathbf{B}(t)$ are now time-varying.

### *The Nonlinear Case*

The same procedure may be attempted, although the elimination step may prove to be intractable. The state equation will be of the general form

$$\dot{\mathbf{x}} = \mathbf{f}(\mathbf{x}, \mathbf{u}, t)$$

### *Solution of the State Equation*

The general methods available for the numerical (i.e. computer) solution of sets of first-order differential equations will not be taken up here. However, if the equations are linear and time-invariant and we are given $\mathbf{A}$, $\mathbf{B}$, $\mathbf{u}\,(t)$, and the initial state $\mathbf{x}(0)$, then the solution is given by

$$\mathbf{x}(t) = e^{\mathbf{A}t}\mathbf{x}(0) + \int_0^t e^{\mathbf{A}(t-\tau)}\mathbf{B}\mathbf{u}(\tau)\,d\tau \qquad (t \geq 0)$$

$$= e^{\mathbf{A}t}\mathbf{x}(0) + e^{\mathbf{A}t}\int_0^t e^{-\mathbf{A}\tau}\mathbf{B}\mathbf{u}(\tau)\,d\tau = \mathbf{x}_{\mathrm{IC}}(t) + \mathbf{x}_{\mathrm{F}}(t)$$

where the first term is the response due to the initial conditions, and the second is the forced response due to the inputs. The second term will also be recognized as the (matrix) convolution

$$e^{\mathbf{A}t} * \mathbf{Bu}(t)$$

so the matrix exponential $e^{\mathbf{A}t}$ may be identified as the matrix impulse response of the system.

In fact, the solution is of the same form as the solution of a single first-order equation, except that the solution is now a matrix. The matrix exponential is also of the same form as its scalar equivalent,

$$e^{\mathbf{A}t} = \mathbf{I} + \mathbf{A}t + \frac{\mathbf{A}^2}{2!}t^2 + \frac{\mathbf{A}^3}{3!}t^3 + \cdots$$

# ILLUSTRATIVE PROBLEMS

## *Oriented Graphs, Network Matrices, and Matrix Equations*

**14.1** (*a*) Draw an oriented graph for the circuit of Fig. 14-5.
(*b*) Choose and label a tree for this graph.
(*c*) Write the augmented incidence matrix $\mathbf{A}_a$ and, after selecting a reference node, determine the incidence matrix $\mathbf{A}$.

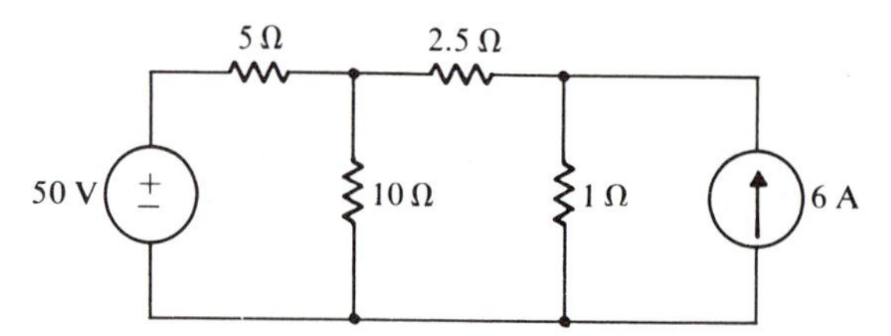


**Fig. 14-5**

(*a*) It is always advisable to reduce the circuit to a form in which all branches are of the "standard" form of Fig. 14-4 *with no branch impedances of zero or infinity*. Thus the voltage source in Fig. 14-5 with its 5-Ω series resistor should be regarded as one branch, and the current source with its parallel 1-Ω resistor as another.

The oriented graph of Fig. 14-6*a* then follows. The reference directions may be assigned arbitrarily, but it is natural to choose the directions in the active branches to coincide with the source direction as we have done here.

(*b*) In Fig. 14-6*b* we have chosen a tree, and in line with the normal convention we have numbered the *links* first, before the tree-branches.

(*c*) Here the augmented incidence matrix is

$$\mathbf{A}_a = \underbrace{\begin{bmatrix} 1 & 1 & -1 & 0 \\ 0 & -1 & 1 & 1 \\ -1 & 0 & 0 & -1 \end{bmatrix}}_{b \text{ branches}} \Bigg\} n_t \text{ nodes}$$

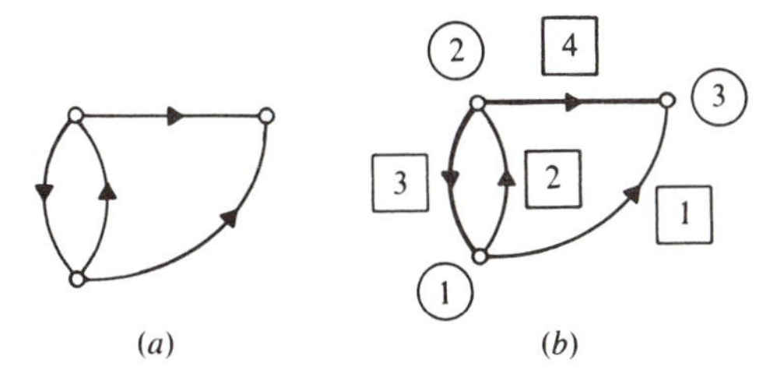


**Fig. 14-6**

And, choosing node ① as the reference, the reduced incidence matrix is

$$\mathbf{A} = \begin{bmatrix} 0 & -1 & 1 & 1 \\ -1 & 0 & 0 & -1 \end{bmatrix}$$

**14.2** The circuit in the previous problem (see Fig. 14-5) is purely resistive, and we may therefore work in the time-domain to find:

(*a*) The branch admittance (conductance) matrix $\mathbf{Y}_b$.

(*b*) The voltage source matrix $\mathbf{V}_s$ and the current source matrix $\mathbf{I}_s$.

(*c*) The node voltage vector $\mathbf{V}_n$, which is the solution of the nodal equation (*14.9*),

$$\{\mathbf{A}\mathbf{Y}_b\mathbf{A}'\}\mathbf{V}_n = \mathbf{A}\mathbf{Y}_b\mathbf{V}_s - \mathbf{A}\mathbf{I}_s \quad \text{or} \quad \mathbf{Y}_n\mathbf{V}_n = \mathbf{I}_{sn}$$

(*a*) The branch admittance matrix is a diagonal matrix in which each diagonal element is the admittance (in this case, the conductance) of the corresponding branch,

$$\mathbf{Y}_b = \begin{bmatrix} 1 & 0 & 0 & 0 \\ 0 & 0.1 & 0 & 0 \\ 0 & 0 & 0.2 & 0 \\ 0 & 0 & 0 & 0.4 \end{bmatrix}$$

(*b*) The voltage source matrix $\mathbf{V}_s$ and the current source matrix $\mathbf{I}_s$ are

$$\mathbf{V}_s = \begin{bmatrix} 0 \\ 0 \\ 50 \\ 0 \end{bmatrix} \quad \text{and} \quad \mathbf{I}_s = \begin{bmatrix} 6 \\ 0 \\ 0 \\ 0 \end{bmatrix}$$

(*c*) Although the solution of the nodal equation (*14.9*) is best done by a computer, it is instructive to carry out the calculations at least once by hand.

First we compute the nodal admittance matrix $\mathbf{Y}_n = \mathbf{A}\mathbf{Y}_b\mathbf{A}'$ as follows:

$$\mathbf{Y}_n = \begin{bmatrix} 0 & -1 & 1 & 1 \\ -1 & 0 & 0 & -1 \end{bmatrix} \begin{bmatrix} 1 & 0 & 0 & 0 \\ 0 & 0.1 & 0 & 0 \\ 0 & 0 & 0.2 & 0 \\ 0 & 0 & 0 & 0.4 \end{bmatrix} \begin{bmatrix} 0 & -1 \\ -1 & 0 \\ 1 & 0 \\ 1 & -1 \end{bmatrix}$$

$$= \begin{bmatrix} 0 & -0.1 & 0.2 & 0.4 \\ -1 & 0 & 0 & -0.4 \end{bmatrix} \begin{bmatrix} 0 & -1 \\ -1 & 0 \\ 1 & 0 \\ 1 & -1 \end{bmatrix} = \begin{bmatrix} 0.7 & -0.4 \\ -0.4 & 1.4 \end{bmatrix}$$

Next we compute the nodal current source matrix $\mathbf{I}_{sn} = \mathbf{A}\mathbf{Y}_b\mathbf{V}_s - \mathbf{A}\mathbf{I}_s$,

$$\mathbf{I}_{sn} = \begin{bmatrix} 0 & -1 & 1 & 1 \\ -1 & 0 & 0 & -1 \end{bmatrix} \begin{bmatrix} 1 & 0 & 0 & 0 \\ 0 & 0.1 & 0 & 0 \\ 0 & 0 & 0.2 & 0 \\ 0 & 0 & 0 & 0.4 \end{bmatrix} \begin{bmatrix} 0 \\ 0 \\ 50 \\ 0 \end{bmatrix}$$

$$- \begin{bmatrix} 0 & -1 & 1 & 1 \\ -1 & 0 & 0 & -1 \end{bmatrix} \begin{bmatrix} 6 \\ 0 \\ 0 \\ 0 \end{bmatrix} = \begin{bmatrix} 10 \\ 0 \end{bmatrix} - \begin{bmatrix} 0 \\ -6 \end{bmatrix} = \begin{bmatrix} 10 \\ 6 \end{bmatrix}$$

That is, the nodal equation $\mathbf{Y}_n \mathbf{V}_n = \mathbf{I}_{sn}$ can now be written

$$\begin{bmatrix} 0.7 & -0.4 \\ -0.4 & 1.4 \end{bmatrix} \mathbf{V}_n = \begin{bmatrix} 10 \\ 6 \end{bmatrix}$$

And, solving for the nodal voltage matrix $\mathbf{V}_n$,

$$\mathbf{V}_n = \begin{bmatrix} 0.7 & -0.4 \\ -0.4 & 1.4 \end{bmatrix}^{-1} \begin{bmatrix} 10 \\ 6 \end{bmatrix} = \frac{1}{0.82} \begin{bmatrix} 1.4 & 0.4 \\ 0.4 & 0.7 \end{bmatrix} \begin{bmatrix} 10 \\ 6 \end{bmatrix} = \begin{bmatrix} 20 \\ 10 \end{bmatrix}$$

Since node ① was the reference,

$$\mathbf{V}_n = \begin{bmatrix} V_2 \\ V_3 \end{bmatrix} = \begin{bmatrix} 20 \\ 10 \end{bmatrix}$$

**14.3** Complete the analysis of the circuit of Fig. 14-5 by finding all the *branch* currents and voltages.

From equation (*14.5*), $\mathbf{V}_b = \mathbf{A}'\mathbf{V}_n$ and, substituting the results of Problems 14.1 and 14.2,

$$\mathbf{V}_b = \begin{bmatrix} 0 & -1 \\ -1 & 0 \\ 1 & 0 \\ 1 & -1 \end{bmatrix} \begin{bmatrix} 20 \\ 10 \end{bmatrix} = \begin{bmatrix} -10 \\ -20 \\ 20 \\ 10 \end{bmatrix}$$

Finally, solving for the branch currents, using equation (*14.8*),

$$\mathbf{I}_b = \mathbf{Y}_b \mathbf{V}_b + \mathbf{I}_s - \mathbf{Y}_b \mathbf{V}_s$$

$$= \begin{bmatrix} 1 & 0 & 0 & 0 \\ 0 & 0.1 & 0 & 0 \\ 0 & 0 & 0.2 & 0 \\ 0 & 0 & 0 & 0.4 \end{bmatrix} \begin{bmatrix} -10 \\ -20 \\ 20 \\ 10 \end{bmatrix} + \begin{bmatrix} 6 \\ 0 \\ 0 \\ 0 \end{bmatrix} - \begin{bmatrix} 1 & 0 & 0 & 0 \\ 0 & 0.1 & 0 & 0 \\ 0 & 0 & 0.2 & 0 \\ 0 & 0 & 0 & 0.4 \end{bmatrix} \begin{bmatrix} 0 \\ 0 \\ 50 \\ 0 \end{bmatrix}$$

$$= \begin{bmatrix} -10 \\ -2 \\ 4 \\ 4 \end{bmatrix} + \begin{bmatrix} 6 \\ 0 \\ 0 \\ 0 \end{bmatrix} - \begin{bmatrix} 0 \\ 0 \\ 10 \\ 0 \end{bmatrix} = \begin{bmatrix} -4 \\ -2 \\ -6 \\ 4 \end{bmatrix}$$

We can use Tellegen's theorem as a partial check on these results. Thus

$$\sum V_k I_k = (-10)(-4) + (-20)(-2) + (20)(-6) + (10)(4) = 0$$

### *Summary of the Matrix Method of Nodal Analysis*

The procedure for solving a circuit problem by matrix nodal analysis is as follows:

1. Draw an oriented graph for the network, making certain that sources do not appear alone in any branch.
2. Select a reference node and write down the incidence matrix $\mathbf{A}$.
3. Write the branch admittance matrix $\mathbf{Y}_b(s)$. This may be done by inspection if the network contains no mutual coupling and no dependent sources.
4. Determine the voltage source matrix $\mathbf{V}_s(s)$ and the current source matrix $\mathbf{I}_s(s)$ using the conventions of the standard branch (see Fig. 14-4) to establish the proper signs.
5. Solve for the nodal voltages $\mathbf{V}_n(s)$ in the equation
$$\{\mathbf{A}\mathbf{Y}_b(s)\mathbf{A}'\}\mathbf{V}_n(s) = \mathbf{A}\mathbf{Y}_b(s)\mathbf{V}_s(s) - \mathbf{A}\mathbf{I}_s(s)$$
6. Finally, $\mathbf{V}_b(s)$ and $\mathbf{I}_b(s)$ may be found from
$$\mathbf{V}_b(s) = \mathbf{A}'\mathbf{V}_n(s) \quad \text{and} \quad \mathbf{I}_b(s) = \mathbf{Y}_b(s)\mathbf{V}_b(s) + \mathbf{I}_s(s) - \mathbf{Y}_b(s)\mathbf{V}_s(s)$$

**14.4** Use the results of Problem 14.3 to demonstrate that the matrix equation $\mathbf{A}\mathbf{I}_b = \mathbf{0}$ is simply Kirchhoff's current law in matrix form.

The results of Problem 14.3 are shown on the graph of Fig. 14-7. Then from KCL,

$$\text{at node ②:} \quad -(-2) - 6 + 4 = 0$$
$$\text{at node ③:} \quad -(-4) - 4 = 0$$

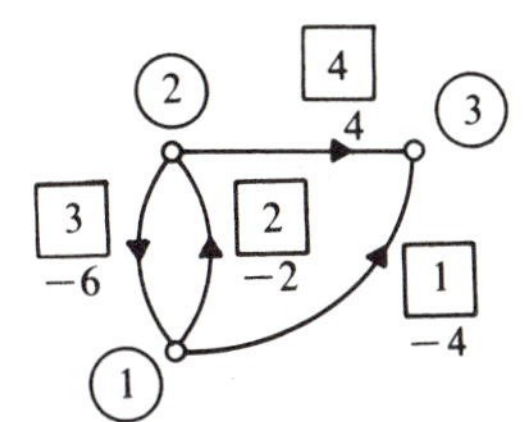


**Fig. 14-7**

Looking now at $\mathbf{A}\mathbf{I}_b$ we have

$$\begin{bmatrix} 0 & -1 & 1 & 1 \\ -1 & 0 & 0 & -1 \end{bmatrix} \begin{bmatrix} -4 \\ -2 \\ -6 \\ 4 \end{bmatrix} = \begin{bmatrix} -(-2) - 6 + 4 \\ -(-4) - 4 \end{bmatrix} = \begin{bmatrix} \Sigma\, i \ \text{ at node ②} \\ \Sigma\, i \ \text{ at node ③} \end{bmatrix} = \mathbf{0}$$

**14.5** The circuit of Fig. 14-8 is in the sinusoidal steady-state, with $i(t) = 7 \cos t$ and $e(t) = 17 \cos(t + 76°)$. Set up the appropriate matrices for nodal analysis in

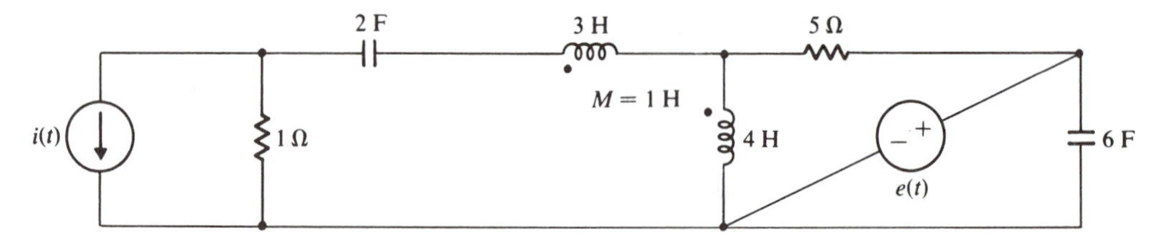


**Fig. 14-8**

the $j\omega$-domain. That is, find:

(*a*) The incidence matrix **A**, choosing the bottom connection as the reference node.

(*b*) The branch admittance matrix $\mathbf{Y}_b(j\omega)$.

(*c*) The phasor source matrices $[\mathbf{V}_s]$ and $[\mathbf{I}_s]$.

(*a*) Refering to the "standard" branch of Fig. 14-4, we see that an impedance in parallel with a voltage source is not permissible. Further, the 6-F capacitor in Fig. 14-8 can have no effect on the circuit, other than its known contribution to the current through the source $e(t)$. We call such an element *redundant,* and *omit it from our calculations*. At the end of the analysis we can easily modify the calculated current through the source $e(t)$ by the appropriate amount.

One possible oriented graph is shown in Fig. 14-9. Branch [1] consists of $e(t)$ and the 5-Ω resistor in series; branch [2] is the 4-H inductor; and branch [3] consists of $i(t)$ in parallel with the 1-Ω resistor.

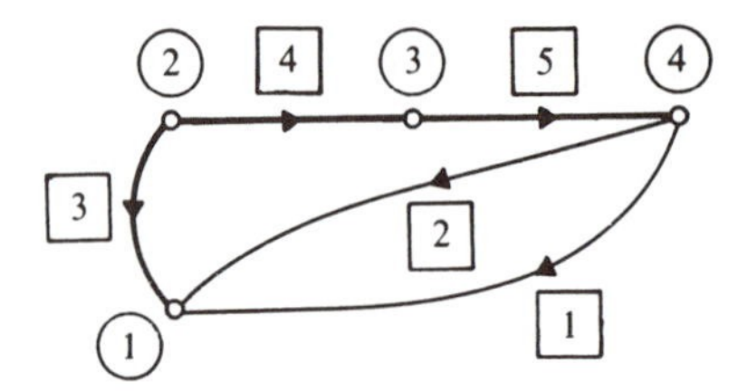


**Fig. 14-9**

The corresponding incidence matrix, choosing node (1) as reference, is

$$\mathbf{A} = \underbrace{\begin{bmatrix} 0 & 0 & 1 & 1 & 0 \\ 0 & 0 & 0 & -1 & 1 \\ 1 & 1 & 0 & 0 & -1 \end{bmatrix}}_{b \text{ branches}} \Bigg\} n \text{ nodes}$$

(*b*) The branch admittance matrix is straightforward for all branches except [2] and [5], where mutual inductance (coupling) is present. For these two branches we can write the $v$-$i$ relations by inspection,

$$\begin{aligned} \mathbf{V}_2 &= j4\mathbf{I}_2 + j\mathbf{I}_5 \\ \mathbf{V}_5 &= j\mathbf{I}_2 + j3\mathbf{I}_5 \end{aligned} \quad \text{or} \quad \begin{bmatrix} j4 & j \\ j & j3 \end{bmatrix} \begin{bmatrix} \mathbf{I}_2 \\ \mathbf{I}_5 \end{bmatrix} = \begin{bmatrix} \mathbf{V}_2 \\ \mathbf{V}_5 \end{bmatrix}$$

To find the admittance matrix for these two elements, we must solve for $\begin{bmatrix} \mathbf{I}_2 \\ \mathbf{I}_5 \end{bmatrix}$. Thus

$$\begin{bmatrix} \mathbf{I}_2 \\ \mathbf{I}_5 \end{bmatrix} = \begin{bmatrix} j4 & j \\ j & j3 \end{bmatrix}^{-1} \begin{bmatrix} \mathbf{V}_2 \\ \mathbf{V}_5 \end{bmatrix} = \frac{1}{-11} \begin{bmatrix} j3 & -j \\ -j & j4 \end{bmatrix} \begin{bmatrix} \mathbf{V}_2 \\ \mathbf{V}_5 \end{bmatrix}$$

Combining this result with the admittance matrix of the uncoupled branches we obtain the overall branch admittance matrix.

$$\mathbf{Y}_b(j\omega) = \begin{bmatrix} 0.2 & 0 & 0 & 0 & 0 \\ 0 & -j\dfrac{3}{11} & 0 & 0 & j\dfrac{1}{11} \\ 0 & 0 & 1 & 0 & 0 \\ 0 & 0 & 0 & j2 & 0 \\ 0 & j\dfrac{1}{11} & 0 & 0 & -j\dfrac{4}{11} \end{bmatrix}$$

(*c*) From Figs. 14-8 and 14-9,

$$[\mathbf{V}_s] = \begin{bmatrix} 17e^{j(t+76°)} \\ 0 \\ 0 \\ 0 \\ 0 \end{bmatrix} \quad \text{and} \quad [\mathbf{I}_s] = \begin{bmatrix} 0 \\ 0 \\ 7e^{jt} \\ 0 \\ 0 \end{bmatrix}$$

The reference directions for branches [1] and [3] in the graph were chosen to coincide with the source directions. Therefore the source phasors appear positively in $[\mathbf{V}_s]$ and $[\mathbf{I}_s]$.

**14.6** The resistive network of Fig. 14-10 is to be analyzed on a mesh current basis. As the first move, modify the circuit so that each branch is in the standard form of Fig. 14-4.

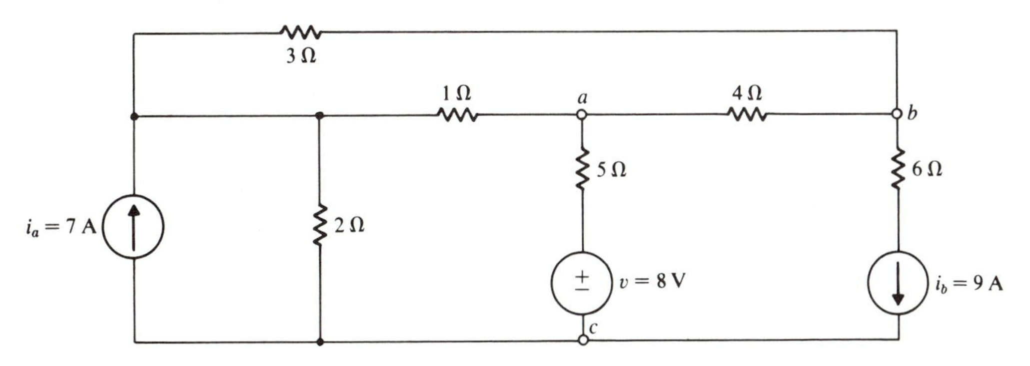


**Fig. 14-10**

An impedance in series with a current source is not permitted in a standard branch (see Fig. 14-4). Here the 6-Ω resistor in series with the current source $i_b$ affects only the voltage across that current source and is therefore omitted as *redundant*. Its known effect on the voltage across the source $i_b$ can be added in at the end of the analysis.

*Now* the current source $i_b$ is in a branch by itself, which also is not permissible! Therefore the circuit must be put into the equivalent form of Fig. 14-11. This *is* equivalent since the source current injected into nodes $b$ and $c$ has not been changed and since no net current is injected into node $a$.

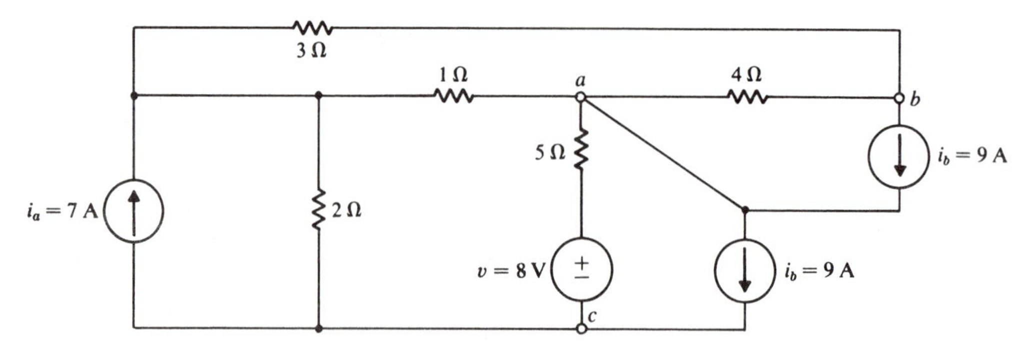


**Fig. 14-11**

The 4-Ω resistor and the parallel source $i_b$ may be taken as one branch, the two sources $v$ and $i_b$ together with the 5-Ω resistor as another, and the source $i_a$ in parallel with 2-Ω as a third.

**14.7** Draw an oriented graph for the network of Fig. 14-11, and obtain the $\mathbf{M}$, $\mathbf{Z}_b$, $\mathbf{V}_s$, and $\mathbf{I}_s$ matrices. Then solve for the branch currents using the mesh matrix method.

An oriented graph is shown in Fig. 14-12 from which we can write by inspection the reduced mesh matrix,

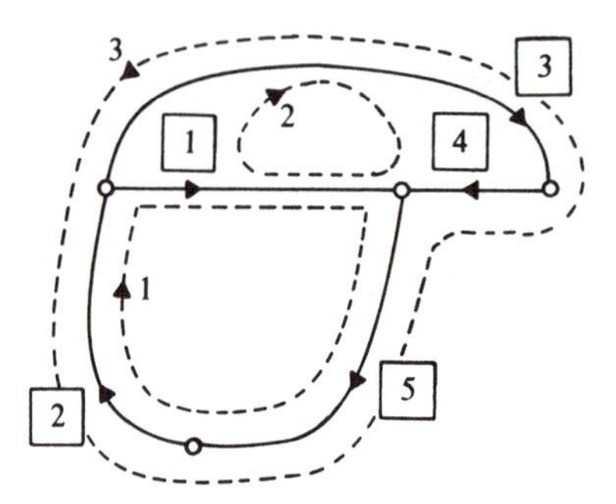


**Fig. 14-12**

$$\mathbf{M} = \begin{bmatrix} 1 & 1 & 0 & 0 & 1 \\ -1 & 0 & 1 & 1 & 0 \end{bmatrix}$$

Next, from Figs. 14-11 and 14-12 we can write

$$\mathbf{Z}_b = \begin{bmatrix} 1 & & & & 0 \\ & 2 & & & \\ & & 3 & & \\ & 0 & & 4 & \\ & & & & 5 \end{bmatrix} \qquad \mathbf{V}_s = \begin{bmatrix} 0 \\ 0 \\ 0 \\ 0 \\ 8 \end{bmatrix} \qquad \mathbf{I}_s = \begin{bmatrix} 0 \\ 7 \\ 0 \\ 9 \\ 9 \end{bmatrix}$$

The mesh equation (*14.10*) is

$$\{\mathbf{M}\mathbf{Z}_b\mathbf{M}'\}\mathbf{I}_m = \mathbf{M}\mathbf{Z}_b\mathbf{I}_s - \mathbf{M}\mathbf{V}_s \quad \text{or} \quad \mathbf{Z}_m\mathbf{I}_m = \mathbf{V}_{sm}$$

where by substitution,

$$\mathbf{Z}_m = \begin{bmatrix} 1 & 1 & 0 & 0 & 1 \\ -1 & 0 & 1 & 1 & 0 \end{bmatrix} \begin{bmatrix} 1 & & & & 0 \\ & 2 & & & \\ & & 3 & & \\ & 0 & & 4 & \\ & & & & 5 \end{bmatrix} \begin{bmatrix} 1 & -1 \\ 1 & 0 \\ 0 & 1 \\ 0 & 1 \\ 1 & 0 \end{bmatrix}$$

$$= \begin{bmatrix} 1 & 2 & 0 & 0 & 5 \\ -1 & 0 & 3 & 4 & 0 \end{bmatrix} \begin{bmatrix} 1 & -1 \\ 1 & 0 \\ 0 & 1 \\ 0 & 1 \\ 1 & 0 \end{bmatrix} = \begin{bmatrix} 8 & -1 \\ -1 & 8 \end{bmatrix}$$

and

$$\mathbf{V}_{sm} = \begin{bmatrix} 1 & 2 & 0 & 0 & 5 \\ -1 & 0 & 3 & 4 & 0 \end{bmatrix} \begin{bmatrix} 0 \\ 7 \\ 0 \\ 9 \\ 9 \end{bmatrix} - \begin{bmatrix} 1 & 1 & 0 & 0 & 1 \\ -1 & 0 & 1 & 1 & 0 \end{bmatrix} \begin{bmatrix} 0 \\ 0 \\ 0 \\ 0 \\ 8 \end{bmatrix}$$

$$= \begin{bmatrix} 59 \\ 36 \end{bmatrix} - \begin{bmatrix} 8 \\ 0 \end{bmatrix} = \begin{bmatrix} 51 \\ 36 \end{bmatrix}$$

That is, the matrix mesh equation is

$$\begin{bmatrix} 8 & -1 \\ -1 & 8 \end{bmatrix} \begin{bmatrix} I_{m1} \\ I_{m2} \end{bmatrix} = \begin{bmatrix} 51 \\ 36 \end{bmatrix}$$

and its solution for the mesh currents follows as

$$\begin{bmatrix} I_{m1} \\ I_{m2} \end{bmatrix} = \begin{bmatrix} 8 & -1 \\ -1 & 8 \end{bmatrix}^{-1} \begin{bmatrix} 51 \\ 36 \end{bmatrix} = \frac{1}{63} \begin{bmatrix} 8 & 1 \\ 1 & 8 \end{bmatrix} \begin{bmatrix} 51 \\ 36 \end{bmatrix} = \begin{bmatrix} 444/63 \\ 339/63 \end{bmatrix}$$

Finally, from equation (*14.5*) we find the branch currents to be

$$\mathbf{I}_b = \mathbf{M}'\mathbf{I}_m = \begin{bmatrix} 1 & -1 \\ 1 & 0 \\ 0 & 1 \\ 0 & 1 \\ 1 & 0 \end{bmatrix} \begin{bmatrix} 444/63 \\ 339/63 \end{bmatrix} = \begin{bmatrix} 105/63 \\ 444/63 \\ 339/63 \\ 339/63 \\ 444/63 \end{bmatrix}$$

Note that the current through the 3-Ω resistor from left to right is $I_3 = 339/63$ or $113/21$ A. This same result was obtained in Problem 2.24 and Problem 2.29.

**14.8** Suppose that we change the circuit of Fig. 14-10 so that $i_b(t)$ is a *dependent* current source equal to $9I_{R_2}$, where $I_{R_2}$ is the downward current through the 2-Ω resistor. Find the corresponding **M**, $\mathbf{Z}_b$, $\mathbf{V}_s$, and $\mathbf{I}_s$ matrices.

The circuit may still be modified as in Fig. 14-11, and the graph of Fig. 14-12 is therefore unchanged, as is the reduced mesh matrix **M**. The first three rows of $\mathbf{Z}_b$, $\mathbf{V}_s$, and $\mathbf{I}_s$ are also unchanged, but the last two rows must be reconsidered.

With reference to Figs. 14-11 and 14-12,

$$I_2 = i_a - I_{R_2} \quad \text{or} \quad I_{R_2} = i_a - I_2$$

where $I_2$ is the current in branch [2], the combination of $i_a$ and 2 Ω in parallel. Hence

$$i_b = 9I_{R_2} = 9i_a - 9I_2 = 63 - 9I_2$$

Then for branch [4],

$$V_4 = 4\{I_4 - i_b\} = 4\{I_4 - 63 + 9I_2\} = 36I_2 + 4I_4 - 252$$

and similarly for branch [5],

$$V_5 = 8 + 5\{I_5 - i_b\} = 45I_2 + 5I_5 - 307$$

Now the $\mathbf{Z}_b$, $\mathbf{V}_s$, and $\mathbf{I}_s$ matrices can be completed, yielding

$$\mathbf{Z}_b = \begin{bmatrix} 1 & 0 & 0 & 0 & 0 \\ 0 & 2 & 0 & 0 & 0 \\ 0 & 0 & 3 & 0 & 0 \\ 0 & 36 & 0 & 4 & 0 \\ 0 & 45 & 0 & 0 & 5 \end{bmatrix} \qquad \mathbf{V}_s = \begin{bmatrix} 0 \\ 0 \\ 0 \\ -252 \\ -307 \end{bmatrix} \qquad \mathbf{I}_s = \begin{bmatrix} 0 \\ 7 \\ 0 \\ 0 \\ 0 \end{bmatrix}$$

**14.9** Find at least three cut set matrices $\mathbf{Q}$ for the oriented graph of Fig. 14-13.

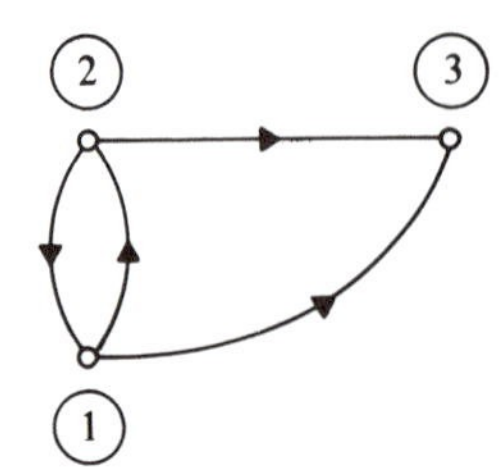


**Fig. 14-13**

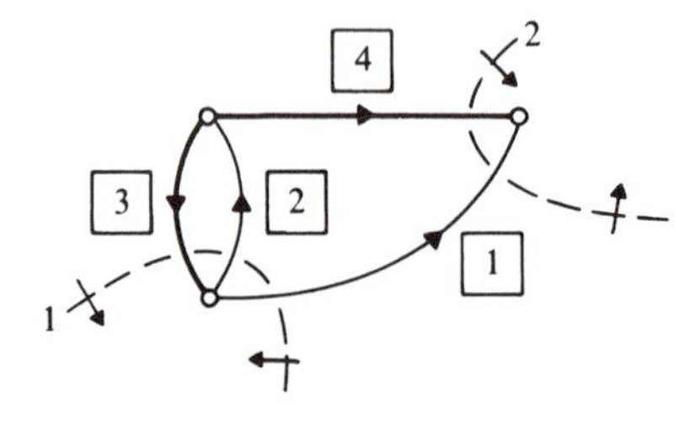


**Fig. 14-14a**

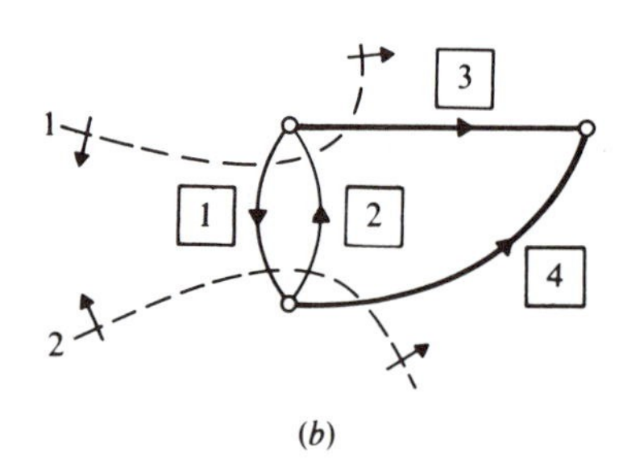


(*b*)

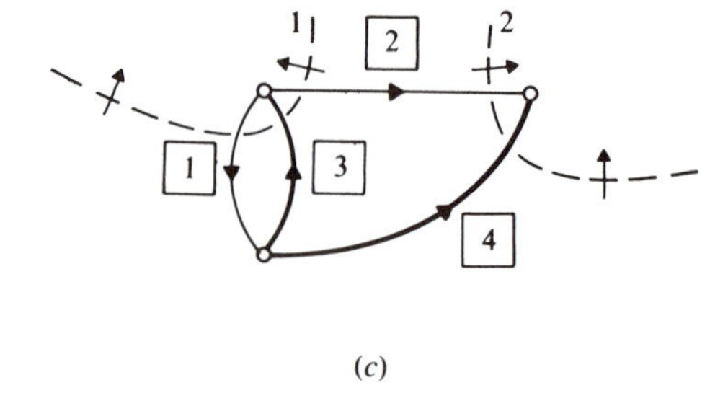


(*c*)

**Fig. 14-14b,c**

First, choosing the tree of Fig. 14-14*a* and numbering the links first,

$$\mathbf{Q}_1 = \begin{bmatrix} -1 & -1 & 1 & 0 \\ 1 & 0 & 0 & 1 \end{bmatrix}$$

Similarly, corresponding to Figs. 14-14*b* and *c*, we have

$$\mathbf{Q}_2 = \begin{bmatrix} 1 & -1 & 1 & 0 \\ -1 & 1 & 0 & 1 \end{bmatrix} \quad \text{and} \quad \mathbf{Q}_3 = \begin{bmatrix} -1 & -1 & 1 & 0 \\ 0 & 1 & 0 & 1 \end{bmatrix}$$

Note that in each case the *last* two columns make up an identity matrix **I**.

**14.10** Write the fundamental loop matrices **B** corresponding to each of the graphs in Fig. 14-14.

Corresponding to Figs. 14-14*a*, *b*, and *c*, respectively,

$$\mathbf{B}_1 = \begin{bmatrix} 1 & 0 & 1 & -1 \\ 0 & 1 & 1 & 0 \end{bmatrix} \quad \mathbf{B}_2 = \begin{bmatrix} 1 & 0 & -1 & 1 \\ 0 & 1 & 1 & -1 \end{bmatrix} \quad \mathbf{B}_3 = \begin{bmatrix} 1 & 0 & 1 & 0 \\ 0 & 1 & 1 & -1 \end{bmatrix}$$

Note that in each case the *first* two columns make up an identity matrix **I**.

**14.11** The **Q** and **B** matrices for the graph of Fig. 14-14*a* are

$$\mathbf{Q} = \begin{bmatrix} -1 & -1 & 1 & 0 \\ 1 & 0 & 0 & 1 \end{bmatrix} \quad \text{and} \quad \mathbf{B} = \begin{bmatrix} 1 & 0 & 1 & -1 \\ 0 & 1 & 1 & 0 \end{bmatrix}$$

Evaluate $\mathbf{BQ}'$.

$$\mathbf{BQ}' = \begin{bmatrix} 1 & 0 & 1 & -1 \\ 0 & 1 & 1 & 0 \end{bmatrix} \begin{bmatrix} -1 & 1 \\ -1 & 0 \\ 1 & 0 \\ 0 & 1 \end{bmatrix} = \begin{bmatrix} 0 & 0 \\ 0 & 0 \end{bmatrix} = \mathbf{0}$$

*Comment:* This is a general result.

**14.12** The fundamental loop matrix **B** corresponding to Fig. 14-14*a* may be partitioned as follows:

$$\mathbf{B} = \left[\begin{array}{cc:cc} 1 & 0 & 1 & -1 \\ 0 & 1 & 1 & 0 \end{array}\right] = [\mathbf{I} \mid \mathbf{F}]$$

while the fundamental cut set matrix **Q** for the same graph may be written

$$\mathbf{Q} = \left[\begin{array}{cc:cc} -1 & -1 & 1 & 0 \\ 1 & 0 & 0 & 1 \end{array}\right] = [\mathbf{E} \mid \mathbf{I}]$$

Show that $\mathbf{E} = -\mathbf{F}'$ or $\mathbf{F} = -\mathbf{E}'$.

Since $\mathbf{F} = \begin{bmatrix} 1 & -1 \\ 1 & 0 \end{bmatrix}$, $\mathbf{F}' = \begin{bmatrix} 1 & 1 \\ -1 & 0 \end{bmatrix}$. But $\mathbf{E} = \begin{bmatrix} -1 & -1 \\ 1 & 0 \end{bmatrix}$ and so $\mathbf{E} = -\mathbf{F}'$.

*Comment:* Again this is a general result. That is,

$$\mathbf{B} = [\mathbf{I} \mid \mathbf{F}] \quad \text{and} \quad \mathbf{Q} = [\mathbf{E} \mid \mathbf{I}]$$

where $\mathbf{E} = -\mathbf{F}'$ or $\mathbf{F} = -\mathbf{E}'$.

**14.13** Determine the fundamental cut set matrix **Q** and the fundamental loop matrix

**B** for the network of Fig. 14-15$a$ using the oriented graph and tree defined in Fig. 14-15$b$.

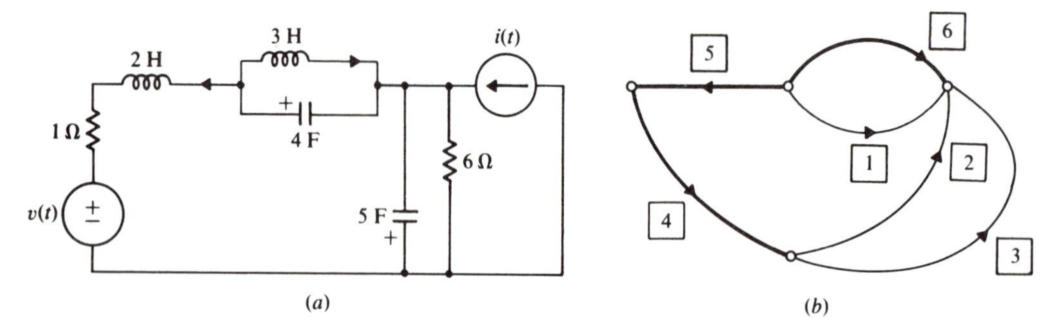


**Fig. 14-15**

Referring to Fig. 14-16$a$, the cut set matrix is

$$\mathbf{Q} = \left[\begin{array}{ccc|ccc} 0 & -1 & -1 & 1 & 0 & 0 \\ 0 & -1 & -1 & 0 & 1 & 0 \\ 1 & 1 & 1 & 0 & 0 & 1 \end{array}\right] = [\mathbf{E} \mid \mathbf{I}]$$

It then follows from the general result of Problem 14.12 that

$$\mathbf{B} = [\mathbf{I} \mid \mathbf{F}] = [\mathbf{I} \mid -\mathbf{E}'] = \left[\begin{array}{ccc|ccc} 1 & 0 & 0 & 0 & 0 & -1 \\ 0 & 1 & 0 & 1 & 1 & -1 \\ 0 & 0 & 1 & 1 & 1 & -1 \end{array}\right]$$

This is easily checked against Fig. 14-16$b$.

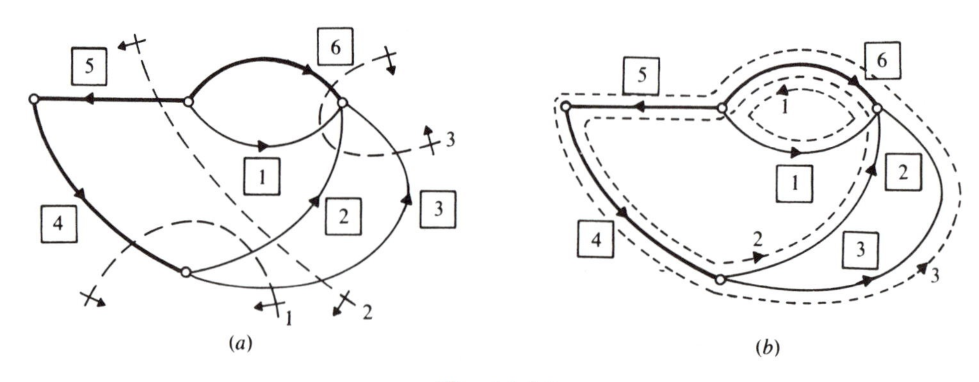


**Fig. 14-16**

**14.14** Show that the branch voltages $\mathbf{v}_b$ and the tree-branch voltages $\mathbf{v}_t$ are related by the matrix equation

$$\mathbf{v}_b = \mathbf{Q}'\mathbf{v}_t$$

for the circuit of Fig. 14-15$a$ and the graph of Fig. 14-15$b$.

The tree-branch vector is by definition

$$\mathbf{v}_t = \begin{bmatrix} v_4 \\ v_5 \\ v_6 \end{bmatrix}$$

Therefore substituting $\mathbf{Q}'$ from Problem 14.13,

$$\mathbf{Q}'\mathbf{v}_t = \begin{bmatrix} 0 & 0 & 1 \\ -1 & -1 & 1 \\ -1 & -1 & 1 \\ 1 & 0 & 0 \\ 0 & 1 & 0 \\ 0 & 0 & 1 \end{bmatrix} \begin{bmatrix} v_4 \\ v_5 \\ v_6 \end{bmatrix} = \begin{bmatrix} v_6 \\ -v_4 - v_5 + v_6 \\ -v_4 - v_5 + v_6 \\ v_4 \\ v_5 \\ v_6 \end{bmatrix} = \begin{bmatrix} v_1 \\ v_2 \\ v_3 \\ v_4 \\ v_5 \\ v_6 \end{bmatrix} = \mathbf{v}_b$$

since from Fig. 14-15 and KVL.

$$v_6 = v_1, \qquad -v_4 - v_5 + v_6 = v_2$$

and

$$-v_4 - v_5 + v_6 = v_3$$

*Comment:* This is also a general result which was stated earlier as equation (*14.5*), page 489.

**14.15** Show that $\mathbf{Qi}_b = \mathbf{0}$ for the graph of Fig. 14-15$b$.

Using $\mathbf{Q}$ from Problem 14.13,

$$\mathbf{Qi}_b = \begin{bmatrix} 0 & -1 & -1 & 1 & 0 & 0 \\ 0 & -1 & -1 & 0 & 1 & 0 \\ 1 & 1 & 1 & 0 & 0 & 1 \end{bmatrix} \begin{bmatrix} i_1 \\ i_2 \\ i_3 \\ i_4 \\ i_5 \\ i_6 \end{bmatrix} = \begin{bmatrix} -i_2 - i_3 + i_4 \\ -i_2 - i_3 + i_5 \\ i_1 + i_2 + i_3 + i_6 \end{bmatrix}$$

Inspection of the graph in relation to KCL shows that each row in the final matrix adds to zero. That is,

$$\mathbf{Qi}_b = \begin{bmatrix} 0 \\ 0 \\ 0 \end{bmatrix} = \mathbf{0}$$

*Comment:* This was stated earlier as equation (*14.6*), page 489, and is in fact simply a matrix statement of Kirchhoff's current law.

**14.16** Suppose that the circuit of Fig. 14-15$a$ is to be analyzed by the cut set method in the $s$-domain. Obtain the necessary matrices.

To find $\mathbf{V}_t(s)$ by the cut set method, we must solve the cut set equation (*14.11*)

$$\{\mathbf{QY}_b(s)\mathbf{Q}'\}\mathbf{V}_t(s) = \mathbf{QY}_b(s)\mathbf{V}_s(s) - \mathbf{QI}_s(s)$$

for the tree-branch voltage $\mathbf{V}_t(s)$. Then the branch voltages and currents follow from equations (14.5) and (14.8),

$$\mathbf{V}_b(s) = \mathbf{Q}'\mathbf{V}_t(s) \quad \text{and} \quad \mathbf{I}_b(s) = \mathbf{Y}_b(s)\mathbf{V}_b(s) + \mathbf{I}_s(s) - \mathbf{Y}_b(s)\mathbf{V}_s(s)$$

Thus our objective here must be to find $\mathbf{Q}$, $\mathbf{Y}_b(s)$, $\mathbf{V}_s(s)$, and $\mathbf{I}_s(s)$. The cut set matrix $\mathbf{Q}$ has already been obtained in Problem 14.13. And by inspection of Fig. 14-15,

$$\mathbf{Y}_b(s) = \begin{bmatrix} 4s & & & & & \\ & 5s & & & 0 & \\ & & \frac{1}{6} & & & \\ & & & 1 & & \\ & 0 & & & \frac{1}{2s} & \\ & & & & & \frac{1}{3s} \end{bmatrix}$$

Now the source matrices must include the initial conditions as equivalent sources. *One* possible set of $s$-domain equivalents for the energy-storage branches is shown in Fig. 14-17.

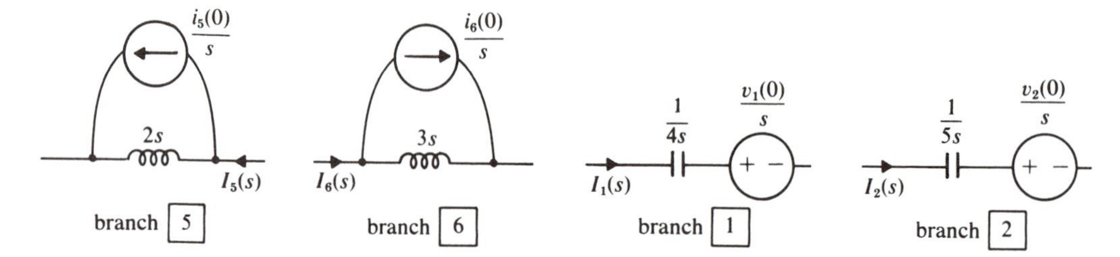


**Fig. 14-17**

The corresponding source matrices are

$$\mathbf{V}_s(s) = \begin{bmatrix} v_1(0)/s \\ v_2(0)/s \\ 0 \\ V(s) \\ 0 \\ 0 \end{bmatrix} \quad \text{and} \quad \mathbf{I}_s(s) = \begin{bmatrix} 0 \\ 0 \\ I(s) \\ 0 \\ i_5(0)/s \\ i_6(0)/s \end{bmatrix}$$

**14.17** The fundamental loop matrix $\mathbf{B}$ has already been determined for the graph of Fig. 14-15$b$ in Problem 14.13. Find the rest of the matrices necessary for the loop analysis of this circuit.

We would find the fundamental loop currents $\mathbf{I}_l(s)$ from equation (14.12),

$$\{\mathbf{B}\mathbf{Z}_b(s)\mathbf{B}'\}\mathbf{I}_l(s) = \mathbf{B}\mathbf{Z}_b(s)\mathbf{I}_s(s) - \mathbf{B}\mathbf{V}_s(s)$$

Then the branch currents and voltages follow from

$$\mathbf{I}_b(s) = \mathbf{B}'\mathbf{I}_l(s) \quad \text{and} \quad \mathbf{V}_b(s) = \mathbf{Z}_b(s)\mathbf{I}_b(s) + \mathbf{V}_s(s) - \mathbf{Z}_b(s)\mathbf{I}_s(s)$$

We must therefore find $\mathbf{Z}_b(s)$, $\mathbf{I}_s(s)$, and $\mathbf{V}_s(s)$. First, by inspection of Fig. 14-15,

$$\mathbf{Z}_b(s) = \begin{bmatrix} \frac{1}{4s} & & & & & \\ & \frac{1}{5s} & & & 0 & \\ & & 6 & & & \\ & & & 1 & & \\ & 0 & & & 2s & \\ & & & & & 3s \end{bmatrix}$$

And $\mathbf{I}_s(s)$ and $\mathbf{V}_s(s)$ have already been found in Problem 14.16.

### *The State Equation*

**14.18** Write the state and output equations for the circuit of Fig. 14-18 in the form

$$\dot{\mathbf{x}} = \mathbf{Ax} + \mathbf{Bu} \qquad \text{where } \mathbf{y} = \begin{bmatrix} v_{C_2} \\ i_{R_1} \end{bmatrix}$$
$$\mathbf{y} = \mathbf{Cx} + \mathbf{Du}$$

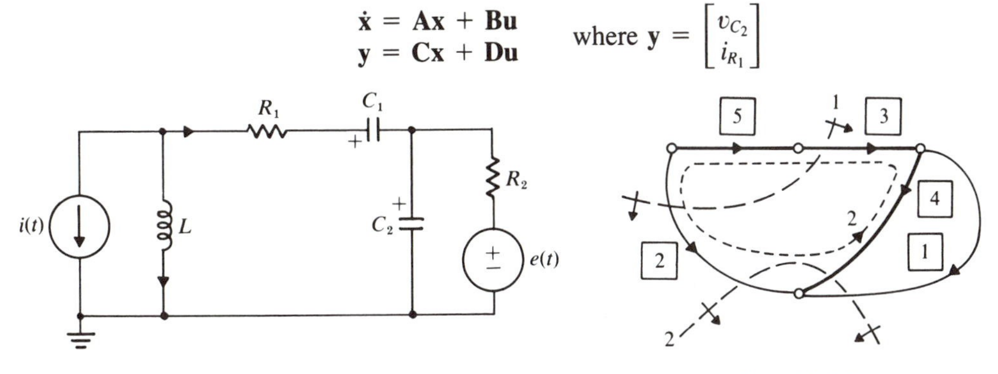


**Fig. 14-18** **Fig. 14-19**

An oriented graph, together with its *proper tree,* is shown in Fig. 14-19. Note that the proper tree contains *all* of the capacitors and *none* of the inductors.

Next we choose the capacitor voltages $v_{C_1}$ and $v_{C_2}$ and the inductor current $i_L$ as the state variables, and write a fundamental cut set equation for each capacitor and the fundamental loop equation for the inductor.

The cut set equation for $C_1$ is $i_3 + i_2 = 0$ or

$$C_1 \frac{dv_{C_1}}{dt} + i(t) + i_L = 0 \qquad (1)$$

Similarly, for $C_2$,

$$C_2 \frac{dv_{C_2}}{dt} + \frac{v_{C_2} - e(t)}{R_2} + i(t) + i_L = 0 \qquad (2)$$

The loop equation for $L$ is $v_2 - v_4 - v_3 - v_5 = 0$, where $v_2 = L(di_L/dt)$, $v_3 = v_{C_1}$, $v_4 = v_{C_2}$, and $v_5 = R_1 i_5 = R_1(-i_2) = -R_1\{i(t) + i_L\}$. Thus

$$L \frac{di_L}{dt} - v_{C_2} - v_{C_1} + R_1\{i(t) + i_L\} = 0 \qquad (3)$$

Now rearranging these three equations to put the derivatives of the state variables on the left and the states and inputs on the right,

$$\frac{dv_{C1}}{dt} = -\frac{1}{C_1}i_L - \frac{1}{C_1}i(t)$$

$$\frac{dv_{C2}}{dt} = -\frac{1}{R_2C_2}v_{C2} - \frac{1}{C_2}i_L + \frac{1}{R_2C_2}e(t) - \frac{1}{C_2}i(t)$$

$$\frac{di_L}{dt} = \frac{1}{L}v_{C1} + \frac{1}{L}v_{C2} - \frac{R_1}{L}i_L - \frac{R_1}{L}i(t)$$

Or, in the matrix form $\dot{\mathbf{x}} = \mathbf{Ax} + \mathbf{Bu}$,

$$\begin{bmatrix} \dot{v}_{C1} \\ \dot{v}_{C2} \\ \dot{i}_L \end{bmatrix} = \begin{bmatrix} 0 & 0 & -\frac{1}{C_1} \\ 0 & -\frac{1}{R_2C_2} & -\frac{1}{C_2} \\ \frac{1}{L} & \frac{1}{L} & -\frac{R_1}{L} \end{bmatrix} \begin{bmatrix} v_{C1} \\ v_{C2} \\ i_L \end{bmatrix} + \begin{bmatrix} -\frac{1}{C_1} & 0 \\ -\frac{1}{C_2} & \frac{1}{R_2C_2} \\ -\frac{R}{L} & 0 \end{bmatrix} \begin{bmatrix} i(t) \\ e(t) \end{bmatrix}$$

We obtained this result somewhat less formally in Problem 2.40.

Now we have defined the outputs to be

$$\mathbf{y} = \begin{bmatrix} y_1 \\ y_2 \end{bmatrix} = \begin{bmatrix} v_{C2} \\ i_{R1} \end{bmatrix}$$

and it can be noted that $y_2 = -\{i(t) + i_L\}$. Therefore in the form $\mathbf{y} = \mathbf{Cx} + \mathbf{Du}$ the output equation becomes

$$\begin{bmatrix} v_{C2} \\ i_{R1} \end{bmatrix} = \begin{bmatrix} 0 & 1 & 0 \\ 0 & 0 & -1 \end{bmatrix} \begin{bmatrix} v_{C1} \\ v_{C2} \\ i_L \end{bmatrix} + \begin{bmatrix} 0 & 0 \\ -1 & 0 \end{bmatrix} \begin{bmatrix} i(t) \\ e(t) \end{bmatrix}$$

**14.19** Write the state and output equations for the circuit of Fig. 14-20, where the state and output matrices are defined by

$$\mathbf{x}' = [v_C \quad i_L] \quad \text{and} \quad \mathbf{y}' = [v_0 \quad i_1]$$

See Fig. 14-21 for the graph and the proper tree.

The fundamental cut set equation for the one capacitor is $i_3 + i_1 + i_2 = 0$ or

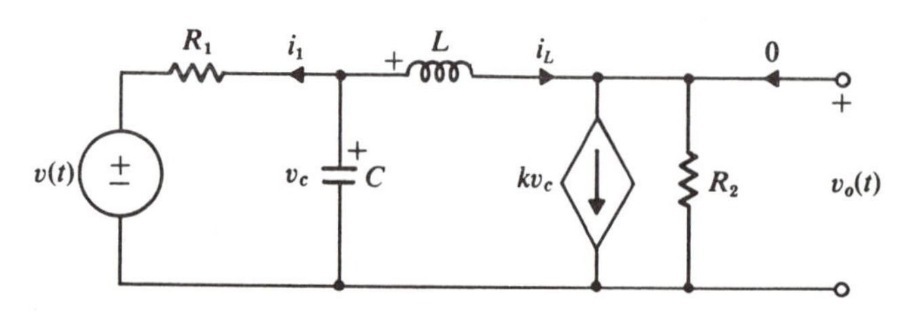


Fig. 14-20

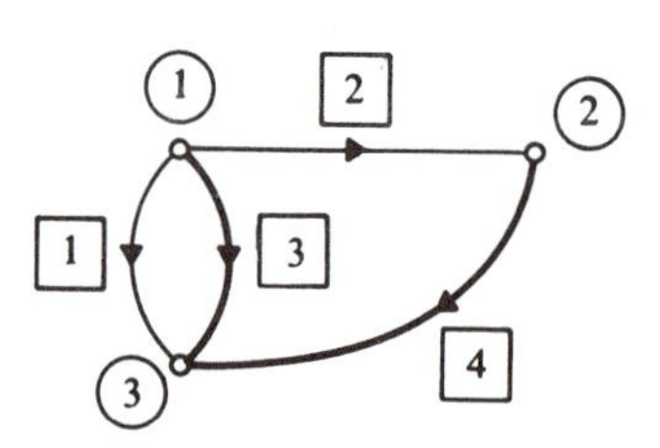


Fig. 14-21

$$C\frac{dv_C}{dt} + \frac{v_C - v(t)}{R_1} + i_L = 0$$

And the fundamental loop equation for the one inductor is $v_2 + v_4 - v_3 = 0$ or

$$L\frac{di_L}{dt} + R_2(i_L - kv_C) - v_C = 0$$

Thus the two state equations are

$$\frac{dv_C}{dt} = -\frac{1}{R_1 C}v_C - \frac{1}{C}i_L + \frac{1}{R_1 C}v(t)$$

$$\frac{di_L}{dt} = \left(\frac{kR_2}{L} + \frac{1}{L}\right)v_C - \frac{R_2}{L}i_L$$

or in standard matrix form,

$$\dot{\mathbf{x}} = \begin{bmatrix} -\frac{1}{R_1 C} & -\frac{1}{C} \\ \frac{kR_2}{L} + \frac{1}{L} & -\frac{R_2}{L} \end{bmatrix}\mathbf{x} + \begin{bmatrix} \frac{1}{R_1 C} \\ 0 \end{bmatrix} u_1$$

where $\mathbf{u} = u_1 = v(t)$.

The two output equations are

$$v_0 = v_4 = R_2(i_L - kv_C) \quad \text{and} \quad i_1 = \frac{v_C - v(t)}{R_1}$$

and so, in standard matrix form,

$$\mathbf{y} = \begin{bmatrix} -kR_2 & R_2 \\ \frac{1}{R_1} & 0 \end{bmatrix}\mathbf{x} + \begin{bmatrix} 0 \\ -\frac{1}{R_1} \end{bmatrix} u_1$$

**14.20** Develop the state equation for the nonlinear network of Fig. 14-22. Choose $q$ and $\lambda$ as the state variables.

The proper tree for this network is shown in Fig. 14-23.

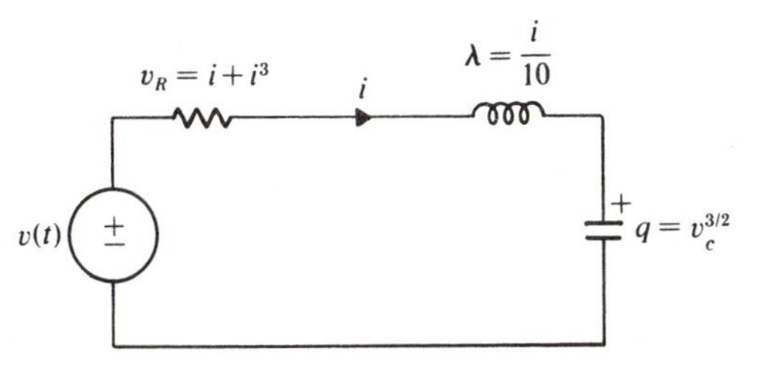


**Fig. 14-22**

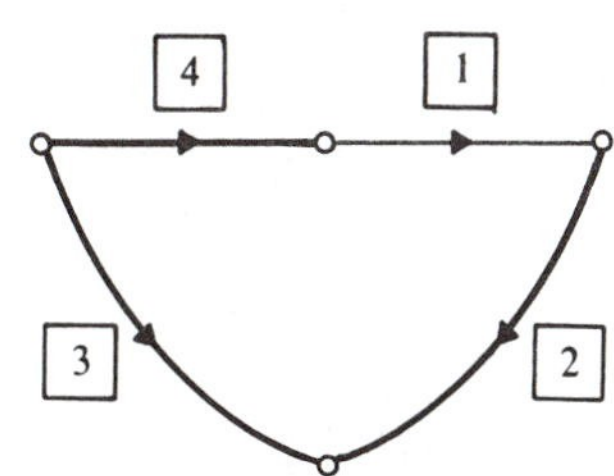


**Fig. 14-23**

The cut set equation is $i_2 - i_1 = 0$ or

$$\frac{dq}{dt} = 10\lambda$$

And the fundamental loop equation is $v_1 + v_2 - v_3 + v_4 = 0$ or

$$\frac{d\lambda}{dt} + q^{2/3} - v(t) + 10\lambda + (10\lambda)^3 = 0$$

Summarizing, the state equation is

$$\begin{bmatrix} \dfrac{dq}{dt} \\ \dfrac{d\lambda}{dt} \end{bmatrix} = \begin{bmatrix} 10\lambda \\ -q^{2/3} - 10\lambda - (10\lambda)^3 + v(t) \end{bmatrix}$$

### *Comments*

1. The state equation can no longer be written $\dot{\mathbf{x}} = \mathbf{Ax} + \mathbf{Bu}$, but has taken on the general form $\dot{\mathbf{x}} = \mathbf{f}(\mathbf{x}, \mathbf{u}, t)$.
2. The inductor is linear. That is, $d\lambda/dt = v_L = 0.1\, di/dt$, which is the equation of a linear inductor with $L = 0.1$ H.

**14.21** Write the state equations for the circuit of Fig. 14-24 for the following three cases:

(*a*) All the passive elements are linear and time-invariant, with $R_1 = 2\ \Omega$ and $R_2 = 3\ \Omega$.

(*b*) All the energy-storage elements are nonlinear but time-invariant, and $R_1 = 2\ \Omega$ and $R_2 = 3\ \Omega$.

(*c*) All the passive elements are linear and time-invariant except for $R_1 = 2 \sin \omega t$ and $R_2 = 3 \cos \omega t$.

A proper tree that may be used for all parts of the problem is shown in Fig. 14-25. From this the fundamental cut set equation for the capacitors is $i_3 + i_2 - i_1 = 0$ while the fundamental loop equations for the inductors are $v_1 + v_3 + v_4 + v_5 = 0$ and $v_2 - v_3 = 0$.

(*a*) For the linear time-invariant case, let

$$\lambda_1 = k_1 i_1, \quad \lambda_2 = k_2 i_2, \quad \text{and} \quad q = k_3 v_C$$

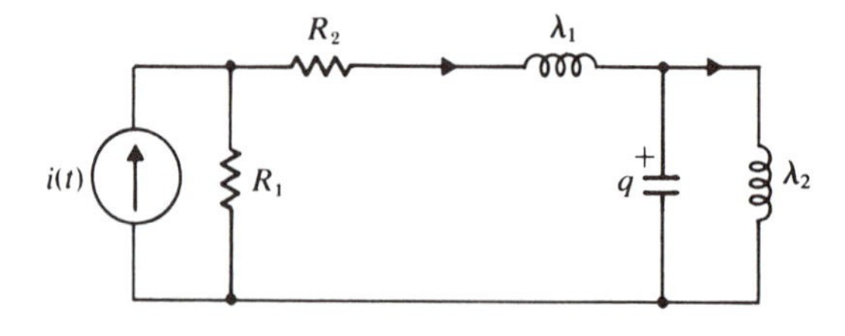


**Fig. 14-24**

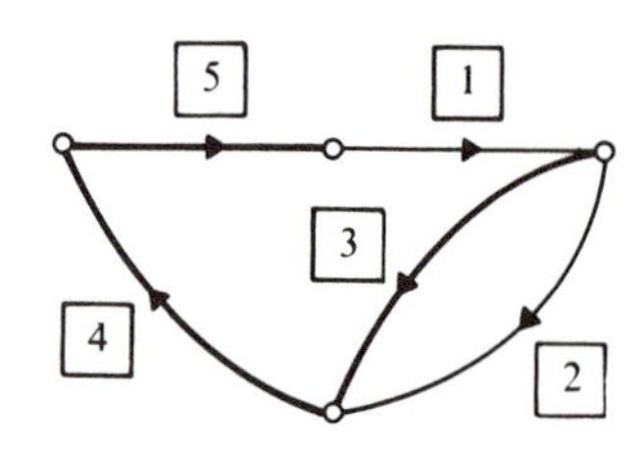


**Fig. 14-25**

where $k_1$, $k_2$, and $k_3$ are constants. Thus the cut set equation becomes

$$\dot{q} + \frac{\lambda_2}{k_2} - \frac{\lambda_1}{k_1} = 0$$

And for the two fundamental loop equations we now have

$$\dot{\lambda}_1 + \frac{q}{k_3} + 2\left\{\frac{\lambda_1}{k_1} - i(t)\right\} + 3\frac{\lambda_1}{k_1} = 0$$

$$\dot{\lambda}_2 - \frac{q}{k_3} = 0$$

Therefore in the standard form $\dot{\mathbf{x}} = \mathbf{Ax} + \mathbf{Bu}$,

$$\begin{bmatrix} \dot{q} \\ \dot{\lambda}_1 \\ \dot{\lambda}_2 \end{bmatrix} = \begin{bmatrix} 0 & \frac{1}{k_1} & -\frac{1}{k_2} \\ -\frac{1}{k_3} & -\frac{5}{k_1} & 0 \\ \frac{1}{k_3} & 0 & 0 \end{bmatrix} \begin{bmatrix} q \\ \lambda_1 \\ \lambda_2 \end{bmatrix} + \begin{bmatrix} 0 \\ 2 \\ 0 \end{bmatrix} i(t)$$

(*b*) For the nonlinear situation, let

$$i_1 = f_1(\lambda_1), \qquad i_2 = f_2(\lambda_2)$$

and

$$v_3 = f_3(q)$$

Then substituting into the original cut set and fundamental loop equations,

$$\dot{q} + f_2(\lambda_2) - f_1(\lambda_1) = 0$$

$$\dot{\lambda}_1 + f_3(q) + 2\{f_1(\lambda_1) - i(t)\} + 3f_1(\lambda_1) = 0$$

$$\dot{\lambda}_2 - f_3(q) = 0$$

or, in the standard form $\dot{\mathbf{x}} = \mathbf{f}(\mathbf{x}, \mathbf{u}, t)$,

$$\begin{bmatrix} \dot{q} \\ \dot{\lambda}_1 \\ \dot{\lambda}_2 \end{bmatrix} = \begin{bmatrix} f_1(\lambda_1) - f_2(\lambda_2) \\ -5f_1(\lambda_1) - f_3(q) + 2i(t) \\ f_3(q) \end{bmatrix}$$

(*c*) In this instance,

$$v_4 = 2 \sin \omega t\{i_1 - i(t)\} = 2 \sin \omega t\left\{\frac{\lambda_1}{k_1} - i(t)\right\}$$

and

$$v_5 = 3 \cos \omega t\{i_1\} = 3\cos \omega t\left\{\frac{\lambda_1}{k_1}\right\}$$

Thus the basic cut set and loop equations become

$$\dot{q} + \frac{\lambda_2}{k_2} - \frac{\lambda_1}{k_1} = 0$$

$$\dot{\lambda}_1 + \frac{q}{k_3} + 2\sin\omega t\left\{\frac{\lambda_1}{k_1} - i(t)\right\} + 3\cos\omega t\left\{\frac{\lambda_1}{k_1}\right\} = 0$$

$$\dot{\lambda}_2 - \frac{q}{k_3} = 0$$

Or, in the matrix form $\dot{\mathbf{x}} = \mathbf{A}(t)\mathbf{x} + \mathbf{B}(t)\mathbf{u}$,

$$\begin{bmatrix} \dot{q} \\ \dot{\lambda}_1 \\ \dot{\lambda}_2 \end{bmatrix} = \begin{bmatrix} 0 & \dfrac{1}{k_1} & -\dfrac{1}{k_2} \\ -\dfrac{1}{k_3} & -\dfrac{1}{k_1}\{2\sin\omega t + 3\cos\omega t\} & 0 \\ \dfrac{1}{k_3} & 0 & 0 \end{bmatrix} \begin{bmatrix} q \\ \lambda_1 \\ \lambda_2 \end{bmatrix} + \begin{bmatrix} 0 \\ 2\sin\omega t \\ 0 \end{bmatrix} i(t)$$

# PROBLEMS

**14.22** Find the augmented incidence matrix $\mathbf{A}_a$ for the oriented graph of Fig. 14-26.

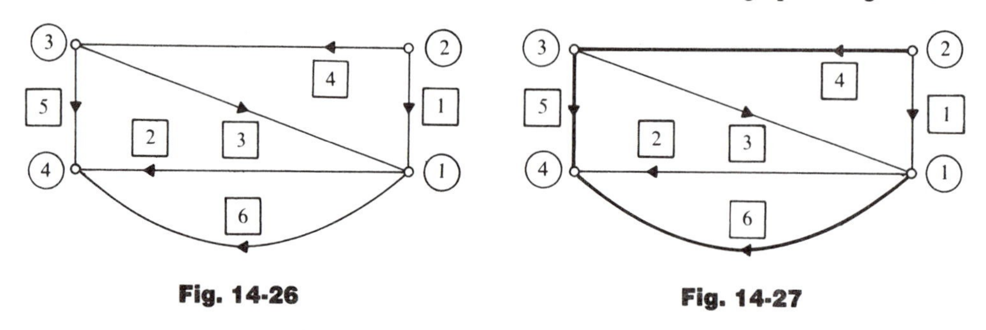


**Fig. 14-26** **Fig. 14-27**

**14.23** Find the incidence matrix **A** for the oriented graph of Fig. 14-26 when node ④ is chosen as the reference.

**14.24** Show that the fundamental loop matrix for the graph of Fig. 14-27 is

$$\mathbf{B} = \underbrace{\begin{bmatrix} 1 & 0 & 0 & -1 & -1 & 1 \\ 0 & 1 & 0 & 0 & 0 & -1 \\ 0 & 0 & 1 & 0 & -1 & 1 \end{bmatrix}}_{b \text{ branches}} = [\underset{\uparrow l \text{ links}}{\mathbf{I}} \;\vdots\; \underset{\uparrow n \text{ tree branches}}{\mathbf{F}}] \}l \text{ loops}$$

**14.25** Write the fundamental cut set matrix **Q** for the graph of Fig. 14-27.

**14.26** Determine the augmented mesh matrix $\mathbf{M}_a$ and the reduced mesh matrix **M** for the graph of Fig. 14-27.

**14.27** Given a circuit's fundamental loop matrix

$$\mathbf{B} = \begin{bmatrix} 1 & 0 & 0 & 0 & 0 & -1 \\ 0 & 1 & 0 & 1 & 1 & -1 \\ 0 & 0 & 1 & 1 & 1 & -1 \end{bmatrix}$$

(*a*) Find the corresponding cut set matrix. (*b*) Draw the oriented graph.

**14.28** A network in the sinusoidal steady-state is shown in Fig. 14.28*a*, while an oriented graph and a chosen tree for this circuit appears in Fig. 14-28*b*. Find: (*a*) $\mathbf{B}$, $\mathbf{Z}_b(j\omega)$, $[\mathbf{V}_s]$ and $[\mathbf{I}_s]$, and (*b*) the matrix loop equation $\mathbf{Z}_l(j\omega)[\mathbf{I}_l] = [\mathbf{V}_{sl}]$.

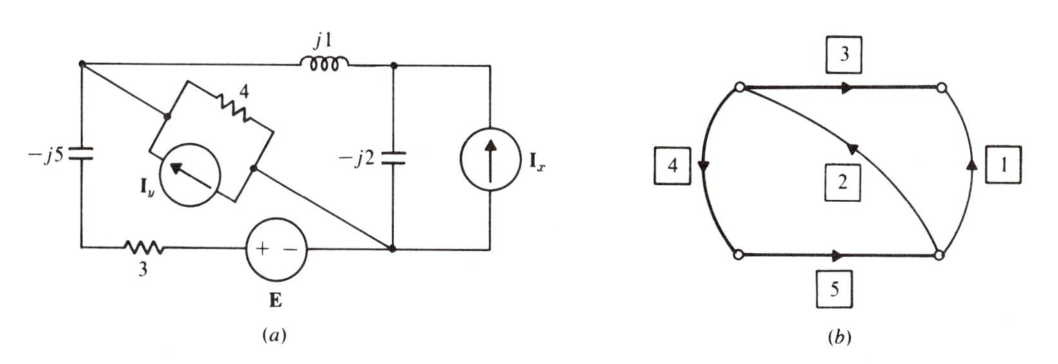


**Fig. 14-28**

**14.29** For the circuit and graph of Fig. 14-28, find
(*a*) **Q** and $\mathbf{Y}_b(j\omega)$, and (*b*) the matrix cut set equation $\mathbf{Y}_t(j\omega)[\mathbf{V}_t] = [\mathbf{I}_{st}]$.

**14.30** Show that the following state equation describes the circuit of Fig. 14-29:

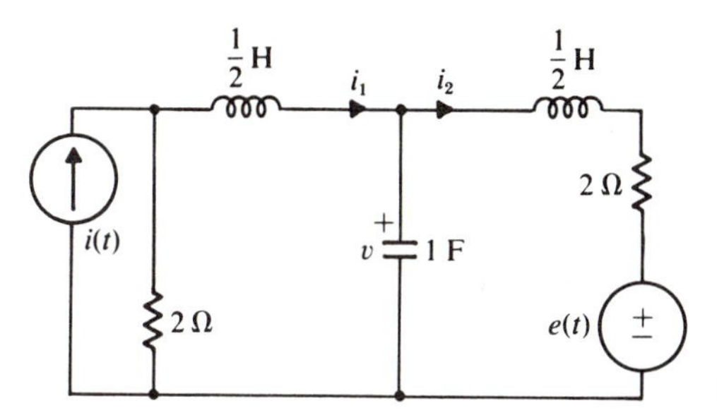


**Fig. 14-29**

$$\begin{bmatrix} di_1/dt \\ di_2/dt \\ dv/dt \end{bmatrix} = \begin{bmatrix} -4 & 0 & -2 \\ 0 & -4 & 2 \\ 1 & -1 & 0 \end{bmatrix} \begin{bmatrix} i_1 \\ i_2 \\ v \end{bmatrix} + \begin{bmatrix} 4 & 0 \\ 0 & -2 \\ 0 & 0 \end{bmatrix} \begin{bmatrix} i(t) \\ e(t) \end{bmatrix}$$

**14.31** Determine the equations

$$\dot{\mathbf{x}} = \mathbf{Ax} + \mathbf{Bu} \qquad \mathbf{y} = \mathbf{Cx} + \mathbf{Du}$$

for the circuit of Fig. 14-30, given

$$\mathbf{x} = \begin{bmatrix} v \\ i_L \end{bmatrix} \quad \mathbf{u} = \begin{bmatrix} i(t) \\ e(t) \end{bmatrix} \quad \mathbf{y} = \begin{bmatrix} v \\ v_a \end{bmatrix}$$

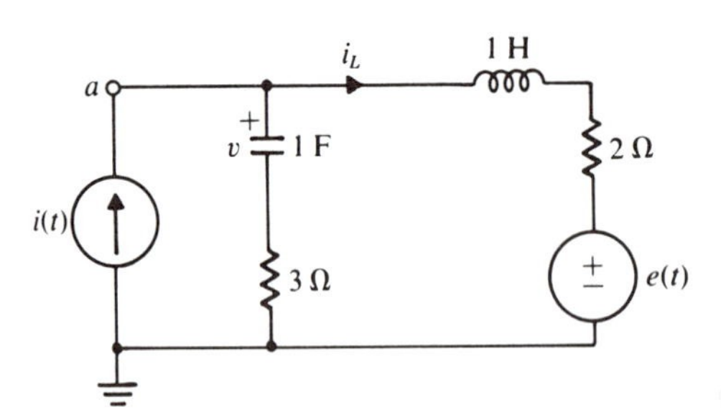


**Fig. 14-30**

**14.32** Write the state equation $\dot{\mathbf{x}} = \mathbf{f}(\mathbf{x}, \mathbf{u}, t)$ for the network of Fig. 14-31 where

$$i = 2 \tanh\left(\tfrac{3}{4}\lambda\right) \quad \text{and} \quad v = q - q^{1/2}.$$

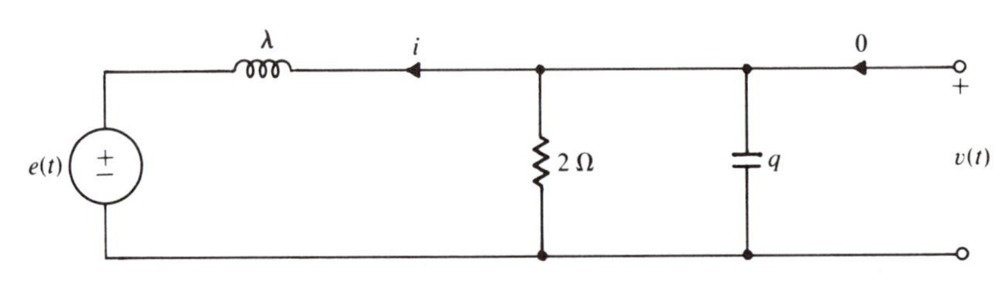


**Fig. 14-31**

# SOLUTIONS TO PROBLEMS

## CHAPTER 1

**1.35** (*a*) See Fig. A-1; (*b*) 250 hertz; (*c*) 500 hertz.

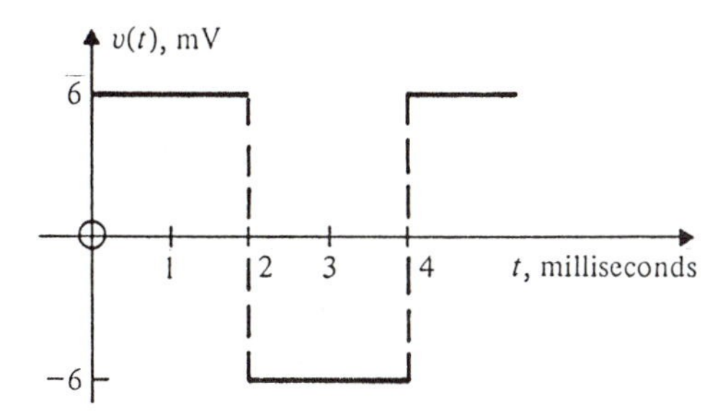

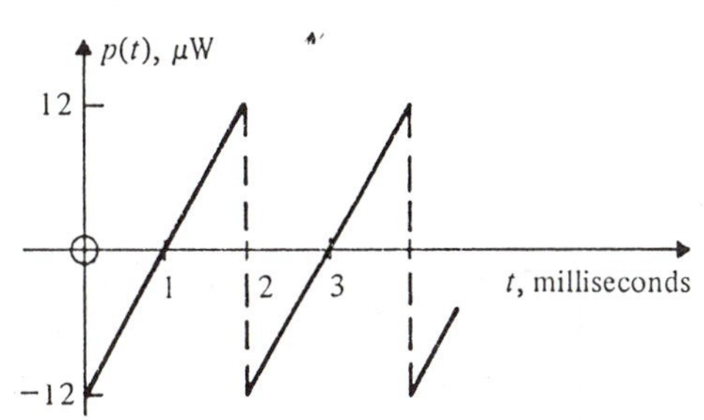


**Fig. A-1**

**1.36** See Fig. A-2.

**1.38** (*a*) $-10M$; (*b*) $-20Me^{-2t}$

**1.39** $p(t) = \frac{1}{2}AB^{3/2}e^{(3/2)t}$

**1.40** $i = 5$ amperes.

**1.41** An inductor of 0.15 henry.

**1.42** (*a*) $Ce^{-t}$; (*b*) No

**1.43** A function *multiplied* by a gate function is "allowed through the gate" only when $p_g(t)$ is *on*. This is illustrated in Fig. A-3.

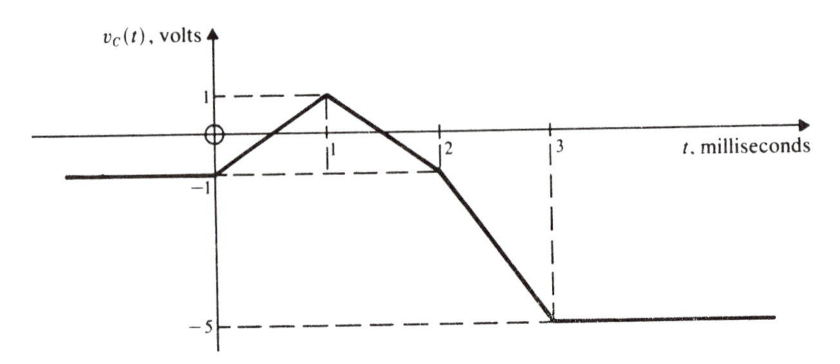


Fig. A-2

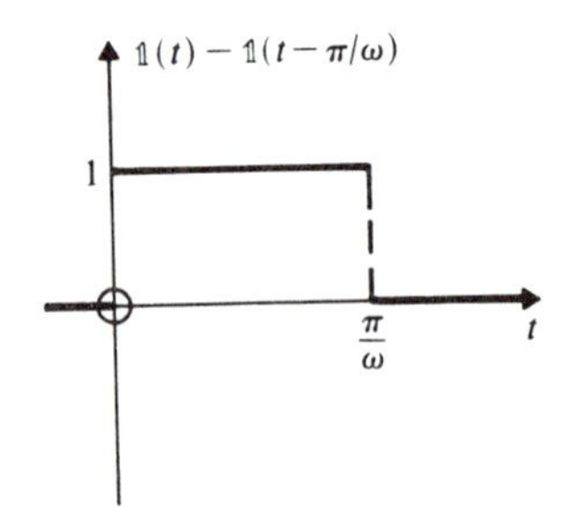


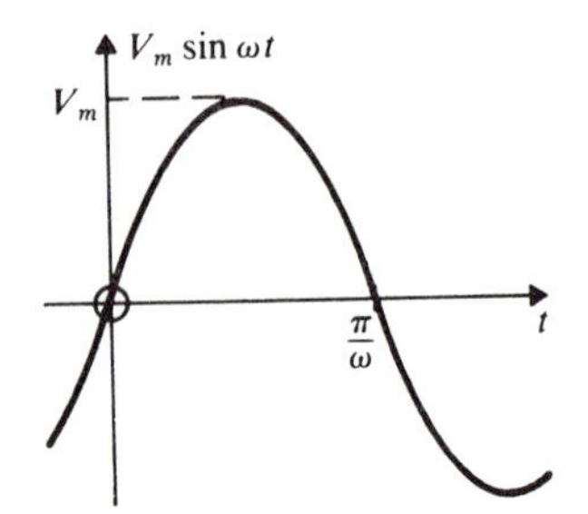


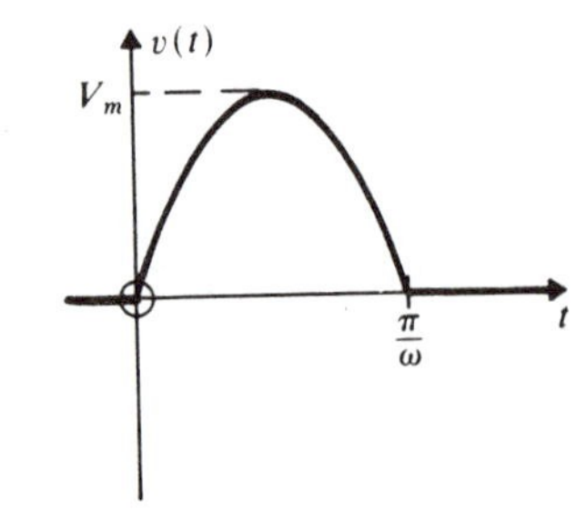


Fig. A-3

**1.44** $i(t) = \frac{7}{L}\mathbb{1}(t)$

**1.45** (*a*) $A/\pi$ volts; (*b*) $A/2$ volts; (*c*) $0.632A$ volts.

**1.46** (*a*) $A/\sqrt{2}$ volts; (*b*) $A/\sqrt{2}$ volts; (*c*) $\frac{1}{2}A\sqrt{1 + 3k}$ volts.

**1.47** 10 amperes.

**1.49** See Fig. A-4.

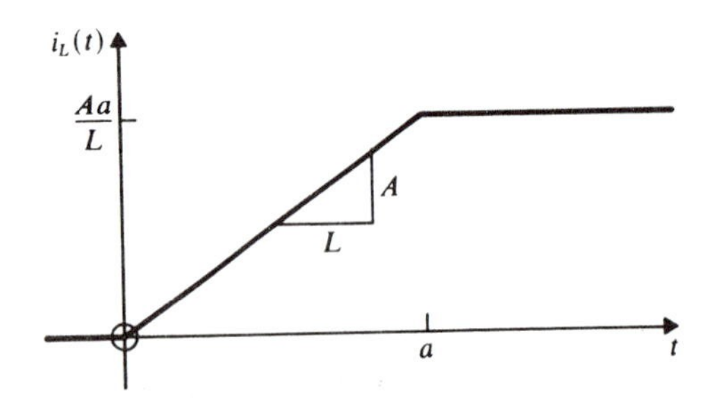


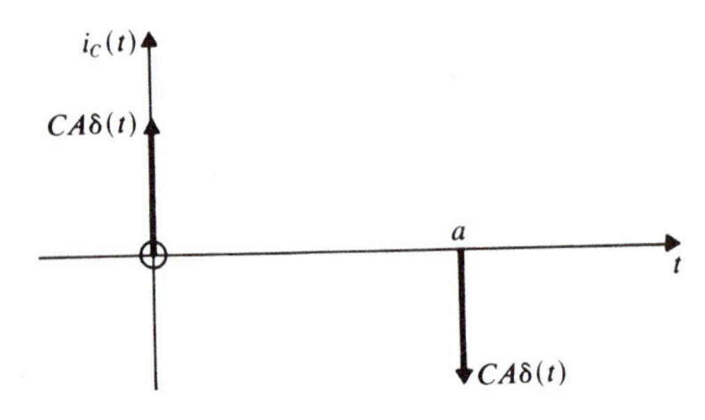


Fig. A-4

## CHAPTER 2

**2.45** For $0 < t < \pi$, $v_4(t) = 20 \cos t - 10/\pi$.
For $\pi < t < 2\pi$, $v_4(t) = 20 \cos t + 10/\pi$.

**2.50** $R = 0.6\ \Omega$

**2.54** Yes

**2.55** The $v$-$i$ characteristic is a straight line cutting the $v$ axis at $V$, and the $i$ axis at $V/r$. These are the open and short circuit conditions, respectively.

For the ideal source, as $r \to 0$, the $v$-$i$ characteristic becomes a horizontal line, cutting the $v$ axis at $V$. The open circuit conditions are unchanged, but the short circuit current becomes infinite.

**2.56** $M = 1$ mH

**2.58** $V = -0.688I + 1.5$; $V_{\text{Th}} = 1.5$ V; and $R_{\text{Th}} = 0.688\ \Omega$; (*a*) $I = 0.89$ A; (*b*) $I = 0.73$ A; (*c*) $I = -1$ A

**2.59** $I = -1.45V + 2.18$; $V_{\text{Th}} = 1.5$ V; and $R_{\text{TH}} = 0.688\ \Omega$

**2.60** $V_{oc} = 1.5$ V; $I_{sc} = 2.18$ A; $V_{\text{Th}} = 1.5$ V; $R_{\text{Th}} = 0.688\ \Omega$

**2.61** $R_{\text{Th}} = 0.688\ \Omega$

**2.63** Both $V_{oc}$ and $I_{sc}$ are zero, so that $R_{\text{Th}} = V_{oc}/I_{sc}$ is indeterminate.

**2.64** There are no resistors in series or parallel, so this network resists your efforts to reduce it to a single equivalent resistance, $R_{\text{Th}}$.

**2.65** For Fig. 2-86*a*,

$$\frac{dv_C}{dt} = -\frac{1}{R_2C}v_C + \frac{1}{C}i_L + \frac{1}{R_2C}v(t)$$

$$\frac{di_L}{dt} = -\frac{1}{L}v_C - \frac{R_1}{L}i_L + \frac{R_1}{L}i(t)$$

For Fig. 2-86*b*,

$$\frac{dv_C}{dt} = -3v_C + i_L - 3v_2(t)$$

$$\frac{di_L}{dt} = -2v_C + 2v_1(t) - 2v_2(t)$$

**2.66** See Fig. A-5, page 518.

## CHAPTER 3

**3.16** $i(t) = -\frac{1}{3}e^{-0.5t}$

**3.17** $\tau = 3.33$ s

**3.18** $di(0)/dt = \frac{1}{3}$ A/s and $dv_C(0)/dt = 2$ V/s

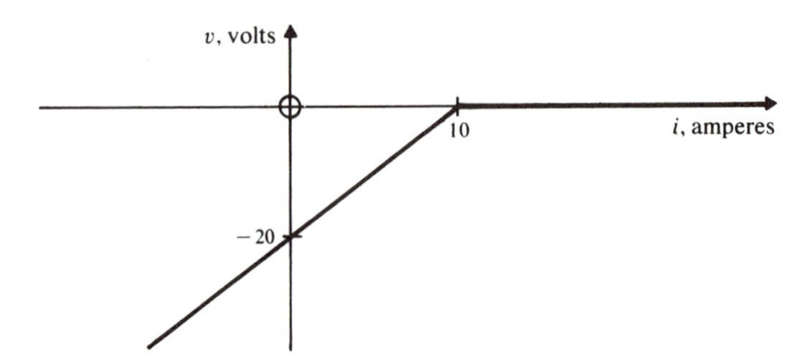


**Fig. A-5**

**3.20** $v(t) = \frac{8}{3}e^{-t} - \frac{4}{3}e^{-2t}$

**3.21** $\omega_n = \sqrt{2}$ and $\zeta = 3\sqrt{2}/4$

**3.22** $v_C(t) = -2e^{-t} + 4e^{-2t}$

**3.23** $i_L(t) = e^{-t} - 4e^{-2t}$

**3.24** $i_{R_1}(0) = 0,\ i_{R_2}(0) = V/R_2,\ di_L(0)/dt = V/L$

## CHAPTER 4

**4.16** $v(t) = 15e^{-15t} + 10e^{-5t}$

**4.18** $v_{ss}(t) = 16$ V

**4.19** $v_{sss}(t) = -20 \cos 10t + 20 \sin 10t = 20\sqrt{2} \sin (10t - \pi/4)$

**4.20** $v(t) = 14e^{-3t} - 2 \cos 3t + 2 \sin 3t = 14e^{-3t} + 2\sqrt{2} \sin (3t - \pi/4)$

**4.21** $v(t) = 5 + e^{-0.5t}(-5 \cos 0.866t + 8.66 \sin 0.866t)$
$= 5 + 10e^{-0.5t} \cos (0.866t - 120°)$

**4.22** $v(t) = -8e^{-t} + 5t^2 - 10t + 10$

**4.23** $v(t) = 10te^{-t} + 2e^{-t}$

## CHAPTER 5

**5.25** $i(t) = -1 + t + 1.11 \cos (2t - 26.6°), \qquad t \geq 0$

**5.26** $v_L(t) = \dfrac{4}{\sqrt{5}} \cos (2t - 18.4°) \qquad \text{for } t \to \infty$

**5.28** $v(t) = -2\sqrt{2}e^{-t} \cos (t - 45°), \qquad t \geq 0$

**5.29** $v(t) = -4e^{-t} \sin t, \qquad t \geq 0$

**5.32** $V(s) = \dfrac{6I(s) + 2E(s) + (s + 3)v(0) + 2i_L(0)}{s^2 + 3s + 2}$

**5.33**
$$\begin{bmatrix} \frac{1}{C_1 s} + 2Ls & -Ls & -Ls \\ -Ls & Ls + \frac{1}{C_2 s} & -\frac{1}{C_2 s} \\ -Ls & -\frac{1}{C_2 s} & R + Ls + \frac{1}{C_2 s} \end{bmatrix} \begin{bmatrix} I_a(s) \\ I_b(s) \\ I_c(s) \end{bmatrix} = \begin{bmatrix} -\frac{v_{C_1}(0)}{s} - 2Li_L(0) \\ V(s) + Li_L(0) - \frac{v_{C_2}(0)}{s} \\ Li_L(0) + \frac{v_{C_2}(0)}{s} \end{bmatrix}$$

**5.34** $\mathbf{x}_F(0.1) = \begin{bmatrix} 0.015 \\ 0.27 \\ 0.46 \end{bmatrix}$

## CHAPTER 6

**6.38** $Z(s) = \dfrac{s^3 + 6s^2 + 11s + 6}{2(s^2 + 2s + 2)}$

**6.39** $\dfrac{I(s)}{V(s)} = \dfrac{1}{R_1 LCs^2 + (R_1 R_2 C + L)s + R_1 + R_2}$

**6.43** $\dfrac{V_b(s)}{I(s)} = \dfrac{R_1 R_2 LCs^2}{(R_1 LC + R_2 LC)s^2 + (L + R_1 R_2 C)s + R_2}$

**6.44** (*a*) $V(s) = \dfrac{s+5}{s+10}\dfrac{10}{s}$ (*c*) $V(s) = \dfrac{10(s+5)}{(s+10)^2}$

(*b*) $V(s) = \dfrac{10}{s+10}$ (*d*) $V(s) = \dfrac{10(s+5)}{s+10}$

**6.45** $Z_o(s) = \dfrac{21s^4 + 30s^3 + 32s^2 + 28s + 4}{21s^3 + 30s^2 + 11s + 1}$

**6.47** $\dfrac{V(s)}{E(s)} = \dfrac{1}{4s^3 + 5s^2 + 7s + 3}$

**6.48** $Z_o(s) = \dfrac{V_2(s)}{I_2(s)} = \dfrac{4s^2 + s + 2}{4s^3 + 5s^2 + 7s + 3}$

**6.49** (*a*) Poles: $s = -2$, $s = \pm j2$. Zero: $s = 0$ (and two zeros at infinity).

(*b*) $K_1 = \dfrac{5}{2}e^{j\pi} = -\dfrac{5}{2}$, $K_2 = \dfrac{5}{2\sqrt{2}}e^{-j45^\circ}$

(*c*) $v(t) = -\dfrac{5}{2}e^{-2t} + \dfrac{5}{\sqrt{2}}\cos(2t - 45^\circ), \quad t \geq 0$

**6.50** (*a*) No restrictions on $b_2$, $b_1$, or $b_0$.

(*b*) The Routh array is

$$\begin{array}{cc} a_3 & a_1 \\ a_2 & a_0 \\ \dfrac{a_1 a_2 - a_3 a_0}{a_2} & 0 \\ a_0 & 0 \end{array}$$

Therefore for stability we require $a_3, a_2, a_1, a_0 > 0$ and $a_1 a_2 > a_3 a_0$.

**6.51** $\dfrac{V_2(s)}{I_1(s)} = \frac{1}{2}\{Z_b(s) - Z_a(s)\}$

**6.52** 3 poles: $s = 0$, $s = \pm j1$ (and one pole at infinity)
4 zeros: $s_{1,2} = +0.5 \pm j0.866$
$s_{3,4} = -0.5 \pm j0.866$

No, the network is not strictly stable, since the poles are on the $j\omega$ axis. This is a case of *marginal stability*.

**6.55** $Z_{\text{Th}}(s) = Z_N(s) = R_1 s(s + 1/R_1 C)/\{s^2 + (R_1/L)s + 1/LC\}$

$$V_{\text{Th}}(s) = \frac{R_1 s^2 I_a(s) - R_1(s + 1/R_1 C)i_L(0)}{s^2 + \dfrac{R_1}{L}s + \dfrac{1}{LC}}$$

where $I_a(s)$ is the transform source current. And $I_N(s) = sI_a(s)/(s + 1/R_1 C) - i_L(0)/s$.

## CHAPTER 7

**7.41** $i_{2,\text{sss}}(t) = 8.96 \sin(2t + \phi - 153.4°)$

**7.42** $X(\omega) = -\dfrac{19}{25}$

**7.44** $v_{a,\text{sss}}(t) = \dfrac{\sqrt{2}}{5}\cos(5t + 31°)$

**7.47** $Z(j\omega) = 2.61e^{-j58.6°}$

**7.48** $\omega = 0.866$ rad/s or $f = 0.138$ Hz

**7.49** $H(j1) = 0.632e^{-j71.6°}$

**7.51** $e(t) = 1.17 \sin(t + 64°)$

**7.53** $Z_0(j3) = 3 - j0.175$

**7.54** $i_{\text{sss}}(t) = 4\cos(3t + 15°)$

**7.55** $Z(j10) = 1.55 - j10.6$, which is equivalent to $R = 1.55\ \Omega$ and $C = 0.0095$ F in series.

**7.56** $v_{\text{sss}}(t) = 87.5\cos(3t - 25°)$

**7.57** $Z_0(j3) = 2.31 + j1.54$

**7.58** $\mathbf{V}_L$ leads $\mathbf{V}_R$ by 127°, as shown in Fig. A-6.

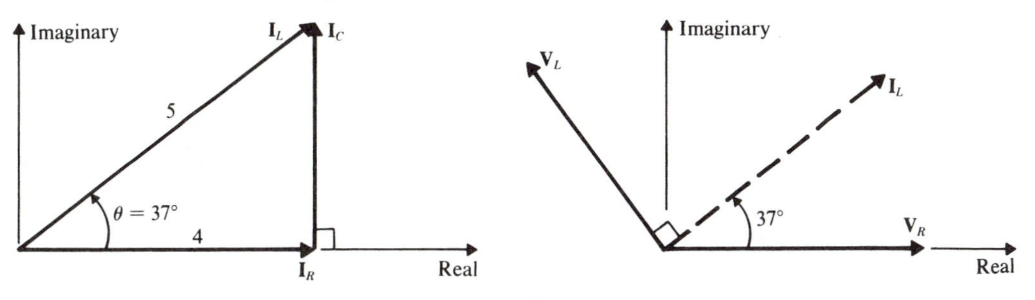


**Fig. A-6**

**7.59** $Z_{\text{Th}}(j1) = 1.414e^{j8.1°} = 1.40 + j0.20$ and $V_{\text{Th}} = 10e^{j(t-36.9°)}$

## CHAPTER 8

**8.25** $H(s) = 2/(s^2 + 3s + 2)$. As a check on the polar plot,

| $\omega$ | $\lvert H(j\omega)\rvert$ | $\underline{/H(j\omega)}$ |
|---|---|---|
| 0 | 1 | 0 |
| 1 | 0.632 | $-71.6°$ |
| 2 | 0.318 | $-108.4°$ |
| $\infty$ | 0 | $-180°$ |

and $\text{Re}[H(j\omega)] = 0$ when $\omega = 1.41$ rad/s.

**8.26** $H(s) = (s + 10)^2/(s + 1)^2$

**8.27** (*a*) 0.1 or $-20$ dB
(*b*) $+1$ or $+20$ dB/decade or $+6$ db/octave
(*c*) 0
(*d*) 4 or 12 dB
(*e*) 70°
(*f*) 0°

**8.30** $\omega_c = 1$ rad/s

**8.35** One

**8.36** (*c*) $Z_{ab}(j\omega)$ for $0 \le \omega \le \infty$

**8.37** See Fig. A-7.

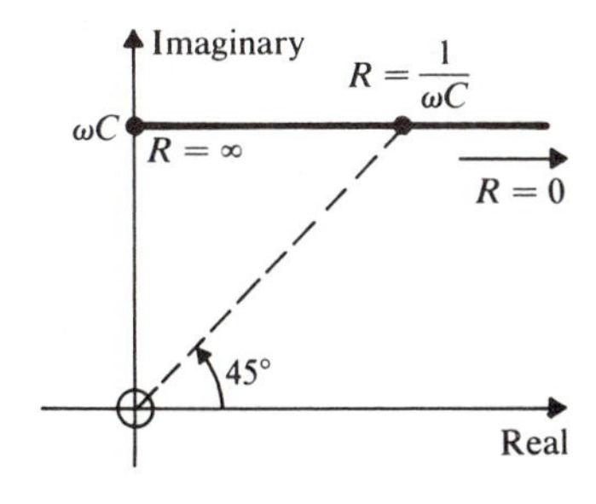


**Fig. A-7**

## CHAPTER 9

**9.24** 200 V, RMS

**9.25** A capacitor of value $C = 0.00125$ F

**9.26** $P$ increases to 6667 W.

**9.27** 394 kVAR

**9.28** (*a*) A 50-Ω resistor and a 0.1-H inductor, (*b*) $P = 100$ W and $Q = 173$ VAR

**9.29** $C = 10\ \mu$F

**9.30** $v_{sss}(t) = \dfrac{5}{\sqrt{2}} \sin(t - 135°)$

**9.33** $P = 2590$ W and $Q = 3460$ VAR

**9.34** $I_L = 36$ A, RMS and $P = 7770$ W

**9.35** $C' = 3C$

**9.36** $R = 17.3\ \Omega$

**9.37** 19.2 A, RMS

**9.38** 7.3 A, RMS

**9.39** 27.3 A, RMS

**9.40** $W_a = 2080$ W and $W_c = 320$ W

## CHAPTER 10

**10.35** $i(t) = 4 + 3e^{-t} - 2t$

**10.36** $v(t) = 6(1 - 2e^{-t} + e^{-2t})$

**10.37** See Fig. A-8.

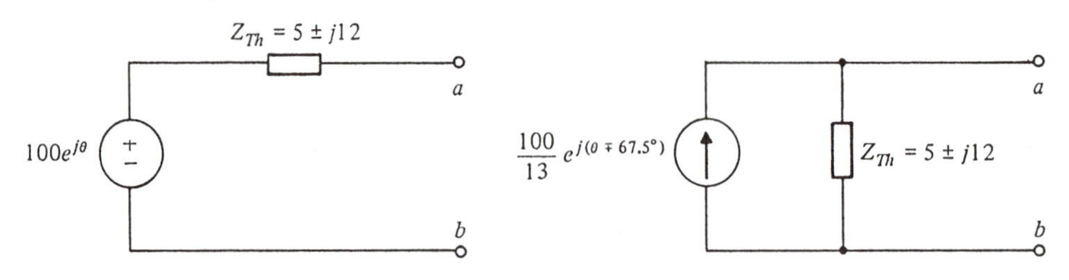


**Fig. A-8**

**10.38** $R = 2/7\ \Omega$ and $P_{max} = 8/7$ W

**10.39** $Z_L(j\omega) = 26.7 + j20$ and $P_{max} = 1.63$ W

**10.40** $\omega_r = 1$ rad/s, $R = 600\ \Omega$, $C = 1/240\ \mu$F and $L = 1.5$ mH

**10.41** $i'(t) = di(t)/dt$

**10.46** See Figs. A-9*a*, *b*, and *c*.

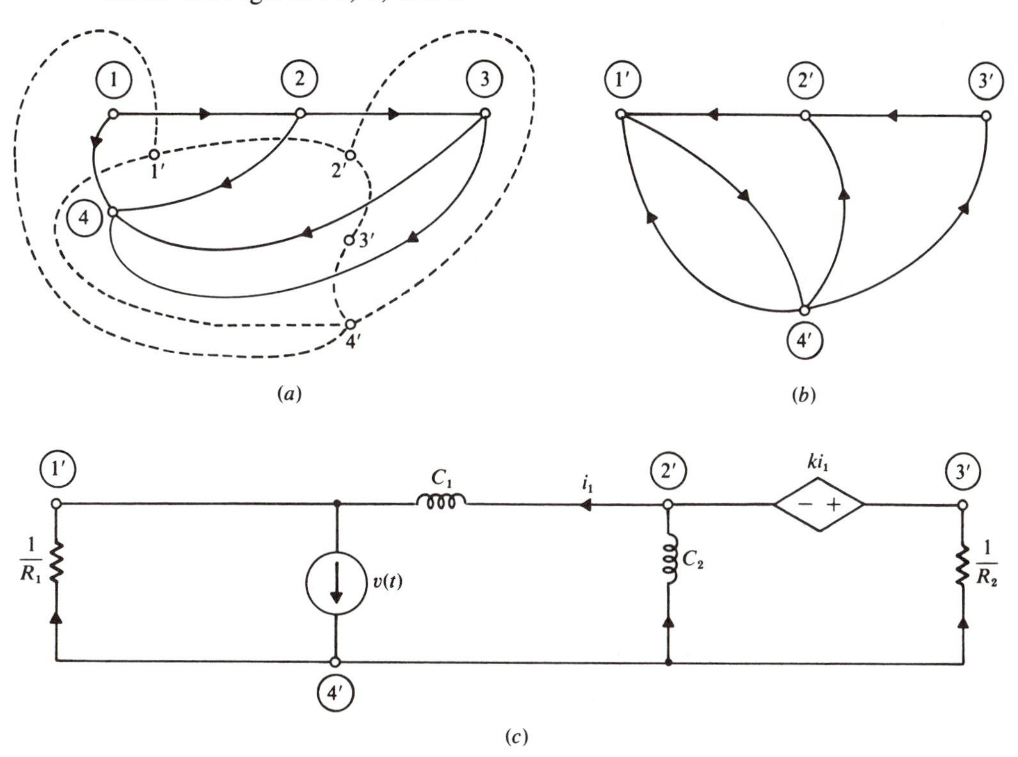


**Fig. A-9**

**10.47** $Z(j\omega) = 0.5 + j0.5$

## CHAPTER 11

**11.30** $\mathbf{h} = \begin{bmatrix} \dfrac{1}{2j\omega} & \dfrac{1}{2} \\ -\dfrac{1}{2} & \dfrac{3j\omega}{2} \end{bmatrix}$

**11.31** $\mathbf{z} = \begin{bmatrix} 1 & 1 \\ 1 & 1 \end{bmatrix}, \mathbf{h} = \begin{bmatrix} 0 & 1 \\ -1 & 1 \end{bmatrix}$

The *y*-parameters do not exist.

**11.32** $h_{21} = -\dfrac{2 + j\omega}{1 + j\omega}$

**11.33** $\mathbf{z} = \begin{bmatrix} 2j\omega L & j\omega L/2 \\ j\omega L/2 & j\omega L \end{bmatrix}$

**11.34** $\dfrac{\mathbf{V}_2}{\mathbf{I}} = \dfrac{R_1R_2z_{21}(j\omega)}{R_2z_{11}(j\omega) + R_1R_2 + R_1z_{22}(j\omega) + \Delta_z}$

## CHAPTER 12

**12.29** $\mathbf{C}_k = \begin{cases} -j\dfrac{4A}{\pi^2k^2}\sin\dfrac{k\pi}{2} & \text{for } k > 0 \\ 0 & \text{for } k = 0 \\ j\dfrac{4A}{\pi^2k^2}\sin\dfrac{k\pi}{2} & \text{for } k < 0 \end{cases}$

$$i(t) = \frac{8A}{\pi^2}\left\{\sin\frac{2\pi}{T}t - \frac{1}{9}\sin\frac{6\pi}{T}t + \frac{1}{25}\sin\frac{10\pi}{T}t - \cdots\right\}$$

**12.30** The dc term is zero because $H(j0) = 0$. The next two terms are

$$\frac{A}{6\pi^2}\cos(2\pi t + 180°) + \frac{100A}{2\pi}\cos(4\pi t + 90°)$$

**12.31** $P = 10\cos 45° + 4\cos 30° \doteq 10.5$ W

**12.32** $y_{ss}(t) = \displaystyle\sum_{k=-\infty}^{\infty} \frac{e^{jkt}}{k(k-j)}$

**12.33** For $g(t) = (-\sin\omega_0 t)\mathbb{1}(-t)$,

$$\mathbf{G}(\omega) = \frac{\pi}{2j}\{\delta(\omega+\omega_0) - \delta(\omega-\omega_0)\} + \frac{1}{2}\left\{\frac{1}{\omega+\omega_0} - \frac{1}{\omega-\omega_0}\right\}$$

**12.34** $\mathbf{V}(\omega) = -\dfrac{2}{j\omega}\cos\dfrac{\omega T}{2} + \dfrac{4}{j\omega^2T}\sin\dfrac{\omega T}{2}$

**12.35** percentage = 29.7

**12.39** $\mathbf{G}(f) = \dfrac{3}{3 + j2\pi f} + \dfrac{40}{16 + 4\pi^2f^2}$

**12.40** $\mathbf{S}(f) = \dfrac{2\sin 2\pi fT}{\pi f}\cos 4\pi fT,\quad \mathbf{S}(\omega) = \dfrac{4\sin\omega T}{\omega}\cos 2\omega T$

**12.41** $g(t) = \begin{cases} \frac{1}{4}e^{-t} & \text{for } t > 0 \\ \frac{1}{2}e^{t} + \frac{5}{4}e^{3t} & \text{for } t < 0 \end{cases}$

## CHAPTER 13

**13.19** See Fig. A-10.

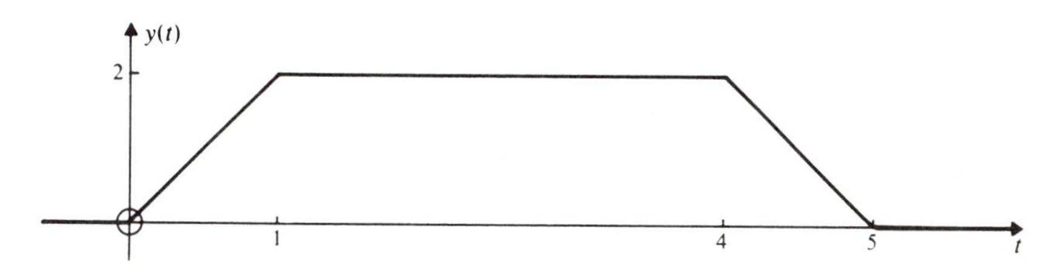


**Fig. A-10**

**13.20** (*a*) $y(t) = \frac{3}{2}\{e^{-(t-5)} + e^{-3(t-5)}\}\mathbb{1}(t-5)$
(*b*) $y(t) = \{4 - 3e^{-t} - e^{-3t}\}\mathbb{1}(t)$

**13.21** $\mathbf{G}(\omega) = \dfrac{\pi}{2j}\{\delta(\omega + \omega_0) - \delta(\omega - \omega_0)\} + \dfrac{1}{2}\left\{\dfrac{1}{\omega + \omega_0} - \dfrac{1}{\omega - \omega_0}\right\}$

**13.22**

| $t$ | $-3$ | $-2$ | $-1$ | $0$ |
|---|---|---|---|---|
| $f_1 * f_2$ | $0$ | $-4$ | $-8$ | $-4$ |

**13.23** $h(t) = \begin{cases} 0 & \text{for } t < 0 \\ 5\delta(t) - 2.5 & \text{for } 0 \le t \le 2 \\ 0 & \text{for } t > 2 \end{cases}$

**13.24** See Fig. A-11.

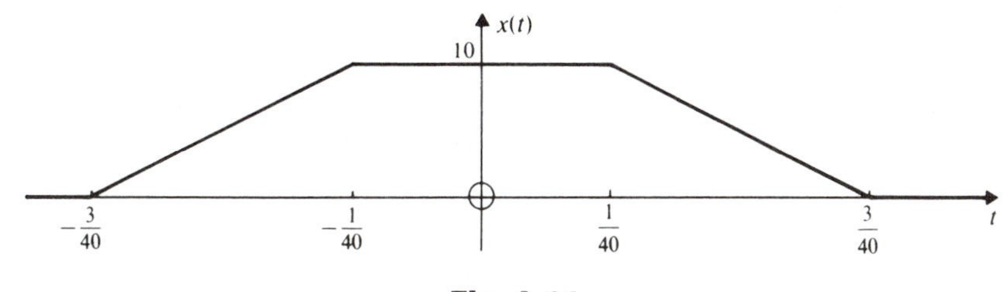


**Fig. A-11**

**13.25** $u(t) * h(t) \doteq 5.5$ at both $t = \pm 1$

## CHAPTER 14

**14.22** $\mathbf{A}_a = \begin{bmatrix} -1 & 1 & -1 & 0 & 0 & 1 \\ 1 & 0 & 0 & 1 & 0 & 0 \\ 0 & 0 & 1 & -1 & 1 & 0 \\ 0 & -1 & 0 & 0 & -1 & -1 \end{bmatrix}$

**14.23** $\mathbf{A} = \begin{bmatrix} -1 & 1 & -1 & 0 & 0 & 1 \\ 1 & 0 & 0 & 1 & 0 & 0 \\ 0 & 0 & 1 & -1 & 1 & 0 \end{bmatrix}$

**14.25** $\mathbf{Q} = \begin{bmatrix} 1 & 0 & 0 & 1 & 0 & 0 \\ 0 & 0 & 1 & 0 & 1 & 0 \\ -1 & 1 & -1 & 0 & 0 & 1 \end{bmatrix} = [\underset{\uparrow\, l \text{ links}}{\mathbf{E}} \;\vdots\; \underset{\uparrow\, n \text{ tree branches}}{\mathbf{I}}]\}n \text{ cut sets}$

**14.26** $\mathbf{M}_a = \begin{bmatrix} -1 & -1 & 0 & 1 & 1 & 0 \\ 0 & 1 & 0 & 0 & 0 & -1 \\ 0 & 0 & 1 & 0 & -1 & 1 \\ 1 & 0 & -1 & -1 & 0 & 0 \end{bmatrix}$

$\mathbf{M} = \begin{bmatrix} -1 & -1 & 0 & 1 & 1 & 0 \\ 0 & 1 & 0 & 0 & 0 & -1 \\ 0 & 0 & 1 & 0 & -1 & 1 \end{bmatrix}$

**14.27** (*a*) $\mathbf{Q} = \begin{bmatrix} 0 & -1 & -1 & 1 & 0 & 0 \\ 0 & -1 & -1 & 0 & 1 & 0 \\ 1 & 1 & 1 & 0 & 0 & 1 \end{bmatrix}$

(*b*) See Fig. A-12.

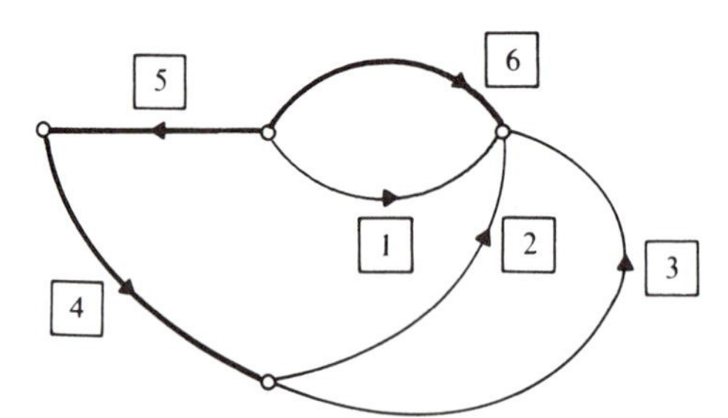


**Fig. A-12**

**14.28** (*a*) $\mathbf{B} = \begin{bmatrix} 1 & 0 & -1 & 1 & 1 \\ 0 & 1 & 0 & 1 & 1 \end{bmatrix}$ $\quad \mathbf{Z}_b(j\omega) = \begin{bmatrix} -j2 & & & & 0 \\ & 4 & & & \\ & & j1 & & \\ & & & -j5 & \\ 0 & & & & 3 \end{bmatrix}$

$$[\mathbf{I}_s] = \begin{bmatrix} \mathbf{I}_x \\ \mathbf{I}_y \\ 0 \\ 0 \\ 0 \end{bmatrix} \qquad [\mathbf{V}_s] = \begin{bmatrix} 0 \\ 0 \\ 0 \\ 0 \\ \mathbf{E} \end{bmatrix}$$

(*b*) $$\begin{bmatrix} 3 - j6 & 3 - j5 \\ 3 - j5 & 7 - j5 \end{bmatrix} \begin{bmatrix} \mathbf{I}_1 \\ \mathbf{I}_2 \end{bmatrix} = \begin{bmatrix} 2\mathbf{I}_x e^{-j90^\circ} - \mathbf{E} \\ 4\mathbf{I}_y - \mathbf{E} \end{bmatrix}$$

**14.29** (*a*) $$\mathbf{Q} = \begin{bmatrix} 1 & 0 & 1 & 0 & 0 \\ -1 & -1 & 0 & 1 & 0 \\ -1 & -1 & 0 & 0 & 1 \end{bmatrix} \qquad \mathbf{Y}_b(j\omega) = \begin{bmatrix} j/2 & & & & 0 \\ & 1/4 & & & \\ & & -j & & \\ & & & j/5 & \\ 0 & & & & j/3 \end{bmatrix}$$

(*b*) $$\begin{bmatrix} -j0.5 & -j0.5 & -j0.5 \\ -j0.5 & 0.25 + j0.7 & 0.25 + j0.5 \\ -j0.5 & 0.25 + j0.5 & 0.583 + j0.5 \end{bmatrix} \begin{bmatrix} \mathbf{V}_3 \\ \mathbf{V}_4 \\ \mathbf{V}_5 \end{bmatrix} = \begin{bmatrix} -\mathbf{I}_x \\ \mathbf{I}_x + \mathbf{I}_y \\ \mathbf{I}_x + \mathbf{I}_y + \frac{1}{3}\mathbf{E} \end{bmatrix}$$

**14.31** $$\dot{\mathbf{x}} = \begin{bmatrix} 0 & -1 \\ 1 & -5 \end{bmatrix}\mathbf{x} + \begin{bmatrix} 1 & 0 \\ 3 & -1 \end{bmatrix}\mathbf{u} \qquad \mathbf{y} = \begin{bmatrix} 1 & 0 \\ 1 & -3 \end{bmatrix}\mathbf{x} + \begin{bmatrix} 0 & 0 \\ 3 & 0 \end{bmatrix}\mathbf{u}$$

**14.32** $$\begin{bmatrix} \dot{q} \\ \dot{\lambda} \end{bmatrix} = \begin{bmatrix} -\frac{1}{2}(q - q^{1/2}) - 2\tanh\left(\frac{3}{4}\lambda\right) \\ q - q^{1/2} - e(t) \end{bmatrix}$$

# INDEX

## G



## H



## I



## J



## K



## L